Advances in Experimental Medicine and Biology

Volume 696

Editorial Board:

NATHAN BACK, *State University of New York at Buffalo*
IRUN R. COHEN, *The Weizmann Institute of Science*
ABEL LAJTHA, *N.S. Kline Institute for Psychiatric Research*
JOHN D. LAMBRIS, *University of Pennsylvania*
RODOLFO PAOLETTI, *University of Milan*

For further volumes:
http://www.springer.com/series/5584

Hamid R. Arabnia · Quoc-Nam Tran
Editors

Software Tools and Algorithms for Biological Systems

 Springer

Editors
Hamid R. Arabnia
Department of Computer Science
415 Boyd Graduate Studies
  Research Centre
University of Georgia
Athens, GA, USA
hra@cs.uga.edu

Quoc-Nam Tran
Department of Computer Science
Lamar University
Beaumont, TX 77710, USA
qntran@buchberger.cs.lamar.edu

ISSN 0065-2598
ISBN 978-1-4419-7045-9   e-ISBN 978-1-4419-7046-6
DOI 10.1007/978-1-4419-7046-6
Springer New York Dordrecht Heidelberg London

Library of Congress Control Number: 2011921718

© Springer Science+Business Media, LLC 2011
All rights reserved. This work may not be translated or copied in whole or in part without the written permission of the publisher (Springer Science+Business Media, LLC, 233 Spring Street, New York, NY 10013, USA), except for brief excerpts in connection with reviews or scholarly analysis. Use in connection with any form of information storage and retrieval, electronic adaptation, computer software, or by similar or dissimilar methodology now known or hereafter developed is forbidden.
The use in this publication of trade names, trademarks, service marks, and similar terms, even if they are not identified as such, is not to be taken as an expression of opinion as to whether or not they are subject to proprietary rights.

Printed on acid-free paper

Springer is part of Springer Science+Business Media (www.springer.com)

Contents

Part I Computational Methods for Microarray, Gene Expression Analysis, and Gene Regulatory Networks

1. **A Technical Platform for Generating Reproducible Expression Data from *Streptomyces coelicolor* Batch Cultivations**.. 3
   F. Battke, A. Herbig, A. Wentzel, Ø.M. Jakobsen, M. Bonin, D.A. Hodgson, W. Wohlleben, T.E. Ellingsen, the STREAM Consortium, and K. Nieselt

2. **MiRNA Recognition with the *yasMiR* System: The Quest for Further Improvements**... 17
   Daniel Pasailă, Andrei Sucilă, Irina Mohorianu, Ştefan Panţiru, and Liviu Ciortuz

3. **Top Scoring Pair Decision Tree for Gene Expression Data Analysis**.. 27
   Marcin Czajkowski and Marek Krętowski

4. **Predictive Minimum Description Length Principle Approach to Inferring Gene Regulatory Networks**..................... 37
   Vijender Chaitankar, Chaoyang Zhang, Preetam Ghosh, Ping Gong, Edward J. Perkins, and Youping Deng

5. **Parsimonious Selection of Useful Genes in Microarray Gene Expression Data**.. 45
   Félix F. González-Navarro and Lluís A. Belanche-Muñoz

6. **Hierarchical Signature Clustering for Time Series Microarray Data**... 57
   Lars Koenig and Eunseog Youn

7	Microarray Database Mining and Cell Differentiation Defects in Schizophrenia 67

Aurelian Radu, Gabriela Hristescu, Pavel Katsel, Vahram Haroutunian, and Kenneth L. Davis

8	miRNA Prediction Using Computational Approach 75

A.K. Mishra and D.K. Lobiyal

9	Improving the Accuracy of Gene Expression Profile Classification with Lorenz Curves and Gini Ratios 83

Quoc-Nam Tran

10	Feature Selection in Gene Expression Data Using Principal Component Analysis and Rough Set Theory 91

Debahuti Mishra, Rajashree Dash, Amiya Kumar Rath, and Milu Acharya

11	Dramatically Reduced Precision in Microarray Analysis Retains Quantitative Properties and Provides Additional Benefits 101

William C. Ray

12	Algebraic Model Checking for Boolean Gene Regulatory Networks 113

Quoc-Nam Tran

13	Comparative Advantages of Novel Algorithms Using MSR Threshold and MSR Difference Threshold for Biclustering Gene Expression Data 123

Shyama Das and Sumam Mary Idicula

14	Performance Comparison of SLFN Training Algorithms for DNA Microarray Classification 135

Hieu Trung Huynh, Jung-Ja Kim, and Yonggwan Won

15	Clustering Microarray Data to Determine Normalization Method 145

Marie Vendettuoli, Erin Doyle, and Heike Hofmann

Part II Bioinformatics Databases, Data Mining, and Pattern Discovery Techniques

16	Estimation, Modeling, and Simulation of Patterned Growth in Extreme Environments 157

B. Strader, K.E. Schubert, M. Quintana, E. Gomez, J. Curnutt, and P. Boston

17 Performance of Univariate Forecasting on Seasonal Diseases: The Case of Tuberculosis ... 171
Adhistya Erna Permanasari, Dayang Rohaya Awang Rambli, and P. Dhanapal Durai Dominic

18 Predicting Individual Affect of Health Interventions to Reduce HPV Prevalence ... 181
Courtney D. Corley, Rada Mihalcea, Armin R. Mikler, and Antonio P. Sanfilippo

19 Decision Tree and Ensemble Learning Algorithms with Their Applications in Bioinformatics ... 191
Dongsheng Che, Qi Liu, Khaled Rasheed, and Xiuping Tao

20 Pattern Recognition of Surface EMG Biological Signals by Means of Hilbert Spectrum and Fuzzy Clustering ... 201
Ruben-Dario Pinzon-Morales, Katherine-Andrea Baquero-Duarte, Alvaro-Angel Orozco-Gutierrez, and Victor-Hugo Grisales-Palacio

21 Rotation of Random Forests for Genomic and Proteomic Classification Problems ... 211
Gregor Stiglic, Juan J. Rodriguez, and Peter Kokol

22 Improved Prediction of MHC Class I Binders/Non-Binders Peptides Through Artificial Neural Network Using Variable Learning Rate: SARS Corona Virus, a Case Study ... 223
Sudhir Singh Soam, Bharat Bhasker, and Bhartendu Nath Mishra

Part III Protein Classification and Structure Prediction, and Computational Structural Biology

23 Fast Three-Dimensional Noise Reduction for Real-Time Electron Tomography ... 233
José Antonio Martínez and José Jesús Fernández

24 Prediction of Chemical-Protein Binding Activity Using Contrast Graph Patterns ... 243
Andrzej Dominik, Zbigniew Walczak, and Jacek Wojciechowski

25 Topological Constraint in High-Density Cells' Tracking of Image Sequences ... 255
Chunming Tang, Ling Ma, and Dongbin Xu

26 **STRIKE: A Protein–Protein Interaction Classification Approach** 263
Nazar Zaki, Wassim El-Hajj, Hesham M. Kamel, and Fadi Sibai

27 **Cooperativity of Protein Binding to Vesicles** 271
Francisco Torrens and Gloria Castellano

28 **The Role of Independent Test Set in Modeling of Protein Folding Kinetics** 279
Nikola Štambuk and Paško Konjevoda

Part IV Comparative Sequence, Genome Analysis, Genome Assembly, and Genome Scale Computational Methods

29 **Branch-and-Bound Approach for Parsimonious Inference of a Species Tree from a Set of Gene Family Trees** 287
Jean-Philippe Doyon and Cedric Chauve

30 **Sequence-Specific Sequence Comparison Using Pairwise Statistical Significance** 297
Ankit Agrawal, Alok Choudhary, and Xiaoqiu Huang

31 **Modelling Short Time Series in Metabolomics: A Functional Data Analysis Approach** 307
Giovanni Montana, Maurice Berk, and Tim Ebbels

32 **Modeling of Gene Therapy for Regenerative Cells Using Intelligent Agents** 317
Aya Sedky Adly, Amal Elsayed Aboutabl, and M. Shaarawy Ibrahim

33 **Biomarkers Discovery in Medical Genomics Data** 327
A. Benis and M. Courtine

34 **Computer Simulation on Disease Vector Population Replacement Driven by the Maternal Effect Dominant Embryonic Arrest** 335
Mauricio Guevara-Souza and Edgar E. Vallejo

35 **Leukocytes Segmentation Using Markov Random Fields** 345
C. Reta, L. Altamirano, J.A. Gonzalez, R. Diaz, and J.S. Guichard

Part V Experimental Medicine and Analysis Tools

36 Ontology-Based Knowledge Discovery in Pharmacogenomics 357
Adrien Coulet, Malika Smaïl-Tabbone, Amedeo Napoli, and Marie-Dominique Devignes

37 Enabling Heterogeneous Data Integration and Biomedical Event Prediction Through ICT: The Test Case of Cancer Reoccurrence 367
Marco Picone, Sebastian Steger, Konstantinos Exarchos, Marco De Fazio, Yorgos Goletsis, Dimitrios I. Fotiadis, Elena Martinelli, and Diego Ardigò

38 Complexity and High-End Computing in Biology and Medicine 377
Dimitri Perrin

39 Molecular Modeling Study of Interaction of Anthracenedione Class of Drug Mitoxantrone and Its Analogs with DNA Tetrameric Sequences 385
Pamita Awasthi, Shilpa Dogra, Lalit K. Awasthi, and Ritu Barthwal

40 A Monte Carlo Analysis of Peritoneal Antimicrobial Pharmacokinetics 401
Sanjukta Hota, Philip Crooke, and John Hotchkiss

Part VI Computational Methods for Filtering, Noise Cancellation, and Signal and Image Processing

41 Histopathology Tissue Segmentation by Combining Fuzzy Clustering with Multiphase Vector Level Sets 413
Filiz Bunyak, Adel Hafiane, and Kannappan Palaniappan

42 A Dynamically Masked Gaussian Can Efficiently Approximate a Distance Calculation for Image Segmentation 425
Shareef M. Dabdoub, Sheryl S. Justice, and William C. Ray

43 Automatic and Robust System for Correcting Microarray Images' Rotations and Isolating Spots 433
Anlei Wang, Naima Kaabouch, and Wen-Chen Hu

44 **Multimodality Medical Image Registration and Fusion Techniques Using Mutual Information and Genetic Algorithm-Based Approaches** 441
Mahua Bhattacharya and Arpita Das

45 **Microcalcifications Detection Using Fisher's Linear Discriminant and Breast Density** 451
G.A. Rodriguez, J.A. Gonzalez, L. Altamirano, J.S. Guichard, and R. Diaz

46 **Enhanced Optical Flow Field of Left Ventricular Motion Using Quasi-Gaussian DCT Filter** 461
Slamet Riyadi, Mohd. Marzuki Mustafa, Aini Hussain, Oteh Maskon, and Ika Faizura Mohd. Nor

47 **An Efficient Algorithm for Denoising MR and CT Images Using Digital Curvelet Transform** 471
S. Hyder Ali and R. Sukanesh

48 **On the Use of Collinear and Triangle Equation for Automatic Segmentation and Boundary Detection of Cardiac Cavity Images** 481
Riyanto Sigit, Mohd. Marzuki Mustafa, Aini Hussain, Oteh Maskon, and Ika Faizura Mohd. Nor

49 **The Electromagnetic-Trait Imaging Computation of Traveling Wave Method in Breast Tumor Microwave Sensor System** 489
Zhi-fu Tao, Zhong-ling Han, and Meng Yao

50 **Medical Image Processing Using Novel Wavelet Filters Based on Atomic Functions: Optimal Medical Image Compression** 497
Cristina Juarez Landin, Magally Martinez Reyes, Anabelem Soberanes Martin, Rosa Maria Valdovinos Rosas, Jose Luis Sanchez Ramirez, Volodymyr Ponomaryov, and Maria Dolores Torres Soto

51 **Cancellation of Artifacts in ECG Signals Using Block Adaptive Filtering Techniques** 505
Mohammad Zia Ur Rahman, Rafi Ahamed Shaik, and D.V. Rama Koti Reddy

52 **Segmentation of Medical Image Sequence by Parallel Active Contour** 515
Abdelkader Fekir and Nacéra Benamrane

53 Computerized Decision Support System for Mass
    Identification in Breast Using Digital Mammogram:
    A Study on GA-Based Neuro-Fuzzy Approaches 523
    Arpita Das and Mahua Bhattacharya

Part VII Computer-Based Medical Systems

54 Optimization-Based Technique for Separation
    and Detection of Saccadic Movements and Eye-Blinking
    in Electrooculography Biosignals 537
    Robert Krupiński and Przemysław Mazurek

55 A Framework for Lipoprotein Ontology 547
    Meifania Chen and Maja Hadzic

56 Verbal Decision Analysis Applied on the Optimization
    of Alzheimer's Disease Diagnosis: A Case Study Based
    on Neuroimaging 555
    Isabelle Tamanini, Ana Karoline de Castro, Plácido Rogério
    Pinheiro, and Mirian Calíope Dantas Pinheiro

57 Asynchronous Brain Machine Interface-Based Control
    of a Wheelchair 565
    C.R. Hema, M.P. Paulraj, Sazali Yaacob, Abdul Hamid Adom,
    and R. Nagarajan

58 Toward an Application to Psychological Disorders
    Diagnosis 573
    Luciano Comin Nunes, Plácido Rogério Pinheiro,
    Tarcísio Cavalcante Pequeno, and Mirian Calíope Dantas
    Pinheiro

59 Enhancing Medical Research Efficiency by Using Concept
    Maps 581
    Varadraj P. Gurupur, Amit S. Kamdi, Tolga Tuncer,
    Murat M. Tanik, and Murat N. Tanju

60 Analysis of Neural Sources of P300 Event-Related
    Potential in Normal and Schizophrenic Participants 589
    Malihe Sabeti, Ehsan Moradi, and Serajeddin Katebi

61 Design and Development of a Tele-Healthcare Information
    System Based on Web Services and HL7 Standards 599
    Ean-Wen Huang, Rui-Suan Hung, Shwu-Fen Chiou,
    Fei-Ying Liu, and Der-Ming Liou

62	**Fuzzy Logic Based Expert System for the Treatment of Mobile Tooth** 607
	Vijay Kumar Mago, Anjali Mago, Poonam Sharma, and Jagmohan Mago

63	**A Microcomputer FES System for Wrist Moving Control** 615
	Li Cao, Jin-Sheng Yang, Zhi-Long Geng, and Gang Cao

64	**Computer-Aided Decision System for the Clubfeet Deformities** 623
	Tien Tuan Dao, Frédéric Marin, Henri Bensahel, and Marie Christine Ho Ba Tho

65	**A Framework for Specifying Safe Behavior of the CIIP Medical System** 637
	Seyed Morteza Babamir

Part VIII Software Packages and Other Computational Topics in Bioinformatics

66	**Lotka–Volterra System with Volterra Multiplier** 647
	Klaus Gürlebeck and Xinhua Ji

67	**A Biological Compression Model and Its Applications** 657
	Minh Duc Cao, Trevor I. Dix, and Lloyd Allison

68	**Open Source Clinical Portals: A Model for Healthcare Information Systems to Support Care Processes and Feed Clinical Research** 667
	Paolo Locatelli, Emanuele Baj, Nicola Restifo, Gianni Origgi, and Silvia Bragagia

69	**Analysis and Clustering of MicroRNA Array: A New Efficient and Reliable Computational Method** 679
	Luca Sterpone, Federica Collino, Giovanni Camussi, and Claudio Loconsole

70	**Stochastic Simulations of Mixed-Lipid Compartments: From Self-Assembling Vesicles to Self-Producing Protocells** 689
	Kepa Ruiz-Mirazo, Gabriel Piedrafita, Fulvio Ciriaco, and Fabio Mavelli

71	**A New Genetic Algorithm for Polygonal Approximation** 697
	Cecilia Di Ruberto and Andrea Morgera

72 Challenges When Using Real-World Bio-data to Calibrate Simulation Systems 709
Elaine M. Blount, Stacie I. Ringleb, and Andreas Tolk

73 Credibility of Digital Content in a Healthcare Collaborative Community 717
Wail M. Omar, Dinesh K. Saini, and Mustafa Hasan

74 Using Standardized Numerical Scores for the Display and Interpretation of Biomedical Data 725
Robert A. Warner

75 ImagCell: A Computer Tool for Cell Culture Image Processing Applications in Bioimpedance Measurements 733
Alberto Yúfera, Estefanía Gallego, and Javier Molina

76 From Ontology Selection and Semantic Web to an Integrated Information System for Food-borne Diseases and Food Safety 741
Xianghe Yan, Yun Peng, Jianghong Meng, Juliana Ruzante, Pina M. Fratamico, Lihan Huang, Vijay Juneja, and David S. Needleman

77 Algebraic Analysis of Social Networks for Bio-surveillance: The Cases of SARS-Beijing-2003 and AH1N1 Influenza-México-2009 751
Doracelly Hincapié and Juan Ospina

Index 763

Contributors

Amal Elsayed Aboutabl
Computer Science Department, Faculty of Computers and Information,
Helwan University, Cairo, Egypt
aaboutabl@helwan.edu.eg

Milu Acharya
Department of Computer Science and Engineering, Institute of Technical
Education and Research, Bhubaneswar, Orissa, India
ayasedky@helwan.edu.eg; ayasedky@yahoo.com

Aya Sedky Adly
Computer Science Department, Faculty of Computers and Information,
Helwan University, Cairo, Egypt
ayasedky@helwan.edu.eg; ayasedky@yahoo.com

Abdul Hamid Adom
School of Mechatronic Engineering, University Malaysia Perlis, 02600,
Pauh, Perlis, Malaysia

Ankit Agrawal
Department of Electrical Engineering and Computer Science, Northwestern
University, 2145 Sheridan Road, Evanston, IL 60208, USA
ankitag@eecs.northwestern.edu; ankit108@gmail.com

S. Hyder Ali
Research Scholar, Anna University, Chennai, Tamil Nadu, India

Lloyd Allison
National ICT Australia, Victorian Research Laboratory, University of Melbourne,
Parkville, VIC 3052, Australia

L. Altamirano
National Institute for Astrophysics, Optics, and Electronics, Luis Enrique Erro
No. 1, Puebla, 72840, Mexico
robles@inaoep.mx

Diego Ardigò
MultiMed s.r.l, Cremona, Italy
diego.ardigo@multi-med.it

Lalit K. Awasthi
Department of Computer Science and Engineering, National Institute of Technology, Hamirpur, Himachal Pradesh, India

Pamita Awasthi
Department of Chemistry, National Institute of Technology, Hamirpur, Himachal Pradesh, India
pamita@nitham.ac.in; p_awasthi@rediff.com; pamitawasthi@gmail.com

Seyed Morteza Babamir
University of Kashan, Kashan, Iran
babamir@kashanu.ac.ir

Emanuele Baj
Fondazione Politecnico di Milano, Piazza Leonardo da Vinci 32, 20133 Milan, Italy
Emanuele.Baj@fondazione.polimi.it

Katherine-Andrea Baquero-Duarte
Laboratory for Automation, Microelectronics and Computational Intelligence (LAMIC), Faculty of Engineering, Universidad Distrital Francisco Jose de Caldas, Bogotá, Colombia
kabaquerod@correo.udistrital.edu.co

Ritu Barthwal
Department of Biotechnology, Indian Institute of Technology, Roorkee, India

F. Battke
Department of Information and Cognitive Sciences, Center for Bioinformatics Tübingen, University of Tübingen, Sand 14, 72076 Tübingen, Germany

Lluís A. Belanche-Muñoz
Departament de Llenguatges i Sistemes Informàtics, Universitat Politècnica de Catalunya, Omega Building, North Campus, 08034 Barcelona, Spain
belanche@lsi.upc.edu

Nacéra Benamrane
Computer Sciences Department, USTO Oran, B.P 1505 El 'mnaouer 31000, Oran, Algeria
nabenamrane@yahoo.com

A. Benis
LIM&Bio – Laboratoire d'Informatique Médicale et de Bioinformatique – E.A.3969, Université Paris Nord, 74 rue Marcel Cachin, 93017 Bobigny, Cedex, France
benis.arriel@gmail.com

Contributors

Henri Bensahel
Service de Chirurgie Infantile, Hôpital de Robert Debré, Paris, France
henriben@noos.fr

Maurice Berk
Mathematics, Imperial College, London, UK
maurice.berk01@imperial.ac.uk

Bharat Bhasker
Indian Institute of Management, Lucknow, India
bhasker@iiml.ac.in

Mahua Bhattacharya
Indian Institute of Information Technology and Management, Morena Link Road, Gwalior 474010, India
mb@iiitm.ac.in; bmahua@hotmail.com

Elaine M. Blount
Old Dominion University, Norfolk, VA, USA

M. Bonin
Microarray Facility Tübingen, Calwer Straße 7, 72076 Tübingen, Germany

P. Boston
New Mexico Tech, Socorro, NM 87801, USA
pboston@nmt.edu

Silvia Bragagia
A.O. Ospedale Niguarda Ca'Granda, Piazza Ospedale Maggiore 3, 20162 Milan, Italy
Silvia.Bragagia@ospedaleniguarda.it

Filiz Bunyak
Department of Computer Science, University of Missouri-Columbia, Columbia, MO 65211, USA
bunyak@missouri.edu

Giovanni Camussi
Dipartimento di Medicina Interna, Molecular Biotechnology Center, Università di Torino, Turin, Italy

Gang Cao
Department of Stomatology, Jinling Hospital, Nanjing 210002, China
caogangfmmu@yahoo.com.cn

Li Cao
College of Civil Aviation, Nanjing University of Aeronautics and Astronautics, Nanjing 210016, China
caoli@nuaa.edu.cn

Minh Duc Cao
Clayton School of Information Technology, Monash University, Clayton,
VIC 3800, Australia
minhduc@monash.edu

Gloria Castellano
Departamento de Ciencias Experimentales y Matemáticas, Universidad Católica
de Valencia San Vicente Mártir, Guillem de Castro-94, 46003 València, Spain
gloria.castellano@ucv.es

Vijender Chaitankar
School of Computing, The University of Southern Mississippi, Hattiesburg,
MS 39402, USA

Cedric Chauve
Department of Mathematics, Simon Fraser University, Burnaby, BC, Canada
cedric.chauve@sfu.ca

Dongsheng Che
Department of Computer Science, East Stroudsburg University, East Stroudsburg,
PA 18301, USA
dche@po-box.esu.edu; dongshengche@gmail.com

Meifania Chen
Digital Ecosystems and Business Intelligence Institute, Curtin University
of Technology, Enterprise Unit 4, De Laeter Way, Technology Park, Bentley,
WA 6102, Australia
m.chen@cbs.curtin.edu.au

Shwu-Fen Chiou
Department of Information Management, National Taipei College of Nursing,
Taipei, Taiwan, ROC

Alok Choudhary
Department of Electrical Engineering and Computer Science, Northwestern
University, 2145 Sheridan Road, Evanston, IL 60208, USA
choudhar@eecs.northwestern.edu

Liviu Ciortuz
Department of Computer Science, "Alexandru Ioan Cuza" University of Iaşi, Iaşi,
Romania
ciortuz@info.uaic.ro

Fulvio Ciriaco
Chemistry Department, University of Bari, Bari, Italy

Federica Collino
Dipartimento di Medicina Interna, Molecular Biotechnology Center, Università
di Torino, Turin, Italy

Courtney D. Corley
Pacific Northwest National Laboratory, Richland, WA, USA
court@pnl.gov

Adrien Coulet
Department of Medicine, Stanford University, Stanford, CA, USA
and
LORIA (CNRS UMR7503, INRIA Nancy Grand-Est, Nancy Université), Campus scientifique, 54506 Vandoeuvre-lès-Nancy, France
adrien.coulet@loria.fr

M. Courtine
LIM&Bio – Laboratoire d'Informatique Médicale et de Bioinformatique, E.A.3969 Université Paris Nord, 74 rue Marcel Cachin, 93017 Bobigny Cedex, France
courtine@limbio-paris13.org

Philip Crooke
Department of Mathematics, Vanderbilt University, Nashville, TN 37240, USA

J. Curnutt
California State University, San Bernardino, San Bernardino, CA 92407, USA
jcurnutt@r2labs.org

Marcin Czajkowski
Faculty of Computer Science, Bialystok University of Technology, Bialystok, Poland
m.czajkowski@pb.edu.pl

Shareef M. Dabdoub
The Biophysics Program, The Ohio State University, Columbus, OH 43210, USA
dabdoub.2@buckeyemail.osu.edu

Tien Tuan Dao
UTC – CNRS UMR, 6600 Biomécanique et Bioingénierie, Compiègne, France
tien-tuan.dao@utc.fr

Arpita Das
Department of Radio Physics and Electronics, University of Calcutta,
92 A.P.C. Road, Kolkata-700009, India
arpita.rpe@caluniv.ac.in

Shyama Das
Department of Computer Science, Cochin University of Science and Technology, Kochin, Kerala, India
shyamadas777@gmail.com

Rajashree Dash
Department of Computer Science and Engineering, Institute of Technical Education and Research, Bhubaneswar, Orissa, India
rajashree_dash@yahoo.co.in

Kenneth L. Davis
Department of Psychiatry, Mount Sinai School of Medicine, One Gustave L. Levy Place, NY 10029, USA
kenneth.davis@mssm.edu

Ana Karoline de Castro
Graduate Program in Applied Computer Sciences, University of Fortaleza (UNIFOR), Av. Washington Soares, 1321, Bl J Sl 30, 60.811-905, Fortaleza, Brazil
akcastrog@gmail.com

Marco De Fazio
STMicroelectronics, Milano, Italy
marco.de-fazio@st.com

Youping Deng
SpecPro Inc., 3909 Halls Ferry Rd, Vicksburg, MS 39180, USA

Marie-Dominique Devignes
LORIA (CNRS UMR7503, INRIA Nancy Grand-Est, Nancy Université), Campus Scientifique, 54506 Vandoeuvre-lès-Nancy, France
devignes@loria.fr

Cecilia Di Ruberto
Department of Mathematics and Computer Science, University of Cagliari, Cagliari, Italy
dirubert@unica.it

R. Diaz
National Institute for Astrophysics, Optics, and Electronics, Luis Enrique Erro No. 1, Puebla, 72840, Mexico
raqueld@inaoep.mx

Trevor I. Dix
Victorian Bioinformatics Consortium, Clayton, VIC 3800, Australia
trevor@infotech.monash.edu.au

Shilpa Dogra
Department of Chemistry, National Institute of Technology, Hamirpur, Himachal Pradesh, India

P. Dhanapal Durai Dominic
Department of Computer and Information Science, Universiti Teknonologi PETRONAS, Bandar Seri Iskandar, 31750 Tronoh, Perak, Malaysia
dhanapal_d@petronas.com.my

Andrzej Dominik
Institute of Radioelectronics, Warsaw University of Technology, Nowowiejska 15/19, 00-665 Warsaw, Poland
a.dominik@elka.pw.edu.pl

Erin Doyle
Bioinformatics and Computational Biology Program, Iowa State University, Ames, IA 50010, USA
and
Department of Plant Pathology, Iowa State University, Ames, IA 50010, USA
edoyle@iastate.edu

Jean-Philippe Doyon
LIRMM, Université Montpellier 2 and CNRS, Montpellier, France
Jean-philippe.Doyon@lirmm.fr; doyonjea@iro.umontreal.ca

Tim Ebbels
Biomolecular Medicine, Imperial College, London, UK
t.ebbels@imperial.ac.uk

Wassim El-Hajj
Bioinformatics Lab, Department of Intelligent Systems, College of Information Technology, UAE University, 17551 Al-Ain, UAE
welhajj@uaeu.ac.ae

T.E. Ellingsen
Department of Biotechnology, SINTEF Materials and Chemistry, Sem Sælands vei 2a, 7465 Trondheim, Norway

Konstantinos Exarchos
Unit of Medical Technology and Intelligent Information Systems, Department of Materials Science and Engineering, University of Ioannina, Ioannina, Greece
and
Department of Medical Physics, Medical School, University of Ioannina, Ioannina, Greece
kexarcho@cc.uoi.gr

Abdelkader Fekir
Mathematics and Computer Science Department, Mascara University, BP 763, Mamounia Route, 29000, Mascara, Algeria
aekfekir@gmail.com; aekfekir@univ-mascara.dz

José Jesús Fernández
Centro Nacional de Biotecnologia (CSIC), Campus UAM, Cantoblanco, 28049 Madrid, Spain
JJ.Fernandez@cnb.csic.es

Dimitrios I. Fotiadis
Unit of Medical Technology and Intelligent Information Systems, Department of Materials Science and Engineering, University of Ioannina, Ioannina, Greece
fotiadis@cs.uoi.gr

Pina M. Fratamico
U.S. Department of Agriculture, Agricultural Research Service, Eastern Regional Research Center, Wyndmoor, PA 19038, USA

Estefanía Gallego
Electronic Technology Department, Computer Engineering School, Seville University, Av. Reina Mercedes s/n, 41012, Sevilla, Spain

Zhi-Long Geng
Department of Anesthesiology, Jincheng Hospital, Lanzhou 730050, China
zlgch@sina.com

Preetam Ghosh
School of Computing, The University of Southern Mississippi, Hattiesburg, MS 39402, USA

Yorgos Goletsis
Department of Economics, University of Ioannina, Ioannina, Greece
goletsis@cc.uoi.gr

E. Gomez
California State University, San Bernardino, San Bernardino, CA 92407, USA
ernestog@csusb.edu

Ping Gong
SpecPro Inc., 3909 Halls Ferry Rd, Vicksburg, MS 39180, USA

J.A. Gonzalez
National Institute for Astrophysics, Optics, and Electronics, Luis Enrique Erro No. 1, Puebla, 72840, Mexico
jagonzalez@inaoep.mx

Félix F. González-Navarro
Dept. de Llenguatges i Sistemes Informàtics, Universitat Politècnica de Catalunya, W-Building, North Campus, 08034 Barcelona, Spain
fgonzalez@lsi.upc.edu

Victor-Hugo Grisales-Palacio
Faculty of Engineering, Laboratory for Automation, Microelectronics
and
Computational Intelligence (LAMIC), Universidad Distrital Francisco Jose de Caldas, Colombia
vhgrisales@udistrital.edu.co

Mauricio Guevara-Souza
Computer Science Department, ITESM CEM, Carretera Lago de Guadalupe Km.3.5, Atizapan de Zaragoza, 52926, Mexico
A00456476@hotmail.com; guevara_mauricio@hotmail.com

J.S. Guichard
National Institute for Astrophysics, Optics, and Electronics, Luis Enrique Erro No. 1, Puebla, 72840, Mexico
jguichard@inaoep.mx

Klaus Gürlebeck
Institut Mathematik/Physik, Bauhaus-Universität Weimar, Coudraystr. 13, 99421
Weimar, Germany
klaus.guerlebeck@uni-weimar.de

Varadraj P. Gurupur
Department of Electrical and Computer Engineering, University of Alabama
at Birmingham, Birmingham, AL 35294-1150, USA
varad@uab.edu

Maja Hadzic
Digital Ecosystems and Business Intelligence Institute, Curtin University
of Technology, GPO Box U1987 Perth, Western Australia 6845, Australia
m.hadzic@curtin.edu.au

Adel Hafiane
ENSI de Bourges, Institut PRISME UPRES EA 4229, 88 boulevard Lahitolle,
18020 Bourges Cedex, France
adel.hafiane@ensi-bourges.fr

Zhong-ling Han
East China Normal University, 3663 North Zhong-Shan Rd, Shanghai 200062,
P.R. China

Vahram Haroutunian
Mental Illness Research, Education and Clinical Centers, Bronx Veterans Affairs
Medical Center, 130 West Kingsbridge Road, Bronx, NY 10468, USA
and
Department of Psychiatry, Mount Sinai School of Medicine, One Gustave L. Levy
Place, NY 10029, USA
vahram.haroutunian@mssm.edu

Mustafa Hasan
Faculty of Computing and Information Technology, Sohar University, P.O. Box 44,
P.C. 311 Sohar, Sultanate of Oman
m.hasan@soharuni.edu.om

C.R. Hema
School of Mechatronic Engineering, University Malaysia Perlis, 02600 Pauh,
Perlis, Malaysia
hema@unimap.edu.my

A. Herbig
Center for Bioinformatics Tübingen, Department of Information and Cognitive
Sciences, University of Tübingen, Sand 14, 72076 Tübingen, Germany

Doracelly Hincapié
Epidemiology Group, National School of Public Health, University of Antioquia,
Medellín, Colombia
doracely@guajiros.udea.edu.co

Marie Christine Ho Ba Tho
UTC – CNRS UMR 6600 Biomécanique et Bioingénierie, Compiègne, France
hobatho@utc.fr

D.A. Hodgson
Department of Biological Science, University of Warwick, Gibbet Hill Road, Coventry CV47AL, UK

Heike Hofmann
Bioinformatics and Computational Biology Program, Iowa State University, Ames, IA 50010, USA
and
Department of Statistics, Iowa State University, Ames, IA 50010, USA
hofmann@iastate.edu

Sanjukta Hota
Department of Mathematics, Fisk University, Nashville, TN 37208, USA
sanjuktahota@gmail.com

John Hotchkiss
Department of Medicine, University of Pittsburgh, Pittsburgh, PA 15261, USA

Gabriela Hristescu
Computer Science Department, Rowan University, 201 Mullica Hill Road, Glassboro, NJ 08028, USA
hristescu@rowan.edu

Wen-Chen Hu
Computer Science Department, University of North Dakota, ND 58202, USA

Ean-Wen Huang
Department of Information Management, National Taipei Universiy of Nursing and Health Sciences, 365, Ming-te-Road, Peitou District, Taipei, Taiwan, ROC
huang@ntunhs.edu.tw

Lihan Huang
U.S. Department of Agriculture, Agricultural Research Service, Eastern Regional Research Center, Wyndmoor, PA 19038, USA

Xiaoqiu Huang
Department of Computer Science, Iowa State University, 226 Atanasoff Hall, Ames, IA 50011, USA
xqhuang@cs.iastate.edu

Rui-Suan Hung
Department of Information Management, National Taipei College of Nursing, Taipei, Taiwan, ROC

Contributors

Aini Hussain
Department of Electrical, Electronic and Systems Engineering, Faculty of Engineering and Built Environment, Universiti Kebangsaan Malaysia, Bangi, 43600 Selangor, Malaysia

Hieu Trung Huynh
Nguyen Tat Thanh College, University of Industry, Ho Chi Minh City, Vietnam
hthieu@hcmut.edu.vn; hieuhtvn@yahoo.com

M. Shaarawy Ibrahim
Computer Science Department, Faculty of Computers and Information, Helwan University, Cairo, Egypt
mhmshaarawy@helwan.edu.eg

Sumam Mary Idicula
Department of Computer Science, Cochin University of Science and Technology, Kochin, Kerala, India
sumam@cusat.ac.in

Ø.M. Jakobsen
Department of Biotechnology, SINTEF Materials and Chemistry, Sem Sælands vei 2a, 7465 Trondheim, Norway

Xinhua Ji
Institute of Mathematics, Academy of Mathematics and Systems Science (AMSS), Chinese Academy of Sciences, Beijing 100190, China
xhji@math.ac.cn

Vijay Juneja
U.S. Department of Agriculture, Agricultural Research Service, Eastern Regional Research Center, Wyndmoor, PA 19038, USA

Sheryl S. Justice
The Center for Microbial Pathogenesis, The Research Institute at Nationwide Children's Hospital, Columbus, OH 43205, USA
sheryl.justice@nationwidechildrens.org

Naima Kaabouch
Electrical Engineering Department, University of North Dakota, ND 58202, USA
naimakaabouch@mail.und.edu

Amit S. Kamdi
UAB School of Public Health, University of Alabama at Birmingham, Birmingham, AL 35294-1150, USA
dramit99@gmail.com

Hesham M. Kamel
Bioinformatics Lab, Department of Intelligent Systems, College of Information Technology, UAE University, 17551 Al-Ain, UAE
hesham@uaeu.ac.ae

Serajeddin Katebi
Department of Computer Science and Engineering, School of Engineering, Shiraz University, Shiraz, Iran

Pavel Katsel
Mental Illness Research, Education and Clinical Centers, Bronx Veterans Affairs Medical Center, 130 West Kingsbridge Road, Bronx, NY 10468, USA
and
Department of Psychiatry, Mount Sinai School of Medicine, One Gustave L. Levy Place, NY 10029, USA
pavel.katsel@mssm.edu

Jung-Ja Kim
Chonbuk National University, Jeonju, Jeollabuk-do, Korea
jungjakim@jbnu.ac.kr

Lars Koenig
Department of Computer Science, Texas Tech University, Lubbock, TX 79409, USA
lars.koenig@ttu.edu

Peter Kokol
Faculty of Health Sciences, University of Maribor, Zitna ulica 15, 2000 Maribor, Slovenia
and
Faculty of Electrical Engineering and Computer Science, University of Maribor, Smetanova 17, 2000 Maribor, Slovenia

Paško Konjevoda
NMR Center, Ruđer Bošković Institute, Bijenička cesta 54, 10002 Zagreb, Croatia
pkonjev@irb.hr

Marek Krętowski
Faculty of Computer Science, Bialystok University of Technology, Bialystok, Poland
m.kretowski@pb.edu.pl

Robert Krupiński
Department of Signal Processing and Multimedia Engineering, West Pomeranian University of Technology in Szczecin, 26-Kwietnia 10 Str., 71-126 Szczecin, Poland
robert.krupinski@zut.edu.pl

Cristina Juarez Landin
Autonomous University of Mexico State, Hermenegildo Galena No.3, Col. Ma. Isabel, Valle de Chalco, Mexico State, Mexico
cjlandin@yahoo.com.mx

Der-Ming Liou
Institute of Biomedical Informatics, National Yang-Ming University Taipei, Taiwan, ROC

Fei-Ying Liu
Division of Mathematics, Daan Junior High School, Taipei, Taiwan, ROC

Qi Liu
College of Life Science and Biotechnology, Tongji University, Shanghai 200092, P.R. China
emailliuqizju@gmail.com

D.K. Lobiyal
Jawaharlal Nehru University, New Delhi, India
dkl@mail.jnu.ac.in

Paolo Locatelli
Fondazione Politecnico di Milano, Piazza Leonardo da Vinci 32, 20133 Milan, Italy
Paolo.Locatelli@fondazione.polimi.it

Claudio Loconsole
Perceptual Robotics Laboratory, Scuola Superiore Sant'Anna, Pisa, Italy

Ling Ma
School of Information and Communication Engineering, Harbin Engineering University, Harbin 150001, China

Anjali Mago
Mago Dental Clinic, Jalandhar, India

Jagmohan Mago
Apeejay College of Fine Arts, Jalandhar, India

Vijay Kumar Mago
DAV College, Jalandhar, India
vijay.mago@gmail.com

Frédéric Marin
UTC – CNRS UMR 6600 Biomécanique et Bioingénierie, Compiègne, France
frederic.marin@utc.fr

Anabelem Soberanes Martin
Autonomous University of Mexico State, Hermenegildo Galena No.3, Col. Ma. Isabel, Valle de Chalco, Mexico State, Mexico
belemsoberanes@yahoo.com.mx

Elena Martinelli
Azienda Ospedaliero-Universitaria di Parma – UO Chirurgia Maxillo-facciale, Parma, Italy
coordinator@neomark.eu

José Antonio Martínez
Department of Computer Architecture, University of Almeria, 04120 Almeria, Spain

Oteh Maskon
Cardiology Care Unit, Universiti Kebangsaan Malaysia Medical Center,
Kuala Lumpur 56000, Malaysia

Fabio Mavelli
Chemistry Department, University of Bari, Bari, Italy
mavelli@chimica.uniba.it

Przemysław Mazurek
Department of Signal Processing and Multimedia Engineering, West Pomeranian
University of Technology in Szczecin, 26-Kwietnia 10 St., 71-126 Szczecin, Poland
przemyslaw.mazurek@zut.edu.pl

Jianghong Meng
Joint Institute for Food Safety and Applied Nutrition (JIFSAN), University
of Maryland, College Park, MD 20742, USA

Rada Mihalcea
Department of Computer Science and Engineering, University of North Texas,
Denton, TX, USA
rada@cs.unt.edu

Armin R. Mikler
Department of Computer Science and Engineering, University of North Texas,
Denton, TX, USA
mikler@unt.edu

A.K. Mishra
Jawaharlal Nehru University, New Delhi, India
and
U.S.I., Indian Agricultural Research Institute, New Delhi, India
akmishra@iari.res.in; misamr@rediffmail.com

Bhartendu Nath Mishra
Department of Biotechnology, Institute of Engineering and Technology,
UP Technical University, Lucknow, India
profbnmishra@gmail.com

Debahuti Mishra
Department of Computer Science and Engineering, Institute of Technical Education
and Research, Siksha O Anusandhan University, Bhubaneswar, Orissa, India
debahuti@iter.ac.in

Irina Mohorianu
Department of Computer Science, "Alexandru Ioan Cuza" University of Iaşi, Iaşi,
Romania
irina.mohorianu@info.uaic.ro

Javier Molina
Electronic Technology Department, Computer Engineering School, Seville
University, Av. Reina Mercedes s/n, 41012, Sevilla, Spain

Giovanni Montana
Mathematics, Imperial College, London, UK
giovanni.montana@imperial.ac.uk

Ehsan Moradi
Department of Neurosurgery, Shiraz Medical School, Shiraz University of Medical Sciences, Shiraz, Iran

Andrea Morgera
Department of Mathematics and Computer Science, University of Cagliari, Cagliari, Italy
andrea.morgera@unica.it

Mohd. Marzuki Mustafa
Department of Electrical, Electronic and Systems Engineering, Faculty of Engineering and Built Environment, Universiti Kebangsaan Malaysia, Bangi, 43600 Selangor, Malaysia
marzuki@eng.ukm.my

R. Nagarajan
School of Mechatronic Engineering, University Malaysia Perlis, 02600, Pauh, Perlis, Malaysia

Amedeo Napoli
LORIA (CNRS UMR7503, INRIA Nancy Grand-Est, Nancy Université), Campus Scientifique, 54506 Vandoeuvre-lès-Nancy, France
napoli@loria.fr

David S. Needleman
U.S. Department of Agriculture, Agricultural Research Service, Eastern Regional Research Center, Wyndmoor, PA 19038, USA

K. Nieselt
Faculty of Science, Center for Bioinformatics Tübingen, Universiy of Tübingen, Sand 14, 72076 Tübingen, Germany
kay.nieselt@uni-tuebingen.de

Ika Faizura Mohd. Nor
Cardiology Care Unit, Universiti Kebangsaan Malaysia Medical Center, Kuala Lumpur 56000, Malaysia

Luciano Comin Nunes
Graduate Program in Applied Computer Sciences, University of Fortaleza, Av. Washington Soares, 1321 – Bl J, Sl 30, CEP: 60811-905, Fortaleza, Ceará, Brazil
lcominn@bnb.gov.br; lcominn@uol.com.br

Wail M. Omar
Faculty of Computing and Information Technology, Sohar University, P.O. Box 44, P.C. 311, Sohar, Sultanate of Oman
w.omar@soharuni.edu.om

Gianni Origgi
A.O. Ospedale Niguarda Ca'Granda, Piazza Ospedale Maggiore 3, 20162 Milan, Italy
Gianni.Origgi@ospedaleniguarda.it

Alvaro-Angel Orozco-Gutierrez
Faculty of Engineering, Research Group on Control and Instrumentation, Technological University of Pereira, Pereira, Colombia
aaog@utp.edu.co

Juan Ospina
Logic and Computation Group, Physical Engineering Program, School of Sciences and Humanities, EAFIT University Medellín, Colombia
jospina@eafit.edu.co

Kannappan Palaniappan
Department of Computer Science, University of Missouri-Columbia, Columbia, MO 65211, USA
palaniappank@missouri.edu

Ștefan Panțiru
Department of Computer Science, "Alexandru Ioan Cuza" University of Iași, Iași, Romania
stefan.pantiru@info.uaic.ro

Daniel Pasailă
Department of Computer Science, "Alexandru Ioan Cuza" University of Iași, Iași, Romania
daniel.pasaila@info.uaic.ro

M.P. Paulraj
School of Mechatronic Engineering, University Malaysia Perlis, 02600, Pauh, Perlis, Malaysia

Yun Peng
Department of Computer Science and Electrical Engineering, University of Maryland, Baltimore, MD 21250, USA

Tarcísio Cavalcante Pequeno
Graduate Program in Applied Computer Sciences, University of Fortaleza, Av. Washington Soares, 1321 – Bl J, Sl 30, CEP: 60811-905, Fortaleza, Ceará, Brazil
tpequeno@unifor.br

Edward J. Perkins
Environmental Laboratory, U.S. Army Engineer Research and Development Center, 3909 Halls Ferry Rd, Vicksburg, MS 39180, USA

Contributors

Adhistya Erna Permanasari
Department of Computer and Information Science, Universiti Teknonologi PETRONAS, Bandar Seri Iskandar, 31750 Tronoh, Perak, Malaysia
and
Electrical Engineering Department, Gadjah Mada University, Jl. Grafika No. 2, Yogyakarta 55281, Indonesia
astya_00@yahoo.com

Dimitri Perrin
Centre for Scientific Computing and Complex Systems Modelling, Dublin City University, Glasnevin, Dublin 9, Ireland
dperrin@computing.dcu.ie

Marco Picone
MultiMed s.r.l, Cremona, Italy
marco.picone@multi-med.it

Gabriel Piedrafita
Department of Biochemistry and Molecular Biology, University Complutense of Madrid, Madrid, Spain

Mirian Calíope Dantas Pinheiro
Graduate Program in Applied Computer Sciences, University of Fortaleza, Av. Washington Soares, 1321 – Bl J, Sl 30, CEP: 60811-905, Fortaleza, Ceará, Brazil
caliope@unifor.br

Plácido Rogério Pinheiro
Graduate Program in Applied Computer Sciences, University of Fortaleza, Av. Washington Soares, 1321 – Bl J, Sl 30, CEP: 60811-905, Fortaleza, Ceará, Brazil
placido@unifor.br

Ruben-Dario Pinzon-Morales
Faculty of Engineering, Research Group on Control and Instrumentation, Technological University of Pereira, Pereira, Colombia
rdpinzonm@utp.edu.co

Volodymyr Ponomaryov
Mechanical and Electrical Engineering Higher School, National Polytechnic Institute of Mexico, Santana Avenue # 1000, San Francisco Culhuacan, Mexico D.F.

M. Quintana
California State University, San Bernardino, CA 92407, USA
quintanm@csusb.edu

Aurelian Radu
Department of Developmental and Regenerative Biology, Mount Sinai School of Medicine, One Gustave L. Levy Place, New York, NY 10029, USA
aurelian.radu@mssm.edu

Dayang Rohaya Awang Rambli
Department of Computer and Information Science, Universiti Teknonologi PETRONAS, Bandar Seri Iskandar, 31750 Tronoh, Perak, Malaysia
roharam@petronas.com.my

Jose Luis Sanchez Ramirez
Mechanical and Electrical Engineering Higher School, National Polytechnic Institute of Mexico, Santana Avenue # 1000, San Francisco Culhuacan, Mexico D.F.
jluissar@yahoo.com.mx

Khaled Rasheed
Department of Computer Science, The University of Georgia, Athens, GA 30602, USA
khaled@cs.uga.edu

Amiya Kumar Rath
Department of Computer Science and Engineering, College of Engineering Bhubaneswar, Orissa, India
amiyaamiya@rediffmail.com

William C. Ray
The Battelle Center for Mathematical Medicine, The Research Institute at Nationwide Children's Hospital, Columbus, OH 43205, USA
ray.29@osu.edu

D.V. Rama Koti Reddy
Instrumentation Engineering, Andhra University, Visakhapatnam-530 003, India

Nicola Restifo
Fondazione Politecnico di Milano, via Durando 38/a, 20158 Milan, Italy
Nicola.Restifo@fondazione.polimi.it; restifo@fondazionepolitecnico.it

C. Reta
National Institute for Astrophysics, Optics, and Electronics, Luis Enrique Erro No. 1, Puebla, 72840, Mexico
creta@inaoep.mx; creta@ccc.inaoep.mx

Magally Martinez Reyes
Autonomous University of Mexico State, Hermenegildo Galena No.3, Col. Ma. Isabel, Valle de Chalco, Mexico State, Mexico
mmreyes@hotmail.com

Stacie I. Ringleb
Old Dominion University, Norfolk, VA, USA
sringleb@odu.edu

Slamet Riyadi
Department of Electrical, Electronic and Systems Engineering, Faculty of Engineering and Built Environment, Universiti Kebangsaan Malaysia, Bangi, 43600 Selangor, Malaysia
riyadi@vlsi.eng.ukm.my

G.A. Rodriguez
National Institute for Astrophysics, Optics, and Electronics, Luis Enrique Erro No. 1, Puebla, 72840, Mexico
g_rodriguez@inaoep.mx

Juan J. Rodriguez
University of Burgos, c/ Francisco de Vitoria s/n, 09006 Burgos, Spain

Rosa Maria Valdovinos Rosas
Autonomous University of Mexico State, Hermenegildo Galena No.3, Col. Ma. Isabel, Valle de Chalco, Mexico State, Mexico

Kepa Ruiz-Mirazo
Department of Logic and Philosophy of Science and Biophysics Unit (CSIC-UPV/EHU), University of the Basque Country, Donostia-San Sebastian and Bilbao, Spain
kepa.ruiz-mirazo@ehu.es

Juliana Ruzante
Joint Institute for Food Safety and Applied Nutrition (JIFSAN), University of Maryland, College Park, MD 20742, USA

Malihe Sabeti
Department of Computer Science and Engineering, School of Engineering, Shiraz University, Shiraz, Iran
sabeti@shirazu.ac.ir

Dinesh K. Saini
Faculty of Computing and Information Technology, Sohar University, P.O. Box 44, P.C. 311 Sohar, Sultanate of Oman
dinesh@soharuni.edu.om

Antonio P. Sanfilippo
Pacific Northwest National Laboratory, Richland, WA, USA
antonio.sanfilippo@pnl.gov

K.E. Schubert
California State University, San Bernardino, CA 92407, USA
schubert@csusb.edu

Rafi Ahamed Shaik
E.C.E., Indian Institute of Technology, Guwahathi-781039, India
rafiahamed@iitg.ernet.in

Poonam Sharma
DAV College, Jalandhar, India

Fadi Sibai
Bioinformatics Lab, Department of Intelligent Systems, College of Information Technology, UAE University, 17551 Al-Ain, UAE
fadi.sibai@uaeu.ac.ae

Riyanto Sigit
Department of Electrical, Electronic and Systems Engineering, Faculty of Engineering and Built Environment, Universiti Kebangsaan Malaysia, Bangi, Malaysia
riyanto@eepis-its.edu

Malika Smaïl-Tabbone
LORIA (CNRS UMR7503, INRIA Nancy Grand-Est, Nancy Université), Campus Scientifique, 54506 Vandoeuvre-lès-Nancy, France
malika@loria.fr

Sudhir Singh Soam
Department of Computer Science and Engineering, Institute of Engineering and Technology, A Constituent College of UP Technical University, Lucknow, India
sssoam@gmail.com

Maria Dolores Torres Soto
Autonomous University of Aguascalientes, University Avenue # 940, University City, Aguascalientes, Mexico
mdtorres@correo.uaa.mx

Nikola Štambuk
NMR Center, Ruđer Bošković Institute, Bijenička cesta 54, HR-10002 Zagreb, Croatia
stambuk@irb.hr

Sebastian Steger
Fraunhofer IGD, Darmstadt, Germany
sebastian.steger@igd.fraunhofer.de

Luca Sterpone
Dipartimento di Automatica e Informatica, Politecnico di Torino, Turin, Italy
luca.sterpone@polito.it

Gregor Stiglic
Faculty of Health Sciences, University of Maribor, Zitna ulica 15, 2000 Maribor, Slovenia
and
Faculty of Electrical Engineering and Computer Science, University of Maribor, Smetanova 17, 2000 Maribor, Slovenia
gregor.stiglic@uni-mb.si

B. Strader
California State University, San Bernardino, CA 92407, USA
bstrader@csusb.edu

STREAM Consortium
https://www.wsbc.warwick.ac.uk/groups/sysmopublic/

Andrei Sucilă
Department of Computer Science, "Alexandru Ioan Cuza" University of Iaşi, Iaşi, Romania
andrei.sucila@info.uaic.ro

R. Sukanesh
Department of ECE, Thiyagarajar College of Engineering, Madurai, Tamil Nadu, India

Isabelle Tamanini
Graduate Program in Applied Computer Sciences, University of Fortaleza (UNIFOR), Av. Washington Soares, 1321, Bl J Sl 30, 60.811-905, Fortaleza, Brazil
isabelle.tamanini@gmail.com

Chunming Tang
School of Information and Communication Engineering, Harbin Engineering University, Harbin 150001, China
tangchunming@hrbeu.edu.cn; tangchunminga@hotmail.com

Murat M. Tanik
Department of Electrical and Computer Engineering, University of Alabama at Birmingham, Birmingham, AL 35294-1150, USA
mtanik@uab.edu

Murat N. Tanju
UAB School of Business, University of Alabama at Birmingham, Birmingham, AL 35294-1150, USA
mtanju@uab.edu

Xiuping Tao
Department of Chemistry, Winston-Salem State University, Winston-Salem, NC 27110, USA
taoxi@wssu.edu

Zhi-fu Tao
East China Normal University, 3663 North Zhong-Shan Rd, Shanghai 200062, P.R. China

Andreas Tolk
Old Dominion University, Norfolk, VA, USA

Francisco Torrens
Institut Universitari de Ciència Molecular, Universitat de València, Edifici d'Instituts de Paterna, P.O. Box 22085, 46071 València, Spain
torrens@uv.es; francisco.torrens@uv.es

Quoc-Nam Tran
Department of Computer Science, Lamar University, Beaumont, TX 77710, USA
qntran@lamar.edu; qntran@buchberger.cs.lamar.edu

Tolga Tuncer
Department of Hematology/Oncology, University of Alabama at Birmingham, Birmingham, AL 35294-1150, USA
ttuncer@azcc.arizona.edu

Edgar E. Vallejo
Computer Science Department, ITESM CEM, Carretera Lago de Guadalupe Km.3.5, Atizapan de Zaragoza, 52926, Mexico
vallejo@itesm.mx

Marie Vendettuoli
Bioinformatics and Computational Biology Program, Iowa State University, Ames, IA 50010, USA
and
Department of Statistics, Iowa State University, Ames, IA 50010, USA
mariev@iastate.edu; mariecv26@gmail.com

Zbigniew Walczak
Institute of Radioelectronics, Warsaw University of Technology, Nowowiejska 15/19, 00-665 Warsaw, Poland
Z.Walczak@elka.pw.edu.pl

Anlei Wang
Electrical Engineering Department, University of North Dakota, ND 58202, USA

Robert A. Warner
Laboratory for Logic and Experimental Philosophy, Simon Fraser University, Burnaby, BC, Canada
hillwarner@frontier.com

A. Wentzel
Department of Biotechnology, SINTEF Materials and Chemistry, Sem Sælands vei 2a, 7465Trondheim, Norway
and
Department of Biotechnology, Norwegian University of Science and Technology (NTNU), Sem Sælands vei 6-8, 7491 Trondheim, Norway

W. Wohlleben
Department of Microbiology/Biotechnology, University of Tübingen, Auf der Morgenstelle 28, 72076 Tübingen, Germany

Jacek Wojciechowski
Institute of Radioelectronics, Warsaw University of Technology, Nowowiejska 15/19, 00-665 Warsaw, Poland
J.Wojciechowski@elka.pw.edu.pl

Yonggwan Won
Chonnam National University, Gwangju, Chonnam, Korea
ykwon@chonnam.ac.kr

Dongbin Xu
School of Information and Communication Engineering, Harbin Engineering University, Harbin 150001, China

Sazali Yaacob
School of Mechatronic Engineering, University Malaysia Perlis, 02600, Pauh, Perlis, Malaysia

Xianghe Yan
U.S. Department of Agriculture, Agricultural Research Service, Eastern Regional Research Center, Wyndmoor, PA 19038, USA
xianghe.yan@ars.usda.gov

Jin-sheng Yang
Department of Neurology, Jincheng Hospital, Lanzhou 730050, China
yangjinsh@163.com

Meng Yao
East China Normal University, 3663 North Zhong-Shan Rd, Shanghai 200062, P.R. China
myao@ee.ecnu.edu.cn

Eunseog Youn
Department of Computer Science, Texas Tech University, Lubbock, TX 79409, USA
eun.youn@ttu.edu

Alberto Yúfera
Electronic Technology Department, Computer Engineering School, Seville University, Av. Reina Mercedes s/n, 41012, Spain
yufera@us.es; yufera@imse-cnm.csic.es

Nazar Zaki
Bioinformatics Laboratory, Department of Intelligent Systems, College of Information Technology, UAE University, 17551 Al-Ain, UAE
nzaki@uaeu.ac.ae

Chaoyang Zhang
School of Computing, The University of Southern Mississippi, MS 39402, USA
chaoyang.zhang@usm.edu

Mohammad Zia Ur Rahman
Instrumentation Engineering, Andhra University, Visakhapatnam-530003, India
mdzr55@gmail.com; mdzr_5@yahoo.com

Preface: Software Tools and Algorithms for Biological Systems

It gives us great pleasure to introduce the book entitled "Software Tools and Algorithms for Biological Systems." This book is composed of a collection of papers received in response to an announcement that was widely distributed to academicians and practitioners in the broad area of computational biology and software tools. Also, selected authors of accepted papers of BIOCOMP'09 proceedings (International Conference on Bioinformatics and Computational Biology: July 13–16, 2009; Las Vegas, NV, USA) were invited to submit the extended versions of their papers for evaluation.

The collection of papers in Part 1 presents computational methods for microarray, gene expression analysis, and gene regulatory networks. Bioinformatics databases, data mining, and pattern discovery techniques are presented in Part 2, followed by Part 3 that is composed of a collection of papers in the areas of protein classification and structure prediction, and computational structural biology. Part 4 presents a set of chapters that discuss comparative sequence, genome analysis, genome assembly, and genome scale computational methods. Experimental medicine and analysis tools are presented in Part 5. The collection of papers in Part 6 presents computational methods for filtering, noise cancellation, and signal and image processing. Important topics in the area of computer-based medical systems appear in Part 7. Finally, Part 8 of the book presents various software packages and novel methods targeted at problems in bioinformatics.

We are very grateful to the many colleagues who helped in making this project possible. In particular, we would like to thank the members of the BIOCOMP'09 (http://www.world-academy-of-science.org/) Program Committee who provided their editorial services for this book project. The BIOCOMP'09 Program Committee members were:

- Dr Niloofar Arshadi, University of Toronto, ON, Canada
- Prof. Ruzena Bajcsy, Member, National Academy of Engineering; IEEE Fellow; ACM Fellow; University of California, Berkeley, CA, USA
- Prof. Alessandro Brawerman, CEO, Globaltalk, Brazil and Positivo University, Brazil
- Dr Dongsheng Che, East Stroudsburg University, PA, USA
- Dr Chuming Chen, Delaware Biotechnology Institute, University of Delaware, DE, USA

- Prof. Victor A. Clincy, MSACS Director, Kennesaw State University, Kennesaw, GA, USA
- Prof. Kevin Daimi, Director CS Programs, University of Detroit Mercy, MI, USA
- Dr Youping Deng (Vice Chair), University of Southern Mississippi, MS, USA
- Prof. Madjid Fathi, Institute and Center Director, University of Siegen, Germany
- Dr George A. Gravvanis, Democritus University of Thrace, Greece
- Dr Debraj GuhaThakurta, Merck & Co. (Rosetta Inpharmatics), Seattle, WA, USA
- Prof. Ray Hashemi, Yamacraw Professor of CS & Coordinator of Graduate Program, Armstrong Atlantic State University, GA, USA
- Dr Haibo He, Graduate Program Director, Stevens Institute of Technology, NJ, USA
- Dr Seddik Khemaissia, Riyadh College of Technology, Riyadh, Saudi Arabia
- Dr Whe Dar Lin, The Overseas Chinese Institute of Technology, Taiwan, R.O.C.
- Prof. Chien-Tsai Liu, Taipei Medical University, Taipei, Taiwan
- Dr Weidong Mao, Virginia State University, VA, USA
- Sara Moein, Multimedia University, Cyberjaya, Malaysia
- Prof. Frederick I. Moxley, Director of Research for Network Science, United States Military Academy, West Point, NY, USA
- Prof. Juan J. Nieto, Director of the Institute of Mathematics, University of Santiago de Compostela, Spain
- Dr Mehdi Pirooznia, The Johns Hopkins University, School of Medicine, MD, USA
- Dr Jianhua Ruan, University of Texas at San Antonio, TX, USA
- Prof. Abdel-Badeeh M. Salem, Head of Medical Informatics & Knowledge Engineering Research Unit, Ain Shams University, Cairo, Egypt
- Avinash Shankaranarayanan, Ritsumeikan Asia Pacific University (APU), Japan and Institute for Applied Materials Flow Management (Ifas), Trier University, Birkenfeld, Germany
- Ashu M. G. Solo (Publicity Chair), Principal/R&D Eng., Maverick Technologies America Inc.
- Dr Quoc-Nam Tran, Lamar University, TX, USA
- Prof. Alicia Troncoso, Head of CS Dept., Pablo de Olavide University, Seville, Spain
- Dr Mary Qu Yang (Vice Chair), National Institutes of Health, MD, USA
- Prof. Lotfi A. Zadeh, Member, National Academy of Engineering; IEEE Fellow, ACM Fellow, AAAS Fellow, AAAI Fellow, IFSA Fellow; Director, BISC; University of California, Berkeley, CA, USA
- Prof. Yanqing Zhang, Georgia State University, Atlanta, GA, USA
- Dr Leming Zhou, University of Pittsburgh, Pittsburgh, PA, USA
- Members of WORLDCOMP Task Force for Artificial Intelligence
- Members of WORLDCOMP Task Force for Computational Biology
- Members of WORLDCOMP Task Force for High-Performance Computing
- Members of WORLDCOMP Task Force for Pattern Recognition

Last but not least, we would like to thank the editorial staff at Springer; in particular, we are indebted to Ms Melanie Wilichinsky (the editorial manager) who provided excellent professional service to us to make this project feasible.

Hamid R. Arabnia, PhD
Professor, Computer Science,
Fellow, Int'l Society of Intelligent Biological Medicine (ISIBM), Editor-in-Chief,
The Journal of Supercomputing (Springer),
Co-Editor/Board, Journal of Computational Science (Elsevier),
Member, Advisory Board, IEEE Technical Committee on Scalable Computing (TCSC),
The University of Georgia,
Department of Computer Science,
Athens, GA 30602-7404, USA.

and

Quoc-Nam Tran, PhD
Associate Professor, Computer Science,
Vice-Chair, BIOCOMP 2010,
Lamar University,
Department of Computer Science,
Beaumont, TX 77710, USA.

Part I
Computational Methods for Microarray, Gene Expression Analysis, and Gene Regulatory Networks

Chapter 1
A Technical Platform for Generating Reproducible Expression Data from *Streptomyces coelicolor* Batch Cultivations

F. Battke, A. Herbig, A. Wentzel, Ø.M. Jakobsen, M. Bonin, D.A. Hodgson, W. Wohlleben, T.E. Ellingsen, the STREAM Consortium, and K. Nieselt

Abstract *Streptomyces coelicolor*, the model species of the genus *Streptomyces*, presents a complex life cycle of successive morphological and biochemical changes involving the formation of substrate and aerial mycelium, sporulation and the production of antibiotics. The switch from primary to secondary metabolism can be triggered by nutrient starvation and is of particular interest as some of the secondary metabolites produced by related *Streptomycetes* are commercially relevant. To understand these events on a molecular basis, a reliable technical platform encompassing reproducible fermentation as well as generation of coherent transcriptomic data is required. Here, we investigate the technical basis of a previous study as reported by Nieselt et al. (BMC Genomics 11:10, 2010) in more detail, based on the same samples and focusing on the validation of the custom-designed microarray as well as on the reproducibility of the data generated from biological replicates. We show that the protocols developed result in highly coherent transcriptomic measurements. Furthermore, we use the data to predict chromosomal gene clusters, extending previously known clusters as well as predicting interesting new clusters with consistent functional annotations.

Keywords Batch fermentation · Chromosomal gene clusters · Microarray design · *Streptomyces coelicolor*

1 Introduction

The filamentous soil bacterium *Streptomyces coelicolor* is the model organism for the *Streptomyces* genus. It presents strikingly complex morphological differentiation and biochemical changes during its life cycle, involving the formation of

---

K. Nieselt (✉)
Faculty of Science, Center for Bioinformatics Tübingen,
University of Tübingen, Sand 14, 72076 Tübingen, Germany
e-mail: kay.nieselt@uni-tuebingen.de

substrate and aerial mycelium, sporulation, and the production of antibiotics, some of which are coloured. The switch from primary metabolism (growth phase) to secondary metabolism (production phase) can be triggered by nutrient starvation, as in most microorganisms, and is of particular interest in the genus *Streptomyces*, as some of the secondary metabolites (antibiotics) are commercially relevant.

In a previous study, transcriptomic time-series data of unprecedented resolution were used to study the metabolic switch of *S. coelicolor* and precisely profile expression changes and allocate them to specific points of time during growth [8]. The foundation for this work was the achievement to grow *S. coelicolor* under highly reproducible conditions in submerged batch fermentations as well as the development of a new custom-designed high-density oligonucleotide DNA microarray.

Using basically the same samples and dataset as in [8], we investigate here this technical basis in more detail focusing on the validation of the array design and on the reproducibility of the data generated from biological replicates. We show that the protocols developed result in highly coherent transcriptomic measurements. Furthermore, we use the same array data of this deepest resolved transcriptomic data ever produced for *S. coelicolor* to algorithmically predict chromosomal gene clusters. We are able to extend previously known clusters as well as to predict interesting new clusters with consistent functional annotations.

2 Materials and Methods

The results presented in this work and in [8] are based on the same set of data. As a consequence, the materials and methods described here are partly identical to those described before. We include them as supplementary material since for the description of the technological platform detailed method descriptions are of particular interest.

2.1 *Array Design and Samples*

An Affymetrix GeneChip® CustomExpress™ was designed based on the RefSeq releases of the genome and the two plasmids of *S. coelicolor* (suppl. Table 1). Altogether 226,576 perfect match probes interrogating 8,205 coding sequences, 10,834 intergenic regions and 3,671 predicted ncRNAs were designed. For details, see the supplementary material. Array data were produced from three biological replicates: 32 samples (hourly from 20 to 44 h, every 2 h from 46 to 60 h after inoculation) of fermenter F199 were processed as well as eight samples of fermenters F201 and F202 (24, 28, 32, 36, 40, 44, 48 and 60 h).

2.2 Data Pre-processing

We used R [9] for data pre-processing and array quality analyses. The robust multi-array average (RMA) method [2] as provided in the `affy` package [5] was used for normalization. We added annotations from StrepDB, SCOcyc [1], TIGR (now JCVI), UniProt [14], Pfam [4], as well as Sanger Protein Classifications [11].

2.3 Reproducibility

As a statistical approach for the inter-fermenter comparison, the mean correlation of the gene expression profiles was calculated for genes within known chromosomal clusters. We defined the mean pairwise between fermenter correlation fcor(C) of a complete chromosomal cluster C as:

$$\text{fcor}(C) = \sum_{g \in C} \text{fcor}(g)/|C| \quad \text{with} \quad \text{fcor}(g) = \sum_{i \in F} \sum_{j \in F, j \neq i} \frac{2 \cdot \text{cor}(f_i(g), f_j(g))}{(|F|^2 - |F|)} \quad (1)$$

where $f_i(g)$ is the expression profile of gene g in fermenter i and F is the set of all fermenters, i.e. here $F = \{F199, F201, F202\}$.

2.4 Clustering of Variant Genes

From the normalized data of fermenter F199, we computed the regularized variance value, $v_r(\mathbf{g})$, for each gene $\mathbf{g}$, defined as the variance divided by the minimum of the gene's expression values. The set of variant genes was determined using only genes with a regularized variance $v_r(\mathbf{g}) > 0.1$. The variant genes were clustered using the QT clustering method [6], a deterministic method ensuring that all clusters fulfill predefined quality criteria. The basis for all clusterings were pairwise Pearson Correlation distances computed as $1 - \text{cor}(\mathbf{g}_i, \mathbf{g}_j)$. The diameter for QT was set to 0.25, and the minimum cluster size was four. All analyses were performed using MAYDAY [3].

2.5 Prediction of Chromosomal Gene Clusters

We developed an unsupervised clustering method in R that predicts clusters of genomically consecutive genes showing highly correlated expression profiles. Neighbouring genes belong to the same chromosomal cluster if the correlation of their expression profiles exceeds 0.6 and their variance exceeds 0.01. All protein coding genes are processed in the order of their genomic positions. The first two neighbouring genes that fulfill the criteria form a preliminary cluster which is extended until a processed gene does not fulfill the criteria. During a second step, consecutive clusters are joined if they are separated by less than six genes and if they show a between cluster correlation exceeding 0.55 defined as the mean pairwise correlation (*mpc*) of expression profiles. In a quality assessment step, clusters with an *mpc* below 0.55 are discarded. All thresholds were empirically determined using known chromosomal clusters as training set. During a fourth step, gaps in the predicted clusters are filled. The analysis of known chromosomal clusters revealed that in many clusters some genes do not show a high correlation with the others. Therefore, we decided to fill the gaps with the genes they contain. This results in improved sensitivity when applied to known clusters without losing specificity.

3 Results and Discussion

3.1 Array Design Validation

Our array design was validated using both in silico as well as *post-hybridization* methods. In silico tests showed a median GC content of 0.64 (95% between 0.52 and 0.80) for probes in genes and non-coding RNA, and 0.68 (95% between 0.52 and 0.84) for intergenic probes.

All probes were checked for cross-hybridization partners in the whole *S. coelicolor* genome. About 99% of the 8,205 probe sets in coding regions contain 13 probes. Of these 8,205 probe sets, 8,060 (98.2%) consist only of unique probes, and another 30 have one non-unique probe. Eleven probe sets have two to four non-unique probes, and 104 probe sets have five or more non-unique probes. The latter probe sets cover tRNA and rRNA genes (39), duplicated genes (37), insertion element transposases (12), plasmid genes (2) and hypothetical proteins (12). Not all of the intergenic region tiling probes could be designed with unique sequences due to the repetitive nature of the intergenic regions. However, of the 64,914 probes interrogating intergenic regions, 61,915 (95.4%) are unique.

After normalization and log transformation, expression value ranges, medians and variances were computed for five groups of probe sets (coding regions, intergenic regions, ncRNAs, negative controls and spike-in controls, suppl. Table 2). The results for the spike-in controls show a strong linear dependency of log spike-in concentrations and expression values ($R^2 = 0.997$, $p < 0.01$).

3.2 Fermentations

The organism was grown in a defined medium based on glucose and glutamate as carbon and glutamate as the sole nitrogen sources (see supplementary material). The initial phosphate concentration was adjusted so that phosphate became the

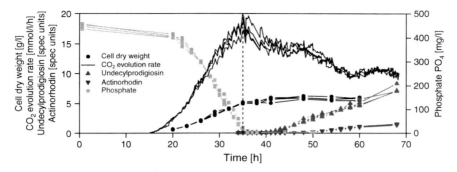

Fig. 1 Online and offline measurements for fermenters F199, F200, F201 and F202. For each measured parameter, results of all four replicates have been overlaid. Depletion of the phosphate in the growth medium of fermenter F199 is indicated by the *dotted vertical line* at 35 h

growth-limiting factor during cultivation, and thus the trigger responsible for the culture's transition from growth to antibiotic production phase. Phosphate depletion occurred at 35–36 h after inoculation for the four biological replicates, at an average biomass concentration of 5.1 g/l dry cell weight, and was followed by a drop in respiration (Fig. 1). The antibiotics undecylprodigiosin and actinorhodin were first detected after 41 and 46 h, respectively.

3.3 Reproducibility

Four biological replicates were cultivated in parallel (Fig. 1). The biomass and the remaining phosphate concentrations in the medium showed an average standard deviation of less than 6%, while the average standard deviation for the levels of undecylprodigiosin and actinorhodin was less than 10% for the four parallel cultivations.

We compared the expression profiles of genes at the time points available for fermenters F199, F201 and F202 (24, 28, 32, 36, 40, 44, 48 and 60 h after inoculation). The mean pairwise correlation between any pair of time points in one fermenter is 0.930, and the mean pairwise correlation between different fermentations *at the same time point* is 0.954, showing that the differences between time points are more significant ($p < 0.001$) than those between synchronized fermenters. To compare the fermentations with respect to the coherence of the offline measurements and the transcriptomics data, we analysed groups of genes whose major expression change coincided with the time of phosphate depletion, e.g. genes reported to be regulated by PhoP [10]. The time of phosphate depletion was delayed about 1 h in fermenter F202. The strong up-regulation of PhoP regulated genes is also delayed about 1 h (suppl. Fig. 1). This verifies that the dramatic shift in expression levels directly corresponds to phosphate depletion.

We analysed the expression profiles of chromosomal gene clusters that show significant variation in more detail. For most clusters the expression profiles of the contained genes show a good correlation between the fermenters (fcor(C), see Sect. 2) with a minimum of 0.48, an average of 0.72 and a maximum of 0.94 (suppl. Table 3), which reveals that the expression changes of most known chromosomal clusters occur highly synchronously during the different fermentation runs.

3.4 Variant Genes and Expression Profile-Based Clustering

After RMA normalization of the 32 samples in fermenter F199, we computed the regularized variance of each gene, which after filtering with a threshold of 0.1 resulted in a set of 322 variant genes. A distribution chart of annotations of the genes based on main role and sub role categories in the Sanger Protein Classification Scheme is shown in Fig. 2. Categories comprised of genes involved in metabolic activity as well as ribosomal proteins are significantly enriched (see Table 1).

Next, we computed a QT clustering of the most variant genes. We observe very tight clusters (Fig. 3, suppl. Table 4), and genes in the same chromosomal cluster were always found clustered together by QT clustering. Detailed analyses of expression-based clustering results revealed that expression clusters consist of

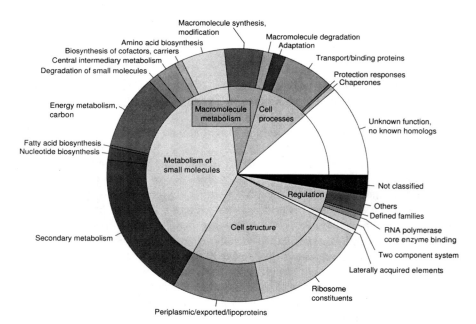

Fig. 2 Distribution of protein classes based on main role (*inner circle*) and sub role categories (*outer circle*) in the Sanger Protein Classification of the most variant genes (mean reg. variance greater 0.1)

Table 1 Categories with significant enrichment ($p < 0.05$) based on the set of variant genes

Category	Description	Enrichment	p value
4.2.0	Ribosome constituents	16.49	<0.0001
3.8.0	Secondary metabolism	5.23	<0.0001
3.7.0	Nucleotide biosynthesis	4.09	0.0061
3.5.0	Energy metabolism, carbon	3.92	<0.0001
3.1.0	Amino acid biosynthesis	3.39	<0.0001
1.6.0	Adaptation	2.92	0.0265

groups of genes with coherent functional annotation [8] with the most significant expression change of genes coinciding with phosphate depletion (Fig. 3d) and whose cluster includes genes that have been previously reported to be activated by PhoP [10, 12, 13].

3.5 Prediction of Chromosomal Gene Clusters

Many genes in a common cellular process are organized in chromosomal clusters. Therefore, we used a supervised approach to assess the coherence of expression by analysing the expression profiles of chromosomal gene clusters from the primary and secondary metabolism [1]. Most of the genes from the known chromosomal clusters were not found to be differentially regulated during the time-series. Those that are differentially regulated, however, show highly correlated expression and partly unexpected dynamics of expression patterns (Fig. 4, suppl. Table 5).

Our densely resolved time-series expression data revealed that genes organized in chromosomal clusters exhibit very similar and highly correlated expression profiles. Therefore, we used these data to predict new and/or uncharacterized chromosomal clusters based on the time-series expression data (see Sect. 2). We validated our method by comparing the predictions with the set of known chromosomal clusters whose mean regularized variance exceeds 0.01. Of the 21 clusters we found all, and the majority of the clusters was precisely delineated (suppl. Table 6). Interestingly, for some of the known clusters our method predicted additional genes. The annotations of these genes were in most cases consistent with that of the cluster, e.g. we found additional membrane proteins to be contained in a cell division cluster and additional ribosomal proteins in the ribosomal protein cluster. A full list of all predicted chromosomal clusters (pcc) is provided as supplementary material. Expression profile plots of the pcc which we describe and discuss in more detail below are shown in Fig. 5.

pcc23 contains the *bioBAD* genes (SCO1244-1246) involved in biotin biosynthesis, which supports the hypothesis that they form a putative operon. Their behaviour is very similar to that of the CPK genes, showing strong up-regulation during early growth (23–24 h) and staying at a low expression level from 26 h onwards.

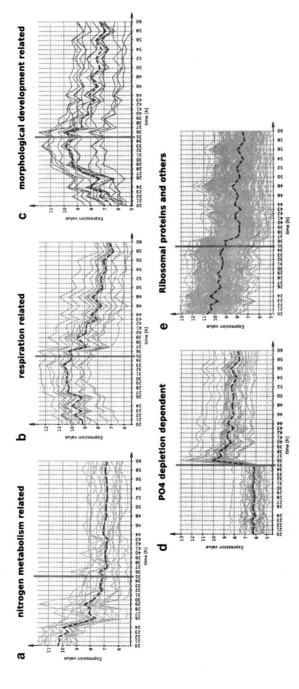

Fig. 3 Expression profiles of clusters resulting from expression profile-based QT clustering. The cluster median is indicated as a *dashed line*. The time point of phosphate depletion is indicated by a *grey vertical line*

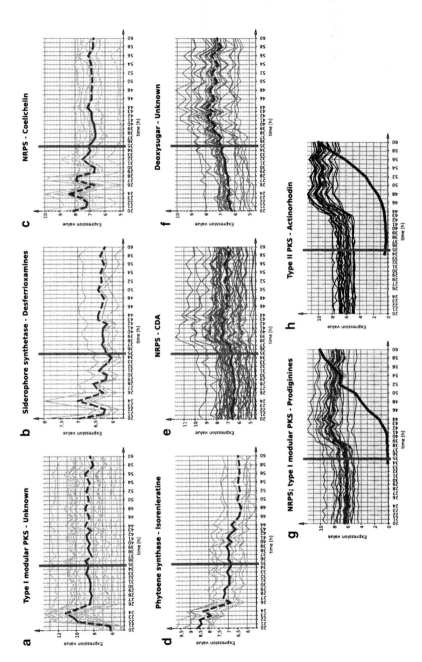

Fig. 4 Expression profiles of chromosomal gene clusters (secondary metabolism). The cluster median is indicated as a *dashed line*. The time point of phosphate depletion is indicated by a *grey vertical line*. The offline measurements for the metabolites of the Actinorhodin and Prodiginines clusters are drawn as *thick lines* in **g** and **h**. Their values were scaled to the interval [0, 10]

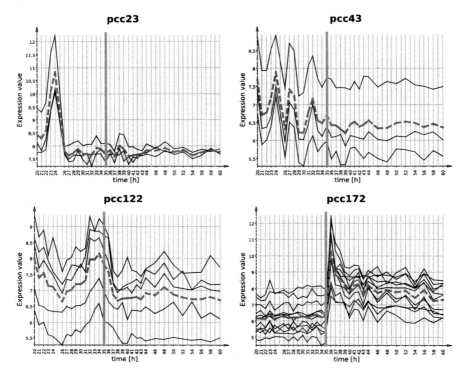

Fig. 5 Expression profiles of predicted chromosomal gene clusters. The cluster median expression profile is indicated as a *dashed line*

The three genes SCO1785–SCO1787 (pcc43) show a periodic expression until the time of phosphate depletion, similar to the coelichelin cluster. They encode transmembrane components involved in iron siderophore import after iron has been chelated.

The five genes (SCO3716–SCO3720) in pcc122 are involved in potassium transport. Their transcription is slightly down-regulated at 26 h and then up-regulated at 34 h, followed by another down-regulation until 38 h after which expression remains unchanged. *kdpABC* (SCO3716–SCO3718) encode subunits of a potassium-transporting ATPase. During response to salt stress, sodium is exported, while the less toxic potassium is imported.

One larger predicted cluster (pcc172) contains 11 genes (SCO4872-SCO4882) whose expression is strongly up-regulated at the time of phosphate depletion showing the same dynamic behaviour as most of the PhoP regulated genes. These genes appear to encode cell wall-modifying enzymes. SCO4876 is a GtrA-like protein. GtrA is predicted to be an integral membrane protein with four transmembrane spans. It may play a role in translocation of undecaprenyl phosphate-linked glucose (UndP-Glc) across the cytoplasmic membrane. This family also includes the teichoic acid glycosylation protein, GtcA, which is essential for decoration of cell wall teichoic acids with glucose and galactose. SCO4875, as an undecaprenyl phosphate 4-deoxy-4-formamido-L-arabinose transferase, may also be involved in modification of teichoic acids. Teichoic acid is a phosphate-rich component of Gram positive cell walls. In other Gram positive bacteria, it acts as a phosphate store and in times of phosphate limitation teichoic acid is catabolised and teichuronic acid, a phosphate-free substitute, is synthesised [7]. Teichoic acids have also been shown to function as a phosphorous storage in *Streptomyces* [15] and maybe the genes in this cluster are involved in this process. Their expression pattern suggests that their transcription regulation is indeed phosphate dependent.

4 Conclusion

Our custom GeneChip® array is a powerful tool for monitoring transcriptional regulation of *S. coelicolor* and generates data of very high quality. Due to the combination of the gene-based and tiling array approaches, it can be used for a wide range of studies. The probe sets for intergenic regions will be of particular use for ChIP-chip experiments. Furthermore, this is the first microarray that allows a genome-wide monitoring of non-coding RNAs. First results of the transcriptomic data suggest that several key proteins of *S. coelicolor* are regulated by non-coding RNAs (data not shown), and these will be subject to future investigations.

Producing reproducible data is of great importance for systems biology research of *S. coelicolor*. The extremely high correlation between the transcriptomics data as well as online/offline measurements produced from distinct fermentations (biological replicates) have shown that the protocols introduced in [8] can be used for highly reproducible submerged batch fermentations and result in very coherent replicate transcriptome data.

We used our visual analytics framework MAYDAY for all our analyses [3]. Its tight integration of many analysis tools and visualizations and the ability to study differential regulation over time as well as between different fermenters or conditions proved essential for a systems biology project like this. MAYDAY also allows to integrate heterogeneous data as well as meta information to support modelling the complex regulatory mechanisms in this organism.

With our algorithms, we could extend known chromosomal clusters by neighbouring genes at the same locus and predicted a number of previously undescribed chromosomal clusters. The successful prediction of known clusters supports the high quality of the data and strengthens our trust in predicted clusters. Furthermore, we were able to detect novel clusters of genes that show interesting expression dynamics, like the cell wall-modifying enzymes of cluster pcc172, which seem to be involved in a highly specific response to phosphate depletion.

The data presented in [8] can be the basis for approaches to (mathematically) model the metabolic switches in *S. coelicolor*. Other types of "omics" data such as metabolite and protein concentrations are currently generated to gain a deeper understanding of this complex system. The next important step will be to determine the regulatory mechanism by performing successive rounds of statistical causal inference combined with experiments of targeted gene disruption. Using one microarray platform and highly standardized growth protocols allows to easily compare and share data. The reference experiments established with our reliable platform can be accessed for comparisons of various conditions and time points. We are currently examining the metabolic switch in response to alternative nutrient limitations (e.g. depletion in nitrogen) during cultivation to reveal similarities and differences to the phosphate limitation described in this study. Integrating all these datasets and network inference will eventually lead to generate systems biology models for *S. coelicolor*.

Supplementary material can be found online at http://www-ps.informatik.uni-tuebingen.de/supplement/.

Acknowledgements We are grateful to Mervyn Bibb for helpful discussion concerning the Affymetrix custom microarray design. We acknowledge the excellent technical help of K. Klein, S. Poths and M. Walter at the Microarray Facility Tübingen, and Anders Øverby and Elin Hansen at SINTEF Materials and Chemistry, Department of Biotechnology. This project was supported by grants of the ERA-NET SySMO Project (GEN2006-27745-E/SYS) and the Research Council of Norway (project no. 181840/I30).

References

1. Bentley, S.D., et al.: Complete genome sequence of the model actinomycete Streptomyces coelicolor A3(2). Nature **417**(6885), 141–147 (2002)
2. Bolstad, B.M., et al.: A comparison of normalization methods for high density oligonucleotide array data based on variance and bias. Bioinformatics **19**(2), 185–193 (2003)
3. Dietzsch, J., et al.: Mayday–a microarray data analysis workbench. Bioinformatics **22**(8), 1010–1012 (2006)
4. Finn, R.D., et al.: The Pfam protein families database. Nucleic Acids Research **36**(suppl_1), D281–288 (2008)
5. Gautier, L., et al.: affy–analysis of Affymetrix GeneChip data at the probe level. Bioinformatics **20**(3), 307–315 (2004)
6. Heyer, L.J., et al.: Exploring expression data: identification and analysis of coexpressed genes. Genome Research **9**(11), 1106–1115 (1999)
7. Liu, W., et al.: Analysis of bacillus subtilis tagab and tagdef expression during phosphate starvation identifies a repressor role for phop-p. Journal of Bacteriology **180**(3), 753–758 (1998)
8. Nieselt, K., et al.: The dynamic architecture of the metabolic switch in Streptomyces coelicolor. BMC Genomics **11:10** (2010)
9. R Development Core Team: R: A Language and Environment for Statistical Computing. R Foundation for Statistical Computing, Vienna, Austria (2006)
10. Rodríguez-García, A., et al.: Genome-wide transcriptomic and proteomic analysis of the primary response to phosphate limitation in Streptomyces coelicolor M145 and in a DeltaphoP mutant. Proteomics **7**(14), 2410–2429 (2007)
11. Sanger Institute: Protein classification scheme. http://www.sanger.ac.uk/Projects/S_coelicolor/classwise.html

12. Sola-Landa, A., et al.: Binding of PhoP to promoters of phosphate-regulated genes in Streptomyces coelicolor: identification of PHO boxes. Molecular Microbiology **56**(5), 1373–1385 (2005)
13. Sola-Landa, A., et al.: Target genes and structure of the direct repeats in the DNA-binding sequences of the response regulator PhoP in Streptomyces coelicolor. Nucleic Acids Research **36**(4), 1358–1368 (2008)
14. UniProt Consortium: The universal protein resource (uniprot). Nucleic Acids Research **36**(Database issue), 190–195 (2008)
15. Voelker, F., et al.: Nitrogen source governs the patterns of growth and pristinamycin production in 'streptomyces pristinaespiralis'. Microbiology **147**(Pt 9), 2447–2459 (2001)

Chapter 2
MiRNA Recognition with the *yasMiR* System: The Quest for Further Improvements

Daniel Pasailă, Andrei Sucilă, Irina Mohorianu, Ştefan Panţiru, and Liviu Ciortuz

Abstract The paper "Using Base Pairing Probabilities for MiRNA Recognition" by Daniel Pasailă, Irina Mohorianu, and Liviu Ciortuz, that has been published in Proceedings of the International Symposium on Symbolic and Numeric Algorithms for Scientific Computing (SYNASC) 2008, IEEE Computer Society, pp. 519–525, has introduced a new SVM for microRNA identification, whose novelty is twofolded: first, many of its features incorporate the base-pairing probabilities provided by Mc-Caskill's algorithm, and second the classification performance is improved using a certain similarity ("profile"-based) measure between the training and test microR-NAs and a set of carefully chosen ("pivot") RNA sequences. Comparisons with some of the best existing SVMs for microRNA identification proved that our SVM obtains truly competitive results.

Here we add several significant extensions to the work reported in Daniel Pasailă et al. Proceedings of the International (SYNASC) 2008, pp. 519–525: testing this classifier on a more recent version of miRBase (12.0), evaluating the effect of using probabilistic patterns instead of non-probabilistic ones, analysing the discriminative power of different categories of features we used, and automatically searching for good pivot RNA sequences, which are critical for classification in our approach.

1 *yasMiR*: The Approach and Main Results

MicroRNAs (henceforth abbreviated miRNAs) are short RNA molecules that play important gene regulatory roles [1]. In the paper [3],[1] we proposed a miRNA recognition system based on a support vector machine [8], which was subsequently named *yasMiR*.[2] It was mainly built upon features using the base-pair binding probabilities

---

[1] Daniel Pasailă and Liviu Ciortuz are joint first authors of both paper [3] and the present paper.
[2] *yasMiR* is an abbreviation for *yet another* **SVM** *for* **miRNA** *recognition*.

L. Ciortuz (✉)
Department of Computer Science, "Alexandru Ioan Cuza" University of Iaşi, Iaşi, Romania
e-mail: ciortuz@info.uaic.ro

Table 1 Categories of features for *yasMiR* SVM. The *rightmost column* gives the number of features in the respective (sub)category

A	– Alignment scores against pivot sequences, where n is the number of pivots used	n
B	– The probabilistic mean for the number of occurrences for each triplet pattern	32
C	– The mean base-pairing distance	1
	– The overall non base-pairing probability	1
	– The non-pairing probability for each nucleotide	4
	– The sum of pairing probabilities for each pair of nucleotides a and b	10
	– The folding minimum free energy (MFE)	1
	– Dinucleotide frequencies	16
	– The average frequency for each nucleotide	4

provided by McCaskill's algorithm [6], supplemented with some other simple features. *yasMiR*'s features are summarized in Table 1: profile similarity scores against "pivot" RNA sequences (A), means of probabilistic triplet patterns (B), and finally other probabilistic and non-probabilistic features (C).

The two remaining parts of this section will summarize the results that we have obtained when comparing *yasMiR* to Triplet-SVM and miPred (previously detailed in [3]) and when testing this classifier on miRBase 12.0. Section 2 will present the effect of using probabilistic patterns instead of non-probabilistic ones in *yasMiR*, and then it will analyse the discriminative power of different categories of features we used. Section 3 will detail our work on the automatic search for good pivot RNA sequences, which are critical for miRNA classification in our approach.

1.1 Comparisons with Triplet-SVM and miPred

We compared *yasMiR* first to the Triplet-SVM classifier [2], after having trained our SVM on the same dataset as Triplet-SVM. The training set included 163 human pre-miRNAs from miRBase registry version 5.0 and 168 pseudo pre-miRNA like hairpins as negative examples. A fivefold cross-validation accuracy of 96.07% was obtained on this training set. On the test datasets created by the authors of Triplet-SVM, our SVM obtained significantly higher prediction results.

Then we made comparative tests with the miPred classifier [7], the best SVM-based miRNA classifier up to our knowledge. Here, the training set included 200 human pre-miRNAs from miRBase version 8.2 as positive examples, and 400 pseudo pre-miRNA hairpins as negative examples. We obtained at fivefold cross-validation an accuracy of 93.66% on this training set, compared to miPred's 93.50%. Running the same tests as miPred, our SVM obtained similar and sometimes significantly better specificity than miPred. Compared to miPred, one of the advantages of our approach is that it makes no use of so-called normalized features which are based on sequence shuffling; in turn it enables the feature computation in our approach to be much less time consuming.

We also checked whether the Random Forests machine learning algorithm is able to obtain comparable results to SVM (as suggested by MiPred, another SVM for miRNA recognition) when using our set of features. While on many test datasets that we used the answer was positive, the overall conclusion is that RF is not a good enough candidate to replace SVM for pre-miRNA identification using our set of features.

1.2 Results on miRBase 12.0

We have also tested our SVM on sequences from miRBase 12.0 (released in October 2008). For the training set, this time all 678 human miRNAs from miRBase 11.0 were used as positive examples, and also 1,256 sequences from the CODING dataset as negative examples. The testing set includes 3,651 positive examples from miRBase 12.0 and 7,198 negative examples from the CODING dataset. The set of 3,651 positives was obtained by removing the positive training sequences from miRBase 12.0, and using the clustering algorithm presented in the supplementary material of the miPred paper [7] for the removal of similar sequences. First, all the sequences were sorted in decreasing length order, and the first one became the representative of the first cluster. Then, each of the remaining sequences was compared with the existing representatives and added into a cluster if the similarity measure with any representative is above 90%. The remaining set of 3,651 sequences is the final set of representatives, using the above algorithm on miRBase 12.0 (after the training positives have been removed). The BLAST system was used for sequence comparison. On this dataset, the yasMiR system obtained 89.64% sensitivity and 97.37% specificity, with the resulting accuracy of 94.77%.

2 yasMiR's Features Analysis

Since *yasMiR* uses features which are a probabilistic (McCaskill) version of the features used by Triplet-SVM, one would question whether our design decision is indeed justified. Therefore, we made a test in which we replaced the features related to the probabilistic triplet patterns with those taken from the Triplet-SVM package. We used the same procedure as for the comparative test between *yasMiR* and Triplet-SVM. The results we obtained for *yasMiR* (Table 2, first column) are usually slightly (and even significantly) better than the ones we obtained with non-probabilistic features computed for triplet patterns (Table 2, second column). This is especially true for the TE-C (human) and CONSERVED-HAIRPIN datasets.

To further analyse *yasMiR*'s set of features, we also investigated what prediction results are obtained when removing each one of the different categories of features defined for our system (see Table 1).

Table 2 Prediction accuracy (%) results obtained by *yasMiR* on the Triplet-SVM datasets when the features for probabilistic triplet patterns were replaced with their non-probabilistic (Triplet-SVM) counterpart

Test	yasMiR	Using non-probabilistic triplet patterns
TE-C: Human pre-miRNAs	100	96.67 (29/30)
TE-C: Pseudo pre-miRNAs	96.2	95.9 (959/1,000)
UPDATED	94.9	94.9 (37/39)
CROSS-SPECIES	95.2	95.87 (557/581)
CONSERVED-HAIRPIN	94.23	93.09 (2,275/2,444)

Table 3 Prediction results for *yasMiR* on miPred datasets (column 1), when removing one category (A, B or C) of its features (columns 2–4). *Bold faces* designate values which are better than those in column 1

	yasMiR			B∪C			A∪C			A∪B		
Test	Sen.	Acc.	Spec.	Sen.	Acc.	Spec.	Sen.	Acc.	Spec.	Acc.	Sen.	Spec.
TE-H	87.80	93.77	96.74	83.73	93.22	**97.96**	89.43	**94.30**	**96.74**	81.30	91.32	96.34
IE-NH	90.35	94.11	95.99	88.58	92.64	94.68	**93.32**	**94.26**	94.73	84.04	92.26	**96.37**
IE-NC		82.95			78.94			59.84			**91.77**	
IE-M		100			100			6.45			100	

Using the same datasets as miPred, we investigated the effect on accuracy, sensitivity and specificity when removing one of the three categories A, B, or C. It can be easily seen in Table 3 that the prediction results with the complete feature set (found in column 1 of Table 3) are in many cases significantly better than those that have been obtained when a category of features is removed. This is especially true for the IE-NC and IE-M datasets. Going into more details, one can see the following facts:

- Retracting the category A of attributes (see column 2 in Table 3) slightly improves the specificity on TE-H (from 96.74% to 97.96%) at the significant cost of sensitivity (from 87.80% down to 83.73%)
- Retracting the category B of attributes (see column 3 in Table 3) slightly improves some of the statistics we obtained previously for *yasMiR* on TE-H and IE-NH but drastically affects the performance on IE-NC (from 82.95% down to 59.84%) and especially on IE-M (from 100% down to 6.45%)
- Retracting the category C of attributes (see column 4 in Table 3) improves the specificity on IE-NC (from 82.95% up to 91.77%) and on IE-NH (from 95.99% to 96.37%), but significantly affects the sensitivity on TE-H (from 87.80% down to 81.30%) and IE-NH (from 90.35% down to 84.04%).

The above analysis implies that each of these categories of features has its own contribution towards the overall good classification results produced by *yasMiR*.

It is also interesting to note that the categories A and C of attributes are more suitable for the TE-H and IE-NH datasets, while B is indispensable for the IE-NC and IE-M datasets. These facts suggest that there are slightly specialized contributions of these categories of features towards discriminating among different categories of RNA sequences.

For expressing the quality of the ith feature we used the $F1$ and $F2$ scores, defined by the following expressions:

$$F1 = \frac{|\mu_i^+ - \mu_i^-|}{|\sigma_i^+ + \sigma_i^-|}, \quad F2 = \frac{(\mu_i^+ - \bar{\mu}_i)^2 + (\mu_i^- - \bar{\mu}_i)^2}{(\sigma_i^+)^2 + (\sigma_i^-)^2},$$

where μ_i^+/μ_i^-, σ_i^+/σ_i^- denote the means and standard deviations of the positive and negative training datasets for the ith feature. After sorting the features in descending order according to the $F1$ and $F2$ scores, we identified the first three features, and they proved to be the same for both sorting measures:

- Feature D: the overall non base-pairing probability ($F1 = 1.21$ and $F2 = 1.64$)
- Feature E: the folding minimum free energy ($F1 = 0.95$ and $F2 = 0.99$)
- Feature F: the probabilistic feature corresponding to the triplet pattern "..." with the nucleotide C on the middle position ($F1 = 0.93$ and $F2 = 0.90$)

The effects on *yasMiR* when each of these three features is removed are shown in Table 4. It is interesting to note that removal of features D and E has a big impact on the IE-NC and IE-M datasets, while feature F seems to be only slightly affecting the result on the IE-NC dataset. Our opinion is that this last feature is made almost redundant by other features.

We therefore tried feature selection applying the Kolmogorov–Smirnov filter [5] for redundancy elimination on the full set of *yasMiR*'s 169 features including the 100 randomly chosen pivots used so far. The Kolmogorov–Smirnov filtering procedure goes as follows: first we rank and sort the features according to the Symmetrical Uncertainty (SU) score which is a normalized version of the mutual information statistics, and then, starting from the top ranking feature that has not yet been filtered, we eliminate all features of lower rank which are redundant to it, according to the Kolmogorov–Smirnov test, up to a certain confidence level.

Using a 0.95 confidence level, the number of features gets reduced to 144 – remarkably, all but one of the 26 eliminated features are pivots – most of the classification statistics on the miPred's test datasets get improved, as shown in Table 5. At 0.90 confidence, things do not go so well, and unfortunately a 5.55% specificity/accuracy loss is reported on the IE-NC dataset (from 82.95% down to 77.20%). However, it is worth noting that this time 31 pivots got eliminated, together with six non-pivot features.

Table 4 Prediction results for *yasMiR* on miPred datasets when removing one of the features D, E or F. Bold faces designate values which are better than those in the first column of Table 3

Test	$A \cup B \cup C \backslash \{D\}$			$A \cup B \cup C \backslash \{E\}$			$A \cup B \cup C \backslash \{F\}$		
	Sen.	Acc.	Spec.	Sen.	Acc.	Spec.	Sen.	Acc.	Spec.
TE-H	86.17	93.76	97.56	**86.17**	93.49	**97.15**	**87.80**	93.49	96.34
IE-NH	**91.24**	**94.99**	**96.87**	90.45	94.40	96.37	90.45	94.14	95.98
IE-NC		67.68			61.95			79.74	
IE-M		19.35			22.58			100	

Table 5 Prediction results of *yasMiR* on miPred's test datasets using 144 features and respectively 132 features selected from the whole set of 169 features via Kolmogorov–Smirnov redundancy filtering. *Bold faces* designate values that are better than those in the first column of Table 3

Test	0.95 Confidence			0.90 Confidence		
	Sensitivity	Accuracy	Specificity	Sensitivity	Accuracy	Specificity
TE-H	87.80	94.30	97.50	87.80	93.76	**96.74**
IE-NH	90.14	94.07	**96.03**	**91.08**	93.39	94.55
IE-NC		**83.28**			77.20	
IE-M		**100**			**100**	

3 Automatically Choosing the Pivots

Until now we performed several runs with *yasMiR* using different sets of randomly generated pivots, and we retained the results for the set of pivots that produced the best overall results on the Triplet-SVM and the miPred test datasets. However, one could ask whether we could get better results by automatically selecting (or improving) the set of pivots.

3.1 Using Clustering

Here we report on using choosing "representative" pivots among a pool of candidates, using clustering and the Euclidean distance between the vectors associated with pivots. For each candidate pivot, its vector was obtained by computing the profile similarity measure between the pivot and each of the sequences in the training set (e.g. TR-H).

The left column in Table 6 shows the results we obtained for 200 pivots automatically selected from a pool of 2,000 randomly generated sequences. The k-means clustering algorithm was used to get those 2,000 sequences grouped into 50 clusters, and then we randomly selected four pivots from each cluster. The results show that the obtained specificity for *yasMiR* SVM's is slightly lower than that obtained with the manually chosen pivots on miRBase 8.2 (on TE-H: from 96.74% to 95.53%, and on IE-NH: from 95.99% to 94.97%), while the sensitivity decreased significantly (TE-H: from 87.80% to 85.37%, and IE-NH: from 90.35% to 83.58%). Remarkably, the specificity/accuracy of *yasMiR* SVM was dramatically improved for IE-NC (from 82.95% to 93.61%, while miPred reported only 68.68%), and for IE-M the specificity/accuracy was kept at 100%.

These results make us conclude that automatically searching for better pivots is worth further working on. In the following section, we will use the Kolmogorov–Smirnov filter for searching among a large pool of randomly generated pivots.

Table 6 *Left column*: prediction results of *yasMiR* on miPred's test datasets using 200 pivots selected via clustering from a pool of 2,000 randomly generated pivots. *Right column*: prediction results of *yasMiR* on miPred's test datasets using the best 13 pivots selected via Kolmogorov–Smirnov filtering from 10,000 randomly generated pivots. *Bold faces* designate values which are better than those in the first column of Table 3

	SVM			KS		
Test	Sensitivity	Accuracy	Specificity	Sensitivity	Accuracy	Specificity
TE-H	85.37	92.14	95.53	85.37	92.53	**96.74**
IE-NH	83.58	91.17	94.97	86.24	91.35	93.90
IE-NC		93.61			87.44	
IE-M		100			100	

3.2 Using the Kolmogorov–Smirnov Filter

The probabilistic alignment scores to pivots used in describing sequences lead to a distance-based description. It is clear that the pivots need not be chosen from positive or negative examples, but at a correct distance from members of these classes. As such, we have tried to implement a non-linear feature selection algorithm to choose a better set of pivots. Such a method is the Kolmogorov–Smirnov filter, which has been reported to work well in conjunction with SVM's. We have implemented such a procedure following directions from [5].

The Kolmogorov–Smirnov filter is divided into two parts. The first part is concerned with ranking the features according to a mutual information measure, and the second part recursively eliminates redundant features. We used a confidence level of 95% for determining whether two features were redundant.

The right column in Table 6 shows that when using the best 13 features selected by the Kolmogorov–Smirnov filter from a large pool of randomly generated pivots, the results obtained by *yasMiR* on the miPred datasets are comparable with those reported in the previous section. On the TE-H dataset, we got a better specificity (96.74%) compared to the one produced via clusterization, while on the IE-NH dataset, the sensitivity improved (from 83.58% to 86.24%) but it still remained significantly lower than the one obtained with hand-chosen pivots (90.35%). On IE-NC, the specificity/accuracy is now at midway between the one obtained via clusterization (93.61%) and the original one, produced by hand-chosen pivots (82.95%). On IE-M, the specificity/accuracy stayed at 100%.

Finally, we would suggest that this method would be best used in conjunction with another feature selection method, where the initial bulk of features would be removed by the Kolmogorov–Smirnov filter, and the final features would be selected by the other, more complex method.

4 Conclusions and Further Work

We proved that the base-pairing probabilities provided by McCaskill's algorithm combined with some other, simple statistical measures make an SVM classifier achieve high pre-miRNA prediction accuracy rates, comparable to the best published results up to our knowledge.

We plan to make direct comparisons with a quite recent kNN-based classifier for non-coding RNAs [9]. Its results seem to be very competitive, due to the use of certain topological features. We will see whether those features could be generalized using again the base-pairing probabilities computed by McCaskill's algorithm. If so, we will check whether adopting them into *yasMiR*'s feature set will make it further improve the quality of pre-miRNA prediction.

It will also be interesting to see whether a more recent work for identifying miRNAs [10], which also used the Triplet-SVM patterns but replaced the automate classifier with a ranking algorithm, will improve its results when replacing the simple triplet features with their enhanced counterpart McCaskill's probabilities.

Acknowledgments LC thanks Dr. Mihaela Zavolan from Biozentrum, University of Basel, Dr. Hélène Touzet from the University of Lille, and Dr. Marti Tammi from the National University of Singapore for useful discussions on miRNA identification.

Availability The source code of our system and the datasets we used can be found at the address www.info.uaic.ro/~ciortuz/yasmir. The technical report [4] offers a unified view on the past [3] and present work on *yasMiR*.

References

1. Andrew Fire, Siqun Xu, Mary Montgomery, Steven Kostas, Samuel Driver, and Craig Mello. Potent and specific genetic interference by double-stranded RNA in Caenorhabditis elegans. *Nature*, 391(6669):806–811, 1998
2. Chenghai Xue, Fei Li, Tao He, Guoping Liu, Yanda Li, and Xuegong Zhang. Classification of real and pseudo microRNA precursors using local structure-sequence features and support vector machine. *BMC Bioinformatics*, 6(310), 2005
3. Daniel Pasailă, Irina Mohorianu, and Liviu Ciortuz. Using base pairing probabilities for MiRNA recognition. In *SYNASC '08: Proceedings of the 2008 10th International Symposium on Symbolic and Numeric Algorithms for Scientific Computing*, pages 519–525, 2008
4. Daniel Pasailă, Irina Mohorianu, Andrei Sucilă, Ștefan Panțiru, and Liviu Ciortuz. Yet another SVM for miRNA recognition: yasMiR, 2010. Technical Report TR-10-01, Faculty of Computer Science, University of Iasi, Romania
5. Jacek Biesiada and Wlodzislaw Duch. Feature selection for high-dimensional data: A Kolmogorov–Smirnov correlation-based filter. *Computer Recognition Systems*, 30:95–103, 2005
6. John S. McCaskill. The equilibrium partition function and base pair binding probabilities for RNA secondary structures. *Biopolymers*, 29:1105–1119, 1990
7. Kwang Loong Stanley Ng and Santosh Mishra. De novo SVM classification of precursor microRNAs from genomic pseudo hairpins using global and intrinsic folding measures. *Bioinformatics*, 23(11):1321–1330, 2007

8. Nello Cristianini and John Shawe-Taylor. *An introduction to Support Vector Machines and other kernel-based learning methods*. Cambridge University Press, New York, NY, USA, 2000
9. Wenjie Shu, Xiaochen Bo, Zhiqiang Zheng, and Shengqi Wang. A novel representation of RNA secondary structure based on element-contact graphs. *BMC Bioinformatics*, 9(1):188, 2008
10. Yunpen Xu, Xuefeng Zhou, and Weixiong Zhang. MicroRNA prediction with a novel ranking algorithm based on random walks. *Bioinformatics*, 24(13), 2008

Chapter 3
Top Scoring Pair Decision Tree for Gene Expression Data Analysis

Marcin Czajkowski and Marek Krętowski

Abstract Classification problems of microarray data may be successfully performed with approaches by human experts which are easy to understand and interpret, like decision trees or Top Scoring Pairs algorithms. In this chapter, we propose a hybrid solution that combines the above-mentioned methods. An application of presented decision trees, which splits instances based on pairwise comparisons of the gene expression values, may have considerable potential for genomic research and scientific modeling of underlying processes. We have compared proposed solution with the TSP-family methods and decision trees on 11 public domain microarray datasets and the results are promising.

1 Introduction

A powerful tool for structural and functional analysis of genomes may be developed from DNA chips [6, 21]. The entire set of genes of an organism can be microarrayed on an area not greater than $1\,cm^2$ and makes possible processing of thousands of expression levels simultaneously in a single experiment [12]. Nowadays, DNA chips may be used to assist diagnosis and to discriminate cancer samples from normal ones [2, 10]. Extracting accurate and simple decision rules that contain marker genes is of great interest for biomedical applications. However, finding a meaningful and robust classification rule is a real challenge, since in different studies of the same cancer, diverse genes consider to be marked [25].

Typical statistical problems that often occur with microarray analysis are dimensionality and redundancy. In particular, we are faced with the *"small N, large P problem"* [27, 28] of statistical learning because the number of samples (denoted by N) comparing to the number of genes (P) remains quite small as N usually does not exceeded one or two hundreds where P is usually several thousands. The high ratio

---

M. Czajkowski (✉)
Faculty of Computer Science, Bialystok University of Technology, Bialystok, Poland
e-mail: m.czajkowski@pb.edu.pl

of features/observations may influence the model complexity [16] and can cause the classifier to overfit the training data. Furthermore, most of the genes are known to be irrelevant for an accurate classification, so gene selection prior the classification should be considered to simplify calculations, decrease model complexity, and often to improve accuracy of the following classification [22].

Recently, many new approaches including hierarchical clustering [1], machine learning [38], methods based on Support Vector Machine [23, 39], neural networks [7], and much more are applied in microarray classification. Usually they are "blackbox" approaches, concentrated mostly on the improvement of accuracy. However, they generate very complex models that are difficult to interpret from medical point of view. Nonlinear models may achieve high accuracy but do not provide any significant and easily understood rules. Simple models like decision trees or rule extraction systems, however, may help in understanding underlying processes. Causal relationships between specific genes can be influenced by other causal relationships which can be identified in the model only if such a model is easy to understand. The restriction in considering the model interpretability in classification system often affects performance in classification. This may be called a trade-off between credibility and comprehensibility of the classifiers [30].

In this chapter, we propose a hybrid solution denoted as TSPDT that applies TSP-family algorithm and decision trees. The rest of the chapter is organized as follows. In the next section, TSP-family algorithms and decision tree classifiers for gene expression analysis are briefly recalled. At the end of this section, TSPDT hybrid solution is presented. In Sect. 3, the proposed approach is experimentally validated on 11 real microarray datasets. At the end, the chapter is concluded and possible future works are suggested.

2 Methods

2.1 A Family of TSP Algorithms

Top Scoring Pair (TSP)-family classifiers are one of the most promising techniques that classify gene expression data using a simple decision rules. These statistical methods are widely used to identify marker genes in microarray datasets [36]. Strong point of the TSP is its parameter-free, data-driven learning property and invariance to any simple transformation of data like normalization or standardization. TSP methods are also used in other algorithms as for example feature selection in SVM classifiers [37]. In this section, we will be focused on the original TSP method [14], its ensemble counterpart k-TSP [30], and our extension Weight k-TSP [8]. There is an additional restriction that k should not exceed 10 in the original article due to computational limitations. Computational cost was reduced by omitting cross-validation procedure for automatic determining parameter k (for all algorithms), and comparison was performed on all possible odd number of pairs manually like in [37].

TSP method was presented by Donald Geman [14] and is based on pairwise comparisons of gene expression values. Despite its simplicity to other methods, classification rates for TSP are comparable or even exceed other classifiers. Discrimination between two classes depends on finding pairs of genes that achieve the highest ranking value called "score".

A k-Top Scoring Pairs (k-TSP) classifier proposed by Aik Choon Tan [30] is a simple extension of the original TSP algorithm. The main feature that differs from those two methods is the number of TSPs included in final prediction. In the TSP method, there can be only one pair of genes, and in k-TSP classifier the upper bound denoted as k can be set up before the classification. The parameter k is determined by a cross-validation and in any prediction the k-TSP classifier uses not more than k top scoring disjoint gene pairs that have the highest score.

In classification Weight k-TSP proposed by us [8], selection of TSPs and prediction were changed comparing to TSP and k-TSP. Ranking modification results from limitation in finding optimal TSPs by TSP. Extended prediction was also proposed to improve accuracy for different types of datasets.

2.2 TSP Decision Tree

There have been several attempts to use decision trees for the classification analysis of the gene expression data. Dudoit et al. [11] compare some classification principles, among which there is the CART system [3]. Tan et al. [29] present the application of C4.5, bagged and boosted decision trees. Usage of the ensemble scheme for classification can be also found in Valentini et al. [34], where bagged ensembles of SVM are presented. The committee of individual classifiers is also presented in [18], where ensembles of cascading trees are applied to the classification of the gene expression data. Evolutionary algorithms with decision trees have also been applied in some classification algorithms for microarray datasets [15]. In [37], authors compare decision trees with SVMs on gene expression data and conclude that bagging and boosting decision trees perform as well as or close to SVM. However, ensemble methods alike decision trees with complex multivariate tests based on linear or nonlinear combination splits are more a "black-box" approach. They are difficult to understand or interpret by human experts and do not have potential for genomic research and scientific modeling of underlying processes. Thus, our goal is to improve classification accuracy of the decision trees in a way which will make them still easy to understand. Better decision trees of this kind imply more informative analysis of gene expression data. By combining the strength of the TSP-family algorithms with decision trees, we believe to receive simple decision rules and competitive classier with applications to microarray data.

Decision trees (also known as classification trees) [24] represent one of the main techniques for classification analysis in data mining and knowledge discovery. TSP Decision Tree like many popular decision trees is based on top-down greedy search [26]. First, the test attribute (and the threshold in the case of continuous attributes) is decided for the root node. Then, the data are separated according to the splitting rule in the current node, and then each subset goes to the corresponding branch. In our research, we have tested splitting rules based on the TSP, k-TSP, and Weight k-TSP algorithms. We have adapted solutions presented in previous section to find the best pair (or pairs) of genes that will separate the data. Root node of the decision tree has identical splitting rule to analogous TSP method. Differences occur in lower parts of the tree where new rules are generated based on the instances in each node. The process is recursively repeated for each branch until leaf node is reached. C4.5-like pessimistic pruning was applied [31] to prevent data overfit. Illustration of the TSPDT is enclosed in the next section, in Fig. 1.

3 Results and Discussions

Performance of the above-mentioned classifiers was investigated on public available microarray datasets summarized in Table 1. Datasets came from Kent Ridge Bio-medical Dataset Repository [19] and are related to the studies of human cancer, including: leukemia, colon tumor, prostate cancer, lung cancer, breast cancer, ovarian cancer, etc. Typical tenfolds cross-validation was applied for datasets that were not arbitrarily divided into the training and the testing sets. To ensure stable results, for all datasets average score of 10 runs is shown.

We have implemented and analyzed TSP-family algorithms, that is TSP, k-TSP, Weight TSP, Weight k-TSP, and proposed hybrid solution, which use this method as splitting criterion, adequately: TSPDT, k-TSPDT, Weight TSPDT, and Weight k-TSPDT. Maximum number of gene pairs k used in all algorithms was default (equal 9) through all datasets. We also included classification results for the traditional decision trees such as J48 (pruned C4.5), ensembles: Bagging, Adaboost, Random Forest, and popular rule learner JRip (RIPPER). Data mining tool called Weka [33] was used for the performance experiments on the above-mentioned methods.

Table 1 Kent Ridge Biomedical gene expression datasets

	Datasets	Abbreviation	Attributes	Training set	Testing set
1	Breast cancer	BC	24,481	34/44	12/7
2	Central nervous system	CNS	7,129	21/39	–
3	Colon tumor	CT	6,500	40/22	–
4	DLBCL vs follicular lymphoma	DF	6,817	58/19	–
5	Leukemia ALL vs AML	LA	7,129	27/11	20/14
6	Lung cancer – Brigham	LCB	12,533	16/16	15/134
7	Lung cancer – University of Michigan	LCM	7,129	86/10	–
8	Lung cancer – Toronto, Ontario	LCT	2,880	24/15	–
9	Ovarian cancer	OC	15,154	91/162	–
10	Prostate cancer	PC	12,600	52/50	27/8
11	Prostate cancer outcome	PCO	12,600	8/21	–

Table 2 Comparison of TSP decision trees accuracy and model size with original methods. The highest classifiers accuracy for each dataset was bolded

Datasets	Classifiers accuracy and size							
	TSP	TSPDT	k-TSP	k-TSPDT	W. TSP	W. TSPDT	W. k-TSP	W. k-TSPDT
1. BC	52.63	73.68	68.42	**78.95**	63.16	57.89	47.37	57.89
	2.00	4.00[a]	18.00	3.00[a]	2.00	7.00[a]	18.00	6.00[a]
2. CNS	49.00	64.17	58.50	63.00	52.17	63.83	50.83	**65.50**
	2.00	6.62[a]	18.00	3.99[a]	2.00	7.09[a]	18.00	3.39[a]
3. CT	83.64	80.29	**88.93**	84.88	85.86	77.71	87.33	80.86
	2.00	4.68[a]	18.00	3.00[a]	2.00	4.79[a]	18.00	4.49[a]
4. DF	72.75	87.54	87.82	**95.25**	88.52	88.52	88.04	90.71
	2.00	6.02[a]	18.00	2.60[a]	2.00	3.17[a]	18.00	2.6[a]
5. LA	73.53	76.47	**91.18**	**91.18**	76.47	76.47	**91.18**	**91.18**
	2.00	2.00[a]	18.00	2.00[a]	2.00	2.00[a]	18.00	1.00[a]
6. LCB	76.51	78.52	83.89	83.89	**97.32**	**97.32**	96.64	94.63
	2.00	2.00[a]	18.00	2.00[a]	2.00	1.00[a]	18.00	1.00[a]
7. LCM	95.87	**98.94**	95.23	97.77	93.02	97.69	97.50	98.40
	2.00	2.90[a]	18.00	2.05[a]	2.00	1.94[a]	18.00	1.89[a]
8. LCT	50.92	58.50	58.42	55.33	55.17	62.42	**72.92**	71.25
	2.00	7.16[a]	18.00	3.38[a]	2.00	4.73[a]	18.00	4.17[a]
9. OC	99.77	**100.00**	**100.00**	**100.00**	97.24	96.45	98.62	98.03
	2.00	1.00[a]	18.00	1.00[a]	2.00	3.07[a]	18.00	3.41[a]
10. PC	76.47	85.29	91.18	94.12	91.18	76.47	**100.00**	**100.00**
	2.00	3.00[a]	18.00	3.00[a]	2.00	8.00[a]	18.00	3.00[a]
11. PCO	72.67	74.17	58.83	59.17	50.17	59.67	88.00	**97.00**
	2.00	1.01[a]	18.00	1.02[a]	2.00	1.38[a]	18.00	1.00[a]
Average	73.07	79.78	80.22	82.14	77.30	77.68	83.49	85.95
	2.00	3.67[a]	18.00	2.46[a]	2.00	4.02[a]	18.00	2.90[a]

[a] Value represents the number of execution TSP algorithm in the decision tree. To calculate the maximum pessimistic number of attributes used in decision tree model, multiply this value by adequate size of TSP (or k-TSP, W.k-TSP) method

Classification was performed with default parameters through all datasets for all algorithms. All classifications were preceded by a step known as feature selection where a subset of relevant features is identified. We decided to use popular for microarray analysis method Relief-F [20] with default number of neighbors (equal 10) and 1,000 features subset size. Experimental results on tested datasets are confronted in Table 2.

3.1 Comparison of TSP-Family Algorithms Methods

Table 2 summarizes classification performance for TSP decision trees and traditional TSP-family methods. First row for each dataset represents classification accuracy and the second row classification size. Results show that for over two-thirds of datasets, classification accuracy increased (or did not change) when hybrid decision tree with TSP methods was applied. Average accuracy for all datasets also increased, from nonsignificant 0.38% for Weight TSPDT to 6.71% for TSPDT. For some datasets, however, like colon tumor, we can observe lower credibility of the decision tree solution. In this case, proposed method overfit to the training data.

However, it should be noted that one of the factor that stands for promising or (with respect to colon tumor dataset) worse results for the decision trees with TSP-family methods is the higher number of features in decision tree model. Maximum pessimistic number of attributes used in the whole decision tree model is from 2.46 to 4.02 times higher to TSP methods. We can observe that average model size of the proposed approach is larger to analogues TSP methods. In case of the decision trees, however, comprehensibility of the generated rules do not have to be always lower to the compared TSP methods.

Let us compare decision rules for TSPDT with k-TSP methods on one of our tested datasets – breast cancer (*LCB*) [35]. In the Fig. 1, we can observe generated rules and trees for analyzed methods and in Table 3 selected gene names. The k-TSP method is an ensemble TSP algorithm where prediction depends on simple majority voting. For every tested instance, there can be different combination of pairs of genes affecting the right decision which may cause difficulties in finding and

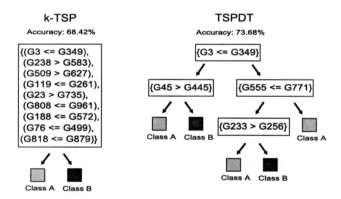

Fig. 1 Outcome for breast cancer dataset for k-TSP and TSPDT method

Table 3 Gene labels and names from Fig. 1

ID	Name	ID	Name	ID	Name	ID	Name
G3	Contig37376 RC	G23	NM 001661	G45	NM 003163	G76	Contig32185 RC
G119	NM 020120	G188	Contig54742 RC	G233	Contig42059 RC	G238	Contig31000 RC
G256	NM 005217	G261	NM 005243	G349	NM 004798	G445	Contig40635 RC
G499	U90911	G509	U82987f	G555	Contig51943 RC	G572	Contig65439
G583	Contig38288 RC	G627	Contig6089 RC	G735	Contig40557 RC	G771	NM 000017
G808	AF161553	G818	NM 019028	G879	NM 018457	G961	Contig32125 RC

extracting significant rules. In our opinion, the TSPDT, alike TSP method, generates much more simpler rules to k-TSP. In the longest path, there have to be tested three pairs of genes; however, no voting procedure is involved. Accuracy for this datasets is also higher for TSPDT method than to k-TSP. If we compare general credibility of the TSPDT and k-TSP algorithm, the results are similar; however, in our opinion proposed solution is much more easier to analyze and interpret by human experts.

3.2 Comparison of Traditional Decision Trees and Rule Classifiers

Performance of the proposed hybrid solution was also compared with popular decision trees and rule classifiers on datasets presented in Table 1. In our research, we have confronted the J48 Tree (pruned C4.5 decision tree) [32], Adaboost (boosting algorithm using Adaboost M1 method) [13], Bagging (reducing variance meta classifier) [4], Random Forest (algorithm constructing a forest of random trees) [5], and rule learner JRip (RIPPER: Repeated Incremental Pruning to Produce Error Reduction) [9]. Results enclosed in Table 4 confirm suggestions [17] that above-mentioned ensemble classification algorithms often outperforms a single classification methods (like C4.5), especially in classifying gene expression data. On the other side, ensemble methods may be rather "black-box" solutions that do not extract simple decision rules. Accuracy results for most of the classifiers are higher from the original TSP algorithm, but they are lower to their extensions. TSPDT solutions also have higher average credibility to those methods.

Table 4 Comparison of TSP decision trees accuracy with original methods. The highest classifiers accuracy for each dataset was bolded

Datasets	Classifiers accuracy				
	J48	Adaboost	Bagging	Random Forest	JRip
1. BC	52.63	57.89	63.15	68.42	73.68
2. CNS	56.66	75.00	71.66	73.33	65.00
3. CT	85.48	79.03	79.03	77.41	74.19
4. DF	79.22	90.90	85.71	88.31	77.92
5. LA	91.17	91.17	94.11	82.35	94.11
6. LCB	81.87	81.80	82.55	93.28	95.97
7. LCM	98.95	96.87	97.91	98.95	93.75
8. LCT	58.97	69.23	61.53	66.66	64.10
9. OC	97.23	99.20	97.62	98.41	98.81
10. PC	29.41	41.17	41.17	29.41	32.35
11. PCO	42.85	47.61	61.90	71.42	42.85
Average	70.40	75.44	76.03	77.09	73.88

4 Conclusion

Performed experiments suggest that proposed hybrid solution may successfully compete with decision trees and popular TSP algorithms for solving classification problems of microarray data. Results suggest that TSPDT may improve credibility and for some cases comprehensibility of the prediction model generated from TSP-family methods.

Furthermore, many possible improvements for decision trees with TSP algorithm still exist. One of the interesting field of endeavor is the tree pruning and the number of gene pairs tested in a single tree node. The problem of overfitting the tree model to the datasets (like in case colon tumor) is still not fully resolved, and we believe that the improvement on this field would reduce the tree size and increase classifier accuracy.

Acknowledgements This work was supported by the grant W/WI/5/08 from Białystok Technical University.

References

1. Alon, U., Barkai, N.: Broad patterns of gene expression revealed by clustering analysis of tumor and normal colon tissues probed by oligonucleotide arrays. Proceedings of the National Academy of Sciences of the USA, 96(12):6745–6750 (1999)
2. Bittner, M., Meltzer, P.: Molecular classification of cutaneous malignant melanoma by gene expression profiling. Nature, 406:536–540 (2000)
3. Breiman, L., Friedman, J.: Classification and Regression Trees, Wadsworth International Group, Belmont, CA, USA (1984)
4. Breiman, L.: Bagging predictors. Machine Learning, 24(2):123–140 (1996)
5. Breiman, L.: Random forests. Machine Learning, 45(1):5–32 (2001)
6. Brown, P.O., Botstein, D.: Exploring the new world of the genome with DNA microarrays. Nature Genetics, 21:33–37 (1999)
7. Cho, H.S., Kim, T.S.: cDNA microarray data based classification of cancers using neural networks and genetic algorithms. Nanotechnology, 1:28–31 (2003)
8. Czajkowski, M., Krętowski, M.: Novel extension of k-TSP algorithm for micro-array classification. Lecture Notes in Artificial Intelligence, 5027:456–465 (2008)
9. Cohen, W.W.: Fast Effective Rule Induction, Twelfth International Conference on Machine Learning, Morgan Kaufmann, San Francisco, CA, USA, 115–123 (1995)
10. Dhanasekaran, S.M.: Delineation of prognostic biomarkers in prostate cancer. Nature, 412:822–826 (2001)
11. Dudoit, S.J., Fridlyand, J.: Comparison of discrimination methods for the classification of tumors using gene expression data. Journal of the American Statistical Association, 97:77–87 (2002)
12. Duggan, D.J., Bittner, M.: Expression profiling using cDNA microarrays. Nature Genetics, 21(suppl 1):10–14 (1999)
13. Freund, Y., Schapire, R.E.: Experiments with a new boosting algorithm, Thirteenth International Conference on Machine Learning, San Francisco, CA, USA, 148–156 (1996)
14. Geman, D., dAvignon, C.: Classifying gene expression profiles from pairwise mRNA comparisons. Statistical Applications in Genetics and Molecular Biology, 3(1):19 (2007)
15. Grześ, M., Krętowski, M.: Decision tree approach to microarray data analysis. Biocybernetics and Biomedical Engineering, 27(3):29–42 (2007)

16. Hastie, T., Tibshirani, R.: The Elements of Statistical Learning. Springer, New York (2001)
17. Hu, H., Li, J.: A Maximally Diversified Multiple Decision Tree Algorithm for Microarray Data Classification, Workshop on Intelligent Systems for Bioinformatics, Hobart, Australia (2006)
18. Jinyan. L., Huiqing, L.: Ensembles of cascading trees, Proceedings of the Third IEEE International Conference on Data Mining, 585–588 (2003)
19. Kent Ridge Bio-medical Dataset Repository: http://datam.i2r.a-star.edu.sg/datasets/index.html
20. Kononenko, I.: Estimating Attributes: Analysis and Extensions of RELIEF. In: European Conference on Machine Learning, Catania, Italy, 171–182 (1994)
21. Lockhart, D.J., Winzeler, E.A.: Genomics, gene expression and DNA arrays. Nature, 405:827–836 (2000)
22. Lu, Y., Han, J.: Cancer classification using gene expression data. Information Systems, 28(4):243–268 (2003)
23. Mao, Y., Zhou, X.: Multiclass cancer classification by using fuzzy support vector machine and binary decision tree with gene selection. Journal of Biomedicine and Biotechnology, 2:160–171 (2005)
24. Murthy, S.: Automatic construction of decision trees from data: A multi-disciplinary survey. Data Mining and Knowledge Discovery, 2:345–389 (1998)
25. Nelson, P.S.: Predicting prostate cancer behavior using transcript profiles. Journal of Urology, 172:28–32 (2004)
26. Rokach, L., Maimon, O.: Top-down induction of decision trees classifiers - A survey. IEEE Transactions on Systems, Man, and Cybernetics - Part C, 35(4):476–487 (2005)
27. Sebastiani, P., Gussoni, E.: Statistical challenges in functional genomics. Statistical Science, 18(1):33–70 (2003)
28. Simon, R., Radmacher, M.D.: Pitfalls in the use of DNA microarray data for diagnostic and prognostic classification. Journal of the National Cancer Institute, 95:14–18 (2003)
29. Tan, A.C., Gilbert, D.: Ensemble machine learning on gene expression data for cancer classification. Applied Bioinformatics, 2:75–83 (2003)
30. Tan, A.C., Naiman, D.Q.: Simple decision rules for classifying human cancers from gene expression profiles. Bioinformatics, 21:3896–3904 (2005)
31. Quinlan, R.: Inductive knowledge acquisition: A case study. Addison-Wesley, Boston, MA, USA, chapt. 9, 157–173 (1987)
32. Quinlan, R.: C4.5: Programs for Machine Learning. Morgan Kaufmann, San Mateo, CA, USA (1993)
33. Witten, I.H., Frank, E.: Data Mining: Practical machine learning tools and techniques, 2nd edn. Morgan Kaufmann, San Francisco, CA, USA (2005)
34. Valentini, G., Muselli, M.: Bagged Ensembles of SVMs for Gene Expression Data Analysis, International Joint Conference on Neural Networks 2003, Portland, OR, USA (2003)
35. Veer, L. J., Dai, H.: Gene expression profiling predicts clinical outcome of breast cancer. Nature, 415:530–536 (2002)
36. Xu, L., Tan, A.C.: Robust prostate cancer marker genes emerge from direct integration of inter-study microarray data. Bioinformatics, 21(20):3905–3911 (2005)
37. Yoon, S., Kim, S.: k-Top scoring pair algorithm for feature selection in SVM with applications to microarray data classification. Soft Computing - A Fusion of Foundations, Methodologies and Applications, 14(2):151–159 (2009)
38. Zhang. H., Yu, C.Y.: Recursive partitioning for tumor classification with gene expression microarray data. Proceedings of the National Academy of Sciences of the USA, 98(12):6730–6735 (2001)
39. Zhang, C., Li, P.: Parallelization of multicategory support vector machines (PMC-SVM) from classifying microarray data. BMC Bioinformatics, 7(Suppl 4):S15 (2006)

Chapter 4
Predictive Minimum Description Length Principle Approach to Inferring Gene Regulatory Networks

Vijender Chaitankar, Chaoyang Zhang, Preetam Ghosh, Ping Gong, Edward J. Perkins, and Youping Deng

Abstract Reverse engineering of gene regulatory networks using information theory models has received much attention due to its simplicity, low computational cost, and capability of inferring large networks. One of the major problems with information theory models is to determine the threshold that defines the regulatory relationships between genes. The minimum description length (MDL) principle has been implemented to overcome this problem. The description length of the MDL principle is the sum of model length and data encoding length. A user-specified fine tuning parameter is used as control mechanism between model and data encoding, but it is difficult to find the optimal parameter. In this work, we propose a new inference algorithm that incorporates mutual information (MI), conditional mutual information (CMI), and predictive minimum description length (PMDL) principle to infer gene regulatory networks from DNA microarray data. In this algorithm, the information theoretic quantities MI and CMI determine the regulatory relationships between genes and the PMDL principle method attempts to determine the best MI threshold without the need of a user-specified fine tuning parameter. The performance of the proposed algorithm is evaluated using both synthetic time series data sets and a biological time series data set (*Saccharomyces cerevisiae*). The results show that the proposed algorithm produced fewer false edges and significantly improved the precision when compared to existing MDL algorithm.

1 Introduction

Reverse engineering of gene regulatory networks remains a major issue and area of interest in the field of bioinformatics and systems biology. A survey paper [1] discusses a number of models related to this area: viz. Bayesian Networks, Dynamic Bayesian Networks, Boolean Networks, Probabilistic Boolean Networks, Differential Equation Models, and Information Theory Models.

C. Zhang (✉)
School of Computing, The University of Southern Mississippi, MS 39402, USA
e-mail: chaoyang.zhang@usm.edu

Information theoretic models gained much attention due to their simplicity and low computational cost. Because of their low data requirements, they are suitable to infer even large-scale networks. Thus, they can be used to study global properties of large-scale regulatory systems [1].

A number of algorithms that implement information theoretic approaches have been proposed in the past [2–6]. The regulatory relationships between genes are derived based on mutual information (MI) in all these algorithms. MI measures the amount of information that can be obtained about one random variable by observing another one. Compared with the correlation coefficient-based metric, the MI is suitable for nonlinear relations and represents a good metric for evaluating the dependency between two random variables [7].

The minimum description length (MDL) principle has been implemented in [3, 4] to estimate the MI threshold. Various implementations of the MDL principle have been studied extensively in [8, 9]. The algorithm proposed in [4] often yields good results, but it does so with an ad hoc coding scheme that requires a user-specified tuning parameter. In the proposed algorithm, we implement the predictive minimum description length (PMDL) principle, which is better suited to time series data and combine it with the conditional mutual information (CMI) metric to reduce the false edges further. In particular, our scheme requires only one threshold parameter as against two threshold values that need to be specified in the scheme proposed in [2].

1.1 Contributions

Our major contribution is the implementation of PMDL principle, which eliminates the need of a fine-tuning parameter. Our work combines the PMDL principle with CMI for the first time to achieve better performance, CMI was used in the past but our scheme adds direction using an ad hoc time delay. We reported the threshold sensitivity of gene regulatory network inference schemes for the first time as it gives the users an estimate of the range of thresholds, which should be used.

The paper is organized as follows. Section 2 covers the Network Model, the information theoretic quantities (MI, CMI), the PMDL principle, and the proposed algorithm. Section 3 demonstrates the performance of the proposed algorithm over synthetic networks and biological time series data and finally we conclude in Sect. 4. A preliminary version of this work has appeared in [10].

2 Methods

The description length of the two-part MDL principle involves calculation of the model length and the data length. As the length can vary for various models, the method is in danger of being biased toward the length of the model [9]. The Normalized Maximum Likelihood Model has been implemented in [3] to overcome this issue. Another such model based on universal code length is the PMDL principle.

We chose to implement the PMDL principle as it suits time series data [11]. The concept of PMDL principle model was proposed in [12].

The description length for a model in PMDL [11] is given as

$$L_D = -\sum_{t=0}^{m-1} \log p(x_{t+1}|x_T), \qquad (1)$$

where $p(x_{t+1}|x_t)$ is the conditional probability or density function. We calculate the description length as data length given in [4]. The network MDL description is sum to model length and data length as given in (1). The model length in the network MDL principle implementation is the amount of memory used by the model in system.

Given the time series data, the data is first preprocessed, which involves filling missing values and quantizing the data. A connectivity matrix is maintained, which has two entries: a 0 and a 1. An entry of 0 indicates that no regulatory relationship exists between genes, but an entry of 1 indicates that a regulatory relationship exists between genes. After preprocessing, the MI matrix is generated. In an n gene network, we will have a connectivity matrix and MI matrix of order $n \times n$. Using every MI value in the matrix, a model is obtained as follows: the MI value is compared with every other value of the MI matrix. If the value is greater than or equal to the selected MI value, the corresponding entry in connectivity matrix is put as 1 else as 0. This way n^2 models are obtained their description lengths are computed. The model with minimum description length is then selected as the best model. After selecting the best model for every valid regulatory connection in the connectivity matrix, the CMI of the genes with every other gene is evaluated, and if any of these values is below the user-specified threshold, the connection is then deleted.

3 Results

3.1 Simulation on Random Synthetic Networks

The proposed algorithm is compared with the one proposed in [4] on synthetic random networks. Benchmark measures such as recall (R) and precision (P) are used to evaluate the performance of inference algorithms. While different definitions for recall and precision exist, in this paper, R is defined as $C_e/(C_e + M_e)$ and P is defined as $C_e/(C_e + F_e)$, where C_e denotes the edges that exist in the true network and in the inferred network, M_e are the edges that exist in the true network but not in the inferred network, and F_e are edges that do not exist in the true network but do exist in the inferred network.

For a specific size of the network, both the algorithms are run for different threshold values 30 times each and the average of P and R is calculated. The algorithms are run for 20, 30, 40, and 50 numbers of genes. The P and R values for each of these networks with different threshold values are given in Table 1.

Table 1 Precision and recall values for synthetic networks

Network size/method	PMDL		MDL	
	Precision	Recall	Precision	Recall
20	0.7900	0.4756	0.6939	0.5133
	0.9464	0.3733	0.8214	0.4451
	0.9944	0.2556	0.8834	0.4
30	0.7224	0.4785	0.6741	0.5015
	0.9044	0.3489	0.7897	0.4385
	0.9808	0.2474	0.8770	0.4163
40	0.6906	0.4806	0.6608	0.4567
	0.8746	0.3417	0.7658	0.3978
	0.9471	0.2528	0.8432	0.3867
50	0.6335	0.4716	0.6394	0.4764
	0.8590	0.3467	0.7401	0.4103
	0.9696	0.2591	0.8266	0.3920

Zhao et al. in [4] reported 0.2–0.4 as suitable values for the tuning parameter, and hence we use the values 0.2, 0.3, and 0.4 to build the networks. Based on the simulations of the proposed algorithm, we found that the threshold for CMI worked best for values in the range of 0.1–0.2. Thus, 0.1, 0.15, and 0.2 threshold values were used to build the networks.

It is observed that the P of the proposed algorithm is higher and R is lower in most of the cases as compared to the algorithm proposed in [4], referred as MDL in Table 1. The number of false negatives is less in the proposed algorithm, and as most biologists are interested in true positives, our results look promising.

3.2 Threshold Sensitivity

Here, we report the performance of our scheme based on different values of the user-specified threshold. For threshold values of 0.15 and 0.2, high precision (>90% in most cases) was observed but the recall for these thresholds was low (from 25% to 30%) as compared to a threshold value of 0.1, which had a fair recall (>47%) and good precision (63–79%) performance. It is observed that as the threshold value is increased, precision increases and at the same time the recall decreases. The simulation experiments show that 0.1 is the optimal threshold value.

3.3 Performance on Saccharomyces cerevisiae Data Set

The time series DNA microarray data from Spellman et al. [13] was used to infer gene regulatory networks. The Spellman experiment was chosen because it provides a comprehensive series of gene expression data sets for Yeast cell cycle. Four time series expression data sets were generated using four different cell synchronization methods: cdc15, cdc28, alpha-factor, and elutriation with 24, 17, 18, and 14 time points, respectively. The alpha-factor data set contained more time points than cdc28 and Elutriation data sets with fewer missing values than cdc15. Therefore, we choose to use time series expression data from alpha-factor method to infer the gene regulatory networks.

As mentioned earlier, preprocessing plays an important part in reverse engineering process. We found that log transformation and normalization of the data had negative effect on the results. Thus, we used the same preprocessing steps as in [4].

Initially, the data is quantized to 0 or 1. To quantize the expression values of every gene, they are sorted in ascending order and the first and last values of the sorted list are discarded as outliers, then the upper 50% are converted to 1 and the lower 50% is converted to 0. Any missing time points are set to the mean of their respective neighbors. If the missing time point is the first or the last one, it is set to the nearest time point value.

The true biological network used for comparison purposes is derived from the yeast cell cycle pathway [14]. A total of six networks were reverse engineered of which three were a result of our proposed method and remaining three were results of [4], the best of three in each case are used for comparison.

The true network and the networks reverse engineered using our proposed algorithm and in [4] are given in Fig. 1. Of the 30 edges inferred by our approach, nine are correctly inferred edges. The method proposed in [4] inferred a total of

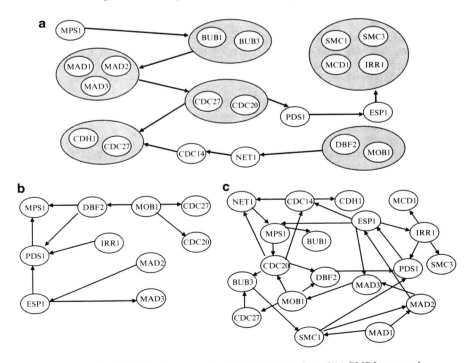

Fig. 1 Networks for (**a**) Biological network, (**b**) MDL approach, and (**c**) PMDL approach

nine edges of which only one is correctly inferred edge. The results favor our approach.

4 Conclusion

We have proposed a new algorithm that implements the PMDL principle for inferring gene regulatory networks from time series DNA microarray data and that eliminates the need of a fine-tuning parameter.

The evaluation results obtained from both synthetic and actual biological data sets show that the PMDL principle is effective in determining the MI threshold and the developed algorithm improves precision of gene regulatory network inference. Based on the sensitivity analysis of all tested cases, an optimal CMI threshold value has been identified. In the future, we will take into account prior knowledge to improve the recall of network inference and use this algorithm to infer gene regulatory networks related to neurological functions using Earthworm time series gene expression data sets with 31 time points generated in the Environmental Laboratory, U.S. Army Engineer Research and Development Center.

References

1. Hecker M, Lambeck S, Toepfer S, van Someren E, Guthke R (2009) Gene regulatory network inference: Data integration in dynamic models – A review. *Bio Systems*, 96, 1, 86–103.
2. Zhao W, Serpedin E, Dougherty ER (2008) Inferring connectivity of genetic regulatory networks using information-theoretic criteria. *IEEE/ACM Transactions on Computational Biology and Bioinformatics*, 5, 2, 262–274.
3. John D, Tabus I, Astola J (2008) Inference of gene regulatory networks based on a universal minimum description length. *EURASIP Journal on Bioinformatics and Systems Biology* (published online April 15, 2008).
4. Zhao W, Serpedin E, Dougherty ER (2006) Inferring gene regulatory networks from time series data using the minimum description length principle. *Bioinformatics*, 22, 17, 2129–2135.
5. Margolin AA, Nemenman I, Basso K, Wiggins C, Stolovitzky G, Dalla Favera R, Califano A (2006) ARACNE: An algorithm for reconstruction of genetic networks in a mammalian cellular context. *BMC Bioinformatics*, 7(Suppl 1), S7.
6. Liang S (1998) Reveal, A general reverse engineering algorithm for inference of genetic network architectures. *Pacific Symposium on Biocomputing*, 3, 18–29.
7. Cover TM, Thomas JA. (1991) Elements of information theory. Wiley-Interscience, New York.
8. Grünwald PD, Myung IJ, Pitt MA (2005) Advances in minimum description length (Theory and Applications). The MIT Press, Cambridge, MA.
9. Hansen MH, Yu B (2001) Model selection and the principle of minimum description length. *Journal of the American Statistical Association*, 96, 454, 746–774.
10. Chaitankar V, Zhang C, Ghosh P, Perkins EJ, Gong P, Deng Y (2009) Gene regulatory network inference using predictive minimum description length principle and conditional mutual information. *Proceedings of International Joint Conference on Bioinformatics, Systems Biology and Intelligent Computing*, 487–490.

11. Rissanen J (2006) An introduction to the MDL principle. Helsinki Institute for Information Technology, Tampere and Helsinki Universities of Technology, Finland, and University of London, England. (www.mdl-research.org/jorma.rissanen/pub/Intro.pdf).
12. Rissanen J (1984) Universal coding, information, prediction and estimation. *IEEE Transactions on Information Theory*, 30, 4, 629–636.
13. Spellman PT, et al. (1998) Comprehensive identification of cell cycle-regulated genes of the yeast *Saccharomyces cerevisiae* by microarray hybridization. *Molecular Biology of the Cell*, 9, 3273–3297.
14. Kanehisa M, et al. (2008) KEGG for linking genomes to life and the environment. *Nucleic Acids Research*, 36, D480–D484.

Chapter 5
Parsimonious Selection of Useful Genes in Microarray Gene Expression Data

Félix F. González-Navarro and Lluís A. Belanche-Muñoz

Abstract Machine learning methods have of late made significant efforts to solving multidisciplinary problems in the field of cancer classification in microarray gene expression data. These tasks are characterized by a large number of features and a few observations, making the modeling a nontrivial undertaking. In this study, we apply entropic filter methods for gene selection, in combination with several off-the-shelf classifiers. The introduction of bootstrap resampling techniques permits the achievement of more stable performance estimates. Our findings show that the proposed methodology permits a drastic reduction in dimension, offering attractive solutions in terms of both prediction accuracy and number of explanatory genes; a dimensionality reduction technique preserving discrimination capabilities is used for visualization of the selected genes.

Keywords Biological data mining and knowledge discovery · Cancer informatics · Gene expression analysis · Tools and methods for computational biology and bioinformatics

1 Introduction

In cancer diagnosis, classification of the different tumor types is of great importance. Traditional methods of tackling the distinction between different types of cancer are primarily based on morphological characteristics of tumorous tissue [8]. Machine learning methods are now extensively used for this task [3]. Typically, a gene expression dataset may consist of dozens of observations with thousands or even tens of thousands of genes. Classifying cancer types using this very high ratio between number of variables and number of observations is a delicate process, because of the high risk of overfitting the data. As a result, dimensionality reduction

L.A. Belanche-Muñoz (✉)
Departament de Llenguatges i Sistemes Informàtics, Universitat Politècnica de Catalunya, Omega Building, North Campus, 08034 Barcelona, Spain
e-mail: belanche@lsi.upc.edu

and in particular *feature selection* (FS) techniques may be very useful. The finding of small subsets of very relevant genes among a huge quantity could derive in much specific and efficient treatments. However, in a FS scenario, gene expression data analysis may entail a heavy computational consumption of resources, due to the extreme sparseness compared to standard datasets in classification tasks [34]. For these reasons, in this work we rely on *filter* measures for feature selection (that are classifier-independent) to keep the computational cost within reasonable bounds. We are also concerned with increasing the *reliability* of the FS process, in the sense of reducing the inherent instability caused by the particular choice of data sample. In addition, to further reduce the chance of overfitting the data, we take the decision of using low-complexity classifiers (five of the eight used classifiers are linear or quadratic) together with a small subset of highly relevant genes.

Of special importance in a practical medical setting is the *interpretability* of the solutions in terms of the obtained gene subsets. A dimensionality reduction technique that provides a data projection – while preserving the class discrimination achieved by a classifier – is also used in our study. Our experimental findings show that the proposed feature selection methodology offers highly competitive solutions both in terms of prediction accuracy and number of explanatory genes. In particular, we report results that offer better performance in both aspects at least for two of the analyzed microarray tasks. In addition, we provide biological evidence for the three most important genes obtained in each microarray dataset.

2 Feature Selection in the Microarray Domain

The bioinformatics community has recognized the FS process as a key issue in gene expression data analysis [25]. In many gene selection methods, a list of the *top ranked* genes based on some merit figure is generated, followed by an inductive step where a classifier is incrementally evaluated on the list [31]. Fisher's criterion [15], the *signal-to-noise* ratio [12], the χ^2 statistic [23], or Wilcoxon's rank sum test [9] are popular choices. However, considering individual contributions only can very likely hinder the discovery of interactions between genes.

Mutual information (MI) has been successfully used in FS for measuring the influence that a feature has over a class or target. Several criteria to evaluate subsets of features use it, mostly in the *bivariate* case, between a feature and the class. A few use a normalized variant defined by $C_{XY} = [I(X;Y)]/H(Y)$, where Y is the class, $I(X;Y) = H(Y) - H(Y|X)$ and H denotes the *entropy*. Note that $I(X;X) = H(X)$, since $H(X|X) = 0$ and $I(X;Y) = I(Y;X)$. The expression for C_{XY}, sometimes referred to as the *coefficient of constraint*, can be better understood by analyzing Fig. 1 (left). It can be seen that by increasing $I(X;Y)$, $H(Y|X)$ decreases; in other words, there is a reduction in the uncertainty of Y due to the action of X. The maximum value that $I(X;Y)$ could take is $H(Y)$. This property has been exploited to create an index to measure subsets of features with respect to a class or target value [35]. Alternatively, both *relevance* and *redundancy* of genes have been assessed using MI, configuring a criterion of minimum redundancy-maximum

5 Parsimonious Selection of Useful Genes in Microarray Gene Expression Data

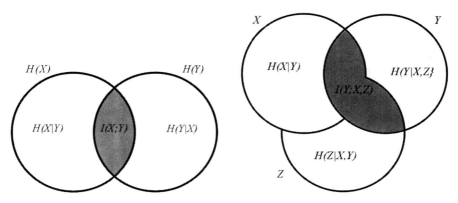

Fig. 1 Entropy concepts. *Left*: basic mutual information. *Right*: mutual information between two variables and class or target variable

relevance (mRMR) [10]. The reader is referred to [25] for a recent compilation on the use of these measures. The computation of MI can be extended from the bivariate to the multivariate case, of a number $n \geq 2$ of variables against another one, as $I(X_1,\ldots,X_n;Y) = \sum_{i=1}^{n} I(X_i;Y|X_1,\ldots,X_{i-1}) = H(Y) - H(Y|X_1,\ldots,X_n)$, whereas conditional MI is expressed in the natural way, as $I(X;Y|Z) = H(Y|Z) - H(Y|X,Z)$.

Instead of by conditioning, in this work we propose to calculate the MI between a class variable Y and two variables X and Z, as shown in Fig. 1 (right). The shaded area represents $I(Y;X,Z) = H(Y) + H(X,Z) - H(X,Y,Z)$, the information that X,Z explain about Z. Given that $I(Y;X,Z) \leq 1$ and that $H(Y)$ acts as the baseline reference, it is wise to normalize it as $[H(Y) + H(X,Z) - H(X,Y,Z)]/H(Y)$, obtaining an index that evaluates the influence of two variables with respect to a class. It takes values between 0 (no relevance) and 1 (maximum relevance).

In order for this expression to be of practical use from the FS point of view, it needs to be extended to the multivariate case. The MI between a subset of variables and the class variable is computed by generating first a "super-feature," obtained considering the concatenation of each combination of possible values of its forming features. In symbols, let $X = \{X_1,\ldots,X_n\}$ be the full feature set and consider a subset $\tau = \{\tau_1,\ldots,\tau_k\} \subseteq X$. A single feature V_τ can be obtained uniquely, whose possible values are the concatenations of all possible values of the features in τ (for completeness, define $V_\emptyset = \emptyset$). An *index of relevance* R of a feature X_i to a class Y with respect to a subset τ is then given by $R(X_i;Y|\tau) = [H(Y) + H(\tau,X_i) - H(\tau,Y,X_i)]/H(Y)$, for $X_i \in X, \tau \subseteq X \setminus X_i$. The use of the super-feature allows a faster implementation in case of discrete variables, whose development is not essential in this context.

To compute the necessary entropies, a discretization process is needed. This change of representation does not often result in a significant loss of accuracy (sometimes significantly improves it [27, 30]); it also offers reductions in learning time [7]. In this work, the CAIM algorithm was selected for two reasons: it is designed to work with supervised data and does not require the user to define a specific number

Algorithm 1: BGS^3 best gene subset search strategy

input : $X = \{X_1,\ldots,X_n\}$: Full gene set; C: Class feature
output: Φ : Best Gene Subset (BGS)

1. $\phi \leftarrow \underset{f \in X}{\operatorname{argmax}} \dfrac{H(C) - H(f,C)}{H(C)}$;
2. $\Phi \leftarrow \{\phi\}$: Current best subset;
3. $R \leftarrow \dfrac{H(C) - H(\Phi,C)}{H(C)}$: Current best relevance;
4. $exit \leftarrow false$;
5. **repeat**
6. $\quad \phi \leftarrow \underset{f \in X \setminus \Phi}{\operatorname{argmax}} \left\{ \dfrac{H(C) + H(\Phi,f) - H(\Phi,C,f)}{H(C)} \right\}$;
7. $\quad$ **if** $|\phi| > 1$ **then** $\phi^+ \leftarrow \underset{f \in \phi}{\operatorname{argmin}} I(\Phi,f)$ **else** $\phi^+ \leftarrow \phi$ **end**;
8. $\quad R^+ \leftarrow \dfrac{H(C) - H(\Phi \cup \phi^+, C)}{H(C)}$;
9. $\quad$ **if** $R^+ > R$ **then** $R \leftarrow R^+$; $\Phi \leftarrow \Phi \cup \phi^+$ **else** $exit \leftarrow true$ **end**
10. **until** $R^+ = 1 \vee exit \vee |\Phi| = n$

of intervals [21]. This way of calculating feature subset relevance is used to evaluate gene subsets, embedding it into a filter *forward-search* strategy, conforming the BGS^3 algorithm standing for *Best Gene Subset Search Strategy*, a supervised filter independent of the search strategy and of the a posteriori inducer, described in the listing below. This algorithm begins by selecting the feature that maximizes its relevance with respect to the class feature (lines 1–3). Then a forward search is conducted: at every step, the feature providing the maximum value of relevance when added to the current subset is selected (line 6). If, at the end of a step, more than one feature renders the same value for relevance (line 7), the feature that produces the *minimum redundancy* of information is chosen. In case the newly added feature brings a benefit, it is added to the current subset (line 9). The algorithm stops when the index of relevance was not improved, its maximum value has been reached, or it has run out of features, whichever comes first (line 10).

3 Experimental Work

The experimental methodology was aimed to achieve results that reflect the true behavior of the system as much as possible, in other words, to obtain *reliably* relevant genes. Bootstrap resampling techniques are used to yield a more stable and thus more reliable measure of predictive ability. The original microarray expression datasets S were used to generate $B = 5,000$ bootstrap samples $S_1,\ldots,S_B$ that play the role of *training sets* in the feature selection process: each relevance value calculated in the algorithm is the *average* across the B bootstrap samples, i.e., the *average behavior* of a feature is used to guide and stabilize the algorithm.

The algorithm is first applied to the discretized bootstrap samples to obtain the best gene subset or BGS (one for each dataset). Then the classifier development stage is conducted using those *original* continuous features that are members of their corresponding BGSs. Eight classifiers were evaluated by means of ten times of tenfold cross validation (10×10cv), a method suitable to handle small-sized datasets: the *k-nearest-neighbors* technique with Euclidean metric (kNN) and parameter $k \in \{1, \ldots, 15\}$, the *Naïve Bayes classifier* (NB), the *linear* and *quadratic discriminant* classifiers (LDC and QDC), *logistic regression* (LR), and the *support vector machine* with linear, quadratic, and radial kernels (lSVM, qSVM and rSVM) and parameter C (regularization constant) (with $C = 2^k$, k running from -7 to 7). The rSVM has the additional smoothing parameter $\sigma = 2^k$, k running from -7 to 7).

Validation of the described approach uses five public-domain microarray gene expression datasets, shortly described as follows: *Colon tumor*: 62 observations of colon tissue, of which 40 are tumorous and 22 normal, 2,000 genes [1]. *Leukemia*: 72 bone marrow observations and 7,129 probes: 6,817 human genes and 312 control genes [12]. The goal is to differentiate acute myeloid leukemia (AML) from acute lymphoblastic leukemia (ALL). *Lung cancer*: distinction between malignant pleural mesothelioma and adenocarcinoma of lung [13]; 181 observations with 12,533 genes. *Prostate cancer*: used in [33] to analyze differences in pathological features of prostate cancer and to identify genes that might anticipate its clinical behavior; 136 observations and 12,600 genes. *Breast cancer*: 97 patients with primary invasive breast carcinoma; 12,600 genes analyzed.

The results of the FS stage are presented in Table 1. Sizes of final BGSs for each dataset are considerable small (as low as three genes in the *Leukemia* problem); remarkably, in all cases the maximum relevance is achieved. Computational times are also reported, ranging from few minutes to several hours – 8 h at most.[1] These times should be judged taking into account that 5,000 resamples for each dataset are

Table 1 Gene subsets selected by BGS^3: $|BGS|$ is their number, R_{max} is the final relevance achieved, and "CPU time" indicates total processing time

Dataset	$\|BGS\|$	R_{max}	CPU time	Gene accession number (GAN) or gene name
Colon tumor	10	1	13 min	M26383 M63391 M76378 X12671 J05032 H40095 H43887 R10066 U09564 H40560
Leukemia	3	1	14 min	M23197_at X95735_at U46499_at
Lung cancer	13	1	4 h	37957_at 1500_at 36536_at 35279_at 33330_at 39643_at 32424_at 40939_at 33757_f_at 33907_at 179_at 39798_at 1585_at
Breast cancer	16	1	4 h	AL080059 NM_003258 NM_003239 NM_005192 Contig7258_RC NM_006115 AL137615 Contig38901_RC AL137514 AF052087 U45975 AF112213 AB037828 NM_005744 NM_018391 NM_003882
Prostate cancer	21	1	8 h	37639_at 37720_at 37366_at 31538_at 37068_at 40436_g_at 39755_at 31527_at 1664_at 34840_at 36495_at 33674_at 39608_at 31545_at 914_g_at 41288_at 37044_at 40071_at 34730_g_at 41732_at 41764_at

---

[1] These figures were obtained in a standard $\times 86$ machine at 2.666 GHz.

Table 2 Final accuracy results with comparison to other references. (**F**) indicates a Filter algorithm, (**W**) a wrapper, and (**FW**) a combination of both. Size of the final gene subset and the used classifier are in brackets

Work	Validation	Colon tumor	Leukemia	Lung cancer	Breast cancer	Prostate cancer
BGS^3(F)	10 × 10cv	89.36	97.89	98.84	83.37	93.43
		(9, 3NN)	(2, NB)	(4, LR)	(12, lSVM)	(3, 10NN)
[5] (F)	200 × 0.632	88.75	98.2	–	–	–
	bootstrap	(14, lSVM)	(23, lSVM)	–	–	–
[31] (W)	10 × 10cv	85.48	93.40	–	–	–
		(3, NB)	(2, NB)	–	–	–
[36] (W)	100 × random	87.31	–	72.20	–	–
	subsampling	(94, SVM)	–	(23, SVM)	–	–
[4] (W)	50 × holdout	77.00	96.00	99.00	79.00	93.00
		(33, rSVM)	(30, rSVM)	(38, rSVM)	(46, rSVM)	(47, rSVM)
[17] (FW)	10 × 10cv	–	–	99.40	–	96.30
		–	–	(135, 5NN)	–	(79, 5NN)
[16] (F)	10 cv	–	98.6	99.45	68.04	91.18
		–	(2, SVM)	(5, SVM)	(8, SVM)	(6, SVM)

being processed. The composition of each BGS is signaled by the gene identifier. These unique IDs will be used to find biological evidence about the significance of the gene in the disease. Recall that the BGSs are constructed adding at every step the gene most informative to the current subset, every new set having more informative power. It seems therefore sensible, in terms of classification performance, to parsimoniously explore the obtained subsets in an incremental fashion, respecting the order in which the genes were found – which is the order reported in Table 1.

The accuracy results are presented in Table 2: shown are the final number of genes obtained in the incremental search from the initial BGSs, giving the best 10 × 10cv accuracy and the used classifier. To be sure, a nonparametric Wilcoxon signed-rank test is used for the (null) hypothesis that the median of the differences between the errors of the *winner* classifiers per dataset and another classifier's error is zero. This hypothesis can be rejected at the 99% level in *all* cases (*p*-values not shown). It is remarkable that in the *Lung Cancer* dataset, as low as four genes are required to get almost 99% of accuracy. On the other hand, the *Breast Cancer* dataset is one of the most difficult problems, followed by *Colon Tumor* (83 and 89%).

It is a common practice to compare to similar works in the literature. Unfortunately, the methodological steps are in general very different, especially concerning resampling techniques, making an accurate comparison a delicate undertaking. Nonetheless, such a comparison is presented in Table 2. Six references which are illustrative of recent work are included. As stated before, the *Colon Tumor* dataset presents difficulties in classification; however, BGS^3 figures are higher than the rest, even with less genes involved and in front of solutions that use a pure wrapper strategy. For the *Leukemia* data, other references achieve better figures, some of them using a much bigger gene subset – 23 or 30 genes. Two results report two genes in their solutions, [31] and [16]. The former does not match the gene subset obtained by our algorithm, and the latter does not give precise information on the obtained genes. The *Lung Cancer* dataset is apparently the easiest to separate. Values as high as 99% are achieved by three of the referenced works, making use of much bigger subset sizes. Incidentally, the solution in [16] agrees in one gene, the 1500_at *WT1-Wilms tumor 1*. The solution offered by BGS^3 in the *Breast Cancer* dataset gives the best result among the references, with almost 4% of difference. The *Prostate Cancer* dataset is well separated by [17], using 79 genes. BGS^3 separates it using only two genes, with a degradation of 3% of accuracy. No information is provided in this reference about which genes are selected. The good performance achieved with low numbers of selected genes are an indication that these are really good ones in separating the classes. However, even if interpretable by mere inspection of the involved genes, the final selection of genes may still provide few clues about the structure of the classes (cancer types). Visualization in a low-dimensional representation space may become extremely important, helping oncologists to gain insights into what is undoubtedly a complex domain. We use in this work a method based on the decomposition of the scatter matrix – arguably a neglected method for dimensionality reduction – with the remarkable property of maximizing the separation between the projections of compact groups of data. This method leads onto the definition of low-dimensional projective spaces with good separation between classes, even when the data covariance matrix is singular; further details about this method can be found in [22]. Such visualization is illustrated by the plots in Fig. 2. These are scatter plots of 2D projections of the classes (using the first two eigenvectors of

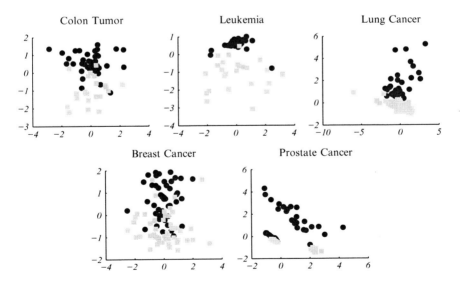

Fig. 2 Visualization of the solutions using the first two eigenvectors of each scatter matrix. In *Colon Tumor* and *Prostate Cancer*, *circles* represent tumorous samples and *squares* indicate normal tissue; in *Leukemia circles* indicate acute lymphoblastic leukemia cells and *squares* acute myeloid leukemia cells; in *Lung Cancer circles* are malignant pleural mesothelioma and *squares* are adenocarcinoma; in *Breast Cancer circles* indicate relapse and *squares* non-relapse

the scatter matrices). It can be seen that separation is in general quite good (although far from perfect); also, the structure itself of the two classes provides clues on the variability of each cancer type.

4 Conclusions

Experimental work in comparison to recent works examining the same datasets reveals that the developed methodology provides very competitive solutions, characterized by small gene subsets and affordable computational demands. The use of resampling methods to stabilize the gene selection process – arguably the most delicate part – has shown to deliver final solutions of very low size and strong relevance. Very noteworthy is the fact that the best classifiers (those that make good use of the gene subsets found in the feature selection phase) are consistently very simple: the *k-nearest-neighbors* technique, *Naïve Bayes classifier*, *logistic regression*, and a linear *support vector machine*. A final concern has been the *practical* use of the results in a medical context, achieved thanks to a dimensionality reduction technique that preserves the class discrimination capabilities. This joint information may become extremely important to help oncologists gain insights into this highly sensitivedomain.

Appendix: Biological Evidence

Biological evidence is assembled in the medical literature studying each specific gene. Only the first (i.e., the more relevant) three genes of each subset are presented, for conciseness.

Colon Tumor: **[M26383]** *IL8-Interleukin 8* encodes a protein member of the CXC chemokine family and is one of the major mediators of the inflammatory response [26], associated with a higher likelihood of developing colon tumor recurrence [24]. **[M63391]** *DES-Desmin* encodes a muscle-specific class III intermediate filament. [19] reported that Interleukin 8 and Desmin act as the central elements in colon cancer susceptibility. **[M76378]** *CSRP1-Cysteine and glycine-rich protein 1* may be involved in regulatory processes important for development and cellular differentiation.

Leukemia: **[M23197_at]** *CD33-CD33* is a putative adhesion molecule of myelomonocytic-derived cells that is expressed on the blast cells in patients with acute myeloid leukemia [29]. **[X95735_at]** *ZYX-ZYXIN* is involved in the spatial control of actin assembly and in the communication between the adhesive membrane and the cell nucleus. This is a gene found in many cancer classification studies (e.g., [6, 9, 12]) and is highly correlated with acute myelogenous leukemia. **[U46499_at]** *MGST1-Microsomal glutathione S-transferase 1* encodes a protein thought to protect membranes from oxidative stress and toxic foreign chemicals and plays an important role in the metabolism of mutagens and carcinogens [26].

Lung Cancer: [37957_at] *ATG4-Autophagy related 4 homolog A*. Autophagy is activated during amino-acid deprivation and has been associated with neurodegenerative diseases, cancer, pathogen infections and myopathies [32]. **[1500_at]** *WT1-Wilms tumor 1* has an essential role in the normal development of the urogenital system; this gene is expressed in several cancer diseases [2]. **[36536_at]** *SCHIP1-Schwannomin interacting protein 1*. The product of the neurofibromatosis type 2 (NF2) tumour suppressor gene, known as schwannomin or merlin, is involved in NF2-associated and sporadic schwannomas and meningiomas [14].

Breast Cancer: [AL080059] *TSPYL5-TSPY like 5*. The gene TSPYL5 encodes testis-specific Y-like protein but its role in human cancer has not been fully understood. Gene expression altered by DNA hypermethylation is often associated with cancers. **[NM_003258]** *TK1-Thymidine kinase 1, soluble* is a cytoplasmic enzyme. Studies have shown that in patients with primary breast cancer, high TK values were shown to be an important risk factor in node-negative patients and seemed to be associated with beneficial effect of adjuvant chemotherapy [28]. **[NM_003239]** *TGFB3-transforming growth factor, beta 3* is a potent growth inhibitor of normal epithelial cells. In established tumor cell systems, however, experimental evidence suggests that TGF-bs can foster tumorhost interactions that indirectly support the progression of cancer cells [11].

Prostate Cancer: [37639_at] *HPN-Hepsin*. Hepsin is a cell surface serine protease and plays an essential role in cell growth and maintenance of cell morphology and it is highly related with prostate cancer and benign prostatic hyperplasia. **[37720_at]** *HSPD1-Heat shock 60kDa protein 1* encodes a member of the chaperonin family; [20] established HSPD1 as a biomarker for prostate malignancy. **[37366_at]** *PDLIM5-PDZ and LIM domain 5*. Although medical evidence for direct implication of PDLIM5 in prostate cancer was not found, it is a gene recurrently found in several works – e.g., [18].

References

1. Alon, U., et al.: Broad patterns of gene expression revealed by clustering analysis of tumor and normal colon tissues probed by oligonucleotide arrays. In: Proceedings of the National Academy of Sciences USA **96**(12) 6745–6750 (1999)
2. Amin, K., et al.: Wilms' tumor 1 susceptibility (wt1) gene products are selectively expressed in malignant mesothelioma. The American Journal of Pathology **146**(2) 344–356 (1995)
3. Duan, K.B., et al.: Multiple SVM-RFE for gene selection in cancer classification with expression data. IEEE/ACM Transactions on Nanobioscience **4**(3) 228–234 (2005)
4. Bu, H.L., et al.: Reducing error of tumor classification by using dimension reduction with feature selection. In: The First International Symposium on Optimization and Systems Biology, Beijing, China, 232–241 (2007)
5. Cai, R., et al.: An efficient gene selection algorithm based on mutual information. Neurocomputing **72** 991–999 (2009)

6. Chakraborty, S.: Simultaneous cancer classification and gene selection with bayesian nearest neighbor method: An integrated approach. Computational Statistics and Data Analysis **53**(4) 1462–1474 (2009)
7. Catlett, J.: On changing continuous attributes into ordered discrete attributes. In: Proceedings of the European working session on Machine learning, Springer, New York, 164–178 (1991)
8. Chu, F., Wang, L.: Applications of support vector machines to cancer classification with microarray data. International Journal of Neural Systems **15**(6) 475–484 (2005)
9. Chu, W., et al.: Biomarker discovery in microarray gene expression data with gaussian processes. Bioinformatics **21**(16) 3385–3393 (June 2005)
10. Ding, C., Peng, H.: Minimum redundancy feature selection from microarray gene expression data. In: Proceedings of IEEE Computational Systems Bioinformatics (2003)
11. Dumont, N., Arteaga, C.: Transforming growth factor-β and breast cancer: Tumor promoting effects of transforming growth factor-β. Breast Cancer Research **2** 125–132 (2000)
12. Golub, T., et al.: Molecular classification of cancer: Class discovery and class prediction by gene expression monitoring. Science **286**(5439) 531–537 (October 1999)
13. Gordon, G.J., et al.: Translation of microarray data into clinically relevant cancer diagnostic tests using gene expression ratios in lung cancer and mesothelioma. Cancer Research **62** 4963–4967 (September 2002)
14. Goutebroze, L., et al.: Cloning and characterization of SCHIP-1, a novel protein interacting specifically with spliced isoforms and naturally occurring mutant NF2 proteins. Molecular and Cellular Biology **20**(5) 1699–1712 (2000)
15. Hedenfalk, I., et al.: Gene-expression profiles in hereditary breast cancer. The New England Journal of Medicine **344** 539–548 (2001)
16. Hewett, R., Kijsanayothin, F.: Tumor classification ranking from microarray data. BMC Genomics **9**(2) (2008)
17. Hong, J.H., Cho, S.B.: Cancer classification with incremental gene selection based on DNA microarray data. In: IEEE/ACM Transactions on Computational Biology and Bioinformatics 70–74 (2008)
18. Hong-Qiang, W., et al.: Extracting gene regulation information for cancer classification. Pattern Recognition **40**(12) 3379–3392 (2007)
19. Jiang, W., et al.: Constructing disease-specific gene networks using pair-wise relevance metric: Application to colon cancer identifies interleukin 8, desmin and enolase 1 as the central elements. BMC Systems Biology **2** (2008)
20. Johansson, B., et al.: The prostate. Proteomic comparison of prostate cancer cell lines LNCaP-FGC and LNCaP-r reveals heatshock protein 60 as a marker for prostate malignancy **66**(12) 1235–1244 (2006)
21. Kurgan, L.A., Cios, K.J.: Caim discretization algorithm. IEEE Transactions on Knowledge and Data Engineering **16**(2) 145–153 (2004)
22. Lisboa, P., et al.: Cluster based visualisation with scatter matrices. Pattern Recognition Letters **29**(13) 1814–1823 (2008)
23. Liu, H., Setiono, R.: Chi2: Feature selection and discretization of numeric atributes. In: IEEE 7th International Conference on Tools with Artificial Intelligence, 338–395 (1995)
24. Lurje, G., et al.: Polymorphisms in VEGF and IL-8 predict tumor recurrence in stage III colon cancer. Annals of Oncology **19** 1734–1741 (2008)
25. Meyer, P.E., Schretter C., Bontempi, G. Information-theoretic feature selection in microarray data using variable complementarity. IEEE Journal of Selected Topics in Signal Processing **2**(3) (2008)
26. National center of biothecnology information. http://www.ncbi.nlm.nih.gov/
27. Ng, M., Chan, L.: Informative gene discovery for cancer classification from microarray expression data. In: IEEE Machine Learning for Signal Processing, 393–398 (2005)
28. Öhrvik, A., et al.: Sensitive nonradiometric method for determining thymidine kinase 1 activity. Clinical Chemistry **50**(9) 1597–1606 (2004)
29. Plesa, C., et al.: Prognostic value of immunophenotyping in elderly patients with acute myeloid leukemia: A single-institution experience. Cancer **112**(3) 572–580 (2007)

30. Potamias, G., et al.: Gene selection via discretized gene-expression profiles and greedy feature-elimination. In: SETN, 256–266 (2004)
31. Ruiz, R., et al.: Incremental wrapper-based gene selection from microarray data for cancer classification. Pattern Recognition **39** 2383–2392 (2006)
32. Scherz-Shouval, R., et al.: Reactive oxygen species are essential for autophagy and specifically regulate the activity of Atg4. The EMBO Journal **26** 1749–1760 (2007)
33. Singh, D., et al.: Gene expression correlates of clinical prostate cancer behavior. Cancer Cell **1** 203–209 (March 2002)
34. Tang, Y., et al.: Development of two-stage svm-rfe gene selection strategy for microarray expression data analysis. IEEE/ACM Transactions on Computational Biology and Bioinformatics **4**(3) 365–381 (2007)
35. Wang, H.: Towards a Unified Framework of Relevance. PhD thesis, University of Ulster (1996)
36. Wang, L., et al.: Hybrid huberized support vector machines for microarray classification and gene selection. Bioinformatics **24**(3) 412–419 (2008)

Chapter 6
Hierarchical Signature Clustering for Time Series Microarray Data

Lars Koenig and Eunseog Youn

Abstract Existing clustering techniques provide clusters from time series microarray data, but the distance metrics used lack interpretability for these types of data. While some previous methods are concerned with matching levels, of interest are genes that behave in the same manner but with varying levels. These are not clustered together using an Euclidean metric, and are indiscernible using a correlation metric, so we propose a more appropriate metric and modified hierarchical clustering method to highlight those genes of interest. Use of hashing and bucket sort allows for fast clustering and the hierarchical dendrogram allows for direct comparison with easily understood meaning of the distance. The method also extends well to use k-means clustering when a desired number of clusters are known.

Keywords Gene pattern discovery and identification · Microarrays

1 Introduction

Clustering microarray data remains the principal tool to extract genes of interest for researchers. Genes of interest include genes that show differential expression between experimental conditions or similar group behavior over the course [1,3]. Clustering and other analysis tools are the prime tools to extract this information from the data quickly, allowing further experiments to be run to test the new set of genes. Traditional clustering algorithms such as k-means [4], hierarchical clustering, and k-medoids are powerful methods to determine the clusters of similar patterns from a variety of data. For any of these methods to produce the desired results, however, they must use an appropriate distance metric to correctly classify similar data. Using a correlation metric [1,5] between pairs of n-dimensional points provides a distance based on the linear dependence, but quantifying the dissimilarity with regard to the

E. Youn (✉)
Department of Computer Science, Texas Tech University, Lubbock, TX 79409, USA
e-mail: eun.youn@ttu.edu

biological aspect is difficult. Two genes that show similar patterns of rise and fall in expression may have the same correlation as two that vary greatly. Such gene pairs would be indistinguishable using a correlation metric and the important similarity of near matches is lost. Fourier transform [6] and statistical analysis [7], each offers different methods of extracting similar patterns from the time series data. Ernst et al. [8] took advantage of the limited combinations for short time series experiments to analyze significant groups with similar behavior patterns. However, the algorithm fails to extend to larger time series due to the exponential growth of potential patterns. Phang et al. [9] introduced trajectory clustering to cluster genes using only their signed changes in expression. For a small number of genes or time points, this is sufficient to create meaningful clusters. When the number of time points grows however, the results of such clustering may degrade into single gene clusters.

Traditional metrics, including Euclidean and Hamming, are capable of clustering microarray data, but these distances between genes in time-series fail to recognize a more important pattern in the data: genes of varying levels that behave similarly. An Euclidean distance may find genes with similar expression levels at each time point, but genes with similar changes in expression at different levels may be considered distant and belong to several different clusters. By analyzing the changes in expression levels of genes throughout an experiment, clusters of genes that behave similarly can be extracted. These genes, more so than genes of similar expression levels, are of interest. The underlying assumption behind this is the same as that for correlated expression levels: those genes are co-regulated. When attempting to determine differentially regulated genes between two samples, genes that change cluster and the new clusters that are formed may establish a subset of genes to investigate further. Also, core clusters, which are found to have identical change patterns, may be of interest without further clustering. The rest of the paper is organized as follows. A new clustering method called Hierarchical Signature Clustering (HSC) is introduced in Sect. 2. Section 3 presents our clustering results on two publicly available time series microarray data and comparison with traditional clustering methods. Section 4 is about further discussion.

2 Methods

The clustering method consists of two traditional parts. A distance metric determines the distance between any two genes' expression patterns and a clustering method creates clusters of genes from this information. The initial treatment of the data involves removing genes that show no or little change over the entire course of the experiment. These genes are assumed to be nondifferentially regulated and are removed from the clustering as in trajectory clustering. If such genes were involved in the activity of interest, they should show significant departure from the standard expression levels. As the expression results are log ratios, an expression level near zero means that the gene showed little change between control and experimental conditions. To filter genes with little change, a near-zero instead of an actual no change can be used herewith the same result. They may be used for postprocessing if they are known to be important regulators, but since they show little change they are not considered in the clustering steps.

Instead of being interested in genes with similar expression levels or ratios, we are interested in genes that show the same pattern of change in expression levels throughout the experiment. Genes with few or no points of difference are grouped together first, with additional genes clustered later. To simplify the signature pattern of each gene, we only consider the sign of the change between each pair of consecutive time points, similar to Ernst et al. [8]. This reduces the number of data points for each gene by one, giving a signature for each gene with $k-1$ signs, where k is the number of time points in the time series microarray experiment. The ith sign of a signature is the sign of the change in the ith and $i+1$th expression levels. While the signature consists of a series of -1, $+1$, and 0, it can be treated as a number in base-3, giving a numerical value for every signature for the time series. This facilitates sorting by allowing a simple bucket sort [10] for every gene. Each signature's numerical representation is hashed, using mod n (the number of genes), to give a target bucket to place the gene. The contents of the bucket are checked to see whether genes already placed there match the signature of the current gene. If not, a linear probe is conducted to find the next available spot. If this collision has occurred before, the similar genes' bucket will be discovered during the probe and the gene will be correctly placed. For long time series, the number of combinations is usually much larger than the number of filtered genes, making such collisions rare. This method provides deterministic results given the same ordering of the data and provides the initial cluster centers. The number of possible signatures increases exponentially with the number of data points, possibly leading to overflow or allocation problems if k is too large, even if just generating empty buckets. Instead, the number of buckets required is defined by n, the number of genes to be clustered: If all n genes are distributed to their own bucket, there can be at most n buckets used. As the table fills, performance may degrade but the effect is limited by the limited number of genes available, usually fewer than 10,000 in the entire dataset and many fewer in the filtered genes.

For cases where the number of time points is small, for example, less than 8, there is likely to be more genes than possible signatures. As a result, some buckets must end up with more than one gene by chance. Further analysis may be necessary to determine whether the clusters are statistically significant as there must be overlap in these cases, but such treatment is ignored by the clustering algorithm itself. Cases that generate buckets with more than the expected number of genes, from either a large number of genes or possible permutations, are treated as more significant and are used in the next step for clustering centers. These genes all share the same behavior at every step during the experiment and may be of further interest themselves. Once the bucket sort is complete, the buckets are sorted based on the number of genes they contain, from largest to smallest. These large buckets form the initial cluster centers, consisting of a group of genes with the same pattern of behavior between every time step. During the clustering step, genes with less similar signatures are merged into these centers.

Once the genes have been sorted into initial clusters, the buckets are merged with each other. Starting with the first and largest target bucket, the Hamming distance from each bucket's signature is calculated. The buckets with $t = 1$ point of difference are removed from the list and their genes are merged into the target bucket, keeping the signature of the original bucket for future merges. After we merge all genes with one point of difference from the genes in the first target bucket, we move to the next bucket in the updated list. By merging downward, there is no chance at some later step that we will need to merge a previous target from the same step since it would have been merged earlier. Since the previous signature is preserved, the distances are only calculated once for all buckets. It is also this merging method that introduces non-deterministic behavior into the algorithm: If the genes are ordered differently, the targets will change, making those merged genes different and the clusters may form differently until even the last step. This does introduce a way to validate clusters, however. By changing the order of genes input, different clusters can be formed and compared to each other after multiple trials to see which constitute a more constant cluster and which are inconstant [11]. The distance between signatures will not change, only the order in which genes are merged and their direction of merging will change. Since we choose signatures with t points of distance from some target, those members joined may have max $(2t, k - 1)$ points of distance from any other signature in the bucket. This process is repeated, shortening the list with each merging step until we have been through the entire list once. Since we put all buckets with the chosen distance into the target bucket, we may add any number of genes during the step if their distance is the same; large buckets may be merged with other large buckets if the distance between them is correct. The process is then repeated from the beginning of the new list for genes with $t = 2$ points of difference up to $k - 1$ points of difference, if necessary. If we have $k - 1$ points in the signature, then between any two genes there can be at most $k - 1$ points of difference, but because of the property that the genes may have $2t$ points of difference, merging may complete after $\lfloor k/2 \rfloor$ steps, or it may continue if the initial target buckets have greater distance than this, although all their members may not. After the final step, all genes belong to the same bucket and the algorithm is complete.

To represent the merging, a dendrogram representation is created using a modified distance metric. The dendrogram accurately reflects the merging process and the distances on the tree remain whole numbers, easily understood as the Hamming distance between targets and mergers. This binary tree representation is created during runtime as merges are carried out. Without maintaining this distance measure in the dendrogram, the tree presented does not match that created by the algorithm itself. By keeping track of the buckets at every step t, we can pick an arbitrary level of the tree and find which genes belonged to clusters at that step. By tracing down the tree and tracking levels, the optimal number of clusters can be found as normal for hierarchical clustering; however, the tree is more intuitive when trying to determine good cutoff values of t.

Normally, HSC generates clusters without any user input. It can be modified to incorporate a k-means clustering when good guesses for values of k are known.

K-means traditionally selects single elements to use as cluster centers and computes the nearest center for each other point in the set. Further expanding on the concept that the largest cluster is of interest, we choose the k largest clusters at some cutoff value for t. We then compute the nearest center for each remaining cluster or singleton and merge them together. By performing the clustering for multiple cutoff values, we can arrive at several potential clusterings. The result is deterministic as before, so by scrambling the initial inputs, different clusterings for each level can be obtained for comparison. This method allows for direct comparison of an existing clustering by guaranteeing that they have the same number of clusters present for comparison.

3 Results

3.1 S. cerevisiae

For initial testing, we chose the *S. cerevisiae* data used by Spellman et al. [1] in both original and updated form. Genes with missing values from the alpha-factor, cdc15, and elutriation trials were removed for initial testing. The gene information was then processed into signatures and clustered using the new method, k-means correlation on the raw data and k-means Hamming on the signatures.

For comparison with the previous findings, we performed the clustering using the full gene dataset without filtering to compare to the findings in the original papers as well as correlation for the single data set. The high number of genes prevents easy visualization of the data, but allows more new combinations of genes to emerge. The original data was the concatenation of eight separate experiments to form a single series with every gene present. The removal of incomplete time series causes the discovered clusters to have missing elements, but allows for correct comparison using the new method. The use of these data with correlation coefficient is unaffected by these missing elements since the covariance and standard deviation can still be calculated. While there are only two pairs of genes which share perfect agreement, the distribution of genes into their clusters is very regular with several hundred small clusters at the intermediate levels. For reference, we use the clusters identified in the Spellman [1] and Eisen [2] papers and look for analogs in the HSC results. While both papers use the same set of data, the Spellman paper focuses on the biological significance of the results while the Eisen paper generated the clustering method used by both. As a good cut-off point, we have chosen the 26th level. At this point, there are 292 clusters from the 3222 genes. At this level, there are 99 clusters of genes with 8 or more members and an additional 73 non-singleton clusters. The top clusters contain 2916 genes which are compared to previous clusters discovered for similarities.

A common cluster identified by both papers is one containing histone-related genes. Of the nine original genes, six are included in the dataset: HHO1, HTA2,

HHF2, HHT2, HTB1, and HTA1. These genes begin to cluster together with 12 out of 81 points of difference. By 14 points, four of six genes are clustered together and by the 22nd point, all genes belong to the same cluster. By the 26th level, the cluster contains 27 members. The genes here cluster tightly despite the differences in the alpha-factor experiment. The original genes from the histone cluster have good correlation, ranging from 0.9586 to 0.7435 for the six genes. HSC forms an initial cluster containing five of six genes with one previously unclustered gene, ERV25, at 20 points of difference. While ERV25 has only 20 points of difference, it has a correlation coefficient of 0.6638 to HTB1, which has the signature used for the entire cluster for comparison. The final merge occurs at 21 points, bringing the single cluster with HHO1 along with a two similar clusters containing seven total genes. The Hamming distance for all the members is no higher than that of HHO1 to HTB1 however, even though the correlation is lower and points of difference are higher. The intercluster correlation of each smaller cluster is much lower than the tight cluster of the histones, and when introduced to the existing cluster, creates correlation coefficients as low as -0.0348. This is due to the algorithm's distance property, which allows for the large variance in intercluster distance, which in turn creates smaller correlation coefficients. The maximum intercluster distance within the Spellman cluster is 24 points of difference with the minimum correlation coefficient of 0.7435. Within the HSC cluster, the maximum distance is 41 points with a minimum correlation of -0.0348. The additional genes show very poor correlation with the original cluster, although similarities can be found in the sharp peaks that exist through the data. The better correlation of the ERV25 gene would suggest that it may actually have related function, while the other genes clustered are false positives for co-regulated genes. The cluster discovered by HSC is shown in Fig. 1. The original cluster's genes are grey with the new genes inserted in black.

Next identified by Spellman is a gene cluster involving the methionine pathway. Eight of 20 genes from the original M cluster are present in the screened data. These genes show very little correlation in contrast to the previous cluster. The genes' correlation coefficients range from 0.8059 for genes MET3 and MET17, but have a lowest correlation coefficient of -0.1013. The original paper describes the cluster as containing 10 genes involved in methionine synthesis while the other members were likely to be involved in methionine metabolism. Fifteen of the genes were found to have similar upstream regions for Met binding to support this. The poor correlation of these genes is seen in the HSC clustering, which has the genes form three clusters for a majority of the algorithm. Three of the five MET genes are clustered by step 20, which is the minimum distance between the three. The distances of the remaining genes are all greater than 30 and as such are not merged until the final steps. This is due, in part, to their similarities to cluster centers over their previously clustered genes and, in part, to the poor correlation they show. By the 30th step, there are two clusters in which each contain three of the eight genes, but the clusters are so large at this point that resolution of similarities is very difficult.

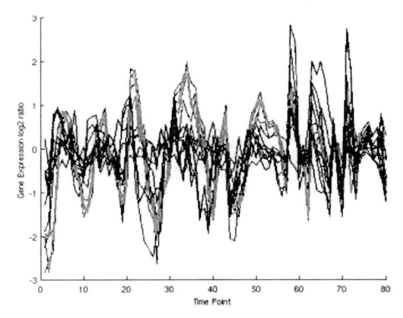

Fig. 1 HSC discovered histone cluster

3.2 G. hirsutum

As a contrasting data set, we chose G. *hirsutum* L. cv. TM1 (cotton) with only six time points [12]. Despite the reduced number of time points, the same filter left 1442 genes from the initial 12063. Due to the smaller number of time points (6), only 243 (3^5) possible combinations exist, leaving a distribution with an expected six genes per initial cluster. While the expectation for multiple genes in a bucket is higher than the previous, more than 15% had at least 9, up to 74 genes in two buckets initially. Both traditional method and HSC find the same initial clusters from the distance metric. From the initial level, standard hierarchical clustering creates one large cluster with six smaller clusters and then merges everything together. While data can be extracted from the initial level, past this it is difficult to compare the 1368 genes placed together in the largest cluster and the entire set at the third level. Between hierarchical and HSC clustering the secondary clusters show noticeable difference in membership beyond the largest cluster. For each step beyond the third merge, however, the members of each cluster may be at maximum distance within-cluster. The highlighted clusters with a distance of 2 must share at least one point of similarity. It is at this level that the standard algorithm creates one cluster containing all genes. This is a product of the small number of time points and would exist for any data set using the standard hierarchical clustering. As the number of time points decreases, traditional clustering yields poorer clusters due to the decreased number of possible combinations and merging steps. Both methods maintain the generalized triangle inequality for each cluster formed at each step, however. HSC clustering has 4 upper levels each with smaller clusters. By merging into a cluster by initial signature, better separation is created and the clusters have more relevance to the initial cluster center. Despite the similar merging criteria, the two methods determine different outlier clusters in addition to the division of the large secondary cluster.

Gene functionality annotation was not available for this experiment, but by comparing the genes for each method's clusters reveals a similar pattern as before with expanded and additional clusters in HSC. The traditional cluster creates a large cluster of more than 1300 genes in the second level, making individual resolution impossible from the initial cluster. HSC takes longer to converge to such a large cluster and has those signatures closer to the root cluster for each merge step.

For additional comparison, we used correlation hierarchical clustering as well as k-means on the original data. As a whole, the data set shows very good correlation. The correlation values for the data have many within 0.9 to as low as 0.13. The largest cluster, which contains 1321 out of the 1360 genes, has correlation better than 0.92. The remaining 24 clusters are largely single genes, which do not create separate clusters, but are individually merged into the large single cluster. The correlation of the data in this set is so high that the differences in correlation coefficients between over 90% of the genes are less than 0.1. Part of this property is due to the fact that there are only five time points in the experiment, creating fewer time points to show disagreement as well as losing some resolution in the data over the entire course of the experiments. In addition, the genes studied may have been selected based on their known or suspected role, giving the set as a whole to have higher than normal correlation for the genes.

4 Discussion

Hierarchical Signature Clustering can be performed very efficiently and provides an easily interpretable distance metric for clustering. It can also be easily modified to satisfy different expression level and distance criteria. The basic sign of change can be converted to use ranges with varying scale instead as done in Ernst et al. [8], giving a larger number of signatures for the same dataset but giving better resolution of genes that change similarly. In scoring, the Hamming distance can be weighted with regard to the positions of points of difference to allow for better scores between groups that have sequential blocks of difference and similarity as this may signify co-regulation for only part of the observed experiment. The method still suffers from the distance criteria of any Hamming metric, but the end results are more easily read and the reasoning behind the clusters more understandable. Clusters containing genes of interest may be placed at the beginning of the bucket list to serve as the initial cluster centers that are merged into since core clusters are determined by the order in which they appear in the initial list, finding genes with similar patterns relative to the signature of the initial bucket.

A final extension includes probing for counter-expressed genes. After initial clusters are created, gene clusters with opposite signatures can be found. Where expression levels of genes from one group increase, levels of genes from the other decrease and vice versa. This pairing gives potential targets for regulation between groups: Members of one cluster may induce the observed change in the other. With the small number of genes involved, focused tests can be run to determine which genes actually participate.

The flexibility gives many points at which the overall method can be augmented to deliver a more tailored result while maintaining the efficient time and space complexity of the general method. The method is comparable to many of the standard clustering algorithms in use and gives similar results with an easily understood explanation for why a cluster is composed as it is.

References

1. Spellman, P.T. et al. (1998) Comprehensive identification of cell cycle-regulated genes of the yeast Saccaromyces cerevisiae by microarray hybridization. *Mol. Biol. Cell*, 9, 3273–3297.
2. Eisen, M.B. et al. (1998) Cluster analysis and display of genome-wide expression patterns. *Proc. Nat'l Acad. Sci. USA*, 95(25):14863-8.
3. Zou, M. and Conzen, S.D. (2005) A new dynamic Bayesian network (DBN) approach for identifying gene regulatory networks from time course microarray data. *Bioinformatics*, 21, 71–79.
4. Hartigan, J.A. and Wong, M.A. (1979) A k-means clustering algorithm. *Appl. Stat.*, 28, 100–108.
5. Bhattacharya, A. and De, R.K. (2008) Divisive Correlation Clustering Algorithm (DCCA) for grouping of genes: detecting varying patterns in expression profiles. *Bioinformatics*, 24, 1359–1366.
6. Kim, J. and Kim H. (2008) Clustering of change patterns using Fourier coefficients. *Bioinformatics*, 24, 184–191.
7. Park, T. et al. (2003) Statistical tests for identifying differentially expressed genes in time-course microarray experiments. *Bioinformatics*, 19, 694–703.
8. Ernst, J. et al. (2005) Clustering short time series gene expression data. *Bioinformatics*, 21, 159–168.
9. Phang T.L., Neville, M.C., Rudolph, M. and Hunter, L. (2003) Trajectory clustering: a nonparametric method for grouping gene expression time courses, with applications to mammary development. *Pacific Symposium on Biocomputing*, 351–362.
10. Dobosiewicz, W. (1978) Sorting by Distributive Partition. *Information Processing Letters*, 7, 1–6.
11. Bréhélin, L., Gascuel1 O. and Martin O. (2008) Using repeated measurements to validate hierarchical gene clusters. *Bioinformatics*, 24, 682–688.
12. Alabady, M.S., Youn, E. and Wilkins, T.A. (2008) Double feature selection and cluster analyses in mining of microarray data from cotton. *BMC Genomics*, 9, 295.

Chapter 7
Microarray Database Mining and Cell Differentiation Defects in Schizophrenia

Aurelian Radu, Gabriela Hristescu, Pavel Katsel, Vahram Haroutunian, and Kenneth L. Davis

Abstract The causes of schizophrenia remain unknown, but a key role of oligodendrocytes and of the myelination process carried out by them has gained increasing support. The adult human brain parenchyma contains a relatively large population of progenitor cells that can generate oligodendrocytes. Defects in these adult oligodendrocyte progenitor cells (OPCs) or in their proliferation/differentiation have received little attention as potential causes of schizophrenia yet. We compared the set of genes whose expression is modified in schizophrenia, as revealed by our microarray studies, with genes specifically expressed in stem cells, as revealed by studies on human embryonic stem cells. We also evaluated the genes that are upregulated when stem cells engage in differentiation programs. These genes can be viewed as fingerprints or signatures for differentiation processes. The comparisons revealed that a substantial fraction of the genes downregulated in the brains of persons with schizophrenia belong to the differentiation signature. A plausible interpretation of our observations is that a cell differentiation process, possibly of adult OPCs to oligodendrocytes, is perturbed in schizophrenia. These observations constitute an incentive for a new direction of study, aimed at investigating the potential role of OPCs in schizophrenia.

1 Introduction

The etiology of schizophrenia (SZ) is so far unknown. A concept that has gained increasing support is that SZ is associated with a deficiency in the myelination process, or more generally in the functioning of the cells responsible for myelination, the oligodendrocytes (ODs). The initial study that introduced this concept [1] was confirmed by subsequent reports from the same and other groups on gene

---

A. Radu (✉)
Department of Developmental and Regenerative Biology, Mount Sinai School of Medicine, One Gustave L. Levy Place, New York, NY 10029, USA
e-mail: aurelian.radu@mssm.edu

expression [2–12] and at protein level [3, 13]. The main evidence that led to this concept was based on genome-wide microarray studies of multiple brain areas. These studies revealed consistent SZ-associated differences in the expression levels of genes involved in myelination or genes specifically expressed in ODs.

ODs, which represent a major constituent of the white matter, are terminally differentiated postmitotic cells, generated during early development from neural progenitor cells [14]. A relatively large population of oligodendrocyte progenitor cells (OPC), which represents 3–9% of the cells, persists, however, in the adult brain parenchyma and most likely continues to generate new ODs during adult life [15–20]. One explanation of the mentioned microarray data is that the pathological processes that generate the SZ-associated differences are initiated in the differentiated ODs. It is, however, possible that the differences occur in fact in the progenitor cells, or in their proliferation/differentiation process, and are inherited by the differentiated ODs. This possibility has not been elaborated and discussed in the literature, with the exception of a recent study that appeared while this work was in progress [7].

A preliminary assessment of the potential role of OPC differentiation in the etiology of SZ can be made in principle by comparing the above-mentioned SZ microarray data with microarray studies that investigated differentiation processes. Unfortunately no genome-wide expression data are available specifically for OPCs differentiation. However, studies of other cell types can yield relevant information, because many cellular processes have a general nature and occur in almost identical form in multiple systems. Due to the recently increased interest in embryonic stem cells (ESCs), several groups have tried to generate a fingerprint or signature of "stemness" using microarrays, namely to identify the genes that are highly expressed in ESCs compared with various differentiated cells. A recent study revealed that careful analysis of three of these reports generates a common and relatively large set of genes that are highly expressed in human ESCs [21]. The study also identified a set of genes that are consistently underexpressed in human ESCs compared to differentiated cells. We compared these sets of genes with the genes whose expression is modified in SZ. The results are compatible with a role of defective OPC differentiation as a causal factor in SZ.

2 Methods

The microarray data used in this study are described in detail in our previous publications [4, 5]. Analysis of 14 brain regions revealed that the temporal and cingulate cortices and the hippocampal formation represent regions of particular abnormality in SZ and may be more susceptible to the disease process(es) [4, 5]. For the investigations described in this report, we used the data for the superior temporal cortex (BA22), which showed the maximal number of altered transcripts in SZ group compared to controls. This dataset contains expression levels for approximately 6,300 genes, from which approximately 900 are downregulated in SZ and approximately 600 are upregulated.

The stem cell database is a combination of datasets generated by three studies aimed at identifying genes that are expressed specifically in human ESCs, and therefore constitute a signature or fingerprint of "stemness" [21]. The analysis revealed that all three studies detected a common and relatively large set of "stemness" genes [21]. Among the 734 genes for which all three studies provided reliable data, 115 were found by all studies to be upregulated and will be referred to as the "stemness" (ST) set of genes. A set of 95 genes was also identified, for which all studies reported higher expression levels in the differentiated cells compared with the ES cells. Given the variety of differentiated cells used in the three studies, these 95 genes can be viewed as a general fingerprint or signature of the differentiated character, and will be referred here to as the "DIF" set of genes. The set of 734 genes, which contains the 115 "stemness" genes and the 95 DIF genes, will be referred to as the stemness/differentiation (S/D) set or database. The operations regarding the gene lists were done using either standard Microsoft Excel 2007 functions and operations or custom application programs written in C++.

3 Results

We compared a set of genes generated by SZ microarray studies [4, 5] with a set presented in a recent publication on stem cell signature [21], referred to here as the stemness/differentiation (S/D) set. The S/D set contains a subset of stemness genes (ST), and a subset that represents a differentiation signature (DIF), as detailed in Sect. 2. Because of the relatively small size of the S/D set, only 321 genes are common between the SZ and S/D datasets and could be compared. As a consequence, all the results presented below are strictly speaking valid only for this common subset of genes. However, except for the fact that the 321 genes were analyzed by all three studies, they are not different in any other way that we are aware of from the remaining 6,300 genes analyzed by the SZ studies. [We found for instance that a possible source of bias, namely different signal strength of the 321 genes compared with the full set, does not occur (data not shown).] Therefore, the observations made by us using the 321 genes are expected to be valid for the full set of genes affected in SZ.

The analysis revealed that a substantial fraction of the genes downregulated in SZ belongs to the DIF set, as detailed below. Among the 321 genes present in both databases, 46 are downregulated in SZ and 44 belong to the DIF set. Fourteen genes belong to both sets, which represent 30.4% of the genes that are downregulated in SZ, and 31.8% of the DIF genes. In order to evaluate whether these relatively high values are statistically meaningful, we generated as controls other sets of 46 genes, extracted randomly from the remaining genes that are not downregulated in SZ (Fig. 1a). The outside left bar represents the genes analyzed in SZ and the subset of genes downregulated in SZ (dark gray), while the outside right bar represents the S/D genes and the DIF genes (light gray). The middle bars represent the subset of genes that are common between the SZ and the S/D databases. The black arrow

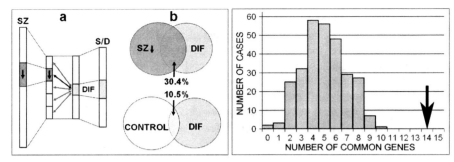

Fig. 1 Comparison of the set of genes downregulated in SZ with the DIF gene set. *Left panel*: (**a**) Schematic representation of the sets of genes that have been compared. (**b**) Venn diagram showing that 30.4% of the genes downregulated in SZ belong to the DIF set, while the corresponding average percentage for the control sets is 10.5%. *Right panel*: the number of genes that are common between the set of genes downregulated in SZ and the DIF set (*arrow*), compared with the distribution of control numbers of common genes. On the horizontal axis – number of common genes between the compared sets; on the vertical axis – number of occurrences. The distribution can be approximated by a normal distribution, which leads to a chance probability value of $p = 2 \times 10^{-6}$ for the occurrence of 14 common genes between SZ and DIF sets

represents the comparison between the set of genes downregulated in SZ and the DIF set, while the gray arrows represent control comparisons. We verified how many DIF genes are present in these control sets. The result is $10.5 \pm 4.3\%$ (standard deviation) (Fig. 1b).

The same procedure was applied in reverse, namely sets of 44 genes have been generated by random extraction from the 321 genes, after removal of the genes that belong to the DIF set. We verified subsequently how many of the genes downregulated in SZ are present in each of these "control" sets. The result is similar to the previous comparison, $11.2 \pm 4.6\%$.

Both types of comparisons (control sets of 46 vs. the 44 DIF set, and control sets of 44 vs. the 46 genes downregulated in SZ) represent comparisons of 44 vs. 46 sets of genes, they are equivalent and they represent instances of the same statistical distribution. The process of random extraction of control sets described above was repeated 24 times for each of the two symmetrical variants, to get a representation of the distribution. The result is presented in Fig. 1, right panel. The horizontal axis represents the number of common genes between the compared sets, and on the vertical axis the number of occurrences. Visual inspection of the distribution suggests that the value of 14 is, with high confidence, different than the control values. In order to substantiate this assertion, we confirmed first that the shape of the distribution is compatible with a normal distribution: the skewness is 0.007 and the kurtosis is -0.47. For a normal distribution with the average of 4.88 and a standard deviation of 2.01, the chance probability value for 14 common genes between SZ and DIF sets is $p = 2 \times 10^{-6}$.

The remaining three comparisons do not reveal any overlap in excess of the control values. Using similar methods, we verified whether the genes downregulated in

Table 1 Percentages of common genes (overlap) between all the combinations of SZ upregulated or downregulated genes and the ST or DIF genes

Overlap Between:		Overlap (%)	Control (%)[a]	p
SZ DOWN and DIF	SZ DOWN	30.4	10.5 ± 4.3	0.000002
	DIF	31.8	11.2 ± 4.5	0.000002
SZ DOWN and ST	SZ DOWN	10.8	9.5 ± 3.2	0.34
	ST	14.2	14.2 ± 4.9	0.50
SZ UP and DIF	SZ UP	10.5	8.1 ± 3.9	0.27
	DIF	4.5	3.8 ± 2.1	0.37
SZ UP and ST	SZ UP	5.2	4.0 ± 2.3	0.30
	ST	2.8	5.4 ± 3.4	0.22

[a]Control values were obtained by substituting repeatedly a set of interest with six sets derived by random extraction of genes that do not belong to the set of interest. The extractions were repeated 24 times and the results were averaged

SZ are enriched in "stemness" genes, and whether the genes that are upregulated in SZ are enriched in "stemness" or DIF genes. As shown in Table 1, none of these comparisons revealed values significantly different than the controls. In conclusion, the only significant finding is that the genes downregulated in SZ are enriched in DIF genes.

4 Discussion

The enrichment of the SZ-downregulated gene set in genes whose expression is upregulated during stem cell differentiation suggests that a defect in a cell differentiation process has a causal role in SZ etiology. Other published information supports this hypothesis. For instance, some of the genes that have been convincingly linked to SZ are known to play important roles in cell differentiation: QKI [10, 22], Sox10 [5, 9, 23, 24], OLIG2 [23–25], and neuregulin [1, 26].

Adult OPCs represent a minor proportion of cells among the parenchymal cells, and this fact is seemingly incompatible with the idea that the SZ microarray data offer information about adult OPCs: it can be argued that the RNAs contributed by OPCs are highly diluted within the parenchymal RNA pool, and differences in their levels are unlikely to be detectable in homogenate-based assays. This view ignores the fact that the expression levels of many genes expressed in ODs are inherited from and reflect expression levels in the OPCs from which they originate. Moreover, the expression levels for many genes in ODs is defined during differentiation and can reflect defects in the process, even if at the time of the microarray analysis only very few cells are engaged in differentiation.

Multiple differentiation processes at various stages of development are involved in generating the cells of the adult CNS. Some of these processes occur in the embryo, others in early postnatal development, while still others generate differentiated cells from progenitors that exist in the adult brain. The question is which differentiation process could be involved in SZ. Three observations appear to be relevant: (1) ODs in the adult brain are involved in SZ; (2) the DIF set of genes is not typical for advanced differentiation processes; and (3) the genes downregulated in SZ are not enriched in brain-specific genes. These considerations suggest that the differentiation process affected in SZ is at an early stage of differentiation of the OPCs. "Early" could mean an early stage in prenatal development, during which defects in gene expression first occur, and are subsequently inherited by the progressively more differentiated cells along the lineage. This interpretation would be in agreement with possible insults to prenatal brain development, which have been the focus of most SZ etiologic research for the past two decades [27].

Less obvious but more interesting, "early" could mean an early stage of differentiation of adult OPCs. This stage could be equivalent to the transition from stem to transit amplifying cells in other stem-based cell systems, in which the cells do not express markers of advanced differentiation yet. This possibility is as legitimate as the prenatal or early postnatal injury hypothesis and has the advantage that it is experimentally verifiable. To our knowledge this possibility has not been explicitly described or explored in the published SZ literature. A recent study examined the expression in SZ of 17 genes expressed in OPCs and in their derivatives [7]. Differences compared to normal controls were found only for four genes expressed after the terminal differentiation of the OPCs. The analyzed set of genes is, however, too small to draw conclusions about defects of lack thereof in the differentiation process of adult OPCs.

In conclusion, our observations suggest that the primary causal defect in SZ may occur initially in adult OPCs, or during their early differentiation.

The involvement of adult OPCs in SZ is compatible with other known facts about SZ and represents an experimentally testable hypothesis. Comparisons of OD abundance in SZ patients vs. controls have revealed lower absolute numbers or percentages, of 20–29% in various brain areas of SZ patients [2, 28, 29]. While these observations are referred to as "reductions" or "decreases" in SZ patients, it is perfectly possible that they represent in fact failures to achieve the increases in OD numbers that occur, as mentioned earlier, in adolescence and early adult life. These data may in fact represent direct confirmations of the hypothesis that the process of adult oligodendrogenesis is defective in SZ.

Another study revealed that the proliferation rate of neural progenitor cells located in the subventricular zone and the subgranular zone of the dentate gyrus of the hippocampus is reduced in SZ patients [30]. Some of the neural progenitors cells from the adult subventricular zone express in fact the OD lineage transcription factor 2 (OLIG2), and primary cultures of these cells give rise to ODs [31]. It is therefore possible that the reduced proliferation found in the subventricular zone of SZ patients represents a particular aspect of a more general situation, namely reduced proliferation of OPCs in many brain areas.

An essential feature of our hypothesis is that it is experimentally verifiable. OPCs and ODs can be isolated from postmortem white matter of SZ patients and can be studied in culture [16, 19, 31]. The properties of these cells as such and during differentiation induced in culture can be compared with cells from control patients.

Agents that could reduce the differences between SZ-derived and control cultures could be identified as leads for potential drug development. In contrast, it is not possible to perform such studies to search for prenatal or early postnatal defects in OPCs or ODs, because at that stage it is not known which donors will develop SZ. It is possible that some SZ cases may occur because of perinatal injury, while others because of adult life deficits. Information obtained by studying the second possibility, which is amenable to experimental investigation, could be useful for the first category, which is not as directly accessible.

In conclusion, we believe that our observations warrant a systematic pursuit of a new direction in SZ research, namely the investigation of OPCs from adult brain white matter.

Acknowledgements We thank Drs Mayte Suárez-Fariñas and Marcelo Magnasco from the Rockefeller University for sharing data, critical reading of the manuscript, and evaluation of the computational/statistical aspects, and Dr Gregory Elder from the Mount Sinai School of Medicine for suggestions and comments. This research was funded by MH45212 (KLD), MH064673 and VA-MIRECC (VH). The generation of the microarray data used here was supported in part by research sponsored by Gene Logic Inc. (Gaithersburg, MD).

References

1. Hakak Y, Walker JR, Li C et al (2001) Genome-wide expression analysis reveals dysregulation of myelination-related genes in chronic schizophrenia. Proc Natl Acad Sci USA 98: 4746–4751.
2. Davis KL, Haroutunian V (2003) Global expression-profiling studies and oligodendrocyte dysfunction in schizophrenia and bipolar disorder. Lancet 362: 758.
3. Dracheva S, Davis KL, Chin B et al (2006) Myelin-associated mRNA and protein expression deficits in the anterior cingulate cortex and hippocampus in elderly schizophrenia patients. Neurobiol Dis 21: 531–540.
4. Katsel P, Davis KL, Haroutunian V (2005) Variations in myelin and oligodendrocyte-related gene expression across multiple brain regions in schizophrenia: a gene ontology study. Schizophr Res 79: 157–173.
5. Katsel P, Davis KL, Haroutunian V (2004) Large-scale microarray studies of gene expression in multiple regions of the brain in schizophrenia and Alzheimers disease. Int Rev Neurobiol 63: 41–81.
6. McCullumsmith RE, Gupta D, Beneyto M et al (2007) Expression of transcripts for myelination-related genes in the anterior cingulate cortex in schizophrenia. Schizophr Res 90(1–3): 15–27.
7. Barley K, Dracheva S, Byne W (2009) Subcortical oligodendrocyte- and astrocyte-associated gene expression in subjects with schizophrenia, major depression and bipolar disorder. Schizophr Res 112(1–3): 54–64.
8. Tkachev D, Mimmack ML, Huffaker SJ et al (2007) Further evidence for altered myelin biosynthesis and glutamatergic dysfunction in schizophrenia. Int J Neuropsychopharmacol 10(4): 557–63.
9. Aston C, Jiang L, Sokolov BP (2004) Microarray analysis of postmortem temporal cortex from patients with schizophrenia. J Neurosci Res 77: 858–866.
10. Aberg K, Saetre P, Jareborg N et al (2006) Human QKI, a potential regulator of mRNA expression of human oligodendrocyte-related genes involved in schizophrenia. Proc Natl Acad Sci USA 103: 7482–7487.

11. Iwamoto K, Bundo M, Yamada K et al (2005) DNA methylation status of SOX10 correlates with its downregulation and oligodendrocyte dysfunction in schizophrenia. J Neurosci 5: 5376–5381.
12. Sugai T, Kawamura M, Iritani S et al (2004) Prefrontal abnormality of schizophrenia revealed by DNA microarray Impact on glial and neurotrophic gene expression. Ann NY Acad Sci 1025: 84–91.
13. Flynn SW, Lang DJ, MacKay AL et al (2003) Abnormalities of myelination in schizophrenia detected in vivo with MRI, and post-mortem with analysis of oligodendrocyte proteins. Mol Psychiatry 8: 811–820.
14. Liu Y, Rao M (2004) Glial progenitors in the CNS and possible lineage relationships among them. Biol Cell 96: 279–290.
15. Polito A, Reynolds R (2005) NG2-expressing cells as oligodendrocyte progenitors in the normal and demyelinated adult central nervous system. J Anat 207: 707–716.
16. Nunes MC, Roy NS, Keyoung HM et al (2003) Identification and isolation of multipotential neural progenitor cells from the subcortical white matter of the adult human brain. Nat Med 9: 439–447.
17. Nishiyama A, Watanabe M, Yang Z et al (2002) Identity, distribution, and development of polydendrocytes: NG2-expressing glial cells. J Neurocytol 31: 437–445.
18. Ruffini F, Arbour N, Blain M et al (2004) Distinctive properties of human adult brain-derived myelin progenitor cells. Am J Pathol 165: 2167–2175.
19. Windrem MS, Nunes MC, Rashbaum WK et al. (2004) Fetal and adult human oligodendrocyte progenitor cell isolates myelinate the congenitally dysmyelinated brain. Nat Med 10: 93–97.
20. Dawson MR, Polito A, Levine JM et al (2003) NG2-expressing glial progenitor cells: an abundant and widespread population of cycling cells in the adult rat CNS. Mol Cell Neurosci 24: 476–488.
21. Suárez-Fariñas M, Noggle S, Heke M et al (2005) Comparing independent microarray studies: the case of human embryonic stem cells. BMC Genomics 6: 99.
22. Haroutunian V, Katsel P, Dracheva S et al (2006) The human homolog of QK1 gene affected in the severe dysmyelination "quaking" mouse phenotype is downregulated in multiple brain regions in schizophrenia. Am J Psychiatry 163: 1–3.
23. Kessaris N, Pringle N, Richardson WD (2001) Ventral neurogenesis and the neuron-glial switch. Neuron 31: 677–680.
24. Stolt CC, Rehberg S, Ader M et al (2002) Terminal differentiation of myelin-forming oligodendrocytes depends on the transcription factor Sox10. Genes Dev 16: 165–170.
25. Georgieva L, Moskvina V, Peirce T et al (2006) Convergent evidence that oligodendrocyte lineage transcription factor 2 (OLIG2) and interacting genes influence susceptibility to schizophrenia. Proc Natl Acad Sci USA 103: 12469–12474.
26. Harrison PJ, Weinberger DR (2005) Schizophrenia genes, gene expression, and neuropathology: on the matter of their convergence. Mol Psychiatry 10: 40–68.
27. Rapoport JL, Addington AM, Frangou S et al (2005) The neurodevelopmental model of schizophrenia: update. Mol Psychiatry 10: 434–449.
28. Hof PR, Haroutunian V, Friedrich VL Jr et al (2003) Loss and altered spatial distribution of oligodendrocytes in the superior frontal gyrus in schizophrenia. Biol Psychiatry 53: 1075–1085.
29. Uranova NA, Vostrikov VM, Orlovskaya DD et al (2004) Oligodendroglial density in the prefrontal cortex in schizophrenia and mood disorders: a study from the Stanley Neuropathology Consortium. Schizophr Res 67: 269–275.
30. Reif A, Fritzen S, Finger M, Strobel A et al (2006) Neural stem cell proliferation is decreased in schizophrenia, but not in depression. Mol Psychiatry 11: 514–522.
31. Menn B, Garcia-Verdugo JM, Yaschine C et al (2006) Origin of oligodendrocytes in the subventricular zone of the adult brain. J Neurosci 26: 7907–7918.

Chapter 8
miRNA Prediction Using Computational Approach

A.K. Mishra and D.K. Lobiyal

Abstract The total *number* of miRNAs in any species even completely sequenced is still an open problem. However, researchers have been using limited techniques for miRNA prediction. Some of the prediction models are developed using different sets of features of pre-miRNA from model organisms. We also focused on exploring dominating set of features using principal component analysis and Info gain in our earlier work, with the assumption that it will reduce complexity of the model and may increase prediction accuracy. In this chapter, we have used attribute relevance analysis techniques for selecting essential attributes based on their relevance. We have explored dominating feature extraction using different machine learning techniques from a set of known mature miRNA sequences. The results are encouraging since the essential attributes selected here are biologically significant. These attributes can be used in deriving rules for miRNA identification.

Keywords Homology · miRNA · Relevance · Prediction

1 Introduction

MicroRNAs (miRNAs) are an evolutionary conserved class of non-coding RNAs of small length approximately 20–25 nucleotides (nt) long and found in diverse organisms, and miRNAs play a very important role in various biological processes. They regulate gene expression at post-transcriptional level by repressing or in-activating target genes [1, 2]. Primary miRNA (pri-miRNA) is several hundred nucleotide transcript processed in the nucleus by Drosha which is a multi-protein complex containing enzymes [3]. It generates ~70nt in case of animals and ~60 to a few hundred nucleotides in case of plants miRNA stem-loop precursor.

A.K. Mishra (✉)
Jawaharlal Nehru University, New Delhi, India
and
U.S.I., Indian Agricultural Research Institute, New Delhi, India
e-mail: akmishra@iari.res.in; misamr@rediffmail.com

The secondary structure plays a vital role for Drosha substrate recognition rather than the primary sequence [4]. miRNA biogenesis is highly associated with stem-loop feature of its precursor's secondary structure. As pre-miRNA secondary structures consisting of stem-loop are highly conserved across different species, extracting informative attributes from secondary structure is a significant step in the identification of miRNA from unknown sequences [5, 6]. There are different approaches to predict the secondary structure of RNA. These can be classified as energy minimization-based, grammar-based, matching-based and evolutionary algorithm-based approaches. Free energy minimization is one of the most popular methods for the prediction of secondary structure of RNA [7, 8].

1.1 Need for Computational Prediction for miRNA Gene Finding

The three fundamental techniques used to isolate miRNA are forward genetics screening, direct cloning and computation analysis through bioinformatics tools. But the most common method is to isolate and clone the miRNA obtained though biological samples, and this method has been widely used in plant miRNA identification. However, miRNAs are difficult to clone due to its short length, very low expression level at specific cell and under specific condition. Hence, it is clear that computational techniques are needed for identification of miRNA genes in any sequenced genome. The same is true for the homology-based search for which there was no clear evolutionary model [9, 10]. Hence, a series of computational tools has been developed for the computational prediction of the miRNA and to find out the homologue between the species and subspecies. In homology-based models, miRNA of known species taken as a query sequence and searching for probable miRNA is done for other species [11, 12]. The present scenario is gaining much preference in the computational identification of the non-coding region of the RNA. However, it is difficult to study the non-coding regions of the RNA as compared to the coding region of the RNA, because it does not possess a strong statistical signal and it lacks generalized algorithm. miRNA and pre-miRNA prediction is still not a widely explored research area in the domain of computational biology. The total number of miRNAs in any species even completely sequenced is still an open problem. However, researchers have been using limited techniques for miRNA prediction.

2 Computational Approaches for miRNA Prediction

Previously, miRNA gene searches were carried out on the basis of both sequence and structure conservation and typical stem loop structure between two closely related species, but now it has become a fact that for a suitable candidate stem-loop should be more conserved than miRNA as well as certain other criteria [13]. Approaches for computational prediction of miRNA can be categorized as follows: filter based, machine learning, target centered, homology based and mixed.

The different tools and techniques have been developed which works on the same principles of finding miRNA with little difference in their approach. Although these tools having similar type of strategy, they differ in conservation patterns and attribute to identify probable candidates.

In *filter-based approach*, first we identify the initial candidates set, the structural criteria is used to further restrict precursor candidates, the conservation criteria is then adopted and at last additional filters may be implemented if needed. The machine learning methods distinguish a positive set of known miRNA and a negative set of the stem-loops which are miRNA precursors. Research community has been using different algorithms based on filter-based approach for identifying potential miRNA in a precursor. miRscan uses a sliding window [14, 15] nucleotides that passes through a precursor for locating probable miRNA. Scores for every window is computed based on the features set derived from phylogenetically known miRNA. MiRseeker [16] is a secondary structure alignment-based tool with certain degree of divergence among the structures. It aligns secondary structures of precursors. The modified version of this tool came in existence in the form of MIRSCAN II [17], which contains certain additional features like conserved motifs, etc. A tool called miRFinder [18] has been developed to predict pre-miRNA that is conserved between two genomes. It uses 18 attributes for miRNA prediction.

The present machine learning algorithms are: SVM, HMM, neural networks and k-nearest neighbor algorithm. Among all the methods, SVM has provided a popular frame to study the distinctive characteristics of miRNA. SVM classifier is trained on a set of positive examples (known miRNA) and a set of negative examples (mRNA, tRNA, rRNA and human and viral genome). ProMiR is a web-based service for the prediction of potential miRNAs in a query sequence of 60–150 nt, using a probabilistic co-learning model [19].

In *target centered approach*, a putative set of miRNA targets were used to find new miRNAs. Target-centered approaches have the benefit of making few assumptions about the structure of miRNA precursors, but are dependent on the identification of highly conserved motifs in 3′-UTRs. FINDMIRNA is a single genome approach. This tool replaces the sieve of cross-species conservation of candidate stem-loops with the detection of potential targets within transcripts of the same species. Based on the comparison of the *Arabidopsis* and *Oryza* genomes, Findmir is a computational approach for finding plant miRNAs [20].

Most of the prediction methods use *homology-based search* as a part of their protocol and even the orthologues search is the integral part of conservation requirement. There are many homology-based approaches, and they are applicable to the original candidates set that usually fail to pass to the filters. These methods are very useful in finding new members of conserved gene clusters of miRNAs and are also useful in finding newly sequenced genomes for homologues of known miRNAs [21] and for validation of miRNA genes in the previously known genomes [22, 23]. As alignment-based methods rely on the sequence conservation, more advanced methods can be developed using structure conservation. ERPIN is a profile-based method [24]. ERPIN uses both sequence and structure conservation and has successfully predicted hundreds of new candidates belonging to different species. Another similar approach is miRAlign [25], which has been proved successful over the conventional search tools such as ERPIN and BLAST.

Mixed approaches are combination of high-throughput experimental data and computational procedures through which large number of miRNA can be revealed. A method called PAL GRADE [26] has been used in detecting the miRNA in *Homo sapiens* based on structural features and thermodynamic stability of the stem-loops. Another method-based probabilistic model used to extract low-expression or tissue-specific miRNAs is miRDeep [27].

Although a number of tools were developed allowing more extensive and sophisticated studies of conservation patterns still a set of rules are lacking that are capable of reducing the false positives. All such tools focus on the conserved sequences, but there are some sequences that are not conserved [28]. Hence, rule-based approach is very useful in the prediction of miRNAs.

3 Methodology

The work reported in this chapter has been carried out in the following five phases: data collection, structure prediction, attribute measurement, attribute relevance analysis and classification of miRNA. There are very limited open and free domain sources of miRNA data available to the research community. MiRBase is one of the highly referred database easily accessible for miRNA research. We downloaded complete dataset of 91 known miRNA sequences of *Bombyx mori* (silk worm) from miRBase sequence database (release 11.0) at http://microrna.sanger.ac.uk [29, 30]. Furthermore, 91 non pre-miRNA sequences were taken from *B. mori* genome data on random basis to make negative set of data.

For prediction, choice of a secondary structure prediction tool was important. We selected RNAfold over Mfold since it is more user-friendly and runs on windows. It gives its output in a dot-bracket form for which we developed a program for further manipulation of the structure. RNAfold software is a small windows-based utility of Vienna RNA secondary structure server available at http://rna.tbi.univie.ac.at/cgi-bin/RNAfold.cgi. RNAfold produces a single structure that has minimum free energy for a given sequence. We produced secondary structures for the each entity, both miRNA and non- miRNA of *B. mori*.

3.1 Measuring Attributes from Pre-miRNA Secondary Structure

In this phase, the values of 9 attributes from 182 secondary structures were determined. This resulted in a relation with 182 tuples of 9 dimensions each. These nine attributes are given below in the set S_A : $S_A = \{$ARM, DFL, NBP, LEN, POP, GCC, MFE, DAS, DAE$\}$.

ARM attribute represents the location of miRNA in pre-miRNA. If miRNA is located in upper arm, then we are taking value 1 and for lower arm 2. In some cases, more than one miRNA occurs in a single pre-miRNA. In such cases, we are setting ARM value to 3. DFL denotes distance from loop. It is the distance between the last nucleotide of hairpin to the first nucleotide in the mature miRNA. NBP denotes the number of base pairs between miRNA and miRNA*. LEN is the total length of the mature miRNA sequence. POP denotes the ratio of number of base pair to the length of miRNA. GCC denotes percentage of GC content in the mature miRNA sequence. MFE is minimum free energy for folding miRNA into secondary structure. DAS is the dominating nucleotide at start ($5'$ UTR) and DAE is the dominating nucleotide at the end ($3'$ UTR). We developed our program in C++ that reads the following output of 3.2 as an input and produces the values of all nine attributes.

3.2 Attribute Relevance Analysis

We have used Weka software (version 3.5.8) [31] for dimension reduction and relevance analysis. Weka takes input in a specific format. Therefore, the data obtained after determining the values of attributes are transformed into the Weka input format (.arff). We applied chi-square attribute evaluation, SVM attribute evaluation, Infogain and consistency subset evaluation methods to determine dominating attributes and for attribute relevance analysis. Chi-square attribute evaluation, SVM attribute evaluation and Infogain methods use rank-based approach for searching, while consistency subset evaluation method is based on greedy stepwise searching strategy.

3.3 Classification of miRNA

We have used a tree-based classifier using j48 (a weka implementation of c 4.5 algorithm) for miRNA prediction [32, 33]. We have trained the classifier with full positive and negative dataset of 91 each and cross validation with fold 10 for *B. mori*. A decision tree was constructed based on the results obtained from classification. Testing of the model was done for the miRNA test set of size 124 of *Apis mellifera*.

4 Results and Discussion

We have derived relevant attributes namely DFL and GCC using chi-square attribute evaluation, Infogain and consistency subset evaluation methods. SVM attribute evaluation gives us the relevant attributes as DFL, POP and GCC. The results are encouraging since the essential attributes selected here are biologically significant.

Most stable RNA structure has larger number of base pairs and GC contents. For classification, we train the classifier using positive and negative samples of 91 sequences each. We have used cross validation with fold 10 and also split the training data with percentage split of 70. After training, we test the model with *A. mellifera* test set of sample size 124. The decision tree drawn after classification is given in Fig. 1 The performance evaluation using weka is given in Table 1.

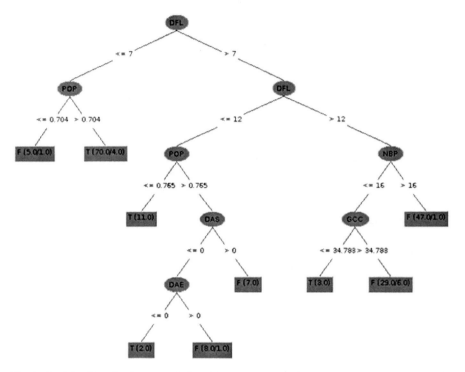

Fig. 1 Decision Tree for *B. mori* with 9 attribute set

Table 1 Performance evaluation using Weka

Dataset	Class	TP Rate	FP Rate	Precision	Recall	F-measure
Full training set	T	0.901	0.044	0.953	0.901	0.927
	F	0.956	0.099	0.906	0.956	0.930
66% split on training set	T	0.808	0.172	0.808	0.808	0.808
	F	0.828	0.192	0.828	0.828	0.828
Cross validation of fold 10	T	0.802	0.088	0.901	0.802	0.849
	F	0.912	0.198	0.822	0.912	0.865
Test dataset	T	0.871	0.524	0.628	0.871	0.730
	F	0.475	0.129	0.784	0.575	0.692

The precision and recall measure on training dataset is quite satisfactory. The model has worked perfectly on dataset of same species. The above results clearly show that our model gives high precision and recall in all cases. However, the model is not able to recall few true cases, and this may be improved by taking a larger training dataset. In the test dataset, the precision and recall values are below expectation. This is because false negative has increased as we have used test dataset of different species. The large training dataset may also help in identifying dominating and essential attributes for increasing the accuracy of prediction model.

5 Conclusion

In this chapter, we have derived miRNA attributes from their secondary structure and used attribute relevance analysis techniques for selecting essential attributes based on their relevance. The results are encouraging since the essential attributes selected here are biologically significant. These attributes are used in deriving rules for miRNA identification. Based on these rules, we can identify miRNAs with a fair accuracy by allowing few mismatches. Incorporating other rule-based mining techniques and scoring criteria can enhance performance of this model. Most computational approaches developed so far make extensive use of evolutionary conservation to predict miRNA genes. Since the cell cannot use the filter of evolutionary conservation to choose among all potential hairpins, we seem to be missing a significant miRNA part. Future developments ought to focus on the need to establish more accurate models for solve these problems using rule-based mining approach.

References

1. Lee RC, Feinbaum RL, et al (1993) The C. elegans heterochronic gene lin-4 encodes small RNAs with antisense complementarity to lin-14. Cell 75: 843–854
2. Lewis BP, Shih IH, Jones-Rhoades MW, et al (2003) Prediction of mammalian microRNA targets. Cell 115: 787–798
3. Lee Y, Ahn C, Han J, et al (2003) The nuclear RNase III Drosha initiates microRNA processing. Nature 425: 415–419
4. Cullen BR (2004) Transcription and processing of human microRNA precursors. Mol. Cell 16: 861–865
5. Bartel DP (2004) MicroRNAs: Genomics, biogenesis, mechanism and function. Cell 116: 281–297
6. Lee, Y. et al (2003) The nuclear RNaseIII Drosha initiates microRNA processing. Nature 425: 415–419
7. Tinoco I Jr, Uhlenbeck X, et al (1971): Estimation of secondary structure in ribonucleic acids. Nature 230: 362–367
8. Zuker M, Stiegler P (1981): Optimal computer folding of large RNA sequences using thermodynamics and auxiliary information. Nucleic Acids Res. 9: 133–148
9. Mette MF, van der Winden J, et al (2002). Short RNAs can identify new candidate transposable element families in Arabidopsis. Plant Physiol. 130: 6–9

10. Sunkar R, Girke T, et al (2005) Cloning and characterization of MicroRNAs from rice. Plant Cell 17:1397–1411
11. Xie Z, Kasschau KD, Carrington JC (2003) Negative feedback regulation of Dicer-Like 1 in Arabidopsis by microRNA-guided mRNA degradation. Curr. Biol. 13: 784–789
12. Jones-Rhoades MW, Bartel DP (2004) Computational identification of plant MicroRNAs and their targets including a stress-induced miRNA. Mol. Cell 14: 787–799
13. Mendes ND, Feeitas AT, Sagot MF (2009) Current tools for the identification of miRNA genes and their target. Nucleic Acids Res. 37: 2419–2433
14. Lim LP, Lau NC, Weinstein EG, et al (2003) The microRNAs of Caenorhabditis elegans. Genes Dev. 17: 991–1008
15. Lim LP, Glasner ME, Yekta S, et al (2003) Vertebrate microRNA genes. Science 299: 1540
16. Lai EC, Tomancak P, et al (2003) Computational identification of Drosophila microRNA genes. Genome Biol. 4(7): R42
17. Ohler U, Yekta S, et al (2004) Patterns of flanking sequence conservation and a characteristic upstream motif for microRNA gene identification. RNA 10: 1309–1322
18. Huang TH, Fan B, et al (2007) MiRFinder: an improved approach and software implementation for genome-wide fast microRNA precursor scans. BMC Bioinformatics 8: 341
19. Nam J-W, Kim JKS-K, et al (2006) ProMiR II: a web server for the probabilistic prediction of clustered, nonclustered, conserved and nonconserved microRNAs. Nucleic Acids Res. 34 (Web Server issue): W455–W458
20. Adai A, Johnson C, et al (2005) Computational prediction of miRNAs in Arabidopsis thaliana. Genome Res. 15: 78–91
21. Xie X, Lu J, Kulbakas E J, Golub TR, et al (2005) Systematic discovery of regulatory motif in human promoters and $3'$ UTR by comparision of several mammals. Nature 434: 338–345
22. Chatterjee R, Chaudhuri K (2006) An approach for the identification of microRNA with an application to Anopheles gambiae. Acta Biochim. Pol. 53: 303–309
23. Weaver D, Anzola J, et al (2007) Computational and transcriptional evidence for microRNAs in the honey bee genome. Genome Biol. 8: R97
24. Legendre M, Lambert A, Gautheret D (2005) Profile-based detection of microRNA precursors in animal genomes. Bioinformatics 21: 841–845
25. Wang X, Zhang J, et al (2005) MicroRNA identification based on sequence and structure alignment. Bioinformatics 21: 3610–3614
26. Bentwich I, Avniel A, et al (2005) Identification of hundreds of conserved and onconserved human microRNAs. Nat. Genet. 37: 766–770
27. Friedlander MR, Chen W, et al (2008) Discovering microRNAs from deep sequencing data using miRDeep. Nat. Biotechnol. 26: 407–415
28. Rose D, Hackermueller J, Washietl S, et al (2007) Computational RNomics of drosophilids. BMC Genomics 8: 406
29. Griffiths-Jones S, Saini HK, van Dongen S, Enright AJ (2008) miRBase: tools for microRNA genomics. Nucleic Acids Res. 36: D154–D158
30. Griffiths-Jones S, Grocock RJ (2006) miRBase: microRNA sequences, targets and gene nomenclature. Nucleic Acids Res. 34: D140–D144
31. Frank E, Hall M, et al (2004) Data mining in bioinformatics using Weka. Bioinformatics 20(15): 2479–2481
32. Quinlan JR (1993) C4.5: Programs for Machine Learning. Morgan Kaufmann Publishers, MA
33. Quinlan JR (1996) Improved use of continuous attributes in c4.5. J Artif Intell Res 4: 77–90

Chapter 9
Improving the Accuracy of Gene Expression Profile Classification with Lorenz Curves and Gini Ratios

Quoc-Nam Tran

Abstract Microarrays are a new technology with great potential to provide accurate medical diagnostics, help to find the right treatment for many diseases such as cancers, and provide a detailed genome-wide molecular portrait of cellular states. In this chapter, we show how Lorenz Curves and Gini Ratios can be modified to improve the accuracy of gene expression profile classification. Experimental results with different classification algorithms using additional techniques and strategies for improving the accuracy such as the principal component analysis, the correlation-based feature subset selection, and the consistency subset evaluation technique for the task of classifying lung adenocarcinomas from gene expression show that our method find more optimal genes than SAM.

Keyword Microarray data mining

1 Introduction

In recent years, microarrays have opened the possibility of creating datasets of molecular information to represent many systems of biological or clinical interest. Gene expression profiles have been used as inputs to large-scale data analysis, for example, to serve as fingerprints to build more accurate molecular classification, to discover hidden taxonomies, or to increase our understanding of normal and disease states. Microarray analysis methodologies developed over the last few years have demonstrated that expression data can be used in a variety of class discovery or class prediction biomedical problems including those relevant to tumor classification [5, 10, 12, 14]. Data mining and statistical techniques applied to gene expression data have been used to address the questions of distinguishing tumor morphology, predicting post treatment outcome, and finding molecular markers for disease. Today, the microarray-based classification of different morphologies, lineages, and cell histologies can be performed successfully in many instances.

Q.-N. Tran (✉)
Department of Computer Science, Lamar University, Beaumont, TX 77710, USA
e-mail: qntran@lamar.edu; qntran@buchberger.cs.lamar.edu

However, gene expression profile presents a big challenge for data mining both in finding differentially expressed genes and in building predictive models, since the datasets are highly multidimensional and contain a small number of records. Although microarray analysis tool can be used as an initial step to extract most relevant features, one has to avoid over-fitting the data and deal with the very large number of dimension of the input. The main types of data analysis needed for biomedical applications include: (a) gene selection for finding the genes most strongly related to a particular class (see, e.g., [3, 9, 13]), and (b) classification for classifying diseases or predicting outcomes based on gene expression patterns, and perhaps even identifying the best treatment for given genetic signature (see [11]).

To improve the prediction accuracy, we modify the Lorenz curves and the Gini coefficients by taking into account the order of classes and the order of gene discretized values and use them for selecting relevant genes. This method selects small subsets of the genes with a high accuracy for classification algorithms. We believe that our method is the first one for gene selection that considers the order of classes and the order of gene discretized values.

We implemented and compared our method with significant analysis of microarray (SAM), one of the most popular gene selection methods [16]. Experimental results with many different classification algorithms for the task of classifying lung adenocarcinomas from gene expression show that: (a) our method is different with SAM in the sense that it finds very different sets of significant genes. There are just four common genes in the sets of 50 most significant genes generated by SAM and our algorithm. Similarly, there are just 38 common genes in the sets of 250 most significant genes generated by SAM and our algorithm. (b) Our method finds better genes for more accurate classification. In [15], our algorithm finds sets of significant genes that can give 12% more accuracy than SAM. (c) Our method finds more optimal genes in the sense that additional techniques and strategies for improving the accuracy such as the principal component analysis (PCA), the correlation-based feature subset (CFS) selection, and the consistency subset evaluation (CSE) technique do not affect the sets of genes selected by our method.

The chapter is organized as follows. In Sect. 2, we present our method for gene selection that was initialized in [15]. In Sect. 3, we present our experimental results with many different classification algorithms using additional techniques and strategies for improving the accuracy for the task of classifying lung adenocarcinomas from gene expression.

2 Lorenz Curves and Gini Ratios for Gene Selection

2.1 First Kind Bias Due to the Order of Classes

We consider a simple example of a gene expression profile in Table 1, where the gene dataset D has d elements and three classes. The gene expression values were discretized into three ranges. When this gene is evaluated by the current

9 A New Algorithm for Gene Selection

Table 1 First kind bias

Range/class	C_1	C_2	C_3
R_1	4	6	30
R_2	6	30	4
R_3	0	4	16

Fig. 1 Lorenz curves

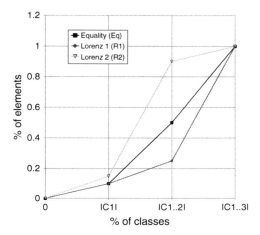

microarray analysis methodologies, for example, by calculating the Gini index $\text{gini}_A(D) = \sum_{i=1}^{m}(|R_i|/d) \cdot \text{gini}(R_i)$, the first two rows contribute equally to the Gini index because $\text{gini}(R_i) = 1 - \sum_{j=1}^{n} p_{i,j}^2$ where $p_{i,j} = (|C_{i,j}|)/|R_i|$ is the relative frequency of class C_j in R_i. That is, when one just considers the probability distribution without taking into account the order of the classes, the first two partitions will be considered the same. Clearly, the two partitions should not be considered the same because partition R_1 says that 75% of patients with gene expression values within this range are classified into Class C_3, while partition R_2 says that 75% of patients with gene expression values within this range are classified into Class C_2. Hence, to have a robust gene selection method, one has to differentiate the partitions with different class orders because they have different amount of information. To solve this problem, we modify the well-known Lorenz curves, a common measure in economics to gauge the inequalities in income and wealth. In Fig. 1, we illustrate how modified Lorenz curves and modified Gini coefficients are calculated.

The equality line (Eq) is defined based on the percentages of elements in $|C_1|$, $|C_{1...2}| = |C_1| + |C_2|$, ..., $|C_{1...n}| = \sum_{i=1}^{n} |C_i|$ at x-coordinates $0, 1/n, 2/n, \ldots, 1$, where n is the number of classes and $|C_1| \leq |C_2| \leq, \ldots, \leq |C_n|$. The Lorenz curve of a partition, say R_j, is defined based on the percentage of elements in $|C_1^j|$, $|C_1^j| + |C_2^j|$, $\ldots, \sum_{i=1}^{n} |C_i^j|$ at x-coordinates $0, 1/n, 2/n, \ldots, 1$. The Gini coefficient of a partition, say R_j, is defined as $(\int_0^1 L(R_j) \cdot dx - \int_0^1 \text{Eq} \cdot dx) / \int_0^1 \text{Eq} \cdot dx$. One can easily see that the partitions with different class orders are now differentiated.

Note that our new method works for any dataset with ≥ 2 classes. For any number of classes, even when the number of classes is equal to 2, the new method is completely different with other microarray analysis methodologies.

Table 2 Second kind bias

Class/range	C_1	C_2	C_3	Class/range	C_1	C_2	C_3
R_1	3	0	0	R_1	3	0	0
R_2	0	100	0	R_2	4	0	0
R_3	4	0	0	R_3	0	100	0
R_4	0	0	5	R_4	0	0	5

2.2 Bias Due to the Order of Gene Expression Values

Another problem with current microarray analysis methodologies comes from the order of discretized gene expression values. We consider another simple example of two gene expression profiles in Table 2 with three classes. The gene expression values were discretized into four ranges. When these genes are evaluated by the current microarray analysis methodologies, for example, by calculating the Gini index of gene A using dataset D $\text{gini}_A(D) = \sum_{i=1}^{m}(|R_i|/d) \cdot \text{gini}(R_i)$ where $d = |D|$, the two genes would have the same rank. Clearly, the gene expression profile on the right-hand side of Table 2 must be ranked higher in comparison with the one on the left because the gene values maintain the order even after being discretized. That is, one has to take into account the gene expression profiles with different value orders because they have different amount of information. To solve this problem, we modify the Gini coefficients by taking into account the splitting status and the Gini ratio. The splitting status of D with respect to the attribute A is calculated as $\text{split}_A(D) = 1 - \sum_{i=1}^{m}[(|R_i|)/d]^2$. The *LorenzGini* value of D with respect to the attribute A is defined as $\Delta\text{gini}(A)/\text{split}_A(D)$, where $\Delta\text{gini}(A) = \text{gini}(D) - \text{gini}_A(D)$.

Furthermore, to take into account the gene expression profiles with different value orders, the Gini coefficient is calculated as $\text{gini}_A(D) = \sum_{i=1}^{m}(|R_i|/d) \cdot \delta(i) \cdot \text{gini}(R_i)$, where $\delta(i)$ is the sum of the normalized distances between the row i and rows $i-1, i+1$. The coefficient $\delta(i)$ is used as a weight to emphasize a row when it is close to its neighbors. The splitting status of the microarray dataset D with respect to a gene A can be calculated as a by-product when the reduction in impurity of D with respect to the attribute A is calculated. Therefore, the time complexity and space complexity of the algorithm are the same as the complexities of Gini index algorithm.

3 Experimentation with Reducing Techniques

Carcinoma of the lung claims more than 150,000 lives every year in the USA, thus exceeding the combined mortality from breast, prostate, and colorectal cancers. The current lung cancer classification is based on clinicopathological features. More fundamental knowledge of the molecular basis and classification of lung carcinomas could aid in the prediction of patient outcome. We use the microarray dataset for lung adenocarcinoma cancer from the Broad Institute for our work on improving the accuracy of microarray classification [1,4]. We use the Dataset A which includes the histologically defined lung adenocarcinomas ($n = 127$), squamous cell lung carcinomas ($n = 21$), pulmonary carcinoids ($n = 20$), and normal lung ($n = 17$).

In this set of experiments, we investigate the optimality of the subsets of genes selected by our new method, called LorenzGini, by checking whether additional reducing techniques such as the PCA, the CFS selection, and the CSE technique can improve the classifying accuracy. We first use the LorenzGini and SAM for selecting most significant genes and create reduced microarray datasets in different sizes.

3.1 Principal Component Analysis

As gene's evolution depends not only on itself but also on the related genes, in this experiment we do not use individual genes for prediction models but their components. A component is a linear combination of related genes. PCA is a vector space transforming method, which functions to reduce multidimensional data to lower dimensions for analysis. PCA transforms a number of possibly correlated genes into a smaller number of uncorrelated variables called principal components. We use the derivation of PCA using the covariance method [7]. The principal components are those linear combinations of the original variables which maximize the variance of the linear combination and which have zero covariance and zero correlation with the previous principal components. The first principal component accounts for as much of the variability in the data as possible, and each succeeding component accounts for as much of the remaining variability as possible. Let $X = (X_1, X_2, \ldots, X_n)$ be an n-dimensional random vector expressed as column vector. Without loss of generality, assume X has zero empirical mean. We want to find a $n \times n$ orthonormal transformation matrix P such that $\mathbf{Y} = \mathbf{P}^\top \mathbf{X}$ with a constraint that cov(Y) is a diagonal matrix and $\mathbf{P}^{-1} = \mathbf{P}^\top$. It can be showed that $\mathbf{P} \text{cov}(\mathbf{Y}) = \text{cov}(\mathbf{X})\mathbf{P}$ where P can be written as $\mathbf{P} = [P_1, P_2, \ldots, P_n]$ and cov($\mathbf{Y}$) can be written as $\begin{bmatrix} \lambda_1 & \cdots & 0 \\ \vdots & \ddots & \vdots \\ 0 & \cdots & \lambda_n \end{bmatrix}$. Substituting into the equation above, we obtain: $[\lambda_1 P_1, \lambda_2 P_2, \ldots, \lambda_n P_n] = [\text{cov}(\mathbf{X})P_1, \text{cov}(\mathbf{X})P_2, \ldots, \text{cov}(\mathbf{X})P_n]$. Note that in $\lambda_i P_i = \text{cov}(\mathbf{X})P_i$, P_i is an eigenvector of the covariance matrix of X. Therefore, by finding the eigenvectors of the covariance matrix of X, we find a projection matrix P that satisfies the original constraints.

It can be proved that there are exactly n such linear combinations. However, typically, the first few of them explain most of the variance in the original data. Hence, instead of working with all the original variables $X_1, X_2, \ldots, X_n$, PCA is first performed and then select only first few principal components to be used depending on the desired cumulative variance. After using SAM and LorenzGini to select most significant genes from a microarray gene expression profile dataset, we further reduce the number of attributes using PCA components for our classification steps. Our goal is for observing whether this additional strategy will bring any improvement in classifying accuracy.

3.2 Correlation-Based Feature Selection

This strategy evaluates the worth of a subset of attributes by considering the individual predictive ability of each feature along with the degree of redundancy between them [6]. Subsets of features that are highly correlated with the class while having low intercorrelation are preferred.

3.3 Consistency Subset Evaluation

This strategy evaluates the worth of a subset of attributes by the level of consistency in the class values when the training instances are projected onto the subset of attributes [8]. Consistency of any subset can never be lower than that of the full set of attributes; hence, the usual practice is to use this subset evaluator in conjunction with a random or exhaustive search which looks for the smallest subset with consistency equal to that of the full set of attributes.

We select two batches of ten different subsets with 50, 100, 150, 200 and 250 most significant genes using LorenzGini and SAM, respectively. We then use the additional reducing techniques to reduce the number of attributes for these subsets of significant genes. For example, for the subsets with 250 most significant genes generated by SAM, the CFS reduced the number of attributes to 67, the CSE reduced the number of attributes to 7, and the PCA reduced the number of attributes to 50. We then use the Bayesian Net classification and libSVM in Weka to check the classifying accuracy [2]. Table 3 shows the accuracy of the gene expression profile classification using the Bayesian Net algorithm for gene selections generated by SAM and LorenzGini. Table 4 shows the accuracy of the gene expression profile classification using the libSVM algorithm. Our conclusion is that the additional

Table 3 Changes in accuracy using Bayesian Net

Number of genes	Bayesian Net					
	SAM			LorenzGini		
	CFS (%)	CSE (%)	PCA (%)	CFS (%)	CSE (%)	PCA (%)
50	5.36	4.77	−1.75	−0.53	−3.15	0.00
100	0.52	−1.07	−0.53	−0.53	−2.10	−1.58
150	1.61	2.69	0.00	1.06	−4.21	−0.53
200	1.63	3.80	−1.64	1.04	−7.33	−1.57
250	1.09	1.09	0.54	0.53	−2.09	−4.19

Table 4 Changes in accuracy using libSVM

Number of genes	libSVM SAM			LorenzGini		
	CFS (%)	CSE (%)	PCA (%)	CFS (%)	CSE (%)	PCA (%)
50	−5.56	−6.12	−12.23	0.05	−3.77	−19.90
100	−0.53	1.09	−14.05	0.05	2.19	−13.11
150	0.00	−3.24	−9.72	2.76	−3.86	−16.58
200	−0.55	−2.72	−8.16	4.44	−0.56	−15.00
250	−0.54	−3.76	−8.60	2.19	−2.19	−16.40

reducing techniques do not help to improve the classifying accuracy for the gene selections generated by LorenzGini. That is, the gene sets generated by LorenzGini are optimal for the reducing techniques with an exception of CFS for ≥ 150 genes using libSVM.

4 Conclusion

We presented a method for gene selection that finds more optimal genes in the sense that additional techniques and strategies for improving the accuracy such as the PCA, the CFS selection, and the CSE technique do not affect the subsets of genes selected.

References

1. http://www.broad.mit.edu/cgi-bin/cancer/datasets.cgi, 2009.
2. http://www.cs.waikato.ac.nz/ml/weka, 2009.
3. BALDI, P., AND LONG, A. D. A bayesian framework for the analysis of microarray expression data: regularized t-test and statistical inferences of gene changes. *Bioinformatics 17* (2001), 509–519.
4. BHATTACHARJEE, A., RICHARDS, W. G., STAUNTON, J., LI, C., MONTI, S., GOLUB, T. R., SUGARBAKER, D. J., AND MEYERSON, M. Classification of human lung carcinomas by mrna expression profiling reveals distinct adenocarcinoma subclasses. *Proc. Natl. Acad. Sci. USA 98*, 24 (2001), 13790–13795.
5. BUTTE, A. The use and analysis of microarray data. *Nat. Rev. Drug Discov. 1*, 12 (2002), 951–960.
6. HALL, M. A. *Correlation-Based Feature Subset Selection*. Hamilton, New Zealand, 1998.
7. JOLLIFFE, I. *Principal Component Analysis*. Springer Series in Statistics. Springer, New York, 2002.
8. LIU, H., AND SETIONO, R. A probabilistic approach to feature selection – a filter solution. In *Proceedings of the 13th International Conference on Machine Learning* (1996), pp. 319–327.
9. MARCHAL, K., ENGELEN, K., BRABANTER, J. D., ZHOU, S., ZHENG, X., WANG, J., AND DELISLE, P. Comparison of different methodologies to identify differentially expressed genes in two-sample cdna microarrays. *J. Biol. Syst. 10* (2002), 409–430.

10. PIATETSKY-SHAPIRO, G., AND TAMAYO, P. Microarray data mining: Facing the challenges. *SIGKDD Explorations 5*, 2 (2003).
11. QUINLAN, J. R. An empirical comparision of genetic and decision-tree classifiers. In *Proceedings of the 5th International Conference on Machine Learning* (Ann Arbor, 1988), pp. 135–141.
12. RAMASWAMY, S., AND GOLUB, T. R. Dna microarrays in clinical oncology. *J. Clin. Oncol. 20* (2002), 1932–1941.
13. STOREY, J. D., AND TIBSHIRANI, R. Statistical significance for genome wide studies. *Proc. Natl. Acad. Sci. USA 100* 16 (2003), 9440–9445.
14. TAMAYO, P., AND RAMASWAMY, S. Cancer genomics and molecular pattern recognition. In *Expression profiling of human tumors: diagnostic and research applications*, M. Ladanyi and W. Gerald, Eds. Humana Press, Clifton, 2003.
15. TRAN, Q.-N. Microarray data mining: A new algorithm for gene selection using Gini ratios. In *Proceedings of IEEE-ITNG 2008 Conference* (Las Vegas, Nevada, 2010).
16. TUSHER, V. G., TIBSHIRANI, R., AND CHU, G. Significance analysis of microarrays applied to the ionizing radiation response. *Proc. Natl. Acad. Sci. USA 98* (2001), 5116–5121.

Chapter 10
Feature Selection in Gene Expression Data Using Principal Component Analysis and Rough Set Theory

Debahuti Mishra, Rajashree Dash, Amiya Kumar Rath, and Milu Acharya

Abstract In many fields such as data mining, machine learning, pattern recognition and signal processing, data sets containing huge number of features are often involved. Feature selection is an essential data preprocessing technique for such high-dimensional data classification tasks. Traditional dimensionality reduction approach falls into two categories: Feature Extraction (FE) and Feature Selection (FS). Principal component analysis is an unsupervised linear FE method for projecting high-dimensional data into a low-dimensional space with minimum loss of information. It discovers the directions of maximal variances in the data. The Rough set approach to feature selection is used to discover the data dependencies and reduction in the number of attributes contained in a data set using the data alone, requiring no additional information. For selecting discriminative features from principal components, the Rough set theory can be applied jointly with PCA, which guarantees that the selected principal components will be the most adequate for classification.

We call this method Rough PCA. The proposed method is successfully applied for choosing the principal features and then applying the Upper and Lower Approximations to find the reduced set of features from a gene expression data.

Keywords Data preprocessing · Feature selection · Principal component analysis · Rough sets · Lower approximation · Upper approximation

1 Introduction

Data mining refers to extracting or mining knowledge from large amounts of data. There are many other terms carrying a similar or slightly different meaning to Data mining, such as knowledge mining from databases, knowledge extraction, data

D. Mishra (✉)
Department of Computer Science & Engineering, Institute of Technical Education
& Research, Siksha O Anusandhan University, Bhubaneswar, Orissa, India
e-mail: debahuti@iter.ac.in

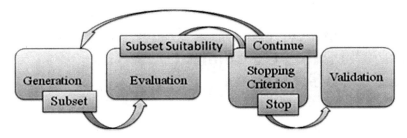

Fig. 1 Feature selection

pattern analysis, data archaeology and data dredging. Data mining treats as synonym for another popularly used term, Knowledge Discovery in Databases (KDD). KDD consists of the following steps to process it such as Data cleaning, Data integration, Data selection, Data transformation, Data mining, Pattern evaluation and Knowledge presentation.

The main aim of feature selection (FS) [1] is to determine a minimal feature from a problem domain while retaining a suitably high accuracy in representing the original features (see Fig. 1). In many real-world problems, FS is must due to the abundance of noisy, irrelevant or misleading features. For instance, by removing these factors, learning from data techniques can benefit greatly.

The usefulness of a feature or feature subset is determined by both its *relevancy* and its *redundancy*. A feature is said to be relevant if it is predictive of the decision feature(s), otherwise it is irrelevant. A feature is considered to be redundant if it is highly correlated with other features. Hence, the search for a good feature subset involves finding those features that are highly correlated with the decision feature(s), but are uncorrelated with each other.

Feature selection algorithms may be classified into two categories based on their evaluation procedure. If an algorithm performs FS independently of any learning algorithm (i.e. it is a completely separate preprocessor), then it is a *filter* approach. In effect, irrelevant attributes are filtered out before induction. If the evaluation procedure is tied to the task of the learning algorithm, the FS algorithm employs the *wrapper* approach. This method searches through the feature subset space using the estimated accuracy from an induction algorithm as a measure of subset suitability.

1.1 Paper Layout

The layout of the paper is as follows: Sect. 1 deals with the introductory concepts of data mining as well as the importance of feature selection in data mining, some recent related works on PCA as well as Rough set theory, then goal of our approach followed by our hybridized proposed model of PCA and Rough Set Theory. Section 2 gives some preliminary concepts regarding PCA and Rough Set Theory.

Section 3 describes our approach in various steps with experimental activities. Discussion on result of our approach is given in Sect. 4 followed by conclusion in Sect. 5.

1.2 Related Work

Jun Yan et al. [2] proposed an effective and efficient dimensionality reduction for large-scale and streaming data preprocessing. This paper finds the optimal solution of a problem in a continuous space, but the computational complexity is more comparative to feature selection algorithm. This algorithm aims at finding out a subset of the most representative features according to some objective function. Michael Davy et al. [3] propose dimensionality reduction for active learning with nearest neighbor classifier. Roman W. Swiniarski [4] proposed a model of Rough Sets Methods in feature reduction and classification. In this paper given a data set with discrete attribute values, it is possible to find a subset (termed reduct) of the original attributes using RST that are the most informative attributes; all other attributes can be removed from the data set with minimal information loss. From the dimensionality reduction perspective, informative features are those that are most predictive of the class attribute. PCA provides feature projection and reduction optimal from the point of view of minimizing the reconstruction error. Hence for selecting discriminative features from principal components, RST is applied to find a reduct of principal components. Consequently, these principal components can describe all the concepts in a data set.

1.3 Goal of the Paper

In this paper, we are concerned on how to construct a heuristic function for feature selection. Our objective is to focus on how to improve the time efficiency of a heuristic feature subset selection algorithm. We employ a new rough set framework hybridized with PCA, which is called positive approximation. The main advantage of this approach stems from the fact that this framework is able to characterize the granulation structure of a rough set using a granulation order. Based on the positive approximation, we develop a common strategy for improving the time efficiency of a heuristic feature selection, which provides a vehicle of making algorithms of rough set-based feature selection techniques faster.

1.4 Proposed Model

The work of this paper, a hybridized Rough Set PCA model for feature selection can be implemented by collecting continuous data sets from data repository and by applying an efficient PCA method, an efficient discritization technique and an efficient

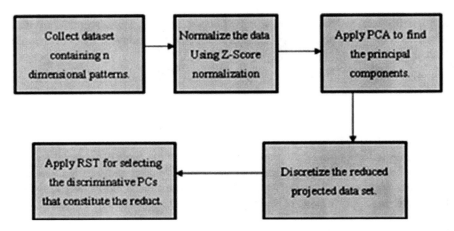

Fig. 2 Hybridized model for feature selection (rough PCA)

reduction algorithm of RST. As a result of which a number of discriminative principal components, more adequate for classification will be obtained. We call this model as Rough PCA (see Fig. 2).

2 Preliminaries

2.1 Principal Component Analysis (PCA)

Principal component analysis [5] is appropriate when you have obtained measures on a number of observed variables and wish to develop a smaller number of artificial variables (called principal components) that will account for most of the variance in the observed variables. The principal components may then be used as predictor or criterion variables in subsequent analysis.

2.1.1 A Variable Reduction Procedure

Principal component analysis is a variable reduction procedure. It is useful when you have obtained data on a number of variables (possibly a large number of variables), and believe that there is some redundancy in those variables, as they are correlated with one another, possibly because they are measuring the same construct. Because of this redundancy, you believe that it should be possible to reduce the observed variables into a smaller number of principal components (artificial variables) that will account for most of the variance in the observed variables.

When correlations among several variables are computed, they are typically summarized in the form of a *covariance matrix*. This is an appropriate opportunity to review just how a covariance matrix is interpreted.

2.1.2 Principal Component

Technically, a *principal component* can be defined as a linear combination of optimally weighted observed variables. To understand the meaning of this definition, it is necessary to first describe how subject scores on a principal component are computed.

Below is the general form for the formula to compute scores on the first component extracted (created) in a principal component analysis:

$$C_1 = b_{11}(X_1) + b_{12}(X_2) + \ldots b_{1p}(X_p), \tag{1}$$

where

$C_1 =$ the subject's score on principal component 1 (the first component extracted)
$b_{1p} =$ the regression coefficient (or weight) for observed variable p, as used in creating principal component 1
$X_p =$ the subject's score on observed variable p.

2.2 Rough Set Theory

Rough set theory [6–8] is a new mathematical approach to imprecision, vagueness and uncertainty. In an information system, every object of the universe is associated with some information. Objects characterized by the same information are indiscernible with respect to the available information about them. Any set of indiscernible objects is called an elementary set. Any union of elementary sets is referred to as a crisp set – otherwise a set is rough (imprecise, vague). Vague concepts cannot be characterized in terms of information about their elements. A rough set is the approximation of a vague concept by a pair of precise concepts, called lower and upper approximations. The lower approximation is a description of the domain objects, which are known with certainty to belong to the subset of interest, whereas the upper approximation is a description of the objects, which possibly belong to the subset. Relative to a given set of attributes, a set is rough if its lower and upper approximations are not equal.

The main advantage of rough set analysis is that it requires no additional knowledge except for the supplied data. Rough sets perform feature selection using only the granularity structure of the data [5].

3 An Interactive Exploration Approach

To find the best subset, we use the structure of the rows to first find the subsets of features that are highly correlated and follow to choose one feature from each subset. The chosen features represent each group optimally in terms of high spread

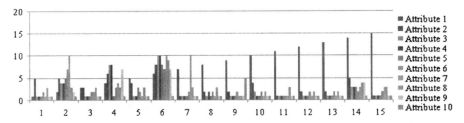

Fig. 3 Plotting of original data set having ten attribute values corresponding to 15 data sets

in the lower dimension, reconstruction and insensitivity to noise. The original data matrix on a synthetic data set has shown given in Fig. 3.

The algorithm can be summarized in the following five steps:

Step 1: Normalization of the Original data set

Using the normalization process, the initial attribute data values are scaled so as to fall within a small specified range. It ensures that any attribute with larger domain will not dominate attributes with smaller domain. Here, we have preferred to use Z-score normalization as it reduces square min error of approximating the input data. An attribute value V of an attribute A is normalized to V' using Z-score normalization by applying the following formula:

$$V' = V - \text{mean}(A)/\text{std}(A). \qquad (2)$$

We have applied the Z-Score normalization method to our Yeast and Breast Cancer data set from the UCI repository [9] data set and also synthetic data set as explained in our exploratory approach.

After applying the above normalization, let the data set be Z.

Step 2: Transforming the Data matrix Z to the new Principal Component axis

PCA is a linear transformation applied on highly correlated multidimensional data set. The input dimensions are transformed to a new co-ordinate system in which the produced dimensions called PCs (Principal Components) are in the descending order of their variance.

PCA has two algebraic solutions:

- Eigen vector decomposition of covariance or correlation of a given data matrix.
- Singular Value decomposition (SVD) of a given data matrix.

The SVD method is used for numerical accuracy.

The basic steps of the transformation using eigenvector decomposition are (see Fig. 4):

1. Calculate the (see Table 1)
   - Covariance matrix or Correlation matrix
2. Calculate the Eigen vector decomposition of the covariance or correlation matrix, which produces the PCs with corresponding Eigen values.
3. Sort the variances, which are represented by the Eigen values in decreasing order.

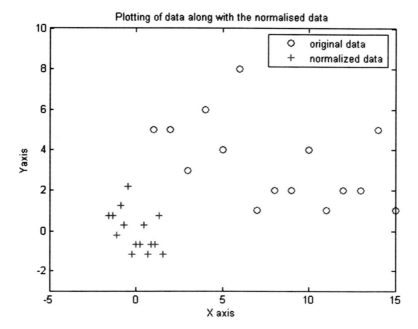

Fig. 4 Plotting of data along with the normalized data after normalization

Table 1 The variances, variances in percentage, and cumulative variances in percentage corresponding to the PCs

PCs	Variances	Variances in %	Cumulative variances in %
pc1	6.210578	62.10578	62.10578
pc2	1.054022	10.54022	72.646
pc3	1.016014	10.16014	82.80614
pc4	0.86546	8.654603	91.46075
pc5	0.455458	4.554576	96.01532
pc6	0.24649	2.464901	98.48022
pc7	0.108508	1.085079	99.5653
pc8	0.030248	0.302483	99.86779
pc9	0.010731	0.10731	99.9751
pc10	0.00249	0.024904	100

Step 3: Determining the number of meaningful components to retain

The transformation of the data set to the new principal component axis produces the number of PCs equivalent to the number of original variables. But for many data sets, the first several PCs explain most of the variance, so the rest can be disregarded with minimal loss of information. The various criteria used to determine how many principal components should be retained for the interpretation are as follows (see Fig. 5):

a. Using Scree diagram plots the variances in percentage corresponding to the PCs symmetrically.

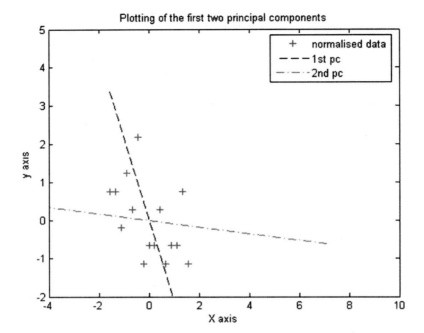

Fig. 5 Plotting of first two principal components using the normalized data

b. Fixing a threshold value of variance, so that PCs having variance more than the given threshold value will be retained rejecting others.
c. Eliminate the components having Eigen values smaller than a fraction of the Eigen values.

Step 4: Form the Transformation matrix with reduced PCs

The transformation matrix with reduced PCs is formed and this transformation matrix is applied to the normalized data set Z to produce the new reduced projected data set Y having reduced attributes.

Step 5: Discretizing the data set

The data set Y is discretized using un-supervised discretization method i.e. Binning by equal frequency. Let the discretized data set be D.

Step 6: Finding the reduct of the Principal Components

A decision table D_T has been formed from the data matrix D, from this D_T the Upper and Lower approximation value has be found. The reduct D_{red} has been composed by finding the boundary region out of the Upper and Lower approximation values. The reduced attribute data set has been constructed containing the columns from the data set D that corresponds to the selected feature set.

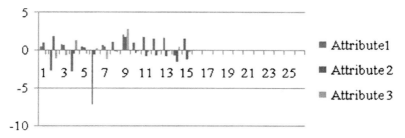

Fig. 6 The plotting of projected reduced data set Y with three attribute values corresponding to the 15 data objects

4 Results and Discussion

The data set generated is given to a synthetic data set described as earlier with 15 data objects and ten attributes and in almost ten iterations. The attribute set was then reduced by the proposed method and fed into the Rough Set Theory model. The graphs are obtained for reductions of the attributes space using the above discussed method (Fig. 6). From the graph obtained, we conclude that the attribute reductions result in the optimum efficiency. Thus the total ten attributes are reduced to three and the efficiency obtained is still similar to that of original approach.

5 Conclusion

In this paper, a hybridized model using rough set and PCA for feature selection has been implemented on some synthetic data set as well as Yeast and Breast Cancer data set by collecting continuous data sets from data repository and by applying an efficient PCA method, an efficient discritization technique and reduction algorithm of RST. As a result of which a no of discriminative principal components, more adequate for classification has been obtained and which gives rise to reduced set of features by retaining its original property.

References

1. Jensen R, "Performing feature selection with ACO. Swarm intelligence and data mining", Abraham A, Grosan C and Ramos V (eds.), Studies in Computational Intelligence, 34, 2006, 45–73.
2. Yan J, Zhang B, Liu N, Yan S, Cheng Q, Fan W, Yang Q, Xi W, Chen Z, "Effective and efficient dimensionality reduction for large-scale and streaming data preprocessing", IEEE transactions on knowledge and data engineering, 18, 3, 2006, 320–333.

3. Davy M, Luz S, "Dimensionality reduction for active learning with nearest neighbour classifier in text categorisation problems", Sixth International Conference on Machine Learning and Applications, 2007.
4. Swiniarski RW, "Rough sets methods in feature reduction and classification" International Journal of Applied Mathematics and Computer Science, 11, 3, 2001, 565–582.
5. Jollie IT, "Principal Component Analysis", Springer-Verlag, New York, 1986.
6. Pawlak Z, "Rough Sets: Theoretical Aspects of Reasoning About Data", Kluwer Academic Publishing, Dordrecht, 1991.
7. Polkowski L, Lin TY, Tsumoto S (Eds), "Rough set methods and applications: new developments in knowledge discovery in information systems", Vol. 56. Studies in Fuzziness and Soft Computing, Physica-Verlag, Heidelberg, 2000.
8. Pawlak Z, "Roughsets" International Journal of Computer and Information Sciences, 11, 1982, 341–356.
9. UCI Repository for Machine Learning Databases retrieved from the World Wide Web: http://www.ics.uci.edu
10. Han J, Kamber M, Data Mining: Concepts and Techniques, Morgan Kaufmann, 2001 279–325.

Chapter 11
Dramatically Reduced Precision in Microarray Analysis Retains Quantitative Properties and Provides Additional Benefits

William C. Ray

Abstract Microarray technology is a primary tool for elucidating the differences between similar populations of nucleic acid molecules. It is frequently used to detect mRNA population shifts that result from perturbations such as environmental changes, host–pathogen interactions, or the shift from therapeutic to toxic drug doses. Unfortunately, current microarray analysis methods provide only undirected discovery tools and cannot be directed to a particular hypothesis. To address this issue, we demonstrate that biologically relevant aspects of expression profiles may be captured by precision-reduced descriptions that are fully human readable, and that biologically relevant relationships may be captured by applying familiar pattern searching techniques to these precision-reduced descriptions. Even expression profiles that are reduced to only a single bit of precision, retain the surprising ability to reproduce the clustering results from the full 16-bit precision original data. We also illustrate that simple verbal descriptions ("my gene's expression went up briefly at timepoint 10") of expression profiles are quantitative entities fully compatible with clustering and searching within the precision reduced data.

1 Introduction

Microarray technology is a primary tool for elucidating the differences between similar populations of nucleic acid molecules. It is frequently used to detect mRNA population shifts that result from perturbations such as environmental changes, host–pathogen interactions, or the shift from therapeutic to toxic drug doses. As such, the use of microarrays permeates areas studying such sequence populations.

Microarray data analysis is often performed by clustering expression profiles into groups with related features. Unfortunately, current clustering methods provide only undirected discovery tools and cannot be directed to a particular hypothesis. Clusters

W.C. Ray (✉)
The Battelle Center for Mathematical Medicine, The Research Institute at Nationwide Children's Hospital, Columbus, OH 43205, USA
e-mail: ray.29@osu.edu

themselves cannot be described other than by enumeration of their members and cannot be used as queries to find genes with similar behaviors. Both of these issues impede microarray expression analysis, especially in the area of expression-pattern databases and intercomparison of experiments.

When observing actual microarray analysis results, one is struck by the often significant variation in signal levels derived for what should theoretically be identically responding probes. In considering this variation, we realized that it was not only a confounding source of error that explains much of the ongoing difficulty that clustering experiments have had in producing results which agree with biological expectations, but also a suggestion that the typical precision used for microarray analysis was far beyond the precision warranted by the data. We hypothesize that biologically relevant aspects of expression profiles may be captured by precision-reduced descriptions that are fully human readable, and that biologically relevant relationships may be captured by applying familiar pattern searching techniques to these precision-reduced descriptions. In this chapter, we demonstrate that expression profiles that are reduced to only a single bit of precision retain the surprising ability to reproduce the clustering results from the full 16-bit precision original data. We also illustrate that simple verbal descriptions ("my gene's expression went up briefly at timepoint 10") of expression profiles are quantitative entities fully compatible with clustering and searching within the precision reduced data.

1.1 Motivation

Microarray technology is an evolutionary descendent of nucleic acid hybridization techniques such as Southern hybridization, scaled to genome-scale numbers of immobilized probes, and micron-scale inter-probe spacing on the substrate [8]. These scales allow (potentially) every sequence in an organism to be simultaneously used as a probe for hybridization to an unknown population of NA sequences, in an attempt to determine the composition of that population.

Probes are localized by a variety of methods into arrays of "spots" with known locations on a specially treated microscope slide, or similar substrate [5]. Spotted slide technology can fit several replicates of the entire probe set for a typical bacterial genome on a single slide. Affymetrix GeneChips can contain gene-walking replicates for human-genome sized probe sets.

The primary interest in determining population compositions is in comparing the composition of differing populations [7]. These populations are frequently bulk cellular RNA prepared from cells grown under differing environmental conditions. However, applications to other populations are being developed, such as the analysis of bulk genomic DNA, or the analysis of mass microbial populations.

Initial processing produces a series of ratios detailing the (probable) relative abundance of the target for each probe, in each population [1]. The null hypothesis is that there is no difference between the populations, implying that the hybridization results would be identical and the observed ratios ≈ 1, for both the experimental and control populations. An array of normalization techniques, embedded controls, and sophisticated applications of statistics are regularly brought to bear on the values [3], yet it is inevitable that some level of error will have to be tolerated in the analysis regardless of the techniques used to minimize it.

It is with this data environment and with its challenges that sophisticated analyses of microarray data, such as the prediction of whole-genome regulatory networks, are faced. This does not imply that it is pointless to attempt sophisticated analyses, but rather that there is no benefit to be found in maintaining precision beyond the basic accuracy of the data, and that methods that aid researcher comprehension may be more advantageous than ones that more elaborately address the data.

Simplification of the data makes it easier to visualize and comprehend for the human researcher attempting to analyze it, and the experimental results detailed in this chapter suggest that surprisingly simplistic proxy descriptions maintain a high level of fidelity to the original data.

1.2 Rationale

Due to the impediments mentioned above, microarray analyses and analysis techniques are so far limited to data-mining, and pattern-discovery tools for microarray data are a serious impediment for the field. The ability to direct queries against expression data would enable exploration of directed hypotheses regarding expression relationships.

Many clustering methods papers note that detecting clusters related to some root pattern is important [9]. However, this is not a common capability of microarray clustering algorithms.

Moreover, the lack of progress may be partially related to the effects described by [6], which show that recent and sophisticated clustering techniques perform poorly compared to older, statistically simpler methods. There is room for interpretation regarding the effects Lichtenberg observes, but it is clear that statistically complex methods are not necessary to produce quality clustering results. Combined with the published success of several simplified-data clustering methods, and our data demonstrating the resilience of MA clustering results to dramatic precision reductions, it is appealing to consider whether simplified descriptions of events observed in the expression patterns – descriptions that are exactly analogous to those already used by researchers when discussing and comparing results – are sufficiently accurate descriptions to allow quantitation, and whether such quantitation can be accomplished using pattern-searching tools that are already familiar to the researcher.

2 Methods

The research reported here grows out of a series of studies we have performed in the generation and evaluation of bit-strings as representations of expression time-series data. The underlying clustering algorithm (Threshold Clustering – TC) and early results have been presented at BIBE04 (BioInformatics and BioEngineering conference of IEEE) and are published in the BIBE04 transactions [4]. In studying this algorithm's performance, we have focused on three key areas:

1. Demonstrating that similar expression magnitudes cannot be expected from biological systems, even when the measurements are from the same transcript.
2. Demonstrating that dramatically simplified descriptions of expression trajectories can contain sufficient information to allow biologically relevant clustering.
3. Demonstrating that similar-subsequence queries against microarray expression databases extract biologically and statistically meaningful results.

We have investigated the first of these through studies of cAMP-induced expression changes in an adenylate cyclase-deficient *Haemophilus influenzae* mutant developed by a collaborator. The second and third have been studied by further characterizing the performance of our previously developed Threshold Clustering (TC) algorithm on existing microarray timeseries data, and on synthetic datasets that reflect biologically reasonable perturbations, and, to eliminate the clustering-algorithm effects, by examining the effect of reduced-precision data on hierarchical clustering results.

It should be noted that the results presented for TC are neither optimized, *nor is TC claimed to be the optimal clustering solution for minimal-data descriptions*. TC functions as a sort of dimensionless analog to the construction of a Voronoi partition, with the distances between points (patterns) calculated based on the longest common subsequence (LCS) shared by the patterns. As such, what we intend by these results is a demonstration of the feasibility of approaching clustering, and cluster consensus with reduced precision, pattern-based clustering algorithms. The results presented here for both the precision reduction and the clustering algorithm should be taken as representative minimum reasonable bounds on performance and quality that can be expected from algorithms of this type.

2.1 Measured Expression Variation

The initial observation that introduced the question of whether the measured fold-change magnitude was meaningful for clustering was a group of *H.i.* genes that appeared to be in an operon, and produced markedly different fold-change values. One of the most basic questions that one might ask of a clustering algorithm is "what genes are co-regulated?" One of the most basic requirements for an answer is that genes that are co-transcribed should co-cluster. Significant variation in the fold change measured from the same transcript makes it difficult to meet this requirement. We therefore wanted to know the extent of variation that might be observed in our *Haemophilus* system, as produced by our microarray core. To assay this variation for cAMP supplementation of our adenylate-cyclase mutant, we conducted a short timeseries (three timepoints, three technical replicates/slide, two slides/dye, dye-flipped) experiment. Probable operon membership and probable co-transcribed genes were predicted for all genes showing significant variation. These analyses suggested that separate assays of the same transcript (as conducted by examining separate genes in an operon) could vary by more than 2,100% in their measured fold change (*comA* at 23.6-fold to *comF* at 1.1-fold at the 125 min timepoint).

To validate that the expression products measured are produced as single transcripts, reverse transcriptase (RT) PCR was used to detect the presence of transcript bridging each intergenic region in the supposed operon. Primers were chosen approximately 200 bp before the stop of the upstream gene and 200 bp after the start of the downstream gene. The intergenic region between HI0053 and HI0054, which are divergent, was included as a negative control. The results clearly indicate that RNA is produced for the region spanning from ∼200 bp upstream of each stop codon, to ∼200 bp into the next gene. This is a strong indication that for each of these putative operons, a full-length transcript is produced. Significant and repeatable differences are observed between the fold-change values for the majority of genes in each of these operons. It is imperative that a clustering algorithm co-cluster the results for probes that target the same RNA molecule. If it cannot co-cluster results from different probes to the same RNA, it can hardly be expected to illuminate more complex relationships. Therefore, when considering questions such as co-regulation, the level of variation displayed by these experiments must not be considered significant for calculating clustering distances. The apparent lower abundance of a PCR product for the HI0050–HI0049 intergenic region brings into question whether HI0049 is regulated *identically* to HI0050. This question actually exists for each gene pair. The PCR product for this intergenic region is clear in the original scan of the gel. Its lower abundance suggests that there may be a weak terminator in this region. This does not, however, demand that the expression pattern for HI0049 should be universally segregated from those of the other members of the operon. Despite any potential internal termination or alternative start sites, full-length transcript is produced for this operon. It may be appropriate to segregate HI0049 if the clustering question involves differential regulation effects, but if it involves the promoter at the 5' end of the full-length transcript, HI0049 must cluster with the other members of its operon.

2.2 *Simplified Data Descriptions and Subsequence-Based Clustering*

Early experiments have led us to standardize on the Yeast cdc28 dataset from [2] for testing. This dataset contains ∼1400 genes that have levels recorded for each of the dataset's 17 timepoints. This subset (cds28) of the entire cdc28 dataset has been used to eliminate complications with differential edge effects.

Examination of cds28 using standard MA clustering tools yields several interesting features that are ideal for comparing the results of different clustering methods. Primary among these are two highly distinct expression behaviors (hereafter **TESTGENES** and **TESTGENES2**) that are each followed by a reasonable number

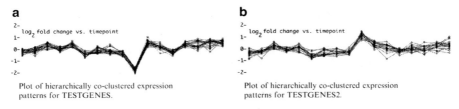

Fig. 1 Two highly distinct expression behaviors in the cds28 dataset. We have extensively characterized the TESTGENES behavior for these results

of genes. In each of several clustering methods [hierarchical, k-means, SOM using Acuity 4.0 software(Axon Software)], these behaviors elicit different, but compositionally similar clusters.

The minimum correlation coefficients (ρ) for TESTGENES and TESTGENES2 are 0.845 and 0.805, respectively. The average ρ for hierarchical clustering using Euclidian distances on cds28 is approximately 0.37 (estimated on nearest linkages from five random subsets of 27, with a standard deviation of 0.042). This value reflects the overall non-random character of the biological data and must be taken into account when comparing correlation coefficients for cluster members. The expression patterns for these genes are shown in Fig. 1a,b. (clustering and associated statistics calculated with Acuity 4.0).

Because of the tightly clustered and intuitively describable nature of the expression events seen in TESTGENES and TESTGENES2, these make useful clusters for determining the impact that reductions in precision have upon the quality of clustering and for testing the application of substring-based algorithms as avenues leading to directed queries using expression profiles.

If our clustering algorithm is resulting in random assignments of patterns to clusters, we would expect to see the expression patterns from a group such as TESTGENES, spread (approximately) uniformly across all clusters predicted by the algorithm. Likewise, independent of the clustering algorithm, if our data reduction is disposing of the biologically important information, the TESTGENES patterns would become randomly distributed, or co-cluster into a large group of data-less patterns. What we observe instead is that applying TC at 90% similarity to a single-bit representation of the data, 20 members of TESTGENES co-cluster in a cluster of size 43 (**cluster 18**). The remaining 7 co-cluster in a cluster of size 107 (**cluster 19**). Fourteen members of TESTGENES2 also co-cluster in cluster 19. The probability of a cluster such as cluster 18 occurring randomly is less than 1 in 2.5×10^{27}. Even the co-occurrence of the remaining seven members of TESTGENES in cluster 19 is improbable.

While these results could indicate a significant increase in false positives, here such an increase does not appear to be the case – nor is this behavior unexpected or undesirable in the context of the expanded latitude of the subpattern-based clustering intent. The majority of the new patterns accumulated in cluster 18 differ from TESTGENES patterns only in magnitude, a variable that we demonstrate has

11 Reduced Precision in Microarray Analysis   107

little meaning later in these results. Since hierarchical clustering's Euclidian measure handles scaled patterns inappropriately, these are likely to be false negatives in the hierarchical TESTGENES result.

Biologically, TESTGENES members are primarily ribosomal protein genes and display a single negative expression pertubation at the cell-cycle juncture between cytokinesis and the appearance of new daughter buds. The 23 extra patterns adopted by cluster 18 show an interesting mix of unknown/hypothetical genes with expression patterns that coincide exactly with the TESTGENES pattern in form and timing, but differ in magnitude (11 genes), regulatory genes with essentially the same pattern and differing magnitudes (3), regulatory genes with single-timepoint perturbations at other timepoints (3), a protein synthesis gene with a very low signal that may be spurious noise, and most interestingly, five additional ribosomal protein genes that are almost certainly influenced by the same regulatory network as the bulk of ribosomal proteins in TESTGENES, but that are shifted by 1 timepoint and therefore are falsely excluded from TESTGENES by hierarchical clustering. The seven members of TESTGENES that are ejected from cluster 18 are ejected only due to the instability of edge cases that is inherent in all clustering algorithms that produce distinct clusters.

Because this comparison of TESTGENES and cluster 18 is performed at only a single point in the respective algorithm's and data reduction's parameter spaces, these results should only be interpreted as a validation of the statistically significant behavior of the pattern-based clustering over the single-bit proxy data representation.

To further investigate the properties of precision reduction, separate from the clustering algorithm, we generated several precision-reduced versions of cds28 and clustered them with hierarchical clustering. The pairwise differences between correlation coefficients were calculated using the full precision data and those calculated using various levels of reduced precision. For $\rho >= 0.9$, with 8 or 4 bits of precision, the residuals are indistinguishable from zero across the entire dataset. Even with a proxy containing only 2-bits of precision, there are less than a dozen residuals that exceed 0.1. The residual error for values of ρ less than 0.9 are even smaller. Paired t-tests at the 95% confidence level confirm that the 8-bit proxy is indistinguishable from the original data, with a confidence interval of $[-0.000036, 0.000028]$ for the residual. The 95% confidence intervals of the 2-bit residuals for $\rho \geq 0.9$, and $\rho \geq 0.7$ are $[0.025739, 0.031624]$ and $[0.007801, 0.008905]$, respectively, while the 95% confidence interval for all ρ is vanishingly small at $[0.000340, 0.000550]$. While clearly statistically distinct from the original data at levels of less than 8-bits precision, these data also demonstrate that large correlations remain large, and small ones remain small (and overall, quite near their original values), even with data that have been reduced to indicate nothing more than the occurrence of a change and its direction. *Clearly, the information content necessary for clustering is resilient to rather extreme reductions in precision.*

The strength of the TC performance is, in fact, somewhat understated based on the minimalist comparisons presented here. For example, the TESTGENES set can be isolated in the hierarchical results primarily by visual means. It is a 27-member cluster with ρ in the range of 0.85–0.95. Separating it from the remainder of cds28, which displays similar correlation coefficients between many patterns, is possible only because the relationship between the TESTGENES cluster and the next closest cds28 pattern is relatively distant with ρ of 0.752. TC, on the other hand, isolates TESTGENES into 2 of only 29 described patterns (at threshold 90%).

2.3 Application of Simplified Descriptions and Subsequence Methods as Directed Queries

Precision reduction would have little real use if the only benefit gained was simplification of expression pattern descriptions. However, the availability of simplified descriptions opens the door to a wide range of new applications and analysis procedures for expression data. Among these are the use of individual patterns, which since the descriptions are simplified, can be conveyed in essentially plain English, as queries against expression pattern databases. The (non) availability of this seemingly simple analysis capability is a serious roadblock to the full utilization of microarray technology. The situation is analogous to where the sequence-analysis community would be if the only capability available was multiple-sequence alignment, rather than directed queries using a single target sequence. The simple ability to ask "is there another pattern like this one?" type queries is transformative. It would allow the researcher to issue hypotheses in terms of expression patterns. Using simplified descriptions to issue these queries, the task is made both tractable and extremely familiar to the bench researcher.

The TC algorithm makes an interesting test algorithm for demonstrating the potential for this type of directed array querying. Because the algorithm is essentially an aglomerative neighborhood collector, it may be seeded by automatically detected "best" cluster seeds, as shown in our earlier comparison to other methods. Alternatively, it may be seeded by researcher-provided patterns, in a simple and direct translation from plain English queries. To demonstrate the feasibility of such an approach, one may consider the expression patterns displayed for TESTGENES. An examination of these patterns leads to the conclusion that they display, as a common element, a dramatic change in expression level at timepoint 10, an equally dramatic change at timepoint 11, and are essentially zero-fold-change over the rest of their timepoints. This expression pattern can be encoded as a string containing zeros in positions (timepoints) of no change, and ones in positions where there is change (0000000001100000). Using this pattern as a seed for TC, we detect 32 genes in cds28.

Not coincidentally, these 32 patterns are all members of cluster 18 as described earlier and have the same relationship to TESTGENES as the overlap between cluster 18 and TESTGENES, plus those elements from cluster 18 that include perturbations only at timepoint 10. They exclude those patterns from cluster 18 that show perturbations at other timepoints, due to the query disallowing variation outside timepoints 10 and 11. Slight variation in the positioning of the perturbation could be allowed in the query string, causing it to more closely capture the cluster 18 set, simply by deleting no-variation positions from the ends of the query (notably, this expression pattern was generated from TESTGENES, rather than cluster 18 – a query generated from cluster 18 would, by definition, exactly recapitulate cluster 18). Of the retrieved patterns, 20 are exactly the members of TESTGENES that co-clustered in cluster 18 of the experiment discussed above. The probability for this occurrence is less than 1 in 6×10^{31}, but of course, being defined to match, it is unsurprising. What may be more surprising is the limited number of additional patterns picked up by the search. These 12 additional patterns are *not* grouped by hierarchical clustering with TESTGENES, yet they display significant (in this case twofold or more between timepoints) changes at, and only at, the timepoints identical to those where the TESTGENES cluster displays changes. The reason hierarchical clustering fails to detect these related patterns is not obvious, but appears to be due to spurious correlations in the noise outside the region of significant change overwhelming the hierarchical correlation calculation.

We have performed numerous other experiments of this nature using query strings extracted from clustered patterns, and the results have been uniformly similar. The query string, matched against even a binary discretized description, can universally select the set of patterns for which it was designed, and additionally finds other patterns that are highly related, but that fail to be clustered by other methods due to lower interest similarities that overwhelm the clustering algorithm.

3 Conclusion and Discussion

Our results demonstrate: that significant variability can be expected between fold-change measurements for genes subject to the same regulatory effects in real experiments; that even signals devoid of all precision information other than the presence of a change still maintain meaningful information for clustering; and that simplified descriptions, amenable to construction and conversation by wet-bench researchers, can be quantitatively applied as queries, returning statistically and biologically meaningful results. All these observations support the applicability of reduced-precision descriptions to improving microarray analysis technology.

Several strong conclusions can be drawn from the data. If we eliminate differences of clustering algorithm from the comparison, and look only at the effect of precision reduction, the primary observation from this study is that the correlation coefficient is surprisingly resiliant to precision reduction. At a reduction to only 2 bits, out of a nominally 32-bit signal, the 95% confidence interval for a paired t-test of the difference between the correlation coefficients of the reduced and original data is [0.0078,0.0089] for correlation coefficients ≥ 0.7. The residual error is even smaller for less significant correlations. The membership of the TESTGENES cluster remains constant when the input data for hierarchical clustering is reduced to only 3 bits and a sign. These results clearly indicate that precision reductions into the single-character-per-observation range maintain a high degree of fidelity to the original information content of the full-precision data, and that they are a reasonable basis for simplified proxy descriptions.

An evaluation of whether subpattern-based clustering warrants further consideration requires analysis of the relationship between our cluster 18 and hierarchical clustering's TESTGENES cluster. The key observation is that examination of the biological context of TESTGENES and cluster 18 suggests that while some of the additional genes detected in cluster 18 are potentially false positives, a significant subset are actually true positives in cluster 18, and were false negatives, incorrectly excluded from TESTGENES by hierarchical clustering. A survey of the yeast data shows approximately 180 annotated ribosomal proteins. The cell's probable stoichiometric requirement for ribosomal proteins suggests that both TESTGENES and our cluster 18 are underestimates of genes involved in the same regulatory scheme.

Finally, we would like to reiterate that these results have been produced using extreme simplifications and both trivial and unoptimized algorithms. They should not, in any way, be construed as a suggestion that a simplistic, single-bit simplification, or a simple Longest-Common-Substring clustering algorithm is the zenith of expression analysis capability. We report results from simplifications and algorithms at the limit of triviality, not to claim their supremacy, but to demonstrate that dramatic simplifications retain surprising analytical power, and that such simplifications enable a type of analysis not feasible with the current state-of-the-art description of expression trajectories. By these results, we wish to encourage the exploration of more powerful methods that retain descriptions within the realm of human comprehension and encoding.

Supplemental information, data, and figures are available from the author's web page at http://www.stickwrld.org/reduced_precision/.

References

1. Chen Y, Kamat V, Dougherty ER, Bittner ML, Meltzer PS, Trent JM (2002) Ratio statistics of gene expression levels and applications to microarray data analysis. Bioinformatics 18(9):1207–1215
2. Cho RJ, Campbell MJ, Winzeler EA, Steinmetz L, Conway A, Wodicka L, Wolfsberg TG, Gabreilian AE, Landsman D, Lockhart DK, Davis RW (1998) A genome-wide transcriptional analysis of the mitotic cell cycle. Molecular Cell 2:65–73
3. Cui X, Churchill GA (2003) Statistical tests for differential expression in cDNA microarray experiments. Genome Biology 4(4):210
4. Erdal S, Ozturk O, Armbruster D, Ferhatosmanoglu H, Ray WC (2004) A time series analysis of microarray data. Transactions of the IEEE Fourth International Symposium on Bioinformatics and Bioengineering BIBE 2004, pp. 366–378
5. Guo Z, Guilfoyle RA, Thiel AJ, Wang R, Smith LM (1994) Direct flourescence analysis of genetic polymorphisms by hybridization with oligonucleotide arrays on glass supports. Nucleic Acids Research 22:5456–5465
6. de Lichtenberg U, Jensen LJ, Fausbøll A, Jensen TS, Bork P, Brunak S (2005) Comparison of computational methods for the identification of cell cycle-regulated genes. Bioinformatics 21(7):1164–1171
7. Rhodes DR, Miller JC, Haab BB, Furge KA (2002) CIT: Identification of differentially expressed clusters of genes from microarray data. Bioinformatics 18(1):205–206

8. Schena M, Shalon D, Heller R, Chai A, Brown PO, Davis RW (1996) Parallel human genome analysis: Microarray-based expression monitoring of 1000 genes. Proceedings of the National Academy of Sciences USA 93(20):10,614–10,619
9. Vlachos M, Kollios G, Gunopulos D (2002) Discovering similar multidimensional trajectories. In: Proceedings of the 18th International Conference on Data Engineering, pp. 673–684, URL http://citeseer.nj.nec.com/vlachos02discovering.html

Chapter 12
Algebraic Model Checking for Boolean Gene Regulatory Networks*

Quoc-Nam Tran

Abstract We present a computational method in which modular and Groebner bases (GB) computation in Boolean rings are used for solving problems in Boolean gene regulatory networks (BN). In contrast to other known algebraic approaches, the degree of intermediate polynomials during the calculation of Groebner bases using our method will never grow resulting in a significant improvement in running time and memory space consumption. We also show how calculation in temporal logic for model checking can be done by means of our direct and efficient Groebner basis computation in Boolean rings. We present our experimental results in finding attractors and control strategies of Boolean networks to illustrate our theoretical arguments. The results are promising. Our algebraic approach is more efficient than the state-of-the-art model checker NuSMV on BNs. More importantly, our approach finds all solutions for the BN problems.

Keywords Gene regulatory networks · Groebner basis · Model checking

1 Introduction

Model checking is an established research direction aimed at increasing the productivity of system designers in which mathematical techniques are used to guarantee the correctness of a design with respect to some specified behavior [1,12]. It has the ability to discover subtle flaws resulting from improbable events. In fact, it is one of the most successful applications of automated reason in computer science. Model checking has several attractive features. Once the Kripke model of the system and the temporal logic specifications have been defined, the process is fully automated. If a computation that violates the specification is found in the Kripke

---

*A preliminary version of this paper appeared in BIOCOMP'09.

Q.-N. Tran (✉)
Department of Computer Science, Lamar University, Beaumont, TX 77710, USA
e-mail: qntran@lamar.edu; qntran@buchberger.cs.lamar.edu

model, it can be displayed to the designer to aid the debugging process. In addition, the model checker can formally verify whether the specification holds [10].

Since temporal logic can assert how the behavior of a system evolves over times, model checking can be used for many finite-state concurrent systems including game designs, Petri nets, and Boolean networks [15]. Model checking are often classified into branching-time or Computational Tree Logic (CTL) model checking and Linear Temporal Logic (LTL) model checking depending on whether time is assumed to have a linear or a branching structure. Typically, computation is carried over Boolean algebras using binary decision diagrams (BDDs) or satisfiability (SAT) solvers.

Researchers have been using BDDs and CTL model checking on Boolean networks. Previous works also showed that BDDs blow up more frequently on random networks than on sequential circuits. In this chapter, we show that LTL model checking may be a better choice for solving the problems in bioinformatics using Boolean networks because LTL is more suitable for this class of problems and LTL gives a shorter coding for the verification process.

The chapter is organized as follows: in the next section we summarize some results from LTL model checking, which are necessary for specifying time-variant behaviors. In Sect. 3, we explain why LTL temporal logic is more suitable for Boolean networks. In Sect. 4, we present our alternative way in which we use algebraic computation for checking temporal systems. In Sect. 5, we present our experimental results in finding attractors and control strategies of Boolean networks to illustrate our theoretical arguments. We compare our work with the state-of-the-art traditional model checkers.

2 Linear Temporal Logic and Model Checking

The first step in formal verification is the representation of formal specification of the design consisting of a description of the desired behavior. As sequential systems capture time-variant behavior, it is not possible to describe their properties completely in the framework of conventional propositional formulas. In a temporal logic, temporal operators are additionally provided, which can be used to express time-variant dependencies. A widely used specification language for designs is Temporal Logic [13], which is a modal logic (i.e., logic to reason about many possible worlds with their semantics based on Kripke structures).

In LTL, time is treated as if each moment has a unique possible future. (An alternative approach is to use branching time (CTL) [5]. However, the Boolean networks evolve over time in a linear and deterministic manner.) Linear temporal formulas are constructed from a set $\wp$ of atomic propositions using usual Boolean connectives as well as temporal connectives **X** (next), **G** (always), **F** (eventually), and **U** (until).

The semantics of LTL formulas are based on an appropriate Kripke structure of the form $K = (\mathbb{N}, \leq, \pi)$, where $\mathbb{N}$ is the set of natural numbers, $\leq \subseteq \mathbb{N}^2$ is the standard linear order, and $\pi : \mathbb{N} \to 2^{\wp}$ is a function that defines what propositions are true at a certain time instant. An LTL formula is interpreted over *computations* viewed as infinite sequences of truth assignments to the atomic propositions in Σ^ω where $\Sigma = 2^\wp$ as follows:

1. $\pi, i \models \varphi$ means that formula φ holds at the time i of the computation π
2. $\pi, i \models p$ iff $p \in \pi(i)$
3. $\pi, i \models \mathbf{X}\varphi$ iff $\pi, i+1 \models \varphi$
4. $\pi, i \models \mathbf{G}\varphi$ iff $\pi, j \models \varphi \; \forall j \geq i$
5. $\pi, i \models \mathbf{F}\varphi$ iff $\pi, j \models \varphi$ for some $j \geq i$
6. $\pi, i \models \varphi \mathbf{U} \psi$ iff $\exists j \geq i$ such that $\pi, j \models \psi$ and $\pi, k \models \varphi \; \forall j > k \geq i$

Note that operators $\mathbf{G}$ and $\mathbf{F}$ can be derived from $\mathbf{X}$ and $\mathbf{U}$ as $\mathbf{F}\varphi \equiv true \, \mathbf{U}\varphi (true \equiv p \vee \neg p)$ and $\mathbf{G}\varphi \equiv \neg \mathbf{F} \neg \varphi$.

In LTL model checking, we assume that the specification is given in terms of properties expressed by LTL formulas. For example the formula $\mathbf{G}(request \rightarrow \mathbf{F} \, grant)$, which refers to the atomic propositions *request* and *grant*, specifies that every state in the computation in which *request* holds is followed by some state in the future in which *grant* holds.

Note that LTL logic and CTL logic have different expressive powers. For example, there is no CTL formula that is equivalent to the LTL formula $\mathbf{A}(\mathbf{FG} \, grant)$. This formula expresses the property that along every path, there is some state from which *grant* will be granted forever. Similarly, there is no LTL formula that is equivalent to the CTL formula $\mathbf{AG}(\mathbf{EF} \, grant)$, where $\mathbf{A}$ and $\mathbf{E}$ are two path quantifiers in CTL logic.

In model checking, the specification is expressed in temporal logic and the system is modeled as a finite state machine. For realistic designs, the number of states of the system can be very large, 10^{20} or higher, and hence explicit traversal of the state space becomes infeasible. Perhaps, the biggest hurdle for the practical use of model checking is the state explosion problem [16]. To relieve the state explosion problem, many approaches have been proposed.

Symbolic model checking using ordered BDDs is a successful and widely used techniques for verifying properties of concurrent hardware and software systems. The underlying model for BDD was studied in the 1950s and 1970s. This basic data structure has been widely used in many areas of computer-aided VLSI design after Bryant's substantial improvement [3] which added some ordering restrictions as well as a sophisticated reduction mechanism to the model. In symbolic model checking, the state space is represented implicitly using symbolic means, and the propositional logic formulas are manipulated using BDDs [9]. Symbolic model checking succeeded in checking systems with unprecedented large state spaces. However, previous works also showed that BDDs blow up more frequently on random Boolean networks than on sequential circuits.

Bounded model checking uses the same basic idea as symbolic model checking using BDD in which the state space of the system is represented implicitly by Boolean formulas. However, instead of manipulating the Boolean formulas using BDDs, the bounded model checker transforms the model and property specifications into a propositional satisfiability (SAT) problem. Given a system M, a temporal logic formula ψ, and a bound k, a Boolean formula is constructed, which is satisfiable if and only if M has a counterexample of length k to ψ. A SAT solver such as SATO or zChaff is used to perform the query. In BMC, the growth of the size of the Boolean formulas can be known in advanced, but predicting the running times of the SAT solver is difficult. BMC is particularly good at finding shallow counterexamples, and some industrial applications have been successful of BMC [2].

Other approaches to relieve the state explosion problem include the partial order methods, complete finite prefixes, compositional methods, symmetry reduction, and abstraction [6].

3 Boolean Network as a Linear Time System

In this section, we review a straightforward way to translate Boolean network's problems into problems in a linear time system and use model checking techniques to solve the problems. A BN is represented by a set of nodes and a set of regulation rules for nodes, where each node corresponds to a gene. At any time t, each node has the value of 1 or 0 corresponding to whether the gene is expressed or not, respectively. A regulation rule for a node decides the state of the node at the next time $t+1$ and is given in the form of a Boolean function. The states of nodes change *synchronously*. Dynamics of a BN is well described by a state transition table and a state transition diagram with $O(2^n)$ entities, where n is the number of nodes on the BN.

3.1 LTL Specifications for Boolean Networks

One of the extensively studied topics for BN is to identify the attractors, the directed cycles in the state transition diagram [7, 11]. Clearly, one can use Tarjan's algorithm for finding the strongly connected components (SCCs) of the state transition diagram with the cost of $O(2^n)$, where n is the number of nodes on the BN. Furthermore, it has been showed that even finding a singleton attractor is NP hard. The problem of identifying the attractors can be easily translated into an LTL model checking problem. For example, to find an attractor of length 4 without a self loop of the BN in Fig. 1, one can use the following LTL formula: LTLSPEC !F(((X X X X(v1) <-> v1) & (X X X X(v2) <-> v2) & (X X X X(v3) <-> v3)) & !((X(v1) <-> v1) & (X(v2) <-> v2) & (X(v3) <-> v3))).

Another extensively studied topic for BN is to find control strategies for a network. In this problem, we are given a BN with n internal nodes and m control nodes, an initial state of the network for internal nodes v^0, and a desired state of the network v^t.

We need to find a sequence of values for the control nodes so that the network would reach the desired state if such a sequence exists. Otherwise, the result is "None". For example, we can add two control nodes into the BN in Fig. 1, where

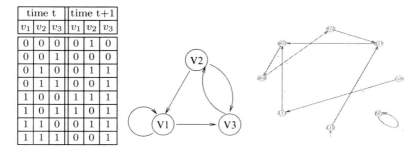

Fig. 1 Network with $v_1(t+1) = v_1(t) \wedge \neg v_2(t)$, $v_2(t+1) = \neg v_3(t)$, $v_3(t+1) = v_1(t) \vee v_2(t)$

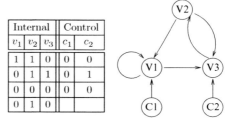

Fig. 2 Control of network with
$v_1(t+1) = v_1(t) \wedge \neg v_2(t) \wedge c_1$,
$v_2(t+1) = \neg v_3(t)$,
$v_3(t+1) = (v_1(t) \vee v_2(t)) \wedge \neg c_2$

$v_1(t+1) = v_1(t) \wedge \neg v_2(t) \wedge c_1$, $v_2(t+1) = \neg v_3(t)$, and $v_3(t+1) = (v_1(t) \vee v_2(t)) \wedge \neg c_2$. To find a control strategy for this BN with the initial state of $[1,1,0]$ and desired state of $[0,1,0]$, one can use the following LTL formula: LTLSPEC !F((v1 <-> 0) & (v2 <-> 1) & (v3 <-> 0)). A result is shown in Fig. 2.

3.2 Limitation of Model Checking for Boolean Networks

We have extensively studied temporal model checking [14, 15] and created a variation of random Boolean networks following this approach. Even though it is straightforward to translate Boolean network's problems into a temporal problem and use model checking techniques to solve the problems, model checking has many limitation for actually solving the problems. One of the biggest problems with traditional model checkers using BDDs or SAT solvers is that model checkers can only automatically find one attractor or solution for the Boolean network. Furthermore, in Sect. 5 we will show that the current model checking systems cannot solve even small size random problems in Boolean network (BN) due to time and memory consumption.

4 Boolean Groebner Bases for Model Checking

We now offer an alternative approach in which we use algebraic computation instead of BDDs and SAT solvers. Once the Boolean formulas have been converted into an equivalent system of polynomials in the corresponding Boolean ring, one can use the results from symbolic computation to perform calculation on the polynomial system. We refer to [4, 14, 15] for an introduction to basic facts on admissible term orders and the method of Groebner bases. It is well known that polynomial reduction is the costliest part for any practical implementation of Buchberger's algorithm for Groebner basis computation. We follow Buchberger's algorithm in a novel way in which extremely efficient bit operations are used for polynomial reduction. We use bit operations for polynomial reduction as an alternative way for SAT and BDD.

From the theory of Groebner bases, we know that a Boolean formula is not satisfiable if and only if the Groebner basis of the polynomials has 1 as a member of the basis [4]. The crux of our approach is that we do not use the costly polynomial reductions during the calculation of the Groebner basis nor in the last polynomial reduction as usual. Instead, we take into account the specific structure of model checking problem using linear algebra [14].

Let $F = \{f_1, \ldots f_s\}$ be a set of polynomials in a Boolean ring $K[X]$, and $\prec$ be a term order on $[X]$. Even though we do not know the reduced Groebner basis of F with respect to the term order $\prec$ yet, the existence and uniqueness of such a Groebner basis are guaranteed. Therefore, for every polynomial p there exists a unique normal form of p with respect to the reduced Groebner basis. Since $p \to^*_{GB(F,\prec)} \mathrm{nf}(p)$, $p - \mathrm{nf}(p)$ is in $I = \langle F \rangle$ and hence

$$p - \mathrm{nf}(p) = \sum_{i=1}^{s} f_i \cdot h_i$$

for some f_i in F and h_i in $K[X]$. In other words, $\mathrm{nf}(p)$ is the smallest monic polynomial with respect to the term order $\prec$ in the I-coset of p. Alternatively, [8] showed that finding the normal form of a polynomial can be transformed into solving a linear algebra system of size $2^{O(n)} \times 2^{O(n)}$ without knowing the reduced Groebner basis of I with respect to the term order $\prec$. If we expand all polynomials (including the unknown polynomials h_i and $\mathrm{nf}(p)$) to sums of monomials: $h_i = \sum_{x \in [X], \deg(x) \leq n} h_{i,x} \cdot x$, $f_i = \sum_{x \in [X], \deg(x) \leq n} f_{i,x} \cdot x$, and $\mathrm{nf}(p) = \sum_{x \in [X], \deg(x) \leq n} r_x \cdot x$ where the $h_{i,x}$ and r_x are unknown coefficients, we have:

$$\begin{aligned} p &= \sum_{x \in [X], \deg(x) \leq n} r_x \cdot x + \sum_{i=1}^{s} \left(\sum_{x \in [X], \deg(x) \leq n} f_{i,x} \cdot x \right) \\ &\quad \cdot \left(\sum_{x \in [X], \deg(x) \leq n} h_{i,x} \cdot x \right) \\ &= \sum_{x \in [X], \deg(x) \leq n} \left(r_x + \sum_{i=1}^{s} \sum_{u,v \in [X], u \cdot v = x} f_{i,u} \cdot h_{i,v} \right) \cdot x \\ &= M.b \end{aligned} \quad (1)$$

where $b = (h_{1,1}, \ldots, h_{1,x}, \ldots, h_{s,1}, \ldots, h_{s,x}, r_1, \ldots, r_x, \ldots)^T$, and M is a matrix of Boolean values (i.e., 0 and 1). The rows of matrix M correspond to terms

$1,\ldots,x_1,\ldots,x_1 \cdot x_2 \cdots x_n$ and the columns correspond to the unknowns $h_{i,x}$s and r_xs for all monomial x from 1 to $x_1 \cdot x_2 \cdots x_n$. Matrix M is free of rows and columns with all zeros, and the rows are rearranged with respect to the order of the monomials. For the columns, we arrange the columns correspond to h_xs before the columns correspond to r_xs. Also, column r_x corresponding to term x will come before column r_y corresponding to term y if $x \prec y$. Finding nf(p) can be done using the following algorithm:

Algorithm 1: [Normal Form]

Given   a set of polynomials F, a term order $\prec$ and a polynomial p.
Find   the normal form nf(p) of p with respect to $I = \langle F \rangle$ and $\prec$.

    Step_1   Build M and b as in (1)
    Step_2   Find a full row rank sub-matrix
        1a. Add the first non-zero row of M into an empty matrix M^+
        1b. For *row* from 2 to the last row of M
        If rank$(M^+ \cup row) \neq$ rank(M^+),
        add *row* into M^+
    Step_3   Find a full column rank sub-matrix
        1a. Add the first non-zero column of M^+ into an empty matrix M'
        Add the corresponding element of vector b into an empty vector b'
        1b. For *col* from 2 to the last column of M^+
        If rank$(M' \cup col) \neq$ rank(M'),
        add *col* into M'
        add the corresponding element of vector b into b'.
    Return   the solution of $p = M'.b'$

It is shown in [14] that Algorithm "Normal Form" always terminates and returns the normal form of p with respect to the given ideal I and term order $\prec$. We derive an algorithm to solve the decision problem for Groebner bases computation in Boolean rings using linear algebra as follows:

Algorithm 2: [GB using Linear Algebra]

Given   a set of polynomials F and a term order $\prec$.
Find   the reduced Groebner basis of $I = \langle F \rangle$ with respect to $\prec$.

    Step_1   Set $G' = \emptyset$; Build matrix M and vector b as in [14].
    Step_2   For all monomial m, $1 \not\prec m \prec x_1 \cdot x_2 \cdots x_n$ do
        If $1 = m + nf(m)$ then stop and return $\{1\}$;
        Add $m + nf(m)$ into G' when m is minimal reducible.
    Step_3   return G'.

It is shown in [14] that Algorithm "GB using Linear Algebra" is P-SPACE. In our approach, temporal formulas are translated into polynomials over the corresponding Boolean ring. Given an LTL formula $\mathbf{E}\phi$ and a Kripke structure M, we construct a Groebner basis whose solutions are exactly the states satisfying the temporal formula as follows: We construct a tableau T for the path formula ϕ. T is a Kripke structure and includes every path that satisfies ϕ. By composing T with M, we find a set of paths that appear in both T and M. A state in M will satisfy $\mathbf{E}\phi$ if and only if it is the start of a path in the composition that satisfies ϕ.

The transition relation for M can be represented by polynomials. In order to represent the transition relation for T, we associate with each elementary formula g a state variable v_g. If g is an atomic proposition, then v_g is just g itself. We describe the transition relation R_T as a Boolean formula in terms of two copies $\bar{v}$ and $\bar{v}'$ of the state variables. The Boolean formula is converted into polynomials to obtain a concise representation of the tableau.

Algorithm 3: Model Checking

Given   a finite state system or program P and an LTL formula φ that specifies the legal computations of the program.
Verify   whether all computations of the system are legal.

- Build a Groebner basis G_P for the system P with respect to an efficient term order and a set of Boolean formulas G_φ for the LTL formula φ.
- Check if $V(G_P) \subset V(G_\varphi)$ or $f \to_{G_P} 0, \forall f \in G_\varphi$.

Algorithm 4: Validity Checking

Given   an LTL formula φ.
Verify   whether the formula φ is valid.

- Build a Groebner basis G_φ for the formula φ with respect to an efficient term order.
- Check if $V(\overline{G_\varphi}) = \emptyset$ or $1 \in \overline{G_\varphi}$.

5 Experimental Results

We have implemented our algebraic model checking approach in Maple. In Fig. 3, we report the running time of our algorithms for finding *all* singleton attractors of randomly generated Boolean networks with different number of genes and in-degree of 3, 4, 5, and 6. Each result is the average running time of 20 randomly generated Boolean networks with the same number of genes and in-degree.

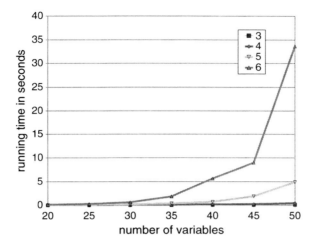

Fig. 3 Finding singleton attractors in BNs

As we mentioned before, even though it is straightforward to translate Boolean network's problems into a temporal problem and use model checking techniques to solve the problems, traditional model checking has many limitation for actually solving the problems. The biggest problems with traditional model checkers using BDDs or SAT solvers are that (a) model checkers can only automatically find one attractor or solution for the Boolean network, (b) model checkers using BDDs consume a lot of memory space for almost all random Boolean networks we have, and (c) model checkers using SAT solvers consume a lot of CPU time without giving a conclusive answer. For experimenting with the problem of identifying the attractors for BN, we generated 3,600 random networks with 25, 30, 35, 40, 45, and 50 internal nodes; in-degree of 3, 4, 5, 6, 7, and 8; cycle length of 1, 2, 3, 4, and 5. For experimenting with the problem of finding the control strategies for BN, we generated 2,160 random networks with 25, 30, 35, 40, 45, and 50 internal nodes; in-degree of 3, 4, 5, 6, 7, and 8; and 5, 8, and 10 control nodes. We wrote some Maple programs to automatically convert the BN problems into NuSMV programs. The NuSMV programs for these random BNs can be downloaded from http://buchberger.cs.lamar.edu/qntran/BN. We use the state-of-the-art techniques and software (NuSMV 2.4.0) in model checking to solve these problems. Experiments were carried on a Dell PowerEdge SC440 with Intel DualCore 2160, 4 Gb RAM running Linux Ubuntu 7.10 Server OS. Our results show that when the symbolic model checking using BDD approach is used, for almost all of the problems, the BDDs were blown up very fast and the system crashed very soon, especially for BNs with more than 30 nodes and in-degree of more than 3. When bounded model checking is used with zChaff SAT solver, for almost all of the problems, memory use was reasonable but NuSMV failed to find a counter-example after 10,000s with a bound of at most 30. At the same time, our algebraic model checking approach using modular solvers found all solutions for the BN problems in a reasonable amount of time and memory consumption.

6 Conclusion

We presented an algebraic model checking approach based on our P-SPACE algorithm for Groebner basis computation in Boolean rings and use it for finding attractors and control strategies of Boolean networks. Our algebraic model checking approach found all solutions for the medium size BN problems in a reasonable amount of time and memory consumption, while other BDD-based and SAT-based model checking approaches failed. We are working on improvements to our methods for dealing with BNs with 100–300 variables.

References

1. A. Biere, A. Cimatti, E. M. Clarke, M. Fujita, and Y. Zhu. Symbolic model checking using SAT procedures instead of BDDs. In *Proceedings of Design Automation Conference (DAC'99)*, 1999
2. P. Bjesse, T. Leonard, and A. Mokkedem. Finding bugs in an alpha microprocessor using satisfiability solvers. In *Proceedings of the 13th International Conference on Computer Aided Verification, CAV 2001*, volume 2102 of *LNCS*, pages 454–464, Paris, France. Springer, 2001
3. R. E. Bryant. Symbolic boolean manipulation with ordered binary decision diagrams. *ACM Computing Surveys*, 24(3):293–318, 1992
4. B. Buchberger. *An Algorithm for Finding a Basis for the Residue Class Ring of a Zero-dimensional Polynomial Ideal (in German)*. PhD thesis, Institute of Mathematics, University of Innsbruck, Innsbruck, Austria, 1965
5. E. Clarke, O. Grumberg, and D. Peled. *Model Checking*. MIT, 1999
6. E. M. Clarke, O. Grumberg, and D. E. Long. Model checking and abstraction. *ACM Transactions on Programming Languages and Systems*, 16(5):1512–1542, 1994
7. E. Dubrova, M. Teslenko, and A. Martinelli. Kauffman networks: analysis and applications. In *ICCAD '05: Proceedings of the 2005 IEEE/ACM International Conference on Computer-Aided Design*, pages 479–484, Washington, DC, USA. IEEE, 2005
8. G. Hermann. Die Frage der endlich vielen Schritte in der Theorie der Polynomideale. *Mathematische Annalen*, 95:736–788, 1925
9. J.R. Burch, E.M. Clarke, K.L. McMillan, D.L. Dill, and L.J. Hwang. Symbolic model checking: 10^{20} states and beyond. In *Proceedings of the Fifth Annual IEEE Symposium on Logic in Computer Science*, pages 1–33, Washington, DC, USA. IEEE, 1990
10. O. Kupferman and M. Vardi. From complementation to certification. In *Proceedings of the 10th International Conference on Tools and Algorithms for the Construction and Analysis of Systems*, volume 2988 of *LNCS*, pages 591–606. Springer, 2004
11. R. Laubenbacher and B. Stigler. A computational algebra approach to the reverse-engineering of gene regulatory networks. *Journal of Theoretical Biology*, 229:523–537, 2004
12. K. L. McMillan. *Symbolic Model Checking*. Kluwer Academic, 1993
13. A. Pnueli. The temporal logic of programs. In *Proceedings of 18th IEEE Symposium on Foundation of Computer Science*, pages 46–57, 1977
14. Q.-N. Tran. Groebner bases computation in boolean rings is P-SPACE. *International Journal of Applied Mathematics and Computer Sciences*, 5(2):109–114
15. Q.-N. Tran and M. Y. Vardi. Groebner bases computation in Boolean rings for symbolic model checking. In *Proceedings of IASTED 2007 Conference on Modeling and Simulation*, Montreal, Canada, 2007
16. A. Valmari. The state explosion problem. In W. Reisig and G. Rozenberg, editors, *Lectures on Petri Nets I: Basic Models, Advances in Petri Nets*, volume 1491 of *LNCS*, pages 429–528. Springer, 1998

Chapter 13
Comparative Advantages of Novel Algorithms Using MSR Threshold and MSR Difference Threshold for Biclustering Gene Expression Data

Shyama Das and Sumam Mary Idicula

Abstract The goal of biclustering in gene expression data matrix is to find a submatrix such that the genes in the submatrix show highly correlated activities across all conditions in the submatrix. A measure called mean squared residue (MSR) is used to simultaneously evaluate the coherence of rows and columns within the submatrix. MSR difference is the incremental increase in MSR when a gene or condition is added to the bicluster. In this chapter, three biclustering algorithms using MSR threshold (MSRT) and MSR difference threshold (MSRDT) are experimented and compared. All these methods use seeds generated from K-Means clustering algorithm. Then these seeds are enlarged by adding more genes and conditions. The first algorithm makes use of MSRT alone. Both the second and third algorithms make use of MSRT and the newly introduced concept of MSRDT. Highly coherent biclusters are obtained using this concept. In the third algorithm, a different method is used to calculate the MSRDT. The results obtained on bench mark datasets prove that these algorithms are better than many of the metaheuristic algorithms.

1 Introduction

Biclustering is the clustering applied along the row and column dimensions simultaneously. Cheng and Church were the first to apply biclustering to gene expression data [1]. There are different types of biclusters. Biclusters with coherent values are biologically more relevant than biclusters with constant values. Hence, in this work biclusters with coherent values are identified. In the case of biclusters with coherent values, the similarity among the genes is measured as the mean squared residue

S. Das (✉)
Department of Computer Science, Cochin University of Science and Technology,
Kochin, Kerala, India
e-mail: shyamadas777@gmail.com

(MSR) score. If the similarity measure (MSR score) of a matrix satisfies certain threshold, it is a bicluster. The degree of coherence of an element in a bicluster with the other elements of the bicluster is revealed by residue score. The residue score of an element bij in a submatrix B is defined as: $RS(bij) = bij - bIj - biJ + bIJ$. Hence, from the value of residue score, the quality of the bicluster can be evaluated by computing the MSR. That is, Hscore or MSR score of the bicluster B is MSR $(B) = \frac{1}{|I||J|} \sum i \in I, j \in J [RS(bij)]^2$. Here I denotes the row set, J denotes the column set, bij denotes the element in a submatrix, biJ denotes the ith row mean, bIj denotes the jth column mean and bIJ denotes the mean of the whole bicluster. A bicluster B is called a δ bicluster if MSR $(B) < \delta$ for some $\delta > 0$, i.e. δ is the MSR threshold (MSRT). If the MSR value of the submatrix is low, it means that the data are correlated. The value of δ depends on the dataset. For yeast dataset, the value of δ is 300 and for lymphoma dataset the value of δ is 1,200. The volume of a bicluster or bicluster size is the product of the number of rows and the number of columns in the bicluster. The larger the volume and smaller the MSR or Hscore of the bicluster, the better is the quality of the bicluster.

2 Datasets Used

Experiments are conducted on the yeast and lymphoma datasets. Yeast dataset is based on Tavazoie et al. [2]. The preprocessed datasets are obtained from http://arep.med.harvard.edu/biclustering. Yeast dataset consists of 2,884 genes and 17 conditions. The values in this dataset are integers in the range 0–600. Missing values are represented by −1. Human B-cell lymphoma expression data contain 4,026 genes and 96 conditions. The dataset was downloaded from the website for supplementary information for the article by Alizadeh et al. [3]. The values in the lymphoma dataset are integers in the range −750–650. There are 47,639 (12.3%) missing values in the lymphoma dataset. Missing values were represented by 999. Missing data in the lymphoma dataset are replaced with random numbers uniformly distributed between −800 and 800. Experiments are also conducted by filtering out genes with small variance over the conditions from both the datasets. The algorithms are implemented in MATLAB.

3 Seed Finding Using K-Means Clustering Algorithm

The gene expression dataset is partitioned into n gene clusters and m sample clusters using the K-Means clustering algorithm. In order to get maximum ten genes per cluster, those gene clusters with more than ten genes are further divided according to the cosine angle distance from the cluster centre. Similarly each sample cluster is further divided into sets of five samples. The number of gene clusters having maximum ten close genes is p and number of sample clusters having maximum five conditions is q. Thus, the initial gene expression data matrix is partitioned into $p*q$ submatrices. From these submatrices, those with MSR value below a certain threshold are selected as seeds [4].

4 Seed-Growing Algorithms

In seed-growing phase, each seed is grown separately by adding genes or conditions one element at a time. List of genes and conditions not included in the seed is maintained separately. Search for the next element to be added to the bicluster starts from condition list followed by gene list. When a gene or condition is added to the bicluster, the MSR increases or decreases. Added node is removed if necessary depending on the constraints used by the algorithm. The entire gene list and condition list are searched for checking the possibility of adding a new element. The three algorithms for biclustering experimented here are the seeds growing using MSRT [5], MSR difference threshold (MSRDT), and iterative search with incremental MSR difference threshold (ISIMSRDT) [6]. All these algorithms are deterministic. Some of the biclustering techniques are giving more importance for producing biclusters with high row variance. Cheng and Church used row variance as an accompanying score to find out trivial biclusters. There is no threshold value for row variance to consider a bicluster as trivial. They considered bicluster with row variance 39.33 as trivial bicluster. In SEBI [7], they were trying to identify biclusters with high row variance by adjusting the fitness function. The minimum value of row variance they obtained for the biclusters in yeast dataset was 317.23. In this study, all biclusters obtained are with row variance above 317.23. Even though no specific steps were taken here to produce biclusters with high row variance, biclusters with row variance above 1,000 were obtained for yeast dataset, and biclusters with the row variance above 5,700 were obtained for lymphoma dataset. But the maximum row variance obtained for lymphoma data in SEBI is only 5691.07.

5 Seed Growing Using MSRT

In MSRT algorithm, after adding one gene or condition, the MSR value of the resulting bicluster is checked to verify whether it exceeds the given MSRT. If it exceeds the given threshold, then it is removed from the bicluster. This step continues till the last gene or condition is verified for inclusion in the bicluster. Compared to metahueristic methods, the algorithm is very fast.

5.1 Bicluster Plots for Yeast and Lymphoma Datasets

In Fig. 1, biclusters obtained by MSRT algorithm on both datasets are shown. Using this algorithm, biclusters with all 17 conditions are identified for the yeast dataset.

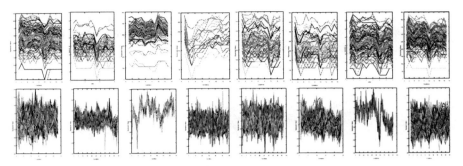

Fig. 1 Biclusters obtained by MSRT algorithm

The first row shows eight biclusters of the yeast dataset and the second row shows eight biclusters of the lymphoma dataset. The value of δ is set to 200. The details about the biclusters are reported in the following format (label, number of genes, number of conditions, volume, MSR, row variance) (a1, 132, 16, 2,112, 199.4098, 483.11), (b1, 69, 15, 1,035, 199.3542, 578.91), (c1, 124, 13, 1,612, 198.9403, 601), (d1, 26, 14, 364, 199.0273, 611.65) (e1, 67, 13, 871, 199.0506, 490.14), (f1, 49, 16, 784, 199.4639, 513.27), (g1, 114, 14, 1,596, 199.52, 508.76), (h1, 117, 17, 1,989, 199.9365, 472), (p1, 586, 28, 16,408, 1179.4, 1,205), (q1, 164, 49, 8,036, 1188.7, 1691.3), (r1, 11, 67, 737, 1186.7, 6222.4), (s1, 171, 59, 10089, 1199.7, 1279.1), (t1, 279, 47, 13,113, 1192.3, 1248.7), (u1, 386, 24, 9,264, 1187.3, 1661.8) (v1, 10, 91, 910, 1190.5, 5308.5), (w1, 1,136, 20, 22,720, 1,194, 1205.2).

5.2 Advantages of MSRT Algorithm

This is fastest among the three algorithms. As no other constraint is used for reducing the MSR score except MSRT, different seeds will result in different biclusters with a few exceptions. Some of the biclusters are with high row variance (more than 1,000 for the yeast dataset and more than 5,700 for lymphoma dataset) even though no measures are taken to get biclusters of high row variance. This method is especially suitable for lymphoma dataset for obtaining biclusters with large size. A bicluster with 91 conditions is obtained for lymphoma dataset. This bicluster is shown in Fig. 1 with label v1.

6 Seed Growing Using MSRDT

The seeds obtained from K-Means algorithm is enlarged by adding more conditions and genes. MSR difference of a gene or condition is the incremental increase in MSR after adding the same to the bicluster. In this method [8], a gene or condition (node) is added to the bicluster. After adding the node if the incremental increase in MSR is greater than MSRDT or if the MSR of the resulting bicluster is greater than δ, then the added node is removed from the bicluster. MSRDT is different for gene list as well as condition list, and it varies depending on the dataset also. The identification of suitable value needs experimentation. It is observed that if MSRDT for condition list is set to 30, then it is possible to get biclusters with all 17 conditions for the yeast dataset. For gene list, the MSRDT is set to 10. By properly adjusting the MSRDT, biclusters of high quality can be obtained. While experimenting, it is found that reducing the MSRDT for condition list eliminates conditions in which expression level changes significantly from the bicluster, whereas reducing the MSRDT for gene list increases coherence. This means condition difference threshold should be large and gene difference threshold should be small.

6.1 Bicluster Plots for Yeast and Lymphoma Datasets

First row of Fig. 2 contains nine biclusters obtained by the algorithm on yeast dataset. Six of the nine biclusters contain all 17 conditions. All biclusters are with MSR less than 200 and row variance above 450. For yeast dataset, all conditions are obtained when the MSRDT for condition lists is set to 30. For gene list, MSRDT is set to 10. Second row shows the nine biclusters obtained by the MSRDT algorithm on Lymphoma dataset. One bicluster (label u2) is having the size 10*91 and row variance above 5,700 even though in SEBI the maximum value of row variance for lymphoma dataset is only 5691.07 and the maximum number of conditions obtained is only 72.

The details about the biclusters are reported in the following format (label, number of genes, number of conditions, volume, MSR, row variance) (a2, 65,17,1105, 198.8756, 619.3479), (b2, 86,171,462, 198.3953, 526.8160), (c2, 74, 17, 1,258, 199.6859, 508.7565), (d2, 77, 16, 1,232, 199.5544, 533.1660), (e2, 81, 17, 1,377, 199.9548, 551.3923), (f2, 140, 16, 2,240, 199.6735, 458.1247), (g2, 55, 17, 935, 199.4912, 534.4627), (h2, 16, 17, 272, 199.3662), (i2, 119, 17, 2,023, 199.5356,

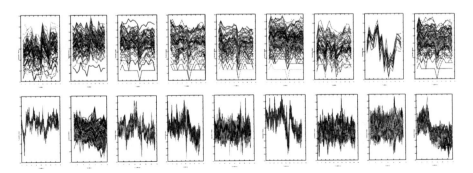

Fig. 2 Biclusters obtained by MSRDT algorithm

518.8431), (p2, 10, 77, 770, 1188.2, 5439.2), (q2, 910, 12, 10,920, 1199.0, 1419.3), (r2, 18, 67, 1,206, 1189.2, 3430.8), (s2, 30, 73, 2,190, 1197.4, 3,902), (t2, 64, 73, 4,672, 1199.6, 1325.5), (u2, 10, 91, 910, 1183.1, 5,702), (v2, 135, 47, 6,345, 1199.3, 1321.1), (w2, 690, 28, 19,320, 1199.1, 1232.3), (x2, 72, 35,2520, 1183.1, 3,959).

6.2 Advantages of MSRDT Algorithm

In this algorithm, a bicluster (label u2) with 91 conditions is obtained for lymphoma dataset, and the MSR is less than that of the bicluster obtained by MSRT algorithm with 91 conditions. In MSRDT algorithm, more genes and conditions can be accommodated compared to MSRT algorithm. It is found that reducing the difference threshold genes in MSRDT algorithm eliminates the possibility of adding inverted rows or mirror images into the bicluster. Formation of mirror images can be eliminated using the MSRDT. This is due to the fact that the genes that form mirror images will have high values for incremental increase in MSR. In Fig. 3, two biclusters with inverted images and the same biclusters with inverted images eliminated are shown.

In the bicluster labelled (m1), there are 105 genes and 13 conditions with MSR value 215.2878. The gene that forms the mirror image when added causes the incremental increase in MSR of 16.8646. All other genes result in incremental increase in MSR less than 2. Similarly in the case of bicluster (m2), there are 73 genes and 17 conditions and MSR value is 182.74. The gene that causes the inverted image when added to the bicluster results in an incremental increase in MSR of 25.8368. While for all other genes, the incremental increase in MSR is less than 2.5. Inverted rows can be obtained by any algorithm which makes use of MSRT alone. Even metaheuristic optimization algorithms with fitness function for minimizing MSR value will result in such biclusters. If MSRDT is applied with a difference threshold of value 10 for the gene list these genes will have to be removed. It is found that decreasing the MSRDT for condition list eliminates conditions which make significant change in the expression level from the bicluster, whereas decreasing MSRDT for gene list increases coherence.

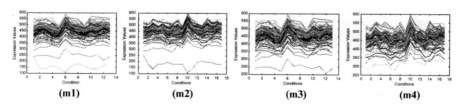

Fig. 3 (**m1**) and (**m2**) show inverted images formed when MSR threshold alone is applied. In (**m3**) and (**m4**), inverted images are removed by MSR difference threshold

7 Seed Growing Using ISIMSRDT

In this algorithm, after adding a gene or a condition if the incremental value of MSR is greater than MSRDT or if the MSR of the resulting bicluster is greater than δ, the added node is removed from the bicluster. In ISIMSRDT algorithm, MSRDT is initialized with a small value and incremented after each iteration in fixed steps until it reaches a final value. Hence, in ISIMSRDT there are three different parameters such as the initial value of MSRDT, the amount by which it is incremented after each iteration and the final value of MSRDT. These three parameters apply for both the gene list and condition list. By properly adjusting the MSRDT parameters, biclusters of high quality can be obtained.

7.1 ISIMSRDT Algorithm

```
Algorithm Iterative_MSRdifference(seed,δ, condorgenethreshinitial,
condorgenethreshincrement, condorgenethreshfinal,)
//Condition list and gene list are maintained separately and these steps should be
//implemented for both the lists separately.
bicluster:= seed; previous=MSR(seed); j := 1;
msrdiffcondorgenethresh=condorgenethreshinitial;
while (msrdiffcondorgenethresh<condorgenethreshfinal)
      While (j <= total _no_conditionsorgenes)
      If conditionorgene[j] is not included in bicluster
        Changed=1; Add conditionorgene[j] to the bicluster; present= MSR(bicluster)
        if (present> δ) or (present-previous)>msrdiffcondorgenethresh
          remove elements of conditionorgene[j] from bicluster; changed=0;
        endif
          if changed==1  previous=present;  endif
        endif
j := j+1
end(while)
msrdiffcondorgenethresh=msrdiffcondorgenethresh+condorgenethreshincrement
end(while)
return bicluster
end(Iterative_MSRdifference)
```

7.2 Bicluster Plots for Yeast and Lymphoma Datasets

For yeast dataset, biclusters are found by setting the initial value of MSRDT for condition list as five. After each iteration, it is incremented by five and the final

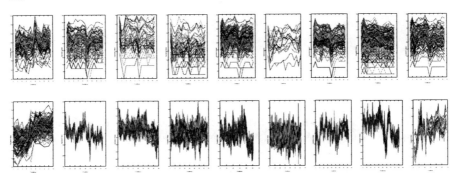

Fig. 4 Biclusters obtained by ISIMSRDT algorithm

value of MSRDT is set to 30. Initial value of MSRDT for gene list is set to 1, and it is incremented by one and the final value is set to 10. For lymphoma dataset, the initial value of MSRDT for condition list is set to 30, and after each iteration it is incremented by 30 and the final value is set to 90. For the gene list, the initial value of MSRDT is set to 50, and it is incremented by 50 after each iteration and final value is set to 150. Experiments are conducted using other values also.

The first row of Fig. 4 shows nine biclusters of the yeast dataset and the second row shows nine biclusters of the lymphoma dataset. The details about the biclusters are reported in the following format (label, number of genes, number of conditions, volume, MSR, row variance) (a3, 98, 17, 1,666, 199.9381, 591.9217), (b3, 107, 17, 1819, 199.982,6, 486.3663), (c3, 43, 17, 731, 199.8613, 550.3640), (d3, 50,17, 850, 199.5999, 511.3709), (e3, 127, 17, 2,159, 199.9656, 471.1995), (f3, 19, 16, 304, 199.9141, 564.4940), (g3, 99, 17, 1,683, 199.9524, 419.2172), (h3, 188, 13, 2,444, 199.9713, 353.8271), (i3, 110, 17, 1,870, 199.9499, 515.1427), (p3, 280,10, 2,800, 1001.40, 2200.1), (q3, 10, 74, 740, 1199.00, 4208.8), (r3, 86, 40, 3,440, 999.88, 2021.6), (s3, 155, 39, 6,045, 999.92, 1102.4), (t3, 51, 51, 2,601, 999.93, 3139.5), (u3, 172, 62, 10,664, 1199.80, 1342.3) (v3, 10, 83, 830, 1194.90, 5082.6), (w3, 10, 92, 920, 1197.40, 5760.1), (x3, 20, 29, 580, 987.80, 4318.2).

7.3 Advantages of ISIMSRDT Algorithm

Since MSRDT is applied in ISIMSRDT algorithm, there is more restriction on the incremental value of MSR. This means that the elements in the biclusters are more tightly packed. This will result in biclusters of larger size and low MSR score. Hence, ISIMSRDT method can produce better biclusters compared with other algorithms like MSRT. The ISIMSRDT algorithm gives the possibility of getting more genes and conditions compared to MSRDT algorithm. In MSRDT algorithm, there is the disadvantage of finding a suitable value for MSR difference threshold. If a small value is assigned, bicluster will be of small size. On the other hand, if a big value is assigned the elements of the resulting bicluster will not be tightly co-regulated. This disadvantage of MSRDT algorithm can be overcome using ISIMSRDT. There is another advantage of using iterative search in ISIMSRDT algorithm. The incremental increase in MSR of a condition not included in a bicluster will vary as the size of the bicluster changes. For example, in the case of bicluster labelled (w3) in Fig. 4, the MSR value of the bicluster when condition 95 is added is 1200.6. Since this is greater than MSRT for lymphoma dataset (1,200), condition 95 is removed from the bicluster. If condition 95 is added after adding condition 96, the MSR of the resulting bicluster is only 1191.9. This is less than 2,000, and hence if 95 is added after adding 96 it is not removed. Since conditions and genes are searched sequentially in all these algorithms, this is possible only if there is an iterative search. This is another option in iterative search for accommodating more conditions. That means apart from finding a suitable value for MSRDT, iterative search has got another advantage of selecting the $(n - k)$th condition whose incremental increase in MSR value reduces after adding the nth condition. In lymphoma dataset, a bicluster (label w3) with 92 conditions is obtained.

8 Comparison

8.1 Comparison of Biclusters Produced by ISIMSRDT, MSRDT and MSRT Algorithms from the Same Seed

A comparison of these three algorithms is given on the basis of size of biclusters obtained and their MSR value starting with the same seed. In terms of bicluster size, ISIMSRDT is always better than the other two algorithms. MSRDT is better than MSRT for all seeds except for seed 3. Difference between biclusters obtained by the three algorithms starting from the same seed is shown in the following format (seed number, bicluster size and MSR of IS1MSRDT, bicluster size and MSR of MSRDT, bicluster size and MSR of MSRT) (1, 110*17, 199.95, 78*16, 199.96, 75*17, 199.95), (2, 93*17, 199.79, 65*17, 198.88, 57*17, 199.09), (3, 99*17, 199.95, 74*17, 199.69, 92*17, 199.71), (4, 96*17, 199.69, 86*17, 198.39, 79*17, 198.96), (5, 125*17, 199.91, 119*17, 199.54, 117*17, 199.94), (6, 19*16, 199.91, 17*16, 199.37, 15*15, 198.58). From the details given, it is clear that ISIMSRDT produces large size biclusters compared to MSRDT. Hence, ISIMSRDT is always better than assigning a single value for MSRDT. Figure 5 displays three biclusters obtained by the three algorithms from the same seed.

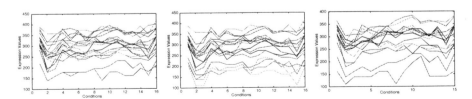

Fig. 5 Biclusters from the same seed for the ISIMSRDT, MSRDT and MSRT algorithms

In the case of lymphoma dataset starting from the same seed, MSRT and MSRDT algorithms produced biclusters with 91 conditions. Even though both are of size 10*91, the bicluster produced by MSRDT is better in terms of MSR value (low) and row variance (high). But for ISIMSRDT, it should be noticed that this algorithm produced bicluster with 92 conditions and higher row variance from the same seed (label w3 of Fig. 4).

8.2 Comparison with Other Algorithms

The comparison of results of MSRT, MSRDT and ISIMSRDT with that of CC [1], SEBI [7], FLOC [9] and DBF [10] for the yeast and the lymphoma datasets is given in Table 1. Sequential evolutionary biclustering (SEBI) is based on evolutionary algorithms. CC algorithm used greedy strategy to find biclusters from the gene expression data matrix. Yang et al. developed a probabilistic algorithm FLOC that can discover a set of possibly overlapping biclusters simultaneously. Zhang et al. presented deterministic biclustering with frequent pattern mining (DBF).

Usually multi-objective algorithms will produce biclusters of larger size. But in the case of multi-objective evolutionary computation [11], the maximum number of conditions obtained is only 11 for the yeast dataset and 40 for the lymphoma dataset. But in these methods, there are biclusters with all 17 and 91 conditions for yeast and lymphoma datasets, respectively. For the yeast dataset, the maximum number of genes obtained for these algorithms in all the 17 conditions is 127 with MSR value 199.9656. Moreover, the three algorithms provide better performance in terms of speed compared to all the metaheuristic and evolutionary algorithms. In the

Table 1 Performance comparison between algorithms

Algorithm	Dataset	Avg. MSR	Avg. volume	Avg. gene number	Avg. condition number	Largest bicluster size
MSRT	Yeast	199.78	1422.87	94.75	14.75	2,112
	Lymphoma	1192.43	14455.30	741.10	38.50	–
MSRDT	Yeast	199.63	2264.80	170.16	14.83	4,400
	Lymphoma	1194.44	5791.63	233.38	58.50	–
ISIMSRDT	Yeast	199.96	1954.20	123.80	16.20	2,444
	Lymphoma	923.47	3458.62	98.00	48.63	–
CC	Yeast	204.29	1576.98	166.71	12.09	4,485
	Lymphoma	850.04	4595.98	269.22	24.50	–
SEBI	Yeast	205.18	209.92	13.61	15.25	1,394
	Lymphoma	1028.84	615.84	14.07	43.57	–
FLOC	Yeast	187.54	1825.78	195.00	12.80	2,000
DBF	Yeast	114.70	1627.20	188.00	11.00	4,000

multi-objective PSO [12], the maximum number of conditions obtained for lymphoma dataset is 84. But for ISIMSRDT algorithm, a bicluster with 92 conditions is obtained.

9 Conclusion

In this chapter, three biclustering algorithms are analysed based on the quality of biclusters produced by them. MSRT algorithm uses MSRT alone. The MSRDT and ISIMSRDT algorithms use both MSRT and MSRDT. All these three algorithms enlarge the seeds produced by the K-Means clustering algorithm. The algorithms are implemented on the yeast and human lymphoma datasets. Comparisons are also given based on the same seed as well as overall performance to demonstrate the effectiveness of the proposed methods. Reducing the difference threshold for genes will eliminate the formation of mirror images from the biclusters of gene expression data. It is found that decreasing the MSRDT for the condition list eliminates conditions which make significant change in the expression level from the bicluster, whereas reducing MSRDT for the gene list increases coherence. Another research finding is in the case of iterative search. Iterative search has got the advantage of selecting the $(n-k)$th condition whose incremental increase in MSR value got reduced after adding the nth condition.

References

1. Cheng Y and Church G M, "Biclustering of expression data", Proceedings of the 8th International Conference on Intelligent Systems for Molecular Biology, pp. 93–103, 2000.
2. Tavazoie S, Hughes J D, Campbell M J, Cho R J and Church G M, "Systematic determination of genetic network architecture", Nature Genetics, Vol. 22, no. 3 pp. 281–285, 1999.
3. Alizadeh A A et al., "Distinct types of diffuse large B-cell lymphoma identified by gene expression profiling", Nature Vol. 43, no. 6769, pp. 503–511, 2000.
4. Chakraborty A and Maka H, "Biclustering of gene expression data using genetic algorithm" Proceedings of Computation Intelligence in Bioinformatics and Computational Biology CIBCB, pp. 1–8, 2005.
5. Das S and Idicula S M, "A novel approach in greedy search algorithm for biclustering gene expression data" International Conference on Bioinformatics, Computational and Systems Biology (ICBCSB), Singapore, Aug 27–29, 2009.
6. Das S and Idicula S M, "Iterative search with incremental MSR difference threshold for biclustering gene expression data", International Journal of Computer Applications, Vol. I, pp. 35–43, 2010.
7. Divina F and Aguilar-Ruize J S, "Biclustering of expression data with evolutionary computation", IEEE Transactions on Knowledge and Data Engineering, Vol. 18, pp. 590–602, 2006.
8. Das S and Idicula S M, "Biclustering gene expression data using MSR difference threshold", International Proceedings of IEEE INDICON, pp. 1–4, 2009.
9. Yang J, Wang H, Wang W and Yu P, "Enhanced biclustering on expression data", Proceedings of the Third IEEE Symposium on BioInformatics and BioEngineering (BIBE'03), pp. 321–327, 2003.

10. Zhang Z, Teo A, Ooi B C and Tan K L, "Mining deterministic biclusters in gene expression data", Proceedings of the Fourth IEEE Symposium on Bioinformatics and Bioengineering (BIBE'04), pp. 283–292, 2004.
11. Banka H and Mitra S, "Multi-objective evolutionary biclustering of gene expression data", Journal of Pattern Recognition, Vol. 39, pp. 2464–2477, 2006.
12. Liu J, Lia Z and Liu F, "Multi-objective particle swarm optimization biclustering of microarray data", IEEE International Conference on Bioinformatics and Biomedicine, pp. 363–366, 2008.

Chapter 14
Performance Comparison of SLFN Training Algorithms for DNA Microarray Classification

Hieu Trung Huynh, Jung-Ja Kim, and Yonggwan Won

Abstract The classification of biological samples measured by DNA microarrays has been a major topic of interest in the last decade, and several approaches to this topic have been investigated. However, till now, classifying the high-dimensional data of microarrays still presents a challenge to researchers. In this chapter, we focus on evaluating the performance of the training algorithms of the single hidden layer feedforward neural networks (SLFNs) to classify DNA microarrays. The training algorithms consist of backpropagation (BP), extreme learning machine (ELM) and regularized least squares ELM (RLS-ELM), and an effective algorithm called neural-SVD has recently been proposed. We also compare the performance of the neural network approaches with popular classifiers such as support vector machine (SVM), principle component analysis (PCA) and fisher discriminant analysis (FDA).

1 Introduction

In recent years, DNA microarray technology has been rapidly developing and providing powerful tools to the medical and biological applications. It enables the monitoring of the expression level of thousands of genes with a single experiment and allows establishing well-defined diagnosis and prognosis that may lead to properly choosing treatments and therapies. In general, many of its applications require classification methods that typically use machine learning techniques, of which the main task is to build a classifier from the identified patterns of expressed genes, and it can then be used to predict the class membership of new patterns.

Many methods have been proposed for achieving classification tasks; traditional methods include the support vector machine (SVM), neural networks and statistical techniques. The special characteristic of microarray datasets is that the number of

---

H.T. Huynh (✉)
Nguyen Tat Thanh College, University of Industry, Ho Chi Minh City, Vietnam
e-mail: hthieu@hcmut.edu.vn; hieuhtvn@yahoo.com

patterns is very small, while the number of the features is extremely large. Therefore, the statistical techniques are not suitable for microarray classification. Also, the SVM approach may take a long period of time to select its model. Several neural network architectures have been proposed for achieving classification tasks; however, it has been shown that the single hidden layer feedforward neural networks (SLFNs) can form boundaries with arbitrary shapes if the activation function is properly chosen.

Traditionally, neural-network training is based on the gradient descent algorithms which are generally slow and may get stuck in the local minima. Those problems have been mitigated via the extreme learning machine (ELM) algorithm that was proposed by Huang et al. [1]. The advantage of this algorithm is that the network parameters are determined via non-iterative steps; first, randomly assign the input weights and hidden layer biases, and then determine the output weights by the pseudo-inverse of the hidden layer output matrix. This ELM algorithm can achieve better performance with faster learning speed in many applications. However, it often requires a large number of hidden nodes that slows down the trained network response to new input patterns. This problem has been overcome by improved ELM algorithms such as the evolutionary extreme learning machine (E-ELM), the least squares extreme learning machine (LS-ELM) [2] and the regularized least squares extreme leaning machine (RLS-ELM) [3]. We have also recently proposed a new approach for training SLFNs with the *tansig* activation function where the network parameters are determined by singular value decomposition (SVD) [4]. This algorithm is simple, does not require iterative processes and can obtain better performance with fast learning speed.

In this chapter, we compare the proposed SVD-neural approach with neural-network training algorithms in terms of performance in microarray classification. Moreover, we present a performance comparison between the neural-network-based approaches and other popular classifiers of microarray data such as SVM, principle component analysis (PCA) and fisher discriminant analysis (FDA).

2 Training Algorithms for SLFNs

In this section, we review training algorithms for SLFNs including BP, ELM and RLS-ELM and present an effective classifier called *SVD-Neural classifier* which is proposed recently [4]. Assume that an SLFN consists of P nodes in input layer, N nodes in hidden layer and C nodes in output layer.

2.1 BP and ELM Training Algorithm

Let $\mathbf{w}_m = [w_{m1}, w_{m2}, \ldots, w_{mP}]^T$ be the input weights connecting from the input layer to the m-th hidden node, and b_m be its bias. The hidden layer output vector corresponding to the input pattern $\mathbf{x}_j$ is given by:

$$\mathbf{h}_j = \left[f(\mathbf{w}_1\mathbf{x}_j + b_1), f(\mathbf{w}_2\mathbf{x}_j + b_2), \ldots, f(\mathbf{w}_N\mathbf{x}_j + b_N)\right]^T, \quad (1)$$

and the i-th output is given by:

$$o_{ji} = \mathbf{h}_j \cdot \boldsymbol{\alpha}_i, \quad (2)$$

where $\boldsymbol{\alpha}_i = [\alpha_{i1}, \alpha_{i2}, \ldots, \alpha_{iN}]^T$ is the weight vector connecting from the hidden nodes to the i-th output node. Assume that there are n profiles or arrays in the training set; each profile consists of P genes, and let $\mathbf{t}_j = [t_{j1}, t_{j2}, \ldots, t_{jC}]^T$ be the desired output corresponding to the input $\mathbf{x}_j$. We have to find parameters $\mathbf{G}$ consisting of input weights $\mathbf{w}$, biases b and output weights $\boldsymbol{\alpha}$ that minimize the error function defined by:

$$E = \sum_{j=1}^{n} (\mathbf{o}_j - \mathbf{t}_j)^2 = \sum_{j=1}^{n}\sum_{i=1}^{C} (\mathbf{h}_j \cdot \boldsymbol{\alpha}_i - t_{ji})^2. \quad (3)$$

This minimization procedure has been traditionally based on the gradient-descent algorithm, in which parameter vector $\mathbf{G}$ is iteratively adjusted as follows:

$$\mathbf{G}_k = \mathbf{G}_{k-1} - \eta \frac{\partial E}{\partial \mathbf{G}}, \quad (4)$$

where η is a learning rate. In the feedforward neural networks, a popular learning algorithm based on gradient descent is back-propagation (BP) algorithm. The gradients can be calculated and the vectors $\mathbf{G}$ can be adjusted via error propagation from the output layer to the input layer. This learning algorithm has several problems such as local minima, learning rate and training speed. Although the BP algorithm has undergone many improvements, it still suffers from the aforementioned problems in most applications.

Huang et al. have proposed an efficient learning algorithm for SLFNs called ELM [1]. This algorithm can overcome problems of gradient-descent-based algorithms. In ELM, the minimization procedure of (3) is restated as finding network parameters that minimize the error of a linear system defined by:

$$\mathbf{HA} = \mathbf{T}, \quad (5)$$

where $\mathbf{H}$ is called the hidden layer output matrix and defined by:

$$\mathbf{H} = [\mathbf{h}_1, \mathbf{h}_2, \ldots, \mathbf{h}_n]^T, \quad (6)$$
$$\mathbf{T} = [\mathbf{t}_1, \mathbf{t}_2, \ldots, \mathbf{t}_n]^T \quad (7)$$

and

$$\mathbf{A} = [\boldsymbol{\alpha}_1, \boldsymbol{\alpha}_2, \ldots, \boldsymbol{\alpha}_C]. \quad (8)$$

The ELM algorithm is based on the basic principle that the output weight matrix $\mathbf{A}$ can be determined by the pseudo-inverse of $\mathbf{H}$ with the small nonzero training error. The ELM algorithm can be summarized as follows:

1. Randomly assign the input weights $\mathbf{w}_m$ and biases b_m, $m = 1, 2, \ldots, N$
2. Determine the output matrix $\mathbf{H}$ of the hidden layer by (6)
3. Determine the output weight matrix $\mathbf{A}$ by:

$$\hat{\mathbf{A}} = \mathbf{H}^\dagger \mathbf{T}, \qquad (9)$$

where $\mathbf{H}^\dagger$ is the pseudo-inverse matrix of $\mathbf{H}$. This algorithm can reduce the learning time by avoiding iteration for tuning the input weights and hidden-layer biases. However, it often requires more hidden nodes than those required by conventional algorithms, and thus the application of trained networks to the classification of new input pattern takes longer time. An approach called regularized least-squares extreme learning machine (RLS-ELM) can overcome these problems [3].

2.2 RLS-ELM Training Algorithm

Unlike ELM, the input weights and hidden layer biases are determined using a linear model defined by:

$$\mathbf{XW} = \mathbf{TQ}, \qquad (10)$$

where $\mathbf{Q} \in \mathbb{R}^{C \times N}$ consists of randomly chosen values, $\mathbf{X} = \begin{bmatrix} \mathbf{x}_1 & \mathbf{x}_2 & \cdots & \mathbf{x}_n \\ 1 & 1 & \cdots & 1 \end{bmatrix}^T$ and $\mathbf{W} = \begin{bmatrix} \mathbf{w}_1 & \mathbf{w}_2 & \cdots & \mathbf{w}_N \\ b_1 & b_2 & \cdots & b_N \end{bmatrix}$. In order to avoid ill-posed problems, the Tikhonov regularization is used, in which the solution for $\mathbf{W}$ of (10) can be replaced by seeking $\mathbf{W}$ that minimizes:

$$||\mathbf{XW} - \mathbf{TQ}||^2 + \lambda ||\mathbf{W}||^2, \qquad (11)$$

where $||\cdot||$ is the Euclidean norm and λ is a positive constant. The solution for $\mathbf{W}$ is given by:

$$\hat{\mathbf{W}} = (\mathbf{X}^T \mathbf{X} + \lambda \mathbf{I})^{-1} \mathbf{X}^T \mathbf{TQ}. \qquad (12)$$

In microarray classification, the size of $\mathbf{X}^T \mathbf{X}$ is $(P+1) \times (P+1)$ which is very huge. Therefore, to reduce the computational cost of inverse matrix, (12) should have the following format:

$$\hat{\mathbf{W}} = \frac{1}{\lambda} \mathbf{X}^T \mathbf{TP} - \frac{1}{\lambda} \mathbf{X}^T (\mathbf{XX}^T + \lambda \mathbf{I})^{-1} \mathbf{XX}^T \mathbf{TQ}. \qquad (13)$$

Thus, RLS-ELM algorithm can be summarized as follows:
1. Randomly assign the values for the matrix $\mathbf{Q}$
2. Estimate the input weights $\mathbf{w}_m$ and biases b_m using (13)
3. Calculate the hidden-layer output matrix $\mathbf{H}$ using (6)
4. Determine the output weight matrix $\mathbf{A}$ using (9)

This algorithm can obtain good performance with small number of hidden nodes which results in extremely high speed for both training and testing [3].

2.3 SVD-Neural Classifier

We have recently introduced the *SVD-Neural classifier* that is an SLFN with the hidden layer activation function *tansig*, and its parameters are determined by non-iterative simple steps based on SVD [4]. From (5), we can see that the matrix **T** is composed of the multiplication of two matrices $\mathbf{H} \in \mathbf{R}^{n \times N}$ and $\mathbf{A} \in \mathbf{R}^{N \times C}$. Thus, if we can reasonably decompose the matrix **T** into two matrices with sizes of $n \times N$ and $N \times C$, then parameters **G** of classifier can be determined simply.

In single value decomposition (SVD) of **T**, there exists an unitary matrix $\mathbf{U} \in \mathbf{R}^{n \times n}$, a diagonal matrix with non-negative real numbers $\mathbf{D} \in \mathbf{R}^{n \times C}$, and an unitary matrix $\mathbf{V} \in \mathbf{R}^{C \times C}$ so that:

$$\mathbf{T} = \mathbf{U}\mathbf{D}\mathbf{V}^\mathrm{T}. \tag{14}$$

Let $\mathbf{U} = [\mathbf{u}_1, \mathbf{u}_2, \ldots, \mathbf{u}_N, \ldots, \mathbf{u}_n]$, $\mathbf{D} = [\sigma_1, \sigma_2, \ldots, \sigma_n]^\mathrm{T}$, and $\mathbf{V} = [\mathbf{v}_1, \mathbf{v}_2, \ldots, \mathbf{v}_C]^\mathrm{T}$. Based on the properties of SVD decomposition, matrix **T** can be approximated by:

$$\mathbf{T} \approx \mathbf{U}_N \mathbf{D}_N \mathbf{V}^\mathrm{T}, \tag{15}$$

where $\mathbf{U}_N = [\mathbf{u}_1, \mathbf{u}_2, \ldots, \mathbf{u}_N]$ and $\mathbf{D}_N = [\sigma_1, \sigma_2, \ldots, \sigma_N]^\mathrm{T}$. Note that $\mathbf{U}_N \in \mathbf{R}^{n \times N}$ and $\mathbf{D}_N \mathbf{V}^\mathrm{T} \in \mathbf{R}^{N \times C}$. Therefore, matrices **H** and **A** can be determined by:

$$\mathbf{H} = \mathbf{U}_N \tag{16}$$

and

$$\mathbf{A} = \mathbf{D}_N \mathbf{V}^\mathrm{T}. \tag{17}$$

Next, we have to determine the input weights and hidden layer biases. From (6) and (16), we have

$$\begin{bmatrix} f(\mathbf{w}_1 \mathbf{x}_1 + b_1) & \cdots & f(\mathbf{w}_N \mathbf{x}_1 + b_N) \\ \vdots & \ddots & \vdots \\ f(\mathbf{w}_1 \mathbf{x}_n + b_1) & \cdots & f(\mathbf{w}_N \mathbf{x}_n + b_N) \end{bmatrix} = \mathbf{U}_N. \tag{18}$$

The input weights and hidden layer biases can be determined using the linear system:

$$\mathbf{X}\mathbf{W} = f^{-1}[\mathbf{U}_N]. \tag{19}$$

Note that $f^{-1}[\mathbf{U}_N]_{ij} = f^{-1}([\mathbf{U}_N]_{ij})$.

Finally, the minimum norm solution for **W** among all possible solutions is given by:

$$\hat{\mathbf{W}} = \mathbf{X}^\dagger f^{-1}[\mathbf{U}_N], \tag{20}$$

where $\mathbf{X}^{\dagger}$ is the pseudo-inverse matrix of $\mathbf{X}$. In summary, the training algorithm for Neural-SVD classifier can be described as follows:

Given a training set $\mathbf{S} = \{(\mathbf{x}_j, \mathbf{t}_j) | j = 1, \ldots, n\}$, and the number of hidden nodes N,

1. Determine SVD of $\mathbf{T}$, and then calculate $\mathbf{U}_N$ and $\mathbf{D}_N$
2. Determine the input weights and hidden layer biases using (20)
3. Determine the output weights using (17)

The SVD of matrix $\mathbf{T}$ is typically determined by a two-step procedure. In the first step, the matrix is reduced to bi-diagonal form which takes $O(nC^2)$ operations, and then the second step is to compute SVD of the bi-diagonal matrix which has computational cost about $O(n)$. Since $C \ll n$ for most classification problems, where C is the number of classes and n is the number of patterns, the overall cost for computing SVD of matrix $\mathbf{T}$ is $O(n)$ flops [5]. Thus, determining parameters of classifier is simple and has low computational complexity. This classifier can obtain a compact network with small number of hidden nodes, and it performs well in classifying microarray data.

Note that SVD has been used with feedforward neural network in some applications [6]. However, in these applications, SVD is used only for feature reduction, and neural networks were still trained by traditional BP algorithms. In this chapter, we focus on training algorithms of neural networks in microarray classification, and the mentioned SVD is a new approach for training SLFNs.

3 Experimental Results

In this section, we evaluate the performance of the classifiers on the microarray datasets used for the benchmark problems that consist of three binary cancer classification problems: leukemia data set, prostate cancer dataset and colon cancer dataset. The initial leukemia data set consisted of 38 bone marrow samples obtained from adult acute leukemia patients at the time of diagnosis, of which 11 suffer from acute myeloid leukemia (AML) and 27 suffer from acute lymphoblastic leukemia (ALL). An independent collection of 34 leukemia samples contained a broader range of samples: the specimens consisted of 24 bone marrow samples and 10 peripheral blood samples derived from both adults and children. The number of input features was 7,129. The objective of this evaluation is to separate the AML samples from the ALL samples. The training set consisted of 38 patterns, and 34 patterns were used for testing.

In the prostate cancer data set, the oligonucleotide microarray contained probes for approximately 12,600 genes and the expressed sequence tags (ESTs). The goal of this study is to classify tumor and non-tumor samples. The training set contained high-quality expression profiles derived from 52 prostate tumors and 50 non-tumor samples obtained from patients undergoing surgery. The test set consisted of 34 tis-

Table 1 Performance comparison with other neural-network approaches (%)

Dataset	Methods	Training	Testing	# Hidden nodes
Leukemia	SVD-Neural	100	95.90	2
	RLS-ELM	100	95.60	2
	ELM	91.35	67.65	20
	BP	98.85	88.52	2
Prostate	SVD-Neural	100	90.85	2
	RLS-ELM	100	90.87	2
	ELM	78.70	60.17	30
	BP	95.07	83.24	2
Colon	SVD-Neural	100	83.63	2
	RLS-ELM	99.75	83.27	2
	ELM	88.35	64.18	20
	BP	95.70	80.27	2

sues, of which 9 were normal and 25 were tumor samples. The raw data are available at http://www-genome.wi.mit.edu/MPR/prostate.

The colon cancer data set contains the expression of the 2,000 genes with highest minimal intensity across the 62 tissues derived from 40 tumor and 22 normal colon tissue samples. The gene expression was analyzed with an Affymetrix (Santa Clara, CA, USA) oligonucleotide array complementary to more than 6,500 human genes. The gene intensity has been derived from about 20 feature pairs that correspond to the gene on the DNA microarray chip using a filtering process. Details for data collection methods and procedures are described in [7], and the data set is available from the website http://microarray.princeton.edu/oncology/.

The average results of 50 trials on three data sets are shown in Table 1, which shows comparison results of neural-network approaches including BP, ELM, RLS-ELM and SVD-Neural classifier. Note that evolutionary ELM (E-ELM) algorithm could not be completed due to "out of memory", even though it was run with two hidden nodes only. It shows that the number of hidden nodes for SVD-Neural approach was two for three data sets which is equal to that of RLS-ELM and BP algorithm, and is about ten times smaller than that of ELM algorithm. The accuracy of SVD-Neural approach is compatible with RLS-ELM on prostate dataset while better than that of ELM and BP. On datasets of leukemia and colon, the SVD-Neural approach outperforms RLS-ELM, ELM and BP algorithms.

In comparison with other popular algorithms of microarray classification, it has shown that classification accuracy of the SVD-Neural classifier was always better than the highest accuracy of the methods based on least squares support vector machines (LS-SVM) and those based on FDA [8] as shown in Table 2. The reported results of FDA were obtained with dimensionality reduction using classical principal component analysis (PCA) as well as kernel principal component analysis (kPCA).

Table 2 Performance comparison with popular classification methods (%)

Dataset	Method	Training Mean	SD	Testing Mean	SD
Leukemia	SVD-Neural	100	0.00	95.93	5.11
	LS-SVM linear kernel	100	0.00	92.86	4.12
	LS-SVM RBF kernel	100	0.00	93.56	4.12
	LS-SVM linear kernel (gamma = inf)	93.61	15.93	87.39	14.61
	PCA + FDA	99.50	1.31	94.40	3.84
	kPCA lin + FDA	99.50	1.31	94.40	3.84
	kPCA RBF + FDA	99.62	0.92	92.02	6.36
Prostate	SVD-Neural	100	0.00	90.85	3.68
	LS-SVM linear kernel	100	0.00	84.31	13.16
	LS-SVM RBF kernel	99.95	0.21	88.10	4.93
	LS-SVM linear kernel (no regul.)	51.45	7.03	48.18	10.25
	PCA + FDA	97.62	1.95	83.89	13.63
	kPCA lin + FDA	97.57	1.90	85.01	9.07
	kPCA RBF + FDA	98.97	1.75	85.01	11.00
Colon	SVD-Neural	100	0.00	83.63	6.15
	LS-SVM linear kernel	99.64	0.87	82.03	7.49
	LS-SVM RBF kernel	98.33	2.36	81.39	9.19
	LS-SVM linear kernel (gamma = inf)	49.40	8.93	51.73	12.19
	PCA + FDA	90.95	5.32	80.30	9.65
	kPCA lin + FDA	95.24	5.32	80.30	9.65
	kPCA RBF + FDA	87.86	11.24	75.11	15.02

4 Conclusion

DNA microarray technology has been attracting research activities in recent years. Many applications and approaches have been proposed for microarray data. One of the typical applications using the machine learning approaches is microarray classification that can help identify the patterns of expressed genes from which class membership for new patterns can be predicted. In this chapter, we evaluated the performance of neural network training algorithms for microarray classification, especially the SVD approach. This is a new training algorithm that has recently been proposed for SLFNs. This algorithm is simple and has low computational complexity.

The performance of the training algorithms on SLFNs including BP, ELM, RLS-ELM and SVD approach was compared in terms of classification accuracy and on the number of hidden nodes on the popular benchmarking data sets. The data sets used for this study were three binary cancer datasets of DNA microarray: leukemia, prostate and colon cancers. The number of hidden units required for SVD, RLS-ELM and BP algorithms are the same, and they are about ten times smaller than that of the ELM algorithm. The classification accuracy of the SVD approach and RLS-ELM is higher than that of ELM and BP.

References

1. Huang, G-B et al (2006) Extreme learning machine: Theory and applications. Neurocomputing 70:489–501.
2. Huynh, HT and Won, Y (2008) Small number of hidden units for ELM with two-stage linear model. IEICE Transactions on Information and Systems E91-D:1042–1049.
3. Huynh, HT et al (2008) An improvement of extreme learning machine for compact single-hidden-layer feedforward neural networks. International Journal of Neural Systems 18:433–441.
4. Huynh, HT and Won, Y (2009) Training single hidden layer feedforward neural networks by singular value decomposition. ICCIT '09 proceedings of the 2009 International conference on computer sciences and convergence information technology. doi: 10.1109/ICCIT.2009.170.
5. Trefethen, LN and David Bau, III (1997) Numerical linear algebra. Philadelphia, Society for Industrial and Applied Mathematics.
6. Wu, C et al (2004) Neural networks for full-scale protein sequence classification: Sequence encoding with singular value decomposition. Machine Learning 21:177–193.
7. Alon U et al (1999) Broad patterns of gene expression revealed by clustering analysis of tumor and normal colon tissues probed by oligonucleotide array. Proceedings of the National Academy of Sciences of the United States of America 96:6745–6750.
8. Pochet, N et al (2004) Systematic benchmarking of microarray data classification: assessing the role of nonlinearity and dimensionality reduction. Bioinformatics 20:3185–3195.

Chapter 15
Clustering Microarray Data to Determine Normalization Method

Marie Vendettuoli, Erin Doyle, and Heike Hofmann

Abstract Most of the scientific journals require published microarray experiments to meet Minimum Information About a Microarray Experiment (MIAME) standards. This ensures that other researchers have the necessary information to interpret the results or reproduce them. Required MIAME information includes raw experimental data, processed data, and data processing procedures. However, the normalization method is often reported inaccurately or not at all. It may be that the scaling factor is not even known except to experienced users of the normalization software. We propose that using a seeded clustering algorithm, researchers can identify or verify previously unknown or doubtful normalization information. For that, we generate descriptive statistics (mean, variance, quantiles, and moments) for normalized expression data from gene chip experiments available in the ArrayExpress database and cluster chips based on these statistics. To verify that clustering grouped chips by normalization method, we normalize raw data for chips chosen from experiments in ArrayExpress using multiple methods. We then generate the same descriptive statistics for the normalized data and cluster the chips using these statistics. We use this dataset of known pedigree as seeding data to identify normalization methods used in unknown or doubtful situations.

Keywords Microarray Data · Clustering · Normalization · Data Processing · Cluster Analysis

---

M. Vendettuoli (✉)
Bioinformatics and Computational Biology Program, Iowa State University,
Ames, IA 50010, USA
and
Department of Statistics, Iowa State University, Ames, IA 50010, USA
e-mail: mariev@iastate.edu; mariecv26@gmail.com

1 Background

ArrayExpress is a publicly accessible database of transcriptomics microarray data maintained by the European Bioinformatics Institute (EBI) and consists now of over 250,000 microarray assays. One major objective of EBI is to make data freely available to the scientific community, under the guidelines of the Microarray and Gene Expression Data (MGED) Society, which has defined standards for Minimum Information About a Microarray Experiment (MIAME). Three of the criteria required by MIAME are (a) the distribution of raw data, (b) the distribution of normalized data, and (c) annotation regarding the normalization protocol. Because normalization methods are an essential component of data collection and processing, it is imperative that annotation is accurate. Additionally, the process of normalization is time-consuming and computationally intensive, motivating researchers to identify efficient approaches beyond repetition to ensure veracity of publicly available data.

In this chapter, we use cluster analysis to determine methodology used for normalization of microarray data. We evaluate this technique against normalized data retrieved from ArrayExpress and against raw data normalized by the authors using the most common normalization methods of Affymetrix Analysis Suite v5.0 (Mas5) and Robust Multichip Average (RMA). As this question first arose to researchers focused on *Arabidopsis thaliana*, the data used for this analysis will be drawn from experiments using this model organism.

1.1 Normalization

Converting raw data from microarray experiments to gene expression values requires a considerable amount of data processing, including normalization. Normalization ensures that all of the chips are comparable to each other and removes systemic effects of experimentation [1]. Data may be log transformed during normalization so that absolute changes in value represent a fold change in expression. Popular algorithms RMA and Mas5 perform normalization as part of a three-step process flanked by background subtraction and summarization, respectively [1].

RMA assumes a common mean background, using perfect match data. After subtracting this value, intensities are adjusted for identical distributions: data undergo logarithmic transformation followed by quantile scaling [2]. In contrast, Mas5 performs background subtraction using a localized value generated from the lowest 2% of data in a zone, with weighted averages based on distance as a smoothing factor. Summaries are calculated by first subtracting ideal mismatch intensities, log transforming the result, and then scaling by a trimmed mean. This expression value is further modified by plotting intensities for each chip's probes against a baseline (also scaled using trimmed mean), fitting a regression line to the middle 98%, and modifying the experimental probeset's intensities so that this regression line becomes the identity $x = y$. Due to the complexity of this process, there is slight variation in the results between Mas5 calculations performed on different platforms and even between using different compilers [3].

1.2 Clustering

Clustering is used to divide data objects into meaningful groups. Ideally, objects that are most closely related to each other are placed in the same cluster, and objects that are dissimilar are placed in different clusters.

Agglomerative hierarchical clustering begins with each object in an individual cluster. At each step of the algorithm, the closest or most similar clusters are joined [4]. Results are displayed as a dendrogram that shows order in which objects were joined. Similarity is determined by a distance measure, with Euclidean most commonly used. As objects are joined into clusters, this distance matrix must be updated, requiring a method for defining the distance between clusters, in addition to individuals. For average-linkage clustering, all pair-wise distances between two clusters are calculated and the average is used as the distance [5]. Ward's method aims to maximize homogeneity within each group by minimizing variation [6].

2 Data Processing Methods

All processing methods were initially performed using R version 2.9.2 [7]. RMA and Mas5 normalization was performed using "affy" package version 1.2 [1]. Packages "cluster" [8] and "pvclust" [9] were used to calculate and annotate dendrograms. Visualization makes use of package "ggplot2" [10]. Data, both raw and pre-normalized, were obtained from ArrayExpress and contained both Affymetrix chips and samples with no array design information available.

2.1 Pre-normalized Dataset

Normalized gene expression data for 50 chips was downloaded from ArrayExpress. Each chip was taken from a different experiment. Most chips listed Affymetrix MA software for the normalization, but not version or scaling factor. For each chip, the following descriptive statistics were calculated from normalized expression values: mean, min, max, variance, skewness, kurtosis, and 10th through 90th quantiles.

2.2 Author-Normalized Dataset

Raw data from 11 randomly selected experiments were downloaded from ArrayExpress. Five chips from each experiment were selected to be normalized. The chips were normalized in groups of five with all chips in a group coming from the same experiment. Additionally, the number of chips selected for each normalization process varied to mitigate impact of experimental design symmetry on results. We chose to group and normalize the chips in this way to reflect a batch process that researchers are likely to encounter.

2.2.1 Mas5

A total of 55 chips (taken from 11 experiments) were normalized twice, once using scale factor 500 and again using scale factor 250. In addition, 25 of these chips were randomly selected and normalized with a scale factor 100.

2.2.2 RMA

Normalization starts with quantile, cyclic loess, or contrast normalization followed by summarization in log base two. We chose quantile normalization because it is the most rapid of the three options [11] minimizing the need for excessive computing. Fifty-five chips were normalized using this method in groups of 5 as described previously, with 25 of these chips randomly selected from those already included in the Mas5 group. Descriptive statistics were computed for each author-normalized chip as described above.

2.3 Clustering

Chips were clustered based on the computed descriptive statistics using iterative calculations of distances according to Ward's method. Elements of the tree are organized with the tightest cluster (smallest distance) to the left. It is important to note that with different datasets, it is expected that new relationships (due to additional chips) will change the hierarchical ordering. By applying a p-value filter to the tree, we were able to designate clusters that the data strongly supported. p-values were obtained by multiscale bootstrap resampling with $n = 1{,}000$.

3 Observations

3.1 Author-Normalized Dataset (Training)

3.1.1 Descriptive Statistics

We used a parallel coordinate plot to examine the frequency of specific values for each descriptive statistic. For some descriptors, such as variance, a casual observer cannot identify which normalization method corresponds to a specific value. Other

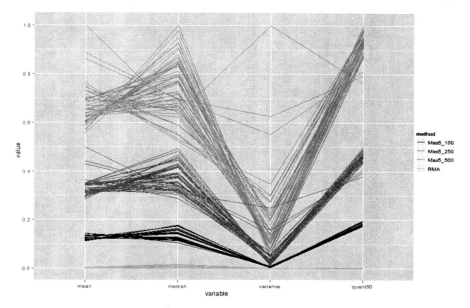

Fig. 1 Distribution of mean, median, variance, and 90th quantile values obtained from author-normalized data

descriptors – mean, median – show obvious grouping by normalization method; however, it is difficult to distinguish boundaries. After examining all descriptive statistics (data not shown), we determined that the distribution of 90th quantile gave the greatest separation between normalization methods (Fig. 1). For all descriptive statistics, RMA normalization shows the least variation in values. The increasing distance between normalization methods for higher level quantiles reflects the range to which each approach scales the chips. A greater scaling factor magnifies variation in observed values.

The shape and confidence interval for descriptive statistics are supported by the central limit theorem: normalized data have an approximately normal distribution for the mean. For quantile values (median, 90th quantile), distribution is normal for the middle quantiles, with increasing right-side skew as one increases past the 75th quantile [12, 13].

3.1.2 Cluster Analysis

The dendrogram shown in Fig. 2 lists a subset of the results of clustering all author-normalized chips. Visual inspection reveals a perfect grouping of microarrays into separate clusters along normalization method. A more robust analysis, by designating a p-value cutoff, generates a list of clusters. The benefit of visual analysis is that relationships are immediately apparent for all chips clustered. However, when

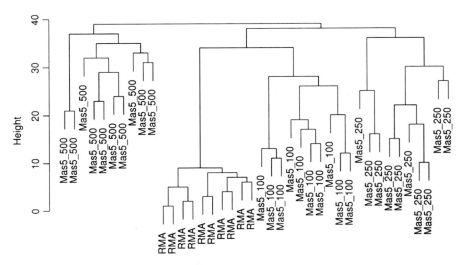

Fig. 2 Dendrogram showing a subset of all author-normalized chips. All distances have p-value of 0.95. Splits are plotted at equally based heights

examining datasets from more than ∼50 chips, the display may become too crowded to distinguish individual labels. Screening clusters by p-values ensures that all selected clusters meet confidence criteria for calculated distances. As number of chips involved in analysis increases, so does confidence level for the overall tree (from $p = 0.70$ for 20 chips to $p = 0.95$ for 40 chips).

3.2 Pre-normalized Dataset (Testing)

3.2.1 Descriptive Statistics

Plots of descriptive statistics for pre-normalized data show that most of these values fall in the same range as the statistics for the author-normalized data, with minimal outliers (Fig. 3). Additional values are expected, as it is possible that data from the ArrayExpress database were normalized by methods not covered above. Documentation from ArrayExpress indicates that the four outliers were subject to ANOVA normalization.

In order to determine the normalization method for each grouping, we seeded pre-normalized data with values generated for use in the author-normalized section. After excluding the outliers, pre-normalized data align well with author-normalized values, especially when looking at the 90th quantile.

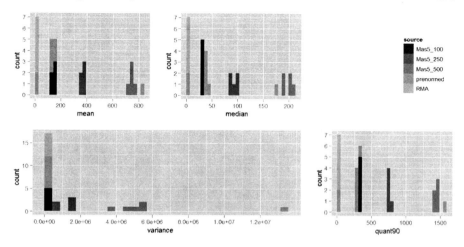

Fig. 3 Descriptive statistics from subset of pre-normalized data seeded with that of author-normalized data

3.2.2 Cluster Analysis

While inspecting descriptive statistic plots allows us to qualitatively define relationships between pre-normalized data and author-normalized methods (Fig. 4), performing cluster analysis as described above provides a quantitative way to measure the strength of the predicted groupings. The tree shown in Fig. 4 indicates that pre-normalized chips 1 and 8 have author-processed, RMA-normalized chips as nearest neighbors. ArrayExpress documentation confirms these chips as being RMA normalized. Remaining pre-normalized chips cluster with Mas5 author-normalized values with various scaling factors. These clusterings also agreed with documentation in ArrayExpress. Additionally, it was observed that clustering is by normalization method irrespective of source data.

3.2.3 Mas5 Scaling Factors

Although ArrayExpress allows researchers to specify that chips were normalized using Mas5, it does not provide a field to report the scaling factor used. Our results show that Mas5-normalized chips with different scaling factors tend to cluster by scaling factor. Additionally, researchers may need to perform analysis across sets of chips normalized using different scaling factors. Therefore, it is useful to be able to convert Mas5 normalized data from one scaling factor to another.

Plotting values from the same dataset normalized with different Mas5 scaling factors show a strong linear relationship (not shown). Therefore, Mas5 normalized data can be transformed from one scaling factor to another much more rapidly than

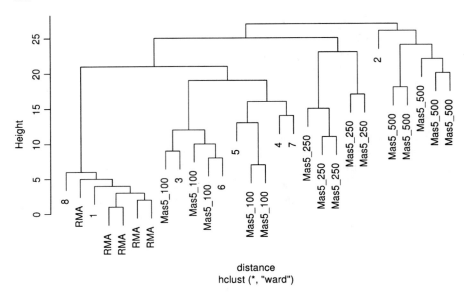

Fig. 4 Subset of seeded data. Pre-normalized values are numbered 1 through 8, and author-normalized values are labeled by normalization method and scaling factor, if applicable. All distances have p-value of 0.95. Splits are plotted at equally based heights

the 5–10 min needed to renormalize each Cel file in R. Transformation is performed simply by multiplying the ratios of each scaling factor. To convert from scaling factor 250 to 500, we multiply by $(500/250) = 2$.

4 Conclusion

We demonstrated that different normalization processes are distinguishable by specific descriptive statistics at both visual and quantitative levels. Incorporation of data with known normalization alongside data of unknown normalization (seeding) is a straightforward and robust technique for a researcher to identify unknown normalization method. Transformation between Mas5 normalizations scaled to different values requires one to simply multiply by a constant value.

These actions (visualization, seeding, and transformation) together define a workflow for researchers to accurately and confidently identify specific normalization methods used to generate a set microarray data, without spending time replicating the normalization process themselves. This not only facilitates easier reproducing of previously published results, but will also mediate novel discoveries as researchers will be able to use previously unusable datasets in ArrayExpress and other databases.

References

1. Gautier, L., Cope, L., Bolstad, B., Irizarry, R.: Affy–analysis of Affymetrix GeneChip data at the probelevel. Bioinformatics. 20, 307–215 (2004)
2. Irizarry, L., Hobbs, B., Collin, F., Speed, T.P.: Exploration, normalization, and summaries of high density oligonucleotide array probe level data. Biostatistics. 4, 240–264 (2003)
3. Lim, W.K., Wang, K., Lefebvre, C., Califano, A.: Comparative analysis of microarray normalization procedures: effects on reverse engineering gene networks. Bioinformatics. 13, 282–288 (2007)
4. Tan, P., Steinbach, M., Kumar, V.: Introduction to Data Mining. Addison-Wesley, Boston (2005)
5. Pevsner, J.: Bioinformatics and Functional Genomics. Wiley-Blackwell, New Jersey (2009)
6. Härdle, W., Simar, L.: Applied Multivariate Statistical Analysis. Springer, Berlin (2003)
7. R Development Core Team.: R: A Language and Environment for Statistical Computing. R Foundation for Statistical Computing. Vienna, Austria (2008)
8. Maechler, M., Rouseeuw, P., Struyf, A., Hubert, M.: Cluster Analysis Basics and Extensions. Unpublished (2005)
9. Suzuki, R., Shimodaira, H.: pvclust: Hierarchical Clustering with p-Values via Multiscale Boostrapp Resampling. R Package Version 1.2-1 (2009)
10. Wickham, H.: ggplot2: An Implementation of the Grammar of Graphics. R Package Version 0.8.3. http://had.co.nz/ggplot2/ (2009)
11. Bolsted, B., Irizarry, R., Åstrand, M., Speed, T.: A comparison of normalization methods for high density oglinucleotide data based on variance and bias. Bioinformatics. 19, 185–193 (2003)
12. David, H., Nagaraja, H.: Order Statistics. Wiley, New York (2003)
13. Sitter, R., Wu, C.: A note on Woodruff confidence intervals for quantiles. Statistics & Probability Letters. 52, 353–358 (2001)

Part II
Bioinformatics Databases, Data Mining, and Pattern Discovery Techniques

Chapter 16
Estimation, Modeling, and Simulation of Patterned Growth in Extreme Environments

B. Strader, K.E. Schubert, M. Quintana, E. Gomez, J. Curnutt, and P. Boston

Abstract In the search for life on Mars and other extraterrestrial bodies or in our attempts to identify biological traces in the most ancient rock record of Earth, one of the biggest problems facing us is how to recognize life or the remains of ancient life in a context very different from our planet's modern biological examples. Specific chemistries or biological properties may well be inapplicable to extraterrestrial conditions or ancient Earth environments. Thus, we need to develop an arsenal of techniques that are of broader applicability. The notion of patterning created in some fashion by biological processes and properties may provide such a generalized property of biological systems no matter what the incidentals of chemistry or environmental conditions. One approach to recognizing these kinds of patterns is to look at apparently organized arrangements created and left by life in extreme environments here on Earth, especially at various spatial scales, different geologies, and biogeochemical circumstances.

1 Introduction

A key aspect of planning a space mission is to set scientific mission objectives with the ability to adapt them based on observations and mission situations. The search for extraterrestrial life is a major scientific objective, but the exact nature of that life and how to confirm it constitute a major debate [9]. A further problem in the search for extraterrestrial life is how to select specific areas that we want to investigate more intensively? Photo surveys, geology, and knowledge of biology here on Earth can take us to a likely general area, but, unlike a dynamic process like that which produced the Martian dust devils [2], subtle biological clues have no motion to point us in the right direction.

K.E. Schubert (✉)
California State University, San Bernardino, CA 92407, USA
e-mail: schubert@csusb.edu

We use life in extreme environments on Earth as analogs for the kinds of life that we could encounter in space [22]. We propose using biopatterns found in some of these analogs to create a series of templates that could be used to indicate areas potentially worthy of deeper investigation. In resource-limited environments (often the case with extreme environments), organisms grow in patterns that are self-enforcing and exhibit hysteresis [1, 14, 21] which could be used to recognize them and their fossil remains at a distance. Particularly on Mars, as the environment apparently became less hospitable over geologic time, extremophiles, similar to that in Earth, were likely the last to survive and should be the easiest to find at least in environments protected from the weathering effects of the harsh surface climate. Obviously, caves would provide such protection [6]. Among the techniques that have been used to model these patterns are evolutionarily stable strategies in game theory and differential equations [8, 14, 15, 21, 29].

While good results have been generated using differential equations, they require tuning of the parameters and experience in mathematical and numerical techniques to obtain valid results. In this work, we developed cellular automata that produce similar predictions to the differential equation models, while preserving the rapid modeling and hypothesis testing of cellular automata. Similar models can also be applicable to group animal behavior [11, 16, 19, 23]. Our method for deriving rules for cellular automata from observed data in organism growth patterns accounts for soil nutrients, water, root growth patterns, and geology allowing scientists to easily examine the effects of modifying conditions without damaging the environment.

We apply this model to identify factors affecting patterning with respect to growth, die-out, and stabilization in extreme environments. We compare the results of our model with biovermiculation microbial mats growing in acid caves, and cyanobacteria-dominated desert crust growth in Zzyzx, CA. These models could be used to rapidly check data from space missions to rate the potential of various locations to offer biosignatures.

2 Cellular Automata

In the 1940s, John von Neumann developed the first cellular automaton, while working on the self-replicating systems biological problem [31]. Physicist, Conrad Zuse, published a book, Calculating Space, in 1969, which proposed that the universe was the output of a gigantic cellular automaton. In 1970, John Conway developed his Game of Life, which is a two-dimensional cellular automaton [13]. In 1983, Stephen Wolfram published the first of many papers on cellular automata culminating in 2002 with the publication of his book, A New Kind of Science [30, 31].

A cellular automaton (CA) is a computational model that is discrete in both space and time. Essentially, we divide space into boxes called cells, and only calculate their values at discrete time using a set of fixed rules. A state could represent anything, and in our case it represents the amount of water, nutrients, and the biomass. The rules describe how the organism grows or dies in the presence of the water,

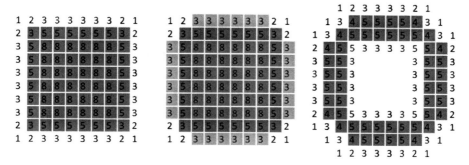

Fig. 1 First transition of a cellular automaton starting with a square of "living" cells

nutrients, and competition from other organisms. CA rules are not usually expressed as formulas, rather they are visual, such as drawing pictures of the neighboring cells and then labeling the next state. Cellular automata have been used to study growth and patterns in forests, arid desert environments, predator–prey problems, and sea shells. It has also been used to study areas as diverse as epidemiology and linguistics [11, 16, 19, 23]. To avoid tedium, we group similar cell patterns using heuristics discerned from discussions with biologists and born out in practice.

Typically CA with the same starting state are deterministic, and we do not use the current value of the state in the calculation of the next state, and use only one state variable. (John von Neumann proposed 29 states in each cell for his self-replicating CA.) Some starting states are so stable that the CA never grow. We consider a generalization of this in which these constraints are relaxed. For this chapter, we only relax the determinism constraint, as this allows for more realistic looking results.

As an example, we will show a simple CA that is calculated with a radius of 1. The only values used to calculate the next state are the values in the eight cells that surround the target cell. The next state of the cell stays the same, unless the sum of the neighbors is 3 (grow) or 7,8 (die).

This can be seen in Fig. 1, where the initial condition is on the right (starts as the blue square). The numbers are the sum of the neighbors. In the middle figure, the cells that are going to die off are colored red, and the ones about to grow are colored light blue. The final figure on the right shows the next state, with the new neighbor sums. This process would then be repeated.

3 Cellular Automata and Partial Differential Equations

Various methods have been researched for directly transforming spatial partial differential equations into cellular automata models through approximation techniques [27, 28]. In Sect. 3.1, two methods of transforming a general partial differential equation into a cellular automata rule are shown. In Sect. 3.2, the methods are then analyzed using the Z-transform to find the theoretical constraints of stability. Section 3.3 illustrates convergence maps of the cellular automata models created from multiple simulation runs. From this information, a set of guidelines was created that can be used to create faster simulations that convergence to stable values.

3.1 Transformation Using Euler's Methods

Two methods were developed for directly transforming differential equations into cellular automata. First, a generalized partial differential equation was derived from surveying biological partial differential equations such as desert vegetation patterns ((1) and (2)) [14, 21],

$$\frac{\partial n}{\partial t} = \frac{yw}{1+\sigma w}n - n^2 - \mu n + \nabla^2 n \tag{1}$$

$$\frac{\partial w}{\partial t} = p - (1-\rho n)w - w^2 n + \delta \nabla^2 (w - \beta n) - v\frac{\partial(w - \alpha n)}{\partial x} \tag{2}$$

Fick's law on population density (3) [26],

$$\frac{\partial P}{\partial t} = f(t, x, P) + d\nabla_x^2 P \tag{3}$$

and predator prey models [24]. The following is the general differential equation derived:

$$f(u_{i,j}) = \frac{\partial u}{\partial t} = m(u_{i,j}) + \nabla_x^2 n(u_{i,j}) + \nabla_x o(u_{i,j}) \tag{4}$$

In this equation, u will be the values that are simulated. The subscript i is the time index of u and j is the space index. Most of the surveyed differential equations could then be broken up into three groupings of terms based on whether the Laplacian or gradient operator was applied to them. The terms can be substituted into the $m(u_{i,j})$, $n(u_{i,j})$, and $o(u_{i,j})$ functions.

Once a differential equation is described in this general form, it can easily be transformed into a cellular automata by two different methods. The three point formula is first applied to the general form eliminating the ∇_x^2 and ∇_x operators. The Forward Euler's method, which is the simpler of the two methods used, can then be applied producing the following formula:

$$u_{i+1,j} = u_{i,j} + h_t \left[m(u_{i,j}) + \frac{n(u_{i,j+1}) - 2n(u_{i,j}) + n(u_{i,j-1})}{h_x^2} \right.$$
$$\left. + \frac{o(u_{i,j+1}) - o(u_{i,j-1})}{2h_x} \right] \tag{5}$$

The Backward Euler's method, which is more stable than the Forward Euler's method, can be applied instead giving the formula:

$$u_{i+1,j} = u_{i,j} + \frac{h_t}{1 - h_t \frac{\partial m(u)}{\partial u}\big|_{i,j}} \left[m(u_{i,j}) + \frac{n(u_{i,j+1}) - 2n(u_{i,j}) + n(u_{i,j-1})}{h_x^2} \right.$$

$$\left. + \frac{o(u_{i,j+1}) - o(u_{i,j-1})}{2h_x} \right] \quad (6)$$

Equations (5) and (6) are in fact cellular automata rules because they state that the value of u at space j for the next time period is equal to u's current value plus or minus some factors of its neighboring values $u_{i,j+1}$ and $u_{i,j-1}$. It is now simple to transform a general differential equation into a cellular automata rule by substituting $m(u_{i,j})$, $n(u_{i,j})$, and $o(u_{i,j})$ in the general differential equation form into either the Forward or Backward formulas. It should also be noted that these equations contain two new important variables as the result of the approximations. The variable h_x is the step size for space and h_t is the step size for time.

3.2 Theoretical Stability Constraints

Equations (5) and (6) were analyzed using the Z-transform to create constraints on stability. The analysis was done on a subset of the general form that used linear terms. The following equation was used to represent this general linear form:

$$f(u) = a_1 u + b_1 + \nabla_x^2(a_2 u) + \nabla_x(a_3 u) \quad (7)$$

One can assign parts of the equations to the following functions of (4): $m(u) = a_1 u + b_1$, $n(u) = a_2 u$, and $o(u) = a_3 u$. Here, the a and b terms are simply coefficients. After the Z-transform was used on Forward and Backward Euler transformed versions of (7), the equation was solved for U_j. The i index is eliminated in the Z-transform. According to Z-transform theory, if the pole and zero values of z are within the unit circle, then the equation should remain stable. Therefore, constraints of stability can be found by setting the poles and zeros values to less than one. The zero constraint created was the same for both the Forward and Backward Euler's formulas:

$$1 > \left| \frac{-1}{2b_1 h_x^2} [(2a_2 - a_3 h_x)(U_{j-1}) + (2a_2 + a_3 h_x)(U_{j+1})] \right| \quad (8)$$

The poles constraint for the Forward Euler's was:

$$1 > \left| 1 + a_1 h_t - \frac{2a_2 h_t}{h_x^2} \right| \quad (9)$$

The poles constraint for the Backward Euler's was slightly different due to the dividing factor that is part of the Backward Euler's equation:

$$1 > \left| 1 + \frac{a_1 h_t}{1 - a_1 h_t} - \frac{2 a_2 h_t}{(1 - a_1 h_t) h_x^2} \right| \qquad (10)$$

The pole constraints demonstrate that there needs to be a balancing between h_t and h_x. If one is too big relative to the other term, the right-hand side would become larger than one, which would mean the associated equation, either (5) or (6), was unstable.

3.3 Convergence Maps and Optimum Convergence

Simulations were run for the Forward and Backward Euler's formulas using Scilab and a one-dimensional cellular automata. The results were combined and graphed onto convergence maps shown in Fig. 2. In this figure simulations were plotted based on the h_x and h_t values used for the Backward Euler's equation. A blue dot represents

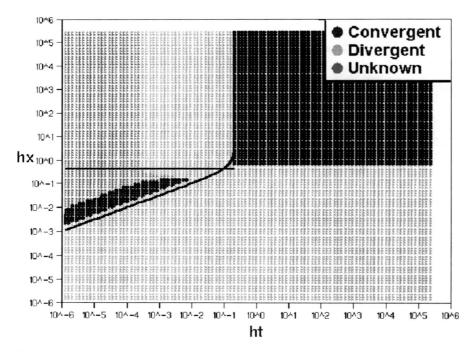

Fig. 2 Convergence map is composed of 10,000 simulations with varying h_t and h_x values for the Backward Euler's function. The *black lines* represent the pole constraints

16 Patterned Growth in Extreme Environments

a simulation that ends in convergence, a green dot represents a simulation that ended in divergence, and a red dot is a simulation that hit the maximum number of iterations allotted without ending in convergence or divergence.

Both the Forward and Backward Euler's convergence maps had a small stripe of convergence in the lower left area. The Backward Euler's map also had an additional area of convergence in the top right, but this area appeared to be an artifact of the Backward Euler's dividing factor. The striped area of convergence was of interest because the convergence values are nearly the same, given the same h_x value, even with different h_t values. In other words, the manner in which space is discretized affects the outcome of the convergence values but how time is discretized does not.

The convergence map in Fig. 2 also shows the pole constraints graphed in black. Through multiple simulations it was found that the area of convergence closely matched the area created by the pole constraints, but not exactly. The difference between the actual area of convergence and the pole constraints appeared to be a constant. Tests also showed that the bottom of the area of convergence converged about ten times faster than the top, therefore making it desirable to use h_x and h_t values close to the bottom pole constraint, but within the area of convergence.

The convergence map in Fig. 3 contains modified zero constraints in magenta. These constraints required substitutions to be made for the U_{j+1} and U_{j-1} terms, which are variables and therefore could not be graphed in the convergence maps.

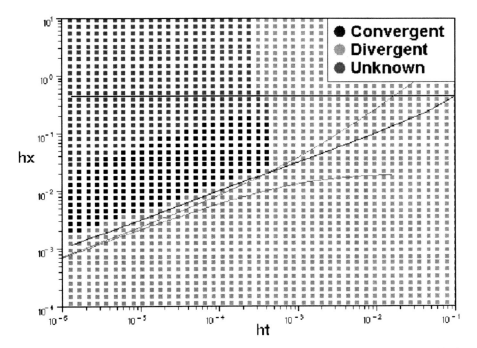

Fig. 3 Convergence map for the Backward Euler's function shows the pole constraints in *black* and the modified zeros constraints in magenta

Simulations showed that most of the time the intersection of the lower pole constraint with the upper zero constraint could be used as an approximation for the maximum h_t value for convergence. This information led to a set of guidelines that used the intersection for finding the maximum h_t value. A lesser h_t value could then be used in the lower pole constraint formula to find an h_x value. This would produce an h_x and h_t pair that would converge quickly because it was a simulation at the bottom of the area of convergence. Hopefully, the direct transformation methods and convergence guidelines can eventually be used in a software tool that will help preform pattern simulations for biologists.

4 Extremophiles

Some organisms form patterns based on a combination of intrinsic growth geometries [18], potentially other biological properties [3], and the environmental conditions in which they exist. It has been suggested that there may be significant benefit in such collective community-level behavior, and that such organism aggregations behave with a remarkable degree of plasticity and adaptability that allows them enhanced capability to meet changing and challenging growth conditions [20]. Obviously such plasticity with respect to the environment would be particularly pertinent to extremophile organisms living in an array of especially challenging physical and chemical conditions.

The CA models are adapted to the biological and geological conditions found on Earth that are most likely to match those in extreme environments from Earth to Mars to Europa. Indeed, if such biopatterning is as universal as we suspect, it could probably work on any life-bearing planet in the galaxy. As test case models of astrobiological significance, cyanobacteria in desert crusts and biovermiculations (hieroglyphic-patterned microbial mats on cave walls) have been identified. This has grown from our group's participation in NASA's Spaceward Bound Project, a teacher science outreach effort that involves extremophile environment research across a number of scientific and engineering disciplines [10].

4.1 Cyanobacteria

Cyanobacteria are photosynthetic bacteria relatives that live in a staggering variety of extreme and non-extreme environments across the planet [25]. They are notable for many reasons, including probably being the oldest fossils, the presumptive original producers of free atmospheric oxygen on Earth, the source of much of our oil, and an ability to grow in an amazing array of extreme environments (including Antarctica). The cyanobacteria in extensive desert crusts at Zzyzx Research Station in the Mojave Desert National Preserve, CA, display the structured growth we are considering. Given their pivotal and early role on the Earth, similar organisms

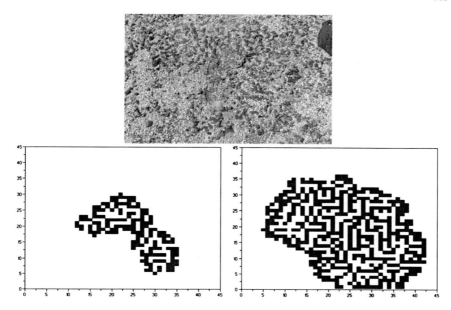

Fig. 4 *Left* picture is of cyanobacteria-rich desert crust patterns from Zzyzx, CA. *Right* graphics are simulated cyanobacteria at $t = 5$ (*top*) and $t = 40$ (*bottom*)

are conceivable on early Mars, and their identification is important for biological objectives on Mars and back into the antique history of Earth.

Figure 4 shows a typical cyanobacterially dominated desert crust (aka cryptogamic crust). These patterns are fairly dense, and so we used a neighborhood with a range of 1 square (1 square in all directions for a total of eight neighbors). Cyanobacteria often grow in blooms when conditions become favorable as we have demonstrated with watering experiments on these aridland, water-limited communities (P. Boston, unpublished results); so we have modeled this using growth on having three neighbors only. We also allowed random death with probability of 10%. The resulting simulation is shown on the right in Fig. 4 and shows much of the same structure as is seen in the picture on left in Fig. 4.

4.2 Biovermiculations

Biovermiculations are microbial mats composed of bacteria, extracellular polysaccharide slime, embedded clay and other particles, in situ precipitated minerals (e.g., sulfur, gypsum, carbonate), and even some small invertebrates such as mites and nematodes. They exist in a wide variety of chemical and physical subsurface settings including caves, mines, lavatubes, and even human structures. Most recently we have identified a photosynthetic variety on the undersides of translucent rocks in the central Australian deserts (P. Boston, unpublished results). Investigators identified them originally from sulfuric acid caves, see [17]. More recently, observers have begun to see them in a wide variety of chemical and physical subsurface settings including acidic mines, cold carbonate caves, both tropical and alpine lavatubes, and even Mayan ruins.

These structures are interesting because of their intrinsically intriguing biology and geochemistry, the distinctive growth patterns that they exhibit, and also because they may be a highly distinctive biosignature that could be interpretable in extraterrestrial settings based on gross morphology [7]. Interest in the biovermiculations has grown as better methods of studying such structures have become available. They also provide a model system of a biomat that might occur on the interiors of various cave types on Mars. Lavatubes have been identified on Mars [4], and more recently confirmed in a more elaborate study [12]. Mechanisms to create solutional caves in evaporite mineral deposits on Mars have also been proposed [5]. Such potential subsurface habitats could conceivably house or have housed microbial populations on Mars and left traces similar to those found in geomicrobiological communities in Earth's subsurface.

The patterning of biovermiculation growth is still a mystery. To understand it will require simulations that test different sets of rules enabling us to arrive at a good pattern match for the microbial mat growth that we are observing in nature. Such simulations will then be correlated with actual pattern examples from cave walls and other occurrences. Figure 5 shows an area that has biovermiculations so thick that they become the solid mat, but there are weird uninhabited areas that follow the rock curvatures. These are not simple water pathways that have prevented growth, and so their origin is unknown. In Fig. 5, no nutrient or water differentiation was induced, only crowding rules, and an ability to pull nutrients from surrounding cells. The result was the depleted region in the middle. Note the shape has many indents, like the actual system in Fig. 5.

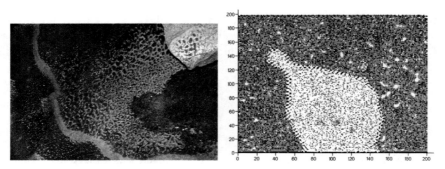

Fig. 5 *Left* picture is biovermiculation with discontinuities. *Right* graphic is simulated biovermiculation growth

5 Identifying Rules from Pictures

Cellular automata use the neighboring cells (within some radius of effect) to determine the next state of the central cell. If we assume the pattern has been around for a long time, and thus has settled into a relatively stable pattern, then we can assume that the next state of any cell is the same as its current state. This implies that most of the rules can be determined for a stable pattern from a picture by making histograms of the number of neighbors around live and dead centers. We say most because

- Relatively stable still implies there will be some cells which will change state in violation of the fundamental assumption
- Not every rule might be active in the picture
- The scale of the cells must still be determined
- The radius of effect must still be determined
- The effects of uneven nutrient distribution is unaccounted for
- Randomness is unaccounted for

Even with these serious issues, a very good first approximation of the rules can be readily obtained using histograms on each cell's neighborhood. The histogram of the live centers (center has value of 1) gives us the stay alive rules, see Fig. 6. The growing rules can also be seen on the graphs where there are live centers but no dead centers (center has value of 0), as the growth rules turn dead centers to live, and it is theoretically the mechanism that caused this, see Fig. 6. In practical histograms, either the left or right side of the ideal histogram will be clear, but the other will be malformed or missing due to which rules were active in the system at that time. A predominantly increasing cellular automaton will have a good right side, while a predominantly decreasing one will have a clear left side.

Consider the graphic of the cellular automaton's state in Fig. 7. It is not obvious what the rules were that generated it by simply looking at the picture, other than the radius of effect is likely to be the thickness of a typical line, which is 3 in

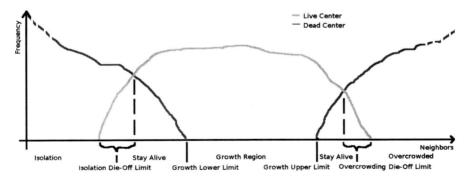

Fig. 6 Idealized histograms of the number of neighbors for both living and dead centers, which show how to read cellular automaton rules

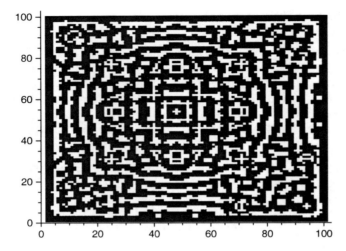

Fig. 7 This graphic is the state of a cellular automaton on a 100 by 100 grid, after 100 time cycles. The cellular automaton used a radius of three cells, the center grew if the neighbors numbered 13–24, it kept its state if the neighbors numbered 8–36, in any other case it died. The initial condition was a box of cells in the center

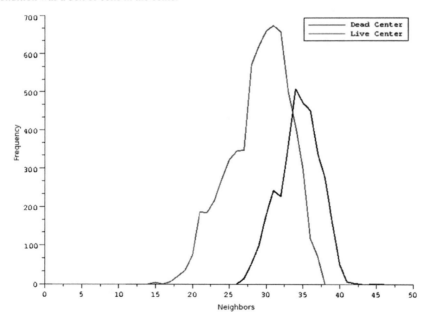

Fig. 8 Dead center and live center histograms of the cellular automaton in Fig. 7

this case. Indeed the radius of effect for this cellular automaton was 3. Now examine its histogram in Fig. 8, which was calculated for neighborhoods of radius 3. The overcrowding limit (upper range of the stay alive rules) should lie between the cross over of the dead center and live center histogram lines and the end of the non-zero portion of the live center histogram, in this case between 34 and 38. The actual upper limit was 36, dead center in the range. The growth range should include the region where the live center histogram is non-zero but the dead center one is not. It can extend below this, but not above. In this case it should include the region 15–26, perhaps a bit lower. The actual growth region was 13–24, again we have a very good approximation. The isolation die-off limit (lower range of the stay alive rules) will be very hard to determine for this cellular automaton because the state in the picture is fairly dense, and thus there are not many of the lower end of the survival spectrum to determine that limit in this case.

The method is not fool proof, but it does provide a good first approximation of the rules from a picture and gets very close to most of the four salient numeric quantifiers.

6 Conclusions and Future Work

The patterns of microbial extremophile growth, developed using cellular automata, can provide starting templates for the types of patterns we may see in lavatubes and caves on Mars, other planets, and in Earth's own rock record of life. These patterns could provide indicators of life similar to the patterns that indicated water flow and possible "ponds" on the surface of Mars. These templates and the ability to rapidly identify high probability areas for detecting life will be vital to planning future Mars missions. The ability to determine correct simulation parameters is crucial to successfully using these models, and thus the optimal step parameters and histogram rule extraction technique are extremely important. We are continuing our work on both the histogram techniques and biological modeling and validation. In particular, we are working on an automated rule generator for field photographs, as well as both field and lab tests of the validity of the cellular automata models with NASA Ames Research Center colleagues.

References

1. Special issue on banded vegetation. Catena vol. 37 (1999)
2. Press release images: Spirit. http://marsrovers.nasa.gov/gallery/press/spirit/20050819a.html (2005)
3. Bassler, B.L.: Small talk: cell-to-cell communication in bacteria. Cell **109**, 421–424 (2002)
4. Boston, P.: Extraterrestrial caves. In: Encyclopedia of cave and karst science, pp. 355–358. Fitzroy-Dearborn Publishers, Ltd., London, UK (2004)
5. Boston, P., Hose, L., Northup, D., Spilde, M.: The microbial communities of sulfur caves: A newly appreciated geologically driven system on Earth and potential model for Mars. In: R. Harmon (ed.) Karst geomorphology, hydrology, & geochemistry, pp. 331–344. Geological Society of America (2006)

6. Boston, P., Ivanov, M., McKay, C.: On the possibility of chemosynthetic ecosystems in subsurface habitats on mars. Icarus **95**, 300–308 (1992)
7. Boston, P., Spilde, M., Northup, D., Melim, L., Soroka, D., Kleina, L., Lavoie, K., Hose, L., Mallory, L., Dahm, C., Crossey, L., Schelble, R.: Cave biosignature suites: Microbes, minerals and mars. Astrobiology Journal **1**(1), 25–55 (2001)
8. CA, K.: Regular and irregular patterns in semiarid vegetation. Science **284**, 1826–1828 (1999)
9. Cady, S.: Formation and preservation of bona fide microfossils. In: Signs of life: A report based on the april 2000 workshop on life detection techniques, committee on the origins and evolution of life, pp. 149–155. National Research Council, The National Academies, Washington, DC (2001)
10. Conrad, L.: Spaceward bound project. http://quest.nasa.gov/projects/spacewardbound/ (2009)
11. Couzin, I., Krause, J., James, R., Ruxton, G., Franks, N.: Collective memory and spatial sorting in animal groups. Journal of Theoretical Biology **218**, 1–11 (2002)
12. Cushing, G., Titus, T., Wynne, J., Christensen, P.: Themis observes possible cave skylights on mars. Geophysical Research Letters **34**(L17201) (2007)
13. Gardner, M.: The fantastic combinations of john conway's new solitaire game 'life'. Scientific American **223**, 120–123 (1970)
14. von Hardenberg, J., Meron, E., Shachak, M., Zarmi, Y.: Diversity of vegetation patterns and desertifcation. Physical Review Letters **87**(19):(198101-14) (2001)
15. HilleRisLambers, R., Rietkerk, M., van den Bosch, F., Prins, H., de Kroon, H.: Vegetation pattern formation in semi-arid grazing systems. Ecology **82**, 50–62 (2001)
16. Hoare, D., Couzin, I., Godin, J.G., Krause, J.: Context-dependent group size choice in fish. Animal Behavior **67**, 155–164 (2004)
17. Hose, L., Palmer, A., Palmer, M., Northup, D., Boston, P., Duchene, H.: Microbiology and geochemistry in a hydrogen sulphide-rich karst environment. Chemical Geology **169**, 399–423 (2000)
18. Jacob, E.B., Aharonov, Y., Shapira, Y.: Bacteria harnessing complexity. Biofilms **1**, 239–263 (2004)
19. Krause, J., Tegeder, R.: The mechanism of aggregation behavior in fish shoals: individuals minimize approach time to neighbours. Animal Behavior **48**, 353–359 (1994)
20. Levine, H., Jacob, E.B.: Physical schemata underlying biological pattern formation-examples, issues and strategies. Journal of Physical Biology **1**, 14–22 (2004)
21. Meron, E., Gilad, E., von Hardenberg, J., Shachak, M., Zarmi, Y.: Vegetation patterns along a rainfall gradient. Chaos, Solutions and Fractals **19**, 367–376 (2004)
22. Nealson, K., Conrad, P.: Life: past, present and future. Philosophical Transactions of the Royal Society of London Series B-Biological Science **354**(1392), 1923–1939 (1999)
23. Pitcher, T., Misund, O., Fernö, A., Totland, B., Melle, V.: Adaptive behaviour of herring schools in the norwegian sea as revealed by high-resolution sonar. ICES Journal of Marine Science **53**, 449–452 (1996)
24. Savill, N.J., Hogeweg, P.: Competition and dispersal in predator-prey waves. Theoretical Population Biology **56**, 243–263 (1999)
25. Seckbach, J.: Algae and cyanobacteria in extreme environments. In: Cellular origin, life in extreme habitats and astrobiology, p. 814. Springer, Dordrecht, Netherlands (2007)
26. Shi, J.: Partial differential equations and mathematical biology. http://www.resnet.wm.edu/~jxshix/math490/lecture-chap1.pdf (2008)
27. Strader, B.: Simulating spatial partial differential equations with cellular automata. M. cs., CSU San Bernardino, San Bernardino, CA, USA (2008)
28. Strader, B., Schubert, K., Gomez, E., Curnutt, J., Boston, P.: Simulating spatial partial differential equations with cellular automata. In: H.R. Arabnia, M.Q. Yang (eds.) Proceedings of the 2009 International Conference on Bioinformatics and Computational Biology, vol. 2, pp. 503–509 (2009)
29. Thiéry, J., d'Herbès, J., Valentin, C.: A model simulating the genesis of banded vegetation patterns in niger. The Journal of Ecology **83**, 497–507 (1995)
30. Wolfram, S.: Twenty problems in the theory of cellular automata. Physica Scripta **T9**, 170–183 (1985)
31. Wolfram, S.: A New Kind of Science. Wolfram Media Inc. (2002)

Chapter 17
Performance of Univariate Forecasting on Seasonal Diseases: The Case of Tuberculosis

Adhistya Erna Permanasari, Dayang Rohaya Awang Rambli, and P. Dhanapal Durai Dominic

Abstract The annual disease incident worldwide is desirable to be predicted for taking appropriate policy to prevent disease outbreak. This chapter considers the performance of different forecasting method to predict the future number of disease incidence, especially for seasonal disease. Six forecasting methods, namely linear regression, moving average, decomposition, Holt-Winter's, ARIMA, and artificial neural network (ANN), were used for disease forecasting on tuberculosis monthly data. The model derived met the requirement of time series with seasonality pattern and downward trend. The forecasting performance was compared using similar error measure in the base of the last 5 years forecast result. The findings indicate that ARIMA model was the most appropriate model since it obtained the less relatively error than the other model.

1 Introduction

Predicting future number of disease incidence is important in planning and managing the suitable policy for reducing number of case. Generally, information systems play a central role in developing an effective comprehensive approach to prevent, detect, respond to, and manage infectious disease outbreaks in human [1]. Some work related to managing disease incidence have been done in several years [2–9]. Hence, different forecasting methods are used in purpose of disease prediction. However, few published studies have compared different forecasting techniques on identical historical data of disease. In many studies, the performance of individual forecasting techniques was reported, including multivariate Markov

---

A.E. Permanasari (✉)
Department of Computer and Information Science, Universiti Teknologi PETRONAS, Bandar Seri Iskandar, 31750 Tronoh, Perak, Malaysia
and
Electrical Engineering Department, Gadjah Mada University, Jl. Grafika No. 2, Yogyakarta 55281, Indonesia
e-mail: astya_00@yahoo.com

chain model to project the number of tuberculosis (TB) incidence in the USA from 1980 to 2010 [10], exponential smoothing to forecast the number of human incidence of Schistosoma haematobium at Mali [11], ARIMA model to forecast the SARS epidemic in China [12], a Bayesian dynamic model to monitor the influenza surveillance as one factor of SARS epidemic [13], seasonal autoregressive models to analyze cutaneous leishmaniasis (CL) incidence in Costa Rica from 1991 to 2001 [14], and the application of decomposition method to predict number of Salmonellosis human incidence [15].

In this chapter, six different methods, namely linear regression, moving average, decomposition, Holt-Winter's, ARIMA, and artificial neural network (ANN), were applied for seasonal disease forecasting that involves a case study for tuberculosis incidence in the USA. Even though the computational techniques differ, the accuracy can be compared through the statistical result of forecast errors.

The remainder of the chapter is structured as follows. Section 2 introduces time series used for model development. Section 3 describes process of model development for each forecasting method. Section 4 presents the result of forecasting. Section 5 reports forecasting performance. Finally, Sect. 5 presents the conclusion of the study and directions for future work.

2 Application Case

The seasonal disease that is having an annual pattern is studied in this chapter. Tuberculosis was selected as the case study. The simulation study used time-series data of tuberculosis incidence in USA for the 168-month period from January 1993 to December 2006. It is obtained from the summary of notifiable diseases in USA from the Morbidity and Mortality Weekly Report (MMWR) published by Centers for Disease Control and Prevention (CDC). The seasonal variation of the original data is shown in Fig. 1.

Refer to Fig. 1 for the maximum indices that are reached in December and minimum indices that are reached in January. Observation of the data pattern indicates the existence of seasonality and the presence of a trend. The underlying pattern of

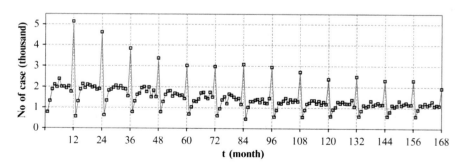

Fig. 1 Time series data for tuberculosis data in USA 1993–2006

the data follows a relatively decreasing trend. The data exhibit seasonal behavior related to the changes of climate. Then, the development of forecast model for tuberculosis incidence must satisfy the following two objectives: (1) can represent the data seasonality; (2) considering the downward trend in the forecasting calculation.

3 Model Development

Based on the time series presented in Sect. 2, the following section presents the development of forecasting model.

3.1 Linear Regression

Regression analysis is a model that develops relationships between a dependent variable and one or more independent variables, in which every component is supposed to be in numerical form.

Time-series data of tuberculosis was formed in monthly data, and then it had 12 seasonal components. Seasonal factor was modeled by applying dummy variables. Thus, 12 ordinal variables might be defined. A month that corresponds to the actual values is assigned as 1, while the rest are assigned into 0. Twelve seasonal components were determined from the monthly data. The 12th month could not provide any information like the first 11, because it is used as a baseline for comparison. In this situation, a monthly trend variable is applied.

The regression formula is:

$$\begin{aligned} y_t &= TR_t + SN_t + \varepsilon_t \\ &= 3653.683 - 6.3886t - 2521.35x_{s1,t} - 2085.35x_{s2,t} - 1732.35x_{s3,t} \\ &\quad - 1657.68x_{s4,t} - 1626.22x_{s5,t} - 1491.9x_{s6,t} - 1603.23x_{s7,t} - 1558.34x_{s8,t} \\ &\quad - 1654.52x_{s9,t} - 1585.71x_{s10,t} - 1709.82x_{s11,t}. \end{aligned} \quad (1)$$

3.2 Moving Average

Moving average is a method that analyzes the average of the time-series fluctuations to direct the time-series changing. Using this method, the future values are obtained by calculating the average of the most recent k data histories.

In order to calculate moving average in the seasonal time series, the existing dataset needs to be deseasonalized first by dividing the actual data with the seasonal factor. After the average calculation, the result is multiplied by the seasonal factor to get the final prediction values. In this chapter, the number of terms in each moving average was chosen as 12.

Based on the calculation, the seasonal factor for the existing historical data are January $= 0.175$, February $= 0.294$, March $= 0.390$, April $= 0.412$, May $= 0.418$, June $= 0.452$, July $= 0.420$, August $= 0.430$, September $= 0.401$, October $= 0.421$, November $= 0.382$, and December $= 0.858$.

3.3 Decomposition

Decomposition method process forecasts result by break time series into its parts: trend, seasonality, cyclical, and irregular (error). Then, a calculation will be done to obtain the value for each component. These values will be projected forward and they will be reassembled to develop a forecast. Based on the current time series (Fig. 1), the time series exhibits decreasing seasonal variation; hence multiplicative decomposition is used [16]. The model is:

$$y_t = TR_t \times SN_t \times CL_t \times IR_t, \qquad (2)$$

where y_t is the observed value of the time series in time period t; TR_t is the trend component (or factor) in time period t; SN_t is the seasonal component (or factor) in time period t; CL_t is the cyclical component (or factor) in time period t; and IR_t is the irregular component (or factor) in time period t.

Because of no cyclic pattern in tuberculosis data, CL component could be removed from (5), and the estimation of $SN_t \times IR_t$ was sn_t/ir_t. With $L = 12$ (number of period a year), the seasonal factor is:

$$sn_t = \left(L \Big/ \sum_{t=1}^{L} \overline{sn_t} \right) \times \overline{sn_t} = 1.000701 \left(\overline{sn_t} \right). \qquad (3)$$

The next step is to calculate the deseasonalized observation in time period t as $d_t = y_t/sn_t$. The estimation of tr_t for the trend TR_t could be obtained by fitting a regression equation to the deseasonalized data. It came out with the following function of tr_t:

$$tr_t = b_0 + b_1 t = 2021.734 - 6.012t. \qquad (4)$$

3.4 Holt-Winter's Method

Holt-Winter's method is the extended form of the simple exponential smoothing that include the seasonality in the approach. This model involves three smoothing parameters, namely level (α), trend (β), and seasonality (γ). Referring to the decreasing trend at the time series, the multiplicative Holt-Winter's method was selected. Every smoothing constant have values between 0 and 1. In the multiplicative method, it is assumed that the observation result is obtained from the product of seasonal values and a seasonal index for that particular period.

Some iteration was done to find the suitable parameters that give the smallest output, the iteration started from 0.01 and incremented by 0.01. The selected parameter was determined by the smallest MAPE with the result: $\alpha = 0.011$, $\beta = 0.01$, and $\gamma = 0.01$.

3.5 ARIMA

In this chapter, seasonal ARIMA (SARIMA) was used because time series exhibited a seasonal variation. A seasonal autoregressive notation (P) and a seasonal moving average notation (Q) form the multiplicative process of SARIMA as $(p,d,q)(P,D,Q)_s$. The subscripted letter 's' shows the length of seasonal period. For example, in an hourly data time series $s = 7$, in a quarterly data $s = 4$, and in a monthly data $s = 12$. The general SARIMA (p,P,q,Q) model is:

$$\phi_p(B)\phi_P(B^s)z_t = \delta + \theta_q(B)\theta_Q(B^s)a_t, \tag{5}$$

where ϕ is an autoregressive operator, θ is a moving average operator, p is the autoregressive order, q is the moving average order, δ is the constant, B is the backshift operator, and a is random shock [16].

Different ARIMA models were applied to find the best fitting model. Then the most appropriate model can be selected using the Bayesian Information Criterion (BIC) and Akaike Information Criterion (AIC) values. The best model was obtained on model C AR(7) AR(29) SAR(12) MA(7) MA(29) SMA(12) that also can be written as SARIMA$(29,0,29)(12,1,12)_{12}$. From this model, the parameter results were $C = -19.347$, AR(7) $= 0.197$, AR(29) $= 0.374$, SAR(12) $= 0.293$, MA(7) $= -0.945$, MA(29) $= 0.0295$, and SMA(12) $= -0.884$. Parameters obtained from the calculation were AR(9) $= 0.154$, SAR(12) $= -0.513$, MA(14) $= 0.255$, and SMA(24) $= -0.8604$. The final model that is expressed is as follows:

$$(1 - 0.197B^7 - 0.374B^{29} - 0.293B^{12} - 0.058B^{19} + 0.110B^{41})z_t$$
$$= -19.347 + (1 - 0.945B^7 + 0.029B^{29} - 0.884B^{12} + 0.835B^{19} - 0.026B^{41})a_t. \tag{6}$$

3.6 Artificial Neural Network

Neural networks are techniques to drive a data model that inspired from the function of brain and nervous system. Typically, neural network consists of an input layer, hidden layers, and an output layer. There are several neural network types: single layer perceptron, linear neuron, multilayer perceptron, competitive networks, self-organizing feature map, and recurrent network.

In this work, multilayer perceptron (MLP) was chosen as the most commonly used neural networks. Several MLP networks with different combination of hidden layers and neuron were generated and tested to obtain the best fitted network. The three layer neural network obtained the best network in iteration that is having 12 input nodes, 3 hidden nodes, and 1 output node.

4 Forecasting Result

The selected model was used to forecast the values of the 12 months. A plot for the whole data point is shown in Fig. 2.

5 Performance of Forecasting

The selected model was used to forecast the values up to 12 months ahead. To perform system validation, forecasting result for each method was compared with actual data to come up with the best univariate model for tuberculosis prediction. Five-year data (2002–2006) were retained as holdout samples to test the performance for each model. The difference between actual data and forecast resulted in error value which was used as the base of the performance error. The graphical view of the error is shown in Fig. 3. It can be observed that the largest error fluctuation is shown by linear regression, while the smallest fluctuation of actual and forecast is given by ARIMA. Furthermore, the calculation of error measure was applied to each method for evaluation and compare forecasting result among them.

Table 1 shows the performance comparison using mean absolute deviation (MAD), root mean square error (RMSE), mean absolute percentage error (MAPE), and Theil's U. The first three measures are categorized as stand-alone accuracy

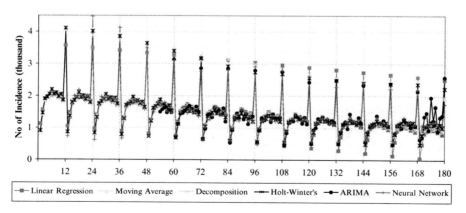

Fig. 2 Forecasting results

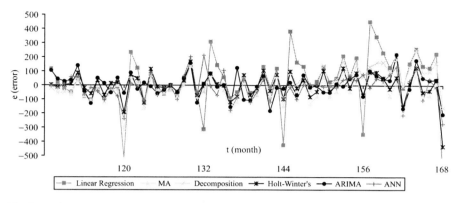

Fig. 3 Performance comparison of forecasting result in terms of error

Table 1 Forecasting measure error

Method	Stand alone measure			Relative measure
	MAD	RMSE	MAPE	Theil's U
Linear regression	142.694	197.557	13.140	0.195
Moving average	78.004	107.389	6.591	0.055
Decomposition	81.667	102.735	7.132	0.055
Holt-Winter's method	65.436	92.549	5.641	0.041
ARIMA	65.677	85.311	6.016	0.033
ANN	69.814	93.874	5.574	0.045

measure. Although these measures are commonly used in some forecasting studies, these have a disadvantage that could lead to certain losses of functions [17]. Therefore, the use of relative measure can solve this problem. Relative measures evaluate the performance of a forecast. Moreover, it can eliminate the bias from trends, seasonal components, and outliers [18].

Holt-Winter's model outperforms others method on MAD values with values 65.436 (Table 1). While, ANN achieves the smallest MAPE among them that giving score of 5.574%. Finally, ARIMA model is superior to both RMSE and Theil's U values. Overall best performance is achieved by ARIMA model, which gives 85.311 of RMSE and relative error on 0.033. Based on the Theil's U-statistics of value 0.033, ARIMA model is highly accurate and presents a close fit.

6 Conclusions

In this chapter, six forecasting methods were applied for seasonal disease forecasting in the case study for tuberculosis. Each method was able to produce accurate future number of tuberculosis incidence. Based on the data, 14 years with a downward trend, a seasonal forecasting method was applied to obtain the optimum result.

Among them, ARIMA model yielded the optimal in terms of simulation performance. Use of this model outperforms the other models with Theil's U value 0.033 that is close to zero. Since these models were developed only in one seasonal disease, further work is still needed to be applied in other disease and also in non-seasonal disease. The contribution made by this chapter is the comparative study of seasonal disease forecasting using univariate forecasting to find the most appropriate model. Finally, given time series data, the recommended model provides good forecast with the relatively minimum error.

References

1. Zeng, D., Chen, H., Tseng, C., Larson, C., Eidson, M., Gotham, I., Lynch, C., and Ascher, M., 'Sharing and visualizing infectious disease datasets using the WNV-BOT portal system', in *Proceedings of the 2004 Annual National Conference on Digital Government Research*, Seattle, 2004, pp. 1–2.
2. Deal, B., Farello, C., Lancaster, M., Kompare, T., and Hannon, B., 'A dynamic model of the spatial spread of an infectious disease: the case Of Fox Rabies in Illinois', *Environmental Modeling and Assessment*, vol. 5, pp. 47–62, 2000.
3. Jinping, L., Qianlu, R., Xi, C., and Jianqin, Y., 'Study on transmission model of avian influenza', in *Proceedings of the International Conference on Information Acquisition 2004*, China, 2004, pp. 54–58.
4. Pfeiffer, D. U. and Hugh-Jones, M., 'Geographical information system as a tool in epidemiological assessment and wildlife disease management', *Revue Scientifique et Technique de l'Office International des Epizooties*, vol. 21, pp. 91–102, 2002.
5. Garner, M. G., Hess, G. D., and Yang, X., 'An integrated modelling approach to assess the risk of wind-borne spread of foot-and-mouth disease virus from infected premises', *Environmental Modelling and Assessment*, vol. 11, pp. 195–207, 2005.
6. Hailu, A., Mudawi Musa, A., Royce, C., and Wasunna, M., 'Visceral leishmaniasis: new health tools are needed', *PLoS Medicine*, vol. 2, pp. 590–594, 2005.
7. Taylor, N., 'Review of the use of models in informing disease control policy development and adjustment', DEFRA, U.K. 26 May 2003.
8. Lees, V. W., 'Learning from outbreaks of bovine tuberculosis near Riding Mountain National Park: applications to a foreign animal disease outbreak', *The Canadian Veterinary Journal*, vol. 45, pp. 28–34, 2004.
9. Stott, A. (2006). Optimisation methods for assisting policy decisions on Endemic diseases [online]. Available at: http://www.sac.ac.uk/mainrep/pdfs/leewp_15_endemic_disease.pdf.
10. Debanne, S. M., Bielefeld, R. A., Cauthen, G. M., Daniel, T. M., and Rowland, D. Y., 'Multivariate Markovian modeling of tuberculosis: Forecast for the United States', *Emerging Infectious Diseases*, vol. 6, pp. 148–157, 2000.
11. Medina, D. C., Findley, S. E., and Doumbia, S., 'State–space forecasting of *Schistosoma haematobium* time-series in Niono, Mali', *PLoS Neglected Tropical Diseases*, vol. 2, pp. 1–12, 2008.
12. Lai, D., 'Monitoring the SARS epidemic in China: a time series analysis', *Journal of Data Science*, vol. 3, pp. 279–293, 2005.
13. Sebastiani, P., Mandl, K. D., Szolovits, P., Kohane, I. S., and Ramoni, M. F., 'A Bayesian dynamic model for influenza surveillance', *Statistics in Medicine*, vol. 25, pp. 1803–1825, 2006.
14. Chaves, L. F. and Pascual, M., 'Climate cycles and forecasts of cutaneous leishmaniasis, a nonstationary vector-borne disease', *PLoS Medicine*, vol. 3 (8), pp. 1320–1328, 2006.
15. Permanasari, A. E., Awang Rambli, D. R., and Dominic, P. D. D., 'Forecasting of zoonosis incidence in human using decomposition method of seasonal time series', in *Proceedings of the NPC 2009*, Tronoh, Malaysia, 2009, pp. 1–7.

16. Bowerman, B. L. and O'Connell, R. T., *Forecasting and Time Series An Applied Approach*, 3rd ed: Pacific Grove, CA: Duxbury Thomson Learning, 1993.
17. Chen, Z. and Yang, Y. (2004). Assessing forecast accuracy measures [Online]. Available at: http://www.stat.iastate.edu/preprint/articles/2004--10.pdf.
18. Permanasari, A. E., Awang Rambli, D. R., and Dominic, P. D. D., 'Prediction of zoonosis incidence in human using Seasonal Auto Regressive Integrated Moving Average (SARIMA)', *International Journal of Computer Science and Information Security (IJCSIS)*, vol. 5, pp. 103–110, 2009.

Chapter 18
Predicting Individual Affect of Health Interventions to Reduce HPV Prevalence

Courtney D. Corley, Rada Mihalcea, Armin R. Mikler, and Antonio P. Sanfilippo

Abstract Recently, human papilloma virus (HPV) has been implicated to cause several throat and oral cancers and HPV is established to cause most cervical cancers. A human papilloma virus vaccine has been proven successful to reduce infection incidence in FDA clinical trials, and it is currently available in the USA. Current intervention policy targets adolescent females for vaccination; however, the expansion of suggested guidelines may extend to other age groups and males as well. This research takes a first step towards automatically predicting personal beliefs, regarding health intervention, on the spread of disease. Using linguistic or statistical approaches, sentiment analysis determines a text's affective content. Self-reported HPV vaccination beliefs published in web and social media are analyzed for affect polarity and leveraged as knowledge inputs to epidemic models. With this in mind, we have developed a discrete-time model to facilitate predicting impact on the reduction of HPV prevalence due to arbitrary age- and gender-targeted vaccination schemes.

Keywords Computational epidemiology · Data mining · Epidemic models · Health informatics · Public health · Sentiment analysis

1 Introduction

This chapter addresses the public health need to assess the population-level disease prevalence impact due to arbitrary interventions, specifically, age- and gender-targeted human papilloma virus (HPV) vaccine schemes. In a January 2009 Science article, HPV is indicated to cause some throat and oral cancers and is well established to cause nearly all cervical cancers [19]. A HPV vaccine has been proven successful to reduce its prevalence in FDA clinical trials, and the vaccine is available

C.D. Corley (✉)
Pacific Northwest National Laboratory, Richland, WA, USA
e-mail: court@pnl.gov

in the USA. Current intervention policy targets pre-teen females for vaccination; however, the expansion of suggested guidelines may extend to other age groups and males as well. Predictive models are important tools in determining disease transmission dynamics and effective vaccination solutions.

With this in mind, we have developed a discrete-time model to facilitate predicting impact on the reduction of HPV prevalence from arbitrary age- and gender-targeted vaccination. Since HPV is highly virulent with more than 30 strains that are sexually transmittable, the age–gender discrete-time model distinguishes a population into different subgroups based on intimate mixing patterns. Population demographics and census data are analyzed to extract demographic parameters and youth, and adult risk behavior surveys are studied to determine the intimate partner contact rates [6, 7].

Recently, HPV has been implicated to cause several oral and throat cancers [19], and it is established that more than 90% of cervical cancers contain HPV DNA [11]. In the USA, 13,000 women are diagnosed with cervical dysplasia and 5,000 die annually, and by the age of 50, 80% of women will have acquired genital HPV infection. Currently, 20 million people are infected with HPV in the USA with 5.5 million new cases annually [16]. Due to the health care and human costs associated with this virus, it is vital to have known impact of vaccination strategies. We realize that 100% vaccination coverage is improbable; hence, vaccination solutions that will have the greatest reduction in the endemic prevalence of cancer-causing HPV types are needed.

2 Related Work

Most work in modeling infectious disease epidemics is mathematically inspired and based on differential equations and SIR/SEIR (Susceptible, Exposed, Infectious, and Recovered) model [2]. Differential equation SIR modeling relies on the assumption of constant population and neglects population demographics [3, 4]. They fail to consider individual interaction processes and assume homogeneous population. Both partial and ordinary differential equation models are deterministic in nature and neglect the stochastic or probabilistic behavior [17]. Nevertheless, these models have been shown to be effective in regions of small population [17]. Modeling sexually transmitted diseases with differential equations has been developed and incorporates sexual activity classes with broad population interaction [1, 9]. Markov models have been developed that are capable of simulating the natural history of HPV and type-specific stages of cervical carcinogenesis [10, 11]. Improved Markov models simulate high- and low-risk HPV infection. They are capable of simulating non-persistent and persistent HPV infections that lead to cervical carcinogenesis [13]. Cost-effectiveness analysis has been performed on the benefits of a HPV vaccine implementing decision and Markov models [16]. The differential equation and Markov models ignore specific demographics and approach modeling at population level [8, 12]. In addition, models have been developed to investigate ethnic inequalities in the incidence of sexually transmitted infections [18].

3 Predicting Individual Affect of Health Interventions

3.1 Data

Spinn3r (www.spinn3r.com) is a web and social media indexing service that conducts real-time indexing of all blogs, with a throughput of over 100,000 new blogs indexed per hour. Blog posts are accessed through an open source Java application-programming interface (API). Metadata available with this dataset include the following (if reported): blog title, blog url, post title, post url, date posted (accurate to seconds), description, full HTML encoded content, subject tags annotated by author, and language.

Data are selected from an arbitrary time period beginning at 1 August 2008 and ending at 28 February 2009 (approximately 130 million English language items). English language Weblog, micro-blog, and mainstream media items are extracted when a lexical match exists to characteristic keywords anywhere in its content (misspellings and synonyms are not included) from web and online social media published in the same time frame. Characteristic keywords include cervical cancer, HPV vaccination, Gardasil, and other keywords relevant to this task. Indexing, parsing, and link extraction code were written in Python, parallelized using pyMPI and executed on a cluster at the University of North Texas Center for Computational Epidemiology and Response Analysis. This compute resource has eight nodes (2.66 GHz Quad Core Xeon processors), 64 core, 256 GB memory, and 30 TB of network storage [14, 15].

This research takes a first step toward predicting the effect of personal beliefs on the spread of disease. One tool developed in computational linguistics is sentiment analysis. Using linguistic or statistical approaches, this tool determines attitude (sentiment) of a text's emotional content. Current state-of-the-art affect labeling methods misclassify one out of five sentiments, and off-the-shelf tools evaluate at 50–75% accuracy. We mine web and social media to retrieve blog posts mentioning vaccination against the HPV. HPV vaccination blog posts from August to September 2008 are manually labeled as objective or subjective (and corresponding polarity). A machine learning sentiment classifier is trained with the positive and negative polarity subjective posts. The following three blog posts demonstrate negative and positive subjective effect and an objective effect.

> **Negative**: I would like to point out that this is exactly the reason why I didn't get the cervical cancer vaccine. Well, that and the fact that it's no longer effective after you turn 25. My point is, I knew this was going to happen. And this was exactly why I was so against it being required for girls to attend school. Yes, let's make an extremely large number of people take a drug that hasn't been on the market for that long. Then let's watch a chunk of them have horrible effects. (Correlates to news story : 4 to 5 times greater risk of Anaphylactic Shock).

> **Positive**: Char! I jogged for another hour straight today! ... granted, I had some pie and ice cream earlier, so that factored in, but... I still did it! I'm training! FOR US! FOR THE MARATHON! :) In other news, I had my first shot in the round for the HPV vaccination today. One less! (Merck Advertising Slogan)
>
> **Objective**: In July, U.S. Citizenship and Immigration Services quietly amended its list of required vaccinations for immigrants applying to become citizens. One of the newest requirements Gardasil, which vaccinates against the HPV.

3.2 Automatic Sentiment Classification

We posit that the domain of training data effects the accuracy of automatic sentiment analysis. To prove this statement, a training dataset has been collected from HPV vaccination weblog items posted in August and September 2008. From the nearly 1,000 English language HPV blog posts, a subset is chosen that clearly demonstrates attitude polarity toward HPV vaccination. This subset is manually annotated with a positive or negative label, in total 93 positive and 64 negative posts. Significant news events that bloggers are discussing in their subjective posts include Gov. Rick Perry (R-TX) mandating vaccination for teenage girls in the state of Texas, notices of adverse side effects, a *New England Journal of Medicine* letter to the editor stating vaccinating women age 25–47 is not justified and announcement by the FDA that Merck's quadrivalent HPV vaccine Gardasil is approved for use in preventing vulvar/vaginal cancer.

To predict subjective polarity, a machine learning classifier is trained on the manually annotated training data. The training data are provided in [5]. A support vector machine (SVM) classifier is chosen to train the classifier, and with fivefold cross-validation it achieves 70% accuracy.[1] Then, 8,656 posts from 1 October 2008 to 28 February 2009 that mention HPV vaccination are labeled by the supervised ML classifier (see Fig. 1). The benefit of using domain-specific training data is significant; using the same SVM classifier trained with L. Lee's 1,800 movie reviews labeled as positive or negative, it achieves 48%. Many of the predicted sentiment magnitude shifts can be correlated to a specific event identified by news releases. A simple news search for stories containing HPV and vaccination keywords returns articles listed in [5]. The magnitude of vaccination beliefs is predicted and knowledge garnered from this analysis is made available as additional features to epidemic models and simulation frameworks (see Fig. 2).

---

[1] Naive Bayes (5-gram character smoothing) and Language-Model-based (8-gram) classifiers where also evaluated; however, they achieved a maximum 69% accuracy with domain-specific training.

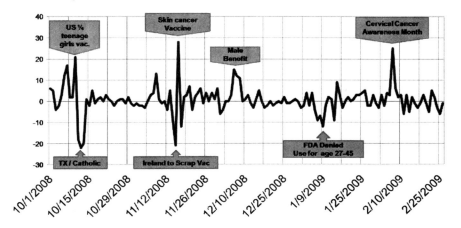

Fig. 1 Predicted human papilloma virus (HPV) vaccination sentiment polarity: 1 October 2008–27 February 2009

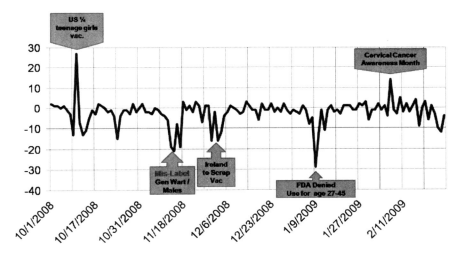

Fig. 2 Predicted Gardasil sentiment polarity: 1 October 2008–27 February 2009

4 Age–Gender Discrete-Time Model

We introduce a age–gender stratified discrete-time model for evaluating the effect of demographic-biased vaccination strategies. This model borrows from previous work in ordinary differential equation models by Hughes, Garnett, Anderson, and Koutsky which predicts the population-level impact of a pre-exposure HPV vaccine [8, 9, 12].

The model developed by Hughes et al. is a SIR-based compartment model that depicts endemic disease prevalence in a given population. Transmission and infection dynamics in this model occur within a sexually active population, and this population is a subset, determined by age, of the total population. Given that individuals generally do not have the same contact rate, the sexually active population is subdivided into disparate groups classified by their average contact rate with the population remaining constant throughout the execution of the model; however, a temporal flow is present with a population proportion aging-out and the same proportion aging-in continuously. The mixing dynamics in the sexually active population subset with constant population size incorporate the concept of homogeneous sexual mixing, in a population, with varying contact rates between individuals. Effective disease transmission is described by varying contact rates between individuals, disease characteristics, and sexual mixing; the homogeneous sexual mixing that occurs is further described by the proportion of contacts that take place within group (assortative) and conversely group-to-group (random). Now, endemic disease prevalence is determined at infection equilibrium, when every infectious individual causes one new infection. The AGe-model calculates the endemic prevalence of *HPV* by continually aging-in new susceptible individuals and aging-out a proportion of each sexual activity class iteratively (see Fig. 3). Each class maintains a constant population size, and the age range in each activity class is uniform. Due to stratification of the population by age, individuals who age-out of each demographic stratum must age-in to the next contiguous age strata. Individuals in the last age stratum exit the sexually active population. Once an individual exceeds the age modeled, they then exit the population from their current state.

5 Results and Discussion

To evaluate the population-level impact of a specific vaccination policy, we measure the relative reduction in endemic prevalence. Endemic prevalence with no vaccination coverage is the baseline to measure the impact of vaccination, and we define the relative reduction in prevalence (RRP, π_p) as change in prevalence due to a vaccine policy (p) compared with the baseline prevalence (θ_0). Our results show a reduction in prevalence within the RRP range of established models, and Hughes et al. cite an RRP of 0.68 with a range of 0.628–0.734; other models such as Sanders and Taira cite a RRP of 0.8 and above [12, 16]. Evaluations are performed in the AGe-model with targeted gender and age demographic settings; both experimental settings use the same disease and vaccine parameters and a demographic composition that corresponds to a particular county in the USA. In each model evaluation, the predicted endemic prevalence is measured and compared, and the potential impact on endemic prevalence by targeting vaccine coverage in certain subgroups is analyzed. For these experiments, vaccine efficacy is 90% and vaccine coverage for each schema is set to 70% and is chosen at this level to represent a median-case intervention coverage/acceptance rate.

18 Predicting Individual Affect of Health Interventions to Reduce HPV Prevalence

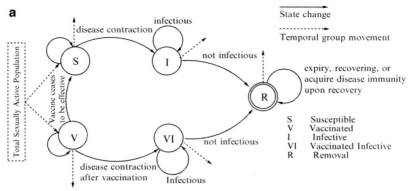

First age stratum (the first n subgroups): The population enters this portion of the model in either the susceptible or vaccinated compartment and a proportion(v) of each compartment will transfer to the next contiguous age stratum.

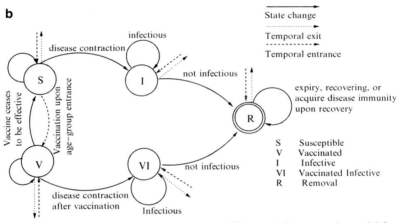

The remaining age strata (remaining $l - n$ subgroups): The population enters the model from the previous age-stratum in to their corresponding compartment and exits to the next contiguous age stratum. in to their corresponding compartment and exits to the next contiguous age stratum.

Fig. 3 AGe-model state diagrams

Although full vaccine coverage in the population would be desirable, costs associated with the vaccine make it infeasible. Table 1 demonstrates that vaccinating only females reduces the prevalence by 47% which is competitive to the impact on prevalence by vaccinating males and females aged 15–19. Clearly, these two policies will have the same impact in the population; however, vaccinating a particular age stratum is more cost-effective than blanketing an entire gender, and this is due to the difference in demographic population levels. Also, the baseline endemic prevalence varies in each experimental setting due to the disparate contact rates and demographic proportions.

The prevalence model evaluates an endemic prevalence of 0.046 and 0.049 for males and females, respectively, across all the counties (Table 1). The low variance

Table 1 Temporal model: endemic prevalence and relative reduction in prevalence per vaccine policy

		OCA	WDC	MFL	FGA	PIA	DTC
Human papilloma virus baseline prevalence (%)							
No vaccine	Males	0.046	0.046	0.046	0.046	0.046	0.046
	Females	0.049	0.049	0.049	0.049	0.049	0.049
	Total	0.048	0.048	0.048	0.048	0.048	0.048
Relative reduction in prevalence per vaccine policy (70% vaccine coverage per target demographic)							
Vaccinate M&F	Males	0.689	0.689	0.689	0.689	0.689	0.689
	Females	0.685	0.685	0.685	0.685	0.685	0.685
	Total	0.687	0.687	0.687	0.687	0.687	0.687
Vaccinate F	Males	0.277	0.277	0.277	0.277	0.277	0.277
	Females	0.465	0.465	0.465	0.465	0.465	0.465
	Total	0.373	0.373	0.371	0.373	0.373	0.373
Vaccinate 15–19 M&F	Males	0.465	0.466	0.464	0.465	0.465	0.465
	Females	0.461	0.462	0.460	0.461	0.461	0.461
	Total	0.463	0.464	0.462	0.463	0.463	0.463
Vaccinate 15–19 F	Males	0.181	0.181	0.180	0.181	0.180	0.181
	Females	0.322	0.322	0.321	0.322	0.321	0.321
	Total	0.253	0.253	0.252	0.253	0.252	0.253
Vaccinate 20–24 M&F	Males	0.367	0.366	0.367	0.367	0.367	0.367
	Females	0.363	0.362	0.363	0.363	0.363	0.363
	Total	0.365	0.364	0.365	0.365	0.365	0.365
Vaccinate 20–24 F	Males	0.138	0.138	0.138	0.138	0.138	0.138
	Females	0.250	0.250	0.250	0.250	0.250	0.250
	Total	0.195	0.195	0.195	0.195	0.195	0.195
Vaccinate 25–29 M&F	Males	0.174	0.174	0.175	0.174	0.175	0.174
	Females	0.172	0.172	0.172	0.172	0.172	0.172
	Total	0.173	0.173	0.173	0.173	0.173	0.173
Vaccinate 25–29 F	Males	0.063	0.063	0.063	0.063	0.063	0.063
	Females	0.123	0.122	0.123	0.122	0.123	0.123
	Total	0.093	0.093	0.094	0.093	0.094	0.094

in endemic prevalence for the evaluations is due to the homogeneous age-strata compositions in the selected counties and the low variance in contact rates per age stratum. The combined effects of these two factors limit the endemic prevalence variability in the temporal model. Measuring the effectiveness of vaccine coverage on a specific age group illustrates a beneficial age to begin vaccination. Table 1 also demonstrates that vaccinating females, aged 15–19 (25.5%), has lower impact than targeting all females in the population (37%); however, this is still a significant reduction in prevalence, and starting vaccination at an earlier age will have the greatest impact with a prevalence reduction compared to vaccinating at a later age. Both experimental settings demonstrated that female vaccination nearly doubled the reduction in prevalence compared to men.

6 Summary

Human papilloma virus, known as HPV, is known to cause nearly all cervical cancers as well as some throat and oral cancers. Primary intervention methods target pre-teen females for vaccination in an effort to reduce the incidence of HPV. Successfully marketing the vaccine will lead to dramatic reduction of the disease. Using linguistic of statistical approaches, sentiment analysis determines a text's affective content. This research automatically predicts personal beliefs, regarding health intervention, on the spread of disease. Moreover, we analyzed self-reported HPV vaccination beliefs published in web and social media for affect polarity and leverage as knowledge inputs to epidemic models and social network simulations. Subjective web and social media items were manually annotated and labeled as "positive" or "negative." The resulting data train an SVM, a machine learning classifier, and then automatically labels web and social media items. This enables the magnitude of vaccination beliefs to be predicted. The knowledge garnered from this analysis can then be made available to epidemic models and simulation frameworks to better inform public health actions.

Acknowledgments This work was partially supported by the Technosocial Predictive Analytics Initiative, part of the Laboratory Directed Research and Development Program at Pacific Northwest National Laboratory (PNNL). PNNL is operated by Battelle for DOE under contract DE-AC05-76RLO 1830.

References

1. Anderson, R., Garnett, G.: Mathematical models of the transmission and control of sexually transmitted diseases. Sexually Transmitted Diseases **27**(10), 636–643 (2000)
2. Bagni, R., Berchi, R., Cariello, P.: A comparison of simulation models applied to epidemics. Journal of Artificial Societies and Social Simulation **5**(3) (2002)
3. Boccara, N., Cheong, K.: Critical behavior of a probabilistic automata network sis model for the spread of an infectious disease in a population of moving individuals. Journal of Physics A : Mathematical and General **26**(5), 3707–3717 (1993)
4. Boccara, N., Cheong, K., Oram, M.: A probabilistic automata network epidemic model with births and deaths exhibiting cyclic behavior. Journal of Physics A : Mathematical and General **27**, 1585–1597 (1994)
5. Corley, C.D.: Mining social media and social network simulation to advance epidemiology. PhD Dissertation, University of North Texas (2009)
6. Douglas, K., Collins, J., Warren, C., Kann, L., Gold, R., Clayton, S., Ross, J., Kolbe, L.: Youth risk behavior surveillance: National college health risk behavior survey–united states, 1995. MMWR CDC Surveillance Summaries : Morbidity and Mortality Weekly Report. CDC Surveillance Summaries / Centers for Disease Control **46**(6), 1–56 (1997)
7. Eaton, D.K., Kann, L., Kinchen, S., Shanklin, S., Ross, J., Hawkins, J., Harris, W.A., Lowry, R., McManus, T., Chyen, D., Lim, C., Brener, N.D., Wechsler, H., for Disease Control, C., (CDC), P.: Youth risk behavior surveillance–united states, 2007. MMWR Surveillance Summaries : Morbidity and Mortality Weekly Report. Surveillance Summaries / CDC **57**(4), 1–131 (2008)
8. Garnett, G.: The geographical and temporal evolution of sexually transmitted disease epidemics. Sexually Transmitted Infections **78**(Suppl I) (2002)

9. Garnett, G., Anderson, R.: Contact tracing and the estimation of sexual mixing patterns: The epidemiology of gonococcal infections. Sexually Transmitted Diseases **20**(4), 181–191 (1993)
10. Goldie, S., Grima, D., Kohli, M., Wright, T., Weinstein, M., Franco, E.: A comprehensive natural history model of hpv infection and cervical cancer to estimate the clinical impact of a prophylactic hpv-16/18 vaccine. International Journal of Cancer **106**, 896–904 (2003)
11. Goldie, S., Kohli, M., Grima, D.: Projected clinical benefits and cost-effectiveness of a humanpapillomavirus 16/18 vaccine. Journal of the National Cancer Institute **96**(8), 604–615 (2004)
12. Hughes, J., Garnett, G., Koutsky, L.: The theoretical population-level impact of a phrophylactic human papilloma virus vaccine. Epidemiology **13**(6), 631–639 (2002)
13. Kulasingam, S., Myers, E.: Potentital health and economic impact of adding a human papillomvirus vaccine to screening programs. Journal of the American Medical Association **290**(6), 781–789 (2003)
14. Miller, P.: pyMPI–an introduction to parallel python using MPI. Livermore National Laboratories (2002). URL https://computing.llnl.gov/code/pdf/pyMPI.pdf
15. Rossum, G.V., Drake, F.: Python language reference. Network Theory Ltd (2003). URL http://www.altaway.com/resources/python/reference.pdf
16. Sanders, G., Taira, A.: Cost effectiveness of a potential vaccine for human papillomavirus. Emerging Infectious Diseases **9**(1), 37–48 (2003)
17. Stefano, D., Fukś, H., Lawniczak, A.: Object-oriented implementation of CA/LGCA modeling applied to the spread of epidemics. In: Proceedings of Canadian Conference on Electrical and Computer Engineering, vol. 1, pp. 26–31 (2000)
18. Turner, K., Garnett, G., Steme, J., Low, N.: Investigating ethnic inequalities in the incidence of sexually transmitted infections: Mathematical modelling study. Sexually Transmitted Infections **80**, 379–385 (2004)
19. Zelkowitz, R.: Cancer. HPV casts a wider shadow. Science **323**(5914), 580–1 (2009). DOI 10.1126/science.323.5914.580

Chapter 19
Decision Tree and Ensemble Learning Algorithms with Their Applications in Bioinformatics

Dongsheng Che, Qi Liu, Khaled Rasheed, and Xiuping Tao

Abstract Machine learning approaches have wide applications in bioinformatics, and decision tree is one of the successful approaches applied in this field. In this chapter, we briefly review decision tree and related ensemble algorithms and show the successful applications of such approaches on solving biological problems. We hope that by learning the algorithms of decision trees and ensemble classifiers, biologists can get the basic ideas of how machine learning algorithms work. On the other hand, by being exposed to the applications of decision trees and ensemble algorithms in bioinformatics, computer scientists can get better ideas of which bioinformatics topics they may work on in their future research directions. We aim to provide a platform to bridge the gap between biologists and computer scientists.

1 Introduction

In the past decade, high-throughput biological technologies have lead to an explosive growth of biological data, including thousands of fully sequenced genomes deposited in the NCBI website, genome-wide microarray expression data deposited in microarray databases, and RNA and protein structures deposited in structure databases, such as Rfam and PDB. The huge amount of accumulated biological data needs to be mined and interpreted to answer fundamental biological questions. For example, we can find out what makes human beings different from chimpanzees by comparing their genomic sequences, or learn how to distinguish healthy people from patients by comparing their cDNA expression profiles.

While it is a challenging task to mine this kind of data, it is rewarding and will probably lead to significant discoveries. Data mining and machine learning techniques have been successfully applied in various areas of bioinformatics, including

---

D. Che (✉)
Department of Computer Science, East Stroudsburg University, East Stroudsburg, PA 18301, USA
e-mail: dche@po-box.esu.edu; dongshengche@gmail.com

H.R. Arabnia and Q.-N. Tran (eds.), *Software Tools and Algorithms for Biological Systems*, Advances in Experimental Medicine and Biology 696,
DOI 10.1007/978-1-4419-7046-6_19, © Springer Science+Business Media, LLC 2011

genomics, proteomics, and system biology. Some supervised approaches are used for classification (i.e., dividing the whole dataset into different classes). The classification algorithms include decision trees, linear discriminant analysis, nave Bayesian classifier, nearest neighbor, neural network, and support vector machines (SVMs). While some other un-supervised approaches are used for clustering, and such clustering approaches include partition clustering (e.g., the K-means algorithm), hierarchical clustering, and bi-clustering. Detailed algorithms and their application in bioinformatics can be found elsewhere [1, 2, 11].

In this chapter, we only focus on the topic of decision tree learning algorithms and ensemble classifiers, and their applications in bioinformatics. Decision tree approaches and ensembles of decision trees have been shown to have wide applications with high performance in solving bioinformatics problems. We hope that this brief review will be useful for both biologists and computational scientists.

2 Decision Trees

A decision tree classification model is represented by a tree-like structure, where each internal node represents a test of a feature, with each branch representing one of the possible test results and each leaf node represents a classification. Depending on which construction algorithms are applied, decision tree models may vary. To understand how decision trees are constructed, we briefly review two of the most common decision tree algorithms, ID3 [14] and Classification and Regression Tree (CART) [3].

2.1 ID3

The ID3 algorithm [6] is the most commonly used decision tree learning approach. It uses a top-down greedy search scheme to search through all possible tree spaces. The algorithm starts with the whole training set and chooses the best feature as the root node. It then splits the set into subsets with the same feature value. If all instances in a subset have the same classification, then the process stops for that branch, and the algorithm returns a leaf node with that classification. If the subset does contain multiple classifications, and there are no more features to test, the algorithm returns a leaf node with the most frequent classification. Otherwise, the algorithm will recursively call itself with the subsets of data.

The key of this algorithm is to find the best feature. One of the popular feature measures is information gain (IG) as follows:

$$\text{IG}(S,A) = E(S) - \sum_{v \in \text{Value}(A)} \frac{|S_v|}{S} E(S_v) \qquad (1)$$

where $E(S)$ is the entropy of training set S, which in turn is defined as:

$$E(S) = -\sum_{i=1}^{C} p_i \log_2 p_i \qquad (2)$$

where p_i is the fraction of class i in all C classes in the whole set S. $Value(A)$ is the set of all possible values for the feature A. S_v is the subset of S for which feature A has the value of v (i.e., $S_v = \{s \in S | A(s) = v\}$).

One main issue with the trees generated by ID3 is the over-fitting problem. The C4.5 algorithm [15] addresses this issue using tree pruning techniques to prune the tree generated by ID3. Basically, C4.5 starts with generating if–then rules for each path from the root to a leaf node of the tree. It then checks the conditions of each rule to see whether removing conditions will actually increase the classification accuracy. If it is the case, the conditions will be removed. The pruned rules are finally sorted based on their classification accuracy and applied in order.

2.2 Classification and Regression Tree

CART [3] is another decision tree learning algorithm. It can be used for both classification and regression, depending on the available information of the dataset. Classification trees are used when we know the class of each instance. The construction of classification trees using CART is similar to that with ID3, with the exception of the information measure. In CART, Gini impurity (G), rather than the entropy, is used as the information measure. G is defined as follows:

$$G(S) = 1 - \sum_{i=1}^{C} (p_i)^2 \qquad (3)$$

In CART, regression trees are used when the instances have continuous target attribute values instead of classes. The selection of the feature is based on the sum of square errors, and the best feature is the one with minimal sum-of-square error.

3 Ensemble Learning

An ensemble classifier is a classifier that combines multiple base classifiers for final classification. Each base classifier can be any kind of supervised classification such as decision trees, neural networks, or SVMs. Recent studies have shown that, in general, ensemble classifiers outperform single classifiers, and there are wide applications of ensemble classification in the bioinformatics field. Here, we briefly review three main categories of ensemble learning algorithms.

3.1 Bagging

Bootstrap aggregating [4], also known as bagging, is one of the earliest ensemble algorithms. Each base classifier (a decision tree or any other classifier) in the ensemble is trained on a subset of the initial training set. The training set of each base classifier is sampled by bootstrap sampling, i.e., randomly selecting a subset of the given dataset with replacement. The classification of a new instance is based on the simple majority voting scheme, i.e., each classifier gives a classification for the instance, and the class that has the maximal number of votes cast by the base classifiers is the final classification. While bagging is one of the most intuitive and simple ensemble classifiers, it is interesting to note that it performed very well in many studies (see Sect. 4).

Random forests [5] is a variant of bagging algorithms whose base classifiers are decision trees. Like bagging, random forests uses the bootstrap sampling and uses un-weighted aggregation of committees for the final classification. However, the construction process of decision trees in the random forest approach is different. Typically, m features out of all M features are randomly selected, and the optimal value of m is usually the square root of M. In addition, each tree in the random forest is fully grown and not pruned. Random forests tend to perform very well, especially for those datasets containing many features.

3.2 Boosting

The boosting algorithm does not create each base classifier independently. Instead, the classifiers are created sequentially where the next base classifier assigns more weights to the mistakes that the previous classifier made. Adaptive boosting (AdaBoost) [9] is one of the most popular boosting algorithms. AdaBoost uses an arbitrary number of base classifiers, with the data sampling and weighted voting scheme. The algorithm starts by building the first base classifier, which is trained on the dataset with equal weights. For the construction of subsequent classifiers, the instances misclassified by the previous classifier are assigned higher weights, while the weights of the instances that are correctly classified remain the same. The weights of all instances in the whole dataset are then normalized and used for sampling for the next classifier. The final classification is based on weighted base classifiers.

3.3 Stacked Generalization

Wolpert [7] proposed the technique called "stacked generalization," or simply "stacking," where the outputs of the base classifiers are the inputs for the second level of classifiers. The purpose of stacking is to estimate and correct the biases of the base classifiers. Thus, the second level classifier may not use the training data that are used for training base classifiers. Stacking is used to learn how the base learners make errors, and the second level classifier tries to overcome the errors.

The major differences between the above-mentioned ensemble algorithms are summarized in Table 1. We also provide a list of some software packages and programs (Table 2), where decision tree algorithms and ensemble learning algorithms are provided and can be used for biologists to classify their biological data.

Table 1 Comparison among ensemble classifiers

Ensemble	Features	Sampling	Voting
AdaBoost	All features	Misclassified data are assigned high weights based on the previous classifier	Weighted majority
Bagging	All features	Random sampling with replacement	Un-weighted majority
Random forest	Partial features	Random sampling with replacement	Un-weighted majority
Stacking	All features	Random sampling with replacement	N/A. Outputs of base classifiers are inputs for the second level of classifier

Table 2 A list of selected tools that contain decision trees and ensemble classifiers

Tools	Content
C5.0/See5	C5.0 is descended from C4.5 and it is used for Unix machines. See5 is a corresponding Windows version (http://www.rulequest.com/see5-info.html)
icsiboost	icsiboost is an open source implementation of AdaBoost over stumps (one-level decision trees) on discrete and continuous attributes (http://code.google.com/p/icsiboost/)
MLC++	MLC++ is a library of C++ classes that contains ID3, C4.5, and OC1 decision tree algorithms. It also contains ensembles such as bagging (http://www.sgi.com/tech/mlc/index.html)
Orange	Orange is machine learning toolbox that contains C4.5 and random forest, and bioinformatics and text mining tools (http://www.ailab.si/orange/)
randomForest	The original random forest was implementation by Leo Breiman and coded in Fortran (http://www.stat.berkeley.edu/ breiman/RandomForests/). The R implementation can be found in http://cran.r-project.org/web/packages/randomForest/index.html
RapidMiner	RapidMiner is an open source toolkit and contains decision tree classifiers and ensembles. RapidMiner is suited for data related to genotyping, proteomics, and mass spectrometry (http://sourceforge.net/projects/yale/)
TMVA	TMVA contains boosting (e.g., AdaBoost), Bagging, random forests (http://tmva.sourceforge.net/)
TunedIT	TunedIT is an integrated system for sharing, evaluation, and comparison of machine learning and data mining algorithms (http://tunedit.org/)
WEKA	Weka is a machine learning package that contains most of decision tree algorithms such as ID3, J48, and ensemble classifiers including AdaBoost, bagging, random fores, and stacking (http://www.cs.waikato.ac.nz/ml/weka)

4 Biological Applications of Decision Tree Methods and Their Ensemble Classifiers

4.1 Cancer Classification

Decision tree approaches and the ensembles of decision tree-based classifiers have been widely applied in cancer classification, including breast cancer, central nervous system embryonic tumor, colon tumor, leukemia, lung cancer, ovarian cancer, pancreatic cancer, and prostate cancer (Table 3). Two major types of data are used

Table 3 A list of cancer classification studies using decision trees and ensemble tools

Group	Cancers	Feature methods used
Tan and Gilbert [13]	Lymphoblastic leukemia, myeloid leukemia, breast cancer, central nervous system embryonal tumor, colon tumor, lung cancer, prostate cancer	Microarray expression data were used. C4.5, bagging and AdaBoost decision trees were tested on datasets. WEKA package was used to perform tests. Bagging constantly performed better than boosting and single C4.5.
Ge and Wong [10]	Pancreatic cancer	Mass-spectrometry data were used. A decision tree algorithm C4.5, and six decision-tree based classifier ensembles (Random forest, Stacked generalization, Bagging, Adaboost, Logitboost and Multiboost) from WEKA were applied on the dataset. Ensemble classifiers outperformed a single decision tree classifier.
Daz-Uriarte and Andrs [8]	Leukemia, breast, NCI 60, adenocarcinoma brain, colon, lymphoma, prostate, Srbct	Microarray expression data were used. Four methods were used, including: random forest, diagonal linear discriminant analysis, K nearest neighbor, and support vector machines. All methods were carried out with R packages. Random forest had the best performance.
Vlahou et al. [19]	Ovarian cancer	Mass spectral data were used. CART was applied with prediction accuracy of 80%.
Qu et al. [13]	Prostate cancer	Mass spectral serum profiles were used. The AdaBoost and Boosted decision stump classfiers were used. Both classifiers had high prediction accuracy.
Wu et al. [20]	Ovarian cancer	Mass spectral data were used. Linear discriminant analysis, quadratic discriminant analysis, k-nearest neighbor classifier, bagging and boosting classification trees, support vector machine, and random forest were used. Random forest outperformed other methods.

for cancer classification, microarray expression data, and mass spectral data. For the microarray expression data, each instance contains thousands of expression values of mRNA. Out of thousands of features, many are actually noisy and thus affect the classification accuracy if they are not filtered out. Thus, the preprocessing step of feature selection (or gene selection) must be carried out before the actual model construction.

Many feature selection methods have been used in cancer classification. Students t test is one of the standard methods for feature selection, and it was used by Ge and Wong [10] and Wu et al. [20]. Another metric, Wilcoxon rank test, similar to Students t test, with the exception of no distribution assumption, was also used in [10]. Stanikov et al. [17] implemented multiple filtering approaches, including backward elimination based on univariate ranking of genes with signal-to-noise ratio and Kruskal–Wallis, random forest-based backward elimination procedure, and SVM-based recursive feature elimination. The detailed techniques of such filtering can be found in [21]. Interestingly, the genetic algorithm approach was also explored for feature selection by Ge and Wong [10].

In general, ensemble learning classifiers outperform single tree classifiers in terms of classification accuracy, and random forests were reported to be the best classifiers as shown in Table 3. There are a few studies showing that SVMs was the best classifiers for cancer classification [17]. The discussion of which approach is truly the best for cancer classification is beyond the scope of this chapter, but we are aware that the use of different datasets, different preprocessing steps, and different performance metrics may lead to different conclusions.

4.2 Genomics Classification

The high-throughput sequencing technology has resulted in thousands of fully sequenced genomes. The wealth of genomic sequence data has provided computational biologists with great opportunities to annotate genomes at different levels, including but not limited to:

- Identifying protein coding regions (or genes) in the genome sequences
- Identifying non-coding RNA genes in the genomic sequences
- Detecting operon structures in the genomic sequences of prokaryotes
- Detecting genomic islands of the genomic sequences of prokaryotes
- Analyzing gene regulation
- Studying genomic structure organization

Classification trees and ensemble learning methods have played important roles in annotating multilevel genomic sequences. For one-dimensional genome annotation, Salzberg et al. [16] used OC1 decision trees for finding genes in DNA sequences. Che et al. [6] developed a J48 decision tree classifier, OperonDT, for operon structure prediction. Beyond one-dimensional genome annotation, computational biologists also study gene–gene interactions using machine learning methods. For example, Middendorf et al. [12] studied gene regulation (i.e., upregulation or downregulation) based on the motif patterns in the upstream regions of the gene, as well as the expression levels of regulators. They built alternating decision trees (ADT) as base classifiers by integrating the features from these two data sources and then built AdaBoost using ADTs.

5 Concluding Remarks

In this chapter, we have reviewed two decision tree algorithms: ID3 and CART. We have also summarized three general ensemble learning algorithms: bagging, boosting, and stacked generalization. A list of related software packages are included for reference. We have also shown the applications of such methods in cancer classification and genomics classification. We hope that this short review can be helpful for both biologists and computational scientists for their future research.

References

1. Baldi, P. and Brunak, S. (2001) Bioinformatics: The Machine Learning Approach (Adaptive Computation and Machine Learning), Second Edition. MIT, Cambridge, MA
2. Bhaskar, H., Hoyle, D.C. and Singh, S. (2006) Machine learning in bioinformatics: A brief survey and recommendations for practitioners, Computers in Biology and Medicine, 36, 1104–1125
3. Breiman, L., Friedman, J., Stone, C. and Olshen, R.A. (1984) Classification and Regression Trees. Chapman & Hall/CRC, New York, NY
4. Brieman, L. (1996) Bagging predictors, Machine Learning, 24, 123–140
5. Brieman, L. (2001) Random forests, Machine Learning, 45, 5–32
6. Che, D., Zhao, J., Cai, L. and Xu, Y. (2007) Operon prediction in microbial genomes using decision tree approach. In Proceedings of CIBCB. Honolulu, 135–142
7. David, H.W. (1992) Stacked generalization, Neural Networks, 5, 241–259
8. Diaz-Uriarte, R. and Alvarez de Andres, S. (2006) Gene selection and classification of microarray data using random forest, BMC Bioinformatics, 7, 3
9. Freund, Y. and Schapire, R. (1995) A decision-theoretic generalization of on-line learning and an application to boosting. In Proceedings of the Second European Conference on Computational Learning Theory. Springer, Berlin, 23–37
10. Ge, G. and Wong, G.W. (2008) Classification of premalignant pancreatic cancer mass-spectrometry data using decision tree ensembles, BMC Bioinformatics, 9, 275
11. Larranaga, P., Calvo, B., Santana, R., Bielza, C., Galdiano, J., Inza, I., Lozano, J.A., Armananzas, R., Santafe, G., Perez, A. and Robles, V. (2006) Machine learning in bioinformatics, Briefings in Bioinformatics, 7, 86–112
12. Middendorf, M., Kundaje, A., Wiggins, C., Freund, Y. and Leslie, C. (2004) Predicting genetic regulatory response using classification, Bioinformatics, 20 Suppl 1, i232–240
13. Qu, Y., Adam, B.L., Yasui, Y., Ward, M.D., Cazares, L.H., Schellhammer, P.F., Feng, Z., Semmes, O.J. and Wright, G.L., Jr. (2002) Boosted decision tree analysis of surface-enhanced laser desorption/ionization mass spectral serum profiles discriminates prostate cancer from noncancer patients, Clinical Chemistry, 48, 1835–1843

14. Quinlan, J.R. (1986) Induction of decision trees, Machine Learning, 1, 81–106
15. Quinlan, J.R. (1993) C4.5: Programs for Machine Learning. Morgan Kaufmann Publishers, San Mateo, CA
16. Salzberg, S., Delcher, A.L., Fasman, K.H. and Henderson, J. (1998) A decision tree system for finding genes in DNA, Journal of Computational Biology, 5, 667–680
17. Statnikov, A., Wang, L. and Aliferis, C.F. (2008) A comprehensive comparison of random forests and support vector machines for microarray-based cancer classification, BMC Bioinformatics, 9, 319
13. Tan, A.C. and Gilbert, D. (2003) Ensemble machine learning on gene expression data for cancer classification, Applied Bioinformatics, 2, S75–83
19. Vlahou, A., Schorge, J.O., Gregory, B.W. and Coleman, R.L. (2003) Diagnosis of ovarian cancer using decision tree classification of mass spectral data, Journal of Biomedicine and Biotechnology, 2003, 308–314
20. Wu, B., Abbott, T., Fishman, D., McMurray, W., Mor, G., Stone, K., Ward, D., Williams, K. and Zhao, H. (2003) Comparison of statistical methods for classification of ovarian cancer using mass spectrometry data, Bioinformatics, 19, 1636–1643

Chapter 20
Pattern Recognition of Surface EMG Biological Signals by Means of Hilbert Spectrum and Fuzzy Clustering

Ruben-Dario Pinzon-Morales, Katherine-Andrea Baquero-Duarte, Alvaro-Angel Orozco-Gutierrez, and Victor-Hugo Grisales-Palacio

Abstract A novel method for hand movement pattern recognition from electromyography (EMG) biological signals is proposed. These signals are recorded by a three-channel data acquisition system using surface electrodes placed over the forearm, and then processed to recognize five hand movements: opening, closing, supination, flexion, and extension. Such method combines the Hilbert–Huang analysis with a fuzzy clustering classifier. A set of metrics, calculated from the time contour of the Hilbert Spectrum, is used to compute a discriminating three-dimensional feature space. The classification task in this feature-space is accomplished by a two-stage procedure where training cases are initially clustered with a fuzzy algorithm, and test cases are then classified applying a nearest-prototype rule. Empirical analysis of the proposed method reveals an average accuracy rate of 96% in the recognition of surface EMG signals.

1 Introduction

Surface electromyographic (SEMG) signal processing has become an important tool in fields such as biological pattern recognition, prosthetic control, clinical diagnosis, and rehabilitation. These signals can be monitored noninvasively using electrodes attached to human skin. They represent the neuromuscular activity from the temporal and spatial aggregation of independent motor unit action potentials (MUAPs) in the muscle [1]. Assiduous efforts have been made in recent years to improve classification accuracy of SEMG signals and to reduce computational complexity of this kind of recognition systems, making them suitable for implementation in real-time applications [3, 6, 9]. Similar to other biological signals, SEMG signals have nonstationary behavior [1] due to the following: metabolic accumulations that

R.-D. Pinzon-Morales (✉)
Faculty of Engineering, Research Group on Control and Instrumentation,
Universidad Tecnologica de Pereira, Pereira, Colombia
e-mail: rdpinzonm@utp.edu.co

produces muscular fatigue, changes in the impedance of the skin-electrode interface, or changes in the MUAP over the time or others conditions. In order to extract discriminative features associated with this kind of behavior, several mathematical transformations have been proposed in the literature. Zardoshti et al. [12], for instance, extracted some time-domain statistics of the SEMG signals such as the integral of the absolute value, and Willison amplitude for movement control of upper limb prosthesis. In some cases, assumptions about local quasi-stationary behavior can be made and the short-time Fourier transform (STFT) can be used to extract discriminative features [4]. However, the uncertainty principle of Heisenberg limits the time-frequency resolution of the STFT, and the dynamics of the SEMG signals cannot be completely described [4]. In this sense, considerable attention has been drawn to Wavelet transform due to its flexibility in time-frequency resolution [11]. The main drawback of this tool lies in the a priori selection of some parameters, namely the choice of the Wavelet function and the decomposition level. In this chapter, a recent technique, suitable for nonstationary and nonlinear signal analysis, called Hilbert–Huang [7] is used. This analysis is based on the empirical mode decomposition (EMD) and the Hilbert Transform (HT). The use of both techniques allows the extractions of the embedded oscillations occurring in the SEMG signals, based on the signal itself, and also the potential extraction of discriminative features [1].

The second part of this pattern recognition paper is related to the classifier. Different supervised techniques have been used in SEMG classification; these include Bayesian classifier, Fisher classifier, Support Vector Machines, etc. However, supervised classification is not always possible due to unlabeled databases or lack of a specialist. Here a fuzzy logic-based unsupervised classifier is used with two main advantages: first, it allows the user to understand the results more easily [8], and second, a labeled database for training is not required. Previous studies have shown encouraging results in EMG signal unsupervised classification [8, 10]. This chapter is organized as follows: in Sect. 2 methods are reviewed. In Sect. 3, the proposed system is introduced in detail. In Sect. 4, results are shown and performance is measured by confusion matrix. Finally, in Sect. 5 conclusions are drawn.

2 Methods

2.1 Hilbert–Huang Analysis

Hilbert–Huang analysis is a novel time-energy-frequency representation of nonlinear and nonstationary signals, which has proved to be a powerful tool in digital signal processing [1,7]. It involves two steps, namely, the EMD, and the construction of a time-frequency space called Hilbert Spectrum (HS), by means of the HT. The purpose of EMD is to decompose a signal into a finite set of well-behaved HT signals, called intrinsic mode functions (IMF), which are nearly band-limited signals. In order to define an orthogonal basis and a meaningful instantaneous frequency,

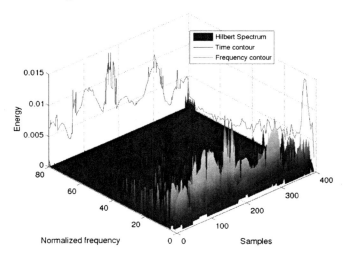

Fig. 1 Hilbert Spectrum (HS). Time contour (*left-hand side plane*), frequency contour (*backplane*). Contours are computed to extract discriminating information from the HS

the IMFs must satisfy two conditions [7]: (1) in the whole dataset, the number of extremes and the number of zero crossings must be either equal or differ by one at most; and (2) at any point, the mean value of the envelope defined by the local maximum and the envelope defined by the local minimum must be zero. The numerical procedure to obtain those IMFs is known as sifting process [7], whose main feature is the adaptive decomposition of the signal exclusively based on the signal's information. Once the IMFs are extracted from the input signal by the sifting process, it is possible to generate the HS from the instantaneous amplitude (IA) and instantaneous frequency (IF) of each IMF, by means of the HT. The HS represents both the variation of frequency and the energy of IMFs over time [7]. A Hilbert Spectrum of a SEMG signal with the time contour and the frequency contour is shown in Fig. 1.

2.2 Fuzzy Clustering Technique

Clustering techniques are mostly unsupervised methods that can be used to organize data into groups based on the similarities among the different data items [2]. The primary task of fuzzy clustering is to split a set of patterns into clusters by means of a performance measure, so that the degree of similarity is higher for data within the same cluster and lower for data in different clusters. For this purpose, the fuzzy clustering Gustafson–Kessel (GK) algorithm [5], which is an extension of the fuzzy c-means where the adaptive distance measure allows the detection of hyperellipsoidal clusters when minimizing the objective function, has been considered. The distance between the data points to be classified and the cluster prototypes (centers) is a Mahalanobis-type metric related to the covariance of the data with respect to each cluster.

The GK algorithm is an iterative process in which the center of clusters **V**, the fuzzy partition matrix **U**, the norm-inducing matrix A_i, and the fuzzy covariance matrix **F** are updated alternatively depending on the behavior of the clusters with the dataset **Z**, until a minimum tolerance is obtained; this process is shown in [2, 5], among others. The structure of the fuzzy covariance matrix **F** contains information about the shape and orientation of each one of the clusters [2], the ratio of the lengths of the clusters axes is given by the ratio of the square roots of the eigenvalues of **F**, and the direction of the axes is given by the eigenvectors of **F**. The following initial parameters must be specified: the number of clusters c, the fuzziness exponent m normally set at 2, the termination tolerance e and cluster volumes ri, simplified at 1 for each cluster. The resulting algorithm may find in general only suboptimal solutions in the convergence, and for this reason it is typically required to run the algorithm several times.

3 The SEMG Recognition System

The SEMG recognition system can be summarized as shown in Fig. 2. It is composed of five steps: first, SEMG signals are acquired from three surface electrodes placed over the forearm in a three-phase configuration. Second, signals collected on each channel are segmented and normalized in time with a 200 ms-width window. Third, the HS is computed for each channel and discriminating features are calculated to build the three-dimensional feature space. Finally, a classifier based on fuzzy clustering is used to recognize the right kind of movement.

3.1 Raw SEMG Preprocessing

Although normal resting muscles show almost no change in their SEMG signals, when a SEMG from a contracting muscle is acquired it shows significant changes in potential. The firing point of the signal is detected by means of energy

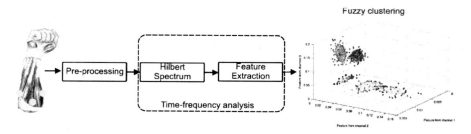

Fig. 2 General structure of the proposed hand movement recognition system

changes. In this study, a firing point detector based on energy changes, similar to the modified integrated EMG (IEMG) method [6], is used. Once the firing point is selected, the signals are segmented up to 200 ms from this point, which is long enough to recognize movements from SEMG signals [4].

3.2 Hilbert Spectrum Generation

After the segmentation, the HS is computed for each channel. First, EMD is used to extract up to seven IMFs, which contain most of the frequencies of interest of the SEMG signal from 5 to 500 Hz. Then, HT is computed for each IMF to obtain the IF and IA, and also the time-frequency plane, the HS.

3.3 Feature Extraction

Out of the time contour computed from the HS, a three-dimensional feature space, with the most discriminating features selected from a library of measures, is constructed. Time contours can be obtained by taking the mean value or the maximum value along the frequency axis of the HS. In this case, the mean value is used. The library of measures used are: the mean value (MV), the standard deviation (SD), the variance (VA), the energy (EN), the Shannon's entropy (SE), the skewness (Sk), the kurtosis (KU), and the mean of the absolute values (MA). These metrics are computed over sliding windows of 40 samples with an overlapping percentage of 50% and then the mean value is taken to obtain just a set of eight features. Once the initial feature space has been extracted from the HS, a set of 24 features (eight for each channel) is obtained.

The most discrimination space (MDS) was selected by trial-and-error and by the highest classification accuracy achieved using a simple Bayesian as a classifier, so that the obtained classification rates will not be enhanced by the fuzzy clustering approach. As a result, the three-dimensional space constructed by the MA of channel 1, the MA of channel 2, and the MA of channel 3 was identified as the MDS. The projection of the movements over the feature space is shown in Fig. 3a.

3.4 Fuzzy Clustering Classification

The three-dimensional space of characteristics is shown in Fig. 3a, where each class could be represented by a center of cluster, and the dispersion of points seems to have an ellipsoidal shape. Because of this, it is useful to use a classifier based on the GK algorithm, as described in Sect. 2.2, which manages prototypes of clusters

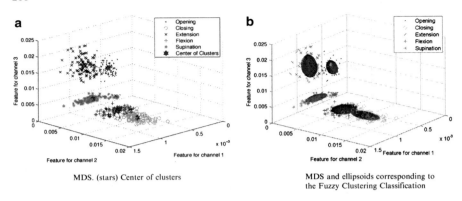

Fig. 3 Projection of the hand movement patterns over the feature space

with ellipsoidal dispersion [6] as appropriate. Two processes can be defined at this point: (1) training by unsupervised clustering of the data by GK fuzzy clustering, and (2) validation of the classifiers decision with the reference classes. The training task begins with the calculation of the reference centers defined by the mean of the labeled data per class. The input parameters of the GK clustering are defined as follows: number of clusters 5 (number of hand movements), fuzziness coefficient 2, and a random initialization of the center of clusters **v**. Finally, the center of clusters **V**, clusters covariance matrix **F**, and fuzzy partition matrix **U** are computed. To find the class label corresponding to each cluster, the Euclidian distance measure between the reference center **K** and the center of clusters **v** is calculated. For each class, the cluster is labeled based on the minimum Euclidean distance of all the reference centers. Each class does have to correspond to one cluster only. If there is any cluster which has a minimum distance with respect to two or more classes, or any cluster which does not have a minimum distance to one class, another iteration of fuzzy clustering has to be run again. When each cluster has been associated with a corresponding class, the classifier is fully set up.

The fuzzy partition matrix **U** is used to calculate the classification accuracy where the fuzzy membership function indicates the degree of membership of each data point to each class. As it is necessary to have the class membership as a crisp data, a defuzzifier based on maximum membership is used to assess the classifier decision. Due to variability in results because of initialization parameters, a sub-training step was taken using 100 different initialization parameters of the fuzzy clustering algorithm. Then the configuration that achieves the minimum least square error is selected as the final classifier.

In the validation step, the fuzzy partition matrix is computed using the Mahalanobis distance for the new data in the feature space and the classifier's decision is then obtained by the maximum defuzzifier.

4 Results and Discussion

4.1 EMG Database

Five hand movements were considered in this study and labeled as follows: closing (CLO), opening (OPE), flexion (FLE), extension (EXT), and supination (SUP) (see Fig. 4). In total, 500 SEMG signals were acquired from the forearm of a non-amputee volunteer using the Delsys Bagnoli-4 system coupled with three parallel-bar active SEMG sensors. For data acquisition, 16-bit per channel resolution, 2 kHz sampling frequency, and 1 000x voltage gain were used. The database was split into two groups using the k-fold cross validation methodology: 70% of the signals were used for the GK training and the 30% left were used for validation. For each case, the HS was computed and the MDS was obtained. This process was repeated 100 times.

This methodology achieved an average accuracy rate of $96.64 \pm 1.6\%$ calculated as: accuracy $= \Sigma(tp)/\text{dat}_{\text{val}}$, where tp corresponds to the true positive results, and dat_{val} is the total amount of validation data. Results are presented in the confusion matrix shown in Table 1. It is important to note that the mean value of the absolute coefficients computed from the time contour of the HS for each channel allows the system to discriminate and disjoin clusters in the three-dimensional feature space. Additionally, it allows the visual review of the results, which is an advantage with respect to high-dimensional feature spaces where graphical representations are not possible.

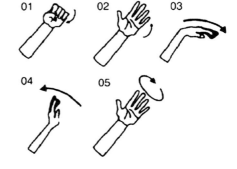

Fig. 4 Hand movements. 01. Closing, 02. Opening, 03. Flexion, 04. Extension, 05. Supination

Table 1 Confusion matrix and classification accuracy

Class	OPE	CLO	EXT	FLE	SUP
OPE	30	0	0	0	0
CLO	0	26	0	4	0
EXT	0	0	30	0	0
FLE	0	0	0	30	0
SUP	0	1	0	0	29
Total	100.0%	86.6%	100%	100.0%	96.6%

5 Conclusions

In this study, a novel method for nonsupervised hand movement pattern recognition from surface electromyographic signals is introduced. The method is based on both the Hilbert transform and a fuzzy clustering classification achieving highly accurate results. Discriminating features extracted from the Hilbert Spectrum proved to be an important source of information about SEMG signals. Such features also allowed the construction of a three-dimensional feature space, which is desirable in real-time implementations. Using the proposed two-stage classification procedure, training cases are initially clustered with the Gustafson–Kessel fuzzy algorithm, which considers the ellipsoidal shape of the clusters. Then, test cases are classified with a nearest-prototype rule to associate the corresponding class for each sample. The proposed method introduces several advantages: (1) assumptions about linearity or stationarity are not needed; (2) previous knowledge about the data in the feature extraction is not required to analyze the signal; and (3) high classification accuracy is achieved. The results of the study reveal a 96% average accuracy rate of recognition for the five movements involved: opening, closing, supination, flexion, and extension. Current research is aimed at improving frequency localization of the IMF signals allowing maximization of the cluster separability. Further study is desirable to increase both the number of movements recognized and the classification accuracy rate, which hopefully will contribute to biomedical research.

References

1. Andrade, A.O., Kyberd, P., Nasuto, S.J.: The application of the hilbert spectrum to the analysis of electromyographic signals. Information Sciences **178**(9), 2176–2193 (2008)
2. Babuška, R., van der Veen, P., Kaymak, U.: Improved covariance estimation for Gustafson–Kessel clustering. In: Proceedings of the 2002 IEEE International Conference on Fuzzy Systems, vol. 2, pp. 1081–1085. Honolulu, Hawaii (2002)
3. Chu, J.U., Moon, I., Mun, M.S.: A real-time emg pattern recognition based on linear-nonlinear feature projection for multifunction myoelectric hand. In: ICORR 2005 9th International Conference on Rehabilitation Robotics, pp. 295–298 (2005)
4. Englehart, K., Hudgins, B., Chan, A.: Short time fourier analysis of the electromyogram: Fast movements and constant contraction. Technology and Disability **15**, 95–103 (2003)
5. Gustafson, D.E., Kessel, W.C.: Fuzzy clustering with a fuzzy covariance matrix. IEEE Conference on Decision and Control including the 17th Symposium on Adaptive Processes (1978)
6. Huang, H.P., Chen, C.Y.: Development of a myoelectric discrimination system for a multi-degree prosthetic hand. In: Proceedings of the 1999 IEEE International Conference on Robotics and Automation, vol. 3, pp. 2392–2397 (1999)
7. Huang, N.E., Shen, Z., Long, S.R., Wu, M.C., Shih, H.H., Zheng, Q., Yen, N.C., Tung, C.C., Liu, H.H.: The empirical mode decomposition and the hilbert spectrum for nonlinear and nonstationary time series analysis. Proceedings of the Royal Society A: Mathematical, Physical and Engineering Sciences **454**, 903–995 (1998)
8. Karlik, B., Tokhi, O.M.: A fuzzy clustering neural network architecture for multifunction upper-limb prosthesis. IEEE Transactions on Biomedical Engineering **50**, 1255–1261 (2003)

9. Marshall, P.W., Murphy, B.A.: Muscle activation changes after exercise rehabilitation for chronic low back pain. Archives of Physical Medicine and Rehabilitation **89**(7), 1305–1313 (2008)
10. Momen, K., Krishnan, S.: Real-time classification of forearm electromyographic signals corresponding to user-selected intentional movements for multifunction prosthesis control. IEEE Transactions on Neural Systems and Rehabilitation Engineering (2007)
11. Wang, G., Wang, Z., Chen, W., Zhuang., J.: Classification of surface emg signals using optimal wavelet packet method based on davies-bouldin criterion. Springer Medical and Biological Engineering **44**(10), 865–872 (2006)
12. Zardoshti-Kermani, M., Wheeler, B., Badie, K., Hashemi, R.: EMG feature evaluation for movement control of upper extremity prostheses. IEEE Transactions on Rehabilitation Engineering **3**(4), 324–333 (1995)

Chapter 21
Rotation of Random Forests for Genomic and Proteomic Classification Problems

Gregor Stiglic, Juan J. Rodriguez, and Peter Kokol

Abstract Random Forests have been recently widely used for different kinds of classification problems. One of them is classification of gene expression samples that is known as a problem with extremely high dimensionality, and therefore demands suited classification techniques. Due to its strong robustness with respect to large feature sets, Random Forests show significant increase of accuracy in comparison to other ensemble-based classifiers that were widely used before its introduction. In this chapter, we present another ensemble of decision trees called Rotation Forest and evaluate its classification performance on different microarray datasets. Rotation Forest can also be applied to different already existing ensembles of classifiers like Random Forest to improve their accuracy and robustness. This study presents evaluation of Rotation Forest classification technique based on decision trees as base classifiers and was evaluated on 14 different datasets with genomic and proteomic data. It is evident that Rotation Forest as well as the proposed rotation of Random Forests outperform most widely used ensembles of classifiers including Random Forests on majority of datasets.

1 Introduction

There have been many supervised classification techniques that were applied to the analysis of microarray and mass spectrometry-based data in recent years. Diaz-Uriarte and Alvarez [1] have shown that ensembles of classifiers can perform very good even in very high-dimensional and noisy gene expression domain, where they can achieve at least as good results as some other advanced machine learning techniques such as support vector machines (SVM) [2, 3], nearest neighbours [4, 5]

G. Stiglic (✉)
Faculty of Health Sciences, University of Maribor, Zitna ulica 15, 2000 Maribor, Slovenia
and
Faculty of Electrical Engineering and Computer Science, University of Maribor, Smetanova 17, 2000 Maribor, Slovenia
e-mail: gregor.stiglic@uni-mb.si

or neural networks [6]. The key idea of building multiple classification models and assembling them in committees of classifiers is the ability of ensembles to increase stability and accuracy compared to a single classifier [7]. This is especially true when classifiers like decision trees or neural networks that are very sensitive to changes in the underlying training set are used. Diversity among the members of an ensemble of classifiers is deemed to be a key issue in classifier combination as it ensures that independent members of an ensemble are built from the same initial dataset.

2 Methods

2.1 Basic Ensemble Building Techniques

One of the first ensemble building techniques was called bagging and was introduced by Breiman in [8]. It is based on random sampling of examples from the training set which is also the basis of bootstrapping [9]. An ensemble of classifiers that was built using bootstrapping is then used to classify an example using the majority vote of the ensemble.

Another basic and at the same time widely used method for building ensembles is called boosting [10]. In our research, boosting is represented by AdaBoost.M1 variant, which is the most commonly used algorithm from boosting family of ensemble building techniques. The main idea of boosting is re-weighting of examples, and it is therefore assumed that each base classifier can handle weighted examples. In cases where this is not possible, a dataset is obtained from a random sample, taking into account the weights distribution. In comparison to bagging, classifiers are built in a sequential process that cannot be parallelized. Although the idea of boosting is very promising and also achieves good accuracy results in practice, there are some drawbacks we should consider when using boosting. One of them is overfitting to the training set examples; although early literature mentions that boosting would not overfit even when running for a large number of iterations. Recent research clearly shows negative overfitting effects when boosting is used on datasets with higher noise content [11].

3 Random Forests

To increase the diversity of classifiers in bagging, Breiman upgraded the basic idea of bagging by combining bootstrapping with random feature selection for decision tree building. It has to be noted that feature randomization that represents an integral part of Random Forests method was introduced earlier by Ho [12] and Amit et al. [13].

Random decision trees created this way are grown by selecting the feature to split on at each node from randomly selected set of features. The number of chosen features is a parameter of the method. In this work, the number of chosen features is set to $\log_2(k+1)$ as in [14], where k is the total number of features.

Random Forests is an ensemble building technique that works well even with noisy content in training dataset and is considered as one of the most competitive and robust methods that can be compared to bagging or boosting [15].

4 Decorate

One of the recently proposed ensemble building techniques that could also be seen as a somehow alternative approach as it significantly differs from the above described techniques is called DECORATE (Diverse Ensemble Creation by Oppositional Relabeling of Artificial Training Examples) [16]. Base classifiers are built using additional artificially constructed training examples. These examples are given outcome labels that disagree with the current decision of the committee, thereby directly increasing diversity of classifiers within the committee.

5 Rotation Forest

One of the most recent classification techniques not only in bioinformatics but also in the machine learning field is called Rotation Forest and was developed by Rodriguez et al. [17]. Rotation Forest classifier was introduced to gene expression classification problems by Stiglic et al. in [18], where it was used as a meta-classifier for meta-classification scheme and applied to 14 different gene expression classification problems. This chapter presents the potential of Rotation Forest in the field of gene expression classification on an even wider scale and additionally presents a novel way to use the rotation of datasets using PCA transformation for building new ensembles of classifiers.

Most ensemble methods can be used with any classification method, but decision trees are one of the most commonly used. There are ensemble methods designed specifically for decision trees, such as Random and Rotation Forests. The latter is based on the sensibility of decision trees to axis rotations; the classifiers obtained with different rotations of a dataset can be very different. This sensibility is usually considered as a disadvantage, but it can be very beneficial when decision trees are used as members of an ensemble. Decision trees obtained from a rotated dataset can still be accurate, because they use all the information available in the dataset, but simultaneously they can be very diverse.

As in Bagging and Random Forests, each member of the ensemble is trained with a different dataset. These datasets are obtained from a random transformation of the original training data. This transformation produces a rotation of the axis.

The transformed dataset has as many examples as the original dataset. All the information that was in the original dataset remains in the transformed dataset, because none of the components is discarded and all the training examples are used for training all classifiers in an ensemble.

Number of features in each group (or number of groups) is a parameter of the method. The optimal value for this parameter depends on the dataset and it could be selected with an internal cross validation. Nevertheless, in this work the default value was used, and groups were formed using three features.

The elimination of classes and examples of the dataset is done because PCA is a deterministic method, and it would not be difficult (especially for big ensembles) that some members of the ensemble had the same (or very similar) grouping of variables. Hence, an additional source of diversity was needed. This elimination is only done for the dataset used to do PCA, while training of ensemble classifiers is done using all examples.

5.1 Rotation of Random Forests

This chapter compares basic and most widely used ensemble building techniques with a novel technique called Rotation of Random Forests (RRF).

Random Forest is based on bagging, using Random Trees as base classifiers. It is also possible to use the Rotation Forest method using Random Trees as base classifiers. We call this method Rotation of Random Forests. The same relationship could be expected between Rotation of Random Forest and Rotation Forest and then between Random Forest and Bagging, that is, the base classifiers will be less accurate but more diverse and this could be beneficial for the ensemble.

On the other hand, one of the advantages of Random Forests over Rotation Forest is that the former are faster. Using Rotation Forest with Random Trees could reduce this time difference, because it is also possible to construct several Random Trees in the same rotated space. This is the equivalent to a Rotation Forest ensemble using Random Forest as base classifiers.

In this chapter, the following configuration is used: ten rotations, with a ten-tree Random Forest classifier in each rotation.

6 Rotation of a Single Decision Tree Example

This section shows an example to illustrate the procedure of single decision tree rotation that represents an integral part of Rotation Forest consisting of several such trees. Decision tree is built using DLBCL-Tumour dataset. Before constructing the forest, a set of features is selected. For this example, ReliefF [19] selection method was used. For simplicity reasons, only six features were selected. The selected features are M14328_s, X02152, X12447, L19686_rna1, J04988 and J03909 that represent gene identifiers from DLBCL-Tumour dataset.

The features are randomly grouped. In our example, for one of the trees in the forest, one of the groups is formed by M14328_s, X12447_at and L19686_rnal; the other group includes the remaining features. Now we have two datasets, one for each group. For each of these datasets, a random proper subset of the classes is selected, and the examples of the classes from the subset are removed from the dataset. For a two-class dataset, a proper subset has zero or one class, and at least the examples of one of the classes will remain in the dataset. From the remaining dataset, a subset of 25% randomly selected examples is removed. Then, PCA is applied to the resulting datasets. The result of PCA is a set of components. For the first group of features, in a particular run, these components were as follows:

$$f_{11} = 0.580 \times M14328\_s + 0.579 \times X12447 + 0.574 \times L19686\_rnal$$
$$f_{12} = 0.814 \times L19686\_rnal - 0.483 \times X12447 - 0.323 \times M14328\_s$$
$$f_{13} = -0.748 \times M14328\_s + 0.657 \times X12447\_at + 0.093 \times L19686\_rnal.$$

And for the second group:

$$f_{21} = 0.583 \times X02152 + 0.581 \times J03909 + 0.568 \times J04988$$
$$f_{22} = -0.822 \times J04988 + 0.438 \times J03909 + 0.365 \times X02152$$
$$f_{23} = 0.726 \times X02152 - 0.686 \times J03909 - 0.043 \times J04988.$$

All these components define a new set of features. The original dataset is then transformed using these components and consists of this new set of features. The transformed dataset will be used to construct the tree. Note that the removal of classes and examples is done only for PCA transformation, while the transformed dataset contains all the training examples. Obtained decision tree is shown in Fig. 1. A Rotation Forest classifier is formed by several decision trees obtained following the procedure described above.

In the case of Rotation of Random Forest method, for each transformed dataset, a Random Forest is constructed.

7 Results

In this section, we extensively compare Rotation Forest and RRF with other methods in the literature on public gene expression datasets. Dimensionality of each dataset is reduced before classification using ReliefF feature selection method. RelieF was recently used in an extensive study by Symons and Nieselt [20], where it was selected as the most effective feature selection method for gene expression classification.

One of the most problematic limitations of supervised classification methods is overfitting to training set of examples. Especially in cases where learning is

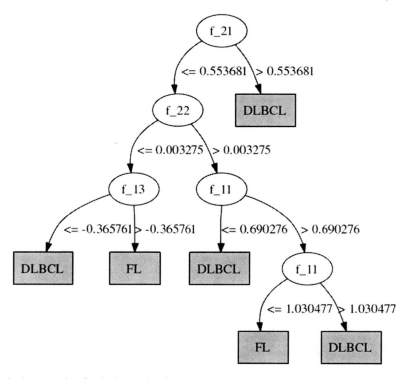

Fig. 1 An example of a single rotation forest decision tree

performed too long (e.g. neural networks) or where training examples are rare, the classifier may adjust to specific random variations of the training data that do not represent true relationships in test data. To solve this problem, we should try to get the most unbiased estimation of classifier accuracy. This is possible by using a so-called external validation, although there are still many papers that do not use that kind of validation and are presenting overly optimistic accuracy rates. The most appropriate methods, according to a study by Ambroise and McLachlan [21], are repeated k-fold cross-validation [22] or suitably defined bootstrap validation methods [23].

Following recommendations from Ambroise and McLachlan, a tenfold cross-validation was used in all experiments. As there were many classifiers that are using randomness in classifier building process, each tenfold cross-validation was repeated 20 times. Instances were randomly shuffled and stratified according to the class values of samples before they were divided into ten subsets, also called folds. The same procedure was done for all tests except the last experiment where 5×10-fold cross-validation was used because of very high computational complexity of testing procedure.

Table 1 Kent ridge repository datasets overview

Dataset	Source	Genes	Patients	Classes
ALL	Yeoh et al. [25]	12,558	327	7
ALLAML	Golub et al. [26]	7,129	72	2
Breast	Van't Veer et al. [27]	24,481	97	2
CNS	Mukherjee et al. [28]	7,129	60	2
Colon	Alon et al. [29]	2,000	62	2
DLBCL	Alizadeh et al. [30]	4,026	47	2
DLBCL-NIH	Rosenwald et al. [31]	7,399	240	2
DLBCL-Tumor	Shipp et al. [32]	6,817	77	2
Lung	Gordon et al. [33]	12,533	181	2
Lung-Harvard	Bhattacharjee et al. [34]	12,600	203	5
Lung-Michigan	Beer et al. [35]	7,129	96	2
MLL	Armstrong et al. [36]	12,582	72	3
Ovarian	Petricoin et al. [37]	15,154	253	2
Prostate	Singh et al. [38]	12,600	102	2

7.1 Data

For effective evaluation of proposed classification methods, a set of 14 gene expression datasets was used. All of them can be downloaded from Kent Ridge Biomedical Data Set Repository [24] and represent different biomarker classification problems originating from microarray or mass spectrometry-based studies. Table 1 contains more details on datasets that can also be found at the above-mentioned repository. Most datasets from this repository are well known, which were used in many studies where classification accuracy was important, and they served for benchmarking the proposed classification methods.

7.2 Classification Accuracy

An important issue in classification performance of gene expression classifiers is the number of selected genes before the classification algorithm is applied. It is well known that feature selection can significantly improve the performance of classification and is an integral part of most gene expression classification schemes. Recent studies [20, 39] show that the most effective classifiers achieve the highest accuracy rates with number of features between 100 and 500 genes. Although feature selection does not play a significant role when comparing classifiers, two different feature selection settings were used in our experiments where 100 and 250 features were selected during cross-validations.

In this study, a set of widely used ensemble building classification methods including Random Forests is compared to two proposed methods – Rotation Forests and RRF. The first one is a novel classification technique in bioinformatics, while the second one proposes an improvement of already proved Random Forests. Compared ensembles of classifiers were built using 100 classifiers inside Weka [40] machine learning framework. Implementations of all methods can be found in Weka environment.

Initially, a comparison of both feature selection methods using 100 and 250 best ranked features was done using average ranks from Friedman test. Results for all six ensemble building methods from 20 × 10-fold cross-validation were used. Rotation Forest, RRF and Random Forests achieve the highest ranks. When comparing feature selection method settings, it can be noted that ReliefF with 250 selected features achieves slightly better results. More detailed results representing averaged 20 × 10-fold cross-validation accuracy rates for ReliefF-250 are presented in Table 2. To directly compare three most accurate ensembles of classifiers, a set of pair-wise Wilcoxon tests was done where Random Forests, Rotation Forest and RRF were compared.

There were no significant differences in accuracy of compared classification methods (using $p < 0.05$) when results from ReliefF-250-based feature selection were used. On the other hand, when compared to the rest of the methods a significant difference can be observed.

Even though there are no statistically significant differences when average accuracy is compared between the three most successful methods, one can observe the superiority of Rotation Forest and RRF by observing the average rank. It can also be observed that Rotation Forest outperforms Random Forest in 9 of 14 datasets when average accuracy is compared. RRF goes even further and can outperform Random Forest in 10 of 14 dataset.

Table 2 Classification accuracy for ReliefF feature extraction using 250 top ranked genes (lower is better)

Dataset	Bagging	Boosting	Decorate	Random Forests	Rotation Forest	RRF
ALL	87.85 ± 5.2	90.86 ± 4.5	86.13 ± 5.2	89.67 ± 4.1	91.38 ± 4.2	88.97 ± 4.4
ALLAML	92.53 ± 9.9	89.67 ± 11.2	94.97 ± 7.8	97.07 ± 6.3	95.55 ± 6.9	97.63 ± 5.2
Breast	64.57 ± 14.4	64.45 ± 14.4	63.74 ± 13.9	68.02 ± 13.6	68.23 ± 13.3	68.42 ± 13.4
CNS	63.67 ± 15.5	60 ± 18.2	59.5 ± 17.9	64.42 ± 13.6	60.58 ± 15.5	65.33 ± 15.8
Colon	83.98 ± 12.3	77.79 ± 14.2	83.54 ± 12.3	83.27 ± 13.3	84.26 ± 12.6	84.24 ± 13.2
DLBCL	89.25 ± 13.4	86.33 ± 16	92.48 ± 12	94.15 ± 9.8	93.58 ± 11.3	94.33 ± 10.3
DLBCL-NIH	61.5 ± 7.8	60.02 ± 8.2	60.77 ± 9	62.21 ± 8.3	62.13 ± 7.9	61.83 ± 8.5
DLBCL-Tumor	89.24 ± 9.6	88.71 ± 10.4	90.08 ± 9.4	92.54 ± 8.8	95.66 ± 7	94.54 ± 7.8
Lung	97.49 ± 3.3	95.58 ± 4.9	99.18 ± 2	99.17 ± 2	99.23 ± 1.9	99.23 ± 1.9
Lung-Harvard	92.43 ± 5.5	93.89 ± 4.9	92.83 ± 5.1	92.79 ± 4.9	93.57 ± 5	92.77 ± 5.1
Lung-Michigan	98.92 ± 3.2	98.92 ± 3.2	97.96 ± 4.1	99.42 ± 1.6	100 ± 0	99.64 ± 1
MLL	92.3 ± 10.1	92.27 ± 10.6	94.92 ± 7.3	94.82 ± 7.8	93.56 ± 8.4	93.38 ± 8.9
Ovarian	98.08 ± 2.3	98.42 ± 2.2	98.68 ± 2	99.31 ± 1.6	99.72 ± 0.8	99.35 ± 1.4
Prostate	90.49 ± 9.8	90.98 ± 8.8	91.59 ± 8.9	93.26 ± 8.1	93.6 ± 7.8	93.46 ± 7.7
Friedman's Avg. Rank	4.44	4.54	4.08	2.88	2.45	2.66

8 Conclusions

This chapter presents a novel Rotation Forest-based classification method for genomic and proteomic classification problems. It also includes experiments comparing leading supervised data mining approaches on a wide range of gene expression classification problems. The results, based on average ranks and average accuracy, indicate that Rotated Random Forest and Rotation Forest can be considered as two of the most accurate ensembles of classifiers in gene expression classification. It is also evident that both methods can improve the performance of Random Forest classifier that was described by Diaz-Uriarte et al. [1] as a "part of the standard tool-box of methods for the analysis of microarray data".

The main motivation for the use of Rotations of Random Forest is that they are faster than Rotation Forests; in this chapter, the used configuration was ten rotations for ensemble, with a ten-tree Random Forest for each rotation. If time is not a cause of main concern, it is possible to construct each Random Tree in a different rotated space. This would improve the diversity among base classifiers without a penalty in their accuracy. It remains to be seen if this configuration would be better than Rotation Forests.

In the near future, when microarray datasets with larger number of samples will become available, Rotated Random Forest can become a rational solution to classification problems, because of low time complexity compared to more accurate Rotation Forest. It should be noted that both methods proved to return more reliable results in comparison to Random Forests in most cases.

References

1. Díaz-Uriarte R, Alvarez de Andrés S (2006) Gene selection and classification of microarray data using random forest. BMC Bioinformatics 7:3
2. Vapnik V (1998) Statistical learning theory. John Wiley and Sons, New York
3. Furey TS, Cristianini N, Duffy N, Bednarski DW, Schummer M, Haussler D (2000) Support vector machine classification and validation of cancer tissue samples using microarray expression data. Bioinformatics 16:906–914
4. Wu W, Xing E, Mian I, Bissell M (2005) Evaluation of normalization methods for cdna microarray data by k-nn classification. BMC Bioinformatics 6(191):1–21
5. Dudoit S, Fridlyand J (2003) Classification in microarray experimentse. In: Speed T (Ed.), Statistical analysis of gene expression microarray data. Interdisciplinary statistics. Chapman & Hall/CRC, Virginia Beach, 93–158
6. Seiffert U, Hammer B, Kaski S, Villmann T (2006) neural networks and machine learning in bioinformatics – theory and applications. In: Proceedings of the 14th European Symposium on Artificial Neural Networks ESANN 2006, 521–532
7. Cunningham, P. (2007) Ensemble Techniques. Technical Report UCD-CSI-2007-5
8. Breiman L (1996) Bagging predictors. Machine Learning 24:123–140
9. Efron B, Tibshirani R (1994) An introduction to the bootstrap. Chapman & Hall/CRC, Virginia Beach
10. Freund Y, Schapire RE (1996) Experiments with a new boosting algorithm. In: Proceedings of the 13th International Conference on Machine Learning, 148–156

11. Rätsch G, Onoda T, Müller KR (2001) Soft margins for AdaBoost. Machine Learning 42(3):287–320
12. Ho, TK (1995) Random decision forest. In: Proceedings of the 3rd Int'l Conf on Document Analysis and Recognition, Montreal, Canada, August 14–18, 1995, 278–282
13. Amit Y, Geman D (1997) Shape quantization and recognition with randomized trees. Neural Computation 9:1545–1588
14. Breiman L (2001) Random forests. Machine Learning 45:5–32
15. Dietterich TG (2002) Ensemble learning. In: Arbib MA (Ed.) The handbook of brain theory and neural networks, 2nd ed. The MIT Press, Cambridge, MA, 405–408
16. Melville P, Mooney RJ (2003) Constructing diverse classifier ensembles using artificial training examples. In: Proceedings of the 18th International Joint Conference on Artificial Intelligence (IJCAI 2003), 505–510
17. Rodríguez JJ, Kuncheva LI, Alonso CJ (2006) Rotation forest: A new classifier ensemble method. IEEE Transactions on Pattern Analysis and Machine Intelligence 28(10):1619–1630
18. Stiglic G, Kokol P (2007) Effectiveness of rotation forest in meta-learning based gene expression classification. In: Proceedings of the 20th IEEE International Symposium on Computer-Based Medical Systems (CBMS 2007), 243–250
19. Robnik-Sikonja M, Kononenko I (1997) An adaptation of relief for attribute estimation in regression. In: Machine Learning: Proceedings of the Fourteenth International Conference (ICML'97), 296–304
20. Symons S, Nieselt K (2006) Data mining microarray data – comprehensive benchmarking of feature selection and classification methods (available at: www.zbit.unituebingen.de/pas/preprints/GCB2006/SymonsNieselt.pdf)
21. Ambroise C, McLachlan GJ (2002) Selection bias in gene extraction on the basis of microarray gene-expression data. Proceedings of the National Academy of Sciences of the United States of America 99:6562–6566
22. Burman P (1989) A comparative study of ordinary cross-validation, v-fold cross-validation and repeated learning-testing methods. Biometrika 76:503–514
23. Efron B, Tibshirani R (1997) Improvements on cross-validation: the. 632 + bootstrap method. Journal of the American Statistical Association 92:548–560
24. Li J, Liu H (2003) Ensembles of cascading trees. In: Proceedings of the IEEE ICDM 2003 Conference 585
25. Yeoh EJ, Ross ME, Shurtleff SA, Williams WK, Patel D, Mahfouz R, Behm FG, Raimondi SC, Relling MV, Patel A, Cheng C, Campana D, Wilkins D, Zhou X, Li J, Liu H, Pui CH, Evans WE, Naeve C, Wong L, Downing JR (2002) Classification, subtype discovery, and prediction of outcome in pediatric acute lymphoblastic leukemia by gene expression profiling. Cancer Cell 1:133–143
26. Golub TR, Slonim DK, Tamayo P, Huard C, Gaasenbeek M, Mesirov JP, Coller H, Loh ML, Downing JR, Caligiuri MA, Bloomfield CD, Lander ES (1999) Molecular classification of cancer: class discovery and class prediction by gene expression monitoring. Science 286:531–537
27. van't Veer LJ, Dai H, van de Vijver MJ, He YD, Hart AA, Mao M, Peterse HL, van der Kooy K, Marton MJ, Witteveen AT, Schreiber GJ, Kerkhoven RM, Roberts C, Linsley PS, Bernards R, Friend SH (2002) Gene expression profiling predicts clinical outcome of breast cancer. Letters to Nature, Nature 415:530–536
28. Pomeroy SL, Tamayo P, Gaasenbeek M, Sturla LM, Angelo M, McLaughlin ME, Kim JY, Goumnerova LC, Black PM, Lau C, Allen JC, Zagzag D, Olson JM, Curran T, Wetmore C, Biegel JA, Poggio T, Mukherjee S, Rifkin R, Califano A, Stolovitzky G, Louis DN, Mesirov JP, Lander ES, Golub TR (2002) Prediction of central nervous system embryonal tumour outcome based on gene expression. Letters to Nature, Nature 415:436–442
29. Alon U, Barkai N, Notterman DA, Gish K, Ybarra S, Mack D, Levine AJ (1999) Broad patterns of gene expression revealed by clustering analysis of tumor and normal colon tissues probed by oligonucleotide arrays. Proceedings of National Academy of Sciences of the United States of America 96:6745–6750

30. Alizadeh AA, Eisen MB, Davis RE, Ma C, Lossos IS, Rosenwald A, Boldrick JC, Sabet H, Tran T, Yu X, Powell JI, Yang L, Marti GE, Moore T, Hudson J Jr, Lu L, Lewis DB, Tibshirani R, Sherlock G, Chan WC, Greiner TC, Weisenburger DD, Armitage JO, Warnke R, Levy R, Wilson W, Grever MR, Byrd JC, Botstein D, Brown PO, Staudt LM (2000) Distinct types of diffuse large B-cell lymphoma identified by gene expression profiling. Nature 403:503–511
31. Rosenwald A, Wright G, Chan W, Connors JM, Campo E, Fisher R, Gascoyne RD, Muller-Hermelink K, Smeland EB, Staut LM (2002) The use of molecular profiling to predict survival after themotherapy for diffuse large-B-cell lymphoma. The New England Journal of Medicine 346:1937–1947
32. Shipp MA, Ross KN, Tamayo P, Weng AP, Kutok JL, Aguiar RC, Gaasenbeek M, Angelo M, Reich M, Pinkus GS, Ray TS, Koval MA, Last KW, Norton A, Lister TA, Mesirov J, Neuberg DS, Lander ES, Aster JC, Golub TR (2002) Diffuse large B-cell lymphoma outcome prediction by gene-expression profiling and supervised machine learning. Nature Medicine 8:68–74
33. Gordon GJ, Jensen RV, Hsiao LL, Gullans SR, Blumenstock JE, Ramaswamy S, Richards WG, Sugarbaker DJ, Bueno R (2002) Translation of microarray data into clinically relevant cancer diagnostic tests using gene expression ratios in lung cancer and mesothelioma. Cancer Research 62(17):4963–4967
34. Bhattacharjee A, Richards WG, Staunton J, Li C, Monti S, Vasa P, Ladd C, Beheshti J, Bueno R, Gillette M, Loda M, Weber G, Mark EJ, Lander ES, Wong W, Johnson BE, Golub TR, Sugarbaker DJ, Meyerson M (2002) Classification of human lung carcinomas by mRNA expression profiling reveals distinct adenocarcinomas subclasses. Proceedings of National Academy of Sciences of the United States of America 98:13790–13795
35. Beer DG, Kardia SL, Huang CC, Giordano TJ, Levin AM, Misek DE, Lin L, Chen G, Gharib TG, Thomas DG, Lizyness ML, Kuick R, Hayasaka S, Taylor JM, Iannettoni MD, Orringer MB, Hanash S (2002) Gene-expression profiles predict survival of patients with lung adenocarcinoma. Nature Medicine 18(8):816–824
36. Armstrong SA, Staunton JE, Silverman LB, Pieters R, den Boer ML, Minden MD, Sallan SE, Lander ES, Golub TR, Korsmeyer SJ (2002) MLL Translocations specify a distinct gene expression profile that distinguishes a unique leukemia. Nature Genetics 30:41–47
37. Petricoin EF, Ardekani AM, Hitt BA, Levine PJ, Fusaro VA, Steinberg SM, Mills GB, Simone C, Fishman DA, Kohn EC, Liotta LA (2002) Use of proteomic patterns in serum to identify ovarian cancer. The Lancet 359:572–577
38. Singh D, Febbo P, Ross K, Jackson DG. Manola J, Ladd C, Tamayo P, Renshaw AA, D'Amico AV, Richie JP, Lander ES, Loda M, Kantoff PW, Golub TR, Sellers WR (2002) Gene expression correlates of clinical prostate cancer behaviour. Cancer Cell 1(2):203–209
39. Statnikov A, Aliferis CF, Tsamardinos I, Hardin D, Levy S (2005) A comprehensive evaluation of multicategory classification methods for microarray gene expression cancer diagnosis. Bioinformatics 21:631–643
40. Witten IH, Frank E (2005) Data mining: practical machine learning tools with Java implementations. Morgan Kaufmann, Massachusetts

Chapter 22
Improved Prediction of MHC Class I Binders/Non-Binders Peptides Through Artificial Neural Network Using Variable Learning Rate: SARS Corona Virus, a Case Study

Sudhir Singh Soam, Bharat Bhasker, and Bhartendu Nath Mishra

Abstract Fundamental step of an adaptive immune response to pathogen or vaccine is the binding of short peptides (also called epitopes) to major histocompatibility complex (MHC) molecules. The various prediction algorithms are being used to capture the MHC peptide binding preference, allowing the rapid scan of entire pathogen proteomes for peptide likely to bind MHC, saving the cost, effort, and time. However, the number of known binders/non-binders (BNB) to a specific MHC molecule is limited in many cases, which still poses a computational challenge for prediction. The training data should be adequate to predict BNB using any machine learning approach. In this study, variable learning rate has been demonstrated for training artificial neural network and predicting BNB for small datasets. The approach can be used for large datasets as well. The dataset for different MHC class I alleles for SARS Corona virus (Tor2 Replicase polyprotein 1ab) has been used for training and prediction of BNB. A total of 90 datasets (nine different MHC class I alleles with tenfold cross validation) have been retrieved from IEDB database for BNB. For fixed learning rate approach, the best value of AROC is 0.65, and in most of the cases it is 0.5, which shows the poor predictions. In case of variable learning rate, of the 90 datasets the value of AROC for 76 datasets is between 0.806 and 1.0 and for 7 datasets the value is between 0.7 and 0.8 and for rest of 7 datasets it is between 0.5 and 0.7, which indicates very good performance in most of the cases.

Keywords Variable learning rate · Artificial neural network · SARS Corona virus · MHC class I binder/non-binder · Epitope prediction · Vaccine designing · T-cell immune response

---

B.N. Mishra (✉)
Department of Biotechnology, Institute of Engineering & Technology,
UP Technical University, Lucknow, India
e-mail: profbnmishra@gmail.com

1 Introduction

Cytotoxic T cells of the immune system monitor cells for infection by viruses, or intracellular bacteria through scanning their surface for peptides bound to MHC class I molecules [1]. The cells that present peptides derived from non-self, e.g., from viruses or bacteria (after binding to MHC molecule), can trigger a T-cell immune response, which leads to the destruction of such cells [2]. T cells do not recognize soluble native antigen but rather recognize antigens that has been processed into antigenic peptides, which are presented in combination with MHC molecules. It has been observed that peptides of nine amino acid residues (9-mers) bind most strongly; peptides of 8–11 residues also bind but generally with lower affinity than nonamers [3, 4]. Binding of a peptide to a MHC molecule is prerequisite for recognition by T cells and, hence is fundamental to understanding the basis of immunity and also for the development of potential vaccines and design of immunotherapy [5].

The SARS coronavirus, sometimes abbreviated to SARS-CoV, is the virus that causes severe acute respiratory syndrome (SARS). On 12 April 2003, scientists working at the Michael Smith Genome Sciences Centre in Vancouver, British Columbia, finished mapping the genetic sequence of a coronavirus believed to be linked to SARS. Passive immunization with convalescent serum has been tested as a way to treat SARS. Control of SARS is most likely to be achieved by vaccination [6]. In this chapter, various MHC class I alleles for SARS coronavirus (Tor2 Replicase polyprotein 1ab) have been used as a case study.

2 Review and Motivation of Present Work

The algorithms for prediction of MHC-binding peptides are based on two concepts: (1) algorithms based on identifying the patterns in sequences of binding peptides, e.g., binding motif, quantitative matrices, and artificial neural networks, and (2) algorithms based on three-dimensional structures for modeling peptide/MHC interactions [7, 8]. The second approach, i.e., based on structures corresponds to techniques with distinct theoretical lineage and includes the use of homology modeling, docking and 3D-threading techniques.

For prediction of T-cell epitope, ANN has been used with genetic algorithms [9, 10] and evolutionary algorithm [11]. Support vector machine has also been used to predict the binding peptides [12, 13]. For improving prediction of MHC class I binding peptides, probability distribution functions have also been used [14]. Threading methods [15] and Gibbs motif sampler [16] approach have also been used for prediction of MHC-binding peptides. In many cases, the number of known binders and non-binders to specific MHC alleles are limited; therefore, the convergence to optimal weights of ANN has to be improved. In current study, the variable learning rate has been used to improve convergence taking various MHC alleles for SARS corona virus as case study.

3 Methodologies

3.1 Variable Learning Rate for ANN Training

The values of learning rate are taken between 0.0 and 1.0. Back propagation network learns using a method of gradient descent to search for a set of weights that can model the given classification problem, so as to minimize the mean squared distance between the network's class prediction and the actual class label of the samples. The learning rate helps to avoid getting stuck at local minimum in the decision space (i.e., where the weights appear to converge, but are not the optimum solution) and encourages finding the global minimum. If the learning rate is too small, then learning will occur at a very slow pace. If the learning rate is too large, then oscillation between inadequate solutions may occur [17, 18]. In gradient descent, learning rate determines the magnitude of the change to be made in the parameters, i.e., weights and a bias of nodes as per (1) and (2); furthermore, the updated weights and biases are given by (3) and (4). The value of error (Err_j) at output node and at internal node is given by (5) and (6), respectively. The input and output to each j-th unit are given by (7) and (8), respectively. For a given training set of input vectors, the learning rate L is kept fixed which leads to poor convergence in case of a small dataset. To improve the convergence, variable learning rate (i.e., the learning rate is updated after each input vector in a given training set of input vectors) has been used as per (9). The error is calculated using (5) after each training vector, and the learning rate is increased by a value a, if the error on the subsequent training vector decreases. It is decreased geometrically by value $b\eta$, if the error on subsequent training vector increases. The value of ΔL has to be calculated after each input vector in the given training set as per (9), and used to update the learning rate, L. The updated learning rate, $L + \Delta L$, has been used for further training to calculate the values of the weights and biases of the nodes as per (1)–(4).

$$\Delta w_{ij} = (L) \text{Err}_j O_i \quad (1)$$

$$\Delta \theta_j = (L) \text{Err}_j \quad (2)$$

$$w_{ij} = w_{ij} + \Delta w_{ij} \quad (3)$$

$$\theta_j = \theta_j + \Delta \theta_j \quad (4)$$

$$\text{Err}_j = O_j(1 - O_j)(T_j - O_j) \quad (5)$$

$$\text{Err}_j = O_j(1 - O_j) \sum_k \text{Err}_k w_{jk} \quad (6)$$

$$I_j = \sum_i w_{ij} O_i + O_j \quad (7)$$

$$O_j = \frac{1}{1 + e^{-I_j}} \quad (8)$$

$$\Delta L = \begin{cases} +a & \text{if } E^{t+\Gamma} \langle E^t \\ -b\eta & \text{otherwise} \end{cases} \quad (9)$$

where Δw_{ij} is the change in the weight w_{ij}, $\Delta \theta_j$ is the change in the bias t_j, W_{ij} is the weight of the connection from unit i to a unit j in the next higher layer, Err_j is the error for unit j at the output layer, Err_j is the error for unit j at hidden layer, I_j is the net input to unit j, O_j is the output of the unit j, T_j is the true output, ΔL is the change in the learning rate, L is the current learning rate, a and b are coefficients E^t is the error at the node in output layer in the previous learning input vector, $E^{t+\Gamma}$ is the error at the node in output layer for current learning input vector.

3.2 Evaluation Parameters

The predictive performance of the model has been evaluated using receiver operating characteristics (ROC) analysis. The area under the ROC curve (AROC) provides a measure of overall prediction accuracy: AROC < 70% for poor, AROC > 80% for good, and AROC > 90% for excellent prediction. The ROC curve is generated by plotting sensitivity (SN) as a function of 1-specificity (SP). The sensitivity, SN = TP/(TP + FN) and SP = TN/(TN + FP), gives percentage of correctly predicted binders and non-binders, respectively. The PPV = [(TP)/(TP + FP)] × 100 and NPV = [(TN)/(FN + TN)] × 100 give the positive probability value, i.e., the probability that a predicted binder will actually be a binder, and negative probability value, i.e., the probability that a predicted non-binder will actually be a non-binder. Tenfold cross validation has been used for training and prediction. The terms TP, FP, TN, and FN related to threshold T are true positive, false positive, true negative, and false negative, respectively. A web-based tool has been used to calculate the area under the ROC curve available at (www.rad.jhmi.edu/jeng/javarad/roc/JROCFITi.html).

4 SARS Corona Virus: A Case Study

4.1 Data Resources

The datasets used for training and testing of (BNB) have been obtained from IEDB Beta 2.0 database (http://www.immuneepitope.org) for HLA-A*0201, HLA-A*0301, HLA-A*1101, HLA-A*0202, HLA-A*0203, HLA-A*0206, HLA-A*2902, HLA-A*3002 and HLA-B*4002 MHC Class I alleles. The strong binders have been retrieved for IC50 < 500. All 9-mers have been filtered after removing the duplicates. For strong non-binders, the records with IC50 > 5000 have been retrieved. The duplicates from binders and non-binders sets have been removed. Furthermore, to keep the ratio of binders and non-binders nearly 1:1, so as to reduce the biasness in learning, the additional 9-mer non-binders were retrieved through EBI-Expasy protein database available at: http://www.expasy.ch. Final sets of binders and non-binders for various alleles have been shown in Table 1.

Table 1 The number of binders and non-binders for various alleles for SARS coronavirus

MHC Alleles	Binders IC < 500			Non-binders IC > 5,000		
	Retrieved	9-mer(s)	Final	Retrieved	9-mer(s)	Final*
A*0201	136	103	103	0	0	104
A*0301	192	116	116	38	23	117
A*1101	202	128	128	16	8	129
A*0202	132	101	101	1	0	101
A*0203	124	96	96	0	0	96
A*0206	135	102	10	0		102
A*2902	178	100	100	80	68	100
A*3002	136	100	77	127	98	77
B*4002	91	36	36	37	9	36

*Additional 9-mer non-binders added

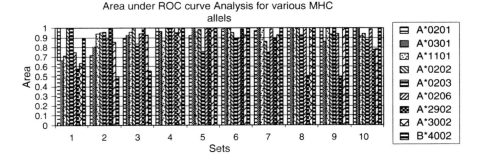

Fig. 1 Analysis of area under ROC curve

5 Results and Discussion

The tenfold cross validation has been used to validate the results. The ANN has been trained ten times for each MHC allele using adaptive learning, each time leaving one of the subsets out of 10, and using the left out subset for prediction. The area under ROC curve has been shown in Fig. 1.

We assembled a dataset of binders and non-binders for various MHC class I alleles to study the impact of the variable learning rate to train the ANN for small datasets. The ten set of binders and non-binders of nearly equal size have been used for tenfold cross validation. The average value of AROC for HLA-A*0201, HLA-A*0301, HLA-A*1101, HLA-A*0202, HLA-A*0203, HLA-A*0206, HLA-A*2902, HLA-A*3002, and HLA-B*4002 MHC Class I alleles is 0.9485, 0.922, 0.9333, 0.9615, 0.8989, 9405, 0.82, 0.906, and 0.8945, respectively, indicating excellent predictions for most of the cases.

The average values for various parameters viz. sensitivity, specificity, accuracy, PPV, and NPV for MHC alleles have been calculated. The average value of sensitivity, i.e., the percent of binders that are correctly predicted as binders, is 91.08.

Higher sensitivity means that almost all of the potential binders will be included in the predicted results. The average specificity, i.e., the percent of correctly predicted as non-binders, is 82.85. The average PPV value is 85.50. It shows that the probability that a predicted binder will actually be a binder is 85.50%. The average NPV is 89.47. It indicates that the probability that a predicted non-binder will actually be a non-binder is 89.47%.

The values used for training the ANN for various MHC alleles are: learning rate L (0.48–0.67), coefficient a (0.3–0.51) and coefficient b (0.0117–0.0725). Area under ROC curve using fixed learning rate for various MHC alleles was found to be 0.5 which indicates very poor prediction.

The modules for the training, classification, and results have been implemented in C using pointers, to improve the efficiency of training and classification through artificial neural network and variable leaning rate.

6 Conclusion

Overall, the study shows that the quality of the prediction of binders and non-binders can be substantially improved using the variable learning rate for artificial neural network training for small datasets. The approach can also be used for various other applications, where the datasets for training are limited. The approach is also useful for the large datasets. The only drawback of the approach is that the value of the parameters is to be adjusted as per the application.

Acknowledgments The authors acknowledge the support of Dr. D. S. Yadav, Assistant Professor in the Department of Computer Science and Engineering, Institute of Engineering and Technology, U P Technical University, Lucknow, and Dr. Firoz Khan, Scientist at CIMAP, Lucknow, India, in reading the entire manuscript and providing the useful suggestions.

References

1. Harriet L Robinson et al. "T cell vaccines for microbial infections", Nature Medicine, vol. 11, no. 4, pp. S25–S32, 2005
2. Anne S De Groot et al. "Genome-derived vaccines", Expert Review of Vaccines, vol. 3, no. 1, pp. 59–76, 2004
3. Anne S De Groot et al. "Immuno-informatics: mining genomes for vaccine components". Immunology and Cell Biology, vol. 80, pp. 255–269, 2002
4. Rino Rappuoli "Reverse vaccinology, a genome based approach to vaccine development", Vaccine, vol. 19, pp. 2688–2691, 2001
5. Tamas G Szabo et al. "Critical role of glycosylation in determining the length and structure of T-cell epitopes – As suggested by a combined in silico systems biology approach", Immunome Research, vol. 5, p. 4, 2009
6. Kathryn V Holmes "SARS coronavirus: a new challenge for prevention and therapy", Journal of Clinical Investigations, vol. 111, pp. 1605–1609, 2003
7. Joo Chuan Tong et al. "Methods and protocols for prediction of immuno-genic epitopes", Briefings in Bioinformatics, vol. 8, no. 2, pp. 96–108, 2006

8. Bing Zhao et al. "MHC-binding peptide prediction for epitope based vaccine design", International Journal of Integrative Biology, vol. 1, no. 2, pp. 127–140, 2007
9. Giuliano Armano et al. "A hybrid genetic-neural system for predicting protein secondary structure", BMC Bioinformatics, vol. 6, no. 4, S3, 2005
10. Yeon-Jin Cho1 et al. "Prediction rule generation of MHC class I binding peptides using ANN and GA", L. ICNC 2005, LNCS 3610, pp. 1009–1016, 2005
11. V Brusic, G Rudy, M Honeyman J Hammer, L Harrison "Prediction of MHC class II-binding peptides using an evolutionary algorithm and artificial neural network", Bioinformatics, vol. 14, no. 2, pp. 121–130, 1998
12. Henning Riedesel et al. "Peptide binding at class I major histo-compatibility complex scored with linear functions and support vector machines", Genome Informatics, vol. 15, no. 1, pp. 198–212, 2004
13. M Bhasin, G P S Raghava "A hybrid approach for predicting promiscuous MHC class I restricted T cell epitopes", Journal of Bioscience, vol. 32, no. 1, pp. 31–42, 2007
14. S S Soam et al. "Prediction of MHC class I binding peptides using probability distribution functions", Bioinformation, vol. 3, no. 9, pp. 403–408, 2009
15. S P Singh et al. "Evaluation of threading based method for prediction of peptides binding to MHC class I alleles", International Journal of Integrative Biology, vol. 4, no. 1, pp. 16–20, 2008
16. S P Singh et al. "Prediction of MHC binding peptides using Gibbs motif sampler, weight matrix and artificial neural network", Bioinformation vol. 3, no. 4, pp. 150–155, 2008
17. T Tollenaere "SuperSAB: fast adaptive back-propagation with good scaling properties", Neural Networks, vol. 3, no. 5, pp. 561–573, 1990
18. Enrique Castillo et al. "A very fast learning method for neural networks based on sensitivity analysis", Journal of Machine Learning Research, vol. 7 pp. 1159–1182, 2006

Part III
Protein Classification and Structure Prediction, and Computational Structural Biology

Chapter 23
Fast Three-Dimensional Noise Reduction for Real-Time Electron Tomography

José Antonio Martínez and José Jesús Fernández

Abstract Electron tomography (ET) allows visualization of the molecular architecture of biological specimens. Real-time ET systems allow scientists to acquire experimental datasets and obtain a preliminary structure of the specimen. This rough structure allows assessment of the quality of the sample and can also be used as a guide to collect more datasets. However, the low signal-to-noise ratio of the ET datasets precludes detailed interpretation and makes their assessment difficult. Therefore, noise reduction methods should be integrated in these real-time ET systems for their full exploitation. However, feature-preserving noise reduction methods are typically computationally intensive, which hinders real-time response. This work proposes and evaluates fast implementations of a sophisticated noise reduction method with capabilities of preservation of biologically relevant features. These implementations are designed to exploit the high performance computing (HPC) capabilities of modern multicore platforms and of graphics processing units. It is shown that the use of HPC on modern platforms makes this noise reduction method able to provide datasets appropriate for assessment in a matter of seconds, thereby making it suitable for integration in current real-time ET systems.

1 Introduction

Electron tomography (ET) has emerged as the leading technique for the structural analysis of unique complex biological specimens [1, 2]. ET has made it possible to directly visualize the molecular architecture of organelles, cells and complex viruses [3, 4]. The principle of ET is based on 3D reconstruction from projections. An individual biological sample is introduced in the electron microscope, and a series of images (so-called tilt series) is recorded by tilting the sample around a single

---

J.J. Fernández (✉)
Centro Nacional de Biotecnologia (CSIC), Campus UAM, Cantoblanco, 28049 Madrid, Spain
e-mail: JJ.Fernandez@cnb.csic.es

axis at different angles and under low-dose conditions to reduce radiation damage. These images are then combined to yield the 3D structure of the specimen by means of tomographic reconstruction algorithms [1, 2].

Real-time ET systems combine computer-assisted image collection with 3D reconstruction and provide the users not only with the acquired tilt series but also with a preliminary structure of the specimen [5, 6]. This rough structure allows the users to easily evaluate the quality of the specimen and decide whether a more time-consuming processing and thorough analysis of the dataset are worthwhile. These systems also guide the users to select further target areas to be imaged. The real-time 3D reconstruction is typically performed on a tilt series with reduced resolution ($2\times$, $4\times$) and using high performance computing (HPC) to obtain the volume in a matter of minutes [5–7].

The current real-time ET systems still have to deal with the challenge of the low signal-to-noise ratio (SNR) of the acquired datasets, which makes the preliminary structure very noisy. Even though this rough reconstructed volume is useful for users to get a first glimpse of the specimen, its low SNR reduces its utility and precludes a detailed inspection. Therefore, for full exploitation of these real-time ET systems, noise reduction methods are necessary. These denoising methods should fulfill a number of requisites: they should be nonlinear to ensure preservation of structural features; there should not be complicated parameters to tune, as these systems are not to run under interactive mode; they should be fast to yield solutions in real-time.

This work proposes and evaluates HPC implementations of a nonlinear noise reduction method based on geometric flow. The method fulfills the requisites to be integrable in current real-time ET systems: nonlinear, with no complicated parameters and fast. The parallel implementations are devised to exploit modern computing platforms, namely multicore computers and graphics processing units (GPUs). The HPC implementations show a significant reduction in the computing time, which definitely makes the denoising method appropriate for real-time ET systems.

2 A Noise Reduction Method Based on Geometric Flow

A noise reduction method with abilities to preserve the structural features of biological interest is necessary. This method should be capable of detecting edges and structures and setting up the strength of the smoothing accordingly. The method should attenuate, or even cancel out, the smoothing in areas where edges are detected, whereas it should apply strong smoothing elsewhere. Anisotropic nonlinear diffusion is currently the most powerful noise reduction technique in ET [8–10]. However, this method is computationally expensive and has a number of free parameters difficult to tune [11], so it is unsuitable for real-time ET.

The Beltrami flow is an efficient noise reduction method based on a geometric diffusion flow approach [12]. It considers images as maps embedded into a higher dimension, that is, a 2D image is considered as a 2-manifold embedded in 3D, i.e. the image $I(x, y)$ is regarded as a surface $S = (x, y, I(x, y))$ in a 3D space (see Fig. 1).

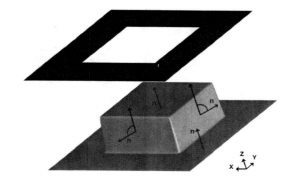

Fig. 1 An image $I(x,y)$ made up of a white square over black background (*top*) is viewed as a surface $S=(x,y,I(x,y))$ in a 3D space (*bottom*). The edges are seen as cliffs in the **Z** direction. At each point of the surface, the projection of the normal **n** (*arrows in gray*) to the **Z** direction (*arrows in black*) acts as an edge indicator

In this work, this idea has been extended to 3D, that is, a 3D volume $I(x,y,z)$ is considered as a 3-manifold embedded in a 4D space $S = (x,y,z,I(x,y,z))$. Embedding the multidimensional image into a higher dimension allows the use of powerful differential geometry operators. In that sense, the Beltrami flow is a geometric flow approach that aims to minimize the area of the image manifold, driving the flow towards a minimal surface solution while preserving edges [12].

The Beltrami flow is formulated as follows [12]:

$$I_t = \frac{1}{\sqrt{g}}\mathrm{div}\left(\frac{\nabla \mathbf{I}}{\sqrt{g}}\right) \quad (1)$$

where $I_t = \delta I/\delta t$ denotes the derivative of the image density I with respect to the time t; $\nabla \mathbf{I}$ is the gradient vector, that is $\nabla \mathbf{I} \equiv (I_x, I_y, I_z)$, being $I_x = \delta I/\delta x$ the derivative of I with respect to x (similar applies for y and z); div is the divergence operator, defined for a vector function $\mathbf{f} = (f_x, f_y, f_z)$ as $\mathrm{div}(\mathbf{f}) = \delta f_x/\delta x + \delta f_y/\delta y + \delta f_z/\delta z$. Finally, g denotes the determinant of the first fundamental form of the surface, which is $g = 1 + |\nabla \mathbf{I}|^2$. The term $1/\sqrt{g}$ in (1) acts as an edge indicator since it is proven to be the projection of the normal-to-the-surface to the vector representing the fourth dimension [12] (see Fig. 1). Therefore, the Beltrami flow is a selective noise filtering method that preserves structural features as minimizes diffusion at and across edges, whereas it applies extensive diffusion elsewhere.

The explicit discretization of the partial differential equation derived from (1) uses an Euler forward difference approximation for I_t and central differences to approximate the spatial derivatives:

$$I^k = I^{k-1} + h_t \frac{I_{xx}(1 + I_y^2 + I_z^2) + I_{yy}(1 + I_x^2 + I_z^2) + I_{zz}(1 + I_x^2 + I_y^2)}{(1 + I_x^2 + I_y^2 + I_z^2)^2}$$
$$- 2h_t \frac{I_x I_y I_{xy} + I_x I_z I_{xz} + I_y I_z I_{yz}}{(1 + I_x^2 + I_y^2 + I_z^2)^2} \quad (2)$$

where I^k is the image in the k-th iteration, h_t is the time step (for stability, the maximum value is the inverse of the squared number of dimensions, i.e. $1/3^2$), I_x is

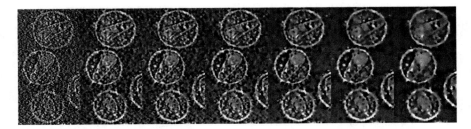

Fig. 2 Noise reduction and feature preservation over a representative ET dataset of human immunodeficiency virions (strain HIV-1) taken from the Electron Microscopy Data Bank (http://emdatabank.org). From *left* to *right*, the original reconstruction and the denoised results at 10, 25, 50, 100, 150 and 200 iterations are shown. The background is progressively flattened with the iterations, whereas the structural features remain sharp. At high number of iterations, some edges begin to look blurred. Only a representative slice of the 3D reconstruction is presented

the derivative with respect to x (similar applies for y and z), I_{xx} is the second-order derivative with respect to x (similar applies for y and z) and I_{xy} is the mixed second order partial derivative with respect to x and y (similar applies for xz and yz).

The noise reduction method based on the Beltrami flow has no complicated parameters to be tuned, as the detection of the edges and estimation of their strength is performed based on g, which is directly computed from the gradient. Nevertheless, the method is solved in an iterative way (2), and hence a number of iterations has to be specified. However, this does not pose a serious inconvenience since a number of 100 iterations typically yield good results for 3D ET datasets [13]. Figure 2 illustrates the performance of this method.

3 HPC Implementations for Modern Platforms

The advent of multicore computers (several independent processors sharing on-chip resources and computer memory) has signalled a historic switch in HPC as the road to higher performance will be via multiple, rather than faster, processors [14]. There are opportunities to obtain exceptional performance by taking advantage of the computing power in commodity multicore computers. These platforms belong to the classical category of tightly coupled shared-memory computers in HPC, made up of a relatively small number of processors sharing a single centralized memory. In these systems, a program can be decomposed into a set of tasks (normally implemented in the form of threads) that are executed in a collaborative fashion, aiming at minimizing the time to solve the global problem. An advantage of these systems is that the data exchange and the synchronization of the threads are greatly facilitated using simple reads/writes through the memory accessible for all the threads.

In this work, the 3D noise reduction method based on the Beltrami flow has been implemented using Pthreads [15] to exploit the power of modern multicore computers. The strategy is based on domain decomposition whereby the 3D volume

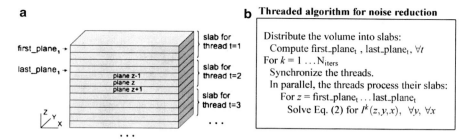

Fig. 3 Multithreaded implementation of the Beltrami flow for multicore computers. (**a**) Scheme showing the distribution of the volume into slabs of Z planes to be processed by the threads in parallel. (**b**) Multithreaded algorithm. t denotes the index of the thread. first_plane$_t$ and last_plane$_t$ are the indices of the first and last Z planes of the volume included in the slab assigned to thread t

is considered as made up of slabs of Z planes (Fig. 3). The volume is thus divided into as many slabs as threads working in parallel (typically the number of threads equals the number of cores in the system). The threads progress independently, each processing its own slab. The processing of a slab basically consists in the solution of (2) for all the Z planes in the slab. There is a synchronization point at the beginning of each iteration to ensure that all the threads start the iteration from the updated volume resulting from the previous iteration. During the processing, there are two buffers that contain the volume in the previous (I^{k-1}) and in the current iteration (I^k). The threads only write on the planes that they are processing in the current volume (I^k), though they can read any location of the previous volume (I^{k-1}) to properly deal with the neighbour dependencies. Taking into account that the spatial derivatives are approximated by means of central differences (2) clearly shows that the processing of a single voxel involves a neighbourhood of the 18 immediate neighbour voxels, in X, Y and Z.

GPUs have emerged as new computing platforms that offer massive parallelism and provide incomparable performance-to-cost ratio for scientific computations [16]. GPUs have hundreds of cores that can collectively run thousands of computing threads. Currently, GPUs architectures can be considered as coprocessors that accelerate the computation with intensive data parallelism. The use of GPUs for general purpose applications has exceptionally increased in the last few years, thanks to the availability of programming interfaces, such as CUDA, that greatly facilitate the development of applications targeted at GPUs [16]. Heterogeneous computing is now supported, where applications use both CPU and GPU. Serial portions of applications are run on the CPU, and parallel portions (kernels) are accelerated on the GPU. In order to develop efficient codes for GPUs with CUDA, the programmer has to take into account several architectural characteristics, such as the topology of the multiprocessors and the management of the memory hierarchy. From the programmer's point of view, the GPU is considered as a set of multiprocessors with shared memory that can run following the Single Program Multiple Data (SPMD) programming model. The GPU architecture allows the host to issue a succession of kernel invocations to the device. Each kernel is executed as a batch of threads organized as a grid of thread blocks. In ET, there is an increasing interest in GPU computing because many image processing procedures are well suited for the type of SPMD parallelism exploited by GPUs [17].

In this work, a CUDA implementation of the 3D noise reduction method has been developed. Different strategies have been tested. The most optimal is one where the CPU sweeps across all the Z planes, and the processing of each Z plane is carried out in the GPU by means of multiple threads working in parallel (Fig. 4). Each thread processes a pixel of the current plane, which essentially consists in the solution of (2) for that pixel. Each thread block is launched on a multiprocessor of the GPU. There is no synchronization among the threads as this is explicitly carried out by the CPU

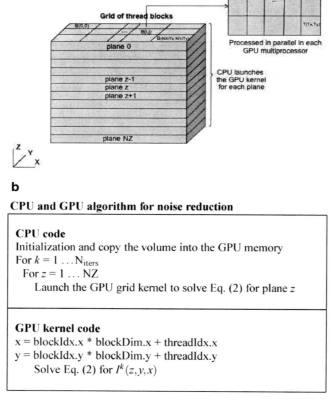

Fig. 4 GPU implementation of the Beltrami flow. (**a**) Scheme showing the distribution of the threads to process the pixels of the current plane in the GPU. NX, NY and NZ denote the dimensions of the volume. Tx and Ty denote the size of the thread block. (**b**) The algorithm running on the CPU and the GPU kernel. The CPU launches the GPU kernel as many times as Z planes

in this plane-by-plane strategy. As happened with the multicore implementation, two buffers are used to contain the volume in the previous (I^{k-1}) and in the current iteration (I^k). Also, the read and write operations and the neighbour dependencies are dealt with similarly.

4 Results

The implementations of the denoising method have been analysed in terms of the processing time and its scalability to exploit modern multicore and GPU platforms, respectively, with the ultimate aim of assessing its potential to be integrated in real-time ET systems. A number of 100 iterations was used to obtain the measures.

To measure the processing time and the scalability, synthetic datasets of different sizes, 256×256×256, 384×384×384, 512×512×512 and 640×640×640 voxels, were used. These sizes are representative of the experimental datasizes (0.5–1.0 GB) currently used in ET. The scalability was assessed based on the speedup factor, which is defined as $S_p = T_1/T_p$ where T_1 is the sequential processing time and T_p is the parallel processing time. The processing time and the speedup for the multithreaded implementation were measured on a state-of-the-art computer with two Intel Xeon E5320 Quad Core (eight cores in total) running at 1.86 GHz under linux. The performance of the GPU implementation was measured on a computer based on Intel Core 2 Duo E8400 at 3.00 GHz using a GPU of a NVIDIA GeForce GTX 295 card, with 30 multiprocessors of eight cores (i.e. a total of 240 cores) at 1.2 GHz, 896 MB of memory and compute capability 1.3.

Figure 5 shows the processing time of the multithreaded implementation to filter the datasets with different sizes used in this work. The graphs show the processing time as a function of the number of threads. There is a significant decrease in the computation time when the potential of parallel processing of modern multicore platforms is exploited. In particular, it is clearly observed that the use of four threads makes the noise reduction process require less than 7 min (400 s) in the largest case. This order of time is similar to that required by the real-time ET reconstruction [6]. Therefore, any modern computer based on a quad-core processor would easily allow the integration of this denoising method in a real-time ET system. Moreover, the use of eight threads further reduces the computing time requirements down to 4 min or less (200 s) in the worst case. Therefore, any state-of-the-art server provided with eight cores would exceed the best expectations.

Figure 6 shows the speedups of the multithreaded implementation as a function of the number of threads and for the different data sizes. All the speedup curves perfectly fit the linear behaviour, which means that the scalability of the parallel strategy used in this work is excellent. In other words, one could expect that, in the general case, the computing time would be reduced by a factor equal to the number of threads used.

Figures 6 and 7 show the speedup and the processing times, respectively, of the GPU implementation as a function of the dataset. Different configurations for the

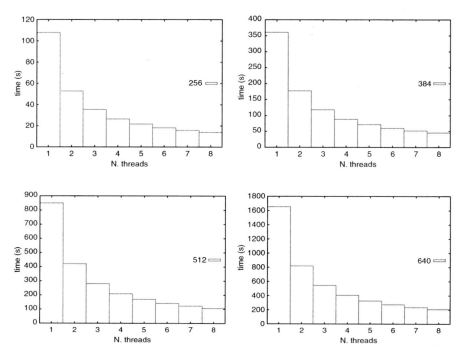

Fig. 5 Processing times of the multithreaded implementation for different data sizes ($256\times256\times256$, $384\times384\times384$, $512\times512\times512$ and $640\times640\times640$) as a function of the number of threads. These times correspond to 100 iterations of the noise reduction method presented in this work. The platform was a state-of-the-art computer with two Intel Xeon E5320 Quad Core (eight cores in total) running at 1.86 Ghz under linux

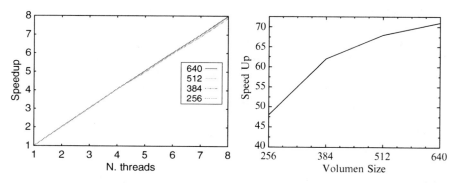

Fig. 6 Speedup factors of the multithreaded (*left*) and GPU (*right*) implementations computed from the processing times shown in Fig. 5, respectively

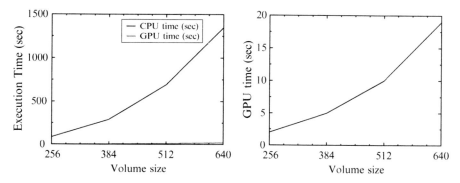

Fig. 7 Processing times of the GPU implementation as a function of data size. A number of 100 iterations of the noise reduction method was used. On the *left*, times on the GPU and on the CPU hosting the GPU are shown in the same plot for comparison. The plot on the *right* shows the processing times on the GPU in a more proper scale. The CPU was an Intel Core 2 Duo E8400 at 3.00 GHz and the GPU belonged to a NVIDIA GeForce GTX 295 card, with 30 multiprocessors of eight cores (i.e. a total of 240 cores) at 1.2 GHz

thread block have been tested. Although the performance is similar for all of them, the best one is achieved with a thread block of size $Tx \times Ty = 16 \times 16$. The GPU implementation achieves an exceptional speedup factor, in the range $[50\times, 70\times]$, and exhibits a monotonically increasing curve with the data size. In view of this curve, it is expected that the GPU implementation will perform even better for larger datasets that will be available in ET in the near future. Even more importantly, the processing time required for denoising the datasets has been reduced to less than 20 s. Therefore, it can be concluded that the exploitation of the computational power within modern GPUs makes real-time noise reduction possible.

5 Conclusion

In this work efficient implementations of a method for noise reduction with potential to be integrated in real-time ET systems have been presented. The denoising method is based on a geometric flow approach, the Beltrami flow, and is equipped with abilities to automatically detect edges. As a consequence, it can properly tune the strength of the smoothing so that biologically relevant information is preserved even though significant noise reduction is carried out elsewhere. Therefore, the resulting datasets can be subjected to detailed inspection and assessment. Furthermore, the method has no complicated parameters to be tuned, which makes it attractive for systems not conceived to be run under interactive mode. The efficient implementations of the method presented here have been intended to exploit the parallel capabilities of modern multicore workstations and of GPUs. It has been demonstrated that good solutions can be obtained in a matter of minutes on standard multicore computers. Moreover, GPUs available in standard graphics cards even outperform these results and are able to provide clean datasets in a few seconds. Therefore, these implementations that take advantage of the computational power in standard computers make this sophisticated noise reduction method suitable for real-time operation. This filtering method is thus expected to be of invaluable help for scientists to assess the quality of their datasets acquired during their ET sessions, and also use this information as guide for subsequent data collection.

Acknowledgements Work supported by grants MCI-TIN2008-01117, JA-P06-TIC1426 and CSIC-PIE-2009201075.

References

1. Fernandez JJ, Sorzano COS, Marabini R, Carazo JM (2006) Image processing and 3D reconstruction in electron microscopy. IEEE Signal Process Mag 23(3):84–94
2. Leis AP, Beck M, Gruska M, Best C, Hegerl R, Baumeister W, Leis JW (2006) Cryo-electron tomography of biological specimens. IEEE Signal Process Mag 23(3):95–103
3. Medalia O, Weber I, Frangakis AS, Nicastro D, Gerisch G, Baumeister W (2002) Macromolecular architecture in eukaryotic cells visualized by cryoelectron tomography. Science 298:1209–1213
4. Cyrklaff M, Risco C, Fernandez JJ, Jimenez MV, Esteban M, Baumeister W, Carrascosa JL (2005) Cryo-electron tomography of vaccinia virus. Proc Natl Acad Sci USA 102:2772–2777
5. Schoenmakers RHM, Perquin RA, Fliervoet TF, Voorhout W, Schirmacher H (2005) New software for high resolution, high throughput electron tomography. Micros Anal 19(4):5–6
6. Zheng SQ, Keszthelyi B, Branlund E, Lyle JM, Braunfeld MB, Sedat JW, Agard DA (2007) UCSF tomography: An integrated software suite for real-time electron microscopic tomographic data collection, alignment, and reconstruction. J Struct Biol 157:138–147
7. Fernandez JJ (2008) High performance computing in structural determination by electron cryomicroscopy. J Struct Biol 164:1–6
8. Frangakis AS, Hegerl R (2001) Noise reduction in electron tomographic reconstructions using nonlinear anisotropic diffusion. J Struct Biol 135:239–250
9. Fernandez JJ, Li S (2003) An improved algorithm for anisotropic nonlinear diffusion for denoising cryo-tomograms. J Struct Biol 144:152–161
10. Fernandez JJ, Li S (2005) Anisotropic nonlinear filtering of cellular structures in cryo-electron tomography. Comput Sci Eng 7(5):54–61
11. Fernandez JJ, Li S, Lucic V (2007) Three-dimensional anisotropic noise reduction with automated parameter tuning: Application to electron cryotomography. Lect Notes Comput Sci 4788:60–69
12. Kimmel R, Malladi R, Sochen NA (2000) Images as embedded maps and minimal surfaces: Movies, color, texture, and volumetric medical images. Int J Comput Vis 39:111–129
13. Fernandez JJ (2009) Tomobflow: Feature-preserving noise filtering for electron tomography. BMC Bioinformatics 10:178
14. Geer D (2005) Chip makers turn to multicore processors. IEEE Comput 38:11–13
15. Butenhof DR (1997) Programming with POSIX(R) Threads. Addison-Wesley Professional, Reading.
16. Nickolls J, Buck I, Garland M, Skadron K (2008) Scalable parallel programming with CUDA. ACM Queue 6:40–53
17. Castano-Diez D, Mueller H, Frangakis AS (2007) Implementation and performance evaluation of reconstruction algorithms on graphics processors. J Struct Biol 157:288–295

Chapter 24
Prediction of Chemical-Protein Binding Activity Using Contrast Graph Patterns

Andrzej Dominik, Zbigniew Walczak, and Jacek Wojciechowski

Abstract The problem of classifying chemical compounds is studied in this chapter. Compounds are represented as two-dimensional topological graphs of atoms (corresponding to nodes) and bonds (corresponding to edges). We use a method called contrast common pattern classifier (CCPC) to predict chemical-protein binding activity. This approach is strongly associated with the classical emerging patterns techniques known from decision tables. It uses two different types of structural patterns (subgraphs): contrast and common. Results show that the considered algorithm outperforms all other existing methods in terms of classification accuracy.

1 Introduction

Computational chemistry and chemical informatics deal with problems (e.g., classification) that can be solved using machine learning and knowledge discovery methods. As the number of large publicly available chemical compounds datasets is increasing, it is necessary to develop effective algorithms that provide accurate classification results. Having in mind that chemical compounds are complex structures, designing of such methods is a challenging task.

The problem of classifying chemical compounds has been recently deeply studied. Two major approaches have been developed: quantitative structure–activity relationships (QSAR) and structural approach. First one concentrates on physicochemical properties derived from compounds. It requires genuine chemical knowledge and it is only limited to chemical applications. Second approach searches directly structure of the compound. It does not require any chemical knowledge and can be used in other domains as well (e.g., web documents classification [14]). Its main idea is based on discovering small, significant substructures

---

A. Dominik (✉)
Institute of Radioelectronics, Warsaw University of Technology, Nowowiejska 15/19, 00-665 Warsaw, Poland
e-mail: a.dominik@elka.pw.edu.pl

(patterns, fragments) within the original structures which discriminate between different classes of structures. Many techniques were used to achieve this goal, like support vector machines (SVM) [12], k-Nearest Neighbours (kNN) [14], or frequent patterns [2].

Recently we proposed a new, universal structural approach (CCPC) for classifying objects represented by graphs [6–10]. This algorithm uses concepts that were originally developed and introduced for data mining (jumping emerging patterns – JEP, and emerging patterns – EP). The CCPC algorithm uses minimal [with respect to size and inclusion (non-isomorphic)] contrast and common connected subgraphs. Contrast subgraphs are patterns that are exclusive for one class of objects, while common subgraphs appear in more than one class of objects. This algorithm was already successfully applied for solving classification problems in different fields of science: computational chemistry, chemical informatics (e.g., detection of mutagenicity, toxicity, and anticancer activity [7, 8]) and information science (e.g., web documents categorization [6]). Results show that it provides at least as high classification accuracy (in most cases classification accuracy is significantly higher [9]) as any current state-of-the-art classifier.

Classification methods, based on the structure of chemical compounds, operate on different representations of molecules (depending on their dimensions and features). Here we concentrate our research on 2D topological graphs. In this representation atoms correspond to labeled vertices (atom symbol is used as vertex label) and bonds to labeled edges (multiplicity of a bond is used as edge label). These graphs are typically quite small (in terms of the numbers of vertices and edges) and the average number of edges per vertex is usually slightly above two.

This chapter discovers the possibility of applying CCPC classifier to predict chemical-protein binding activity (ability of a chemical compound to bind to a particular protein). This is a very important issue during drug design process, because drug's efficiency may be affected by the degree to which it binds to proteins within blood plasma. The less bound a drug is, the more efficiently it can traverse cell membranes or diffuse [17].

This chapter is organized as follows. In Sect. 2, the state of the art in graph mining is briefly described. Preliminary terminology on graph theory and the concept of CCPC classifier are introduced in Sects. 3 and 4, respectively. Section 5 concentrates on experiments. Conclusions and final remarks are in Sect. 6.

2 Related Work

There are numerous types of patterns that can be used to build classifiers, e.g., frequent, common, and contrast. Contrast patterns are substructures that appear in one class of objects and do not appear in other classes. In data mining, patterns that uniquely identify certain class of objects are called jumping emerging patterns (JEP). Patterns common for different classes are called emerging patterns (EP). Concepts of JEP and EP [13] have been deeply researched as a tool for classification purposes in databases. They are reported to provide high classification accuracy results. The concept of contrast subgraphs was studied in [1, 6, 7, 15]. Ting and Bailey proposed an effective algorithm (containing backtracking tree and hypergraph traversal algorithm) for mining all disconnected contrast subgraphs from dataset containing graphs. Common graph is a subgraph isomorphic to at least one graph in at least two different classes of graphs, while frequent graph is subgraph isomorphic to at least as many graphs in a particular class of graphs as specified threshold (minimal support of a graph).

Mining patterns in graphs dataset which fulfil given conditions is a much more challenging task than mining conceptually similar patterns in decision tables (relational databases). The most important difficulty is checking subgraph isomorphism and isomorphism of graphs. Both of these are computationally complex tasks. The first problem is proved to be *NP*-complete, while the complexity of the other one is still not known. All the algorithms for solving the isomorphism problem present in the literature have an exponential time complexity in the worst case, but polynomial solution has not been yet disproved. A universal exhaustive search algorithm for both of these problems was proposed in [16]. It operates on the matrix representation of graphs and tries to find a proper permutation of nodes. Search space can be greatly reduced using nodes (edges) invariants and iterative partitioning [11]. Moreover, multiple graph isomorphism (for a set of graphs determine which of them are isomorphic) problems can be efficiently performed with canonical labeling [11]. Canonical label is a unique representation (code) of a graph such that two isomorphic graphs have the same canonical label.

One of the most popular approaches for graph classification is based on SVM. SVMs have good generalization properties (both theoretically and experimentally) and they operate well in high-dimensional datasets. Numerous different kernels using all three compound representations were designed for this method [5, 12].

3 Preliminary Terminology

In this section, we introduce some basic concepts and definitions [6–8, 10] that will be used in the subsequent sections.

All discussed graphs are assumed to be undirected, connected (any two vertices are linked by a path), labeled (both vertices and edges possess labels), and simple (without self-loops and parallel edges). By the size of a graph, we mean the number of its edges. Capital letters ($G, S, \ldots$) denote single graphs while calligraphic letters ($\mathcal{G}, \mathcal{N}, \mathcal{P}, \ldots$) denote sets of graphs.

3.1 Labeled Graph

Labeled graph G is a quadruple (V, E, α, β), where V is a non-empty finite set of vertices, $E \subseteq V \times V$ is a non-empty finite set of edges, and α, β are functions assigning labels to vertices and edges, respectively.

3.2 Subgraph

A graph $S = (W, F, \alpha, \beta)$ is a subgraph of $G = (V, E, \alpha, \beta)$ (written as $S \subseteq G$) if: (1) $W \subseteq V$ and (2) $F \subseteq E \cap (W \times W)$.

3.3 Isomorphism and Subgraph Isomorphism

Let $\mathcal{G}$ be a set of graphs, $G' = (V', E', \alpha', \beta')$ and $G = (V, E, \alpha, \beta)$. We say that G' is subgraph isomorphic to G (written as $G' \simeq G$) if there is an injective function $f : V' \longrightarrow V$ such that: (1) $\forall e' = (u', v') \in E'\ \exists e = (f(u'), f(v')) \in E$, (2) $\forall u' \in V', \alpha'(u') = \alpha(f(u'))$, and (3) $\forall e' \in E', \beta'(e') = \beta(f(e'))$. If $f : V' \longrightarrow V$ is a bijective function then G' is isomorphic to G (written as $G' = G$). If G' is not subgraph isomorphic to G then we write $G' \not\simeq G$.

A graph G' is $\mathcal{G}$-isomorphic (written as $G' \simeq \mathcal{G}$) if: (1) $\exists G \in \mathcal{G} : G' \simeq G$. A graph G' is not $\mathcal{G}$-isomorphic (written as $G' \not\simeq \mathcal{G}$) if: (1) $\forall G \in \mathcal{G} : G' \not\simeq G$.

3.4 Common Subgraph

Given the sets of graphs $\mathcal{P}$, $\mathcal{N}$ and a graph $M_{\mathcal{P}\mathcal{N}}$. $M_{\mathcal{P}\mathcal{N}}$ is a common subgraph for $\mathcal{P}$ and $\mathcal{N}$ if: (1) $M_{\mathcal{P}\mathcal{N}} \simeq \mathcal{P}$ and (2) $M_{\mathcal{P}\mathcal{N}} \simeq \mathcal{N}$. Set of all common subgraphs for $\mathcal{P}$ and $\mathcal{N}$ will be denoted by $\mathcal{M}_{\mathcal{P}\mathcal{N}}$. Set of all minimal (with respect to the size, i.e., containing only one edge and two vertices) common subgraphs for $\mathcal{P}$ and $\mathcal{N}$ will be denoted as $\mathcal{M}_{\mathcal{P}\mathcal{N}}^{\text{Min}}$.

3.5 Contrast Subgraph

Given the sets of graphs $\mathcal{N}$ and a graph P. A graph $C_{P \to \mathcal{N}}$ is a contrast subgraph of P with respect to $\mathcal{N}$ if: (1) $C_{P \to \mathcal{N}} \simeq P$ and (2) $C_{P \to \mathcal{N}} \not\simeq \mathcal{N}$. It is minimal (with respect to subgraph isomorphism) if all of $C_{P \to \mathcal{N}}$'s strict subgraphs are not contrast subgraphs. Set of all minimal contrast subgraphs of P with respect to $\mathcal{N}$ will be denoted as $\mathcal{C}_{P \to \mathcal{N}}^{\text{Min}}$.

Given the sets of graphs $\mathcal{P} = \{P_1, \ldots, P_n\}$ and $\mathcal{N}$. Let $\mathcal{C}_{P_i \to \mathcal{N}}^{\text{Min}}$ be the set of all minimal contrast subgraphs of P_i with respect to $\mathcal{N}$, where $i \in \langle 1, n \rangle$. $\mathcal{C}_{\mathcal{P} \to \mathcal{N}}^{\text{Min}}$ is a set of all minimal contrast subgraphs of $\mathcal{P}$ with respect to $\mathcal{N}$ if: (1) $\forall C \in \mathcal{C}_{\mathcal{P} \to \mathcal{N}}^{\text{Min}}\ \exists J \in \mathcal{C}_{P_i \to \mathcal{N}}^{\text{Min}} : J \simeq C$, for $i \in \langle 1, n \rangle$, (2) $\forall J_1 \in \mathcal{C}_{\mathcal{P} \to \mathcal{N}}^{\text{Min}}\ \neg \exists J_2 \in \mathcal{C}_{\mathcal{P} \to \mathcal{N}}^{\text{Min}} \setminus J_1 : J_2 \simeq J_1$.

$\mathcal{C}_{\mathcal{P} \to \mathcal{N}}^{\text{Min}}$ contains all minimal subgraphs (patterns) which are present in $\mathcal{P}$ (i.e., each subgraph in $\mathcal{C}_{\mathcal{P} \to \mathcal{N}}^{\text{Min}}$ is subgraph isomorphic to at least one graph from $\mathcal{P}$) and are not present in $\mathcal{N}$ (i.e., each subgraph in $\mathcal{C}_{\mathcal{P} \to \mathcal{N}}^{\text{Min}}$ is not subgraph isomorphic to any graph from $\mathcal{N}$). What is more $\mathcal{C}_{\mathcal{P} \to \mathcal{N}}^{\text{Min}}$ contains only minimal (with respect to size and subgraph isomorphism) subgraphs.

3.6 Support and Growth Rate

Given the sets of graphs $\mathcal{G}, \mathcal{N}, \mathcal{P}$ and a graph G. Let $\mathcal{S} = \{G' \in \mathcal{G} : G \simeq G'\}$. Support of graph G in $\mathcal{G}$ is defined as follows: $\mathrm{supp}_{\mathcal{G}}(G) = \frac{\mathrm{card}(\mathcal{S})}{\mathrm{card}(\mathcal{G})}$, where $\mathrm{card}(\mathcal{G})$ denotes the cardinal number of set $\mathcal{G}$. Growth rate of graph G in favor of $\mathcal{P}$ against $\mathcal{N}$ is expressed as follows: $\rho_{\mathcal{P} \to \mathcal{N}}(G) = \frac{\mathrm{supp}_{\mathcal{P}}(G)}{\mathrm{supp}_{\mathcal{N}}(G)}$.

4 Contrast Common Pattern Classifier

In this section, we briefly describe classification algorithm called CCPC, which is based on the concept of contrast and common graphs. For additional information on CCPC classifier, refer to [6–10].

Measures for classical emerging patterns designed for classification purposes are mainly based on the support of a pattern in different classes of objects. We adapted some classical scoring schemes to be used in conjunction with contrast and common subgraphs.

Let $\mathcal{G}$ be a set of training graphs (graphs used for the learning of a classifier) and G be a test graph (graph to classify). For the sake of convenience, let us assume that $\mathcal{G}$ is divided into two disjoint non-empty decision classes: positive $\mathcal{P}$ and negative $\mathcal{N}$, $\mathcal{G} = \mathcal{N} \cup \mathcal{P}$. Let $\mathcal{C}^{\mathrm{Min}}_{\mathcal{P} \to \mathcal{N}}$ be the set of all minimal contrast subgraphs of $\mathcal{P}$ with respect to graph set $\mathcal{N}$ and $\mathcal{C}^{\mathrm{Min}}_{\mathcal{N} \to \mathcal{P}}$ be the set of all minimal contrast subgraphs of $\mathcal{N}$ with respect to $\mathcal{P}$. Let $\mathcal{M}^{\mathrm{Min}}_{\mathcal{P}\mathcal{N}}$ be the set of all minimal common subgraphs for $\mathcal{P}$ and $\mathcal{N}$.

Let us now define a few score routines used for classification. Score is obtained using contrast subgraphs according to the following equations:

$$\mathrm{scConA}_{\mathcal{P}}(G) = \sum_{K \in \mathcal{K}} \mathrm{supp}_{\mathcal{P}}(K)$$
$$\mathcal{K} = \{K : K \in \mathcal{C}^{\mathrm{Min}}_{\mathcal{P} \to \mathcal{N}} \wedge K \simeq G\} \quad (1)$$

$$\mathrm{scConB}_{\mathcal{P}}(G) = \frac{1}{\lambda_{\mathcal{P}}} * \sum_{K \in \mathcal{K}} \mathrm{supp}_{\mathcal{P}}(K)$$
$$\mathcal{K} = \{K : K \in \mathcal{C}^{\mathrm{Min}}_{\mathcal{P} \to \mathcal{N}} \wedge K \simeq G\} \quad (2)$$

$$\mathrm{scConA}_{\mathcal{N}}(G) = \sum_{K \in \mathcal{K}} \mathrm{supp}_{\mathcal{N}}(K)$$
$$\mathcal{K} = \{K : K \in \mathcal{C}^{\mathrm{Min}}_{\mathcal{N} \to \mathcal{P}} \wedge K \simeq G\} \quad (3)$$

$$\mathrm{scConB}_{\mathcal{N}}(G) = \frac{1}{\lambda_{\mathcal{N}}} * \sum_{K \in \mathcal{K}} \mathrm{supp}_{\mathcal{N}}(K)$$
$$\mathcal{K} = \{K : K \in \mathcal{C}^{\mathrm{Min}}_{\mathcal{N} \to \mathcal{P}} \wedge K \simeq G\} \quad (4)$$

where $\lambda_{\mathcal{P}}, \lambda_{\mathcal{N}}$ are scaling factors. They are median values from statistics of the contrast scores, (1), (3), determined for each graph from both classes: $\mathcal{P}, \mathcal{N}$. $\mathcal{K}$

used in: (1), (2) and (3), (4) is a temporary set, which contains minimal contrast subgraphs of $\mathcal{P}$ ($\mathcal{N}$) with respect to $\mathcal{N}$ ($\mathcal{P}$) that are subgraph isomorphic to test graph G. Scores for both classes are calculated as a sum of supports of all subgraphs from $\mathcal{K}$ in certain class (i.e., $\mathcal{P}, \mathcal{N}$).

Our first aim is to classify all graphs using only contrast patterns (contrast subgraphs). Sometimes knowledge obtained from contrast subgraphs is not sufficient to classify test objects (i.e., test object remains unclassified). In his case, our classifier takes also advantage of other structural patterns: common subgraphs. These patterns play marginal role in the classification procedure and are used only when test object cannot be classified using solely contrast patterns. In such a case scores are calculated according to the following equations:

$$\text{scComA}_{\mathcal{P}}(G) = \sum_{K \in \mathcal{K}} \text{supp}_{\mathcal{P}}(K)$$
$$\mathcal{K} = \{K : K \in \mathcal{M}_{\mathcal{P}\mathcal{N}}^{\text{Min}} \wedge K \simeq G\} \qquad (5)$$

$$\text{scComA}_{\mathcal{N}}(G) = \sum_{K \in \mathcal{K}} \text{supp}_{\mathcal{N}}(K)$$
$$\mathcal{K} = \{K : K \in \mathcal{M}_{\mathcal{P}\mathcal{N}}^{\text{Min}} \wedge K \simeq G\} \qquad (6)$$

$$\text{scComB}_{\mathcal{P}}(G) = \sum_{K \in \mathcal{K}} \rho_{\mathcal{P} \to \mathcal{N}}(K)$$
$$\mathcal{K} = \{K : K \in \mathcal{M}_{\mathcal{P}\mathcal{N}}^{\text{Min}} \wedge K \simeq G\} \qquad (7)$$

$$\text{scComB}_{\mathcal{N}}(G) = \sum_{K \in \mathcal{K}} \rho_{\mathcal{N} \to \mathcal{P}}(K)$$
$$\mathcal{K} = \{K : K \in \mathcal{M}_{\mathcal{P}\mathcal{N}}^{\text{Min}} \wedge K \simeq G\} \qquad (8)$$

In (5) and (6), score depends directly on the support of the minimal common subgraphs, whereas in (7), (8) score depends on the growth rate of certain patterns. $\mathcal{K}$ is a temporary set, which contains minimal common subgraphs for $\mathcal{P}$ and $\mathcal{N}$ that are subgraph isomorphic to test graph G.

4.1 Training and Testing Phase

Classifier train process looks as follows. First, all minimal [with respect to size and inclusion (non-isomorphic)] contrast subgraphs characteristic for negative and positive classes are discovered ($\mathcal{C}_{\mathcal{P} \to \mathcal{N}}, \mathcal{C}_{\mathcal{N} \to \mathcal{P}}$). Then, all minimal [with respect to size and inclusion (non-isomorphic)] common subgraphs for both classes are discovered ($\mathcal{M}_{\mathcal{P}\mathcal{N}}^{\text{Min}}$). Subgraph discovery is performed using Depth First Search (DFS) code generation method, and all necessary isomorphism checking are performed using canonical labeling, nodes invariants, and iterative partitioning methods [11]. Additional speed up (as well as reduced memory consumption) can be achieved by limiting size of discovered subgraphs, i.e., instead of discovering all contrast subgraphs only some of them are investigated up to a given size (number of edges).

Maximal size of discovered subgraph is an additional method parameter denoted as s. This limitation may influence classification accuracy.

Classification process looks as follows. First scores based on contrast subgraphs are calculated for each class. We can choose between presented scoring schemes: scConA [scConA$_\mathcal{P}$(G), scConA$_\mathcal{N}$(G)] from (1), (3) and scConB [scConB$_\mathcal{P}$(G), scConB$_\mathcal{N}$(G)] from (2), (4). The tested graph G is assigned to a class with a higher score. If both scores are equal, then G remains unclassified and scores based on common subgraphs are calculated. Again we can choose one of the two approaches: scComA [scComA$_\mathcal{P}$(G), scComA$_\mathcal{N}$(G)] from (5), (6) and scComB [scComB$_\mathcal{P}$(G), scComB$_\mathcal{N}$(G)] from (7), (8). The tested graph is assigned to a class with a higher score. If both scores are equal, then G remains unclassified.

5 Experiments

5.1 Datasets

We used Jorissen dataset collection from drug virtual screening experiment [3]. This collection contains chemical compounds along with the information whether they have binding affinity to a particular protein. Five different proteins are considered: α-1A adrenoceptor, cyclin-dependent kinase 2, cyclooxygenase 2, coagulation factor Xa, and phosphodiesterase 5. For each protein, 50 binders (chemical structures that bind to the protein) are provided. In addition, there is a list selected by domain expert decoys, i.e., chemical compounds that are very similar to binders but do not bind to target protein.

For our experiments, we created five datasets from original data. Each dataset corresponds to single protein and contains 50 binders and 50 decoys (randomly selected from provided decoys list). Each dataset defines binary classification problem with following decision classes: positive (binder) and negative (non-binder, i.e., decoy). Similar datasets were used in [2, 4].

The characteristics of used datasets and detailed information on the structure (minimal, average, maximal number of atoms and bonds) of chemical compounds in each dataset are shown in Table 1.

Table 1 Characteristics of datasets

Dataset	Protein target	Number of compounds (%)			Minimum		Average		Maximum	
		Total	Negative	Positive	A	B	A	B	A	B
A1A	α-1A adrenoceptor	100	50	50	8	7	26	28	47	51
CDK2	Cyclin-dependent kinase 2	100	50	50	8	7	22	24	46	46
COX2	Cyclooxygenase 2	100	50	50	8	7	26	29	46	46
Fxa	Coagulation factor Xa	100	50	50	8	7	26	29	46	46
PDE5	Phosphodiesterase 5	100	50	50	8	7	25	28	46	49

5.2 Experiment Plan

We concentrated our research on one issue: performance of a classifier. In order to make a good comparison with other available research results (for these datasets), we estimated classification accuracy [ability to assign the correct class to an example (graph)] using tenfold cross-validation procedure [6, 7].

Accuracy of a classifier is expressed as the percentage of correctly classified examples. Additionally we computed precision and recall of classifiers.[1]

We tested all possible four scoring schemas for CCPC [i.e., *(scConA, scComA)*; *(scConA, scComB)*; *(scConB, scComA)*; *(scConB, scComB)*]. We also limited the maximal size of investigated subgraphs for this classifier to 5 (i.e., $s \leq 5$).

5.3 Results

Figure 1 shows the classification accuracy of CCPC classifier using different scoring routines ($s = 5$) for all datasets. Results show that contrast subgraphs play dominant role in classification process, whereas common subgraphs have a very little influence on final results. This can be explained by the fact that almost all of the compounds are classified only by contrast patterns.

Another observation is that methods based on *scConB* scoring scheme provides higher classification results than methods based on *scConA* scoring scheme. The best results were obtained using *(scConB, scComB)* scoring scheme.

Figure 2 shows the classification accuracy of CCPC (*scConB, scComB, s = 5*) for negative and positive class for all datasets. It is worth noticing that algorithm

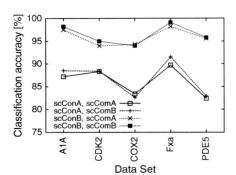

Fig. 1 Classification accuracy of CCPC using different scoring routines ($s = 5$) for all datasets

---

[1] Precision [TP/(TP + FP)] is the number of true positives (i.e., the number of examples correctly labeled as belonging to the positive class) divided by the total number of examples labeled as belonging to the positive class (i.e., the sum of true positives and false positives). Recall [TP/(TP + FN)] is the number of true positives divided by the total number of examples that actually belong to the positive class (i.e., the sum of true positives and false negatives).

Fig. 2 Classification accuracy of CCPC (scConB, scComB, s = 5) for negative and positive class for all datasets

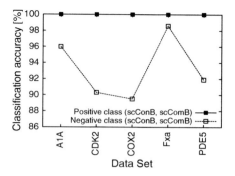

Fig. 3 Classification accuracy for COX2

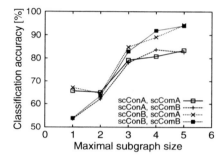

always correctly classified all positive examples (compounds that bind to protein). It only misclassified negative examples – decoys.

Figure 3 shows the classification accuracy of CCPC classifier for COX2 dataset as a function of maximal discovered subgraph size ($s \in <1,5>$). It can be observed that classification quality measure increases with maximal contrast subgraph size.

5.4 Results Comparison

Figure 4 shows the classification accuracy for the following methods: CCPC (scConB, scComB, s = 5) and kernel methods which were reported to provide best classification accuracy [4]. CCPC classifier outperformed all other methods for all evaluated datasets. The difference between CCPC and runner-up varies from 0.6% (Fxa) to 4.2% (A1A).

Table 2 reports precision and recall of investigated methods on all datasets. The best results are bolded. Recall measure for CCPC classifier are always 100% because there are no false-negative examples (classification accuracy for positive class is 100%). As far as precision is considered, GPM provides best results (it outperformed CCPC method in three datasets). On the other hand, recall measures for GPM was significantly lower than for CCPC.

Fig. 4 Classification accuracy comparison

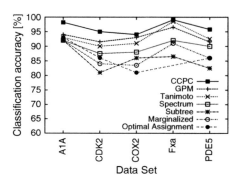

Table 2 Precision (recall) comparison

Method	A1A	CDK2	COX2	Fxa	PDE5
GPM	88.3 (96.7)	**96.0** (88.4)	**96.1** (91.6)	**100.0** (93.6)	92.2 (89.7)
Tanimoto	85.6 (**100.0**)	89.1 (93.7)	89.8 (93.1)	97.5 (**100.0**)	92.7 (93.6)
Spectrum	85.8 (98.0)	86.5 (90.1)	85.9 (93.5)	90.8 (95.2)	86.8 (93.3)
Subtree	87.7 (93.2)	79.2 (84.7)	88.6 (83.7)	87.5 (84.3)	75.9 (86.3)
Marginalized	83.2 (**100.0**)	81.3 (92.1)	83.5 (90.9)	87.4 (96.9)	83.5 (95.0)
Optimal ass.	87.7 (**100.0**)	81.3 (87.9)	82.9 (79.9)	79.8 (93.5)	85.3 (89.6)
CCPC	**96.2** (**100.0**)	90.9 (**100.0**)	90.9 (**100.0**)	98.0 (**100.0**)	**92.8** (**100.0**)

6 Conclusions

In this chapter, we presented a novel approach for predicting chemical-protein binding activity. Results show that proposed solution provides higher classification accuracy than other existing algorithms. What is more, construction and structure of our classifier are simpler than approaches based on kernel methods, e.g., SVM. This feature lets the domain experts to modify the classifier with the professional knowledge by adding new patterns to contrast or common subgraphs sets or by modifying their supports.

The main concept of our classifier is domain independent; hence, it can be used to solve classification problems in other areas as well. It was successfully applied to solving other problems from the area of chemical informatics [7, 8] and information science [6].

References

1. Borgelt, C., Berthold, M.R.: Mining molecular fragments: finding relevant substructures of molecules. In: Proceedings of the 2nd IEEE International Conference on Data Mining, pp. 51–58. IEEE Computer Society, Washington, DC (2002)
2. Fei, H., Huan, J.: Structure feature selection for graph classification. In: CIKM '08, pp. 991–1000. ACM, California (2008)

3. Jorissen, R.N., Gilson, M.K.: Virtual screening of molecular databases using a support vector machine. Journal of Chemical Information and Modeling **45**, 549–561 (2005)
4. Smalter, A., Huan, J., Lushington, G.: CPM: A graph pattern matching kernel with diffusion for accurate graph classification. Information Telecommunication and Technology Center, University of Kansas, Lawrence (2008)
5. Swamidass, S.J., Chen, J., Bruand, J., Phung, P., Ralaivola, L., Baldi, P.: Kernels for small molecules and the prediction of mutagenicity, toxicity and anti-cancer activity. Bioinformatics **21**, 359–368 (2005)
6. Dominik, A., Walczak, Z., Wojciechowski, J.: Classification of web documents using a graph-based model and structural patterns. In: Kok, J.N., Koronacki, J., López de Mántaras, R., Matwin, S., Mladenic D., Skowron, A. (eds.) Knowledge Discovery in Databases: PKDD 2007, pp. 67–78. Springer, Heidelberg (2007)
7. Dominik, A., Walczak, Z., Wojciechowski, J.: Classifying chemical compounds using contrast and common patterns. In: Beliczynski, B., Dzielinski, A., Iwanowski, M., Ribeiro, B. (eds.) Adaptive and Natural Computing Algorithms, pp. 772–781. Springer, Heidelberg (2007)
8. Dominik, A., Walczak, Z., Wojciechowski, J.: Detection of mutagenicity, toxicity and anti-cancer activity using structural contrast graph patterns. In: Rutkowski, L., Tadeusiewicz, R., Zadeh, L.A., Zurada, J. (eds.) Computational Intelligence: Methods and Applications, pp. 255–266. Academic Publishing House EXIT, Warsaw (2008)
9. Dominik, A., Walczak, Z., Wojciechowski, J.: Classification of graph structures. In: Wang, J. (eds.) Encyclopedia of Data Warehousing and Mining, 2nd Ed., pp. 202–207. Information Science Reference, Hershey (2008)
10. Dominik, A.: Graph classification using structural contrast patterns. Warsaw University of Technology, Warsaw (2009)
11. Fortin, S.: The graph isomorphism problem. University of Alberta, Alberta (1996)
12. Ralaivola, L., Swamidass, S.J., Saigo, H., Baldi, P.: Graph kernels for chemical informatics. Neural Networks **18**, 1093–1110 (2005)
13. Ramamohanarao, K., Bailey, J.: Discovery of emerging patterns and their use in classification. In: Gedeon, T.D., Fung, L.C.C. (eds.) Australian Conference on Artificial Intelligence, pp. 1–12. Springer, Heidelberg (2003)
14. Schenker, A., Last, M., Bunke, H., Kandel, A.: Classification of web documents using graph matching. International Journal of Pattern Recognition and Artificial Intelligence **18**, 475–496 (2004)
15. Ting, R.M.H., Bailey, J.: Mining minimal contrast subgraph patterns. In: Ghosh, J., Lambert, D., Skillicorn, D.B., Srivastava, J. (eds.) SDM, pp. 202–207. SIAM, Maryland (2006)
16. Ullmann, J.R.: An algorithm for subgraph isomorphism. JACM **23**, 31–42 (1976)
17. Plasma protein binding – Wikipedia, The Free Encyclopedia (2009)

Chapter 25
Topological Constraint in High-Density Cells' Tracking of Image Sequences

Chunming Tang, Ling Ma, and Dongbin Xu

Abstract The multi-target tracking in cell image sequences is the main difficulty in cells' locomotion study. Aim to study cells' complexity movement in high-density cells' image, this chapter has proposed a system of segmentation and tracking. The proposed tracking algorithm has combined overlapping and topological constraints with track inactive and active cells, respectively. In order to improve performance of algorithm, size factor has been introduced as a new restriction to quantification criterion of similarity based on Zhang's method. And the distance threshold for transforming segmented image into graph is adjusted on considering the local distribution of cells' district in one image. The improved algorithm has been tested in two different image sequences, which have high or low contrast ration separately. Experimental results show that our approach has improved tracking accuracy from 3% to 9% compared with Zhang's algorithm, especially when cells are in high density and cells' splitting occurred frequently. And the final tracking accuracy can reach 90.24% and 77.08%.

Keywords Cells' tracking · Segmentation · High-density cells · Overlapping · Topological constraint · Size factor

1 Introduction

Detection of cells' locomotion plays a key role in many biological processes, such as embryonic development, wound healing, immune response and cancer metastasis. Study on cells' tracking is an essential part of cytology and biology [1]. Main methods for cells' tracking are mostly based on overlapping [2], active contour model [3], level set model [4], mean shift [5] and their improved algorithms. Active contour

---

C. Tang (✉)
School of Information and Communication Engineering, Harbin Engineering University, Harbin 150001, China
e-mail: tangchunminga@hotmail.com

models have been successfully applied in real-time cells' tracking. However, it is difficult to deal with the merging and splitting of target areas, and also difficult to set global optimized parameters because of their sensitivity. Level set model is easier to achieve cells' tracking. However, its performance depends on some parameters, some of which have no meaning and difficult to be predicted. Moreover, when level set is applied in multi-target tracking in frames, average luminance differences of frames will make its performance worse. Mean shift is a fast and robust tracking algorithm. But as it only searches the gray-scale center of targets, recognition efficiency may reduce dramatically when cells move fast.

All the above algorithms are usually suitable for tracking isolated or segmented cells. They are difficult to implement high density cells' tracking [6]. Zhang [7] has built a model based on graph theory, which may identify and track cells by considering topological relationships of cells and their neighborhood globally. This model can track cells even in high density successfully. However, when cells' distribution is dense and cells having been divided simultaneously, it easily leads to tracking error. In order to improve this, our proposed algorithm focuses on tracking high-density cells in image sequences with high or low contrast ratio.

In this chapter, segmentation algorithms of two different image sequences have been explained first. Clustered cells are then separated after global segmentation based on coding of cell shape or local gray threshold. In cells' tracking part, all cells have been classified into two categories: inactive cells and active cells. The former is tracked by overlapping first. The latter is tracked by improved topological constraint method. In order to improve the tracking accuracy of Zhang's algorithm, area term has been introduced into the quantification criterion of similarity and adjusted distance threshold has been set for segmented image.

2 Segmentation

Two image sequences have been tested. Sequence I is fluorescent images, which are characterized by strong gray-scale contrast between cells and background. One of the original images is shown in Fig. 1a. Sequence II is a neuron stem cell image imaged by confocal microscope, which has low image contrast ratio with serious problems of cells' adhesion and cluster in segmentation. One of the original images is shown in Fig. 2a. In sequence I, isolated cells can be segmented by Otsu [8]. After morphological operation and other denoising measurement have been adopted, the binary images can be obtained. Freeman code [9] is then applied for the clustered cells being separated further. One segmentation illustration is shown in Fig. 1. The cells in white rectangle in Fig. 1b are clustered. They are separated successfully in Fig. 1c after Freeman code is applied.

Segmentation is a tough task in sequence II. The detailed algorithm is based on level set algorithm without re-initialization to obtain isolated cells and clustered cells' contour first. After curvature term is added to accelerate convergence, iteration

25 Topological Constraint in High-Density Cells' Tracking of Image Sequences

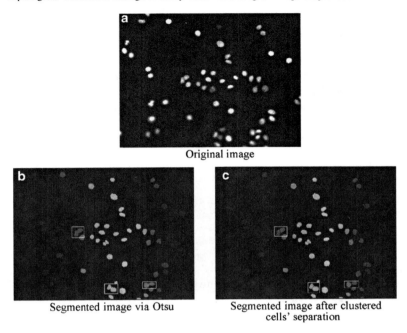

Fig. 1 Segmentation of one frame in sequence I

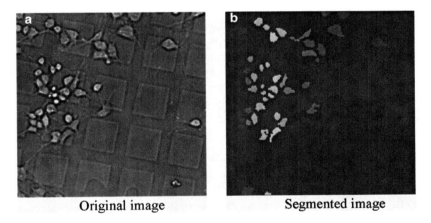

Fig. 2 Segmentation of one frame in sequence II

terminate condition is changed to measure norm energy to decrease complexity and save computation time. After that, local gray threshold is combined with the result of curve evolution. Adhesion and clustered cells are then separated. One segmentation illustration is show in Fig. 2b.

3 Tracking

After the initial frame of one segmented image sequence is scanned, different IDs are assigned to different cells. The purpose of cells' tracking is achieved if all cells in a whole image sequence can keep their original IDs. In order to achieve satisfied tracking results, appropriate algorithms should be chosen to track different cells based on their characteristics of movement and proliferation. Cells are classified into inactive ones and active ones. The inactive cells are defined as those locomotive distances in two adjacent frames are less than average radius of cells. Otherwise they are defined as active cells. The details of tracking will be described in the following part.

3.1 Overlap Method

Overlap method [2] used for inactive cells' tracking is based on cell's overlap region between two successive frames to ensure one cell keeping its unique ID till last frame of one image sequence.

3.2 Improved Topological Constraint

To the active cells, which overlapping cannot track, topological constraint is then applied. This algorithm is based on graph model to describe the topological relationships of cells [7]. It can convert tracking cells' feature to matching vertexes in two similar structured graphs. It combines the advantages of features-linked rule and topological constraint. Therefore, it is able to represent the integral structure of cells' distribution. Tracking can then depend on their neighbourhood relationships, especially when cells are in high density.

3.2.1 Graph Description of Cell Image

Graph G can be expressed as $G = \{V, E\}$, which usually describes the relationships between objects [10]. To cell images, V represents all cells in an image, constituting a vertex set. A cell's size, location and ID are saved in a vertex subset $R(i)$. E represents the adjacency relationship between cells.

According to [7], after a fixed dt is set, degree of a vertex $R(i)$ and its neighbourhood set $N_{F_k,R(i)}$ and directions of neighbourhood distributions $D_{F_k,R(i)}$ are defined.

3.2.2 Improved Topological Constraint

For the active cells, searching regions are built up according to the distance limitation first. Only the cells partitioned in this region are regarded as candidates.

Matched cells are then confirmed via correlation calculation. As only the cell which is similar to cell X in previous frame both in their properties and neighbour constitution, it can be regarded as the matched cell of cell X in current frame, whereas properties of cells refer to cells' sizes and centroids' distance. Neighbour constitution refers to the cell's degree and directions of neighbour distribution. Apart from distance and angle, cells' size is added to be another topological constraint, which can avoid some disturbance from neighbour cells during tracking. Vertex matching is carried out based on the following topological constraints: Degrees of two matched cells should be closest in all possible matching pairs. Size difference between two matched cells should be the smallest in all possible matching pairs. Other cells cannot exist in the direction of the line linked two matched cells' centroids. And direction angle should be minimal between two matched cells in all possible matching pairs.

Parameter dt is important when segmented image is transformed into graph, which can directly determine cell's degree. Higher density district is usually set relative shorter dt. After unmatched cell with each candidate constitutes a pair, similarity of each pair is calculated via (1), which is called quantification criterion of similarity. The pair with the largest Q would be regarded as a matched one. Then the cell's ID can be propagated between two successive frames. Equation (2) calculates similarity degree of the size. Constants δ_S is used to adjust Q's sensitivity to the size factors. It should be set dependently on given image. In (1), $Q_{\text{SSN}}[R(i),R'(j)]$ calculates the similarity of neighbour constitution, d' in [7] has been replaced by nbr. Because d' estimates two centroids' distance of two clustered cells in next frame according to the centroids' distance of these two clustered cells in current frame. Since all clustered cells have been separated before tracking in our presented paper, nbr here represents the number of neighbour cells $R(i)$.

$$Q[R(i),R'(j)] = Q_{\text{Dist}}[R(i),R'(j)] \cdot Q_{\text{SSN}}[R(i),R'(j)] \cdot Q_{\text{size}}[R(i),R'(j)] \quad (1)$$

$$Q_{\text{size}}[R(i),R'(j)] = \exp\left(-\frac{\text{abs}[R_{\text{size}}(i) - R'_{\text{size}}(j)]}{\delta_S}\right) \quad (2)$$

It is known that each pair of vertexes is assigned a weight according to graph theory. These weights should indicate similarity degree of two vertexes. As Q has been computed via (1), which can reflect similarity degree of the two cells, Q is set as weight. Cells' tracking is then transformed into vertexes matching inter-graphs so far. Mapping from a cell image to a graph G is shown in Fig. 3.

3.2.3 Other Cases Happened in Tracking

Main problems of cells' tracking are cells' loss, emergence and cleavage. The first two problems can be solved according to its position and consistence in image. Cleavage cells have been identified by their position, distance and neighbour distribution. Data table is used to record cleavage cells' information.

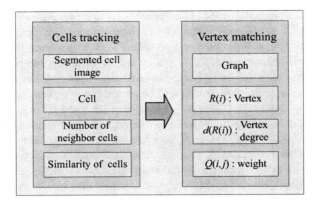

Fig. 3 Mapping from a cell image to a graph

Table 1 Comparison of tracking results of sequence I

Frame number	Number of segmented cells	Cell's number of correct tracking		Accuracy rate of tracking (%)	
		Zhang's	Improved	Zhang's	Improved
2	60	60	60	100	100
20	73	69	73	94.52	100
40	90	81	84	90.00	93.33
60	109	99	101	90.83	92.66
80	132	117	121	88.64	91.67
100	164	142	148	86.59	90.24

Table 2 Comparison of tracking results of sequence II

Frame number	Number of segmented cells	Cells number of correct tracking		Accuracy rate of tracking (%)	
		Zhang's	Improved	Zhang's	Improved
2	36	36	36	100	100
10	37	34	35	91.89	94.59
20	37	30	33	81.08	89.19
30	39	31	34	79.49	87.18
40	43	32	36	74.42	83.72
50	48	33	37	68.75	77.08

4 Experiments and Results

Fluorescence image sequence I: 672*512*100 has been tested. In this sequence, cells' splitting occurred frequently. Cell's number increases from 57 in initial frame to 164 in the final frame. Comparison of tracking result of sequence I between Zhang's and our improved algorithm is listed in Table 1.

Gray-scale image sequence II is 250*250*50, where cells are in high density and much irregularity in shapes. Another comparison result is shown in Table 2. All cells' trajectories in sequences I and II are shown in Fig. 4a, b.

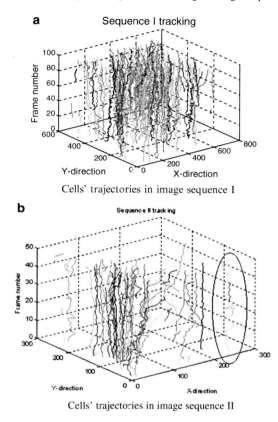

Fig. 4 Cells' trajectories in three dimensions

5 Conclusion and Discussion

Experimental results show our proposed segmentation and tracking system has indeed improved the tracking accuracy from 3% to 9% compared with Zhang's algorithm based on topological constraints, especially when cells are in high density and cells' splitting occurred frequently. Some trajectory segments have been circled in black in Fig. 4b. As errors in segmentation because of low contrast ratio, trajectory of one cell has been broken for three times. The next job we might focus is on the analysis of cells' trajectories to improve tracking accuracy further.

Acknowledgements The authors would like to thank Dr Xiaobo Zhou in Harvard Medical School, Harvard University, for providing the image sequence I and DSS of Chalmers University of Technology for recording the stem cell image sequences II. This project is supported by National Natural Science Foundation of China (NSFC). Grant number: 60875020.

References

1. Gage FH (2000) Mammalian neural stem cell. Science 287:1433–1438
2. Tang C, Bentsson E (2005) Segmentation and tracking of neural stem cell. ICIC 2005, Hefei, China, LNCS 3645. Germany: Springer, pp. 851–859
3. Zimmer C, Labruyere E et al (2002) Segmentation and tracking of migrating cells in videomicroscopy with parametric active contours: a tool for cell-based drug testing. IEEE Transactions on Medical Imaging 21:1212–1221
4. Dzyubachyk O, Niessen W et al (2007) A variational model for level-set based cell tracking in time lapse fluorescence microscopy images. 4th IEEE International Symposium on Biomedical Imaging: From Nano to Macro, 2007. ISBI 2007, pp. 12–15
5. Debeir O, Ham P V et al (2005) Tracking of migrating cells under phase-contrast video microscopy with combined mean-shift processes. IEEE Transactions on Medical Imaging 24:697–711
6. Kirubarajan T, Bar-Shalom Y et al (2001) Multiassignment for tracking a large number of overlapping Objects. IEEE Transactions on Aerospace and Electronic Systems 37:2–19
7. Zhang L, Xiong H et al (2007) Graph theory application in cell nucleus segmentation, tracking and identification. Proceedings of the 7th IEEE International Conference on BIBE 2007: 226–232
8. Otsu N (1979) A threshold selection method from gray-level histograms. IEEE Transactions on Systems, Man and Cybernetics 9:62–66
9. Kaneko T, Okudaira M (1985) Encoding of arbitrary curves based on the chain code representation. IEEE Transactions on Communications 33:697–707
10. Diestel R (2008) Graphy theory, 2nd ed. Beijing: Springer

Chapter 26
STRIKE: A Protein–Protein Interaction Classification Approach

Nazar Zaki, Wassim El-Hajj, Hesham M. Kamel, and Fadi Sibai

Abstract Protein–protein interaction has proven to be a valuable biological knowledge and an initial point for understanding how the cell internally works. In this chapter, we introduce a novel approach termed STRIKE which uses String Kernel to predict protein–protein interaction. STRIKE classifies protein pairs into "interacting" and "non-interacting" sets based solely on amino acid sequence information. The classification is performed by applying the string kernel approach, which has been shown to achieve good performance on text categorization and protein sequence classification. Two proteins are classified as "interacting" if they contain similar substrings of amino acids. Strings' similarity would allow one to infer homology which could lead to a very similar structural relationship. To evaluate the performance of STRIKE, we apply it to classify into "interacting" and "non-interacting" protein pairs. The dataset of the protein pairs are generated from the yeast protein interaction literature. The dataset is supported by different lines of experimental evidence. STRIKE was able to achieve reasonable improvement over the existing protein–protein interaction prediction methods.

Keywords Pattern classification and recognition · Protein–protein interaction · Protein sequence analysis · Amino acid sequencing · Biological data mining and knowledge discovery

1 Introduction

The prediction of protein–protein interaction (PPI) is one of the fundamental problems in computational biology as it can aid significantly in identifying the function of newly discovered proteins. Understanding PPI is crucial for the investigation

N. Zaki (✉)
Bioinformatics Laboratory, Department of Intelligent Systems,
College of Information Technology, UAE University, 17551 Al-Ain, UAE
e-mail: nzaki@uaeu.ac.ae

of intracellular signaling pathways, modeling of protein complex structures and gaining insights into various biochemical processes. To enhance this understanding, many experimental techniques have been developed to predict the proteins' physical interactions which could lead to the identification of the functional relationships between proteins. These experimental techniques are, however, very expensive, significantly time consuming and technically limited which creates a growing need for the development of computational tools that are capable of identifying PPIs. To this end, many impressive computational techniques have been developed. Each of these techniques has its own strengths and weaknesses, especially with regard to the sensitivity and specificity of the method. Some of the techniques such as the Association Method (AM) [1], Maximum Likelihood Estimation (MLE) [2], Maximum Specificity Set Cover (MSSC) [3] and Domain-based Random Forest [4] have used domain knowledge to predict PPI. The motivation of these techniques was that molecular interactions are typically mediated by a great variety of interacting domains. Another method called Protein–Protein Interaction Prediction Engine (PIPE) [5] was developed based on the assumption that some of the interactions between proteins are mediated by a finite number of short polypeptide sequences. These sequences are typically shorter than the classical domains and are used repeatedly in different proteins and contexts within the cell. However, identifying domains or short polypeptide sequences is a long and computationally expensive process. Moreover, these techniques are not universal because their accuracy and reliability are dependent on the domain information of the protein partners.

In this chapter, we introduce a novel approach termed STRIKE which uses string kernel (SK) approach to predict PPI. This has been shown to achieve good performance on text categorization tasks [6] and protein sequence classification [7]. The basic idea of this approach is to compare two protein sequences by looking for common subsequences of a fix-length. The string kernel is built on the kernel method introduced by Haussler et al. [8] and Watkins et al. [9]. The kernel computes similarity scores between protein sequences without ever explicitly extracting the features. The subsequence is any ordered sequence of amino acids occurring in the protein sequence and is not necessarily contiguous. The subsequences are weighted by an exponentially decaying factor of their full length in the sequence, hence emphasizing those occurrences that are close to contiguous.

We understand that the subsequence similarity between two proteins may not necessarily indicate interaction. However, it is an evidence that we cannot ignore. Subsequence similarity would allow one to infer that homology and homologous sequences usually have the same or very similar structural relationships.

A drawback of this approach emerges when the level of similarity between the protein pairs is too low to pick up interaction. The reasonable explanation is that in the case of low similarity sequence, there are always similar patterns of identical amino acid residues which could be seen in the two sequences. The pattern of sequence similarity reflects the similarity between experimentally determined structures of the respective proteins or at least corresponds to the known key elements of one such structure [10]. Structural evidence indicates that interacting pairs of close homologs usually interact in the same way [11]. Other evidences are derived using a combination of some genomic features such as structurally known interacting Pfam domains and sequence homology as a means to assign reliability to the PPIs in *Saccharomyces cerevisiae* [12]. The likelihood ratio in this study expresses the reliability of such genomic feature. In our case, there is no doubt that the SK method will be a good technique of reflecting homology between protein pairs. The intensive comparison between subsequences existing in protein pairs may capture structural domain knowledge or typically subsequences that are shorter than the classical domains and could appear repeatedly in the protein pairs of interest. We are also encouraged by the success of a recently published work using pair-wise alignment as a way to extract meaningful features to predict PPI. The PPI based on Pairwise Similarity (PPI-PS) method consists of a representation of each protein sequence by a vector of pair-wise similarities against large subsequences of amino acids created by a shifting window which slides over concatenated protein training sequences. Each coordinate of this vector is typically the E-value of the Smith–Waterman score [13]. One major drawback of the PPI-PS is that each protein is represented by computing the Smith–Waterman score against large subsequences created by concatenating protein training sequences. However, comparing short sequence to a very long one will result in some potentially valuable alignments to be missed out. The SK, however, tackles this weakness by capturing any match or mismatch existing in the protein sequence of interest.

2 Method

In order to classify protein pairs as "interacting" or "non-interacting", STRIKE performs the following steps: (1) data preparation step in which protein pairs in the dataset are concatenated; (2) the training step in which the support vector machine (SVMs) classifiers are constructed and (3) the testing step which uses SVMs to determine whether the protein pair is "interacting" or "non-interacting". Steps (2) and (3) require the computation of kernel similarity scores between protein sequences. The feature space is generated by all subsequences of bounded length. In order to derive the SK, we start from the features and then compute their inner product. SK maps strings to a feature vector indexed by all k-tuples of amino acids. A k-tuple will have a non-zero entry if it occurs as a subsequence anywhere (not necessarily contiguous) in the string.

In order to derive the string kernel (SK), we start from the features and then compute their inner product. Hence, the criterion of satisfying the Mercer's condition (positive semi-definiteness) automatically applies here. It maps the amino acid sequence (string) to a feature vector indexed by all k-tuples of amino acids. A k-tuple has a non-zero entry if it occurs as a subsequence anywhere (not necessarily contiguous) in the protein sequence of interest. The weighting of the feature will be the sum over the occurrences of the k-tuple in the protein sequence.

Following Cristianini et al. [14] and Lodhi et al. [6], the string kernel can be defined as follows.

Let Σ be a finite set of amino acids. A string is a finite sequence of amino acids from Σ, including the empty sequence. For protein sequences s, t, we denote by $|s|$ the length of the sequence $s = s_1 \ldots s_{|s|}$ and by st the string obtained by concatenating the sequences s and t. The string $s[i:j]$ is the substring $s_i \ldots s_j$ of s. We say that u is a subsequence of s, if there exist indices $i = (i_1, i_2, \ldots, i_{|u|})$, with $1 \leq i_1, i_2, \ldots, i_{|u|} \leq |s|$, such that $u_j = s_{i_j}$ for $j = 1, 2, \ldots, |u|$ or $u = s[i]$ for short. The full length $l(i)$ of the subsequence in s is $i_{|u|} - i_1 + 1$. We denote by Σ^n the set of all finite amino acid sequences of length n and by Σ^* the set of all protein (strings):

$$\Sigma^* = \bigcup_{n=0}^{\infty} \Sigma^n. \tag{1}$$

We now define the feature space $F_n = \Re^{\Sigma^n}$. The feature mapping ϕ for protein sequence s is given by defining the u coordinate $\phi_u(s)$ for each $u \in \Sigma^n$. We define

$$\phi_u(s) = \sum_{i:u=s[j]} \lambda^{l(i)} \tag{2}$$

for some $\lambda \leq 1$. These features measure the number of subsequence occurrences in the protein sequence s weighting them according to their lengths. Hence, the inner product of the feature vectors for two protein sequences s and t gives a sum over all common subsequences weighted according to their frequency of occurrence and lengths:

$$k_n(s,t) = \sum_{u \in \Sigma^n} \langle \phi_u(s) \cdot \phi_u(t) \rangle \tag{3}$$

$$= \sum_{u \in \Sigma^n} \sum_{i:u=s[i]} \lambda^{l(i)} \sum_{j:u=t[j]} \lambda^{l(j)} \tag{4}$$

$$= \sum_{u \in \Sigma^n} \sum_{i:u=s[i]} \sum_{j:u=t[j]} \lambda^{l(i)+l(j)} \tag{5}$$

This technique can be illustrated by a simple example. Consider the following four concatenated protein pairs sequences:

> s1   > s2   > s3   > 4

lql   lqal   hgs   gqsl

In this case, sequences s1 and s2 belong to the interacting set and the sequences s3 and s4 belong to the non-interacting set. For each substring, there is a dimension of feature space, and the value of such coordinate depends on how frequently and compactly such string (such as the ones highlighted in bold) is embedded in the protein sequences of interest. Let us first assume that we are comparing the first two concatenated protein sequences s1 and s2, where there exists one string in each sequence. For simplicity, we set the length of substring to two. In other words, these sequences are implicitly transformed into feature vectors, where each feature

Table 1 Mapping two strings "lql" and "lqal" to six-dimensional feature spaces

	lq	ll	ql	la	qa	al
$\phi(lql)$	λ^2	λ^3	λ^2	0	0	0
$\phi(lqal)$	λ^2	λ^4	λ^3	λ^3	λ^2	λ^2

Table 2 Mapping all the strings to dimensional feature spaces

	S1(lql)	S2(lqal)	S3(hgs)	S4(gqsl)
S1(lql)	–	$k(lql, lqal) = 0.102$	$k((lql), hgs) = 0$	$k(lql, gqsl) = 0.031$
S2(lqal)		–	$k((lqal), hgs) = 0$	$k(lqal, gqsl) = 0.016$
S3(hgs)			–	$k(hgs, gqsl) = 0.031$
S4(gqsl)				–

vector is indexed by the substrings of length two. The six-dimensional feature space, feature vectors $\phi(lql)$, $\phi(lqal)$ and the corresponding kernel are given in Table 1.

In this case, the un-normalized kernel $k(lql, lqal)$ can clearly be computed as $\lambda^4 + \lambda^7 + \lambda^5$. Assuming that the decay factor λ is equal to 0.5, $k(lql, lqal) = 0.102$. Similarly, the remaining kernels are computed and summarized in Table 2.

From Table 2, it is clearly shown that the proposed method is able to capture the potential interaction between the two sequences "lql" and "lqal".

3 Experimental Work

STRIKE is implemented in Perl. However, the SK component is a modified version of Lodhi's [6] approach which was implemented in C++. SK is based on a simple gradient-based implementation of SVM [15] known as "Adatron". Many different techniques have been applied to speed up the computation of the kernel matrix. First, the program reads a single file containing training protein pairs and test protein pairs sets. Second, two files containing the indexes of training set and test set are added.

STRIKE uses as input a file containing the training and testing protein pairs. A second file containing all the protein sequences from the yeast S. cerevisiae is included to retrieve the corresponding protein sequences. As STRIKE runs from the command line, the user enters the name of the file containing the protein pairs and the number of positive training examples, the negative training examples, the positive testing examples and the negative testing examples. The user has the option to change the weighted decay factor (λ) and the size of the substring (n). Both parameters have effects on the generalization performance of an SVM learner that manipulates the information encoded in a string kernel. SK weighs the substrings of the amino acids according to their proximity in the protein sequence. This is the parameter that controls the penalization of the interior gaps in the substrings. STRIKE outputs the classification results which contain the accuracy, precision, recall and the F1 value. The performance of STRIKE is measured by how well it can distinguish between "interacting" and "non-interacting" protein pairs. STRIKE was used to classify between 100 "interacting" protein pairs (157 proteins) and 100 "non-interacting" protein pairs (77 proteins). The datasets were randomly selected by Sylvain et al. [5] and used to evaluate the PIPE's performance. It was generated from the yeast protein interaction literature for which at least three different lines of experimental evidence supported the interaction. The dataset can be downloaded from http://faculty.uaeu.ac.ae/nzaki/STRIKE.htm.

4 Results and Discussion

The performance of STRIKE was measured by how well the system can recognize interacting protein pairs. In order to analyze the evaluation measures in PPI prediction, we first explain the contingency table (Table 3). The entries of the four cells of the contingency table are described as follows:

tp = number of interacting sequences classified as "interacting".
fn = number of non-interacting sequences classified as "interacting".
fp = number of interacting sequences classified as "non-interacting".
tn = number of non-interacting sequences classified as "non-interacting".
$n = tp + fn + fp + tn$ (total number of sequences).

The information encoded in the contingency table is used to calculate the following PPI evaluation measures:

$$\text{Precision}(\text{Pr}) = tp/(tp+fp) \qquad (6)$$

$$\text{Recall}(\text{Re}) = tp/(tp+fn) \qquad (7)$$

$$F1 = 2[(\text{Pr} \times \text{Re})/(\text{Pr}+\text{Re})] \qquad (8)$$

$$\text{Accuracy} = (tp+tn)/n. \qquad (9)$$

To evaluate the performance of STRIKE, we first grouped the 100 "interacted" protein pairs into two sets: A (50 pairs) and B (50 pairs). We also grouped the 100 "non-interacted" protein pairs into two sets: C (50 pairs) and D (50 pairs). We then combine A with C to create a training dataset and B with D to create a testing dataset.

To study the effectiveness of varying weighted decay factor (λ) on the generalization performance, a series of experiments were conducted. The value of the subsequence length (n) was kept fixed to 2. The analysis results are summarized in Fig. 1. Figure 1a shows the relation between different values of λ and the

Table 3 The contingency table

	Interacting sequence	Non-interacting sequence
Classified interacting	True positives (tp)	False negatives (fn)
Classified non-interacting	False positives (fp)	True negatives (tn)

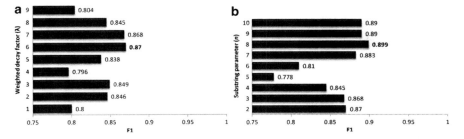

Fig. 1 Performance of STRIKE with a varying weight decay factor (λ) and subsequence length (n) values

corresponding influence of F1. It is interesting to point out that F1 peaks at a value of 0.6.

Further experiments were conducted to study the effectiveness of varying the subsequence length (n). The value of λ was set to 0.6. The analysis results are shown in Fig. 1b. It is apparent in this results that F1 peaks at a subsequence length of 8.

In this case, STRIKE was able to achieve 89% accuracy, 0.831 precision, 0.98 recall and 0.899 F1. The substring parameter (n) and the weighted decay factor (λ) were set to 8 and 0.6, respectively.

The obtained results are superior to PIPE. PIPE performance resulted in 0.61 sensitivity, 0.89 specificity and 0.75 overall accuracy when tested to detect yeast protein interaction pairs. In addition to its superior accuracy, STRIKE has two further advantages when compared to PIPE. First, the PIPE method is computationally intensive and the evaluation of PIPE's performance over the same dataset took around 1,000 h of computation time, compared to only 54.9 min using STRIKE. Second, as indicated by the PIPE authors, their method is expected to be weak if it is used for the detection of novel interactions among genome-wide large-scale datasets which is not the case using STRIKE.

5 Conclusion

In this chapter, we introduced a novel approach termed STRIKE (String Kernel) which uses String Kernel method to predict protein–protein interaction. STRIKE classifies protein pairs into "interacting" and "non-interacting" sets that is based solely on amino acid sequence information. STRIKE makes no use of prior knowledge, yet it has been used with considerable success.

One of STRIKE's strength is that the subsequences are weighed by an exponentially decaying factor of their full length in the sequence, hence emphasizing those occurrences that are close to being contiguous. However, replacing the weighted decaying factor using a protein weight matrix (scoring matrix) will add sensitivity to the comparison as it implicitly represents a particular theory of protein sequence evolution. The parallel processing framework to compute the string kernel presented in this chapter will be implemented and assessed in a large dataset such as the one proposed by Xue-Wen et al. [4].

References

1. Sprinzak E, Margalit H (2001) Correlated sequence-signatures as markers of protein-protein interaction. J Mol Biol 311: 681–692
2. Deng M, Mehta S, Sun F, Cheng T (2002) Inferring domain-domain interactions from protein-protein interactions. Genome Res 12: 1540–1548
3. Huang TW, Tien AC, Huang WS, Lee YC, Peng CL, Tseng HH, Kao CY, Huang CY (2004) POINT: a database for the prediction of protein-protein interactions based on the orthologous interactome. Bioinformatics 20: 3273–3276
4. Xue-Wen C, Mei L (2005) Prediction of protein–protein interactions using random decision forest framework. Bioinformatics 21: 4394–4400
5. Sylvain P, Frank D, Albert C, Jim C, Alex D, Andrew E, Marinella G, Jack G, Mathew J, Nevan K, Xuemei L, Ashkan G (2006) PIPE: a protein-protein interaction prediction engine based on the re-occurring short polypeptide sequences between known interacting protein pairs. BMC Bioinformatics 7: 365
6. Lodhi H, Saunders C, Shawe-Taylor J, Cristianini N, Watkins C (2002) Text classification using string kernels. J Mach Learn Res 2: 419–444
7. Zaki NM, Deris S, Illias RM (2005) Application of string kernels in protein sequence classification. Appl Bioinformatics 4: 45–52
8. Haussler D (1999) Convolution kernels on discrete structures. Technical Report UCSC-CRL-99–10, University of California, Santa Cruz
9. Watkins C (2000) Dynamic Alignment Kernels. Advances in Large Margin Classifiers. Cambridge, MA, MIT Press, 39–50
10. Koonin EV, Galperin MY (2002) Sequence-Evolution-Function: Computational Approaches in Comparative Genomics. Dordrecht, Kluwer Academic Publishers
11. Zaki NM (2009) Protein-protein interaction prediction using homology and inter-domain linker region information. In: Lecture Notes in Electrical Engineering, Vol. 39. New York, Springer, 635–645
12. Patil A, Nakamura H (2005) Filtering high-throughput protein-protein interaction data using a combination of genomic features. BMC Bioinformatics 6: 100
13. Zaki NM, Lazarova-Molnar S, El-Hajj W, Campbell P (2009) Protein-protein interaction based on pairwise similarity. BMC Bioinformatics 10: 150
14. Cristianini N, Shawe-Taylor J (2000) An Introduction to Support Vector Machines. Cambridge, UK: Cambridge, University Press
15. Friess T, Cristianini N, Campbell C (1998) The Kernel-Adatron algorithm: a fast and simple learning procedure for support vector machines. 15th International Conference on Machine Learning. Morgan Kaufmann, CA, 188–196

Chapter 27
Cooperativity of Protein Binding to Vesicles

Francisco Torrens and Gloria Castellano

Abstract Electrostatics role is studied in protein adsorption to phosphatidylcholine (PC) and PC/phosphatidylglycerol (PG) small unilamellar vesicles (SUVs). Protein interaction is monitored vs. PG content at low ionic strength. Adsorption of lysozyme, myoglobin and bovine serum albumin (BSA) isoelectric point (pI) is investigated in SUVs, along with changes in protein fluorescence emission spectra. Partition coefficients and cooperativity parameters are calculated. At pI, binding is maximum while at lower/higher pHs binding drops. In Gouy–Chapman model activity coefficient goes with square charge number, which deviations indicate asymmetric location of anionic lipid in the bilayer inner leaflet, in agreement with experiments and molecular dynamics simulations. Vesicles bind myoglobin anti-cooperatively and lysozyme/BSA cooperatively. Hill coefficient reflects subunit cooperativity of bi/tridomain proteins.

Keywords Molecular dynamics and simulation · Molecular interactions · Tools and methods for computational biology and bioinformatics · Protein modeling · Macromolecular structure prediction

1 Introduction

In earlier publications the binding of vinyl polymers to anionic SUVs [1], polyelectrolytes to cationic micelles [2], melittin to zwitterionic SUVs [3] and lysozyme–myoglobin–bovine serum albumin (BSA) to zwitterionic SUVs [4, 5] were studied by spectrofluorimetry. In the present report, proteins were used as models because they are water-soluble, globular with known three-dimensional structures, covering pI 5–11. The objective is to extend the studies to anionic lipid bilayers.

F. Torrens (✉)
Institut Universitari de Ciència Molecular, Universitat de València,
Edifici d'Instituts de Paterna, P.O. Box 22085, 46071 València, Spain
e-mail: torrens@uv.es; francisco.torrens@uv.es

2 Materials and Methods

The PG, phosphatidylserine (PS), phosphatidylinositol (PI), cardiolipin (CL), hen egg-white lysozyme, horse heart myoglobin and BSA were obtained from Sigma (St. Louis, MO). Egg-yolk PC was obtained from Merck (Darmstad, Germany) and purified. Salt, buffers and reagents were of the highest purity available. The SUVs of PC/PG, SP, PI and CL were prepared by dissolving the lipid in chloroform/methanol. Solvent was evaporated under a stream of $N_{2(g)}$ and lipid was dried under vacuum overnight. The 0.010 mol L^{-1} 3-(*N*-morpholino)-propanesulphonic acid (MOPS)–NaOH pH 7.0, acetate pH 4.0 or glycine buffer pH 9.0, at a given NaCl concentration 0–1 mol L^{-1}, was added to the dry film and the suspension was extensively vortexed. Lipid dispersion was sonicated for 20 min, at a temperature above lipid phase transition temperature, using an ultrasonic generator, with a microtip probe (Vibra-Cell^{TM}, Sonics & Materials, Inc., Danbury, CT), at a power setting 4 and 50% duty cycle. Samples were centrifuged for 15 min, at 35,000 g, to remove probe particles and remaining multilamellar aggregates. Lipid content in SUVs was determined by phosphorus assay. Integrity was controlled by negative-stain electron microscopy. Steady-state fluorescence measurements were recorded using a Perkin Elmer (Beaconsfield, UK) LS-50 spectrofluorometer, with 1.0×1.0 cm quartz cuvette. Excitation–emission bandwidths were 5 nm. Excitation wavelength was set to 280 nm. Spectra were corrected by comparison to quinine sulphate standard. In the binding experiments, proteins fluorescence emission spectra in buffer were monitored 300–440 nm. Titrations were performed adding SUV solution small aliquots to protein at a desired concentration in 1 mL, and data represent independent experiments. Possible weak fluorescence contribution from buffer–lipid was subtracted. In lipid–protein mixtures, emission fluorescence intensity changes at $\lambda_{\text{lysozyme–BSA}} = 345$ or $\lambda_{\text{myoglobin}} = 337$ nm, I^λ, were analysed vs. R_i (lipid/protein molar ratio) and, from intensity increase, fraction of bound protein $\alpha = \left(I^\lambda - I^\lambda_{\text{free}}\right)\left(I^\lambda_{\text{bound}} - I^\lambda_{\text{free}}\right)$ was estimated. The I^λ_{bound} was extrapolated from double-reciprocal plot. Total protein concentration was 1 µmol L^{-1}.

3 Results and Discussion

The partition coefficient for the protein, between the lipid and aqueous phases, is the ratio of the protein activity in the lipid phase a_p^L to that in aqueous phase a_p^A:

$$K_r = \frac{a_p^L}{a_p^A} = \frac{c_p^L \gamma_p^L}{c_p^A \gamma_p^A} \qquad (1)$$

where c_p^L and c_p^A are the protein concentrations in lipid–water, and γ_p^L and γ_p^A the activity coefficients. When lipid volume is negligible with regard to solvent one:

$$\frac{c_p^L}{c_p^A} = \frac{(\alpha/R_i^*)}{(1-\alpha)[P]_T \bar{v}_L} \tag{2}$$

where $\bar{v}_L = 0.785 \, \text{L mol}^{-1}$ is lipid partial molar volume and $(1-\alpha)[P]_T$ aqueous free protein concentration. Since protein has access only from vesicle outside, $\alpha/R_i^* = (\alpha/R_i)/\beta$ is corrected by fraction of lipid in outer leaflet $\beta = 0.65$:

$$\frac{(\alpha/R_i^*)}{(1-\alpha)[P]_T} = \frac{K_r \bar{v}_L}{\gamma} = \frac{\Gamma}{\gamma} \tag{3}$$

where $\gamma = \gamma_p^L / \gamma_p^A$. Parameter Γ is proportional to partition coefficient $\Gamma = K_r \bar{v}_L$:

$$\Gamma = C \bar{v}_L \exp\left(\frac{\Delta \bar{G}_p^{o,A} - \Delta \bar{G}_p^{o,L}}{RT}\right) \tag{4}$$

where constant C depends on molar masses and densities of water–lipid, and $\Delta \bar{G}_p^{o,A}$ and $\Delta \bar{G}_p^{o,L}$ are protein molar free energies in water–lipid. $\Delta \bar{G}_p^{o,A}$ is:

$$\frac{\Delta \bar{G}_p^{o,A}}{RT} = -\frac{N_A (z_p^+)^2 e^2}{2RT(R_p + 2R_w) 4\pi \varepsilon_o} \left(1 - \frac{1}{\varepsilon_w}\right) + \frac{\Delta G_{cav}}{RT}$$
$$- \frac{4N_A z_p^+ e \mu_w}{RT(R_p + R_w)^2 4\pi \varepsilon_o} + \frac{4N_A z_p^+ e \theta_w}{2RT(R_p + R_w)^3 4\pi \varepsilon_o} - \frac{4N_A (z_p^+)^2 e^2 \alpha_w}{2RT(R_p + R_w)^4 4\pi \varepsilon_o} \tag{5}$$

where R_p is protein ion effective radius, z_p^+ its actual charge, N_A Avogadro number, $e = 1.6021892 \times 10^{-19}\,\text{C}$ proton charge, $R = 8.3143\,\text{J mol}^{-1} \cdot \text{K}^{-1}$ gas constant, $T = 293\,\text{K}$ temperature, $\varepsilon_o = 8.854 \times 10^{-12}\,\text{C}^2 \text{N}^{-1} \text{m}^{-2}$ vacuum permittivity, $R_w = 2.8\,\text{Å}$ water solvation effective radius, $\varepsilon_w = 78.5$ its relative permittivity, $\mu_w = 1.86\,\text{D}$ its dipole moment, $\theta_w = 3.9 \times 10^{-26}\,\text{statC} \cdot \text{cm}^2$ its quadrupole moment and $\alpha_w = 1.65 \times 10^{-40}\,\text{C}^2 \text{m}^2 \text{J}^{-1}$ its polarizability. Cavitation energy ΔG_{cav} is:

$$\Delta G_{cav} = \tau A_{cav} + W_o = \tau A_{cav} - RT \ln(1 - V_{cav}\rho_w^n) \tag{6}$$

where τ is water–air surface tension ($435\,\text{J Å}^{-2}$), $A_{cav} = 4\pi R_p^2$ and $V_{cav} = 4\pi R_p^3/3$ cavity surface area and volume, W_o cavitation work for zero-surface solute and ρ_w^n water number density. An expression similar to (5) is obtained for $\Delta \bar{G}_p^{o,L}$:

$$\frac{\Delta \bar{G}_p^{o,L}}{RT} = -\frac{N_A (z_p^+)^2 e^2}{2RT R_p 4\pi \varepsilon_o} \left(1 - \frac{1}{\varepsilon_L}\right) + \frac{N_A z_p^+ z_L e^2}{RT 4\pi \varepsilon_L \varepsilon_o R_p (1 + \kappa R_p)} \tag{7}$$

where $\varepsilon_L = 20$ is the relative permittivity of a lipid membrane,

$$\kappa = \left(\frac{2e^2 N_A I}{\varepsilon_w \varepsilon_o kT}\right)^{1/2} \quad (8)$$

is inverse Debye screening length, z_L lipid head charge and k Boltman constant. Protein effective interfacial charge for membrane-bound state v is:

$$\ln \gamma = 2v \sinh^{-1}\left[vb\left(\alpha/R_i^*\right)\right] \quad (9)$$

$$b = \frac{20e}{\beta A_L (8\varepsilon_w \varepsilon_o RTI)^{1/2}} = 3.10\left(400/I\right)^{1/2} \quad (10)$$

where b depends on ionic strength I and $A_L = 70$ Å^2 lipid head area. One expects:

$$\ln \Gamma \propto \ln \gamma \propto z_L \sinh^{-1}(z_L) \approx z_L^2 \quad (11)$$

Lysozyme charges are 12.0, 8.0 and 6.0 e.u. at pHs 4.0, 7.0 and 9.0 decaying near $pI = 10.7$. Binding curves were analysed with partition equilibrium model, taking z_p^+ as adjustable parameter, assuming lysozyme solvated in both phases. A globular shape for lysozyme ($R_p = 20.3$ Å) was used. Adsorption on PC/PG at pH 7.0 and $I = 0.015$ mol L^{-1}, Fig. 1, increases with PG content, as expected for cationic lysozyme–anionic PG electrostatic interaction. The PC/PG shows high coverage, and v calculation did not converge for fixed z_L because Gouy–Chapman model is

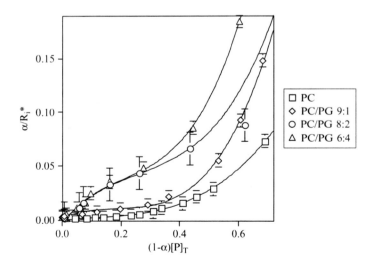

Fig. 1 Effect of vesicle charge on lysozyme–PC/PG (pH 7.0)

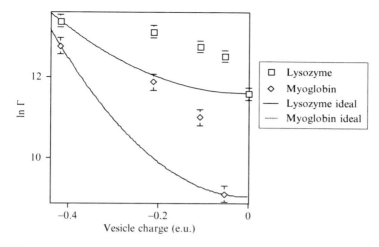

Fig. 2 Effect of vesicle charge on $\ln \Gamma$ for lysozyme (pH 7.0)/myoglobin (pH 4.0)–PC/PG

limited to low coverages. As ad-molecule is cationic $z_L^{eff} > z_L$, which was treated as a fitting parameter. The $z_p^+ < q$ because it decays via counterions.

Cationic lysozyme (pH 7) showed electrostatic selectivity for anionic SUVs, meaning that electrostatic interaction dominates over hydrophobic one, in agreement with Förster resonance energy transfer (FRET) between cationic surfactant protein (SP)-B of pulmonary surfactant and anionic lipids [6]. Myoglobin (pH 4) opposed selectivity confirms that vesicle interaction is dominated by hydrophobic interaction at low PG content, in agreement with FRET between SP-C and anionic lipids. Variation of $\ln \Gamma$ vs. z_L, Fig. 2, shows parabolas representing Gouy–Chapman model, above which data deviation indicates effective charge increment $z_{L,eff} > z_L$, in agreement with that in PC/PG anionic PG is asymmetrically located in the inner leaflet of the bilayer. Partial correlation diagram of the symmetric correlation matrix **R**, with regard to five properties $\{<z_p^+>, \Delta \overline{G}_p^{o,A}, \Delta \overline{G}_p^{o,L}, <\Gamma>, <\nu>\}$, is calculated with program GraphCor [7]; intercorrelations are high ($|r| \geq 0.75$).

A program, written with IMSL subroutine CLINK to perform *cluster analysis* (CA), was applied to lysozyme and myoglobin/PC/PG. *Single-* and *complete-linkage hierarchical* CAs performed a binary taxonomy of the data, which allowed the *dendrogram*, splitting three classes: {lysozyme/PC/PG 80/20,60/40}, {lysozyme/PC/PG 95/5,90/10,myoglobin/PC/PG 60/40} and {lysozyme/PC, myoglobin/PC,myoglobin/PC/PG 95/5,90/10,80/20}. Program SplitsTree permitted analysing CA results by *split decomposition*. Principal component analysis factor F_1 explains 90% of variance, F_{1-2} 97%, F_{1-3} 99.9%, etc. The same classes are obtained as in partial correlation diagram, dendrogram, radial tree and splits graph: ($F_1 < F_2$, Fig. 3, *left*), ($F_1 \approx F_2$, *middle*) and ($F_1 > F_2$, *right*). Adsorption isotherm forms from lysozyme–BSA solutions are *sigmoid*.

Figure 4 shows Scatchard $B/F - B$ plot for lysozyme–BSA (pH 7.0), and myoglobin (pH 4.0) to PC, where B is equilibrium molar concentration of *bound ligand*

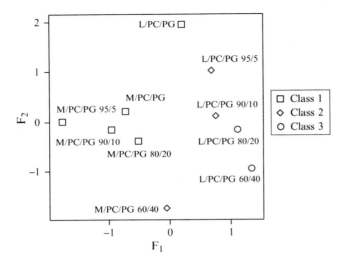

Fig. 3 Principal component analysis F_2 vs. F_1 scores plot of lysozyme–myoglobin/PC/PG

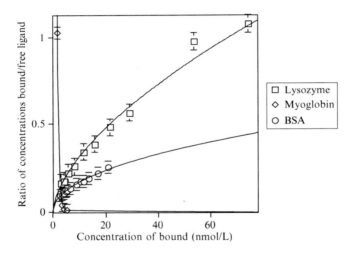

Fig. 4 Scatchard plot of lysozyme/myoglobin/bovine serum albumin–PC

and F that of *free ligand*. Myoglobin concavity indicates that for a given B/F, B is lower than for an ideal model suggesting negative cooperativity. Lysozyme–BSA convexity denotes that for a given B/F, B is greater than that expected for an ideal model signifying positive cooperativity, in agreement with results for BSA–biomaterial adsorption [8].

Figure 5 shows Hill $\log P_B/(1 - P_B) - \log F$ plots, for lysozyme–BSA (pH 7.0) and myoglobin (pH 4.0) to PC, where $P_B = B/B_{max}$ is adsorption degree and B_{max} maximum B. Slopes represent Hill coefficient $h = 1.749$, 0.402 and 1.598

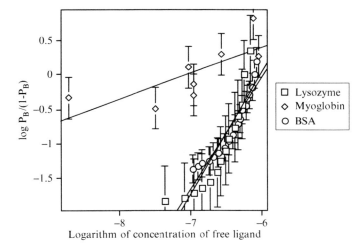

Fig. 5 Hill plot of lysozyme-, myoglobin- and bovine serum albumin-PC

for lysozyme, myoglobin and BSA. The PC bound myoglobin with cooperativity constant $h = 0.402 \leq 1$, meaning negative cooperativity, in agreement with reversible myoglobin–O_2 in water ($h = 1$), and peptides monodansylcadaverine (DNC)-melittin- and melittin-PC ($h = 0.888$ and 0.909). Hill coefficient reflects negative cooperativity of melittin and monodomain myoglobin.

4 Conclusions

From the present results and discussion, the following conclusions can be drawn:

1. For multivalent ions, Gouy–Chapman model is semi-quantitative. The activity coefficient goes with square charge number. For lysozyme- and myoglobin-PC/PG, deviations from the ideal model indicate asymmetric location of anionic lipid, in the inner leaflet of bilayer, in agreement with experiments and molecular dynamics simulations.
2. Myoglobin- and melittin-zwitterionic vesicles were described by a partition model, modulated by electrostatic charging of membrane as protein accumulates at interface. Further binding is difficult because charge repulsion dominates. Lysozyme and albumin followed a positively cooperative model, which represents interaction between protein considered as dipole moment and headgroups as isolated anion. Hill coefficient reflects subunit cooperativity of bi/tridomain proteins.

Acknowledgement The authors dedicate this manuscript to Prof. Dr Agustín Campos, who was greatly interested in this research and would have loved to see its conclusion.

References

1. Torrens F, Campos A, Abad C (2003) Binding of vinyl polymers to anionic model membranes. Cell Mol Biol 49:991–998
2. Torrens F, Abad C, Codoñer A, García-Lopera R, Campos A (2005) Interaction of polyelectrolytes with oppositely charged micelles studied by fluorescence and liquid chromatography. Eur Polym J 41:1439–1452
3. Torrens F, Castellano G, Campos A, Abad C (2007) Negatively cooperative binding of melittin to neutral phospholipid vesicles. J Mol Struct 834–836:216–228
4. Torrens F, Castellano G, Campos A, Abad C (2009) Binding of water-soluble, globular proteins to anionic model membranes. J Mol Struct 924–926:274–284
5. Torrens F, Castellano G. (2009) Comparative analysis of the electrostatics of the binding of cationic proteins to vesicles: Asymmetric location of anionic phospholipids. Anal Chim Acta 654:2–10
6. Pérez-Gil J, Keough KMW (1998) Interfacial properties of surfactant proteins. Biochim Biophys Acta 1408:203–217
7. Torrens F, Castellano G (2006) Periodic classification of local anaesthetics (procaine analogues). Int J Mol Sci 7:12–34
8. Sánchez-Muñoz OL, Nordström EG, Prérez-Hernández E (2003) Electrophoretic and thermodynamic properties for biomaterial particles with bovine serum albumin adsorption. Biomed Mater Eng 13:147–158

Chapter 28
The Role of Independent Test Set in Modeling of Protein Folding Kinetics

Nikola Štambuk and Paško Konjevoda

Abstract The testing of a bioinformatics algorithm on the training set is not the best indicator of its future performance because of the misleadingly optimistic results. The optimal method of testing is the calculation of error rate on an independent dataset (test set). We have tested the validity of the FOLD-RATE method for the prediction of protein folding rate constants $[\ln(k_f)]$ using sequences, structural class information and experimentally verified folding rate constants of the Protein Folding Database (PFD). PFD is a publicly accessible repository of thermodynamic and kinetic data of interest for the researchers of different profiles, standardized by the *International Foldeomics Consortium*. Our results show that when the standardized PFD dataset is used to test a protein fold rate prediction method, the estimation of validity may differ significantly.

1 Introduction

Investigation of the protein folding kinetics is of importance for defining the rules that govern protein folding and for determining its secondary and tertiary structures [1–5]. Prediction of the protein folding rates from the primary amino acid sequence of an individual protein is one of the most important challenges in molecular and computational biology [2]. Computational analysis and molecular modeling often represent a starting/elementary points of the protein structure-function analysis because they are not time consuming, and the costs of the procedures are relatively modest, when compared to the experimental methods (e.g. by X-ray, NMR, circular dichroism spectroscopy) [6]. The accurate prediction of the protein folding rates from the amino acid sequence would be a particularly useful starting point of the protein structure-function analysis, since primary amino acid compositions of proteins are well known and easily available [2, 3, 5].

N. Štambuk (✉)
NMR Center, Ruđer Bošković Institute, Bijenička cesta 54, HR-10002 Zagreb, Croatia
e-mail: stambuk@irb.hr

Until recently, it was not possible to test individual models of the protein folding kinetics on a defined set of relevant protein structures with reported experimental conditions, data analysis methods and adopted standards for the kinetic and thermodynamic data collection and analysis [1, 4, 7]. Due to the efforts of Maxwell et al., more than 35 researchers from eight countries formed the *International Foldeomics Consortium* for the standardization of data reporting, acquisition and analysis [4, 7]. This resulted in a publicly accessible repository of thermodynamic and kinetic data on protein folding and a new database that collects data into a single public resource named Protein Folding Database (PFD) [4, 7]. PFD is of particular importance for the researchers in the field of protein folding kinetics since it enables evaluation of models on an independent dataset [4, 7].

One recently developed method for the prediction of the protein folding rates from their amino acid sequences is a FOLD-RATE method which takes as input amino acid sequence and structural class information [2]. The procedure is based on the multiple regression equations of 49 diverse physical–chemical, energetic and conformational amino acid properties [2, 8, 9]. It was reported to exhibit a strong overall correlation of 0.96–0.99 between predicted and experimental folding rate constants $[\ln(k_f)]$ of 77 proteins [2]. We have tested the validity of the FOLD-RATE using sequences, structural class information and experimentally verified protein folding rate constants of the PFD (Table 1).

2 Results

We determined Pearson correlation coefficients (r) for the folding rate constants $[\ln(k_f)]$ between experimentally obtained PFD and theoretically calculated FOLD-RATE data [10]. The correlations were calculated for all proteins and for particular folding classes (α, β and $\alpha + \beta$). Since FOLD-RATE method enables $\ln(k_f)$ estimation for the proteins of unknown and known class, correlations for both options were calculated. Correlation coefficients of experimentally and theoretically obtained $\ln(k_f)$, presented in Table 2, did not confirm strong correlations reported by Gromiha et al. [2].

The impact of the protein class option in FOLD-RATE prediction was evaluated by a mountain plot, a useful graphical technique for comparison of a new method(s) with a reference method [11]. It is created by computing a percentile for each ranked difference between a new method and a reference method, with the following transformation for all percentiles above 50: percentile = 100 − percentile [11]. The alternative name is *folded empirical cumulative distribution plot*. If two methods are unbiased with respect to each other, the mountain will be centered over zero. Long tails in the plot reflect large differences between the methods, as shown in Fig. 1. The distribution plot shows that FOLD-RATE-based prediction of $\ln(k_f)$ is better if the protein class option is omitted. This is contrary to results of Gromiha et al. [2], who obtained the correlation of 0.87 without structural class information, and the 0.96–0.99 with structural class information.

Table 1 Proteins of the PFD used for validation of the FOLD-RATE method prediction

Protein[a]	PFD experimental ln(k_f)[a]	FOLD-RATE unknown class ln(k_f)	FOLD-RATE known class ln(k_f)	Fold class
Abp1 SH3	2.46	−1.09	−15.20	β
ACBP	6.96	6.75	6.32	α
ADAh2	6.80	−0.64	−11.50	α+β
Apo-azurin	4.91	7.72	11.80	β
CheW	7.44	4.62	21.70	β
C12	5.75	0.03	−3.23	α+β
CTL9	3.27	7.79	17.80	α+β
EC298	9.08	4.73	17.90	α
FKBP12	1.60	2.94	5.13	α+β
Fyn SH3	4.88	3.89	0.08	β
GW1	3.98	5.07	−10.60	β
IM7[a]	7.20	2.34	11.50	α
IM9[a]	7.33	6.19	9.56	α
λ-repressor	10.38	4.77	8.26	α
L23	2.02	2.13	0.97	α+β
mAcP	−1.58	5.90	15.30	α+β
NTL9	6.55	10.70	−17.40	α+β
Protein G	6.30	8.29	28.00	α+β
Protein L	4.10	4.81	−7.58	α+β
Raf RBD	8.36	11.10	−0.03	α+β
S6	6.07	5.07	24.80	α+β
Sho1 SH3	2.11	1.99	−4.76	β
Spectrin SH3	1.05	1.11	−4.79	β
Src SH2	8.74	12.50	8.19	α+β
Src SH3	4.36	2.58	3.18	β
Tm1083	6.85	3.77	14.00	α
U1A	4.62	3.34	11.30	α+β
Ubiquitin	7.33	6.36	6.64	α+β
Urm1	2.58	6.18	12.10	α
VisE	2.03	6.34	1.47	α

Note: [a]Two-state dataset [4]

Correlations between FOLD-RATE folding rate parameter $\ln(k_f)$ calculated with and without protein class option also suggest detrimental role of this parameter (all proteins: $r = 0.2903$, $p = 0.1204$; $\alpha = -0.5018$, $p = 0.2173$; $\beta = 0.6317$, $p = 0.0961$; $\alpha + \beta = 0.1741$, $p = 0.5598$).

FOLD-RATE method is based exclusively on the frequency of amino acids, at the expense of the protein length and the sequence. It calculates the same $\ln(k_f)$ for different modifications of the same protein sequence (Table 3).

Table 2 Correlations between experimental (PFD) and FOLD-RATE calculated folding rate parameter $\ln(k_f)$ – without and with protein class options

Without class option	Pearson's correlation (r)	p-value	Lower boundary (95% CI)	Upper boundary (95% CI)
All proteins (30)	0.3261	0.0786	0.0387	0.6142
α (8)	−0.4138	0.3250	−0.8660	0.4106
β (8)	0.6449	0.0865	−0.1096	0.9279
$\alpha + \beta$ (14)	0.3761	0.1896	−0.1930	0.7559
With class option	Pearson's correlation (r)	p-value	Lower boundary (95% CI)	Upper boundary (95% CI)
All proteins (30)	0.2716	0.2073	−0.1653	0.5282
α (8)	0.4340	0.2986	−0.3899	0.8720
β (8)	0.7924	0.0160	0.1985	0.9606
$\alpha + \beta$ (14)	−0.1819	0.5336	−0.6498	0.3859

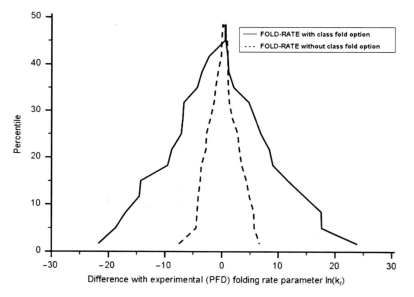

Fig. 1 Mountain plot of differences between experimentally verified folding rate constants of the Protein Folding Database (PFD), and FOLD-RATE-based prediction of $\ln(k_f)$ with and without protein class option

3 Discussion

Gromiha et al. [2] used the back-check and jack-knife tests for testing of the FOLD-RATE method and obtained, respectively, the correlation of 0.99 and 0.96 between experimental and predicted folding rates. However, the testing of bioinformatics algorithms on the training set is not the best indicator of future performance be-

Table 3 FOLD-RATE method calculates the same $\ln(k_f)$ for different modifications of the same protein sequence (values for *without class* option are presented)

	Peptide sequence Abp1 SH3	$\ln(k_f)$
Original	MAPWATAEYDYDAAEDNELTFVENDKIINIEFVDDDWWLGELE KDGSKGLFPSNYVSLGNLEHHHHHH	−1.09
Reverse	HHHHHHELNGLSVYNSPFLGKSGDKELEGLWWDDDVFEINIIK DNEVFTLENDEAADYDYEATAWPAM	−1.09
Alphabetical	AAAAADDDDDDDDEEEEEEEFFFGGGGHHHHHHIIIKKKLLL LLLMNNNNNPPSSSTTVVVWWWYYY	−1.09
Multiple	MAPWATAEYDYDAAEDNELTFVENDKIINIEFVDDDWWLGELE KDGSKGLFPSNYVSLGNLEHHHHHHMAPWATAEYDYDAAEDNE LTFVENDKIINIEFVDDDWWLGELEKDGSKGLFPSNYVSLGNL EHHHHHH	−1.09

cause of the over-optimistic prediction results [12]. The optimal method of testing is calculation of error rate on the independent dataset (test set) [12]. We used independently derived protein folding rate constants of the PFD that was standardized by the *International Foldeomics Consortium* to test predictive validity of the FOLD-RATE method. Our results suggest significantly lower performance of the FOLD-RATE method when using independent test set.

Current research standards require scientific and computational intelligence in documenting and managing data [13]. Databases containing information, i.e. big datasets, are not sufficient by itself. The scientific community needs simple, quick and free access to standardized and specific (often small) datasets, supplemented by open code software packages for data mining and statistics. PFD is a good example from the field of experimental and theoretical research on the mechanisms of protein folding. This database is a publicly accessible repository of thermodynamic and kinetic data of interest for the researchers of different profiles [4, 7]. It follows a strategy and recommendations of the *International Foldeomics Consortium*. However, when the results of standardized PFD dataset are used to test other prediction methods based on unstandardized datasets, the results may differ, as shown in Table 2 and Fig. 1. It seems that the analysis of the protein folding will follow the situation of the climate research [14] where publicly available data and open code of R software contributed to the "community cleverness" [10, 13].

Acknowledgement The support of the Croatian Ministry of Science, Education and Sports is gratefully acknowledged (grant no. 098–0982929–2524).

References

1. Fulton KF, Devlin GL, Jodun RA et al (2005) PFD: a database for the investigation of protein folding kinetics and stability. Nucleic Acids Res 33:D279–D283
2. Gromiha MM, Thangakani AM, Selvaraj S (2006) FOLD-RATE: prediction of protein folding rates from amino acid sequence. Nucleic Acids Res 34:W70–W74

3. Huang K (2005) Lectures on statistical physics and protein folding. World Scientific, New Jersey
4. Maxwell KL, Wildes D, Zarrine-Afsar A et al (2005) Protein folding: defining a "standard" set of experimental conditions and a preliminary kinetic data set of two-state proteins. Prot Sci 14:602–616
5. Nölting B (2006) Protein folding kinetics: biophysical methods. Springer, Berlin
6. Young DC (2009) Computational drug design: a guide for computational and medicinal chemists. Wiley, Hoboken
7. Fulton KF, Bate MA, Faux NG et al (2007) Protein folding database (PFD 2.0): an online environment for the International Foldeomics Consortium. Nucleic Acids Res 35:D304–D307
8. Gromiha MM, Oobatake M, Sarai A (1999) Important amino acid properties for enhanced thermostability from mesophilic to thermophilic proteins. Biophys Chem 82:51–67
9. Gromiha MM, Oobatake M, Kono H, Uedaira H, Sarai A (2000) Importance of surrounding residues for protein stability of partially buried mutations. J Biomol Struct Dyn 18:281–295
10. R Development Core Team (2005). R: A language and environment for statistical computing. R Foundation for Statistical Computing, Vienna, Austria. ISBN 3-900051-07-0. URL: http://www.R-project.org.
11. Krouwer JS, Monti KL (1995) A simple, graphical method to evaluate laboratory assays. Eur J Clin Chem Clin Biochem 33:525–527
12. Witten IH, Frank E (2005) Data mining: practical machine learning tools and techniques. Elsevier, San Francisco
13. Editorial (2008) Community cleverness required. Nature 455:1
14. Pocernich M (2006) R's role in the climate change debate. R News 6:17–18

Part IV
Comparative Sequence, Genome Analysis, Genome Assembly, and Genome Scale Computational Methods

Chapter 29
Branch-and-Bound Approach for Parsimonious Inference of a Species Tree from a Set of Gene Family Trees

Jean-Philippe Doyon and Cedric Chauve

Abstract We describe a Branch-and-Bound algorithm for computing a parsimonious species tree, given a set of gene family trees. Our algorithm can consider three cost measures: number of gene duplications, number of gene losses, and both combined. Moreover, to cope with intrinsic limitations of Branch-and-Bound algorithms for species trees inference regarding the number of taxa that can be considered, our algorithm can naturally take into account predefined relationships between sets of taxa. We test our algorithm on a dataset of eukaryotic gene families spanning 29 taxa.

Keywords Comparative genomics · Evolution and phylogenetics

1 Introduction

Speciation is the fundamental mechanism of genome evolution, especially for eukaryotic genomes. However, other events can happen, which do not result immediately in the creation of new species but act as fundamental evolutionary mechanisms, such as gene duplication and loss. Duplication is the genomic process where one or more genes of a single genome are copied, resulting in two copies of each duplicated gene. Gene duplication allows one copy to possibly develop a new biological function through point mutation, while the other copy often preserves its original role. A gene is considered to be lost when the corresponding sequence has been deleted by a genomic rearrangement or has completely lost any functional role. Genes of contemporary species that evolved from a common ancestor, through speciations and duplications, are said to be homologs [5] and are grouped into a gene family.

J.-P. Doyon (✉)
LIRMM, Université Montpellier 2 and CNRS, Montpellier, France
e-mail: doyonjea@iro.umontreal.ca; Jean-philippe.Doyon@lirmm.fr

The availability of large datasets of gene families makes now possible to perform genome-scale phylogenetic analyses. A widely used approach, named Gene Tree Parsimony (GTP for short), is based on the notion of *reconciliation* between a gene tree and a species tree introduced in [6], and seeks a species tree with a minimum overall reconciliation cost with the whole set of input gene trees. Given a gene tree G and a species tree S for the corresponding taxa, the reconciliation cost is the minimum number of duplications, losses, or mutations (duplications plus losses) that is needed to explain the (possible) discrepancies between G and S. Computing a most parsimonious reconciliation between a given gene tree and a species tree can be done in linear time [13], but inferring a parsimonious species tree is an NP-complete problem for both duplication and mutation criteria [8], although fixed-parameter tractable algorithms have been described in [7]. Hence, in most cases, studies based on GTP use either a brute-force approach when the number of taxa is low [11], greedy heuristics [2], or local search approach with edit operations on species trees [1,9]. Although such heuristics, especially local-search ones, are fast and proved to be effective on large datasets [12], they do not guarantee to infer an optimal species tree.

A Branch-and-Bound approach is a classical method when dealing with hard species tree inference problems, as it implicitly explores the space of species trees and guarantees to find an optimal phylogeny for a given criterion. This method has been used for evolutionary criteria such as Maximum Parsimony and Maximum Likelihood, but it has not been considered up to now for the GTP. In this work, we present a Branch-and-Bound algorithm that guarantees to find a species tree S with the minimum cost for a given gene tree G, and works for the three usual criteria (duplications, losses, and mutations). Our algorithm relies on a new way to explore the space of species trees that allows to update efficiently the cost (for the three considered costs) of a partial species tree.

2 Preliminaries

Except when indicated, any considered tree is rooted, binary, unordered, and leaf-labeled. For simplicity, we consider that each leaf label is an integer. For a given tree T, let $V(T)$, $r(T)$, $L(T)$, $\Lambda(T)$, and $I(T)$, respectively, denote its vertex set, root, leaf set, label set (the set of integers that appear at its leaves), and internal vertex set (i.e., $V(T) \setminus L(T)$). For a vertex u of T, we denote by u_1 and u_2 its children (when $u \notin L(T)$), by $p(u)$ its parent, by $s(u)$ its sibling (when $u \neq r(T)$), and by T_u the subtree of T rooted at u. The distance between two vertices u and v of a tree T, where u is a descendant of v, is denoted $d_T[u,v]$ and is the number of vertices on the path from u to v in T, excluding u and v.

A *species tree* S is a tree such that each element of $\Lambda(S)$ represents an extant species and labels exactly one leaf of S (there is a bijection between $L(S)$ and $\Lambda(S)$). A *gene tree* G is a tree such that $\Lambda(G) \subseteq \Lambda(S)$ (each leaf of G represents an extant gene that belongs to a species of $\Lambda(S)$). A *forest* is a set of trees. A *species forest*

$\mathcal{F}$ is a set of species trees with disjoint label sets. Given two trees S_1 and S_2 of a species forest $\mathcal{F}$, we denote by $(S_1 + S_2)$ the tree obtained by joining S_1 and S_2 under a common (binary) root x (i.e., creating two edges from x to the roots of S_1 and S_2).

A reconciliation between a gene tree G and a species tree S maps each internal vertex of G onto a vertex of S and induces an evolutionary history in terms of gene duplications and losses. The Lowest Common Ancestor mapping defines (see Definition 1 below) the most widely used reconciliation, as it depicts a parsimonious evolutionary process for each of the three usual combinatorial criteria [3]. This parsimonious reconciliation is the one we consider here. Definitions 2–4 below define how to read the different costs associated with this reconciliation between a given gene tree G and a given species tree S.

Definition 1. The LCA mapping between a gene tree G and a species tree S, denoted $M_S : V(G) \to V(S)$, is defined as follows: given a vertex u of G, $M_S(u)$ is the unique vertex x of S such that $\Lambda(G_u) \subseteq \Lambda(S_x)$ and either x is a leaf of S, or $\Lambda(G_u) \not\subseteq \Lambda(S_{x_1})$ and $\Lambda(G_u) \not\subseteq \Lambda(S_{x_2})$.

Definition 2. An internal vertex $u \in I(G)$ is a duplication if $M_S(u) = M_S(u_1)$ and/or $M_S(u) = M_S(u_2)$. The duplication cost between G and S is $d(G,S) = \sum_{u \in I(G)} d(u,S)$, where $d(u,S)$ has value 1 if and only if $u \in I(G)$ is a duplication and 0 otherwise.

Definition 3. The loss cost between G and S is $l(G,S) = \sum_{u \in I(G)} l(u,S)$, where $l(u,S)$ equals (1) 0 if $M_S(u) = M_S(u_1) = M_S(u_2)$; (2) $d_S[M_S(u_i), M_S(u)] + 1$ if $M_S(u_i) \neq M_S(u)$ for either $i = 1$ or $i = 2$; (3) $d_S[M_S(u_1), M_S(u)] + d_S[M_S(u_2), M_S(u)]$, otherwise.

Definition 4. The mutation cost between G and S is $m(G,S) = l(G,S) + d(G,S)$.

The LCA mapping between a gene tree G and a species forest $\mathcal{F}$, denoted $M_\mathcal{F} : V(G) \to V(\mathcal{F})$, is defined similarly to the case of a single species tree (see Definition 1). For a given vertex $u \in V(G)$, $M_\mathcal{F}(u) = M_S(u)$, if there is a species tree S of $\mathcal{F}$ such that $M_S(u)$ is defined. Otherwise, $M_\mathcal{F}(u)$ is said to be undefined (which is denoted $M_\mathcal{F}(u) = \varnothing$ from now).

The goal of this work is the design of an exact method to solve the following optimization problem, given a cost measure c (either d, l, or m) for the reconciliation between a gene tree and a species tree.

MINIMUM C SPECIES TREE PROBLEM
INPUT. A gene tree forest $\mathcal{G} = \{G_1, \ldots, G_k\}$
OUTPUT. A species tree S such that $\sum_{i=1}^{k} c(G_i, S)$ is minimized.

3 A Branch-and-Bound Algorithm for the GTP

Our algorithm is based on a classical Branch-and-Bound scheme, and we describe below its main components: the architecture of the exploration tree and the lower bounds. All the details are in the technical report [4]. We assume that there are n

taxa, denoted by $\{1, 2, \ldots, n\}$, and denote by $\mathcal{K}^n$ the set of all possible species trees on these n taxa. Without loss of generality, we describe our algorithm for a single gene tree G.

The algorithm is based on the exploration of a rooted tree denoted $\mathcal{T}^n$, where each vertex corresponds to a forest of species trees, and such that each internal forest corresponds to an incomplete species tree for n species, and each leaf forest to a complete species tree S of $\mathcal{K}^n$. The Branch-and-Bound explores this tree and each time it visits a forest denoted $\mathcal{F}$, it computes a lower bound on the cost $c(G, S)$ (where $c = l$ or $c = d$) of any species tree $S \in \mathcal{K}^n$ located in $\mathcal{T}^n_{\mathcal{F}}$, that is the subtree of $\mathcal{T}^n$ rooted at $\mathcal{F}$. To ensure the optimality of this approach, such a lower bound has to respect the following definition.

Definition 5. Let $\pi : \mathcal{K}^n \to \mathbb{N}$ be an objective function. A function $\omega : V(\mathcal{T}^n) \to \mathbb{N}$ is a Consistent Lower Bound (CLB) for π if and only if (1) it is non-decreasing along the path that starts at $r(\mathcal{T}^n)$ and ends at any leaf $\{S\}$ of $\mathcal{T}^n$ and (2) $\omega(\{S\}) = \pi(S)$.

Given a CLB, denoted $c(G, \mathcal{F})$, for the considered cost $c(G, S)$, then $\mathcal{T}^n_{\mathcal{F}}$ is explored if and only if $c(G, \mathcal{F}) < c(G, S_{\min})$, where $S_{\min} \in \mathcal{K}^n$ is the best solution found since the beginning of the exploration of $\mathcal{T}^n$ and is updated when a species tree with a lower cost is found. Such a Branch-and-Bound guarantees to find an optimal species tree $S_{\min}$ such that $c(G, S_{\min})$ is minimum.

This section is separated in two parts. First, we formally define the space tree $\mathcal{T}^n$ and give important combinatorial properties that are central in the design of the Branch-and-Bound. Second, we define the two CLBs for the costs $d(G, S)$ and $l(G, S)$.

The main structural feature of $\mathcal{T}^n$ is that a child $\mathcal{F}'$ of an internal forest $\mathcal{F}$ is defined by joining two of its trees under a (new) vertex (thus forming a new clade). This architecture is different from the classical one used in a Branch-and-Bound approach for phylogenetic inference, where the exploration starts with a tree with two leaves and one internal node, an then iteratively add a leaf and an edge until a complete tree is obtained. The advantage of the architecture of $\mathcal{T}^n$ over the classical one is that it is more adapted to efficiently compute the LCA mapping during its traversal, which is essential to rapidly explore the space and solve our problem. Definition 7 formally describes the architecture of $\mathcal{T}^n$, and Property 1 shows that it is an appropriate structure for the exploration of $\mathcal{K}^n$. Below, given a species tree S over $\{1, \ldots, n\}$, $\min[\Lambda(S)]$ denotes the minimum label of the leaves of S. We also define an order on the trees of a forest as follows.

Definition 6. Given two trees S and S' of a forest $\mathcal{F}$, $S \prec_{\mathcal{F}} S'$ (resp. $\preceq_{\mathcal{F}}$) if and only if $min[\Lambda(S)] < min[\Lambda(S')]$ (resp. $\leq$).

Definition 7. $\mathcal{T}^n$ is an ordered and rooted tree where each vertex is an ordered species forest $\mathcal{F}$ on $\{1, \ldots, n\}$, with a distinguished tree called the *branching tree* $\beta(\mathcal{F})$. The branching structure of $\mathcal{T}^n$ is defined below.

1. The root forest $r(\mathcal{T}^n)$ is the forest composed of n trees $\{S_1, \ldots, S_n\}$, where S_i is the tree reduced to a single vertex labeled i, and its branching tree is the tree S_1.

2. Each leaf of T^n is a forest $\mathcal{F}$ containing a single tree that is a species tree from $\mathcal{K}^n$.
3. A forest $\mathcal{F}_x$ is a child of an internal forest $\mathcal{F}$ if and only if there exists two trees S_{x_1} and S_{x_2} in $\mathcal{F}$, with $S_{x_1} \prec_{\mathcal{F}} S_{x_2}$, such that $\mathcal{F}_x = \mathcal{F} - \{S_{x_1}, S_{x_2}\} \cup \{S_x\}$, where $S_x = (S_{x_1} + S_{x_2})$ is the branching tree of $\mathcal{F}_x$, and either $S_{x_2} = \beta(\mathcal{F})$ or $\beta(\mathcal{F}) \preceq_{\mathcal{F}} S_{x_1}$.

Finally, the children of an internal vertex (i.e., forest) $\mathcal{F}$ of T^n are totally ordered as follows: if $\mathcal{F}_x$ and $\mathcal{F}_y$ are two children of $\mathcal{F}$, where the corresponding branching trees are, respectively, $S_x = (S_{x_1} + S_{x_2})$ and $S_y = (S_{y_1} + S_{y_2})$, then $\mathcal{F}_x$ precedes $\mathcal{F}_y$ if and only if either (1) $S_{x_1} \prec_{\mathcal{F}} S_{y_1}$ or (2) $S_{x_1} = S_{y_1}$ and $S_{x_2} \prec_{\mathcal{F}} S_{y_2}$.

Property 1 below follows from the structure of T^n described in Definition 7, and in particular from the order on the trees in a forest, which ensures that no two different paths from the root can lead to the same species forest.

Property 1. The tree T^n is such that (1) there are no two nodes that represent the same species forest, (2) $L(T^n) = \mathcal{K}^n$, (3) its height is $\Theta(n)$, (4) the number of children of each internal vertex is bounded by $O(n^2)$.

The general principle of our Branch-and-Bound algorithm is to visit T^n starting at its root, and then to recursively visit the children of the starting vertex forest $\mathcal{F}$ according to the order described in Definition 7. There are two key points that explain how this exploration can be done in time linear in the size of T^n. First, to efficiently visit the children of an internal vertex $\mathcal{F}$ according to the order defined in Definition 7, it is sufficient that the trees of $\mathcal{F}$ are ordered according to $\prec_{\mathcal{F}}$. Second, after visiting a child forest of $\mathcal{F}$, the order $\prec_{\mathcal{F}}$ is easily recovered in constant time using lists and pointers. Together that the height of T^n is in $\Theta(n)$, this proves the following result.

Proposition 1. *The complete exploration of a subtree T of T^n can be implemented to run in time $\Theta(|V(T)|)$ and space $\Theta(n)$.*

We finally introduce some combinatorial properties on the architecture of T^n that will be used to define a CLB for the cost $l(G, S)$. According to Definition 3, the cost $l(u, S)$ induced by an internal vertex u of G depends on the distance in S between $M_S(u)$ and $M_S(u_1)$ (resp. $M_S(u_2)$). Hence, the main idea behind a CLB for $l(G, S)$ resides on the definition of a CLB for the distance $d_S[M_S(u_1), M_S(u)]$ {resp. $d_S[M_S(u_2), M_S(u)]$}. Formally, considering a forest $\mathcal{F}$ of T^n, a non-root vertex u of G, and any species tree $S \in \mathcal{K}^n$ that is located at a leaf of $T_{\mathcal{F}}^n$, the question is as follows: how can a CLB for $d_S[M_S(u), M_S(p(u))]$ be efficiently computed during the traversal of T^n along the path that connects $r(T^n)$ and $\mathcal{F}$? To define such a CLB, we introduce *incremental forests*.

Definition 8. Let $\mathcal{F}$ be a forest of T^n and $\mathcal{F}_x$ one of its children, whose branching tree is $S_x = (S_{x_1} + S_{x_2})$. Given a non-root vertex u of G, if the mapping $M_{\mathcal{F}}(u)$ is defined in either S_{x_1} or S_{x_2} and $M_{\mathcal{F}_x}(p(u))$ is not defined, then $\mathcal{F}_x$ is said to be an *incremental forest* for u.

Each incremental forest for u located between $r(T^n)$ and $\mathcal{F}$ (including $\mathcal{F}$) corresponds to an increment of one on $d_S[M_S(u), M_S(p(u))]$. If $d_\mathcal{F}(u)$ denotes the number of such incremental forests, then the property below immediately follows from the usual LCA mapping (between G and S) and Definition 7.

Property 2. Given a leaf forest $\{S\}$ of T^n and $u \in V(G) \setminus \{r(G)\}$, $d_S[M_S(u), M_S(p(u))] = d_{\{S\}}(u)$, and $d_\mathcal{F}(u)$ is a CLB for $d_S[M_S(u), M_S(p(u))]$.

For both duplication and loss criteria, we formally define a cost for a forest $\mathcal{F}$ of T^n. Below, $I'(G)$ denotes the subset $\{u \in I(G) : M_\mathcal{F}(u) \neq \varnothing\}$, and when the context is unambiguous, S refers to the species tree of $\mathcal{F}$ where the mapping $M_\mathcal{F}(u)$, of $u \in I'(G)$, is defined.

Definition 9. The duplication cost between a forest $\mathcal{F}$ of T^n and a gene tree G is denoted $d(G, \mathcal{F})$ and is defined by $d(G, \mathcal{F}) = \sum_{u \in I'(G)} d(u, S)$.

Definition 10. The loss cost between a forest $\mathcal{F}$ of T^n and a gene tree G is denoted $l(G, \mathcal{F})$ and is defined as follows: $l(G, \mathcal{F}) = \sum_{u \in I'(G)} l(u, S) + \sum_{u \in I(G) \setminus I'(G)} [d_\mathcal{F}(u_1) + d_\mathcal{F}(u_2)]$.

From Definitions 2, 3, and 5 and Property 2, the following result is immediate: $l(G, \mathcal{F})$ (resp. $d(G, \mathcal{F})$) is a CLB for $l(G, S)$ (resp. $d(G, S)$). The main issue now is to detect as efficiently as possible the vertices of G for which the LCA mapping is undefined in the current visited forest but is defined when one of its children is visited. Definition 11 below describes the smallest forest of subtrees of G that contains all these vertices, and Corollary 1 below gives the complexities to update the considered CLB.

Definition 11. Let $\mathcal{F}$ be an internal forest of T^n and $\mathcal{F}_x$ be one of its child, where $S_x = (S_{x_1} + S_{x_2})$ is the branching tree. $\mathcal{G}_x$ denotes the forest of subtrees of G such that its root set is $\mathcal{M}_{S_x} = \{u \in V(G) \setminus \{r(G)\} : M_{S_x}(u) \neq \varnothing \text{ and } M_{S_x}(s(u)) = \varnothing\}$ and its leaf set is $\mathcal{M}_{S_{x_1}} \cup \mathcal{M}_{S_{x_2}}$.

Corollary 1. *If the loss (resp. duplication) cost $l(G, \mathcal{F})$ (resp. $d(G, \mathcal{F})$) is given, then $l(G, \mathcal{F}_x)$ (resp. $d(G, \mathcal{F}_x)$) can be computed in time $\Theta(|V(\mathcal{G}_x)|)$ and space $O(|V(\mathcal{G}_x)|) + \Theta(|V(\mathcal{F}_x)|)$.*

Proposition 1, together with Corollary 1, gives the fundamental properties of the complexity of our Branch-and-Bound algorithm: (1) the exploration of the visited species forests requires a time linear in the number of these forests, (2) visiting a new species forest while updating the appropriate CLB requires a time linear in the number of vertices of gene forest whose LCA mapping is updated, and (3) the total space complexity is linear in the number of considered taxa. We do not have theoretical properties of the two CLBs we introduced, and we will assess how efficient they are to cut large subspaces of T^n experimentally in Sect. 4.

4 Experimental Results

We considered 1,111 gene family trees that have been manually corrected by experts and contain gene families from 29 eukaryotic genomes. The corresponding reference species tree is denoted by S_0 and corresponds to the NCBI taxonomy tree, except that three nodes of the tree were considered as multifurcations due to different phylogenetic hypothesis regarding the corresponding clades.

First, to gain some insight on the whole space of species trees for a given dataset, we selected $n = 8$ species from the 29 considered genomes, removed from the gene trees all genes from other species, and performed an exhaustive exploration of the 135,135 species trees. The aim is to study the shape of the space $\mathcal{K}^n$ according to the three combinatorial criteria we considered to evaluate their performance for phylogenetic inference. First, we observed that for the loss and mutation cost, the loss cost distribution is similar to a normal and almost symmetrical distribution with a mean located around 0.5, while the distribution for the duplication cost is less smooth. Second, it appears clearly from Table 1 that, on this dataset, the duplication cost seems to be slightly better than the two other criteria, although in terms of normalized cost, the difference is relatively marginal. Finally, we can notice that S_0 is close to the optimal species tree, both in terms of duplication and/or loss events, and in the classical Robinson and Foulds distance between phylogenetic trees [10].

Next, we attacked the problem of computing a parsimonious species tree for the 29 considered genomes. There are about 10^{36} possible species trees, and the Branch-and-Bound starting from the root of $\mathcal{T}^n$ was not completed after a few days of computation. There are two reasons for this problem: first, during the traversal of $\mathcal{T}^n$, the CLB of the newly visited forest is computed in linear time; second, the subtree of $\mathcal{T}^n$ induced by the pruned forests is not small enough for an exhaustive exploration.

We then decided to reduce the number of considered species trees by integrating prior information on the seeked species tree. We followed this hypothesis: the more a clade from the species tree S_0 is respected among the considered gene trees, the more probable it is present in an optimal solution of the GTP problem. Let $\mathcal{K}^n(S)$ denotes the subset of species trees that are consistent with a given tree $S \in \mathcal{K}^n$. We found 19 clades which are respected by a majority of the 1,111 gene trees, and defined four species trees, denoted S_i for $i \in \{1,2,3,4\}$, such that $\mathcal{K}^n(S_i) \subset \mathcal{K}^n(S_{i+1})$, for $0 \leq i \leq 3$. Hence, $\mathcal{K}^n(S_0)$ (resp. $\mathcal{K}^n(S_4)$) is the smallest (resp. largest)

Table 1 Minimum (col. 1) and maximum (col. 2) costs for the loss, duplication, and mutation criteria. Col. 3: costs of S_0, both absolute and normalized. Col. 4: Robinson and Foulds distance between the optimal solution S_{min} and S_0

	$c(G, S_{min})$	$c(G, S_{max})$	$c(G, S_0)$	$RF(S_0, S_{min})$
Loss	3,577	35,072	5,226 (0.05)	3
Dup.	2,229	6,313	2,425 (0.04)	1
Mut.	5,812	41,355	7,651 (0.05)	3

Table 2 For each criterion and each constrained species tree S_i, the *optimal cost* for the GTP problem applied on the set of allowed solutions $\mathcal{K}^n(S_i)$ and the *CPU time* used by the Branch-and-Bound

	Optimal cost in $\mathcal{K}^n(S_i)$		CPU time (s)				
	S_0	$S_1 \ldots S_4$	S_0	S_1	S_2	S_3	S_4
Loss	22,464	21,257	54	3,147	7,897	26,444	59,997
Mut	27,691	26,328	49	3,944	10,697	39,742	94,429
Dup	5,140	4,941	50	7,296	32,117	?	?

For each of the three criteria, the optimal cost in $\mathcal{K}^n(S_i)$, for $i = 1, 2, 3,$ and 4, is the same. The "?" character indicates that the Branch-and-Bound process was not terminated after 4 days, where the optimal solution of $K^n(S_2)$ was the best solution found so far for both processes.

set of species trees. For S_1, S_2, S_3, and S_4, the number of possible species trees is, respectively, 127,575, 893,025, 9,823,275, and 29,469,825. For the three usual criteria, we applied the Branch-and-Bound to solve the GTP problem first on the smallest set $\mathcal{K}^n(S_0)$, and then the optimal solution for $\mathcal{K}^n(S_i)$, with increasing index i from 0 to 3, was then used as the first upper bound for the Branch-and-Bound applied on $\mathcal{K}^n(S_{i+1})$. For each criterion and each constrained species tree, the results are summarized in Table 2 above.

For the duplication criterion, the best solution found in $\mathcal{K}^n(S_1)$ and $\mathcal{K}^n(S_2)$ is the same. For the loss and mutation criteria and the four sets $\mathcal{K}^n(S_i)$, the optimal solution is the same. For both criteria, this means that the optimal solution for the GTP applied on the largest set $\mathcal{K}^n(S_4)$ can be found solely by applying the Branch-and-Bound on the smallest set $\mathcal{K}^n(S_1)$, although that its optimality status requires the use of $\mathcal{K}^n(S_4)$. The Robinson and Foulds distance between the two optimal solutions for loss (i.e., mutation) (in $\mathcal{K}^n(S_4)$) and duplication (in $\mathcal{K}^n(S_2)$) criteria is 4, while their distance with S_0 (i.e., the reference species tree), respectively, is 5 and 3.

References

1. M.S. Bansal, O. Eulenstein, and A. Wehe. The gene-duplication problem: Near-linear time algorithms for nni-based local searches. *IEEE/ACM Transactions on Computational Biology and Bioinformatics*, 6(2):221–231, 2009.
2. C. Chauve, J.P. Doyon, and N. El-Mabrouk. Gene family evolution by duplication, speciation and loss. *Journal of Computational Biology*, 15(8):1043–1062, 2008.
3. C. Chauve and N. El-Mabrouk. New perspectives on gene family evolution: Losses in reconciliation and a link with supertrees. In *Research in Computational Molecular Biology, 13th Annual International Conference, RECOMB 2009, Tucson, AZ, USA, May 18-21, 2009. Proceedings*, volume 5541 of *Lecture Notes in Computer Science*, pages 46–58. Springer, 2009.
4. J.P. Doyon and C. Chauve. Branch-and-bound approach for parsimonious inference of a species tree from a set of gene family trees. Technical report, LIRMM, 2010. URL http://hal-lirmm.ccsd.cnrs.fr/lirmm-00448481/fr/.
5. W.M. Fitch. Homology a personal view on some of the problems. *Trends in Genetics*, 16:227–231, 2000.

6. M. Goodman, J. Czelusniak, G.W. Moore, R.A. Herrera, and G. Matsuda. Fitting the gene lineage into its species lineage, a parsimony strategy illustrated by cladograms constructed from globin sequences. *Systematic Zoology*, 28:132–163, 1979.
7. M.T. Hallett and J. Lagergren. New algorithms for the duplication-loss model. In *Proceedings of the Fourth Annual International Conference on Computational Molecular Biology, RECOMB 2000, April 8-11, 2000, Tokyo, Japan*, pages 138–146. ACM Press, 2000.
8. B. Ma, M. Li, and L. Zhang. From gene trees to species trees. *SIAM Journal on Computing*, 30(3):729–752, 2000.
9. R.D.M. Page. GeneTree: Comparing gene and species phylogenies using reconciled trees. *Bioinformatics*, 14(9):819–820, 1998.
10. D.F. Robinson and L.R. Foulds. Comparison of phylogenetic trees. *Mathematical Biosciences*, 53:131–147, 1981.
11. M. Sanderson and M. Mcmahon. Inferring angiosperm phylogeny from EST data with widespread gene duplication. *BMC Evolutionary Biology*, 7:S3, 2007.
12. A. Wehe, M.S. Bansal, J.G Burleigh, and O. Eulenstein. DupTree: A program for large-scale phylogenetic analyses using gene tree parsimony. *Bioinformatics*, 24(13):1540–1541, 2008.
13. L. Zhang. On a Mirkin-Muchnik-Smith conjecture for comparing molecular phylogenies. *Journal of Computational Biology*, 4(2):177–187, 1997.

Chapter 30
Sequence-Specific Sequence Comparison Using Pairwise Statistical Significance

Ankit Agrawal, Alok Choudhary, and Xiaoqiu Huang

Abstract There has been a deluge of biological sequence data in the public domain, which makes sequence comparison one of the most fundamental computational problems in bioinformatics. The biologists routinely use pairwise alignment programs to identify similar, or more specifically, related sequences (having common ancestor). It is a well-known fact that almost everything in bioinformatics depends on the inter-relationship between sequence, structure, and function (all encapsulated in the term *relatedness*), which is far from being well understood. The potential relatedness of two sequences is better judged by statistical significance of the alignment score rather than by the alignment score alone. This chapter presents a summary of recent advances in accurately estimating statistical significance of pairwise local alignment for the purpose of identifying related sequences, by making the sequence comparison process more sequence specific. Comparison of using pairwise statistical significance to rank database sequences, with well-known database search programs like BLAST, PSI-BLAST, and SSEARCH, is also presented. As expected, the sequence-comparison performance (evaluated in terms of retrieval accuracy) improves significantly as the sequence comparison process is made more and more sequence specific. Shortcomings of currently used approaches and some potentially useful directions for future work are also presented.

1 Introduction

It is a well-known fact that almost everything in bioinformatics depends on the inter-relationship between sequence, structure, and function (all encapsulated in the term "relatedness"), which is far from being well understood. With

---

A. Agrawal (✉)
Department of Electrical Engineering and Computer Science, Northwestern University,
2145 Sheridan Road, Evanston, IL 60208, USA
e-mail: ankitag@eecs.northwestern.edu; ankit108@gmail.com

sequencing becoming more and more easy and affordable, there is a huge amount of sequence data in the public domain for the analysis of which computational sequence comparison techniques would have to play a key role. Pairwise sequence alignment is an extremely important and common application in the analysis of DNA and protein sequences [9, 12, 14, 19, 20, 27]. It forms the basic step of many other bioinformatics applications such as multiple sequence alignment, database search, finding protein function, protein structure, and phylogenetic analysis for making various high level inferences about the DNA and protein sequences.

A typical pairwise alignment program aligns two sequences and constructs an alignment with maximum similarity score. Although related sequences will have high similarity scores, the threshold alignment score T below which the two sequences can be considered unrelated depends on the probability distribution of alignment scores between random, unrelated sequences [16]. Therefore, the biological significance of a pairwise sequence alignment is gauged by the statistical significance rather than the alignment score alone. This means that if an alignment score has a low probability of occurring by chance, the alignment is considered statistically significant and hence biologically significant. However, it must be noted that statistical significance does not necessarily imply biological significance although it may be a good preliminary indicator of biological significance [7, 14, 16, 19].

The alignment score distribution depends on various factors such as alignment program, scoring scheme, sequence lengths, and sequence compositions [16]. Figure 1 shows two alignment score distributions (probability density functions) X and Y. Consider scores x and y in the score distributions X and Y, respectively. Clearly $x < y$, but x is more statistically significant than y, since x lies more in the right tail of the distribution and is less probable to have occurred by chance. The shaded region represents the probability that a score equal or higher could have been obtained by chance. Thus, it is useful to estimate the statistical significance of an alignment score to comment on the relatedness of the two sequences being aligned.

Currently, rigorous statistical theory for alignment score distribution is available only for ungapped alignment, and not even for its simplest extension, i.e., alignment with gaps. Statistics of score distributions from more sophisticated alignment techniques incorporating more biological features, therefore, is not expected to be straightforward. Thus, biologically relevant statistical significance estimates of scores from such programs can be extremely useful.

With the all-pervasive use of sequence alignment methods in bioinformatics making use of ever-increasing sequence data, and with development of more and more sophisticated alignment methods with unknown statistics, it is expected that computational and statistical approaches for accurate estimation of statistical significance of pairwise alignment scores would be very useful for computational biologists.

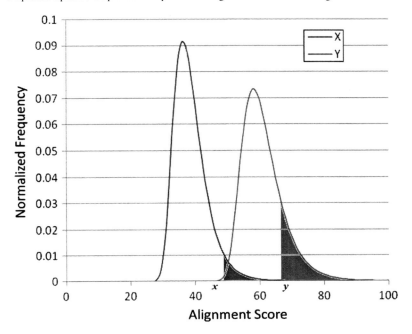

Fig. 1 Two alignment score distributions X and Y depicting the advantage of statistical significance over alignment scores. $x < y$, but x is more statistically significant than y. The *shaded region* represents the probability that a score equal or higher could have arisen by chance

2 Nature of the Problem

For the purpose of homology detection, biologists are interested in finding if there are any biologically relevant targets sharing important structural and functional characteristics, which are common to the two sequences. Thus, pairwise local alignment [26] is commonly used for this purpose, which aligns two sequences using a substitution matrix consisting of pairwise scores of aligning different residues with each other (like BLOSUM62), and give an alignment score for the given sequence-pair. Since the evolution of DNA and proteins in living organisms is influenced by a number of random factors, and any observed patterns may be due to such factors, rather than from the selective pressure maintaining a certain function [14], the question here arises how likely it is that a high local similarity score is obtained by chance, which gives importance to the concept of statistical significance.

Statistical significance of a pairwise alignment score is commonly assessed by its P-value, which denotes the probability that an alignment with this score or higher occurs by chance alone. Here, the test statistic is the alignment score, the null hypothesis is that the aligned sequences are unrelated, and the alternate hypothesis is that the sequences are related (homologous). Therefore, the P-value is the probability of seeing an alignment score as extreme as the observed value, assuming that the null hypothesis is true. In Fig. 1, the shaded region represents the P-value. It is easy to see that estimation of the P-value for an alignment score requires the knowledge of the distribution of the local alignment scores, especially in the right tail region, since the related sequences will (generally) have high scores.

Score distribution for ungapped local alignment is known to follow a Gumbel-type EVD [13], as shown in Fig. 1 with analytically calculable parameters, K and λ. The probability that the optimal local alignment score S exceeds x is given by the P-value:

$$\Pr(S > x) \sim 1 - e^{-E},$$

where E is the E-value and is given by

$$E = Kmne^{-\lambda x},$$

and m and n are the lengths of the two sequences being aligned.

For gapped alignment score distribution, no perfect statistical theory has yet been developed, although there is ample empirical evidence that it also closely follows Gumbel-type EVD [8, 12, 15, 17, 20, 27]. Therefore, the frequently used approach has been to fit the score distribution to an extreme value distribution to get the parameters K and λ. In general, the approximations thus obtained are quite accurate [14].

A good pairwise alignment-based sequence comparison strategy should therefore have the following characteristics:

1. *Sequence-specificity*: Since the distribution of alignment scores and hence the statistical significance depends on various factors such as alignment program, scoring scheme, sequence lengths, and sequence compositions [16], all these factors for the specific sequence-pair being aligned should be taken into account.
2. *Statistical significance accuracy*: The approach should be able to estimate the P-values for high scores in the tail region of the distribution accurately.
3. *Retrieval accuracy*: Most importantly, the approach should assign lower P-values to pairs of related sequences than to pairs of unrelated sequences, which is commonly measured by retrieval accuracy (also known as coverage).
4. *Speed*: The estimation process should be fast enough to be usable in practice.

Some excellent reviews on statistical significance in sequence comparison are available in the literature [14, 16, 19, 22].

3 Database Statistical Significance Versus Pairwise Statistical Significance

Database search programs typically use heuristics to obtain a suboptimal local alignment in less time. Popular database search programs are BLAST [9], FASTA [20, 21], SSEARCH (using full implementation of Smith–Waterman algorithm [26]), and PSI-BLAST [9, 24]. The database statistical significance so obtained for a pairwise comparison is dependent on the size and composition of the database, and will be different for different database, and even for the same database at different times, since the size of the database can change with time. In recent years, there have been significant improvements in the BLAST and PSI-BLAST programs [24, 28, 29], which have been shown to improve the retrieval accuracy of database searches using composition-based statistics and other enhancements.

An alternative method to estimate statistical significance of a pairwise alignment is to estimate pairwise statistical significance, which is database independent and more sequence specific. A study of pairwise statistical significance and its comparison with database statistical significance [1, 2] compared various approaches to estimate pairwise statistical significance such as ARIADNE [15], PRSS [21], censored-maximum-likelihood fitting [10], and linear regression fitting [12] to find that maximum likelihood fitting with censoring left of peak (described as type-I censoring in [10]) is the most accurate method for estimating pairwise statistical significance. Further, this method was compared with database statistical significance in a homology detection experiment. Pairwise statistical significance described in [1, 2] can be understood to be obtainable by the following function:

$$PairwiseStatSig(Seq1, Seq2, SC, N)$$

where $Seq1$ is the first sequence, $Seq2$ is the second sequence, SC is the scoring scheme (substitution matrix, gap opening penalty, gap extension penalty), and N is the number of shuffles. The function $PairwiseStatSig$, therefore, generates a score distribution by aligning $Seq1$ with N shuffled versions of $Seq2$, fits the distribution to an extreme value distribution using censored maximum likelihood fitting to obtain the statistical parameters K and λ, and returns the pairwise statistical significance estimate of the pairwise alignment score between $Seq1$ and $Seq2$ using the parameters K and λ. The scoring scheme SC can be extended to use sequence pair-specific distanced substitution matrices or multiple parameter sets, as used in [3] and [4], respectively. Furthermore, a sequence-specific/position-specific scoring scheme SC_1 specific to one of the sequences (say $Seq1$) can be used to estimate pairwise statistical significance using sequence-specific/position-specific substitution matrices [5].

4 Experimental Results

The performance of sequence comparison at sequence level can be evaluated by homology detection experiments, wherein a set of queries and database sequences is used with prior knowledge of which queries are related to which sequences in the database, and a sequence-comparison strategy is evaluated in terms of its ability to separate related sequence pairs from unrelated ones. The studies on pairwise statistical significance described in the earlier section used the same experiment setup as used in [25]. A non-redundant subset of the CATH 2.3 database (Class, Architecture, Topology, and Hierarchy, [18]) available at ftp://ftp.ebi.ac.uk/ pub/software/

unix/fasta/prot_sci_04/ was selected in [25] to evaluate seven structure comparison programs and two sequence comparison programs. This dataset consists of 2,771 database sequences and 86 query sequences and is considered as a valid benchmark for testing protein comparison algorithms [23].

Following [25], error per query (EPQ) versus coverage plots were used to visualize and compare the results. To create these plots, the list of pairwise comparisons are sorted based on decreasing statistical significance (increasing P-values). While traversing the sorted list from top to bottom, the coverage count is increased by one if the two sequences of the pair are homologs, else the error count is increased by one. At any given point in the list, EPQ is the total number of errors incurred so far, divided by the number of queries; and coverage is the fraction of total homolog pairs so far detected. The ideal curve would go from 0 to 100% coverage, without incurring any errors, which would correspond to a straight line on the x-axis. Therefore, a better curve is one which is more to the right.

Since the EPQ versus coverage curves on the complete dataset can be distorted due to poor performance by one or two queries (if those queries produce many errors at low coverage levels) [25], performance is usually compared using individual queries, following the work in [25]. The coverage of each of the 86 queries at the 1st, 3rd, 10th, 30th, and 100th error is recorded, and the median coverage at each error level is compared across different sequence-comparison methods.

Figure 2 presents the summarized results from [2–5]. There are ten comparisons shown in this figure, ordered according to their relative performance:

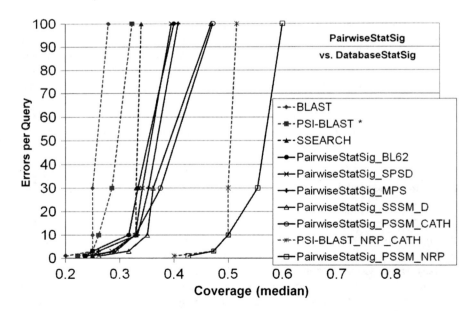

Fig. 2 Comparison of different sequence-comparison approaches in terms of retrieval accuracy. For detailed explanation of the legends and conclusions from the figure, see the text

1. BLAST: The results obtained using BLAST on the test database (subset of CATH database).
2. PSI-BLAST*: The results obtained by directly using PSI-BLAST on the test database without using pre-trained PSSMs.
3. SSEARCH: The results obtained using SSEARCH on the test database.
4. PairwiseStatSig_BL62: The results obtained using pairwise statistical significance with standard substitution matrix BLOSUM62 [2].
5. PairwiseStatSig_SPSD: The results obtained using pairwise statistical significance with sequence-pair-specific distanced substitution matrix (the rate matrix corresponding to BLOSUM62 was used, and to get the sequence-pair-specific distanced matrices, the distances were allowed to range from 150 to 300 [3]).
6. PairwiseStatSig_MPS: The results obtained using pairwise statistical significance with multiple parameter sets (BLOSUM50 and BLOSUM80 were simultaneously used to get the alignment score distribution [4]).
7. PairwiseStatSig_SSSM_D: The results obtained using pairwise statistical significance with sequence-specific substitution matrices for the database sequences constructed using the method described in [5].
8. PairwiseStatSig_PSSM_CATH: The results obtained using pairwise statistical significance with position-specific substitution matrices that were derived using PSI-BLAST searches against the test database [5].
9. PSI-BLAST_NRP_CATH: The results obtained by first using PSI-BLAST on the big non-redundant protein database (NRP) to get PSSMs for the test queries, and then using them as starting PSSMs for database searches against the test database.
10. PairwiseStatSig_PSSM_NRP: The results obtained using pairwise statistical significance with position-specific substitution matrices that were derived using PSI-BLAST searches against the non-redundant protein (NRP) database [5].

There are two major observations that can be made from Fig. 2. First, for all relevant comparisons, pairwise statistical significance performs at least comparable or significantly better than database statistical significance. Second, as the sequence-comparison process is made more and more sequence-specific, the sequence comparison performance improves significantly.

The current implementations of pairwise statistical significance estimation takes about 2 s on an Intel 2.8 GHz processor for average length protein sequences (200–250).

5 Conclusion and Future Work

This chapter presents the summary of the recent advances in sequence-specific sequence comparison using pairwise statistical significance. Current results provide clear empirical evidence that sequence-comparison performance improves as the sequence-comparison process is made more and more sequence specific.

Although the comparison results show that using pairwise statistical significance outperforms the traditional database search programs, currently pairwise statistical significance estimation is nonetheless much computationally expensive to be directly used in a large database search since it involves generation of sequence-specific empirical score distributions and subsequent curve fitting.

The current shortcomings of pairwise statistical significance (low speed) and database statistical significance (low retrieval accuracy) provide for a lot of scope for future work. Combining the strengths of the two approaches should result in an effective sequence comparison strategy. In an existing work [6], pairwise statistical significance was used to reorder the hits returned by BLAST/PSI-BLAST searches and was shown to enhance the database search performance slightly. But since this method cannot recover the hits missed by BLAST/PSI-BLAST, the magnitude of potential improvement is limited. Future work includes designing new strategies to speed up the pairwise statistical significance estimation process by exploiting the potential parallelism available in the estimation procedure, since many alignments need to be computed to generate sequence-specific score distributions, which can be performed in parallel. Acceleration can also be potentially achieved using FPGA- [11] and GPU-based implementations specifically designed for this problem, taking into account the special feature of pairwise statistical significance estimation procedure. Apart from addressing the issue of speed, researchers can also try to further make the sequence-comparison process more sequence-specific, by constructing and using better quality position-specific substitution matrices.

Acknowledgements The authors would like to thank Dr. Sean Eddy for making the C routines of censored maximum likelihood fitting available online, Dr. William R. Pearson for making the benchmark protein comparison database available online, and Dr. Volker Brendel for helpful discussions and providing links to the data. This work was supported in part by NSF grants CNS-0551639, IIS-0536994, NSF HECURA CCF-0621443, and NSF SDCI OCI-0724599, NSF IIS-0905205, DOE FASTOS award number DE-FG02-08ER25848 and DOE SCIDAC-2: Scientific Data Management Center for Enabling Technologies (CET) grant DE-FC02-07ER25808.

References

1. Agrawal, A., Brendel, V., Huang, X.: Pairwise statistical significance versus database statistical significance for local alignment of protein sequences. In: Bioinformatics Research and Applications, *LNCS(LNBI)*, vol. 4983, pp. 50–61. Berlin/Heidelberg: Springer (2008)
2. Agrawal, A., Brendel, V.P., Huang, X.: Pairwise statistical significance and empirical determination of effective gap opening penalties for protein local sequence alignment. International Journal of Computational Biology and Drug Design **1**(4), 347–367 (2008)
3. Agrawal, A., Huang, X.: Pairwise statistical significance of local sequence alignment using substitution matrices with sequence-pair-specific distance. In: Proceedings of International Conference on Information Technology, ICIT, pp. 94–99 (2008)
4. Agrawal, A., Huang, X.: Pairwise statistical significance of local sequence alignment using multiple parameter sets and empirical justification of parameter set change penalty. BMC Bioinformatics **10**(Suppl 3), S1 (2009)

5. Agrawal, A., Huang, X.: Pairwise statistical significance of local sequence alignment using sequence-specific and position-specific substitution matrices. IEEE/ACM Transactions on Computational Biology and Bioinformatics (2009). DOI http://doi.ieeecomputersociety.org/10.1109/TCBB.2009.69. 25 Sept. 2009
6. Agrawal, A., Huang, X.: PSIBLAST_PairwiseStatSig: reordering PSI-BLAST hits using pairwise statistical significance. Bioinformatics **25**(8), 1082–1083 (2009). DOI 10.1093/bioinformatics/btp089. URL http://bioinformatics.oxfordjournals.org/cgi/content/abstract/25/8/1082
7. Altschul, S.F., Boguski, M.S., Gish, W., Wootton, J.C.: Issues in searching molecular sequence databases. Nature Genetics **6**(2), 119–129 (1994)
8. Altschul, S.F., Gish, W.: Local alignment statistics. Methods in Enzymology **266**, 460–80 (1996)
9. Altschul, S.F., Madden, T.L., Schäffer, A.A., Zhang, J., Zhang, Z., Miller, W., Lipman, D.J.: Gapped BLAST and PSI-BLAST: A new generation of protein database search programs. Nucleic Acids Research **25**(17), 3389–3402 (1997). DOI 10.1093/nar/25.17.3389. URL http://dx.doi.org/10.1093/nar/25.17.3389
10. Eddy, S.R.: Maximum-likelihood fitting of extreme value distributions (1997). Available: URL ftp://selab.janelia.org/pub/publications/Eddy97b/Eddy97b-techreport.pdf. Accessed 13 January 2011
11. Honbo, D., Agrawal, A., Choudhary, A.: Efficient pairwise statistical significance estimation using FPGAs. BIOCOMP, pp. 571–577 (2010)
12. Huang, X., Brutlag, D.L.: Dynamic use of multiple parameter sets in sequence alignment. Nucleic Acids Research **35**(2), 678–686 (2007). DOI 10.1093/nar/gkl1063. URL http://nar.oxfordjournals.org/cgi/content/abstract/35/2/678
13. Karlin, S., Altschul, S.F.: Methods for assessing the statistical significance of molecular sequence features by using general scoring schemes. Proceedings of the National Academy of Sciences, USA **87**(6), 2264–2268 (1990). DOI 10.1073/pnas.87.6.2264. URL http://www.pnas.org/cgi/content/abstract/87/6/2264
14. Mitrophanov, A.Y., Borodovsky, M.: Statistical significance in biological sequence analysis. Briefings in Bioinformatics **7**(1), 2–24 (2006). DOI 10.1093/bib/bbk001
15. Mott, R.: Accurate formula for p-values of gapped local sequence and profile alignments. Journal of Molecular Biology **300**, 649–659 (2000)
16. Mott, R.: Alignment: Statistical Significance. Encyclopedia of Life Sciences (2005). URL http://mrw.interscience.wiley.com/emrw/9780470015902/els/article/a0005264/current/abstract
17. Olsen, R., Bundschuh, R., Hwa, T.: Rapid assessment of extremal statistics for gapped local alignment. In: Proceedings of the Seventh International Conference on Intelligent Systems for Molecular Biology, pp. 211–222. AAAI Press (1999)
18. Orengo, C.A., Michie, A.D., Jones, S., Jones, D.T., Swindells, M.B., Thornton, J.M.: CATH - A hierarchic classification of protein domain structures. Structure **28**(1), 1093–1108 (1997)
19. Pagni, M., Jongeneel, C.V.: Making sense of score statistics for sequence alignments. Briefings in Bioinformatics **2**(1), 51–67 (2001). DOI 10.1093/bib/2.1.51
20. Pearson, W.R.: Empirical statistical estimates for sequence similarity searches. Journal of Molecular Biology **276**, 71–84 (1998)
21. Pearson, W.R.: Flexible sequence similarity searching with the FASTA3 program package. Methods in Molecular Biology **132**, 185–219 (2000)
22. Pearson, W.R., Wood, T.C.: Statistical significance in biological sequence comparison. In: D.J. Balding, M. Bishop, C. Cannings (eds.) Handbook of Statistical Genetics, pp. 39–66. Chichester, UK: Wiley (2001)
23. Rocha, J., Rosselló, F., Segura, J.: Compression ratios based on the universal similarity metric still yield protein distances far from cath distances. CoRR abs/q-bio/0603007 (2006)
24. Schäffer, A.A., Aravind, L., Madden, T.L., Shavirin, S., Spouge, J.L., Wolf, Y.I., Koonin, E.V., Altschul, S.F.: Improving the accuracy of PSI-BLAST protein database searches with composition-based statistics and other refinements. Nucleic Acids Research **29**(14), 2994–3005 (2001)

25. Sierk, M.L., Pearson, W.R.: Sensitivity and selectivity in protein structure comparison. Protein Science **13**(3), 773–785 (2004). DOI 10.1110/ps.03328504
26. Smith, T.F., Waterman, M.S.: Identification of common molecular subsequences. Journal of Molecular Biology **147**(1), 195–197 (1981). URL http://view.ncbi.nlm.nih.gov/pubmed/7265238
27. Waterman, M.S., Vingron, M.: Rapid and accurate estimates of statistical significance for sequence database searches. Proceedings of the National Academy of Sciences, USA **91**(11), 4625–4628 (1994). DOI 10.1073/pnas.91.11.4625. URL http://www.pnas.org/cgi/content/abstract/91/11/4625
28. Yu, Y.K., Altschul, S.F.: The construction of amino acid substitution matrices for the comparison of proteins with non-standard compositions. Bioinformatics **21**(7), 902–911 (2005). DOI 10.1093/bioinformatics/bti070
29. Yu, Y.K., Gertz, E.M., Agarwala, R., Schäffer, A.A., Altschul, S.F.: Retrieval accuracy, statistical significance and compositional similarity in protein sequence database searches. Nucleic Acids Research **34**(20), 5966–5973 (2006). DOI 10.1093/nar/gkl731

Chapter 31
Modelling Short Time Series in Metabolomics: A Functional Data Analysis Approach

Giovanni Montana, Maurice Berk, and Tim Ebbels

Abstract Metabolomics is the study of the complement of small molecule metabolites in cells, biofluids and tissues. Many metabolomic experiments are designed to compare changes observed over time under two or more experimental conditions (e.g. a control and drug-treated group), thus producing time course data. Models from traditional time series analysis are often unsuitable because, by design, only very few time points are available and there are a high number of missing values. We propose a functional data analysis approach for modelling short time series arising in metabolomic studies which overcomes these obstacles. Our model assumes that each observed time series is a smooth random curve, and we propose a statistical approach for inferring this curve from repeated measurements taken on the experimental units. A test statistic for detecting differences between temporal profiles associated with two experimental conditions is then presented. The methodology has been applied to NMR spectroscopy data collected in a pre-clinical toxicology study.

1 Introduction

Metabolomics (also metabonomics or metabolic profiling) is the study of the complement of small molecule metabolites in cells, biofluids and tissues [7]. Along with transcriptomics and proteomics, it is an important component of systems biology approaches which often use several such "omics" technologies to report the state of an organism at multiple biomolecular levels. Metabolomics experiments usually use sophisticated analytical techniques such as nuclear magnetic resonance (NMR) spectroscopy or mass spectrometry (MS) to assay metabolite levels, and these instruments produce large volumes of highly complex but information-rich data. An important component of all omics studies, particularly for metabolomics,

---

G. Montana (✉)
Mathematics, Imperial College, London, UK
e-mail: giovanni.montana@imperial.ac.uk

is the statistical and bioinformatic methods that are used to process and model the data. In recent years, there has been much interest in developing algorithms for both low level processing (e.g. peak detection, baseline correction) and higher level modelling such as classification and regression [2]. Time is an extremely important variable in biological experiments, and many metabolomics studies produce time course data. A common experimental design consists of collecting repeated measurements in two or more experimental groups, for instance a control group and a drug-treated group; the scientific interest lies in detecting metabolites whose temporal response appears to be different, compared to the controls, due to treatment. Other related tasks include classification of individual time profiles into categories, prediction of future behaviour based on early time points and identification of the dynamical characteristics of the system. In all these applications, ignoring the time ordering leads to an obvious loss of information. When testing for differential temporal profiles, this loss of information can also result in a loss of power. Despite the need for statistical methods that make full use of the time information, the use of advanced time series analysis methods in metabolomics has not been extensively documented. One reason for this may be the extremely short nature of typical time series; it is rare to find datasets with more than 5–10 time points, and standard techniques from time series analysis typically require several tens of data points to adequately model the time variation. Other difficulties include the high dimensionality of the data (depending on the resolution, an experiment may consist of thousands of short time series), strong degree of collinearity in the repeated measurements, missing data and presence of noise. For these reasons, most analyses of time series data in metabolomics do not explicitly include the time ordering in the model; that is, the same results would be obtained if the time index was permuted.

In an attempt to overcome the complications inherent in modelling short time series arising in metabolonic studies, in this chapter we propose a functional data analysis approach. Each observed time series is seen as the realisation of an underlying stochastic process or smooth curve that needs to be estimated. Such estimated curves are then treated as the basic observational unit in the data analysis, as in [8]. The motivating study and dataset are described in Sect. 2. The suggested methodology, based on functional mixed-effect models, is described in Sect. 3. The experimental results are presented in Sect. 4.

2 A Toxicology Study: NMR Spectroscopy of Rodent Urine

The motivating dataset that originated this work comes from a pre-clinical toxicology study conducted by the Consortium for Metabonomic Toxicology [5]. The objective of the study was to characterise the metabolic response of laboratory rats to the liver toxin hydrazine, as assayed by the ^{1}H NMR spectroscopy of urine. A summary of the experimental setup is provided below and further details can be found in [1]. In the study, 30 Sprague–Dawley rats were randomly assigned to three treatment groups (controls, low dose and high dose) in equal sample sizes of 10. Urine samples were collected at 0–8 and 8–24 h on the day before treatment, and further samples collected at 8, 24, 48, 72, 96, 120, 144 and 168 h (7 days) post-treatment. Five animals in each group were killed at 48 h post-dose for histopathological and clinical chemistry evaluation. All animal studies were carried out in accordance with national legislation and were subject to appropriate local review. Urine samples were buffered in sodium phosphate, combined with an internal reference (TSP) and antibacterial agent (sodium azide). ^{1}H NMR spectra were acquired at 600 MHz with suppression of the water resonance using a standard pre-saturation pulse sequence. An exponential line-broadening factor of 1 Hz was applied to each free induction decay before Fourier transformation using XWinNMR and phasing, baseline correction and referencing carried out automatically using an in-house MATLAB routine. For statistical analysis, each spectrum was segmented by integrating the signal in regions of equal width (0.001 ppm) in the chemical shift ranges δ0.20–10.0, excluding spectral artefacts in the region δ4.50–5.98. All spectra were then normalised to a constant integrated intensity of 100 units to take account of large variations in urine concentration. The final dataset consisted of 8,020 integrated NMR intensities observed at 10 time points for all rats but those that were killed at 48 h post-dose, for which there were only 5 time points. Figure 1 provides an example of the raw data observed for the ten rats in the high-dose group, for one of the variables (3.04 ppm). In what follows, we focus exclusively on the comparison between controls and high dose group.

3 Time Series Analysis Using Functional Mixed-Effects Models

Each observation being modelled is denoted by $y(t_{ij})$ and represents the integrated NMR intensity observed on rat i at time j, where $i = 1,2,\ldots,n_k$, $j = 1,2,\ldots,m_i$, and n_k represents the sample size in group k. In our setting, $n_1 = n_2 = 10$ and m_i can be either 5 or 10. As clearly seen in Fig. 1, these observations present several typical features. First, all time series are short and exhibit a clear serial correlation structure, in which subsequent observations are highly correlated with the previous ones. Second, by design, half of the time series are only observed over the first 5 time points, thus introducing a clear missing data pattern; other data points are also missing (e.g. rats 1 and 9), due to technical issues during the measurement process or because these points were deemed outliers and therefore were removed. Finally, there is a clear individual variability across all temporal profiles in this group: while some rats (e.g. 7 and 9) display a response that increases over time, others (9 and 10) do not show any response at all. In order to properly account for all these features, we suggest to model $y(t_{ij})$ non-parametrically. The suggested model is $y(t_{ij}) = \mu(t_{ij}) + v_i(t_{ij}) + \varepsilon_{ij}$; it postulates that the observed integrated NMR intensity $y(t_{ij})$ can be explained by the additive effect of three components: a mean response $\mu(\cdot)$ observed at time t_{ij}, which is assumed to be a smooth, non-random curve defined over the time range of interest; a rat-specific deviation from that mean curve, $v_i(\cdot)$, which is also assumed to be a smooth, but random curve observed over

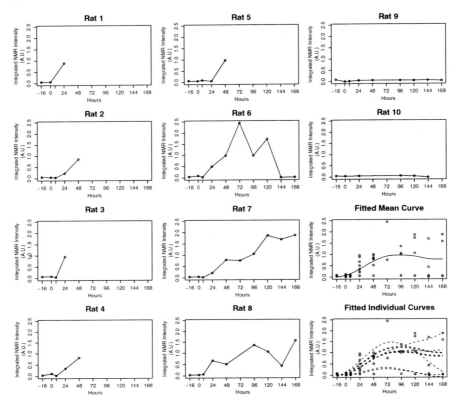

Fig. 1 Short time series observed for one metabolite (creatine, 3.04 ppm) for all ten rats in the high dose group. Shown are also the fitted mean curve and individual curves using the functional mixed-effects model (1)

the same time range; an error term ε_{ij} which accounts for the variability not explained by the first two terms. Formally, we treat each $v_i(t_{ij})$, for $i = 1, 2, \ldots, n_k$, as independent and identically distributed realisations of an underlying stochastic process; specifically, we assume that $v_i(t_{ij})$ is a Gaussian process with zero-mean and covariance function $\gamma(s,t)$, that is $v_i(t_{ij}) \sim \mathrm{GP}(0, \gamma)$. The error terms ε_{ij} are assumed to be independent and normally distributed with zero mean and covariance matrix $\mathbf{R}_i$. Moreover, we do not assume that all rats have been observed at the same design time points, which make the model more flexible and applicable to more general experimental settings. All the distinct design time points are collected together and denoted by $(\tau_1, \tau_2, \ldots, \tau_m)$.

A choice has to be made regarding the specific parameterisation of the smooth curves $\mu(t_{ij})$ and $v_i(t_{ij}), i = 1, 2, \ldots, n$. We suggest to represent the curves using cubic smoothing splines; see, for instance, [3]. The key idea of smoothing splines consists in making full use of all the design time points and then fitting the model by adding a smoothness or roughness constraint; by controlling the size of this constraint, we are then able to avoid curves that appear too wiggly. A natural way of measuring the roughness of a curve is by means of its integrated squared second derivative, assuming that the curve is twice-differentiable. We call $\eta = [\eta(\tau_1), \ldots, \eta(\tau_m)]^T$ the vector containing the values of the mean curve estimated at all design time points, and, analogously, the rat-specific deviations from the mean curve, for rat i, are collected in $\mathbf{v}_i = [v_i(\tau_1), \ldots, v_i(\tau_m)]^T$. With this notation in place, a representation of the mean curve is given by $\mu(t_{ij}) = \mathbf{x}_{ij}^T \eta$, and the individual curves are given by $v_i(t_{ij}) = \mathbf{x}_{ij}^T \mathbf{v}_i, i = 1, 2, \ldots, n$ where $\mathbf{x}_{ij} = (x_{ij1}, \ldots, x_{ijm})^T$ and each element of this vector is defined as:

$$x_{ijr} = \begin{cases} 1 & \text{if } t_{ij} = \tau_r,\ r = 1, \ldots, m \\ 0 & \text{else} \end{cases}$$

The fact that the individual curves are assumed to be Gaussian processes is captured by assuming that the individual deviations are random and follow a zero-centred Gaussian distribution with covariance $\mathbf{D}$, where $\mathbf{D}(r,s) = \gamma(\tau_s, \tau_r)$, $r, s = 1, \ldots, m$. Finally, in matrix form, the suggested model can then be rewritten as:

$$\mathbf{y}_i = \mathbf{X}_i \eta + \mathbf{X}_i \mathbf{v}_i + \varepsilon_i \qquad \mathbf{v}_i \sim N(\mathbf{0}, \mathbf{D}) \qquad \varepsilon_i \sim N(\mathbf{0}, \mathbf{R}_i) \qquad (1)$$

For simplicity, we assume that $\mathbf{R}_i = \sigma^2 \mathbf{I}$. In this form, model (1) is a linear mixed-effects model. Clearly, the model accounts for the fact that, at a given variable, the observed repeated measurements for each rat are correlated. Specifically, under the assumptions above, we have that $\text{Cov}(\mathbf{y}_i) = \mathbf{X}_i \mathbf{D} \mathbf{X}_i^T + \mathbf{R}_i$, which separates out the variability due to individual fluctuations around the mean curve and the variability due to measurement error. A common approach to estimating the unknown parameters of a linear mixed-effects model is by maximum likelihood (ML) estimation. The ML estimators of the mean curve $\mu(\cdot)$ and each individual curve $v_i(\cdot)$ can be obtained by minimising a penalised version of the generalised log-likelihood for model (1), obtained by adding two additional terms that impose a penalty on the roughness of the curves: a term $\lambda \eta^T \mathbf{G} \eta$ for the mean curve and a term $\lambda_v \sum_{i=1}^{n_k} \{\mathbf{v}_i^T \mathbf{G} \mathbf{v}_i\}$ for the individual curves. The matrix $\mathbf{G}$ is the roughness matrix that quantifies the smoothness of the curve, as in [3], and the two scalars λ_v and λ are smoothing parameters controlling the size of the penalties. In principle, n_k distinct individual smoothing parameters can be introduced in the model, but such a choice would incur a great computational cost during the model fitting and model selection process; hence, we prefer to assume that all individual curves observed at each NMR intensity share the same smoothness. For this reason, a unique smoothing parameter λ_v is used to control the smoothness of the rat-specific deviations from the mean curve. After a rearrangement on the terms featuring in the penalised log-likelihood, the model can be rewritten in terms of the regularised covariance matrices $\mathbf{D}_v = (\mathbf{D}^{-1} + \lambda_v \mathbf{G})^{-1}$ and $\mathbf{V}_i = \mathbf{X}_i \mathbf{D}_v \mathbf{X}_i^T + \mathbf{R}_i$. When both these variance components are known, the ML estimators $\hat{\eta}$ and $\hat{v}_i$, for $i = 1, 2, \ldots$, can be derived in closed-form as the minimisers of the penalised generalised log-likelihood. However, the variance components are generally unknown. All parameters can be estimated iteratively using an EM algorithm, which begins with some initial guesses of the variance components (e.g. $\hat{\sigma}^2 = 1$ and $\hat{\mathbf{D}} = \hat{\sigma}^2 \mathbf{I}_M$). The algorithm repeats the following steps until convergence: (a) compute the regularised variance components $\mathbf{D}_v$ and $\mathbf{V}_i$; (b) compute the ML estimates of fixed and random effects, $\hat{\eta}$ and $\hat{\mathbf{v}}_i$, respectively; (c) update $\hat{\sigma}^2$ and $\hat{\mathbf{D}}$ using their conditional expectations $E(\hat{\sigma}^2|\mathbf{y}, \eta = \hat{\eta})$ and $E(\hat{\mathbf{D}}|\mathbf{y}, \eta = \hat{\eta})$. The smoothing parameters λ and λ_v are generally found as those values, in the two-dimensional space $(\Lambda \times \Lambda_v)$, that optimise a given model selection criterion such as the Akaike Information Criterion (AIC), the Bayesian Information Criterion (BIC) or generalised cross-validation. Exhaustive grid-based searches in the parameter space are generally adopted in practical settings, as in [10]. Alternatively, one can incorporate model selection in the estimation procedure by treating the smoothing parameters as variance components, as in [4]. We prefer to disassociate the model selection from the estimation as this is a much more flexible approach. Moreover, we choose to optimise the *corrected* AIC which, unlike the standard AIC, includes a correction for small sample sizes. Based on Monte Carlo simulations, we have found that this criterion provides a better model fit when the sample sizes are small (data not shown). In our implementation, the search for optimal smoothing values $\hat{\lambda}$ and $\hat{\lambda}_v$ is performed using the downhill simplex optimisation algorithm.

We wish to compare the temporal profiles under two experimental groups. After fitting model (1) to the data, independently for each variable, we obtain the estimated mean curves $\hat{\mu}^{(1)}(t)$ describing the response profile in the control group and, similarly, the $\hat{\mu}^{(2)}(t)$ curve describing the response in the drug-treated group. In order to test the null hypothesis of no treatment effect over the entire period, we need to be able to assess whether the assumption that $\mu^{(1)}(t) = \mu^{(2)}(t)$ holds true, that is whether the two curves are equal. One way of measuring the dissimilarity between these two curves consists in computing the L_2 distance between them, that is:

$$D = \int [\mu^{(1)}(t) - \mu^{(2)}(t)]^2 dt \qquad (2)$$

which can be assessed using the estimated (smoothed) curves $\hat{\mu}^{(1)}(t)$ and $\hat{\mu}^{(2)}(t)$, thus yielding the observed distance $\hat{D}$. We use this dissimilarity measure as a test statistic. Since the null distribution of this statistic is not available in closed form, we resort to a non-parametric bootstrap approach to approximate the probability $P(D > \hat{D})$ under the null. The non-parametric bootstrap approach resamples the model fits to simulate samples under the null hypothesis of no effect. For each variable, we fit two models to the data: a null model that ignores the class assignments by pooling all rats together, and an alternative model in which the groupings are preserved. Null samples are constructed by adding resampled (with replacement) within- and between-individual residuals from the alternative model to the mean curve of the null model. Denoting $\hat{\eta}^{(0)}$ as the fitted mean under the null model, and $\hat{\mathbf{v}}_i^{(1)}$ and $\hat{\varepsilon}^{(1)}$ as the between- and within-individual residuals, respectively, under the alternative model, randomly sampled data can be generated as $y_{ij}^* = \mathbf{x}_{ij}\hat{\eta}^{(0)} + \mathbf{x}_{ij}\hat{\mathbf{v}}_{ij}^* + \varepsilon_{ij}^*$. A total of B bootstrap datasets are generated in this way to estimate the null distribution of the test statistic. The procedure is repeated for all the variables and a correction for multiple testing is applied to control the false discovery rate.

4 Experimental Results

In this section, we report on experimental results obtained by applying the methodology described in Sect. 3. We also compare our approach with a simpler procedure that consists in performing a standard t-test at each time point, and then selecting the corresponding minimum p-value across all time points. The resulting 8,020 p-values are then ranked in increasing order of magnitude, after a correction for multiple testing that controls the false discovery rate, analogously to the multivariate functional approach. Figure 1 shows the fitted mean curve as well as all the individual curves, in the high-dose group and for one variable, obtained from model (1). It can be noted that the fitted mean curve captures the trend observed in the raw data, which indicates a rapid increase in integrated NMR intensity over time, shortly after the treatment, followed by a slow time-dependent decrease after approximately 120 h. The fitted individual curves provide additional information on each rat's response and can also be used to impute the missing data, perform predictions or cluster the rats based on their temporal responses. Figure 2 shows the mean integrated NMR intensity as a function of ppm; the colour indicates the magnitude of the associated p-value on a $-\log_{10}$ scale, so that larger values support the alternative hypothesis of a treatment effect. The functional model identified 158 variables as being statistically significant, whereas the t-test approach identified 7,421 variables, of the total 8,020 available variables. Not surprisingly, the

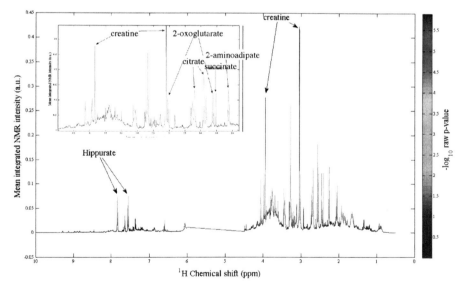

Fig. 2 Mean high dose spectrum showing all the 8,020 variables. Each integrated NMR intensity value is colour coded to reflect the magnitude of the corresponding p-value obtained using model (1) jointly with test statistic (2). The top scoring variables correspond to real peaks, and, in several cases, these peaks can be assigned to known metabolites. The inset shows an expansion of the region 2.1–4.4 ppm

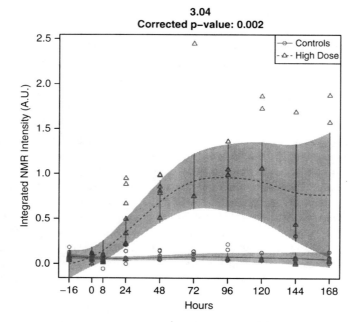

Fig. 3 Fitted mean curves for both control and high dose group for the top scoring variable (creatine, 3.04 ppm), according to the functional mixed-effects model (1). The shaded area represents the 95% confidence bars around the fitted curves

univariate approach suffers from a high rate of false positives. Figure 3 shows the temporal profiles for controls and drug-treated rats corresponding to the top-scoring variable according to the functional model; the shaded areas represent 95% confidence bars and quantify the model uncertainty. The two temporal responses are clearly different, and this aspect has been captured by the suggested test statistic measuring the area between the two smoothed curves. A further level of validation of the functional approach may be obtained by investigating whether the most significant variables can be assigned to metabolites previously reported in the literature as associated with hydrazine toxicity. Of the top 25 variables ranked by the p-value, all could be assigned to one of the metabolites creatine, 2-aminoadipate, 2-oxoglutarate, citrate or succinate. In stark contrast, only 13 of the top 25 variables in the t-test approach could be associated with a clear peak and of those just 8 could be assigned to a known metabolite (results not shown). Note that each metabolite may be represented by more than one peak corresponding to several variables. All variables highlighted by the functional model showed a consistent and smooth response to hydrazine administration across the time course. Differential regulation of all five metabolites mentioned above has previously been observed in urinary studies of hydrazine toxicity in both the rat and mouse [6]. Earlier studies also report changes in these metabolites upon hydrazine administration [9]. In particular, increased excretion of creatine and 2-aminoadipate, together with decreased excretion of 2-oxoglutarate, citrate and succinate were observed. The appearance of 2-aminoadipate is an effect highly specific to the mechanism of action of hydrazine in the rat, and its high significance in the functional approach builds confidence that the results of this method can have meaningful biological interpretations. We note that other metabolites with previously reported associations with hydrazine toxicity are also highlighted by our approach (but at lower significance levels), such as hippurate, taurine and trimethylamine-N-oxide [6]. Overall, the functional model is able to highlight many metabolites previously associated with hydrazine toxicity, in contrast to the univariate approach which results in a large number of false-positive associations.

References

1. M. E. Bollard, H. C. Keun, O. Beckonert, T. M. D. Ebbels, H. Antti, A. W. Nicholls, J. P. Shockcor, G. H. Cantor, G. Stevens, J. C. Lindon, E. Holmes, and J. K. Nicholson. Comparative metabonomics of differential hydrazine toxicity in the rat and mouse. *Toxicology and Applied Pharmacology*, 204(2):135–151, Apr 2005
2. T. Ebbels and R. Cavill. Bioinformatic methods in NMR-based metabolic profiling. *Progress in Nuclear Magnetic Resonance Spectroscopy*, 55:361–374, 2009
3. P. J. Green and B. W. Silverman. *Nonparametric Regression and Generalized Linear Models*. Chapman and Hall, London, 1994
4. W. Guo. Functional mixed effects models. *Biometrics*, 58:121–128, 2002
5. J. C. Lindon, H. C. Keun, T. M. Ebbels, J. M. Pearce, E. Holmes, and J. K. Nicholson. The consortium for metabonomic toxicology (COMET): Aims, activities and achievements. *Pharmacogenomics*, 6(7):691–699, Oct 2005
6. A. W. Nicholls, E. Holmes, J. C. Lindon, J. P. Shockcor, R. D. Farrant, J. N. Haselden, S. J. Damment, C. J. Waterfield, and J. K. Nicholson. Metabonomic investigations into hydrazine toxicity in the rat. *Chemical Research in Toxicology*, 14(8):975–987, Aug 2001
7. L. M. Raamsdonk, B. Teusink, D. Broadhurst, N. Zhang, A. Hayes, M. C. Walsh, J. A. Berden, K. M. Brindle, D. B. Kell, J. J. Rowland, H. V. Westerhoff, K. van Dam, and S. G. Oliver. A functional genomics strategy that uses metabolome data to reveal the phenotype of silent mutations. *Nature Biotechnology*, 19(1):45–50, Jan 2001
8. J. Ramsay and B. Silverman. *Functional Data Analysis*. Springer, New York, 2006
9. S. M. Sanins, J. K. Nicholson, C. Elcombe, and J. A. Timbrell. Hepatotoxin-induced hypertaurinuria: a proton NMR study. *Archives of Toxicology*, 64(5):407–411, 1990
10. H. Wu and J.-T. Zhang. *Nonparametric Regression Methods for Longitudinal Data Analysis*. Wiley, New York, 2006

Chapter 32
Modeling of Gene Therapy for Regenerative Cells Using Intelligent Agents

Aya Sedky Adly, Amal Elsayed Aboutabl, and M. Shaarawy Ibrahim

Abstract Gene therapy is an exciting field that has attracted much interest since the first submission of clinical trials. Preliminary results were very encouraging and prompted many investigators and researchers. However, the ability of stem cells to differentiate into specific cell types holds immense potential for therapeutic use in gene therapy. Realization of this potential depends on efficient and optimized protocols for genetic manipulation of stem cells. It is widely recognized that gain/loss of function approaches using gene therapy are essential for understanding specific genes functions, and such approaches would be particularly valuable in studies involving stem cells. A significant complexity is that the development stage of vectors and their variety are still not sufficient to be efficiently applied in stem cell therapy. The development of scalable computer systems constitutes one step toward understanding dynamics of its potential. Therefore, the primary goal of this work is to develop a computer model that will support investigations of virus' behavior and organization on regenerative tissues including genetically modified stem cells. Different simulation scenarios were implemented, and their results were encouraging compared to ex vivo experiments, where the error rate lies in the range of acceptable values in this domain of application.

Keywords Molecular dynamics and simulation · Molecular interactions · Computational systems biology

1 Introduction

Gene therapy is a novel therapeutic branch of modern medicine. Its emergence is a direct consequence of the revolution heralded by the recombinant DNA methodology using transduction, which is a method of gene transfer using infection of a cell.

A.S. Adly (✉)
Computer Science Department, Faculty of Computers and Information,
Helwan University, Cairo, Egypt
e-mail: ayasedky@helwan.edu.eg; ayasedky@yahoo.com

Gene therapy delivers genetic material to the cell using vectors to generate a therapeutic effect by correcting an existing abnormality or providing a new function. Furthermore, the procedure allows the addition of new functions to cells, such as inducing pluripotent stem cells, which are capable of regenerating a new healthy tissue [1–3]. Although much effort has been directed in the last decade toward improvement of protocols in gene therapy, and in spite of many considerable achievements in basic research, gene delivery systems still need to be optimized to achieve effective therapeutic interventions, especially when dealing with stem cells. In fact, during the past years, only monogenic inherited diseases, such as cystic fibrosis, were considered primary targets for gene therapy, but now after realizing the importance of artificially derived pluripotent cells, it becomes a center of attraction to the entire world. This concept is new compared to the traditional medicine [4–6]. However, there is now a growing awareness of the need to use models and computer simulation to understand its processes and behaviors [7, 8].

2 Applying Gene Therapy in the Regenerative Process

Scientists are one step closer to create a gene therapy/stem cell combination to combat genetic diseases. This research may lead to not only curing the disease, but also repairing the damage left behind [3, 9, 10]. While gene therapy is a burgeoning field that has shown great results in treating genetic disorders, many of those diseases leave behind heavily damaged tissue that the body is unable to repair. So even if the disease is completely eradicated, quality of life may not necessarily improve, and without help, health can still continue to deteriorate. Since stem cell research began, there has been a hope that use of those cells may help alleviate some of the trauma left behind by genetic diseases [11–13]. On the other hand, stem cells intended to treat or cure a disease could end up harming the patient, simply because they are no longer under the control of the clinician. However, gene therapy has the potential to solve this problem, using strategies for genetically modifying stem cells before transplantation to ensure their safety. Accordingly, this new field encompasses many novel challenges, treatments, and restoration of biological function. However, there are many technical hurdles between their promise and the realization of these uses, which will only be overcome by continued intensive stem cell research [10, 14, 15].

3 Proposed Model

This work applies the idea of intelligent agents where its structure is defined as a specific environment in which agents are situated. The behavior of agents is influenced by states and types of agents that are situated in adjacent and at distant sites. The capabilities of agents include both ability to adapt and ability to learn. Adaptation implies sensing the environment and reconfiguring in response. This can be achieved through the choice of alternative problem-solving rules or algorithms, or through the discovery of problem-solving strategies. Learning may proceed through observing actual data, and then it implies a capability of introspection and analysis of behavior and success. Alternatively, learning may proceed by example and generalization, and then it implies a capacity to abstract and generalize. We could group agents used in this work into five classes based on their degree of perceived intelligence and capability: (1) simple reflex agent (acts only on the basis of the current percept), (2) model-based reflex agent (keeps track of the current state of its world using an internal model, before choosing an action), (3) goal-based agent (store information regarding desirable situations, to select the one which reaches a goal state), (4) utility-based agent (distinguish between goal states and non-goal states and can define how desirable a particular state is), and (5) learning agent (can initially operate in unknown environments and to become more competent than its initial knowledge alone might allow) [16].

However, the states or values of the components evolve synchronously in discrete time steps according to identical rules. The value of a particular site is determined by the previous values or states of a neighborhood of sites around it. We tried to create a very general structure that would allow the accurate simulation of most viral/cellular phenomena. This required selecting a set of molecular components and choosing a collection of interaction rules that would be sufficiently diverse to describe the events that typically occur both inside and outside a cell. We developed a simulation model, hoping that it would empower scientists in the process of growing tissues and organs in vitro (in the laboratory) and safely implant them when the body is unable to be prompted into healing itself. Modeling and simulation can play a key role in such systems, as they can be both predictive tools for expansion potential and tools to characterize and measure current characteristics and behavior of regenerative cells.

4 Mathematical Model

To be able to implement such a model on a computer and to analyze its properties by simulation experiments, it is necessary to formulate its mathematical model. This model offers the possibility to follow the fate of each individual cell in the system and investigate the behavior within and between individuals.

To estimate selection coefficients, one conventionally writes simple exponential growth models for the various viral variants. Since viral growth need not be exponential, we first derive a somewhat more realistic model that, however, has to remain sufficiently generic for estimating different viruses under different circumstances. This requires the assumption that the dynamics of free virus particles are much faster than those of productively infected cells. Adsorption of retrovirus particles is diffusion limited. Valentine and Allison used the theory of Brownian motion to develop a model for its adsorption on flat surfaces [17].

The fraction of adsorbed viruses $\beta(t)$ is defined as the ratio of the number of adsorbed viruses a, to the total number of virus particles in solution v, which is the product of virus concentration c, times the volume of the virus-containing medium, vol. Mathematically, the fraction of adsorbed viruses is formulated as:

$$\beta(t) = \frac{a}{v} \qquad (1)$$

$$\beta(t) = \frac{a}{c \times \text{vol}} \qquad (2)$$

A general model for the viral life cycle allows for at least two stages: free virions and infected cells. Infected cells appear when virions infect target cells, and virions appear from infected cells. Let $\tau(t)$ be a function representing target cell availability (and/or other factors limiting viral replication).

Considering number of infectious viral particles v and number of productively infected cells I, we write the mathematical equations needed to compute number of viral particles, the viral replication index R, the change in infectious viral particles $\Delta v/\Delta t$, the change in productively infected cells $\Delta I/\Delta t$, and the transduction efficiency TE as follows:

Number of viral particles can be defined as the ratio of the virus production index γ, to the virus decay index δ, multiplied by number of productively infected cells I.

$$v = \frac{\gamma}{\delta} \times I \qquad (3)$$

Viral replication index can be defined as the ratio of the virus production index γ, to the virus decay index δ, multiplied by the infection index λ.

$$R = \frac{\gamma}{\delta} \times \lambda \qquad (4)$$

Change in infectious viral particles can be defined as the product of the number of productively infected cells I, times virus production index γ, then we subtract from this term the product of the number of viral particles v, times virus decay index δ.

$$\frac{\Delta v}{\Delta t} = \gamma \times I - v \times \delta \qquad (5)$$

Change in productively infected cells can be defined as the product of the number of viral particles v, times the infection index λ, times the available target cells $\tau(t)$, then we subtract from this term the product of the number of productively infected cells I, times its death index ε.

$$\frac{\Delta I}{\Delta t} = v \times \lambda \times \tau(t) - I \times \varepsilon \qquad (6)$$

However, by replacing v with its value $\gamma/\delta \times I$, the equation would be as follows:

$$\frac{\Delta I}{\Delta t} = \frac{\gamma}{\delta} \times \lambda \times I \times \tau(t) - I \times \varepsilon \qquad (7)$$

In addition, by replacing the term $\gamma/\delta \times \lambda$ with its equivalent value R, we obtain the following equation:

$$\frac{\Delta I}{\Delta t} = I \times R \times \tau(t) - I \times \varepsilon \qquad (8)$$

Transduction efficiency can be defined as the ratio of number of gene transfer events T to the number of available target cells $\tau(t)$.

$$\text{TE} = \frac{T}{\tau(t)} \qquad (9)$$

Finally, the expected number of gene transfer events per cell MOI (multiplicity of infection) can be defined as the ratio of the number of genes transfer events T to the number of productively infected cells I.

$$\text{MOI} = \frac{T}{I} \qquad (10)$$

However, by replacing T with its value $I \times \text{MOI}$, the equation would be as follows:

$$\text{TE} = \frac{I \times \text{MOI}}{\tau(t)} \qquad (11)$$

The actual number of viruses that will enter any given cell is a statistical process: some cells may absorb more than one virus particle, while others may not absorb any. The probability that a cell will absorb v virus particles when inoculated with an MOI can be calculated for a given population using a Poisson distribution [18–21].

$$P(v) = \frac{\text{MOI}^v \times e^{-\text{MOI}}}{v!} \qquad (12)$$

where v is the number of virus particles that enter the cells, and $P(v)$ is the probability that a cell will get infected by v virus particles. However, the need for formal specification systems has been noted for years. Formal methods may be used to give a description of the system to be developed, at whatever level(s) of detail desired. This formal description can be used to guide further development activities; additionally, it can be used to verify that the requirements for the system being developed have been completely and accurately specified.

5 Experimental Results

Different simulations for a variety of typical field scenarios were performed, having different results, and different degrees of relevance to the actual experimental measurements. These simulations were compared with actual experiments, and the comparison shows good agreement between them, with error rates varying from

5% to 12%. We will present our simulation results compared to experimental measurements included in the investigations of Chan et al. [22] of the self-renewal and differentiation of human fetal mesenchymal stem cells (hfMSCs) after transduction with onco-retroviral and lentiviral vectors. Retroviral and lentiviral approaches offered the initial methodology that launched the field and established the technological basis of nuclear reprogramming with rapid confirmation across integrating vector systems [23, 24]. We will first present their experimental results, Figs. 1 and 3, and then our simulation results, Figs. 2 and 4, to evaluate and check our system's validity.

Their experiment [22] included hfMSCs that were transduced initially with an onco-retroviral vector encoding the LacZ transgene. Static transduction of three samples of first trimester fetal blood-derived hfMSCs from 9^{+1}, 9^{+6}, and 10^{+6} weeks of gestation (derived from three different donors) revealed a mean efficiency of $18.1 \pm 8.6\%$, $43.0 \pm 2.8\%$, and $45.1 \pm 2.8\%$ after one, two, and three cycles of transduction, respectively (Fig. 1a). However, centrifugational forces can be used to increase the transduction efficiency. It is a process that involves the use of the centrifugal force for the isolation of small volumes of molecules, where more-dense components migrate away from the axis of the centrifuge, while less-dense

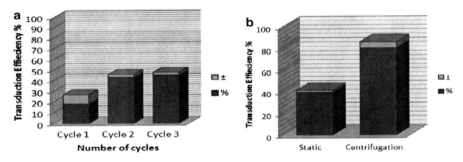

Fig. 1 Chan, O'Donoghue, de la Fuente et al. investigated results: (**a**) effect of transduction cycle number on the efficiency of transduction; (**b**) static transduction vs. centrifugational transduction

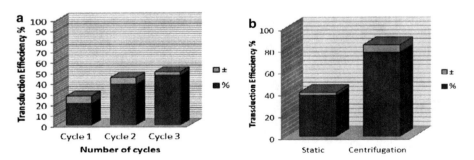

Fig. 2 Simulation results: (**a**) effect of transduction cycle number on the efficiency of transduction, error rate = 11.64%; (**b**) static transduction vs. centrifugational transduction, error rate = 5.22%

components migrate toward it. However, beyond two cycles, there was no observed increase in transduction efficiency with either static or centrifugational transduction. Transduction under centrifugational forces demonstrated roughly a twofold increase in transduction efficiency ($n = 3$) from $39.0 \pm 1.8\%$ to $80.2 \pm 5.4\%$ after two cycles (paired t-test; $p = 0.002$) (Fig. 1b). Using a lentiviral vector, transduction efficiencies of $97.7 \pm 1.4\%$ with mean fluorescence intensities (MFIs) of $2,395 \pm 460$ were achieved with the same three samples (Fig. 3). This was achieved after a single cycle at a multiplicity of infection (MOI) of 11. Raising viral titers was detrimental to the hfMSCs, with massive cell death observed after MOI of >22.

In a similar way, our simulation of static transduction of three samples of first trimester fetal blood-derived hfMSCs from 9^{+1}, 9^{+6}, and 10^{+6} weeks of gestation revealed a mean efficiency of $20.3 \pm 7.2\%$, $38.6 \pm 6.4\%$, and $47 \pm 3.2\%$ after one, two, and three cycles of transduction, respectively, with an error rate $= 11.64\%$ (Fig. 2a). Simulating transduction under centrifugational forces also demonstrated roughly a twofold increase in transduction efficiency ($n = 3$) from $37.8 \pm 2.6\%$ to $77.3 \pm 6.8\%$ after two cycles (paired t-test; $p = .002$) with an error rate $= 5.22\%$ (Fig. 2b). And using a lentiviral vector agent, transduction efficiencies were of

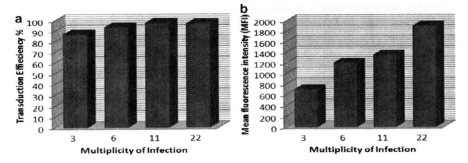

Fig. 3 Chan, O'Donoghue, de la Fuente et al. investigated results: (**a**) percentage of lentiviral-transduced cells with varying MOI; (**b**) mean fluorescence intensity with varying MOI after lentiviral transduction

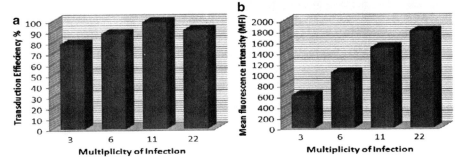

Fig. 4 Simulation results: (**a**) percentage of lentiviral-transduced cells with varying MOI, error rate $= 5.4\%$; (**b**) mean fluorescence intensity with varying MOI after lentiviral transduction, error rate $= 9.8\%$

$98.9 \pm 1.1\%$ having error rate $= 5.4\%$ with MFIs of $2,117 \pm 633$ were achieved with the same three samples and with error rate $= 9.8\%$ (Fig. 4). This was achieved after a single cycle at MOI of 11. Raising viral titers was also detrimental to the hfMSCs, with massive cell death observed after MOI of >22.

References

1. Anna L. David, Donald Peebles (2008) Gene therapy for the fetus: is there a future? Best Practice & Research: Clinical Obstetrics and Gynaecology 22, 1, 203–218
2. Duanqing Pei (2009) Regulation of pluripotency and reprogramming by transcription factors. Journal of Biological Chemistry 284, 6, 3365–3369
3. Hongyan Zhou, Shili Wu, Jin Young Joo et al (2009) Generation of induced pluripotent stem cells using recombinant proteins. Cell Stem Cell, 4, 5, 381–384
4. Bo Feng, Jia-Hui Ng, Jian-Chien Dominic Heng et al (2009) Molecules that promote or enhance reprogramming of somatic cells to induced pluripotent stem cells. Cell Stem Cell, 4, 4, 301–312
5. Prashant Mali, Zhaohui Ye, Holly H. Hommond et al (2008) Improved efficiency and pace of generating induced pluripotent stem cells from human adult and fetal fibroblasts. Stem Cells. doi: 10.1634/stemcells
6. Monya Baker (2007) Adult cells reprogrammed to pluripotency, without tumors. Nature Reports Stem Cells. doi: 10.1038/stemcells.2007.124
7. Anthony Atala (2006) Recent developments in tissue engineering and regenerative medicine. Current Opinion in Pediatriacs 18, 2, 167–171
8. William W. Cohen (2007) A computer scientist's guide to cell biology. Springer Science + Business Media, LLC, e-ISBN 978–0–387–48278–1
9. Anthony Atala, Robert Lanza, James A. Thomson et al (2007) Principles of Regenerative Medicine. Academic Press, ISBN-10: 0123694108
10. Peter A. Horn, Julia C. Morris, Tobias Neff et al (2004) Stem cell gene transfer—efficacy and safety in large animal studies. Molecular Therapy 10, 417–431. doi: 10.1016/j.ymthe.2004.05.017
11. Marieke Bokhoven, Sam L. Stephen, Sean Knight et al (2009) Insertional gene activation by lentiviral and gammaretroviral vectors. Journal of Virology 83, 283–294
12. Youngsuk Yi, Sung Ho Hahm, Kwan Hee Lee (2005) Retroviral gene therapy: safety issues and possible solutions. Current Gene Therapy, 5, 25–35
13. Wilfried Weber, Martin Fussenegger (2006) Pharmacologic transgene control systems for gene therapy. Journal of Gene Medicine 8, 5, 535–556
14. R Gardlík, R Pálffy, J Hodosy et al (2005) Vectors and delivery systems in gene therapy. Medical Science Monitor 11, 4, RA110–RA121. PMID 15795707
15. Sandy B. Primrose, Richard M. Twyman (2006) Principles of Gene Manipulation and Genomics. Malden, MA: Blackwell Publishing. isbn=1405135441
16. Stuart J. Russell, Peter Norvig (2003) Artificial Intelligence: A Modern Approach. Upper Saddle River, NJ: Prentice Hall, 2nd ed. ISBN 0–13–790395–2
17. Stylianos Andreadis, Thomas Lavery, Howard E. Davis et al (2000) Toward a more accurate quantitation of the activity of recombinant retroviruses: alternatives to titer and multiplicity of infection. Journal of Virology 74, 3, 1258–1266
18. Donald E. Knuth (1969) The Art of Computer Programming. Seminumerical Algorithms, Volume 2. Reading, MA: Addison Wesley
19. Joachim H. Ahrens, Ulrich Dieter (1974) Computer methods for sampling from gamma, beta, Poisson and binomial distributions. Computing. doi: 10.1007/BF02293108
20. Joachim H. Ahrens, Ulrich Dieter (1982) Computer generation of Poisson deviates. ACM Transactions on Mathematical Software. doi: 10.1145/355993.355997

21. Ronald J. Evans, J. Boersma, N. M. Blachman et al (1988) The entropy of a Poisson distribution: problem 87–6. SIAM Review. doi: 10.1137/1030059
22. Jerry Chan, Keelin O'Donoghue, Josu de la Fuente et al (2005) Human fetal mesenchymal stem cells as vehicles for gene delivery. Stem Cells 23, 93–102
23. Timothy J. Nelson, Andre Terzic (2009) Induced pluripotent stem cells: reprogrammed without a trace. Regenerative Medicine. doi: 10.2217/rme.09.16
24. Cesar A. Sommer, Matthias Stadtfeld, George J. Murphy et al (2009) Induced pluripotent stem cell generation using a single lentiviral stem cell cassette. Stem Cells. doi: 10.1634/stemcells.2008-1075

Chapter 33
Biomarkers Discovery in Medical Genomics Data

A. Benis and M. Courtine

Abstract This chapter presents a system, called DiscoCini, assisting the biology experts to explore medical genomics data. First, it computes all the correlations (based on ranks) between gene expression and bioclinical data. The amount of generated results is huge. In a second step, we propose an original visual approach to simply and efficiently explore these results. Thanks to sets of data generated during experiments in the field of the obesities genomics, we show how DiscoClini allows easily identification of complex disease biomarkers.

1 Introduction

In biomedical researches, understanding the mechanisms related to gene expression represents a major key challenge. Better knowledge of the gene activation or inhibition is an opening way to numbers of perspectives in therapeutics and diagnosis, and will be one of the fundamentals of a personalized medicine [1]. An increasing number of diseases are studied using functional genomic technologies allowing the exploration of many gene expressions simultaneously. cDNA chips [2] is one of these, and generated data combined with bioclinical measurements data allow biomarker discovery. However, all these data are proned to inter-individual variability and to an inconstant precision (handling and equipment errors, etc.). Classically in Medical Functional Genomics (MFG) to discover and to identify a gene as a biomarker for a physiological status, biologists use the literature and their prior knowledge. This kind of approach allows to select a reduced number of potentially relevant relationships existing between gene expression and bioclinical data. To define a relationship as an interesting one, biologists mainly considered the statistical significance of this relation [3, 4]. However, combining these two

A. Benis (✉)
LIM&Bio – Laboratoire d'Informatique Médicale et de Bioinformatique – E.A.3969,
Université Paris Nord, 74 rue Marcel Cachin, 93017 Bobigny, Cedex, France
e-mail: benis.arriel@gmail.com

steps in an analysis process critically reduces the chances to discover a new and unexpected biomarker. To overcome these limitations, it is important to provide a system wherein all available combinations between all datasets are automatically computed to disclose automatically relevant relationships to experts. We propose a system to support the relevant relationship discovery in a very large number of gene expression values from cDNA chips and several bioclinical data. Hereinafter, we present our system called DiscoClini which is based on correlation discovery. We describe its main steps including computation and visualization. In a third part, we present results based on experiments in the field of obesity research. At the end, we conclude and propose potentially new ways of works.

2 Materials and Methods

A large number of approaches exist to define relationships between two datasets. In biology, the correlation coefficient is currently used. It indicates the intensity and the direction of a "linear" relationship between two variables. Several correlation approaches exist [5]. In MFG, the rough data contain many biases linked to experimental conditions, experimenter habits, equipment calibration, inter-individual differences, etc. The data quality is not constant and varies at each measure. Hence, we use the Spearman's correlation coefficient of ranks [5], denoted ρ_S (or r alternatively). This allows reducing the impact of the "fuzzy" quality of data.

In MGF field, important number of papers disclose and analyze results of bioclinical research protocols. The relationships in MFG data are defined as correlations. These studies are a priori based and are composed of two main steps. First, genes of interest are defined on regard to the literature, the expert interests, and/or prior knowledge and/or computational filtering [5–7]. Second, computation steps are done on the preselected and restricted number of MFG data are done between the selected genes with few bioclinical parameters associated with a specifically studied disease. This kind of approach is expensive, needs time, and reduces the chances of finding new knowledge. Other approaches were proposed to define relationships between genetic and/or genomics and bioclinical data [8] such as SNPs. As an example, ObeLinks [9] is a system using a Galois lattice to classify genomic data and phenotypic to discover relationships between these two types of information. However, this kind of approach is not relevant to discover relationships to predict the progression of the disease status of patients. Nevertheless, in other fields, such as in Physics or in Management, systems were proposed. Natarajakumar [10] examines the quality of the signals transmitted over the oceans in terms of sea level and rainfall. Linear relationships were discovered using genetic algorithms and using several thousands of measurements to compute each relationship. According to this last constraint, this is not appropriate to our scope. In other fields, correlation discovery systems have been proposed. Chiang [11], who developed the concept of Linear Correlation Discovery, proposed a data mining system allowing discovery in large datasets related to Client Relationship Management. His approach consists in automatically coupling all the attributes of a database to detect groups providing information validated using bootstrap. This approach is conceptually close to DiscoClini's first step. Nonetheless, Chiang uses for each parameter few thousands of replicates and expert is not involved in the discovery process.

Interpretation of correlation values is not a trivial task. It is dependent of the application field. Cohen [12] and Hopkins [13] have proposed, respectively, interpretation guides for Psychological and Physical studies. Each one proposed a different segmentation of the [0;|1|] interval. Herein, we propose one fitting with the MGF experts needs. Another issue related to the interpretation of the correlation values is based on considering only their significance test values in view of a predefined threshold. Due to the in vitro experiment's expensive costs, this interpretation is crucial. Traditionally, the biologists consider a relationship as relevant according to its statistical significance. DiscoClini generates a very large number of correlation values and their related significance tests; this implicates a problem of multiplicity of tests. To deal with, one approach is to compute the False Discovery Rate [3, 4]. However, a high absolute correlation value is potentially highly interesting [14]. For these reasons, the FDR is used in DiscoClini only as a filter allowing reducing the number of potentially disclosed relevant.

Information visualization (IV) is essential to assist the user in the exploration of result. In DiscoClini, IV is used to assist the expert in the exploration step of the discovery process. These were chosen according to our Knowledge Discovery process goals to allow interactivity between the user and the system [15]. Methods for displaying correlations are a fundamental element in decision making [16]. Visualizations can disclose and highlight the most interesting correlations to facilitate interpretation, or the user should interact with the system which does not exhaustively point out the most interesting one, or all the computed results are provided to the user without assisting. DiscoClini provides a visualization step is disclosing and highlighting the most interesting correlations and allows interaction with the user.

3 DiscoClini: A System for Computer-Aided Biomarker Discovery

DiscoClini is based on a process assisting the expert in the biomarker discovery process and using known concepts. The general dataflow of DiscoClini is as follows: (1) defining data sources gene expression data from cDNA chips and bioclinical data; (2) extracting from the data sources the individuals included in a correlational study; (3) computing: (3a) for each attribute of each dataset, univariate statistics, and (3b) for each combination of a gene expression and bioclinical set, bivariate statistics; (4) detecting outliers; (5) guided exploration of the results of steps 3 and 4; (6) biological validation of potential biomarker discoveries. The main goal of DiscoClini is to reduce and to optimize the expert working time devoted to explore and to analyze data, to allow him to mainly deal with the biological aspects of his work. In this chapter, we present with further details the DiscoClini's steps (3b) and (5).

Step 3b of DiscoClini consists in combining a gene expression set and a bioclinical one and to compute bivariate statistics such as previously explained (correlation coefficient value, tests of significance and of multiplicity). This computational step involve global and local approaches. "Global" approach calculates the intensity and the direction of each relationship computing ρ_S, p, and q, and other statistical measures none disclosed herein. In the "local" approach, each bivariate dataset is treated to discover subgroups of points. Each correlation is calculated on a subset of adjacent points (known as a segment of points linearly correlated) using a windowing approach, and when the better ones are automatically screened and when they are detected, the same statistics computed with the "global" approach. These algorithms are implemented in R [17]. For example, they are running, both, for 40 individuals, 1 bioclinical parameter, and 39,709 genes in 50 min, on a server under Linux using a processor Intel quad-Xeon 3.6 GHz and 4 GB of RAM.

To allow the expert to easily explore the previously generated results, visualization tools were developed. The first graphical approach in DiscoClini (Fig. 1) proposes to dynamically cluster into nodes of a Hasse diagram (Fig. 1b) the results describing the relationships (between gene expression and a bioclinical parameters selected in a list of these parameters (Fig. 1a1)).

The number of relationships is depending on the threshold values of ρ_S and/or p and/or q and/or n which could be easily updated using for each on horizontal sliders widgets (Fig. 1a2). These four simple statistical parameters were chosen because they can simply be explained to a non-statistician user. Indeed, more the absolute value of ρ_S tends to 1, better the correlation is; more p and q tend to 0, better are the significance values of the relationship; and more n is high more easily the results related to a relationship can be generalized to a large population. These four values defining the static threshold (Fig. 1a3 for minimal values and Fig. 1a4 for maximal one) of the sliders are colored in black. The dynamic values of the thresholds (herein $|r|=0.66$, $p=0.05$, $q=0.05$ and $n=95$) define for ρ_S, p, q, and

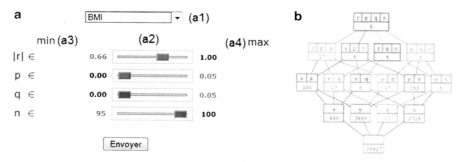

Fig. 1 Screenshot of the Visual Data Mining interface. (**a**) GUI allowing the user to choose which bioclinical parameter he or she wants to study and to dynamically define the thresholds of ρ_S, p, q, and n; (**b**) Hasse diagram of all the relationships computed between BMI values and gene expression data, with the empirical guideline

n can be modified by the expert, using the slider cursors; the values are dynamically colored in red to allow the expert to perceive his action on these sliders. Furthermore and yet to facilitate interesting correlation discovery, nodes colored in dark (denoted "first color") include relationships responding to all the four constraints. Nodes colored in less dark (denoted "second color") include relationships which have a potentially interesting correlation value according to the values defined using the related slider. This interface and more particularly the Hasse diagram (Fig. 1b) is a "gateway" to another "visual" representation. Furthermore, in this Hasse diagram some nodes where colored with the first color ({r, p, q, n} and {r, q, n}) and other ones with the second color ({r, p, q}, {p, q, n}, {r, p}, {p, q}, {q, n}, and {q}). This colorization is done to provide to the user to quickly be aware if relationships are responding to the field expert constraint, the first color nodes being the most interesting one and the second color interesting one. The black ones are nodes that can although be explored with interest; however, these one are less interesting regarding the application field and the importance to reduce the risk of false discovery. At the next step, DiscoClini provides to the expert a table showing the results of the global and local computational steps and related to the constraint related to the node selected on the Hass diagram (Fig. 1). We propose, here, two intuitive graphical languages: one assisting the exploration: DcSymb, and the second for visualizing the relations: DcVisu. DcSymb symbolically reformulates the previously calculated values (ρ_S, p, q, and other statistics). It should be applied to any kind of bidimensional datasets. The interpretation of the DcSymb language is based on the threshold of ρ_S, p, and n defined in the previous interface (Fig. 1). Hence, a positive correlation is represented by "/", a negative one by a "\", and a potential correlation without interest by a "O". When the relation is significant for p, said symbols are in boldface, and when they are related to at least the number of individuals defined by the expert, they are colored in red. The reformulation of the results through DcSymb is used in one of our visualization methods for assisting exploration of the results by the expert. DcSymb is very quickly understandable and increases the efficiency of the exploration process. At the end, the expert can request from the system to obtain a synthesis of the computational and the exploration step. The document is generated in PDF and HTML formats and includes data related to all the DiscoClini process including the descriptions of the correlational study, the sources files, the computational times for each substep of the process, a copy of the Hasse diagram responding to the user constraints, and a list of relevant genes (for the higher and not null node of the lattice). This report is generated to allow the expert to work and to use the results in another context like preparing a paper submission. The second graphical language, DcVisu, assisting the exploration is provided by selecting a node of the Hasse diagram. DcVisu aims to visually disclose a set of relationships simultaneously. A gradient of colors from blue to pink through white is used. These colors were chosen because they are not part of the colors used classically for visualizing cDNA chips data (from green to red). The user can define a number of intervals to discretize to bioclinical data according to the level of details wanted on the relationship's display. For example, with DcVisu, the expression of a gene for each individual is shown on a horizontal axis as a point which color depends on the previous scale of colors.

4 Results

DiscoClini was experimented and was validated at every step of its development, using data from different clinical research protocols related to obesity.

Experiments were done in the context of protocol exploring the impact of the epinephrine on the gene expression in muscles of nine healthy men [18]. We applied DcVisu to provide to the experts a parallel view with relevant relationships. This experiment has helped to define that the epinephrine is involved in inflammatory mechanisms. According, firstly to this discovery and secondly to the interest of the biologists for the results of a systematic computation of correlations and then visualization, the main concepts of DiscoClini were validated. A second experiment was done during a protocol studying the behavior of the adipose tissue under a very low caloric diet and involving 29 individuals. DiscoClini has helped to define that a reduced loss of weight influences the inflammatory profile [19]. This experiment has validated that computing all possible correlations between gene expression and bioclinical data increasing significantly the chances to discover biomarkers. Another contribution of DiscoClini is the discovery of a new marker of the adiposity [20]. Herein, we compared the results obtained by a full human discovery process and by DiscoClini. DiscoClini provides the same results than an expert with a lower human investment.

Another experiment was done on data collected from different clinical research protocols. The data were related to 39 obese women being in a basal status. DiscoClini computed all the possible correlations, using the global and the local approaches, in approximately 24 h for 22 bioclinical parameters and 39 cDNA chips. This is a very short time compared with the few days to few months expert-working time necessary to only define a list of genes of interest and to compute with a priori some correlation values. Hereinafter, we disclose the guided exploration of the results computed between the gene expression data and the BMI values. DiscoClini generated Hasse diagram is shown in Fig. 1. The grouping step was done according to the following thresholds: $|r| \geq 0.66$, $p \leq 0.05$, $q \leq 0.05$, and $n \geq 95\%$. The node {r, p, q, n} disclosing the relationship fitting with all the four constraints includes four potentially interesting biomarkers. These relationships are the same as those in nodes {r, p, n}, {r, q, n}, and {p, q, n}. The node {r, p, q} includes 17 relationships, 12 being "new" one. This means that these relationships are potentially interesting, but including less than 38 individuals. According to the Hasse Diagram, the node {p} relates to a high number of relations (3688), and the related list cannot be efficiently explored by the expert directly; nevertheless according to this high number of significant relationships, we can consider that q is efficiently computed and it is still possible to consider that the 18 relationships included in this node are also potentially relevant one. Selecting node {r, p, q, n} in Hass diagram provided us a table as shown in Fig. 2.

In Fig. 2, the first column gives the gene name. Selection of it allows the user to be redirected to a dedicated webpage on SOURCE. The next two columns are the graphical representation of the relationship using, respectively, DcVisu and Dcsymb.

Gene	DCVisu	DCSymb	n	r	p	q	p/q	v.s.
G1	• • ••• • •• ••• • •• • •	\	38	-0.66444	0.00001	0.03069	0.00018	A
G2	• • •• ••• ••••• •• •	\	39	-0.71205	<1E-005	0.00740	0.00005	A
G3	• •••• •••• • • •• • •	\	38	-0.69133	<1E-005	0.02050	0.00008	A
G4	•• •••••• •• ••••• •	\	39	-0.66734	<1E-005	0.03069	0.00011	A
G5	••••• ••••••• •• • •	\	37	-0.66319	0.00001	0.03088	0.00025	A

Fig. 2 Sample results disclosed using DcVisu obtained after selection of the node $\{\rho_S, p, q, n\}$ of the Hasse diagram presented in Fig. 1

The selection of one of these representations allows showing the related relationship in a two-dimensional plot including the "classic" and the "robust" regression lines. This last representation facilitates the relationship understanding. The fourth column shows n, the number of individuals involved in the related relationship. The next column is the r value. The two next columns correspond to the values of p and q. Column "p/q" indicates the ratio of p and q which indicate when this number is near 0 that the risk of false discovery is very low. The last column indicates the level of "contamination" of the relation by outliers; we use letters from A (clean relationship) to F (highly contaminated relationship). Outliers are detected using a modified version of the k-medoids algorithm. Processing time between selection of a clinical parameter and the redirection to SOURCE website takes few minutes.

5 Perspectives and Conclusions

Hereinbelow, we have presented DiscoClini devoted to assist the biologists to explore MFG data. First, the system computes all the correlations between gene expression and bioclinical data. Second, DiscoClini discloses to the expert interesting relationship using original visual approaches that we propose to simply and efficiently deal with the exploration of the high number of computed relationships. Its advantages were shown on experiments for the identification of biomarkers. DiscoClini provides an objective and *no* a priori-based process. However, limitations exist. The parallel visualization of relationships is limited to a small number of lines to remain usable. Data used, in the MFG, are noisy and incomplete. Therefore, we developed a method to automatically detect suspicious values to improve the quality of the results provided to experts. The number of study subject included in each experiment being currently limited, it is still not possible to automatically discover more complex descriptions of relationships (i.e. mathematical functions). DiscoClini should be efficiently used, according to few modification, to other fields where the number of replicates is low.

References

1. Kennedy, PJ. (2008). To secure the promise of personalized medicine for all Americans by expanding and accelerating genomics research and initiatives to improve the accuracy of disease diagnosis, increase the safety of drugs, and identify novel treatments, and for other purposes. Genomics and Personalized Medicine. H.R.6498
2. Schena, M., D. Shalon, R. Davis, and P. Brown (1995). Quantitative monitoring of gene expression patterns with a complementary DNA microarray. Science 270(5235), 467–470
3. Benjamini, Y. and Y. Hochberg (1995). Controlling the discovery rate: A practical and powerful approach to multiple testing. Journal of the Royal Statistical Society 57, 289–300
4. Storey, J. (2002). A direct approach to false discovery rates. Journal of the Royal Statistical Society: Series B (Statistical Methodology) 64(3), 479–498
5. Nelson, D. (2004). The Penguin Dictionary of Statistics. Penguin Books
6. Tusher, VG., R. Tibshirani, and G. Chu (2001). Significance analysis of microarrays applied to the ionizing radiation response. Proceedings of the National Academy of Sciences of the United States of America 98, 5116–5121
7. Eisen, M., P. Spellman, P. Brown, and D. Botstein (1995). Cluster analysis and display of genome-wide expression patterns. Proceedings of the National Academy of Sciences of the United States of America 95, 14863–14868
8. Ramaswamy, S. and T. Golub (2002). DNA microarrays in clinical oncology. Journal of Clinical Oncology 20(7), 1932–1941
9. Courtine, M. (2002). Regroupement conceptuel de données structurées pour la fouille de données. PhD dissertation, University Paris 6
10. Natarajakumar, B., V. Kurisunkal, R. Moore, and D. Braaten (2004). Rain heights over the oceans: Relation to rain rates. In: U.C.F.T.O. Symposium (Ed.), Cairns, Great Barrier Reef, Australia, pp. 126–132. URSI Commission F Triennium Open Symposium
11. Chiang, R., C. Cecil, and E. Lim (2005). Linear correlation discovery in databases: A data mining approach. Data and Knowledge Engineering 53(3), 311–337
12. Cohen, J. (1988). Statistical Power Analysis for the Behavioral Sciences. Lawrence Erlbaum Associates Inc, Mahwah, NJ
13. Hopkins, W. (2004). A New View of Statistics: Sportscience. Online available: http://www.sportsci.org/resource/stats/index.html
14. Efron, B. (2007). Correlation and large-scale simultaneous significance testing. Journal of the American Statistical Association 102, 93–103
15. Keim, DA., Lee, J., B. Thuraisingham, and C. Wittenbrink (1995). Database issues for data visualization: Supporting interactive database explorartion. In A. Wierse, G. G. Grinstein, and U. Lang (Eds.), Proceedings of the IEEE Visualization Work. Database Issues for Data Visualization, Number 1183, pp. 12–25. Springer, New York
16. Kovalerchuk, B. (2001). Review of Visual Correlation Methods. Online available: http://www.cwu.edu/borisk/visualization/review2b.pdf
17. R Development Core Team (2006). R: A Language and Environment for Statistical Computing. R Foundation for Statistical Computing, Vienna. ISBN 3900051070
18. Viguerie, N., K. Clément, P. Barbe, M. Courtine, A. Benis, D. Larrouy, B. Hanczar, V. Pelloux, C. Poitou, Y. Khalfallah, G. S. Barsh, C. Thalamas, J. D. Zucker, and D. Langin (2004). In vivo epinephrine-mediated regulation of gene expression in human skeletal muscle. Journal of Clinical Endocrinology and Metabolism 89(5), 2000–2014. 0021–972x Journal Article Validation Studies
19. Clément, K., N. Viguerie, C. Poitou, C. Carette, V. Pelloux, C. Curat, A. Sicard, S. Rome, A. Benis, J. Zucker, H. Vidal, M. Laville, G. Barsh, A. Basdevant, V. Stich, R. Cancello, and D. Langin (2004). Weight loss regulates inflammation-related genes in white adipose tissue of obese subjects. FASEB Journal 18(14), 1657–1669
20. Taleb, S., D. Lacasa, J. Bastard, C. Poitou, R. Cancello, V. Pelloux, N. Viguerie, A. Benis, J. Zucker, J. Bouillot, C. Coussieu, A. Basdevant, D. Langin, and K. Clément (2005). Cathepsin S, a novel biomarker of adiposity: Relevance to atherogenesis. FASEB Journal 19(11), 1540–1542

Chapter 34
Computer Simulation on Disease Vector Population Replacement Driven by the Maternal Effect Dominant Embryonic Arrest

Mauricio Guevara-Souza and Edgar E. Vallejo

Abstract In this chapter, we present a series of computer simulations on the genetic modification of disease vectors. We compared the effectiveness of two techniques of genetic modification, transposable elements and maternal effect dominant embryonic arrest (MEDEA). A gene drive mechanism based on MEDEA is introduced in the population to confer immunity to individuals. Experimental results suggested that the genetic maternal effects could be necessary for the effectiveness of a disease control strategy based on the genetic modification of vectors.

Keywords Epidemic models · Medical informatics

1 Introduction

Creating new and effective mechanisms for controlling vector-borne diseases such as malaria and dengue is a major epidemiology concern worldwide. Malaria parasite, *Plasmodium falciparum*, is the pathogen that has killed more people in the history of humanity. Nowadays, malaria kills over two million people per year, mostly children, in Africa alone [1, 2].

The use of insecticides was the primary control strategy of a vector population in the past decades. This type of control was effective at the beginning but as the time passed by, this strategy encouraged the emergence of insecticide-resistant strains. Further drawback of these strategies are the negative effects they have on the environment.

Insecticide is useful to reduce the size of vector populations. However, there are other species that feed from the vector population; hence, reducing its number greatly would produce critical disturbance on ecosystems. Therefore, these disease

---

M. Guevara-Souza (✉)
Computer Science Department, ITESM CEM, Carretera Lago de Guadalupe Km. 3.5,
Atizapan de Zaragoza, 52926, Mexico
e-mail: guevara_mauricio@hotmail.com

control strategies are no longer considered effective, and the use of insecticides has been banned in most countries of the world.

In addition, *P. falciparum* has developed resistance for the most used treatment drugs such as chloroquine, sulfadoxine, and mefloquine.

New discoveries in areas such as molecular biology and bioengineering had contributed to our understanding of many biological mechanisms at the molecular level [3]. Nowadays, it is possible to manipulate biological processes for our convenience. Genetic modification of organisms to confer disease refractoriness to them is now possible and it is a potentially effective approach worth exploring [4].

Several altruistic organizations had provided financial support for research on disease control strategies based on genetic modification to confer disease immunity to the vector population. In effect, the Gates Foundation have provided a considerable funding for the creation of disease-resistant strains to the vector of malaria parasite [5].

The creation of individual disease-resistant strains is possible, in principle. However, for the disease control strategy to be effective, there are other conditions that must be satisfied. To mention one, the refractory gene should confer some fitness advantage to the modified mosquitoes with respect to their wild counterpart or else it would be removed from the gene pool within a few generations as an effect of natural selection. In addition, the refractory gene drive mechanism should not produce a significant alteration on the mosquitoes phenotype; hence, it does not affect its adaptation capabilities [6].

There are some issues to consider when dealing with a genetic modification strategy. First of all, there is much concern on the environmental consequences of the release of genetically engineered mosquitoes. Second, there is much uncertainty on the level of fixation of the refractory gene at the population level required to produce an effective disease control strategy [6].

It is believed that natural selection alone is not likely to produce the rapid fixation of a new gene; hence, complementary molecular mechanisms such as maternal effect dominant embryonic arrest (MEDEA) have been recently proposed to expedite this process [7].

As a result, the implementation of this strategy often includes controlled laboratory experiments. These experiments are typically expensive, but most importantly the controlled environment could lead to wrong conclusions, especially related to the fitness of modified mosquitoes [8].

The lifespan of the malaria vector, *Anopheles gambiae*, is approximately 1 month. This makes very difficult to study multiple scenarios of spread and prevalence of the resistant genes in real populations.

Studying population genetics experimentally is, in general, extremely difficult. Therefore, mathematical and computer simulation models are widely used because they have proved to be useful tools for studying many aspects on the genetics of populations [9, 10].

The study of biology have been increasingly relied upon artificial life simulation models. New computer algorithms that simulate population dynamics have been used to study gene flow in populations, among other applications [11]. The use of computer simulations has proved to be more tractable than mathematical models previously used for studying these fundamental aspects of the biology of organisms [12].

In our previous work, we used transposable elements to introduce a refractory genes in a wild population. The use of transposable elements alone was not enough to fix the transposable elements in the wild population, and so the use of another biological mechanism is needed [13].

In this chapter, we extended the previous work on simulating gene replacement in vector populations to study the effects of the MEDEA mechanisms on the fixation of modified genes as a potential control strategy [13].

Different scenarios for the spread and prevalence of modified genes are simulated. Experimental results show that the MEDEA mechanism can considerably contribute to accelerate the fixation of refractory genes in vector populations. The rapid fixation of these new genes is a necessary condition for the genetic modification of disease vectors.

2 Maternal Effect Dominant Embryonic Arrest

Transposable elements are small DNA sequences that can move around different positions in the genome. They are often called jumping genes. Transposons can modify the length of the genome, multiply themselves, and cause mutations. They are valuable tools to produce genetic modifications [4].

MEDEA is a biological mechanism that is used to favor offspring that possess a collection of genes. This mechanism was first observed in the flour beetle *Tribolium castaneum* [14].

MEDEA consists roughly of three parts: a toxin, an effector, and an antidote. The offspring that inherits the toxin and the effector, but not the antidote, will die [15]. Figure 1a shows the parts of the MEDEA mechanism.

This technique has been proven in the laboratory mainly in female fruit flies. The researchers blocked the production of a very important protein, the Myd88, that is necessary for embryonic development but also inserted a copy of that same protein in a DNA area that is activated early in the development. The offspring that inherits the extra copy of Myd88 survived, and the rest died before hatch [16].

As we can see in Fig. 1b, mosquitoes that inherited at least one copy of the antidote survived, while the mosquitoes that carry two non-MEDEA genes die.

There is much hope that this mechanism helps to enhance the fixation of a desirable gene rapidly in wild vector populations to provide an effective disease control strategy [17].

Laboratory experiments showed that it is possible to spread a gene in a population using the MEDEA mechanism. In these studies, the fixation of the gene in the population was accomplished in about 20 generations with a percentage of insertion of 25% of transgenic mosquitoes [15].

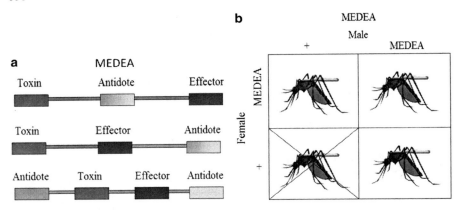

Fig. 1 (a) MEDEA mechanism, (b) MEDEA effect in offspring

Although MEDEA seems very promising, this mechanism possesses some limitations. First, MEDEA females have a lower fecundity due to the offspring that dies because of the MEDEA mechanism [16].

Another negative effect is that MEDEA mechanism has variable degrees of maternal effect lethality; hence, it is difficult to forecast the effect of MEDEA when introduced in a population [17].

Finally, MEDEA, as any transposable element, produces a negative effect on the fitness of the carriers, making more difficult to spread among population [4].

3 Model Description

The computer model developed as part of this research is capable of simulating the most important biological processes that participate in gene flow among populations: genetic mutation and reproduction. In addition, we simulate the MEDEA mechanism. The mosquitoes are haploid and they reproduce sexually using random mating.

To describe the mosquito, we use the following attributes: DNA sequence, age, sex, and fitness.

The genetic mutation process is fairly simple. Mutation consists of a random substitution of a particular locus. We fixed the mutation rate probability to 10%. It is considerably high compared with the actual mutation rate in nature, that is 1%. However, since we are simulating only a fragment of the mosquitoes' genome, we set the mutation rate higher to produce a proportional effect.

The reproduction takes place when a male and a female mosquito mate. The offspring inherit the attributes from its parents with little modifications. The DNA sequence of the offspring is calculated via a random crossover between the DNA sequence of its mother and father, and then we apply mutation to the resulting sequence.

Table 1 MEDEA embryo survival

Male	Female	Embryo survival(%)
M/+	+/+	98.4
M/M	M/M	98.6
+/+	M/+	48.7
M/M	M/+	98.4
M/+	M/+	73.6
M/+	M/M	57.2
+/+	M/M	20.2
M/M	+/+	98.5

M MEDEA allele, + wild allele
Data source [15]

The probability of the offspring to be a male or a female is 50% for each case. The fitness is calculated as an average of the fitness of the mother and the father. The fitness is very important because the number of offspring that a mosquito can produce is proportional to its fitness.

The MEDEA mechanism is simulated using five boolean variables. These variables are: two MEDEA alleles that the mosquito inherit from its parents that determine the embryo survival, a variable that shows whether the mosquito acquired the MEDEA toxin, a variable for the effector, and the last variable is the antidote.

The probability of a mosquito to survive is shown in Table 1.

The computer model has a feature known as turnover. The turnover controls the number of individuals in a population and produces a maximum of them [18]. We use the turnover because of computational limited resources, and in nature the populations do not grow unboundedly due to limited natural resources as well.

3.1 Simulation Scenarios

In all the simulations described in this chapter, we used four meta populations. Each meta population is initialized with a fixed number of mosquitoes that served as genetic seeds. At every time step in the simulation, all the mosquitoes are exposed to the application of the biological processes described above. In all our experiments, 100 generations were simulated. The turnover was fixed to a maximum of 100,000 mosquitoes. This population size was large enough to minimize the genetic drift effect to fix the MEDEA mechanism. All of these parameters were determined empirically from preliminary simulations. At the beginning, we had two populations: a population of native mosquitoes and a population of mosquitoes carrying the MEDEA mechanism. Each population evolved independently, and when the native population reached a certain size, the MEDEA mosquitoes were inserted in the native population to observe the gene flow between both populations. At the end of each generation, we counted the number of mosquitoes without the MEDEA mechanism and the mosquitoes with the MEDEA mechanism, and we calculate the percentage of fixation of MEDEA.

4 Experiments and Results

In this chapter, we report two types of experiments. The first experiment intends to determine the percentage of mosquitoes with the MEDEA mechanism needed to replace the malaria carrying alleles of the entire population using the MEDEA mechanism and with transposons only.

The second experiment is aimed to observe the effect in the percentage of fixation of the MEDEA mechanism when it disturbs the fitness of the mosquitoes.

4.1 Determination of the Percentage of Insertion of MEDEA Mosquitoes

For this experiment, the scenario that we used was the following: the maximum population size was fixed to 100,000 individuals to minimize genetic drift effect. The generations simulated was 100. The percentage of insertion used was 5, 10, and 20%. In this experiment, the MEDEA mechanism has no fitness effect in the carrier mosquito. This experiment was intended to determine the percentage of MEDEA mosquitoes needed to fix the mechanism in the 100% of the native population.

As can be seen in Fig. 2a, the larger the MEDEA mosquito percentage inserted, the fixation of the MEDEA increases. We can also reach a 100% fixation of the mechanism, which is very superior compared with the fixation obtained in our previous experiments.

4.2 MEDEA Versus Transposon

The use of transposons to confer immunity to a vector population was explored in early attempts to achieve an effective strategy for controlling a disease.

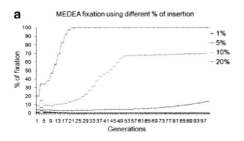
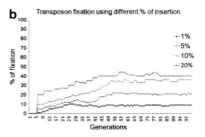

Fig. 2 Results of the experiment of different percentage of transgenic mosquitoes insertion. (**a**) With MEDEA; (**b**) transposon only

The transposon control strategy encounters several difficulties when put into practice. The transposons tend to diminish the fitness of the carrier, and so it is very difficult to fix it on the population in a reasonable time [14].

In a computer model that simulates the introduction of a transposon in a wild population, even with an introduction of 20% of transgenic mosquitoes the percentage of fixation was insufficient. Only 40% of the total population carried the transposon after 100 generations [13]. In this experiment the transposon does not reduce the fitness of the carrier.

In this work with the same 20% introduction of MEDEA mosquitoes in the population, the percentage of fixation was 100%; hence, we can deduce by these results that the MEDEA disease control strategy has more chances to succeed than a strategy based on transposons alone. Figure 2b shows the results of the simulations made with the scenario described above.

4.3 Effect of the MEDEA Mechanism in the Mosquitoes' Fitness

The goal of this experiment is to determine the effects in the population and in the fixation percentage of the MEDEA mechanism when a cost in the fitness of the carrier mosquito is involved.

This is important because a gene that does not confer an advantage is discarded by natural selection. For this experiment, the scenario is very similar to that of the first experiment. The maximum number of individuals in the population is 100,000, and we used a 20% insertion of MEDEA mosquitoes and the fitness decrement caused by the MEDEA mechanism to the carrier is 5, 10, and 20%. The next figure shows the results of the experiment.

The percentage of fixation of the MEDEA mechanism is inversely proportional to the fitness cost. As can be seen in the Fig. 3, even with a decrement in fitness of 5% caused by the MEDEA mechanism, it achieved a 100% of fixation in only 50 generations.

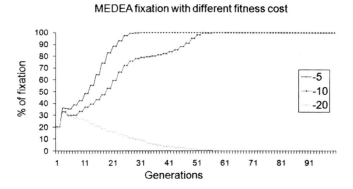

Fig. 3 Fitness effect of the MEDEA machanism

5 Discussion

The disease control strategies based on genetic modification are considered promising approaches for controlling vector populations as an application of the new advances in genomic science. Insect-borne pathogens have had devastating effects all over the world, mainly in third world countries [19].

The disease control strategies based on genetic modification of vectors are promising in principle, but there are still some uncertainties on the effectiveness of such control strategies. The experiments are done in a very controlled environment, but there is no reason to think that MEDEA mechanism will behave the same way in the wild than in the laboratory experiments.

Another concern about genetic modification is that there is no way to measure the environment impact that the introduction of transposable elements can cause. Crucial studies on ethical, social, and legal issues are needed for the realization of such a scientific feat.

Computer simulations are very valuable as a first approach and for studying the potentials of the disease control strategies, and to help for establishing a set of conditions under it would be effective.

There is still a lot of studies and research to close the gap between the reality and the abstractions presented here. Still, we believe that computer simulations are capable of modeling with a good level of accuracy the fundamental aspects of many biological mechanisms and will be increasingly useful for supporting the study of fundamental questions on the biology of organisms [20].

Acknowledgements This work was supported by a Consejo Nacional de Ciencia y Tecnología (CONACYT) Award number SEP-204-C01-47434. We want to thank Charles E. Taylor, Bruce Hay, and Catherine Ward for the feedback and ideas provided for the realization of this work.

References

1. Levy S (2006) Mosquito Modifications: New Approach to Controlling Malaria. BioScience 57:816–821
2. Zhong D (2006) Dynamics of Gene Introgression in the Malaria Vector Anopheles gambiae. Genetics 172:1–27
3. Wang R (2004) Microsatellite markers and Genotyping Procedures for Anopheles gambiae. Parasitology Today 15:33–37.
4. Deceliere G (2005) The Dynamics of Transposable Elements in Structured Populations. Genetic Society of America 169:467–474
5. Kidwell M (1997) Transposable Elements as Source of Variation in Animals and Plants. The National Academy of Sciences 94:7704–7711
6. Deceliere G (2006) TEDS: a Transposable Element Dynamics Simulation Environment. Bioinformatics 22:2702–2703
7. Vernick K, Collins F, Gwadz R (1989). A General System of Resistance to Malaria Infection in Anopheles gambiae Controlled by Two Main Genetic Loci. The American Journal of Tropical Medicine and Hygiene 40:585–592
8. Le Rouzic A (2005) Models of the Population Genetics of Transposable Elements. Genetic Research 85:171–181

9. Clough J (1996) Computer Simulation of Transposable Elements: Random Template and Strict Master Model. Molecular Evolution 42:52–58
10. Manoukis N (2004) Detecting Recurrent Extinction in Metapopulations of Anopheles gambiae: Preliminary Results Using Simulation. WSES Transaction on Systems:1–26
11. Langton C (1997) Artificial Life. MIT Press, USA
12. Wilson W (1997) Simulating Ecological and Evolutionary Systems. Cambridge University Press, UK
13. Guevara M, Vallejo E (2008) A Computer Simulation Model of Gene Replacement in Vector Population. IEEE BIBE 2008:1–6
14. Smith N (1998) The Dynamics of Maternal Effect Selfish Genetic Elements. Journal of Theoretical Biology 191:173–180
15. Chen C, Huang H, Ward C, Su J (2007) A Synthetic Maternal-Effect Self Genetic Element Drives Population Replacement in Drosophila. Science 316:597–600
16. Kambris Z (2003) DmMyD88 controls dorsoventral patterning of the Drosophila embryo. EMBO reports 4:64–69
17. Wade M, Beeman R (1994) The Population Dynamics of Maternal Effect Selfish Genes. Genetics 138:1309–1314
18. Hartl D (1997) Principles of Population Genetics. Sinauer, USA
19. Marshal J, Morikawa K, Manoukis N, Taylor C (2007) Predicting the Effectiveness of Population Replacement Strategy Using Mathematical Modeling. Journal of Visualized Experiments (5):227
20. Krane D, Raymer M (1997) Fundamental Concepts of Bioinformatics. Pearson, USA

Chapter 35
Leukocytes Segmentation Using Markov Random Fields

C. Reta, L. Altamirano, J.A. Gonzalez, R. Diaz, and J.S. Guichard

Abstract The segmentation of leukocytes and their components plays an important role in the extraction of geometric, texture, and morphological characteristics used to diagnose different diseases. This paper presents a novel method to segment leukocytes and their respective nucleus and cytoplasm from microscopic bone marrow leukemia cell images. Our method uses color and texture contextual information of image pixels to extract cellular elements from images, which show heterogeneous color and texture staining and high-cell population. The *CIE $L^*a^*b^*$* color space is used to extract color features, whereas a 2D Wold Decomposition model is applied to extract structural and stochastic texture features. The color and texture contextual information is incorporated into an unsupervised binary Markov Random Field segmentation model. Experimental results show the performance of the proposed method on both synthetic and real leukemia cell images. An average accuracy of 95% was achieved in the segmentation of real cell images by comparing those results with manually segmented cell images.

1 Introduction

In blood cell imaging, the current tendency of segmentation consists in extracting cells from complicated backgrounds and separating them into their morphological components such as nucleus, cytoplasm, holes, and some others [1]. Hence, the segmentation process helps to obtain a better representation of the cells in the image to facilitate its understanding for further steps such as cell identification and classification.

Segmenting the nucleus and cytoplasm of leukocytes from bone marrow images is a very difficult task, as the images show heterogeneous staining and high-cell

---

C. Reta (✉)
National Institute for Astrophysics, Optics, and Electronics,
Luis Enrique Erro No. 1, Puebla, Mexico 72840
e-mail: creta@ccc.inaoep.mx

population. Some segmentation techniques such as thresholding, edge detection, pixel clustering, and growing regions have been combined to extract the nucleus and cytoplasm of leukocytes [2–6]. These techniques could be applied as the images showed uniform backgrounds and high contrast that appropriately defined the objects of interest. Conversely, in our work there are images with low contrast among cell elements and a variety of colors and textures that make cellular elements difficult to distinguish. For this reason, we propose a segmentation algorithm based on color and texture pixel features that can work in bone marrow images showing heterogeneous staining.

This paper is organized as follows. Section 2 describes the proposed method for segmenting the nucleus and cytoplasm of leukocytes. Section 3 shows the segmentation results. Finally, conclusions are presented in Section 4.

2 Cell Segmentation Model

We analyzed the color and texture features of blood cells to design a robust algorithm able to appropriately segment leukocytes and their respective nucleus and cytoplasm from bone marrow images. This is important because in future work we will extract descriptive features of these elements, analyze them, and obtain information to monitor and detect diseases originated in leukocytes.

2.1 Color Analysis

When bone marrow cells are dyed using the Romanowsky method, the color intensities acquired by leukocytes in their nucleus are darker than in their cytoplasm, the nucleus exhibits tonalities of purple, the cytoplasm shows blue tones in lymphocytes and rose tones in myelocytes, and red blood cells get shades of orange and rose. According to this description, we decided to use the *CIE $L^*a^*b^*$* color space because it highlights the visual differences between these colors, and it provides accuracy and a perceptual approach in the color difference calculation. Since our image collection is in the *RGB* color space, a transformation from *RGB* to the *CIE $L^*a^*b^*$* space was done using the formulas presented in [7].

Figure 1 shows the *CIE $L^*a^*b^*$* color channels of a bone marrow cell image. As we can see, the luminosity channel L^* provides an adequate nucleus color representation because it allows us to identify objects with different light reflection. We can also note that channel b^* highlights elements in shades of purple and blue providing a suitable color representation of leukocytes.

Since channels L^* and b^* contain valuable information about the nucleus and the cells, respectively, they were used to create two similar groups in their channel intensity values using the k-means clustering algorithm with k = 2 and k = 3. The first group was established by selecting the cluster that better represents the nucleus

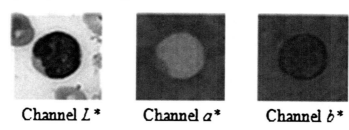

Fig. 1 Blood smear in the *CIE L*a*b** color space

or cell information. All the pixels that do not correspond to the first group fall in the second one. We finally calculated to each group its channel intensity statistics – mean, variance, and standard deviation – to incorporate them as color features for the segmentation model.

2.2 Texture Analysis

An analysis of texture was done by using the Wold decomposition model, which combines the structural and stochastic approach to describe texture. We chose this model because it can be applied to heterogeneous textures found in natural images, and for its invariant properties to translation, rotation, and scale.

Wold 2D-theory interprets the image texture by means of the sum of three mutually orthogonal components: a harmonic field, a generalized evanescent field, and a purely deterministic field [8]. These fields describe the periodicity, directionality, and randomness of the objects texture, respectively. Figure 2 shows a diagram of the orthogonal components of the Wold decomposition for a selected channel.

To parameterize the harmonic field, we used the method proposed by [8]. In this algorithm (see Fig. 3), we first solved the sinusoidals using the discrete Fourier transform (DFT); next, harmonic peaks were located by identifying the largest isolated peaks in the harmonic frequencies. We established a value of 10 as the amplitude threshold, which is sufficient to find all the peaks that were considered harmonic components in the cell images. Finally, the harmonic field parameterization was done by evaluating the amplitude and phase values of the DFT from the frequencies identified as peaks.

In the parameterization of the generalized evanescent field, we used the algorithm proposed by [9]. In this algorithm (see Fig. 4), the DFT without harmonic components is used to find four evanescent lines using a Hough transform. The parameterization of this field is carried out by evaluating the amplitude and phase values of the DFT from the frequencies of the evanescent lines identified.

The texture structural component is the orthogonal sum of the harmonic and generalized evanescent fields. The stochastic component parameterization is done by evaluating the amplitude and phase values of the residual DFT once the structural component is removed.

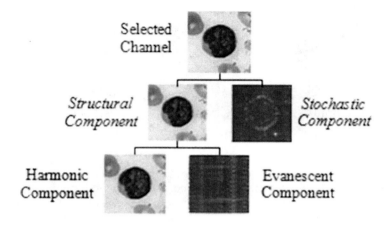

Fig. 2 Wold decomposition texture model

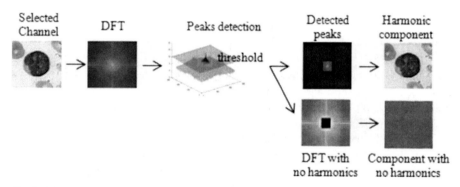

Fig. 3 Harmonic field parameterization

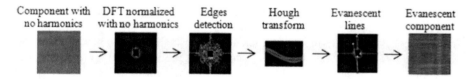

Fig. 4 Generalized evanescent field parameterization

2.3 Color and Texture MRF Model

In this work, we propose a method for segmenting leukocytes from bone marrow images using Markov Random Fields. This method is formulated within the Bayesian approach and incorporates contextual information from color and texture image pixels allowing segmenting cells in images with heterogeneous staining.

The MRF model consists of the definition of a neighborhood and cliques, and the definition of the *a priori* probability and the probability model that allows deriving the posterior probability.

2.3.1 Definition of the a Priori Probability

The color intensity features and the structural and stochastic texture fields are modeled through the definition of the *a priori* energy $U_p(f)$ as follows:

$$U_p(f) = \sum_{s \in S} \left(w_s f_s + \sum_{r \in f_{N(s)}} v_{sr} f_s f_r \right), \quad (1)$$

where f_s is the labeling configuration of state s, f_r is the labeling configuration of neighboring states $N(s)$ of state s, and the coefficients w_s and v_{sr} are the spatial variation parameters of the stochastic W and structural V texture components, respectively. We defined the spatial variation parameters w_s and v_{sr} using clique potentials $V_c(\cdot)$ over all possible cliques C as follows:

$$V_1(f) = w_s \in W \quad V_2(f) = \begin{cases} v_{sr} = |v_s - v_r| & \text{if } s \neq r, s, r \in C, v_s, v_r \in V \\ 0 & \text{otherwise} \end{cases} \quad (2)$$

Thus, the *a priori* probability $P(f)$ can be expressed as:

$$P(f) = \frac{1}{Z} \exp\left[\frac{1}{T} U_p(f)\right] \quad (3)$$

where Z is a standardization constant and T represents the probability distribution form [10].

2.3.2 Definition of the Probability Model

The probability model $P(d \mid f)$ is defined by a Gaussian distribution as follows:

$$P(d \mid f) = \prod_{s \in S} P(d_s \mid f_s) \quad (4)$$

$$P(d_s \mid f_s = k) = \frac{1}{\sqrt{2\pi^n |Cov_k|}} \exp\left[-\frac{1}{2}(d_s - \mu_k)^T Cov_k^{-1}(d_s - \mu_k)\right] \quad (5)$$

where d_s represents the observed color of state s, f_s is the labeling configuration of s, μ_k and Cov_k are the mean and covariance matrix of class k, respectively, and n is the number of image color channels [11].

2.3.3 Posterior Probability

According to [10] to determine the labeling of f, it is necessary to maximize the posterior probability. Considering the definition of the *a priori* probability $P(f)$ (eq. 3) and the probability model $P(d\,|\,f)$ (eq. 4), the posterior probability to maximize $f^* = \arg\max_{f \in F} P(d\,|\,f)P(f)$ is equivalent to:

$$f* = \arg\max_f [\ln P(d\,|\,f) + P(f)]$$

$$f* = \arg\min_f \sum_{s \in S} \left[\ln \sqrt{2\pi^n |Cov_k|} + \frac{1}{2}(d_s - \mu_k)^T Cov_k^{-1}(d_s - \mu_k) \right] \quad (6)$$

$$+ \frac{1}{Z} \left[\sum_{s \in S} w_s f_s + \sum_{s \in S} \sum_{r \in f_{N(s)}} v_{sr} f_s f_r \right] \quad (7)$$

Thus, the posterior energy incorporates the stochastic w_s and structural v_{sr} texture fields of the Wold decomposition into the *a priori* energy $P(f)$ and adds color information by means of the definition of the probability model $P(d\,|\,f)$.

3 Experimental Results

In this work, we used a subset of the IMSS San Jose cells image collection that contains 200 bone marrow leukemia cell images with different color staining. These images were digitalized [12] using a digital camera connected to a Carl Zeiss optical microscope with a 100× objective.

The proposed method showed good qualitative segmentation results allowing the extraction of leukocytes and their respective nucleus. The cytoplasm can be obtained from these by applying the difference-set operation. To measure the accuracy of the segmentation algorithm in a quantitative way, we tested it on a variety of color images, including real and synthetic blood cell images.

The *Real* images evaluation data set corresponds to a subset of the original images collection; it contains 20 leukemia cell images with color and texture variations among them. The *Synthetic* images evaluation data set was created from real images using an image editor; it contains 10 *Real–Synthetic* images and 10 *Synthetic–Synthetic* images. *Real–Synthetic* images were created by using a known shape and the color and texture of real cells. In the case of *Synthetic–Synthetic* images, we know the shape of the cell and we fill it with known colors and textures.

Table 1 shows the results of the evaluation of the algorithm. These results were obtained by comparing our automatic cell segmentation algorithm with cells manually segmented by an expert (or with known shapes in the case of synthetic images).

Table 1 Evaluation of the cells segmentation algorithm for real and synthetic cells images

	Real images		Real–Synthetic		Synthetic–Synthetic	
	Nucleus	Cell	Nucleus	Cell	Nucleus	Cell
Precision	95.87%	95.75%	97.30%	98.07%	99.00%	99.79%

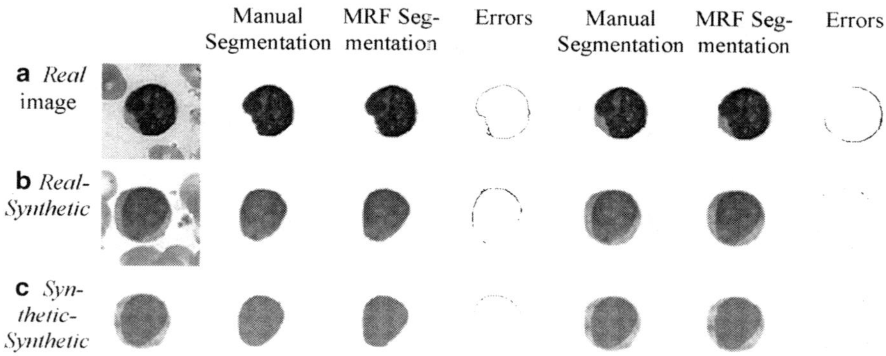

Fig. 5 Evaluation of the cell segmentation algorithm for different kinds of cells images

Figure 5 shows an example of the cell segmentation using *Real*, *Real–Synthetic*, and *Synthetic–Synthetic* cell images.

Experimental results show that the proposed methodology allows the extraction of leukocytes and their nucleus with high accuracy when experimenting with synthetic images, in which the shape of the nucleus and cells is known. The best results are obtained using *Synthetic–Synthetic* (synthetic in shape and texture) images because the image texture was known. The error in the segmentation of *Real–Synthetic* images (synthetic shape images with real texture) was caused by the incorporation of real textures; whereas the error of the segmentation of *Real* images is caused by both, the vagueness of the cell segmentation algorithm and the errors generated during the manual segmentation process.

Figure 6 shows some examples of the results of the segmentation of real cells in images with different staining and cell population.

4 Conclusions

In this paper, we proposed a novel method based on Markov Random Fields that uses color and texture contextual information of image pixels to segment leukocytes and their respective nucleus and cytoplasm. It is important to note that our method can be applied to images that show heterogeneous color and texture staining, a desirable property when working with bone marrow smears. Experimental results show that our method achieves a segmentation accuracy of 95% when it is compared with

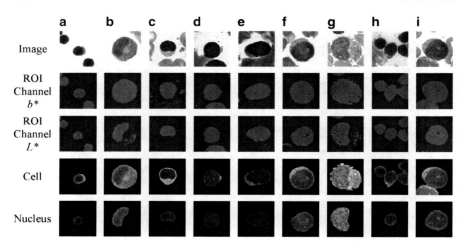

Fig. 6 Examples of the segmentation in images with different staining and cell population

a manual segmentation. However, when we apply our method to images with a high cell population, we are able to correctly segment the cell nucleus, we can also segment leukocytes when they are touched by red globules (see Figs. 6(f) and 6(g)), but when a leukocyte is touching another leukocyte, our method is not able to separate them (see Figs. 6(h) and 6(i)).

Our future work involves the creation of a cell separation algorithm. We will also work on the extraction of descriptive features from the segmented nucleus and cytoplasm of the cells and use that information to classify the cells into different acute leukemia subtypes.

References

1. Kumar B, Joseph D, and Sreenivas T (2002) Teager energy based blood cell segmentation. IEEE 14th Int Conf on Digital Signal Processing 619–622
2. Won C, Nam J, and Choe Y (2004) Segmenting cell images: a deterministic relaxation approach. LNCS 3117:281–291
3. Colantonio S, Gurevich I, and Salvetti O (2007) Automatic fuzzy-neural based segmentation of microscopic cell images. LNCS. doi: 10.1007/978-3-540-76300-0_12
4. Theera-Umpon N (2005) White blood cell segmentation and classification in microscopic bone marrow images. LNCS. doi: 10.1007/11540007_98
5. Dorini L, Minetto R, and Leite N (2007) WBC segmentation using morphological operators and scale-space analysis. IEEE XX Braz Symp on Comput Graphics 294–304
6. Kyungsu K, Jeon J, Choi W et al (2001) Automatic cell classification in human's peripheral blood images based on morphological image processing. LNCS. doi 10.1007/3-540-45656-2
7. Paschos G (2001) Perceptually uniform color spaces for color texture. IEEE Trans on Image Processing 932–937
8. Francos J (1993) An unified texture model based on a 2-D Wold-Like decomposition. IEEE Trans on Signal Processing 2665–2678

9. Liu F and Picard R (1999). A Spectral 2D Wold Decomposition Algorithm for Homogeneous Random Fields. IEEE Int Conf on Acoustics, Speech and Signal Processing 3501–3504
10. Li S (2003) Modeling Image Analysis Problems Using Markov Random Fields. Stochastic Processes: Modelling and Simulation. North-Holland 473–513
11. Lopez E, Altamirano L (2008) A method based on tree-structured Markov random field and a texture energy function for classification of remote sensing images. IEEE 5th Int Conf on Electrical Engineering, Computing Science and Automatic Control 540–544
12. Morales B, Olmos I, Gonzalez JA, Altamirano L, Alonso J, and Lobato R (2001) Bone marrow smears digitalization. Private medical images collection concluded on December 2001 at Lab de Especialidades del Inst Mexicano del Seguro Social. Puebla, Mexico

Part V
Experimental Medicine and Analysis Tools

Chapter 36
Ontology-Based Knowledge Discovery in Pharmacogenomics

Adrien Coulet, Malika Smaïl-Tabbone, Amedeo Napoli, and Marie-Dominique Devignes

Abstract One current challenge in biomedicine is to analyze large amounts of complex biological data for extracting domain knowledge. This work holds on the use of knowledge-based techniques such as knowledge discovery (KD) and knowledge representation (KR) in pharmacogenomics, where knowledge units represent genotype–phenotype relationships in the context of a given treatment. An objective is to design knowledge base (KB, here also mentioned as an ontology) and then to use it in the KD process itself. A method is proposed for dealing with two main tasks: (1) building a KB from heterogeneous data related to genotype, phenotype, and treatment, and (2) applying KD techniques on knowledge assertions for extracting genotype–phenotype relationships. An application was carried out on a clinical trial concerned with the variability of drug response to montelukast treatment. Genotype–genotype and genotype–phenotype associations were retrieved together with new associations, allowing the extension of the initial KB. This experiment shows the potential of KR and KD processes, especially for designing KB, checking KB consistency, and reasoning for problem solving.

1 Introduction

Knowledge discovery in databases (or KDD) is aimed at analyzing large amount of data and extracting interesting and reusable knowledge units [7]. The KDD process is based on three main steps: data preparation, data mining, and interpretation of (interesting) extracted patterns (where extracted patterns become knowledge units). This process is usually guided by an analyst who is a domain expert and whose objective is to identify interesting knowledge units with respect to her or his goal.

A. Coulet (✉)
Department of Medicine, Stanford University, Stanford, CA, USA
and
LORIA (CNRS UMR7503, INRIA Nancy Grand-Est, Nancy Université),
Campus scientifique, 54506 Vandoeuvre-lès-Nancy, France
e-mail: adrien.coulet@loria.fr

In pharmacogenomics, interesting knowledge units are given by associations between *drug treatments, genotypes*, and *phenotypic traits*. One problem in KDD is the large amount of patterns extracted from data that cannot be all interpreted by the analyst. One way to select interesting units is to be guided by domain knowledge.

This is precisely the goal of this chapter to show how domain knowledge can be reused in a computer program for guiding the analysis of large amount of data in pharmacogenomics. For that, domain knowledge has to be represented with a knowledge representation (KR) language into a set of formulas that will constitute what is called a domain ontology. Knowledge units within an ontology can be obtained as results of a KDD process. The present research work goes a step further by presenting a KDD process working at two levels: the first level is related to data and data mining, while the second level consists in mining knowledge units themselves, i.e., assertions lying in an ontology. This is why this method is called *role assertion analysis* (RAA). An experiment was carried out on a clinical trial on the variability of the response to *montelukast* treatment. Associations such as *genotype–genotype* and *genotype–phenotype* were retrieved, but new associations too, which have been reused for extending the initial ontology in pharmacogenomics.

This chapter aims at showing the potential of methods such as KDD methods for extracting interesting knowledge units and KR methods for representing knowledge units in ontologies in the context of biological studies.

2 Materials and Methods

2.1 Representing Domain Knowledge in Pharmacogenomics

An ontology was especially designed for representing domain knowledge in pharmacogenomics called SO-Pharm – *Suggested Ontology for Pharmacogenomics* – [4]. SO-Pharm concepts represent pharmacogenomics relationships between a drug, a genomic variation, and a phenotypic trait. It enables as well to represent a patient her or his clinical features, such as a phenotypic trait or the value of a particular polymorphism (SNP).

Indeed, SO-Pharm can be seen as the articulation of several "sub-ontologies", each of them corresponding to a sub-domain related to pharmacogenomics, e.g., genotypes, phenotypes, drugs, and clinical studies.[1] Relationships represent pharmacogenomics knowledge units, such as the ternary association between a genotype item, a phenotype item, and a drug treatment. SO-Pharm is developed in OWL-DL and is available on the OBO Foundry web site[2] and on the NCBO BioPortal.[3]

---

[1] The ontologies in articulation are listed at http://www.stanford.edu/~coulet/ontologies.jpg.
[2] http://www.obofoundry.org/.
[3] http://bioportal.bioontology.org.

2.2 The Studied Dataset

The dataset analyzed in this chapter (see Sect. 4) is taken from a pharmacogenomics study investigating the diversity of responses to *montelukast*, which is a drug used for the treatment of asthma and of symptoms of seasonal allergies. The first results of this study were published in 2006 by Lima et al. [10] and were produced from both genetic and clinical data observed for a subset of 61 patients. The features observed for these patients are 26 SNP genotypes along with two main phenotypic traits. Genotyped SNPs are localized on five different genes involved in the leukotriene pathway: *ABCC1, ALOX5, CYSLTR1, LTA4H,* and *LTC4S,* localized on chromosomes 16, 10, X, 5, and 12, respectively. The phenotype features are as follows:

- The outcome of an asthma exacerbation during 6 months of treatment is marked as "Exa" for *exacerbation* and can take two values, *Yes* or *No*.
- The change in the percentage of predicted "forced expiratory volume in one second" (FEV1) between baseline and after 6 months of treatment is marked as "Per" for *Percent change in predicted FEV1*. "Per" values range in interval $[-0.16; 1.16]$.

Genotype and phenotype data are publicly accessible in two separate files available in the pharmacogenomics database PharmGKB.[4]

2.3 Knowledge Representation Formalism

2.3.1 Description Logics

Description Logics (DL) [1] are the KR formalism used for representing SO-Pharm. A DL ontology consists in a terminological component or TBox, including concept definitions, and an assertional component or ABox, including assertions about individuals, mainly *concept instantiations* and *role instantiations*. For example:

- Patient(pa01) states that pa01 is an instance of the concept Patient(pa01)
- ClinicalItem(exacerbation_yes) states that exacerbation_yes is an instance of the concept ClinicalItem
- hasClinicalItem(pa01, exacerbation_yes) is a role instantiation stating that pa01 is in relation with exacerbation_yes through the role hasClinicalItem

In this chapter, two specific DL constructors are used. First, the fills constructor is a concept constructor filling the range of a role, denoted by (R : b), meaning that given an individual a instance of a concept to which R is attached, then b is related to a through R, where b is said the filler of role R.

---

[4] http://www.pharmgkb.org/do/serve?objId=PA142628130.

Second, the role composition constructor is denoted by R ∘ S for two roles R and S and represents the composition of the two roles: (R ∘ S)(a,c) means that there exists b such that R(a,b) and S(b,c). For example, the set of individuals having a genotype item which is a variant of *LTC4S* gene is denoted by the expression:
hasGenotypeItem ∘ isVariantIn : LTC4S

2.3.2 Linking Data to a DL Ontology

In the following, the KDD process is guided by an ontology through two transformations: the first is applied on initial data for obtaining assertions of concepts and roles, and the second consists in mining role assertions. The first transformation involves the transformation of attribute-value pairs into assertions of concepts and roles, following the methodology of semantic data integration [3]. As in a data mediation approach, heterogeneous data sources are mapped to a reference vocabulary or data model, which here is an ontology. Then, data extraction is processed by *wrappers* and centralized by a mediation system. The set of mappings defined between each data source and the ontology enables the translation of datasets corresponding to query answers into a reference format. The originality of our approach lies in the translation of data into a set of assertions that are used in sequence to populate the ontology. The original details about mapping between data and assertions can be found in Poggi et al. [11]. Here, this is the way how the SO-Pharm ontology was populated with assertions coming from montelukast study. The 61 patients of the study with their phenotypic and genetic data lead to the creation of 61 assertions of the concept Patient, 162 assertions of the concept ClinicalItem and its sub-concepts, and assertions of the role hasClinicalItem (details in Sect. 3).

2.4 Data Mining Method

Association Rules (AR) are probabilistic data dependencies extracted from a dataset, of the form $B_1 \rightarrow B_2$, where the *antecedent* B_1 and the *consequent* B_2 are sets of attributes. The rule expresses that the presence of attributes in B_1 implies the presence of attributes in B_2 with a *support* and a *confidence*. The support counts the frequency of the set of attributes $B_1 \cup B_2$ in the reference set of objects, while the confidence measures the conditional probability of obtaining B_2 knowing B_1.

The number of extracted AR can be large and reduced sets of AR need to be identified. In this work, formal concept analysis (FCA) [8] was used to extract AR and the special set of *reduced minimal non-redundant rules* or RMNR [9] was chosen. RMNR represents a reduced set of rules with a minimal antecedent and a maximum consequent, from which all other rules can be derived. The set RMNR was computed with the CORON platform [13].

3 Role Assertion Analysis

In the following, we make precise the RAA process, where role assertions lying in an ontology are mined for extending this ontology. The process is characterized by two main original features consisting in (1) the mining of knowledge units such as assertions (and not data), and (2) the effective representation of the extracted patterns as knowledge units in an ontology. It is therefore necessary to prepare and transform ontology assertions into a format compliant with the data mining process, and to transform as well the results of the data mining process into knowledge units. The RAA method is based on the complementarity of DL and FCA that was studied in other contexts [2, 12].

3.1 RAA in Three Steps

RAA is a semi-automatic process taking as input a DL ontology $\mathcal{O} = (\mathcal{T}, \mathcal{A})$, a DL concept description denoted by C_0, and a parameter called maximum depth d_{max}, and returning as output a *refined* version of the original ontology, i.e., $\mathcal{O}$ extended with new concept descriptions, new roles, and new role assertions. Figure 1 shows the three main steps of RAA that can be read with respect to the three KDD steps, namely preparation, mining, and interpretation:

1. Exploration of a set of assertions and transformation into a binary table
2. Identification of RMNR from the binary table
3. Interpretation of RMNR in terms of new DL concepts and roles

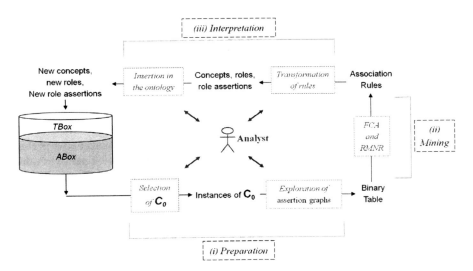

Fig. 1 The different steps of role assertion analysis (RAA). The analyst is a human expert who guides the process

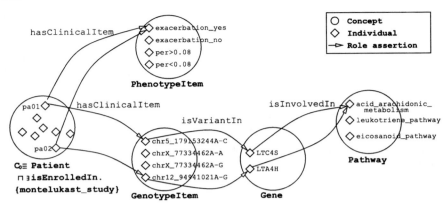

Fig. 2 Assertion graphs of individuals pa01 and pa02 with $d_{max} = 3$

During the first step, the analyst manually defines or selects a concept called C_0 from the ontology $\mathcal{O}$. This allows the analyst to focus on the set of instances of C_0, for analyzing the characteristics of this subset of relevant individuals, and to refine a given part of the ontology. For example, the analyst may be interested in finding regularities in clinical items associated with patients involved in the montelukast study. Then, C_0 is defined as $C_0 \equiv$ Patient ⊓ isEnrolledIn : montelukast_study.

Given C_0, a binary table is built for expressing how instances of C_0 are related to other instances in the ontology (the binary table is a data structure compliant with data mining algorithms). This is done by exploring a so-called assertion graph, i.e., a graph whose vertexes are all instances reached from the instances of C_0 through roles and role compositions. Figure 2 shows the assertion graph of pa01 and pa02.

The exploration process of the assertion graph takes into account the possible cycles and is constrained by a maximal depth of exploration (d_{max} set up by the analyst, e.g., $d_{max} = 3$). A path between two individuals is interpreted as a DL concept where a role is filled, and where this role can be a role composition. For example, the path between pa01 and gene LTC4S is interpreted as pa01 instantiating the concept hasClinicalItem ∘ isVariantIn : LTC4S.

Then all paths are compiled into a binary table Objects × Attributes. A DL role filler expression is transformed into a string used for labeling a new attribute in the table. An instance becomes an object in the table and relations between this object and other objects through role filling are the attributes associated with the reference object. For example, the path hasClinicalItem ∘ isVariantIn : LTC4S gives rise to a line in the table for the object pa01, and to a column of the table for the attribute hasClinicalItem_o_isVariantIn : LTC4S, and the symbol × states this fact at the crossing of the line and the column (see Table 1). This process is done for all considered paths and gives a binary table on which data mining algorithms will be run.

The next step in RAA consists in applying data mining algorithms to the binary table built above to extract AR and more precisely reduced minimal non-redundant

Table 1 An excerpt of a binary table to be mined with two objects and six attributes

Objects \ Attributes	hasClinicalItem: exacerbation_yes	hasClinicalItem: chr5_179153244A-C	hasClinicalItem_o _isVariantIn:LTC4S	hasClinicalItem_o _isVariantIn_o _isInvolvedIn: acid_arachidonic_metabolism	hasClinicalItem: chr12_9494102lA-G	hasClinicalItem_o _isVariantIn:LTA4H	...
pa01	×	×	×	×			
pa02	×			×	×	×	
...							

rules or RMNR (actually the mining process is based on the design of a concept lattice using FCA techniques that cannot be detailed here [5, 6]). An example of RMNR is the following:

{hasClinicalItem : ChrX_77334462A − G} ⟶
  {hasClinicalItem : chrX_77367837A − G,
  hasClinicalItem : Per ≤ 0.08,
  isEnrolledIn_o_isDefinedBy_o_isComposedOf : montelukast_treatment}

The last step in RAA consists in selecting and interpreting some RMNR for defining new concepts and roles and enriching the initial ontology. For example, the above rule will be represented as the following concept expressions:

PatientWithLowChgeInFev ≡ hasClinicalItem: chrX_77334462A − G ⊓
                          hasClinicalItem: chrX_77367837A − G ⊓
                          hasClinicalItem: Per ≤ 0.08 ⊓
    isEnrolledIn ∘ isDefinedBy ∘ isComposedOf : montelukast_treatment

Here, all attributes involved in the rule are selected for defining the new concept which represents patients from the montelukast study showing low changes in FEV1 and two genotypes on the chromosome X (note that the new concept has been named by the analyst). Once a new concept, say C_{new}, is defined and named by the analyst, C_{new} is inserted in the ontology under the control of the *classifier* associated with the ontology. For example, the SO-Pharm classifier will infer the following subsumption relation: PatientWithLowChgeInFev ⊑ Patient, meaning that the first concept is less general than the second.

In addition, selected attributes lead to the creation of new roles. For example, in C_{new}, the relation between the individuals chrX_77334462A − G and montelukast_treatment yields the new role assertion (if it does not already exist): interactsWith(chrX_77334462A − G, montelukast_treatment).

4 Results and Discussion

4.1 Implicit or Hidden Knowledge

With high *minimum support* and *confidence*, the RAA process reveals general characteristics of the dataset which are usually known by the analyst but which are not necessarily formalized in the ontology. For example, the rules extracted with $d_{\max} = 3$, $support = 0.8$, and $confidence = 0.8$ indicate that patients involved in the study have been genotyped for five different genes, namely *ABCC1*, *ALOX5*, *CYSLTR1*, *LTA4H*, and *LTC4S*. Another extracted property was that every patient was white, suggesting some unknown characteristic of the recruitment procedure.

When minimum support and confidence are set to lower values, the number of rules increases and more pertinent associations can be found. For example, *genotype–genotype* associations are relevant for characterizing sub-populations of the study in terms of co-segregated genotypes. In Lima et al., three groups of genotypes associated with *linkage disequilibrium* (LD) are reported. By contrast, the RAA method yields seven groups of genotypes from which three partially correspond to the above reported groups, and two are associated with the *CYSLTR1* gene that has no corresponding group in the previous study.

4.2 Extraction and Classification of New Pharmacogenomics Knowledge Units

Relevant pharmacogenomics knowledge units express associations between genotype items, phenotype items, and treatments. In the present study, the drug treatment is the same for all patients. Therefore, the search for knowledge units can be reduced to the search for *genotype–phenotype* associations by filtering rules that contain at least one genotype and one phenotype attributes.

Table 2 summarizes the associations found in the original study and those found by RAA. Central column of the table shows the results reported by Lima et al. on the basis of χ^2 and *likelihood-ratio* statistical tests. Five genotypes (1–5) were found associated with three different phenotypes. The right column shows results obtained with RAA. From five isolated rules, four single genotypes (3, 4, 6, 8) and one pair of genotypes (7.1 and 7.2) are found associated with three different phenotypes.

Thus, RAA detected two of the five genotypes initially associated with a particular phenotype (3 and 4). Three other genotypes (1, 2, and 5) observed in Lima et al. were not found with RAA because they concern small subgroups of the panel.

By contrast, RAA enables the identification of four new genotypes (6, 7, and 8) associated with two of the four phenotypes considered in the study. Once validated by the analyst, such associations can be considered as new knowledge units and can be inserted in the ontology. In the present case, an example of new concept is PatientsWithLowChangeInFEV1 already seen above (i.e. defining the

Table 2 Genotypes specifically associated with the phenotypes listed in the left column. Central column shows specific genotypes reported in Lima et al. using statistical methods (χ^2 and $likelihood - ratio$) [10]. Right column shows the associated genomic variations found by role assertions analysis

Phenotype	Specific genotypes Lima et al. (χ^2, $likelihood - ratio$)		Specific genotypes Role assertion analysis	
Per="≥ 0.09"	Chr10:45221095G-G	(1)	∅	
	Chr16:15994335C-T	(2)		
Per="≤ 0.08"	∅		Chr10:45211490A-A	(6)
			ChrX:77334462A-G	(7.1)
			ChrX:77367837A-G	(7.2)
Exa="No"	Chr5:179153244A-C	(3)	Chr5:179153244A-C	(3)
			Chr16:161443440C-G	(8)
Exa="Yes"	Chr12:94941021A-G	(4)	Chr12:94941021A-G	(4)
	Chr12:94941021G-G	(5)		

class of patients enrolled in the montelukast study displaying both the phenotype Per $=\leq 0.08$ and the genotype (7) of Table 2). An example of new role assertion is given by:
    interactsWithDrugTreatment(chrX_77334462A − G, montelukast_treatment)

5 Conclusion

RAA is an original approach for KD allowing to analyze data and to extend a domain ontology with the analysis results. RAA involves various knowledge technologies such as DL and FCA formalisms. The application of RAA to pharmacogenomics led to the extraction of new knowledge units either as new concept definitions or new role assertions. Here, new unit has to be understood as not explicitly lying in the initial ontology and pertinent because validated by the analyst. Such units, as part of an ontology, can be shared, disseminated, and reused for various knowledge-base tasks. The experiment on montelukast study shows the influence of several parameters of the RAA process, such as maximal depth parameter, minimum support, and confidence, on the size of the resulting binary table, on the number of extracted rules, and on the efficiency of the whole discovery process. RAA can be extended with alternative mining methods such as extensions of FCA, new algorithms for mining itemsets and rules, and relational learning methods.

RAA is a KD process based on ontologies and can be seen as a symbolic and complementary method for statistics and data analysis in clinical trials. RAA provides a formal and explicit representation of pharmacogenomics domain to be used for organizing the concepts of the domain and for solving problems through reasoning capabilities. This is probably a new way of thinking at biological problems from a computational point of view that is worth investigating more deeply.

Acknowledgements We would like to thank Dr. Pascale Benlian for her help in the interpretation of RAA results.

References

1. F. Baader, D. Calvanese, D.L. McGuinness, D. Nardi, and P.F. Patel-Schneider, editors. In *The Description Logic Handbook: Theory, Implementation, and Applications*. Cambridge University Press, Cambridge, 2003.
2. F. Baader, B. Ganter, B. Sertkaya, and U. Sattler. Completing description logic knowledge bases using formal concept analysis. In M.M. Veloso, editor, *IJCAI*, pages 230–235, 2007.
3. D. Calvanese, G. De Giacomo, M. Lenzerini, D. Nardi, and R. Rosati. Description logic framework for information integration. In *Proceedings of KR 98*, pages 2–13, 1998.
4. A. Coulet, M. Smaïl-Tabbone, A. Napoli, and M.D. Devignes. Suggested Ontology for Pharmacogenomics (SO-Pharm): Modular construction and preliminary testing. In *Proceedings of the International Workshop on Knowledge Systems in Bioinformatics - KSinBIT'06*, volume 4277 of LNCS, pages 648–657, 2006.
5. A. Coulet, M. Smaïl-Tabbone, A. Napoli, and M.D. Devignes. Ontology refinement through Role assertion analysis: Example in pharmacogenomics. In F. Baader, C. Lutz, and B. Motik, editors. *Proceedings of the 21st International Workshop on Description Logics (DL 2008)*, volume 353 of *CEUR Workshop Proceedings*, 2008.
6. A. Coulet. *Construction and Use of a Pharmacogenomics Knowledge Base for Data Integration and Knowledge Discovery*. Thèse en Informatique, Université Henri Poincaré (Nancy 1), France, Oct. 2008 (in French).
7. W.J. Frawley, G. Piatetsky-Shapiro, and C.J. Matheus. Knowledge discovery in databases: An overview. In *Knowledge Discovery in Databases*. The AI Magazine, 13:1–30, 1992.
8. B. Ganter and R. Wille. *Formal Concept Analysis: Mathematical Foundations*. Springer, Berlin, 1999.
9. M. Kryszkiewicz. Concise representations of association rules. In *Proceedings of the ESF Exploratory Workshop on Pattern Detection and Discovery*, pages 92–109, London, UK, 2002. Springer, Berlin, 2002.
10. J.J. Lima, S. Zhang, A. Grant, L. Shao, K.G. Tantisira, H. Allayee, J. Wang, J. Sylvester, J. Holbrook, R. Wise, S.T. Weiss, and K. Barnes. Influence of leukotriene pathway polymorphisms on response to montelukast in asthma. *Am. J. Respir. Crit. Care Med.*, 173(4):379–385, 2006.
11. A. Poggi, D. Lembo, D. Calvanese, G. De Giacomo, M. Lenzerini, and R. Rosati. Linking data to ontologies. *J. Data Semant.*, 10:133–173, 2008.
12. S. Rudolph. *Relational Exploration: Combining Description Logics and Formal Concept Analysis for Knowledge Specification*. Thesis in Computer Science, Technischen Universität – Dresden, Germany, Dec. 2006.
13. L. Szathmary. *Symbolic Data Mining Methods with the Coron Platform*. Thèse en Informatique, Université Henri Poincaré (Nancy 1), France, Nov. 2006.

Chapter 37
Enabling Heterogeneous Data Integration and Biomedical Event Prediction Through ICT: The Test Case of Cancer Reoccurrence

Marco Picone, Sebastian Steger, Konstantinos Exarchos, Marco De Fazio, Yorgos Goletsis, Dimitrios I. Fotiadis, Elena Martinelli, and Diego Ardigò

Abstract Early prediction of cancer reoccurrence constitutes a challenge for oncologists and surgeons. This chapter describes one ongoing experience, the EU-Project NeoMark, where scientists from different medical and biology research fields joined efforts with Information Technology experts to identify methods and algorithms that are able to early predict the reoccurrence risk for one of the most devastating tumors, the oral cavity squamous cell carcinoma (OSCC). The challenge of NeoMark is to develop algorithms able to identify a "signature" or bio-profile of the disease, by integrating multiscale and multivariate data from medical images, genomic profile from tissue and circulating cells RNA, and other medical parameters collected from patients before and after treatment. A limited number of relevant biomarkers will be identified and used in a real-time PCR device for early detection of disease reoccurrence.

Keywords Cancer informatics · Experimental medicine and analysis tools · Image processing in medicine and biological sciences · Biological data mining and knowledge discovery · Biological data integration and visualization

1 Introduction

Malignant neoplasms are – as a whole – the second cause of death in the western world. The main mechanism causing progressive inability and death in cancer patients is represented by the occurrence of loco-regional relapses and distant metastases whose incidence remains still high despite the recent development of several effective treatments. Reoccurrence is mainly due to the persistence of tumor cells after treatment without any clinical, laboratory, and imaging evidence of residual disease. The identification of new, reliable biomarkers of disease, discriminating

M. Picone (✉)
MultiMed s.r.l, Cremona, Italy
e-mail: marco.picone@multi-med.it

which patients are at highest risk of relapses, is therefore of primary importance in cancer research to allow focus follow-up efforts and limit adjuvant chemotherapy only to high-risk patients. In addition to a better risk prediction, the identification of biomarkers for early diagnosis of relapses would have the potentiality to improve patient's survival. Oral squamous cell carcinoma (OSCC) represents about 5% of all cancers and provides a prototypical example of this issue. OSCC has a reoccurrence rate of about 25–50% over a period of 5 years, leading to severe consequences on physical appearance and ability to eat and speak, invasive and disabling surgical interventions, and death [1]. A strict follow-up is usually undertaken, and additional treatments are often planned to reduce the risk of reoccurrence even in the presence of disease remission (an approach called "adjuvant"). Adjuvant chemo- and radiotherapy treatments have important side effects, and a large group of patients would probably not require them to decrease the risk of reoccurrence. However, it is currently almost impossible to identify the high-risk subjects to be candidated to an aggressive treatment.

In this chapter, we describe the prototypical test case of the NeoMark project aiming to provide IT support to candidate biomarker identification from clinical, imaging, and molecular biology data. In NeoMark, we investigate an innovative strategy to identify relevant biomarkers of cancer reoccurrence risk and presence, integrating high-throughput gene expression analysis in tumor and blood cells, and IT-assisted imaging with traditional staging and follow-up protocols, to improve the stratification of reoccurrence risk and the earlier identification of loco-regional relapses. The idea behind NeoMark is that by analyzing a sufficient set of different types of data (clinical, biomedical, genomic, histological, from digital imaging, from surgery evidence, etc.) of patients affected by OSCC before treatment and at the time of remission, a set of relevant biomarkers appearing only in the presence of the disease might be identified. The reoccurrence of the same biomarker phenotype during post-remission follow-up may precede the clinical manifestation of the relapse thus allowing earlier intervention.

To pursue this ambitious aim, we need to collect, store, integrate, and analyze heterogeneous clinical and biomedical data, merge health care and data mining, and introduce molecular biology into clinical practice. The NeoMark integrated platform addresses these issues and enables the medical doctor jointly analyze, revise, and exploit traditional clinical data in conjunction with image analysis tools and gene expression data in better supported clinical decisions during the follow-up of OSCC patients in remission phase. The work presented in this chapter is co-funded by the European Union under the 7th Framework Program, Information and Communication Technologies (EU-FP7-ICT-2007-2-22483-NeoMark).

2 System Description

The versatile user requirements and especially the integration of heterogeneous input data required a careful design of the NeoMark system. Our goal was to integrate as much functionality as possible in a single unified service-oriented

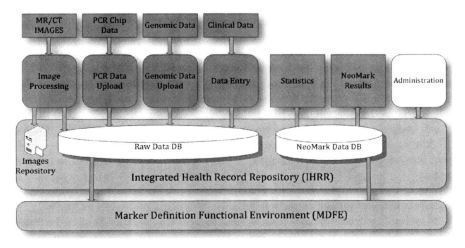

Fig. 1 NeoMark system overview

system, achieving great flexibility and usability. These properties increase the user acceptance and may decrease human error. The basic scheme of the implemented system can be seen in Fig. 1. Most of the user interaction is done via the web interface. The physician can manage patients, enter clinical data, and view all features and the NeoMark results. The clinician can upload genomic data and researchers can view anonymous statistics, which could serve as a base for future research on oral cancer. Furthermore, entering imaging data and uploading PCR chip data can be performed using standalone tools (see Sects. 2.1 and 2.3 for details). Heterogeneous NeoMark data (general information about the patient, Clinical, filtered, and cleaned Genomic and Imaging data as well as OSCC reoccurrence prediction, patient-specific, and disease-specific risk factors) are stored in a single database – the Integrated Health Record Repository (IHRR) – on a central NeoMark server. The participating hospitals are connected to the very same NeoMark server, allowing the data-driven training algorithm to incorporate patient data from all participating hospitals at the same time and enabling users to perform inter-hospital comparisons of patient data.

The NeoMark server contains an Apache-based web server, a MySQL-based database, an FTP server for the storage of medical images, and the *marker definition functional environment (MDFE)*, a data analysis module which is the core of the system. Based on the heterogeneous input data, this module estimates the likelihood of a relapse and identifies OSCC risk factors (for details, see Sect. 3).

In order to protect the patient's right of privacy, none of the data (e.g., the name) that allows to uniquely identify the patient is stored on the central NeoMark server. Only a unique NeoMark ID can identify a patient. Those IDs are connected to the patients via individual databases that are located within each hospital. A JAVA-based tool – *the sensitive data tool* – that can be started directly from the web interface provides a user interface which allows to manage that kind of patient data.

2.1 Image Processing and Image Feature Extraction

Medical images cannot directly be analyzed by the NeoMark MDFE because the amount of information is too high while not being meaningful due to the lack of semantic enrichment. An experienced radiologist uses the *image processing tool* to extract meaningful numeric features of tumors and suspicious lymph nodes from CT, CT with contrast, MR T1 TSE, and MR T2 TSE images.

All images are acquired before treatment and then every 6 month during follow-up. The high resolution (1 mm slice thickness) CT images cover the entire head and neck region, whereas the MR images only cover the tumors and significant lymph nodes.

The further processing steps [2] consist of image registration, tumor/lymph node segmentation, and feature extraction (FE).

In order to extract features from several images jointly, they have to be transformed to a common coordinate system. This process, called image registration, has to be performed on each of the images. Due to the rigidity of the human skull, it is sufficient to only allow translations and rotations (rigid registration). However, to achieve good registration results having only those six degrees of freedom, the position of the head relative to the spine has to be as similar as possible in all images. A coarse prealignment is performed manually by the radiologist. Then the registration is automatically refined by maximizing a suitable similarity measurement. For images from different modalitites (MR/CT), mutual information [3,4] has proven to achieve good results.

In the next step, the regions of interest (ROIs, here tumors and suspicious lymph nodes) need to be segmented. (Semi-) Automated lymph node segmentation has been studied exhaustively [5–8]. The image processing tool uses a mass spring model [9] to segment lymph nodes. This approach balances internal forces (i.e., forces that preserve the shape) and external forces (i.e., forces that push the model toward the lymph node borders) that are applied to an initial spheric lymph node model that has been placed in the center of a lymph node by the user.

Tumors in the oral cavity, however, cannot be segmented automatically with sufficient accuracy because of the high variance in shape and appearance and their similarity to surrounding tissue, which requires the knowledge of an experienced radiologist. Therefore, a user-friendly graphical user interface allows the radiologist to delineate the tumor in each image slice.

Once the boundaries of the ROIs are known, geometric and texture-based features can be extracted. A selection of them is:

- *The volume.* It can be easily computed by simply counting all voxels in the ROI and multiplying that by the cuboid volume of one voxel.
- *The axes.* Assuming that the ROI is convex, the axes of ROI coincide with axes of the enclosing minimal volume bounding box. An approximation of that box can be computed by performing a main component analysis on the points of the surface of the ROI [10]. The directions of the axes are then given by the eigenvectors. This approach for computing the axes is rotation invariant and not affected

by subjectivity in contrast to current clinical practice, where the radiologist delineates the axes manually in the axial plane only.
- *The contrast take up rate.* This texture-based feature is computed by measuring the mean squared error of the CT and the CT with contrast image. This feature, as well as the water content, is extremely sensitive to accurate registration because it directly compares voxel values in different images.
- *The water content.* Water appears bright in T2 MR images and dark in T1 MR images. Therefore, a measure of the water content is the difference between T2 voxels and the average T2 voxel multiplied with the difference between the average T1 voxel and the T1 voxels.

In addition to the automated extracted features, the radiologist enters further properties like the location and the amount of infiltration of the surrounding tissue. This enables a semiautomatic staging [11] of the tumor.

2.2 Genomic Data Cleaning and Filtering

Gene expression data are usually coming from FE files. A FE file is usually a tab-delimited text file that contains all the data extracted from the experiment by the FE software. The file contains the Log2-ratio value as well as (very) raw intensity data, background information, meta-data on the experiment and on the scanning settings, annotation data to identify genes, etc. In order to integrate into a study database, the expression values of each sample should be extracted from FE file and uploaded into the data matrix after being assigned to the correct patient or sample. The relevant information that is stored in the database are Feature Name, Probe Name Gene Name, Systematic Name, Description, and Log2-ratio. For this reason, we have a specific tool that analyzes *control features, duplicate features, filtering of genes based on low data quality,* and *filtering of genes with high number of missing values* and generates as output a cleaned file with a small dimension that contains only relevant information. This cleaned file can be uploaded from a specific page of NeoMark Web Application into the database. For the genomic data, there are two kinds of tables. The first one contains the cleaned relevant information with the Log2-ratio value and in addition also the normalized one calculated with the information that is already stored in the database. The other table is designed to store genomic data expression following the group ontology.

2.3 PCR Tool

A novel qRT-PCR platform is under development in STMicroelectronics to obtain quantitative information about the PCR amplification of the targeted genes. It is a portable, real-time, integrated analytical system based on qRT-PCR performed in an array of silicon microchambers. The small size of the components and its low power requirements make this system an ideal candidate for further miniaturization into a hand-held, point-of-care device. The qRT-PCR lab-on-chip is disposable and relatively inexpensive to make this method of analysis economically viable. The excellent thermal conductivity of silicon makes it ideal in applications requiring rapid cycles of heating and cooling. The silicon core of the qRT-PCR microchip is fabricated using photolithographic techniques: heaters are fabricated directly onto the surface of the chip, along with the thermal sensors monitoring the temperature and providing feedback to the temperature controller, while cooling is achieved via forced air using a fan. Designs for the qRT-PCR chips range from a single reaction chamber to arrays of microchambers of varying sizes and depths for multiple simultaneous reactions. The PCR reaction dynamics can be monitored locally in real-time, using a dedicated optical system recording the fluorescence intensity at each thermal cycle. The real-time PCR portable device analyzes a set of predefined genes (up to 20) and reports their expression value in relation with a housekeeping gene. In the long-term approach, this system could be the low cost alternative to the microarray devices. The purpose of the system is to speed up and lower the cost for the gene expression analysis for a particular set of genes of particular relevance in the oncology of oral cancer. During a first evaluation phase, the returned values of gene expression are correlated to data reported from the microarray tool. Results must be comparable in terms of returned value for gene expression, which is usually correlated to a reference gene of known expression. In the final release, however, the gene expression values could be directly sent to the genomic data repository system of the Neomark Database. The upload is integrated in the instruments' GUI, and reports to the Neomark Database. Any cleaning and filtering of RAW data are done in the system GUI, and uploaded data will be clean and with only significant values. Up to now significant results have been achieved in successfully amplifying both DNA and RNA, with quantitative results comparable with the one from the same sample amplified on a standard lab thermal cycler.

3 Data Analysis

The analysis of the heterogeneous data constitutes the cornerstone of the NeoMark artificial intelligence component. The aim of this component is twofold: (1) to assess the risk of reoccurrence in the very early stages of treatment, i.e., as soon as the patient reaches remission, and (2) to efficiently and effectively model the disease evolution during the whole follow-up period based on a multitude of heterogeneous data, thus monitoring the patient's therapeutic progression. As described in the clinical scenario of the NeoMark project, for each patient who has been diagnosed with oral cancer a wide range of heterogeneous data is collected and analyzed. Specifically, due to the complex nature of the disease, a holistic approach is performed which integrates a great multitude of clinical, imaging, and genomic data to "frame" every possible aspect related to the onset and progression of oral cancer.

3.1 Early Risk Assessment

For this purpose, an initial "snapshot" of all the patient's attributes is acquired which consists of clinical, imaging, and gene expression data. Later, certain basic preprocessing steps (i.e., outlier detection and missing values handling) are used which aim to ameliorate the quality of the available input set. Next, the wrapper feature selection algorithm [12] is used to discard redundant and non-informative features and maintain the most discriminatory ones. The reduced set of attributes, maintained after the feature selection, is then provided as input to a classification algorithm; in the final integrated NeoMark server, the user may choose among a selection of algorithms which have been trained carefully and yield the best results [13]. The objective of this classification is to stratify the patients into two classes, i.e., remittents and non-remittents, based only on their initial clinical profile.

3.2 Disease Evolution Monitoring

In the present study, we use Dynamic Bayesian Networks to early identify potential relapses of the disease, during the follow-up. As described in the clinical scenario, a snapshot of the patient's medical condition is acquired during every predefined follow-up with the doctor. By exploiting the information of history snapshots, we aim to model the progression of the disease in the future. The proposed prognostic model is based on DBNs, which are temporal extensions of Bayesian Networks (BNs) [14]. A BN can be described as where is a directed acyclic graph, where the nodes correspond to a set of random variables $X = \{x_1, x_2, \ldots, x_N\}$, and P is a joint probability distribution of variables in X, which factorizes as:

$$P(X) = \prod_{i=1}^{N} P[x_i | \pi_G(x_i)] \qquad (1)$$

where $\pi_G(x)$ denotes the parents in G. A DBN can be defined as a pair $DB = (B_0, B_{\text{trans}})$ where B_0 is a BN, defining the prior $P(X_0)$, and B_{trans} is a two-slice temporal BN (2TBN) which defines $P(X_t | X_{t-1})$. The semantics of a DBN can be defined by "unrolling" the 2TBN until we have T time-slices. The resulting joint distribution is given by:

$$P(X_1, X_2, \ldots, X_T) = \prod_{i=1}^{T} \prod_{i=1}^{N} P[x_i^t | \pi_G(x_i^t)] \qquad (2)$$

In order to build a model that successfully evaluates the current state or predicts a state in the future (next time slice), we need to finetune both the intra- and the inter-slice dependencies of the DBN network, using both expert knowledge as a prior model and experimental data to get a more accurate posterior model. After the training procedure, we are able to conjecture about the probability of any variable for every time slice, including of course the probability for reoccurrence.

4 Conclusion

We have presented a novel ICT-enabled cancer reoccurrence prediction method and have described the system implementing this idea. In addition to the great innovation of collecting and jointly interpreting such an enormous amount of heterogeneous data, the development of the NeoMark system led to further innovations:

- The data analysis component predicts not only the probability of a relapse overall but also the probability at a given time. All predictions are updated upon retrieval of follow-up input data.
- For the first time genomic data obtained from a PCR chip will eventually replace the expensive and complex laboratory-based genomic data extraction.
- The innovative semiautomatic multimodal image FE algorithms extract imaging features of tumors and lymph nodes that are well suited for further processing by the data analysis component due to their numeric manner and robustness.

Currently the system is in a late stage of development, and the first patient's data are about to be entered. Once enough NeoMark data are collected, the system can be trained and first results can be obtained and evaluated. We believe not only to be able to predict the reoccurrence of OSCC but also to obtain yet unknown risk factors of the disease.

In any case, the system will provide a large database customized to oral cancer which enables to perform a variety of clinical studies. Furthermore, the system may support the radiologist in diagnosing by providing a semantically enriched image database.

References

1. B. Boyle, P. Levin, editor. *World Cancer Report*. International Agency for Research on Cancer, 2008.
2. S. Steger, M. Erdt, G. Chiari, and G. Sakas. Feature extraction from medical images for an oral cancer reoccurrence prediction environment. In *World Congress on Medical Physics and Biomedical Engineering*, 2009.
3. F. Maes, A. Collignon, D. Vandermeulen, G. Marchal, and P. Suetens. Multimodality image registration by maximization of mutual information. *IEEE Trans Med Imaging*, 16(2):187–198, 1997.
4. W. M. Wells, P. Viola, H. Atsumi, S. Nakajima, and R. Kikinis. Multi-modal volume registration by maximization of mutual information. *Med Image Anal*, 1(1):35–51, Mar 1996.
5. D. M. Honea, G. Yaorong, W. E. Snyder, P. F. Hemler, and D. J. Vining. Lymph node segmentation using active contours. volume 3034, pages 265–273. SPIE, 1997.

6. D. Maleike, M. Fabel, R. Tetzlaff, H. von Tengg-Kobligk, T. Heimann, H-P. Meinzer, and I. Wolf. Lymph node segmentation on CT images by a shape model guided deformable surface method. volume 6914, page 69141S. SPIE, 2008.
7. J. Rogowska, K. Batchelder, G. S. Gazelle, E. F. Halpern, W. Connor, and G. L. Wolf. Evaluation of selected two-dimensional segmentation techniques for computed tomography quantitation of lymph nodes. *Invest Radiol*, 31(3):138–145, Mar 1996.
8. G. Unal, G. Slabaugh, A. Ess, A. Yezzi, T. Fang, J. Tyan, M. Requardt, R. Krieg, R. Seethamraju, M. Harisinghani, and R. Weissleder. Semi-automatic lymph node segmentation in In-mri. In *Proc. IEEE International Conference on Image Processing*, pages 77–80, 8–11 Oct. 2006.
9. J. Dornheim, H. Seim, B. Preim, I. Hertel, and G. Strauss. Segmentation of neck lymph nodes in CT datasets with stable 3d mass-spring models segmentation of neck lymph nodes. *Acad Radiol*, 14(11):1389–1399, Nov 2007.
10. G. Barequet and S. Har-peled. Efficiently approximating the minimum-volume bounding box of a point set in three dimensions. *J Algorithms*, 38:82–91.
11. R. V. P. Hutter, M. Klimpfinger, L. H. Sobin, C. Wittekind, F. L. Greene, editor. *TNM Atlas*. Springer, Berlin, 5th edition, 2007.
12. G. H. John and R. Kohavi. Wrappers for feature subset selection. *Artif Intell*, 97:273–324, 1997.
13. V. Kumar, P.-N. Tan, M. Steinbach. Introduction to data mining. Pearson Addison Wesley, Boston, 1st edition, 2006.
14. K. P. Murphy. Dynamic bayesian netoworks: Representation, inference and learning. University of California, 2002.

Chapter 38
Complexity and High-End Computing in Biology and Medicine

Dimitri Perrin

Abstract Biomedical systems involve a large number of entities and intricate interactions between these. Their direct analysis is, therefore, difficult, and it is often necessary to rely on computational models. These models require significant resources and parallel computing solutions. These approaches are particularly suited, given parallel aspects in the nature of biomedical systems. Model hybridisation also permits the integration and simultaneous study of multiple aspects and scales of these systems, thus providing an efficient platform for multidisciplinary research.

Keywords High-performance computing · Computational biology and medicine · Complexity

1 Introduction

Biomedical systems are typically complex, with a large number of intricate entities, involved in non-trivial interactions, at all scales from cell level and interactions between molecules, to population level and social interactions between individuals. This poses difficulties with their formal modelling and definition. Using the right level of description to investigate the phenomena of interest is crucial [13].

While the source of complexity may differ between systems, it is often possible to identify some parallel aspects in system behaviour or in the modelling techniques used to investigate it. This permits efficient use of high-performance architectures, which is facilitated by the wider availability of these.

In this chapter, we identify the complex nature of biomedical systems and categorise their parallel aspects. We also review how several systems have been successfully investigated using parallel computing, before discussing hybrid modelling alternatives for instances where scaling up alone is not efficient.

D. Perrin (✉)
Centre for Scientific Computing & Complex Systems Modelling, Dublin City University, Glasnevin, Dublin 9, Ireland
e-mail: dperrin@computing.dcu.ie

H.R. Arabnia and Q.-N. Tran (eds.), *Software Tools and Algorithms for Biological Systems*, Advances in Experimental Medicine and Biology 696,
DOI 10.1007/978-1-4419-7046-6_38, © Springer Science+Business Media, LLC 2011

2 Biomedical Systems are Typically Complex

2.1 Structural Complexity

A first element of complexity in biological systems results from their components and the way these are structurally organised, e.g. for nucleosome impact on gene regulation (one of the main interests in the field of Epigenetics). Histones, the proteins that form nucleosomes, can be modified by the addition of different functional groups to their tails, and these changes can alter gene expression [1].

Possible changes interact with each other and may have conflicting impacts on gene expression [20]. The number of possible states therefore goes from a few dozens for histone H2A to several millions for histone H3. One gene involves hundreds of nucleosomes, and each nucleosome includes two histones of each type. It is thus clear that its epigenetic evolution is a system where complexity is a direct consequence of the structural organisation and the large number of configurations.

2.2 Scale Complexity

Neuronal functions are typically complex but each neuron can be seen as a relatively simple unit: they integrate input signals to form an output potential, propagated to neighbouring neurons. In that context, the overall function and complexity is a consequence of the number of neurons, estimated at approximately 100 billion in the human brain [33]. This has been extensively used in the field of artificial intelligence, with the development of artificial neural networks [10]. Here, complexity is scale based and independent of the structure of individual system units.

2.3 Adaptive Systems

In a number of biological systems, scale and structure combine and lead to complexity, as can be observed for the immune system. Not only does it involve a large number of cells and various organs (all of which structurally complex), but the interactions between these can also evolve over time, through memory of past immune responses. In that sense, it is both complex and adaptive. Key principles of complex adaptive systems are *emergence*[1] and *self-organisation*.[2]

Given such properties, these systems are difficult to describe fully. In particular, the contribution and importance of low-level unsupervised interactions to the overall evolution process are far from trivial. Consequently, several modelling approaches have been proposed, and many relied on advances in high-performance computing.

---

[1] Patterns of system evolution arising from an abundance of simple, low-level, interactions [7].

[2] Increased complexity obtained without intervention from an outside source [12].

3 Parallelism and Large-Scale Computing

A list of the world's fastest computers, the *TOP500* project [24], highlights hardware advances. Recent notebooks and entry-level supercomputers from late-1990s lists have similar performances. Furthermore, while a single teraflop/s computer ranked first until 2000 (Intel's ASCI Red), current top-ranking systems were benchmarked over 1 petaflop/s (i.e. three orders of magnitude faster).

Costs for top-ranking systems are stable,[3] and commodity clusters are ubiquitous [19]. As a result, use of parallel computing has become more widespread. This trend is also facilitated by the latest developments on software dedicated to parallel computing and the suitability of parallel computing to complex systems modelling.

3.1 Parallelism in Complex Systems

In the context of complex systems, parallelism can be observed as part of either the system itself or the techniques used. This is listed here in decreasing order for feasibility and efficiency of parallelisation (see e.g. [14] for details).

Embarrassingly parallel problems can be broken down into subparts, each completely independent of the others. Given the interactions found in complex systems, these are rarely fully parallel themselves, but some techniques, e.g. Monte Carlo simulations, have such properties [16]. *Regular and synchronous problems* require that the same sequence of instructions (regular algorithm) is applied to all data with synchronous communications, and with each processor finishing its task at the same time. Technical examples include approaches involving Fast Fourier transforms (synchronous), matrix vector products and sorting (loosely synchronous). The immune system can also, under certain conditions, be considered in this category (Sect. 4.2). *Irregular and/or asynchronous problems* are characterised by modelling approaches in which irregular algorithms cannot be implemented efficiently, involving message passing and high communication overhead. Any moving boundary simulation typically falls into this category.

Most complex systems require, at some point in their analysis or modelling, the use of high-end computing. To facilitate this, a number of tools have been developed.

3.2 Dedicated Software and Libraries

Software package R is often used in the analysis of biological systems [11], but is limited by the requirements of such analysis for very large systems. Parallel

---

[3] IBM's ASCI White (fastest computer from 11/2000 to 06/2002) cost $110 million. IBM's Roadrunner (ranked first since 06/2008) $130 million, but delivered 200 times the computing power.

frameworks have, therefore, been developed. In particular, Simple Parallel R INTerface (SPRINT) does not require that users master parallel programming paradigms [17].

If such approaches are common for data analysis, modelling often requires that parallel applications are developed ab initio. In that context, message-passing interface (MPI) has become a de facto standard, available on most current distributed-memory computers. MPI is a communications protocol providing support for point-to-point message passing and collective operations, letting developers specify workload partition and application behaviour on each of the processes used [14, 15]. OpenMP can be seen as an equivalent for shared-memory architectures [4].

The main advantages of MPI are high performance and scalability, as well as portability. The main drawback, however, is that MPI requires a detailed, low-level implementation. To address this, new parallel languages are proposed. Chapel has a multiresolution approach, to combine the benefits of high- and low-level languages [3]. Programmers can use abstract code and global-view data aggregates, as well as specify low-level behaviour. While still under development, it represents a very interesting prospect in terms of simplifying the implementation of parallel programs.

4 High-Performance Computing for Biomedical Systems

In recent years, a number of studies have focused on the application of large-scale, parallel resources to the investigation of complex biological and medical systems, where their nature provided a basis for efficient application of such resources.

4.1 Biological Data Mining

Collection and analysis of large datasets have acquired a crucial role in biology, in particular, for studies investigating genes and their function. Expressed sequence tags (EST) (sub-sequences of transcribed cDNA sequences, obtained through sequencing of cloned mRNA), are particularly useful for gene discovery and functional annotation of putative gene products [26]. EST databases are becoming increasingly large, with millions of entries. To facilitate analysis of such datasets, a number of parallel tools have been implemented, e.g. ParPEST [8].

Single-nucleotide polymorphisms are also commonly studied and correspond to DNA sequence variations, limited to a single nucleotide between members of the same species. These variations can be substitutions, deletions or insertions, and non-synonymous SNPs are possible contributors to complex disease traits [34]. Small-scale genome-wide association studies do not create problems for serial tools. Larger studies require parallel computing and tools such as EPISNPmpi [23].

Another common technology, microarray analysis, is used for large-scale transcriptional profiling, through measurement of expression levels of thousands of genes at the same time, and grouping them based on their expression under multiple conditions time points (biclustering). Several methods have been developed, including some based on classic optimisation techniques such as genetic algorithms. A well-known limitation of this technique, however, is that a large population of sequences is required [18]. Several parallelisation strategies can be used to address this [2], and it is possible to implement a parallel genetic algorithm which significantly outperforms its serial equivalent on large microarray datasets [9].

4.2 Immune System Models

As mentioned above, the immune system is typically complex. It involves a large number of cells, of various types, for which the population size dynamically evolves. Additionally, the interactions between these cells are non-trivial and may be altered over time. As a result, direct analysis through in vivo or in vitro experiments is sometimes limited, and a number of mathematical and computational models have been developed. Because of the complexity of the overall system, these models are often limited to a subset of the immune system, e.g. the innate immune response [32], or a specific context, such as HIV infection.

In the context of HIV research, cellular automata have been developed (e.g. [29]) and are structurally suited to parallelisation. They have been optimised for cluster architectures, and communication overload was identified as the main limitation [16].

Agent-based approaches, reputedly resource-consuming, also require efficient parallelisation. This has been investigated in detail in the context of an immune model based on the lymph network structure [28]. In such an approach, the system consists of a collection of lymph nodes (where the immune response takes place), which are connected through a network (that cells and virions use to move from one node to another). Consequently, each node can be considered as an independent defense unit, and the model is regular and synchronous. Optimised implementation permits simulations a thousand lymph nodes and more than one billion immune cells. This scale and level of detail enabled the inclusion of localised features of the immune system, such as the gastrointestinal tract.

5 Challenges and Perspectives

5.1 Model Hybridisation

As detailed previously, there are a number of examples where, using high-end computing, it is possible to scale up existing approaches and, as a result, gain new insights into the complex systems under investigation. There are, however, cases where scaling up alone does not provide any answer. Complementary solutions must, therefore, be sought, such as model hybridisation.

Dynamics of disease spread within a population are of crucial importance in terms of public health (e.g. monitoring of existing outbreaks, preemptive evaluation of intervention policies). Here, the model has to deal with millions of people live in modern urban environments, each with a refined social behaviour. To date, most approaches have focused on tackling either one aspect or the other.

Network-based models have been used to investigate the impact of social structures on disease spread (see e.g. [21]). This is motivated by the fact that, for most infectious diseases, contact is required for a new infection to occur (sexually transmitted infections being obvious examples). Social structures are represented by a network where nodes correspond to individuals (or groups) and edges correspond to social links between these. Infection spread is then implemented as a stochastic propagation over the network [5].

There are, however, two limitations to this approach. First, networks with million of nodes are difficult to obtain from real data or to generate ab initio. More crucially, because they are based on social structures, these models cannot account for casual contacts between strangers (e.g. in crowded areas and public transports), which are crucial in common infectious diseases such as influenza.

Conversely, agent-based approaches are suited to model such infections between strangers, which *emerge* from individual behaviour. An infection between strangers on a bus is the result of their individual choices to board, and not because of any relationship involving them (as opposed to colleagues who *have* to be in the same office). Such a system can be efficiently parallelised, as described above for the immune model. The main limitation, however, is the lack of formal framework to include social complex structures.

Given the limitations of both paradigms, it becomes apparent that a hybrid model is the most efficient solution, combining elements of both approaches. In particular, network-based concepts can be used to generate socially realistic populations, while the agent basis can simulate an epidemic outbreak within these. This hybrid approach was successfully implemented and optimally parallelised, providing a realistic framework on which to investigate disease outbreaks and related policies [6].

Moving forward, advances in both hardware and computational techniques permit the investigation of larger, and increasingly complex, systems. Model hybridisation offers a powerful method to account for all aspects of these systems. This was exemplified here for epidemic modelling, but also applies to other contexts and methods (see, e.g. shape space hybridisation in immune models [30]).

5.2 Computing as a Platform for Collaboration and Multidisciplinarity

Complex systems usually involves several fields of expertise and multiple research teams. A typical example is Epigenetics, at the interface of three distinct fields:

- Chemistry, with studies investigating how functional groups are added and removed, and how modifying enzymes such as DNMTs interact with each other.

- Biology and Genetics, including analysis of how gene expression patterns are affected by epigenetic modifications.
- Medicine, as epigenetic changes have been identified in a number of diseases, such as cancer [25] and neural disorders [22].

Another limitation is that, while successful in investigating specific phenomena, the scientific community has so far failed to explain the system-wide complex interactions in epigenetic control. Overall system complexity and technical constraints lead most research groups to focus on a single change in one given context. Results are fragmented, limiting both reproducibility and interpretability.

Integrating these partial results is crucial to understanding the overall biomedical system. Computer-based modelling can provide a useful framework to address this need. Several models have been proposed [27]. These must be seen as a complement to in vivo and in vitro experiments, and as a framework on which to integrate scattered results, facilitating collaboration between multiple teams and distinct fields.

Computer applications are frequently developed as a collaborative tool for aspects in everyday life (see e.g. [31]), and experience from these can be transferred into the implementation of a platform for collaboration and multidisciplinary research.

6 Conclusion

Biomedical systems are typically complex, as a consequence of the number of entities they involve and of the intricate interactions between these. Additionally, they are often adaptive: the nature of these interactions can evolve over time.

To investigate these systems, mathematical and computational models have been developed and have moved from phenomenological and descriptive approaches to quantitative and predictive models. This requires significant computing resources and benefits from recent hardware and software advances, as well as from high-performance computing becoming more widely available.

Thanks to parallel aspects of biomedical systems, these computing resources can be efficiently applied to large-scale systems, such as the immune response or biological dataset analysis. In instances where parallel implementation and scale increase alone do not provide sufficient insight into the system under investigation, model hybridisation permits the integration and simultaneous study of multiple aspects and scales of the system. This reinforces the potential of computing as a platform for collaboration and multidisciplinary research.

Acknowledgements The author warmly acknowledges financial support from the Irish Research Council for Science, Engineering and Technology (Embark Initiative, immune modelling), Science Foundation Ireland (Research Frontiers Programme 07/RFP/CMSR724, epigenetic modelling), and Dublin City University (Career Start Award, socio-epidemic modelling). The author also wishes to thank both the SFI/HEA Irish Centre for High-End Computing, and the Centre for Scientific Computing & Complex Systems Modelling, for the provision of computational facilities and support.

References

1. Barski A et al. (2007). Cell 129(4):823–837.
2. Cantu-Paz E (1995). IlliGAL report 95007, University of Illinois (IL).
3. Chamberlain BL et al. (2007). Int J High Perform Comput Appl 21(3):291–312.
4. Chandra R et al. (2000). Parallel Programming in OpenMP. Morgan Kaufmann, San Francisco.
5. Chen Y et al. (2008). Phys Rev Lett 101(5):1–4.
6. Claude B et al. (2009). Proceedings of the International Conference on Computational Aspects of Social Networks, IEEE Computer Society.
7. Crutchfield JP (1994). Physica D 75(1–3):11–54.
8. D'Agostino N et al. (2005). BMC Bioinformatics 2005, 6(Suppl 4):S9.
9. Duhamel C, Perrin D (2008). Technical report RR-08-07, LIMOS.
10. Egmont-Petersen M et al. (2002). Pattern Recognit 35(10):2279–2301.
11. Gentleman R et al. (2005). Bioinformatics and Computational Biology Solutions Using R and Bioconductor. Springer, New York.
12. Goertzel B (1992). J Soc Evol Syst 15(1):7–53.
13. Goldenfeld N, Kadanoff LP (1999). Science 284:87–89.
14. Gropp W et al. (1999). Using MPI: Portable Parallel Programming With the Message-Passing Interface, 2nd edition. MIT Press, Cambridge.
15. Gropp W et al. (1999). Using MPI-2: Advanced Features of the Message Passing Interface. MIT Press, Cambridge.
16. Hecquet D et al. (2007). Comput Biol Med 37(5):691–699.
17. Hill J et al. (2008). BMC Bioinformatics 9:558.
18. Jackson WC, Norgard JD (2008). J Optim Theory Appl 136(3):431–443.
19. Kitchen CA, Guest MF (2009). Adv Comput 75:1–111.
20. Koch CM et al. (2007). Genome Res 17(6):691–707.
21. Kretzschmar M, Wiessing LG (1998). AIDS 12(7):801–811.
22. Kubota T (2008). Environ Health Prev Med 13(1):3–7.
23. Ma L et al. (2008). BMC Bioinformatics 9:315.
24. Meuer HW (2008). Informatik-Spektrum 31(3):203–222.
25. Miyamoto K, Ushijima T (2005). Jpn J Clin Oncol 35(6):293–301.
26. Nagaraj SH et al. (2007). Brief Bioinform 8(1):6–21.
27. Perrin D et al. (2008). ERCIM News 72:46.
28. Perrin D et al. (2009). In silico Biology – Making the most of parallel computing. In Handbook of Research on Biocomputation and Biomedical Informatics (A. Lazakidou, editor). Information Science Reference.
29. Ruskin HJ et al. (2002). Physica A 311:213–220.
30. Ruskin HJ, Burns J (2006). Physica A 365(2):549–555.
31. Sato M, Sumiya K (2006). Proceedings of The Fourth International Conference on Creating, Connecting and Collaborating through Computing (C5 '06). IEEE Computer Society.
32. Vasilescu C et al. (2007). J Crit Care 22(4):342.
33. Williams RW, Herrup K (1988). Annu Rev Neurosci 11:423–453.
34. Yue P, Moult J (2006). J Mol Biol 356(5):1263–1274.

Chapter 39
Molecular Modeling Study of Interaction of Anthracenedione Class of Drug Mitoxantrone and Its Analogs with DNA Tetrameric Sequences

Pamita Awasthi, Shilpa Dogra, Lalit K. Awasthi, and Ritu Barthwal

Abstract Numbers of drugs are being synthesized every year to meet the target of safe and disease-free society. Presently molecular modeling technique is used to unfold the mechanism of action of drugs alone or in conjunction with experimental methodologies. There are a number of drugs which are successfully developed using this methodology. Mitoxantrone (MTX) – 1, 4-dihydroxy-5, 8-bis {[2-(2-hydroxyethyl) amino] amino}-9, 10-anthracenedione is marketed under the name Novantrone, an anticancer drug used in chemotherapy. Its important analog ametantrone and various other analogs differ from one another in the position of side chain or functionalities on the chromophore eventually exhibit varied biological activities. DNA binding is an important phenomenon for anticancer activity of these drugs. In order to understand the interactions of the drug molecules with its receptor site, at atomic level, we have carried out computer simulations of drug and DNA alone and also in complex mode in water as a medium. All the simulations are being carried out using molecular operating environment (MOE) and X3DNA software tools on SUN SOLARIS platform. Interaction energy of all the drug molecules with DNA is determined and compared. Also the structural changes in DNA and drug before and after complex formation are studied extensively.

Keywords Deoxyribose nucleic acid · Mitoxantrone · Molecular operating environment

1 Introduction

Anthracycline antibiotics represent a diverse group of potent chemotherapeutic agents that interact with double-stranded DNA by interfering with the process of replication and transcription. Binding to DNA is believed to be essential for its

---

P. Awasthi (✉)
Department of Chemistry, National Institute of Technology, Hamirpur, Himachal Pradesh, India
e-mail: pamita@nitham.ac.in; p_awasthi@rediff.com; pamitawasthi@gmail.com

anticancer activity [1–5]. It is very well accomplished that DNA adopt marked stable conformation depending upon the arrangements of the base pairs. The conformation stability of backbone is further related to the pseudo-rotation phase angle of sugar molecule and ultimately to base pairs, slight variation in one component sometimes disturbs the conformation of DNA completely.

Nowadays, drug molecules are designed in such a way that they emulate the property of DNA-binding proteins and deviate the cell machinery from its desired act. Majority of the DNA-binding drugs works on this mechanism and ultimately mislead the genetic material from its normal course of action [2–4, 6]. Interaction of various anthracycline agents with DNA has been studied using biochemical as well as physicochemical techniques [7–20]. Among all anthracycline antibiotics, mitoxantrone (MTX) – 1, 4-dihydroxy-5, 8-bis{[2-(2-hydroxyethyl) amino] amino}-9, 10-anthracenedione, named as novatrone, is one of the promising drug used in chemotherapy [21, 22]. It is cytotoxic and has major clinical value in the treatment of certain leukemias as well as solid tumors such as advanced breast cancer and non-Hodgkin lymphoma [22, 23]. Presently mitoxantrone is used in combination with other cytotoxic drugs for the treatment of other cancers such as ovarian, prostate, and lung cancer [24]. Numerous studies on the binding of mitoxantrone to DNA have been undertaken, and all indicate that it exerts its biological effects by intercalation into DNA and by making electrostatic cross links with DNA to stabilize the binding [16, 17]. The drug also inhibits the activity of topII, an enzyme introducing double strand breaks in DNA and forms a cleavable drug–DNA–topII complex which ultimately leads to cell apoptosis [18, 19, 25, 26]. Besides interfering with topII–DNA complex, mitoxantrone also produces the free radical by reduction process in biological system [27]. Large number of analogs of mitoxantrone has been synthesized and biologically evaluated for their anticancer action [28–36]. The pharmacological properties of drug molecules are greatly affected by the substitution groups and side chain present over the chromophore. Simply by incorporating amido group in place of amino group affects the topII-mediated DNA cleavage to a larger extent [29, 35]. The sequence specificity of the drug molecules toward the particular DNA sequences relies on the sequence-dependent deviation from canonical B-DNA. Transcriptional essay determined the sequence specificity of binding of mitoxantrone, and it was observed that 93% of overall blockages were before the pyrimidine $(3'–5')$ purine sequences [20]. Computationally, the preference for the binding site as well as favorable mode of interaction of drug has been reported in literature, but the exact mechanism of action of mitoxantrone at cellular level is still unclear [13, 37].

In this chapter, we have reported the molecular dynamics study of mitoxantrone and its various analogs (Fig. 1) differing from one another on the basis of arrangement of functionalities around the chromophore [21, 29, 31, 33]. Due to structural diversification, they interact differently with the receptor site, i.e., DNA. We have studied the interactions of mitoxantrone and its various analogs with four different tetrameric sequences, i.e., d-5'-ACAG-3'; d-5'-ACGC-3'; d-5'-ACGG-3'; and d-5'-ACGT-3' [26] (Fig. 2). We have tried to give realistic approach to the system by simulating the molecules (drug and DNA alone or in complex format) in water as medium and observed very interesting results which is quite compatible with

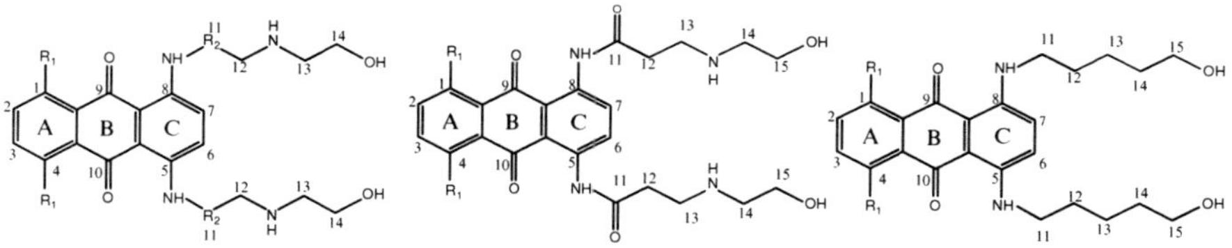

39 Molecular Modeling Study of Interaction of Anthracenedione Class of Drug MTX 387

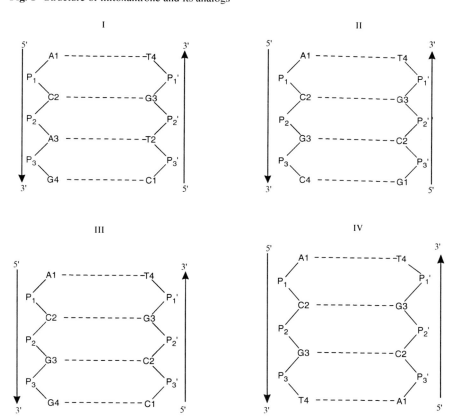

1 R1 – OH, R2 – CH$_2$, Mitoxantrone
2 R1 – H, R2 – CH$_2$, Ametantrone
3 R1 – CH$_3$, R2 – CH$_2$, CH3-analogue
4 R1 – OH, R2 – CO, Amidol analogue
5 R1 – OH, AmidoII analgue
6 R1 – OH, NSC analogue

Fig. 1 Structure of mitoxantrone and its analogs

Fig. 2 Four DNA tetramer sequences chosen for study – d-5'ACAG-3' I; d-5'-ACGC-3' II; d-5'-ACGG-3' III; and d-5'-ACGT-3' IV

experimental work [16, 38]. Structural behavior of drug and DNA before and after binding is observed and interaction energy is being calculated. The aim of this study is to explore the structural and energetic aspects of the interaction of mitoxantrone and its analogs with different DNA tetrameric sequences.

2 Methodology

Molecular operating environment (MOE) software tool is use to carry out molecular modeling studies on drug and DNA alone and in complex form followed by X3DNA software (www.rutchem.rutgers.edu) to unfold the structural fluctuations at backbone in DNA. An initial model of double-helical B-DNA structure was generated using builder module of molecular operating environment (Chemical computing group, Montréal, Canada) on dual processor SunBlade 2500 workstation. Energy of the molecule was minimized using steepest descent and conjugation gradient algorithms to remove any initial strain due to short contact in starting structure using MMFF94X force field. Dielectric constant was fixed as 1 followed by conformational search protocol. Temperature of the medium was kept 300 K throughout the simulation, and molecular dynamics was carried out for 100 ps. A total of 200 structures were saved at regular interval of 0.5 ps. Simulation was carried out in explicit water to impasse realistic condition. All the 200 structures were studied and one extracted out based on their lowest total potential energy. Same protocol was followed for all the drugs and tetramer sequences for comparative study. DNA backbone angles and helicoidal parameters were studied using X3DNA software on Sun Solaris platform. The interaction energy [39] of various complexes is calculated by:

$$\text{Interaction energy} = E\_total - (E\_drug + E\_oligonucleotides)$$

The potential energy term for whole complex is E_total, for its tetranucleotide it is E_oligo, and for mitoxantrone or its analogs it is E_drug. Van der Waal and electrostatic contributors to the interaction energy E_inter between drug and DNA are given as $E\_int\text{-}vdw$. Models of mitoxantrone and DNA tetrameric sequences are build in the building and sequence editing module of MOE, respectively. Mitoxantrone–DNA tetrameric complex is made by simply intercalating the mitoxantrone at the desired d-5'-CpG-3' step and followed by minimization protocol. Same procedure is followed for other analogs in a complex formation with four tetrameric DNA sequences.

3 Results and Discussion

3.1 Energy Analysis

3.1.1 Interaction Energy of the Drug–DNA Tetrameric Complexes

Interaction energy was quite comparable with all the four tetrameric sequences except in case of d-5'-ACGT-3', where slight decrease in interaction energy has been observed (supplementary Table 1). This could be due to the CH_3 group present in the C5 position of thymine residue, exhibiting steric effect. CH3 analog also exhibits

Table 1 Comparison of total energy, interaction energy and other components in kcalmol⁻1

S.no	Complex	Total energy E	Interaction energy	E_Vdw	E_tor
I	ACAG-MTX	−3777.43	1134.844	1958.10	127.046
	ACAG-AMT	−3729.85	1332.924	1902.23	134.031
	ACAG-AMDI Analogue	−3883.62	1102.240	1949.90	140.536
	ACAG-AMDII analogue	−3881.17	1245.071	2016.99	119.966
	ACAG-CH3-Analogue	−3627.76	1252.134	143.0972	1952.2943
	ACAG-NSC	−3552.92	1392.35	1793.32	150.374
II	ACGC-MTX	−3800.34	1086.736	1949.774	130.348
	ACGC-AMT	−3837.19	1312.776	1851.68	131.367
	ACGC-AMDI Analogue	−3783.72	1225.036	1945.67	144.585
	ACGC-AMDII analogue	−3652.42	1448.623	1735.528	127.23
	ACGC-CH3 Analogue	−3937.59	1161.896	135.9657	1956.71
	ACGC-NSC analogue	−3454.75	1465.32	1749.57	170.54
III	ACGG-MTX	−3668.15	1101.095	1877.098	139.7131
	ACGG-AMT	−3726.80	1084.555	1817.479	138.0342
	ACGG-AMD1 Analogue	−3735.63	1059.115	2079.339	155.3092
	ACGG-AMDII analogue	−3560.75	1422.463	1868.097	145.041
	ACGG-CH3 Analogue	−3692.80	799.275	137.2311	2105.95
	ACGG-NSC analogue	−3502.43	1299.81	1918.50	141.00
IV	ACGT-MTX	−3814.92	871.157	1917.32	125.939
	ACGT-AMT	−3920.06	918.517	1939.33	125.609
	ACGT-AMDI	−3542.55	1217.117	1931.01	155.571
	ACGT-AMDII analogue	−3545.92	1354.117	1809.850	143.054
	ACGT-CH3-Analogue	−3839.19	814.507	146.4082	2033.253
	ACGT-NSC analogue	−3454.13	1264.93	1743.03	137.023

the same behavior with d-5′-ACGG-3′ sequence, where the interaction energy is also very less in comparison to other analogs. Large deviation in energy has been found for d-5′-ACAG-3′ and d-5′-ACGC-3′ sequences with amido as well as CH3 analogs. This observation can be due to the change in functionality around 1, 4 positions on the chromophore in CH3 analog and addition of functional group in the side chain of amidoI and amidoII analogs.

3.1.2 Total Energy of the Complex

Not much of variation has been observed in terms of total energy of drug–DNA complexes for all the four tetrameric DNA sequences (supplementary Table 1). In the case of mitoxantrone and ametantrone, the favorable tetrameric sequence in terms of minimum energy is d-5′-ACGT-3′, whereas in the case of amido and CH3 analogs d-5′-ACAG-3′ and d-5′-ACGG-3′, respectively. The observation in terms of energy confirmed that the side chain offers some specific interaction with DNA atoms, which ultimately is responsible for sequence-specific identification by drug at DNA site. While in the case of CH3 analog, bulky CH_3 group is present which exhibits steric hindrance for favorable interaction to be offered by side chain. Additionally, the fragment of total energy, i.e., E_vdw energy is comparable for all the three analogs, whereas in the case of CH3 analog less energy has been observed in comparison to others. Similarly E_tor energy has been observed very high for CH3 analog in comparison to other analogs. This observation support that the OH group at 1, 4 positions played a very important role in sequence-specific identification at DNA and further favorable interaction responsible for anticancer activity.

Comparison of total energy, interaction energy, and further add on components to the total energy like-E_vdw and E_tor of DNA tetramer–drug complexes is shown in supplementary Table 1 for all the drugs – mitoxantrone(MTX), ametantrone(AMT), amidoI(AMDI), amidoII(AMDII) analogs, CH3 analog, and NSC analog. Energy of DNA–drug complexes determined in Kcal mol^{-1}.

3.2 Drug–DNA Interactions

3.2.1 Drug (Mitoxantrone, Ametantrone, AmidoI and II Analogs, CH3 Analog, and NSC Analog) DNA Tetramer Interactions (5′-ACAG-3′, 5′-ACGG-3′, 5′-ACGC-3′, and 5′-ACGT-3′) After Simulations

Hydrogen bonding is the major contributor toward the drug–DNA interactions and plays a critical role in stabilization of complex in addition to electrostatic, hydrophobic forces of attraction among drug–DNA complex. Therefore, polar and nonpolar groups present in the DNA as well as in drug in the presence of water molecules and cation are responsible for these interactions. All the four tetrameric sequences chosen for the study [26] bear a common interaction site d-5′-CpG-3′ with adenine base at $n+1$ position, whereas bases at $n-1$ vary. Much of the variation has been observed in terms of interaction of mitoxantrone with these bases. Drugs show preference for such sequences i.e., C or G at $(n-1)$ terminal and make stable complexes [26]. We have tried to simulate the drug–DNA complex in water and tried to study that the interaction of drug differs from one another in terms of functional groups, on different position around the chromophore and also variation in the side chain.

Case I (Fig. 2)

14OH group of mitoxantrone involved in H-bonding with oxygen atom of phosphate group of P2′ and P2 linkages of respective chain. Ametantrone drug have different behavior, and its 9O atom involved in H-bonding with NH$_2$ group of G3. In amidoI analog, 1OH group involved with oxygen atom of sugar ring of residue A3. In amidoII analog, 15OH group formed H-bond with oxygen atom of phosphate group P3′. In case of CH3 analog, 12NH group of drug involved with oxygen atom of the sugar ring of G3 and 14OH group of side chain involved in H-bonding with oxygen atom of phosphate P2 and P3' linkages. In NSC analog, 1OH group forms H-bond with oxygen atom of sugar ring of residue G3 and 15OH group involved in H-bonding with oxygen atom of phosphate P1 linkage.

Case II (Fig. 2)

OH group of mitoxantrone involved in H-bonding with oxygen atom of phosphate P2' and P2 linkages of respective chains and 12NH group also formed hydrogen bond with N7 of G3 Fig. 3(a). 14OH groups of ametantrone also involved with oxygen atom of phosphate group P2 and P2', respectively. AmidoI analog behaves differently, and in this case 1OH involved in H-bonding with oxygen atom of sugar ring of residue G3 and O9 atom involved with 11NHgroup and 15OH of amidoII forms hydrogen bonding with oxygen atom of phosphate P3'. In CH3 analog, the terminal 14OH group of both the side chains involved with oxygen atom of phosphate P2' and P2 linkage. In NSC analog, 1OH group forms H-bonding with oxygen atom of phosphate P2' linkage and 11NH involved with oxygen atom of sugar ring of residue G3.

Case III (Fig. 2)

14OH groups which are present at the side chain of mitoxantrone involved in hydrogen bonding with oxygen atom of phosphate group P2 and P2' linkages. Ametantrone drug also involved with oxygen atoms of phosphate group of P2' and P2 linkages. In amidoI analog, 12NH group involved in H-bonding with oxygen

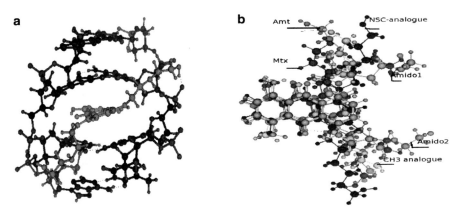

Fig. 3 (**a**) Intercalation of mitoxantrone drug into DNA tetramer sequence d-5'-ACGG-3' (**b**) vertically superimposed drugs after simulation, respectively

atom of phosphate P2' linkage and 11NH group of amidoII form H-bonding with oxygen atom of sugar ring of residue G3. In the case of CH3 analog, 14OH group involved with oxygen atom of phosphate P2 linkage.

Case IV (Fig. 2)

Similarly 14OH groups of mitoxantrone involved with oxygen atom of phosphate P2 linkage. In Ametantrone, 14OH group involved in H-bonding with oxygen atom of P2' and P2 linkages. In amidoI analog, 14OH group involved in H-bonding with oxygen atom of free A1 phosphate group and in case of CH3 analog 14OH group involved H-bonding with P2' and P2 linkages. In NSC analog, 11NH group involved with oxygen atom of the sugar ring of residue G3. All these interactions are summarized in supplementary Table 2.

3.2.2 Conformational Changes at Drug After Intercalation

1, 4 and 5, 8 substitution positions at anthraquinone chromophore are important for a molecule to be active against the cancer cells as they are being designated as pharmacophoric group [36]. Ametantrone and CH3 analog shows almost similar flexibility at the side chain position at 5, 8 like that of mitoxantrone, while in both the cases 1, 4 OH group is replaced by H-atom or CH_3 group, respectively Fig. 3(b). In amidoI and amidoII analogs, restriction in the flexibility of side chain is due to the incorporation of CO and $COCH_2$ group as studied by Zaggatto et al. in place of $11CH_2$ group, which further hinders the favorable interactions with the DNA backbone and ultimately being a reason for restricting the topII-mediated DNA cleavage, and the same observation has been seen for all the four tetrameric sequences.

3.3 Conformational Changes at DNA After Intercalation

3.3.1 Backbone Torsional Angles

Another important effect of mitoxantrone and its analogs binding to DNA tetramer sequences is regarding the conformation of the sugar-base backbone. For a nucleotide in a B-DNA type conformation, the phosphate groups are normally found in gauche, gauche (g,g) conformational position with $\alpha =$ O3'-P-O5'-C5' and $\zeta =$ C3'-O3'-P-O5' angles with -60 and -90, respectively, whereas after interacting with drug molecules they transforms into the gauche, trans (g,t) conformation with $\alpha = -60$ and $\zeta = -180$. This phenomenon is associated with transition from g,g to g,t on intercalation of drug chromophore due to opening of adjacent base pairs at intercalation site. We have calculated the backbone torsional angles for DNA before and after the simulation of the complex from X3DNA program, averaged, compared,

Table 2 Atomic interaction (A°) of drug molecules with all four tetrameric sequences

D1/N1		D2/N1		D3/N1		D4/N1		D5/N1		D6/N1	
14OH-T2P'G3	1.69	9O-G3NH2	1.90	1OH-A3O1'	1.64	14OH-C2P2A3	1.37	15OH-T2P3'C1	1.48	1OH-G3O1'	1.50
14OH-A3P2C2	1.60	–	–	–	–	14OH-G3P2'T2	1.29	–	–	15OH-A1P1C2	2.26
–	–	–	–	–	–	12NH-A3O1'	3.46	–	–	–	–
D1/N2		**D2/N2**		**D3/N2**		**D4/N2**		**D5/N2**		**D6/N2**	
14OH-G3P2C2	1.68	14OH-C2P2'G3	1.35	1OH-G3O1'	1.51	14OH-C2P2'G3	1.44	15OH-G1P3'C2	1.50	1OH-G3P2'C2	2.87
14OH-C2P2'G3	1.63	14OH-G3P2C2	1.41	–	–	14OH-G3P2C2	1.44	–	–	11NH-G3O1'	1.34
12NH-G3N7	2.87	–	–	–	–	12NH-G3N7	2.28	–	–	–	–
12NH-G3N7	2.90	–	–	–	–	–	–	–	–	–	–
D1/N3		**D2/N3**		**D3/N3**		**D4/N3**		**D5/N3**		**D6/N3**	
14OH-C2P2'G3	1.81	14OH-C2P2'G3	1.44	12NH-C2P2'G3	1.93	14OH-G3P2C2	1.42	11NH-G3O1'	2.26	–	–
14OH-G3P2C2	1.56	14OH-G3P2C2	1.48	–	–	–	–	–	–	–	–
D1/N4		**D2/N4**		**D3/N4**		**D4/N4**		**D5/N4**		**D6/N4**	
14OH-G3P2C2	1.55	14OH-C2P2'G3	1.73	14OH-A1P	1.64	14OH-C2P2'G3	1.56	–	–	11NH-G3O1'	2.0
–	–	14OH-G3P2C2	1.76	–	–	14OH-G3P2C2	1.44	–	–	–	–

D1, D2, D3, D4, D5 & D6 are drugs mitoxantrone, ametantrone, amidol-analogue, CH3, amidoII and NSC analogues: N1,N2,N3 & N4 are the tetramer sequences 5'-ACAG-3',5'-ACGC-3',5'-ACGG-3' and 5'-ACGT-3 respectively.

and found some variation in these angles. It is observed that angles α to ζ show variation in all the tetrameric base sequences with six different drug analogs. The (α) angle varies from approximately −60° to −90° for almost all the sequences. This angle has quite comparative value for almost all the complexes except d-5′-ACAG-3′ with CH3 analog and d-5′-ACGG-3′ with amidoII analog as shown in the supplementary Table 3. Large variation has been seen for angle β (beta) where conformation varies from *cis* to *trans*, while the γ does not show much variation from its B-DNA configuration of all the four tetrameric sequences. Torsional angle δ (delta) is in *trans* conformation for all the base sequences, and the value is close to starting B-DNA conformation. Angle ε (epsilon) showed marked variation from its starting B-DNA structure in case of d-5′-ACAG-3′ sequence with ametantrone, amidoI, and CH3 analogs. Torsional angle ζ (zeta) showed large variation and adopt *cis* to *trans* conformation. It is observed that χ (chi) (sugar-base connectivity) values lie in the range −80° to −127° adopting favorable anti-conformation. The value of pseudo-rational phase angle also called glycosidic bond rotation adopts overall favorable S conformation (supplementary Table 3).

3.3.2 Helicoidal Parameters

Average value of helicoidal parameters of tetramer DNA sequences and drug–DNA complex determined from X3DNA program. Intra-base pair parameters shear, stretch, stagger, buckle, propeller, and opening values for all the four tetramer sequences in complex and uncomplex mode with drugs are compared. The observed average value of shear and stretch lies in the range ± 0.4Å and -0.07 to -0.1Å, respectively. Stagger values lie in the range of -0.09 to 0.2Å, while buckle base pair showed large variation from B-DNA conformation in almost all the complexes. Propeller and opening base pair lie in the range of $\sim 3°$ to $\pm 10°$ and $-8°$ to $10°$, respectively. Among inter-base pair parameters, behavior of shift, slide, tilt, roll, and inclination on interaction with drug made a difference in almost all the sequences. The average value of shift and slide varies in the range -0.07 to $+1.4$ for all complexes, and the rise value for all the complexes lies in the range 2.7–4.2Å° except AMDI analog with 5′-ACGG-3′ sequence. Tilt value show large variation while comparing with its B-DNA. The negative value of roll has been observed for all the complexes. But the large negative value is observed for CH3 analog with 5′-ACAG-3′ sequence and MTX with 5-ACGG-3′. This decreased value is compensated by propeller twist value, so as to prevent the destacking of the bases. Not much variation has been observed in twist value for all the complexes. The average value of helicoidal parameters are shown in supplementary Tables 4 and 5.

4 Conclusions

Molecular modeling studies done by Zagotto et al. showed that stereochemistry of mitoxantrone and amido analogs is almost similar, whereas difference was found

Table 3 Average backbone torsion angles of the constructed tetra nucleotides 5' ACAG 3', 5' ACGC 3', 5' ACGG 3', and 5' ACGT 3' with mitoxantrone (MTX), ametantrone (AMT), amidoI (AMDI), amidoII (AMDII) and CH3-analogue

S.No	Structures	α	β	γ	δ	ε	ζ	χ	P
I	ACAG	−69.08	41.07	51.68	133.96	150.48	−56.18	−118.8	121.23
	ACAG-MTX	−45.28	51.63	26.78	130.41	−111.53	−17.03	−120.68	132.10
	ACAG-AMT	−59.01	−8.60	50.48	117.31	−79.56	−113.76	−87.18	107.97
	ACAG-AMDI	−42.31	32.43	30.23	137.60	−145.15	−31.51	−122.97	147.30
	ACAG-AMDII	−81.01	81.20	28.73	136.97	−159.55	117.46	122.48	143.37
	ACAG-CH3-Analogue	92.60	115.01	87.72	132.18	−89.76	−92.76	−91.22	148.25
	ACAG-NSC-Analogue	−60.13	80.07	38.35	120.42	−21.91	−115.01	−121.97	133.60
II	ACGG	−61.46	80.012	52.60	128.96	−152.50	108.83	−125.86	138.05
	ACGG-MTX	−39.68	73.07	26.58	124.77	−125.43	−71.68	−130.25	130.41
	ACGG-AMT	−63.61	33.38	61.12	117.52	−85.58	−61.81	−129.30	121.55
	ACGG-AMDI	−37.70	11.02	25.62	112.55	−25.33	−108.55	−134.25	110.57
	ACGG-AMDII	28.90	−45.86	79.66	128.67	−101.65	109.78	−70.87	141.07
	ACGG-CH3-Analogue	−62.60	123.05	63.45	121.02	−83.68	−115.10	−91.42	156.27
	ACGG-NSC-Analogue	−66.23	63.98	53.85	127.42	−81.76	−66.28	−126	131.75
III	ACGC	−69.35	116.22	56.27	128.38	−143.30	−117.13	−131.77	138.70
	ACGC-MTX	−79.46	73.33	9.95	125.07	−24.01	−35.24	−135.81	130.73
	ACGC-AMT	−78.48	43.46	28.76	118.26	−106.53	−95.26	−122.23	122.85
	ACGC-AMDI	−63.85	38.16	58.12	129.47	−156.13	−113.51	−128.50	137.06
	ACGC-AMDII	−89.51	43.15	25.60	112.95	−91.48	−94.06	−138.61	106.90
	ACGC-CH3-Analogue	−61.91	126.82	52.50	124.05	−21.81	−107.36	−89.32	165.36
	ACGC-NSC-Analogue	−55.35	33.36	91.26	119.17	−96.78	−86.98	−134.72	118.48
	ACGT	−63.03	112.45	47.06	139.65	−149.13	−74.53	−124.46	143.27
IV	ACGT-MTX	−88.28	33.12	23.13	121.45	−32.41	121.45	−127.96	129.58
	ACGT-AMT	−55.36	71.26	55.73	121.02	−121.88	121.02	−91.01	127.82
	ACGT-AMDI	−61.96	65.13	43.92	115.48	−86.06	115.48	−81.56	115.91
	ACGT-AMDII	−65.53	114.57	62.2	125.76	−151.65	−119.35	−120.75	126.35
	ACGT-CH3-Analogue	−100.21	67.07	42.4	114.42	−84.76	−93.41	−84.31	103.28
	ACGT-NSC-Analogue	−64.93	1.05	37.43	124	−34.55	−99.96	−41.17	128.7

while comparing their electrostatic potential. From their study, it was very clear that the amido analogs place itself inside the DNA–topII cleavable complex in different stereochemistry than mitoxantrone.

Our simulation supports these results as in mitoxantrone terminal 14OH group of the side chain forms hydrogen bonding with oxygen atom of phosphorous group at C2pG3 with all tetrameric sequences, i.e., d-5'-ACAG-3'; d-5'-ACGC-3';

Table 4 Average Value of intra base pair parameters for DNA-Drug complexes

S.no	Nucleotide alone and complex	shear	stretch	stagger	buckle	propeller	opening
I	ACAG	0.12	−0.15	−0.83	4.47	3.74	−7.76
	ACAG-MTX	.40	−0.10	−0.41	−7.17	−2.61	−10.48
	ACAG-AMT	0.22	−0.01	−0.58	−6.95	6.54	−4.42
	ACAG-AMDI	0.20	−0.13	−0.43	−1.91	−0.30	−0.56
	ACAG-AMDII	−0.07	0.09	−0.19	−0.40	0.53	−3.66
	ACAG-CH3	0.08	0.01	0.28	20.85	−6.75	−9.60
	ACAG-NSC	0.31	−0.20	0.16	−14.09	−15.20	−12.50
II	ACGG	−0.17	−0.22	0.23	−1.71	−6.77	−5.32
	ACGG-MTX	0.15	0.13	−0.09	0.81	2.76	0.56
	ACGG-AMT	−0.22	−0.03	−0.17	7.29	4.07	−1.28
	ACGG-AMDI	0.06	−0.22	−0.65	−11.40	1.26	−3.81
	ACGG-AMDII	0.13	−0.15	0.80	9.19	−21.39	−3.04
	ACGG-CH3	0	−0.05	−0.06	−4.22	0.38	−0.44
	ACGG-NSC	0.35	0.01	−0.09	2.08	4.92	−3.65
III	ACGC	0.23	−0.10	−0.49	6.64	8.17	2.79
	ACGC-MTX	−0.17	−0.14	−0.12	−3.97	1.31	−1.87
	ACGC-AMT	−0.17	0.10	−0.03	10.48	−15.63	3.13
	ACGC-AMDI	−0.01	−0.14	0.18	−0.10	7.69	−5.73
	ACGC-AMDII	−0.23	−0.11	0.39	8.96	−1.06	1.19
	ACGC-CH3	−0.07	−0.22	−0.21	−.5.78	5.62	−4.86
	ACGC-NSC	0.11	0.04	0.20	2.62	−13.76	−0.83
IV	ACGT	0.02	−0.18	−0.08	1.87	10.10	−3.94
	ACGT-MTX	−0.03	−0.04	−0.27	−12.51	−4.86	−7.12
	ACGT-AMT	−0.18	−0.07	−0.19	−8.70	6.09	−5.04
	ACGT-AMDI	−0.13	0.06	0.09	1.09	−4.61	−10.67
	ACGT-AMDII	−0.30	0.10	−0.04	9.07	−12.56	1.12
	ACGT-CH3	0.12	−0.02	0.11	−15.34	−10.96	−15.98
	ACGT-NSC	0.46	0.05	0.18	2.08	−3.64	8.66

d-5′-ACGG-3′; and d-5′-ACGT-3′, whereas amido analogs fail to do so. This is due to the restriction in the flexibility offered by the amido group introduced in side chain and further restricts the interaction of terminal OH group with phosphate oxygen at the backbone.

Phosphate oxygen at 5′-CpG-3′ was identified as a preferable binding site of topI/II enzyme. TopI/II enzyme cleaves the DNA strand and forms a covalent bond with phosphorous atom of DNA backbone on one side. Tyrosine-723 residue of topo-isomerase enzyme involves in hydrogen bonding with phosphate group of DNA backbone at the time of unwinding process for DNA replication, transcription etc. [40]. 1, 4 OH group also play an important role. It has been observed in number of intercalators of anthracycline series that presence of either –OH or –OCH$_3$ group on the ring A and B is important for anticancer activity (Fig. 1). Similarly, absence or substitution of OH group from 1, 4 positions on the chromophore alters the

Table 5 Average Value of inter base pair parameters for DNA-Drug complexes

S.no	Nucleotide alone and complex	Shift	Slide	Rise	Tilt	Roll	Twist
I	ACAG	0.07	−0.15	3.66	0.5	−14.81	37.51
	ACAG-MTXz	0.03	−0.48	3.68	−4.96	−4.84	34.43
	ACAG-AMT	1.13	−1.17	2.75	−7.11	−4.84	25.01
	ACAG-AMD	0.59	−1.36	3.99	3.20	−10.15	32.18
	ACAG-CH3	1.42	−0.61	3.75	8.83	−23.66	36.58
	ACAG-NSC	0.83	−0.55	4.06	9.73	−8.08	33.31
II	ACGG	0.44	−0.22	3.57	1.80	1.42	35.10
	ACGG-MTX	1.27	−0.22	3.71	3.40	−22.78	33.82
	ACGG-AMT	0.41	−0.70	3.49	2.39	0.42	33.47
	ACGG-AMDI	0.92	−1.42	1.62	0.26	7.23	33.19
	ACGG-AMDII	0.91	0.03	4.1	4.94	−2.79	35.77
	ACGG-CH3	0.72	−1.06	3.83	3.39	−5.92	32.68
	ACGG-NSC	0.01	−0.98	4.27	4.49	−13.35	35.05
III	ACGC	0.31	−0.18	3.82	2.68	−12.37	33.83
	ACGC-MTX	0.49	−0.33	3.86	0.65	−7.45	30.97
	ACGC-AMT	−0.67	−0.79	3.76	−1.51	−5.28	36.21
	ACGC-AMDI	0.40	−0.78	3.69	−2.35	−15.16	30.36
	ACGC-AMDII	1.14	−1.07	4.29	5.44	−11.27	34.92
	ACGC-CH3	−0.24	−0.94	3.59	−3.77	−5.54	30.63
	ACGC-NSC	1.03	−0.46	3.74	1.17	−8.26	37.12
IV	ACGT	−0.1	−0.61	3.79	0.14	−8.21	33.22
	ACGT-MTX	−0.4	−1.09	3.37	−0.96	−6.76	32.05
	ACGT-AMT	0.18	−1.21	3.49	−1.46	−4.26	27.07
	ACGT-AMDI	−0.51	−0.79	4.05	−3.57	−10.18	32.65
	ACGT-AMDII	−0.07	−0.86	3.24	1.03	−6.45	31.11
	ACGT-CH3	−0.76	−0.94	3.64	−2.79	−5.93	37.44
	ACGT-NSC	−0.45	0.19	3.86	−1.47	−15.55	39.35

interaction of drug molecules with DNA which is noted for ametantrone and CH3 analogs. Also the variation in the topology of the side chain at 5, 8 position affects the activity of drugs as reported in the literature, further confirmed by interaction of NSC, amidoI, and amidoII analogs with all the four DNA tetrameric sequences. For effective interactions with the binding site, side chain conformation should be in extended and flexible mode. This behavior was absent from amidoI and amidoII.

Mechanism of action of mitoxantrone at cellular level is still not confirmed, as recent study by Phillips et al. explains the intercalation mode of mitoxantrone binding to CpG site along with the position of flanking sequences. It confirms the previous studies of intercalation mode of binding of drug to its receptor site. It was only d-5'-CpG-3' or 5'-CpA-3' sites where the intercalation is noted for anthracycline derivatives. But very recently structural studies [41, 42] carried out by

Davies et al and Barthwal et al. using P^{31} NMR methodology elucidate the unusual intercalation platform adopted by drug as reported by Wang et al. for ametantrone intercalation [43].

Acknowledgments Research work reported in this manuscript is supported by the research grant under R&D, Ministry of Human Resources and Development, Government of India. The facilities provided by the NMR Centre, India Institute of Technology Roorkee (IIT-R), Uttaranchal, are highly acknowledged. (Tables 1–5 as a supplementary material.)

References

1. Frederick C. A., Williams L. D., Ughetto G., Van der Marel G. A., Van Boom J. H., Rich A. and Wang A. H. J. 1990. Structural comparison anticancer drug–DNA complex: Adriamycin and Daunomycin. Biochemistry 29: 2538–2549
2. Arcamone F. 1981. Doxorubicin: Anticancer Antibiotics. Academic Press, New York
3. Neidle S. and Waring M. J. 1983. Molecular Aspects of Anti-Cancer Drug Action. Macmillan, London
4. Searle M. S.1993. NMR studies of drug-DNA interactions. Prog. NMR Spectrosc. 25: 403–480
5. Lown J. W. 1993. Discovery and development of anthracycline antibiotics. Chem. Soc. Rev. 22: 165–176
6. Di Marco A., Arcamone F. and Zunino F. In "Antibiotics". (Eds. Corcoran J.W. and Hahn I.E.) 1974. Springer-Verlag, Berlin. 101–108
7. Durr F. E., Wallace R. E. and Citarella R.V. 1983. Molecular and biochemical pharamacology of mitoxantrone. Cancer Treat. Rev. 10: 3–11
8. Islam S. A., Neidle S., Gandecha B. M., Partridge M., Patterson L. H. and Brown J. R. 1985. Comparative computer graphics and solution studies of the DNA interaction of substituted anthraquinones based on doxorubicin and mitoxantrone. J. Med. Chem. 28: 857–864
9. Lee B. S. and Dutta P. K. 1989. Optical spectroscopic studies of the antitumor drug 1,4-dihydroxy-5,8-bis [[2-[(2-hydroxyethyl)amino]ethyl]amino]-9,10-anthracenedione (mitoxantrone). J. Phys. Chem, 93: 5665–5672
10. Kolodziejczyk P. and Suillerot A. G. 1987. Circular dichroism study of the interaction of mitoxantrone. Ametantrone and their Pd (II) complexes with deoxyribonucleic acid. Biochem. Biophys. Acta. 926: 249–257
11. Chen K. X., Gresh N. and Pullman B. 1985. A theoretical investigation on the sequence selective binding of mitoxantrone to double–stranded tetranucleotides. J. Biomol. Struct. Dyn. 3: 445–466
12. Gresh N. and Kahn P. H. 1990. Theoretical design of novel, 4 base pair selective derivatives of mitoxantrone. J. Biomol. Struct. Dyn. 7: 1141–1159
13. Chen K. X., Gresh N. and Pullman B. 1986. A theoretical investigation on the sequence selective binding of mitoxantrone to double–stranded tetranucleotides. Nucleic Acids Res. 14: 3799–3812
14. Reidar L., Alison R. and Bengt N. 1992. The CD of ligand-DNA Systems. 2. Poly (dA-dT) B-DNA. Biopolymers 32: 1201–1214
15. Reidar L., Alison R. and Bengt N. 1992. The CD of ligand-DNA Systems. 1. Poly (dG-dC) B-DNA. Biopolymers 31: 1709–1720
16. Lown J. W. and Hanstock C. C. 1985. High field 1H–NMR analysis of the 1:1 intercalation complex of the antitumor agent mitoxantrone and the DNA duplex [d(CpGpCpG)]. J. Biomol. Struc. Dyn. 2: 1097–1106
17. Kotovych G., Lown J. W. and Tong P. K. 1986. High field 1H and 31P NMR studies on the binding of the anticancer agent mitoxantrone to d–[CpGpApTpCpG)2. J. Biomol. Struct. Dyn. 4: 111–125

18. Parker B. S, Cullinane C. and Phillips D. R. 1999. Formation of DNA adducts by formaldehyde–activated mitoxantrone. Nucleic Acids Res. 27: 2918–2923
19. Parker B. S., Cullinane C. and Phillips D. R. 2000. Formaldehyde activation of mitoxantrone yields CpG and CpA specific DNA adducts. Nucleic Acids Res. 28: 983–989
20. Panousis C. and Phillips D. R. 1994. DNA sequence specificity of Mitoxantrone. Nucleic Acids Res. 22: 1342–1345
21. Murdock K. C., Child R. G., Fabio P. F., Angier R. B., Wallace T. E., Durr F. E. and Citarella R. V. 1979. Antitumor agents.1, 4 Bis–[aminoalkyl) amino]–9, 10 anthracenedione. J. Med. Chem. 22: 1024–1030
22. Collier D. A. and Neidle S. 1988. Synthesis, molecular modeling, DNA binding and antitumor properties of some substituted Amidoanthraquinone. J. Med. Chem. 1: 847–857
23. Smith I. E. 1983. Mitoxantrone (novantrone): a review of experimental and early clinical studies. Cancer Treat Rev. 10: 103–115
24. Faulds D., Balfour J. A., Chrisp P. and Langtry H. D. 1991. Mitoxantrone. A review of its pharmacodynamics and pharmacokinetic properties and therapeutic potential in the chemotherapy of cancer. Drugs 41: 440–449
25. Zee-Cheng R. K-Y and Cheng C. C. 1978. Antineoplastic agents. Structural-activity relationship study of bis(substituted amino-alkyloamino)anthraquinone. J. Med. Chem. 21: 291–294
26. Parker B. S., Buley T., Evinson B., Cutts S. M., Neumann G. M., Iskander M. N. and Phillips D. R. 2004. Molecular understandings of mitoxantrone–DNA adduct formation: effect of cytosine methylation and flanking sequences. J. Biol. Chem. 279: 18814–18823
27. Tarasiuk J., Tkaczyk-Gobis K., Stefanska B., Dzieduszycka M., Priebe W., Martelli S. and Borowski E. 1998. The role of structural factors of anthraquinone compounds and their quionone modified analogues in NADH dehydrogenase catalysed oxygen radical formation. Anticancer Drug Des 13: 923–939
28. Bailly C., Rouier S., Bernier J. L. and Waring M. J. 1986. DNA recognition by two mitoxantrone analogues: influence of the hydroxyl groups. FEBS Lett. 379: 269–272
29. Zaggatto G., Moro S., Uriatte E., Ferrazzi E., Palu G. and Palumbo M. 1997. Amido analogues of mitoxantrone: physico chemical properties, molecular modeling cellular effects and antineoplastic potential. Anticancer Drug Des. 12: 99–112
30. Gatto B., Zagatto G., Sissi C., Cera C., Uriate E., Palu G., Caparnici G. and Palumbo M. 1996. Peptidyl anthraquinones as potential antineoplastic drugs: synthesis, DNA binding, redox cycling and biological activity. J. Med. Chem. 39: 3114–3122
31. Horn D. E., Michael S. L., Amy J. F. and Gray E. 2000. Synthesis of symmetrically substituted 1, 4–bis [(aminoalkyl) amino]–5, 8–Dimethylanthracene–9, 10–diones. ARKIVOC 1: 876–881
32. Morier-Teissier E., Boitte N., Helbecque N., Bernier J., Pommery N., Duvalet J., Fournier C., Catteau J. and Honichart J. 1993. Synthesis and antitumor properties of an anthraquionone bisubstituted by the copper chelating peptide Gly-Gly-L-His. J. Med. Chem. 36: 2084–2090
33. Zagotto G., Sissi C., Gatto B. and Palumbo M. 2004. Aminoacyl-analogues of mitoxantrone as novel DNA-damaging cytotoxic agents. ARKIVOC 204–218
34. Sakladanowski A. and Konopa J. 2000. Mitoxantrone and Ametantrone induce interstand cross-links in DNA of tumour cells. Br. J. Cancer 82: 1300–1304
35. Konopa J. 2001. Antitumor acridines with diaminoalkylo pharmacophoric group. Pure Appl. Chem. 73: 1421–1428
36. Krapcho A. P., Getahun Z., Avery K. L., Vargas K. J. and Hacker M. P. 1991. Synthesis and antitumor evaluations of symmetrically and unsymmetrically substituted 1,4-bis [(aminoalkyl)amino]anthracene-9,10-diones and 1,4-bis[(aminoalkyl)amino]-5,8-dihydroxyanthracene-9,10-dione. J. Med. Chem. 34: 2373–2380
37. Mazerski J., Martelli S. and Borowski E. 1998. The geometry of intercalation complex of antitumor mitoxantrone and ametantrone with DNA: molecular dynamics simulations. Acta Biochim. Pol. 45: 1–11
38. Kaur M. 2006. Structural studies of anticancer drug mitoxantrone and its complex with DNA, Ph.D. Thesis

39. Rehn C. and Pindur U. 1996. Model building and molecular mechanics calculation of mitoxantrone–deoxytetranucleotide complexes: molecular foundation of DNA intercalation as cytostatic active principle. Monatsh. Chem. 127: 631–644
40. Stewart L., Redinobo M. R., Qui X., Hol W. G. and Champoux J. J. 1998. A model for the mechanism of human topoisomerase 1. Science 279: 1534–1541
41. Barthwal R., Monica, Awasthi P., Srivastava N., Sharma U., Kaur M. and Govil G. 2003. Structure of DNA hexamer sequence d-CGATCG by two-dimensional nuclear magnetic resonance spectroscopy and restrained molecular dynamics. J. Biomol. Struct. Dyn. 21: 407–423
42. Davies D. B., Eaton R. J., Baranovsky S. F. and Veselkov A. N. 2000. NMR investigation of the complexation of daunomycin with deoxytetranucleotides of different base sequence in aqueous solution. J. Biomol. Struct. Dyn. 17: 887–901
43. Yang X., Robinson H., Gao Y.-G. and Wang A. H. J. 2000. Binding of a macrocyclic bisacridine and ametantrone to CGTACG involves similar unusual intercalation platforms. Biochemistry 39: 10950–10957

Chapter 40
A Monte Carlo Analysis of Peritoneal Antimicrobial Pharmacokinetics

Sanjukta Hota, Philip Crooke, and John Hotchkiss

Abstract Peritoneal dialysis-associated peritonitis (PDAP) can be treated using very different regimens of antimicrobial administration, regimens that result in different pharmacokinetic outcomes and systemic exposure levels. Currently, there is no population-level pharmacokinetic framework germane to the treatment of PDAP. We coupled a differential-equation-based model of antimicrobial kinetics to a Monte Carlo simulation framework, and conducted "in silico" clinical trials to explore the anticipated effects of different antimicrobial dosing regimens on relevant pharmacokinetic parameters (AUC/MIC and time greater than 5 × MIC) and the level of systemic exposure.

Keywords Peritoneal dialysis · In silico clinical trial

1 Introduction

Although cheap, simple, and effective, peritoneal dialysis does have a number of complications, one of which is peritoneal dialysis-associated peritonitis (PDAP; bacteria growing in the peritoneal cavity). When it occurs, PDAP contributes to morbidity and increased cost, may hasten "technique failure" (a patient can no longer be treated with peritoneal dialysis), and can lead to repeated courses of treatment with antibiotics [1–3]. Potentially repetitive courses of antimicrobial chemotherapy for recurrent episodes of PDAP may expose microbial flora in other, non-peritoneal, sites (such as skin or gut) to selection pressure favoring the induction or overgrowth of microorganisms resistant to that antimicrobial agent [4–10]. Similarly, a prolonged single bout of PDAP can have deleterious effects on patient outcome. Fungi, *Clostridia* species, and vancomycin-resistant

S. Hota (✉)
Department of Mathematics, Fisk University, Nashville, TN 37208, USA
e-mail: sanjuktahota@gmail.com

Enterococcus represent clinically relevant potential beneficiaries of selection pressure associated with protracted antimicrobial therapy, each to the detriment of the individual patient. Moreover, frequent or prolonged courses of antimicrobial therapy may render the peritoneal dialysis population an incubator for the development of antimicrobial resistant microorganisms – a public health issue previously identified in the hemodialysis population [11–16]. Recent demonstration of heteroresistance in *Staphylococcus aureus*, with selection of increasingly vancomycin-tolerant microorganisms, highlights another potentially important facet of this issue [15, 16].

At present, there is no well-defined population level framework for optimizing treatment of such infections; the dynamic complexity of the processes involved suggests that in silico clinical trials using mathematical models that emulate "virtual patients" could provide a useful approach. Optimized antibiotic kinetics for treatment of PDAP, by diminishing the burden of antimicrobial exposure, may diminish the adverse ecological consequences of treating this affliction. The main objective of this study was to characterize the predicted intraperitoneal antimicrobial concentration profiles under two different treatment regimens for two first line antimicrobial agents.

2 The Model

A simple two compartment, variable volume pharmacokinetic model was developed to describe the distribution and levels of antibiotics as a function of the dialysis prescription, patient characteristics, and dosing regimen. The model (Fig. 1) incorporated sequential exchanges, transfer of antimicrobial between the peritoneal cavity and the systemic volume of distribution, elimination by residual renal function, extraperitoneal protein binding of antimicrobial, and the modulation of peritoneal volume by lymphatic flow. The governing equations and details of their manipulation can be found in the Appendix.

2.1 Criteria of Pharmacokinetic Adequacy and Systemic Exposure

Two common measures of pharmacokinetic adequacy were examined: (1) the total time during which the intraperitoneal antimicrobial concentration exceeded five times the minimal inhibitory concentration (MIC) of the microorganism ($5 \times$ MIC), and (2) the ratio of the area under the concentration curve over time divided by the MIC (AUC/MIC). Goals for these indices were >400 (for AUC/MIC) and $>50\%$ (for time above $5 \times$ MIC). The average extraperitoneal concentration of the antimicrobial agent over the 4-day treatment course was also computed.

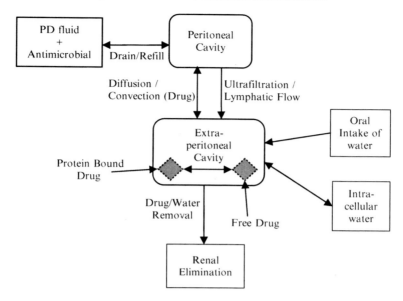

Fig. 1 Schematic of system modeled

2.2 Dialysis Schedule

We modeled two archetypes of peritoneal dialysis: continuous ambulatory peritoneal dialysis (CAPD) and automated peritoneal dialysis (APD). In the former implementation, the patient "fills" and drains approximately four times per day with each cycle lasting $\sim$4 h; the patient then drains and fills for a long ($\sim$8 h) overnight dwell. In APD, a solenoid device (cycler) is used to fill and drain the peritoneal cavity while the patient sleeps. At the end of the nocturnal cycling period, the abdomen is filled and one or two "manual" fill/drain cycles are performed during the daylight hours before the next evening, when the cycler repeats its schedule of multiple fill/drain cycles. We modeled a CAPD regimen of four daytime exchanges with a long night-time dwell and an APD regimen in which there are five 2-h fill/drain cycles at night, followed by two long (7 h) fill/drain cycles during the day.

2.3 Antimicrobial Dosing Regimens

Four dosing regimens were examined for two different antimicrobials: cefazolin and ceftazidime, both first-line agents in the treatment of PDAP:

- CAPD continuous: every bag of dialysate contains a fixed quantity of antimicrobial

- CAPD single daily dose: a large dose of antimicrobial is administered in the long dwell (daytime) bag
- APD single daily dose: a large dose of antimicrobial is instilled with the first daytime exchange each day
- APD continuous: every bag of dialysate contains a fixed quantity of antimicrobial (the same as used in CAPD continuous)

With the exception of the "APD continuous" regimen, we based our simulated dosing on ISPDN guidelines for a 70-kg patient with significant residual renal function [17]. There is less information regarding continuous (antimicrobial in every bag) APD dosing in the guidelines. The protein-binding characteristics of both drugs were taken into consideration for computation.

2.4 Monte Carlo Implementation: Population Construction

A Monte Carlo implementation was used to reflect population-level variability in potentially important model parameters. Briefly, a population of 1000 "virtual patients" was constructed. In each patient, the infecting organism's MIC, the extraperitoneal clearance of antimicrobial, peritoneal mass transfer coefficient for the antimicrobial and residual intraperitoneal volume were drawn randomly from uniform, uncorrelated distributions.

- MIC span range from 1 to 10 µg/ml
- Extraperitoneal clearance corresponded to a clearance of 2–12 ml/min
- Residual volume 0–0.5 l
- Lymphatic flow from 6 to 66 ml/h
- Mass transfer coefficient from 0.004 to 0.012

The range of MTAC was selected to generate population mean values for peritoneal half-lives of cefazolin and ceftazidime matching those determined by Manley (~3.5h and 2.5h, respectively) [18, 19]. Consistent with published data, protein binding of cefazolin was assumed to be ~50%, with a range from 25% to 75%; that for ceftazidime was assumed to be ~25%, with a range from 10% to 30% [20]. Putative urine output was set at 1 l/day for all simulations, and fluid intake was adjusted so that the volume of distribution remained constant (rising or falling $<0.1\%$) between the beginning and end of the simulation ("patients" experienced neither net weight gain nor loss over the 4-day period). Insensible losses (sweat, respiration) were not considered. Incorporation of free (unhindered) convective transport of antimicrobial (sieving coefficient of 1) as a limiting case only slightly changed the quantitative results and did not change the qualitative conclusions; for simplicity it is not described further.

3 Results

Pharmacokinetics was examined over a treatment period of 4 days on each of 1,000 virtual patients undergoing the appropriate dialysis regimen and dosing scheme. Four days were selected because failure of clinical improvement at 4 days is often used as an indication for a change in therapy. Figure 2 presents the anticipated intraperitoneal concentration curves over the 4-day period for each dosing regimen and dialysis mode, averaged over all members of the relevant population.

For both antimicrobials and in both CAPD and APD, continuous dosing consistently satisfied the criterion of maintaining the intraperitoneal antimicrobial concentration greater than $5 \times$ MIC for at least 50% of the treatment time (Table 1). Single daily dose regimens failed to satisfy this criterion in 75–100% of simulated treatment courses (Table 2). For cefazolin, continuous dosing afforded satisfactory AUC/MIC values for both APD and CAPD. In contrast, for ceftazidime, neither continuous nor single daily dosing afforded satisfactory AUC/MIC results for either APD or CAPD. Time averaged extraperitoneal concentrations of antimicrobial were similar between dosing regimens, varying over a ~ 2 fold range between intermittent and continuous dosing for both modes of peritoneal dialysis and both

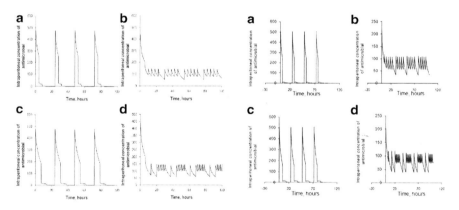

Fig. 2 Concentration profiles. Predicted intraperitoneal concentration profiles for each dialysis mode and dosing regimen with Cefazolin (*left*) and Ceftazidime (*right*) over 96 h. Panel **a**: CAPD, single daily dose; Panel **b**: CAPD, antimicrobial in every bag; Panel **c**: APD, single daily dose; Panel **d**: APD, antimicrobial in every bag

Table 1 Population mean values for percent time above $5 \times$ MIC, AUC/MIC, and extraperitoneal concentration: Cefazolin and Ceftazidime

	CAPD Single daily		CAPD Every bag		APD Single daily		APD Every bag	
	Cefazo	Ceftaz	Cefazo	Ceftaz	Cefazo	Ceftaz	Cefazo	Ceftaz
% Time above $5 \times$ MIC	24	25	99	99	36	377	99	99
Mean AUC/MIC	242	252	460	337	354	359	479	353
Avg. conc. extraperitoneal	9.6	10.6	20	13.5	13.8	14.9	20.7	13.9

Table 2 Probability that AUC/MIC or time above 5 × MIC criteria will be satisfied: Cefazolin and Ceftazidime

	CAPD Single daily		CAPD Every bag		APD Single daily		APD Every bag	
	Cefazo (%)	Ceftaz (%)	Cefazo (%)	Ceftaz (%)	Cefazo (%)	Ceftaz (%)	Cefazo (%)	Ceftaz (%)
Probability AUC/MIC >400	0	0	93	2.7	19	22.3	100	8.7
Probability time >5 × MIC	0	1.7	100	100	18	25	100	100

antimicrobials (Table 1). Changes in residual volume had a relatively modest effect on predicted pharmacokinetic outcomes in which increases or decreases in residual volume did not remarkably change the AUC/MIC or time above 5 × MIC.

4 Discussion

Our results suggest that a continuous dosing of both cefazolin and ceftazidime produces an acceptable fraction of time spent at >5 × MIC in both CAPD and APD. In contrast, in no case does single daily dosing afford "adequate" AUC/MIC ratios. The results for individual "virtual patients" suggest that these results are robust over clinically relevant ranges and combinations of MIC, extraperitoneal clearance (residual renal function), peritoneal mass transfer coefficient for the antimicrobial, lymphatic flows, and residual volumes.

This is, to our knowledge, the first application of Monte Carlo pharmacokinetic techniques and simulated clinical trials to peritoneal dialysis in general and PDAP in particular. We chose this approach because of the significant variation in patient (and microorganism) characteristics, such as MIC, MTAC, and residual renal function. These key pharmacokinetic parameters are generally not known at the time of patient presentation, vary significantly between patients, and might change over the course of treatment. A Monte Carlo implementation allows analysis of anticipated consequences of different dosing regimens across all plausible combinations of patient and microorganism characteristics. The *qualitative* predictions of such analyses (subject to the pharmacokinetic metric selected for adequacy) suggest both the approaches most likely to result in pharmacokinetic adequacy and the ecological costs attending this dosing regimen.

We restricted our analysis to a pair of cell-wall active antimicrobials that display concentration-independent killing, that is, killing is not dependent on peak levels, but the fraction of the treatment time during which the antimicrobial concentration exceeds a level dependent on the MIC of the microorganism. Other antimicrobials, such as aminoglycosides, quinolones, and daptomycin, display different, concentration-dependent killing profiles. For these agents, peak intraperitoneal concentration (and the number of times this is attained) modulates antibacterial activity.

Such agents are readily addressed by the model and approach, but require tracking of different pharmacokinetic outcomes. In addition, for clarity and brevity of presentation, we did not address oral antimicrobial administration.

This work cannot address the relationship between pharmacokinetic metrics of "adequacy" and clinical efficacy, a problem common to all pharmacokinetic analyses. Regimens that yield "inadequate" pharmacokinetics might still provide clinical efficacy, and regimens predicted to be effective on pharmacokinetic grounds might still be ineffective clinically. The role of peritoneal "washout" of transiently stunned/nutrient-deprived microorganisms would be a factor in the former, a potential suggested by reports of effective treatment of cefazolin resistant *S. epidermidis* by cefazolin [21–27]. Failure of pharmacokinetically acceptable regimens could relate to the presence of biofilm or the reduction in killing rate observed when cell wall active agents are deployed against slowly dividing (or non-dividing) bacterial populations. These are issues for further, pharmacodynamic, analyses.

Nonetheless, our results suggest that "in silico" clinical trials based on Monte Carlo pharmacokinetic analyses can identify specific regimens that are likely to provide a higher population level probability of attaining pharmacokinetic adequacy while minimizing systemic exposure. The ranges of patient characteristics and dosing regimens that can be examined would be difficult, if not impossible, to explore in a clinical trial. Moreover, as the induction of collateral resistance is a statistically rare (although clinically very important) event, evaluating the effects of different dosing regimens on the induction of antimicrobial resistant clones using traditional clinical trial approaches would be even more difficult. Pharmacokinetic analyses might provide useful information for constructing rational treatment strategies that minimize this risk.

Acknowledgements This study is supported by NIH 7R21AI055818-02.

Appendix

Using the mass balance principle, two sets of dynamic equations of intraperitoneal pharmacokinetics are developed. One set computes the quantity of the antimicrobial in the peritoneal cavity during an intraperitoneal dwell and other calculates the volumes of the peritoneal cavity and the extraperitoneal volume of distribution.

Equations of Antimicrobial Profile During Dwell

$$\frac{dD_p}{dt} = -(k_{\text{dif1}} + k_{\text{con1}})D_p + (k_{\text{dif2}} + k_{\text{con2}})D_{\text{epf}}$$

$$\frac{dD_{\text{epf}}}{dt} = -(k_{\text{dif2}} + k_{\text{con2}} + k_{\text{fb}})D_{\text{epf}} + (k_{\text{dif1}} + k_{\text{con1}})D_p + k_{\text{bf}}D_{\text{epb}} - \left(\frac{D_{\text{epf}}}{V_{\text{ep}}}\right)Q_u$$

$$\frac{dD_{\text{epb}}}{dt} = k_{\text{fb}}D_{\text{epf}} - k_{\text{bf}}D_{\text{epb}} - \left(\frac{D_{\text{epb}}}{V_{\text{ep}}}\right)Q_u$$

Equations of Volume Profile During Dwell

$$\frac{dV_p}{dt} = Q_o(t) + Q_{ic-in}(t) - Q_{ic-out}(t) - Q_{lym}(t)$$

$$\frac{dV_{ep}}{dt} = \alpha \left(Q_o(t) + Q_{ic-in}(t) - Q_{ic-out}(t) \right) - Q_u(t) + Q_{lym}(t)$$

with initial values $V_p(0) = V_{p0}$ and $V_{ep}(0) = V_{ep0}$.

In the equations D_p, D_{epf}, D_{epb} represent the amount of the antimicrobial in the peritoneal cavity, the amount of free drug in the extraperitoneal volume, and the protein-bound drug in the extraperitoneal volume, respectively. V_p and V_{ep} denote the volume of the peritoneal cavity and the extraperitoneal volume of distribution, respectively. k_{dif}, k_{con}, k_{fb} and k_{bf} are respectively diffusion, convection, and free drug to bound drug and bound drug to free drug constants. $Q_{ic}, Q_o, Q_u,$ and Q_{lym} represent the intracellular flow, oral intake, renal elimination, and lymphatic flow, respectively.

The antimicrobial concentration, $C_p(t)$, in the peritoneum, and the concentration, $C_{ep}(t)$ in the extraperitoneal volume of distribution at any point t during the dwell is computed as:

$$C_p(t) = \frac{D_p(t)}{V_p(t)} \text{ and } C_{ep}(t) = \frac{D_{epf}(t)}{\alpha V_{ep}(t)}$$

where α is the fraction of extracellular water.

Pharmacokinetics During Drain and Subsequent Refilling

When the peritoneal cavity is drained, a residual amount still remains in the peritoneal cavity and is denoted as $V_{residual}$. The quantity of antimicrobial which remains in the peritoneal cavity due to the residual volume is denoted as $D_{residual}$ and is given by:

$$D_{residual} = D_{epf} \frac{V_{residual}}{V_p}$$

This residual amount of drug will be added to the next bag of infused dialysate. The volume of the peritoneum (V_p) and the contained quantity of antimicrobial (D_p) at the beginning of the next cycle will then be

$$V_p = V_{residual} + V_{bag} \text{ and } D_p = D_{residual} + D_{bag}$$

where V_{bag} and D_{bag} represent, respectively, the volume of fresh bag of dialysis fluid and the amount of drug infused into the dialysate. These values in concert with the computed value for V_{ep} are used as the initial values for the pharmacokinetics of the subsequent dwell, and the first set of drug equations are solved to compute the current values for $D_p(t)$ and $D_{ep}(t)$. Concentration equations are used to determine the antimicrobial concentration in the peritoneal cavity and the extraperitoneal volume of distribution.

Computational Framework

For each dosing regimen, a virtual patient population of 1,000 patients was constructed. Each virtual patient had unique (and different) values for microorganism MIC, residual renal function, peritoneal mass transfer coefficient, and peritoneal residual volume. The equations above were appropriately parameterized and numerically evaluated over a 96-h period. The mean population values for intraperitoneal and extraperitoneal antimicrobial concentration were computed and used in Fig. 2. The time above $5 \times$ MIC as a fraction of total time was computed for each patient, as was the final AUC/MIC for each patient. These patient-specific pharmacokinetic outcomes are presented in Tables 1 and 2. The figures related to these data are placed online for convenience. Computation was performed using *Mathematica* and also using a purpose-written Visual Basic program at a time step of 6 s.

References

1. Ota K, Mineshima M, Watanabe N, Naganuma S: Functional deterioration of the peritoneum: does it occur in the absence of peritonitis? *Nephrol Dial Transplant* 2:30–33, 1987
2. Selgas R, Fernandez-Reyes MJ, Bosque E, *et al*: Functional longevity of the human peritoneum: how long is continuous peritoneal dialysis possible? Results of a prospective medium long-term study. *Am J Kidney Dis* 23:64–73, 1994
3. Davies SJ, Bryan J, Phillips L, Russell GI: Longitudinal changes in peritoneal kinetics: the effects of peritoneal dialysis and peritonitis. *Nephrol Dial Transplant* 11:498–506, 1996
4. Finkelstein E, Jekel J, Troidel L, *et al*: Patterns of infection in patients maintained on long-term peritoneal dialysis therapy with multiple episodes of peritonitis. *Am J Kidney Dis* 39:1278–1286, 2002
5. Zelenitsky S, Barns L, Findlay I, *et al*: Analysis of microbiological trends in peritoneal dialysis-related peritonitis from 1991 to 1998. *Am J Kidney Dis* 36:1009–1013, 2000
6. Golper T, Hartstein A: Analysis of the causative pathogens in uncomplicated CAPD-associated peritonitis: duration of therapy, relapses, and prognosis. *Am J Kidney Dis* 7:141–145, 1986
7. West TE, Walshe JJ, Krol CP, Amsterdam D: *Staphylococcal peritonitis* in patients on continuous peritoneal dialysis. *J Clin Microbiol* 23(5):809–812, 1986
8. Holley JL, Bernardini J, Johnston JR, Piraino B: Methicillin-resistant staphylococcal infections in an outpatient peritoneal dialysis program. *Am J Kidney Dis* 16:142–146, 1990
9. Pérez-Fontán M, Rosales M, Rodríguez-Carmona A, García Falcón T, *et al*.: Mupirocin resistance after long-term use for *Staphylococcus aureus* colonization in patients undergoing chronic peritoneal dialysis. *Am J Kidney Dis* 39:337–341, 2002
10. Sieradzki K, Roberts RB, Serur D, Hargrave J, *et al*: Heterogeneously vancomycin-resistant Staphylococcus epidermidis strain causing recurrent peritonitis in a dialysis patient during vancomycin therapy. *J Clin Microbiol* 37:39–44, 1999

11. Tokors, JI, Frank M, Alter, M, Arduino, M: National surveillance of dialysis-associated diseases in the United States, 2000. Available at: www.cdc.gov/ncidod/hip/Dialysis/dialysis. htm; p 29
12. Tokars J: Infections due to antimicrobial-resistant pathogens in the dialysis unit. *Blood Purif* 18:355–360, 2000
13. Tokars JI, Gehr T, Jarvis WR, et al: Vancomycin-resistant enterococci colonization in patients at seven hemodialysis centers. *Kidney Int* 60:1511–1516, 2001
14. D'Agata EM, Green WK, Schulman G, et al: Vancomycin-resistant enterococci among chronic hemodialysis patients: a prospective study of acquisition. *Clin Infect Dis* 32(1):23–29, 2001
15. Atta MG, Eustace JA, Song X, et al: Outpatient vancomycin use and vancomycin-resistant enterococcal colonization in maintenance dialysis patients. *Kidney Int* 59:718–724, 2001
16. Wong S, Ho PL, Woo P, Yuen K: Bacteremia caused by staphylococci with inducible vancomycin heteroresistance. *Clin Infect Dis* 29:760–767, 1999
17. Piraino B, Bailie GR, Bernardini J, Boeschoten E, Gupta A, Holmes C, Kuijper EJ, Li PK, Lye WC, Mujais S, Paterson DL, Fontan MP, Ramos A, Schaefer F, Uttley L: ISPD Ad Hoc Advisory Committee. Peritoneal dialysis-related infections recommendations: 2005 update. *Perit Dial Int* 25(2):107–131, 2005
18. Manley HJ, Bailie GR, Asher RD, Eisele G, Frye RF: Pharmacokinetics of intermittent intraperitoneal cefazolin in continuous ambulatory peritoneal dialysis patients. *Perit Dial Int* 19(1):65–70, 1999
19. Grabe DW, Bailie GR, Eisele G, Frye RF: Pharmacokinetics of intermittent intraperitoneal ceftazidime. *Am J Kidney Dis* 33(1):111–117, 1999
20. Goodman & Gilman's the pharmacological basis of therapeutics, 11th edition, edited by LL Bruton, McGraw-Hill, New York, 2006
21. Macdonald WA, Watts J, Bowmer MI: Factors affecting *Staphylococcus epidermidis* growth in peritoneal dialysis solutions. *J Clin Microbiol* 24:104–107, 1986
22. Watson SP, Clements MO, Foster SJ: Characterization of the starvation-survival response of *Staphylococcus aureus. J Bacteriol* 180:1750–1758, 1998
23. Clements MO, Foster SJ: Starvation recovery of *Staphylococcus aureus* 8325-4. *Microbiology* 144:1755–1763, 1998
24. Hermsen ED, Hovde LB, Hotchkiss JR, Rotschafer JC. Increased killing of staphylococci and streptococci by daptomycin compared with cefazolin and vancomycin in an in vitro peritoneal dialysate model. *Antimicrob Agents Chemother* 47(12):3764–3767, 2003
25. Glancey GR, Cameron JS, Ogg CS: Peritoneal drainage: an important element in host defense against staphylococcal peritonitis in patients on CAPD. *Nephrol Dial Transplant* 7: 627–631, 1992
26. Hotchkiss JR, Hermsen ED, Hovde LB, Simonson DA, Rotschafer JC, Crooke PS: Dynamic analysis of peritoneal dialysis associated peritonitis, *ASAIO J* 50(6):568–576, 2004
27. Ariano RE, Franczuk C, Fine A, Harding GK, Zelenitsky SA: Challenging the current treatment paradigm for methicillin-resistant *Staphylococcus epidermidis* peritonitis in peritoneal dialysis patients. *Perit Dial Int* 22(3):335–338, 2002

Part VI
Computational Methods for Filtering, Noise Cancellation, and Signal and Image Processing

Chapter 41
Histopathology Tissue Segmentation by Combining Fuzzy Clustering with Multiphase Vector Level Sets

Filiz Bunyak, Adel Hafiane, and Kannappan Palaniappan

Abstract High resolution, multispectral, and multimodal imagery of tissue biopsies is an indispensable source of information for diagnosis and prognosis of diseases. Automatic extraction of relevant features from these imagery is a valuable assistance for medical experts. A primary step in computational histology is accurate image segmentation to detect the number and spatial distribution of cell nuclei in the tissue, along with segmenting other structures such as lumen and epithelial regions which together make up a gland structure. This chapter presents an automatic segmentation system for histopathology imaging. Spatial constraint fuzzy C-means provides an unsupervised initialization. An active contour algorithm that combines multispectral edge and region informations through a vector multiphase level set framework and Beltrami color metric tensors refines the segmentation. An improved iterative kernel filtering approach detects individual nuclei centers and decomposes densely clustered nuclei structures. The obtained results show high performances for nuclei detection compared to the human annotation.

1 Introduction

Advances in biomedical imaging systems and staining techniques made high resolution, multispectral, and multimodal imagery of tissue biopsies available for diagnosis and prognosis of diseases. But the analysis of these imagery is a subjective, highly tedious, and time-consuming task that requires great expertise. The aim of the newly developing computer-assisted diagnosis (CAD) approaches is to provide new perspectives to develop algorithms for classification of histological images in a clinical setting, to reduce the high variability between analysts caused by subjectivity of the process, and to help processing of increasingly high volumes

F. Bunyak (✉)
Department of Computer Science, University of Missouri-Columbia,
Columbia, MO 65211, USA
e-mail: bunyak@missouri.edu

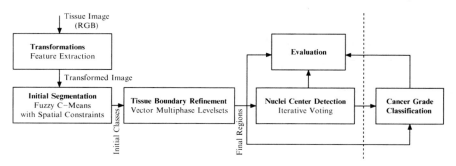

Fig. 1 Process flow for our histopathology image analysis system

of data. Automated quantitative grading of tissue patches is beginning to compare favorably with visual analysis by experts for assigning a Gleason grade to histological imagery. The shape and arrangement of glandular and nuclear structures are related to tissue type. Classification studies use graphs describing the spatial arrangement of the nuclei (i.e., Delaunay triangulation of nuclei centers) [1–6] along with regional intensity and texture features [7–9] to determine tissue type and cancer grade.

This chapter presents a robust image segmentation system for histopathology imagery to be used as a first step to cancer grade classification. The process flow for our histopathology image analysis system is given in Fig. 1. The fuzzy spatial clustering initializes the tissue class, and then the vector-based multiphase level-sets with Beltrami color edge stopping function refine the segmentation. This process allows extraction of nuclei, lumen, and epithelial cytoplasm regions. The detection of individual nuclei centers is performed by an improved iterative voting process. This chapter is extended from our earlier work in [10–12] and is organized as follows. Section 2 describes the fuzzy C-means algorithm with spatial constraint, Sect. 3.1 presents level set-based segmentation refinement process. Section 4 describes nuclei center detection. Results and conclusions are presented in Sects. 5 and 6.

2 Fuzzy C-Means with Spatial Constraint

In this section, we describe the method used to initialize multiclass contours for the multiphase level set image segmentation method. Algorithms based on fuzzy clustering are widely used in color image segmentation. They present good performance in separation of image color classes. It is advantageous to use such kind of methods to initialize level set classes to avoid local minimum convergence and to reduce the number of iterations. The classical fuzzy C-means algorithm is convenient in such cases, but it is not robust to noise, outliers, etc., which could introduce errors in the initialization of level set contours for each class. In [13], we have shown that the spatial correlation reduces noise, outliers, and other artifacts effects. The proposed spatial constraint for fuzzy C-means (SCFCM) method yields coherent regions and classes. It is based on minimizing the following objective function:

$$J_M(U,V) = \sum_{i=1}^{C} \sum_{j=1}^{N} u_{ij}^m \| \mathbf{x_j} - \mathbf{v_i} \|^2 + \alpha \sum_{i=1}^{C} \sum_{j=1}^{N} u_{ij}^m e^{-\Sigma_{k \in \Omega} u_{ik}^m} \quad (1)$$

where $X = \{\mathbf{x_1}, \mathbf{x_2}, \ldots, \mathbf{x_N}\}$ denotes the dataset (pixel feature vector). $V = \{\mathbf{v_1}, \mathbf{v_2}, \ldots, \mathbf{v_C}\}$ represents class centers. $U = [u_{ij}]$ is the partition matrix which satisfies the condition: $\sum_i^C u_{ij} = 1 \quad \forall j$, Ω is a set of neighbors. The parameter α is a weight that controls the influence of the second term (spatial information). The objective function (1) has two components. The first component is same as the Fuzzy C-means, the second is a penalty term. This last component reaches the minimum when the membership value of neighbors in a particular cluster is large. Therefore, the classification of a pixel depends strongly on its neighbors membership to a particular class. The optimization of (1) with respect to U was solved using Lagrange multiplier technique. The obtained membership update function is given by:

$$u_{ij} = \frac{1}{\sum_{p=1}^{C} \left(\frac{\|\mathbf{x_j}-\mathbf{v_i}\|^2 + \alpha e^{-\Sigma_{k\in\Omega} u_{ik}^m}}{\|\mathbf{x_j}-\mathbf{v_p}\|^2 + \alpha e^{-\Sigma_{k\in\Omega} u_{pk}^m}} \right)^{1/(m-1)}} \quad (2)$$

The method preforms the same steps as the original fuzzy C-means algorithm except for the membership function update.

3 Tissue Boundary Refinement Using Active Contours

While clustering-based algorithms, particularly SCFCM presented in the previous section, provide coarse initial segmentations, further refinement of the tissue boundaries are needed for higher accuracy and robustness. In recent years, PDE-based segmentation methods such as active contours have gained a considerable interest in biomedical image analysis. Active contours evolve a curve $\mathscr{C}$, subject to constraints from a given image. They are classified as parametric [14] or geometric [15, 16] according to their representation. Geometric active contours provide advantages such as eliminating the need to re-parameterize the curve and automatically handling topology changes using a level set implementation [17]. In level set-based active contours, a curve $\mathscr{C}$ is represented implicitly via a Lipschitz function $\phi : \Omega \mapsto \mathbb{R}$ by $\mathscr{C} = \{(x,y)|\phi(x,y) = 0\}$, and the evolution of the curve is given by the zero-level curve of the function $\phi(t,x,y)$ [18]. A regularized Heaviside function is used as a numerical indicator function for the points inside and outside of $\mathscr{C}$. In this section, we present two tissue boundary refinement approaches, a region-based and an edge-based, both using level set-based geometric active contours and both initialized by our SCFCM clustering: (1) vector multiphase active contours, (2) geodesic active contours.

3.1 Region-Based Vector Multiphase Active Contours

In [19], Chan and Vese presented a multiphase extension of their two-phase level set image segmentation algorithm [18]. The multiphase approach enables efficient partitioning of an image into n classes using just $log(n)$ level sets without leaving any gaps or having overlaps between level sets. Chan and Vese also extended their two-phase level set image segmentation algorithm for scalar valued images to vector-valued images such as color or multispectral images [20]. We combine the scalar multiphase approach with the two-phase feature vector approach to handle both multiple image classes and vector-valued imagery. In the case of histopathology imaging derived from H&E-stained cancer tissue biopsies, four image classes have been shown to produce good feature sets for image classification-based cancer grading [3]. We segment multichannel histopathology images using vector multiphase level sets (4-phase in this case) with RGB color as the input feature vector $\mathbf{u_0}$. The two level set functions that partition the image domain Ω into four phases are initialized with four classes obtained from SCFCM, the fuzzy clustering with spatial constraint method, described in Sect. 2. The vector multiphase energy functional $F(\Phi)$ can then be defined as:

$$F(\Phi) = \lambda_0 \int_\Omega \| \mathbf{u_0} - \mathbf{c_{00}} \|^2 [1 - H(\phi_1)][1 - H(\phi_2)] d\mathbf{x} + \lambda_1 \int_\Omega \| \mathbf{u_0} - \mathbf{c_{01}} \|^2$$
$$\times [1 - H(\phi_1)] H(\phi_2) \, d\mathbf{x} + \lambda_2 \int_\Omega \| \mathbf{u_0} - \mathbf{c_{10}} \|^2$$
$$\times H(\phi_1)[1 - H(\phi_2)] \, d\mathbf{x} + \lambda_3 \int_\Omega \| \mathbf{u_0} - \mathbf{c_{11}} \|^2 H(\phi_1) H(\phi_2) \, d\mathbf{x}$$
$$+ \mu_1 \int_\Omega g(\nabla \mathbf{u_0}) \, |\nabla H(\phi_1)| d\mathbf{x} + \mu_2 \int_\Omega g(\nabla \mathbf{u_0}) \, |\nabla H(\phi_2)| d\mathbf{x} \qquad (3)$$

In order to exploit edge information between different regions, as in [21] for regularization, we use geodesic length measure, $\sum_{1 \leq i \leq m} \mu_i \int_\Omega g(u_0) |\nabla H(\phi_i)| d\mathbf{x}$ [16], which is basically the length term weighted by an edge stopping function $g(u_0)$. The Euler–Lagrange equations for finding the infimum of 3 can be derived using the calculus of variations as [19]:

$$\frac{\partial \phi_1}{\partial t} = \delta(\phi_1) \left\{ \mu_1 \, g(\nabla \mathbf{u_0}) \, \text{div} \left(\frac{\nabla \phi_1}{|\nabla \phi_1|} \right) + \mu_1 \, \nabla g(\nabla \mathbf{u_0}) \cdot \frac{\nabla \phi_1}{|\nabla \phi_1|} \right.$$
$$- (\lambda_3 \| \mathbf{u_0} - \mathbf{c_{11}} \|^2 - \lambda_1 \| \mathbf{u_0} - \mathbf{c_{01}} \|^2) H(\phi_2)$$
$$\left. - (\lambda_2 \| \mathbf{u_0} - \mathbf{c_{10}} \|^2 - \lambda_0 \| \mathbf{u_0} - \mathbf{c_{00}} \|^2)[1 - H(\phi_2)] \right\},$$

$$\frac{\partial \phi_2}{\partial t} = \delta(\phi_2) \left\{ \mu_2\, g(\nabla \mathbf{u_0})\, \text{div}\left(\frac{\nabla \phi_2}{|\nabla \phi_2|}\right) + \mu_2\, \nabla g(\nabla \mathbf{u_0}) \cdot \frac{\nabla \phi_2}{|\nabla \phi_2|} \right.$$
$$- (\lambda_3 \parallel \mathbf{u_0} - \mathbf{c_{11}} \parallel^2 - \lambda_2 \parallel \mathbf{u_0} - \mathbf{c_{10}} \parallel^2) H(\phi_1)$$
$$\left. - (\lambda_1 \parallel \mathbf{u_0} - \mathbf{c_{01}} \parallel^2 - \lambda_0 \parallel \mathbf{u_0} - \mathbf{c_{00}} \parallel^2)[1 - H(\phi_1)] \right\} \qquad (4)$$

where $\mathbf{c_{ij}}$ is the mean feature vector of all pixels associated with class or phase ij, and $\delta(\phi_k) = H'(\phi_k)$ is the Dirac delta function.

3.2 Edge-Based Geodesic Active Contours

In level set-based geodesic active contours [16], the level set function ϕ is evolved using the speed function:

$$\frac{\partial \phi}{\partial t} = g(\nabla \mathbf{u_0})[F_c + \mathcal{K}(\phi)]|\nabla \phi| + \nabla \phi \cdot \nabla g(\nabla \mathbf{u_0}) \qquad (5)$$

where F_c is a constant velocity, $\mathcal{K} = \text{div}[\nabla \phi/(|\nabla \phi|)]$ is the curvature term (for regularization), and $g(\mathbf{u_0})$ is an edge stopping function. F_c pushes the curve inward or outward depending on its sign. The regularization term $\mathcal{K}$ ensures boundary smoothness. The external image-dependent force $g(\mathbf{u_0})$ is used to stop the curve evolution at object boundaries. The term $\nabla g \cdot \nabla \phi$ is used to increase the basin of attraction for evolving the curve to the boundaries of the objects.

Although the spatial gradients for single channel images lead to well-defined edge operators, edge detection in multichannel images (i.e., color histopathology images) is not straightforward to generalize since gradients in different channels can have inconsistent orientations. Both in geodesic active contours and in vector multiphase active contours (regularization term in 3), we use an edge stopping function g obtained from Beltrami color metric tensor $\mathcal{E}$ (6) [22] that considers the multichannel image as a vector field and computes the tensor gradient.

$$\mathcal{E} = \begin{bmatrix} 1 + \sum_{i=R,G,B} \left(\frac{\partial \mathbf{I}_i}{\partial x}\right)^2 & \sum_{i=R,G,B} \frac{\partial \mathbf{I}_i}{\partial x} \frac{\partial \mathbf{I}_i}{\partial y} \\ \sum_{i=R,G,B} \frac{\partial \mathbf{I}_i}{\partial x} \frac{\partial \mathbf{I}_i}{\partial y} & 1 + \sum_{i=R,G,B} \left(\frac{\partial \mathbf{I}_i}{\partial y}\right)^2 \end{bmatrix} ; \quad g(\nabla \mathbf{I}) = \exp[-\text{abs}(\mathbf{det}(\mathcal{E}))] \qquad (6)$$

Classical geodesic active contours are two phase and can segment an image into only two classes. In order to segment the three-class histopathology images, we use two level sets both initialized by SCFCM segmentation mask (Sect. 2). First level set segments the lumen regions from the rest of the image, and second level set segments the nuclei regions from the rest of the image. Two binary masks one for lumen and one for nuclei classes are produced from the multiclass SCFCM mask (1-for lumen and 0-for everything else, and 1-for nuclei and 0-for everything else, respectively). The masks are dilated with a large enough structuring element to ensure that they fully contain the regions of interests (lumen and nuclei). Geodesic active contours refine the segmentation by moving from the contours of these initial coarse masks inward toward the actual lumen or nuclei boundaries where they stop.

4 Nuclei Center Detection

Segmentation of individual cells or nuclei is an important and challenging necessity for biomedical image analysis. Analysis of cell morphology (shape, structure, color, texture), distribution, motility, and behavior heavily relies on identification of individual cells. In histopathology, image analysis, shape, and arrangement of glandular and nuclear structures are related to tissue type, and graphs describing the spatial arrangement of the nuclei (i.e., Delaunay triangulation of nuclei centers) are used in some automated Gleason grade classification studies [1–3] along with other features. In some tissues, nuclei structures are so densely clustered that staining process cannot visually separate them. In these cases, a cluster separation step is needed for quantification of single nuclei properties. Various approaches have been proposed for cluster decomposition and for individual center or seed point detection in cells and nuclei. These approaches include: watershed segmentation [21,23,24], gradient vector diffusion [25], elliptical model fitting using genetic algorithms [26], regularized centroid transform [27], blob detectors [28], etc. An extensive overview of related work can be found in [29]. Cluster decomposition in histopathology images is particularly challenging because nuclei clusters tend to be tightly fused and large with tens of nuclei.

The module described in this section detects individual nuclei centers from the nuclei mask produced in the previous section. The technique used is based on iterative voting using the oriented kernels approach described in [30, 31]. The approach detects nucleus centers from incomplete boundary information through voting and perceptual grouping. The method applies a series of cone-shaped kernels (Fig. 2a) that vote iteratively along the radial directions [30]. Orientation and shape of the kernel are refined and focused within the iterative process. Center of mass is refined at each iteration until it converges to a focal response. This technique has been chosen because of its noise immunity. Our three additions, such as pre-filtering, improved radial directions, and post-validation, further improve the detection performance.

1. *Pre-filtering* identifies stand-alone nuclei through ellipse fitting. The pre-filtering step have two benefits: (a) by avoiding the unnecessary iterative voting, the nuclei detection process is speeded up; and more importantly, (b) fragmentation (detection of multiple centers for a single nuclei) is considerably reduced. For

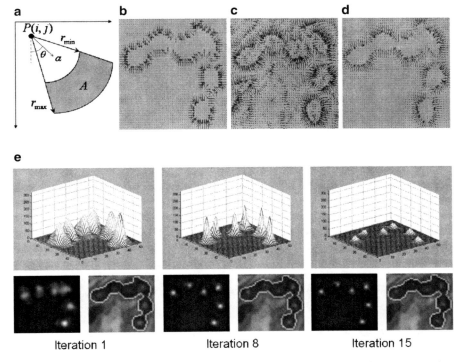

Fig. 2 (a) Cone shaped kernel and the voting area. (b–d) Radial directions: (a) from segmentation mask, (b) from input image, (c) our improved. (e) Evolution of the voting landscape matrix V during nuclei center detection with iterative voting using improved radial directions in (d)

elongated ellipses, votes accumulate on multiple peaks rather than a single peak, which results in fragmentation. In histopathology images, stand-alone nuclei tend to be more elongated than the clustered nuclei and produce most of the fragmentation.

2. *Improved radial directions* we incorporate image gradient information and region segmentation information into the computation of radial directions. Use of gradient information contributes to more smooth and precise directions; incorporation of region segmentation information reduces the effects of nuclei and cytoplasm texture. The fusion (Fig. 2d) results in higher precision, better localization, and robustness to cluster contour inaccuracies. Figure 2e shows the evolution of the voting landscape $V(i;j)$ and the resulting centers for a sample nucleus cluster image.

3. *Post-validation* is a rule-based module that studies each cluster and the associated detected centers, to accept, further split, or merge the centers. The reasoning is done using some statistics on distance between centers, generalized voronoi diagrams, and area of center influence zones.

5 Results and Discussion

The automatic tissue segmentation experiments were evaluated using prostate biopsy histopathology images with various grades of cancer.[1] We applied clustering-based methods [K-means, fuzzy C-means, and SCFCM (Sect. 2)], active contours segmentation (vector multiphase and geodesic in Sect. 3.1), and their combination.

5.1 Region Segmentation Performance

Figure 3 shows a cropped region from segmentation results of the different techniques on a sample Gleason grade 3 tissue image. K-means and FCM algorithms (shown in Fig. 3b) provide similar results. They do not incorporate any spatial information and tend to fragment regions. The SCFCM method (Fig. 3c) reduces this deficiency using spatial constraints. Both of the level set-based approaches geodesic and vector multiphase (Fig. 3d–f) result in higher segmentation accuracy. For histopathology image segmentation, vector multiphase level sets are more reliable and robust compared to geodesic active contours. Geodesic active contours are

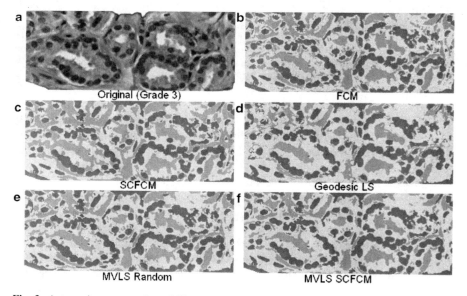

Fig. 3 Automatic segmentation of Gleason grade 3 histopathology image with nuclei shown in dark-gray (*red*), pseudo-lumen in mid-gray (*green*), epithelial cytoplasm in light-gray (*yellow*)

---

[1] Histopathology imagery provided by Michael Feldman (Department of Surgical Pathology, University of Pennsylvania) and ground truth from Anant Madabhushi (Rutgers).

Table 1 Region segmentation measures for classes nuclei (C1), pseudo-lumen (C2), and cytoplasm (C3). Left: confusion matrix, right: M_I, $\mathcal{H}$ criteria

Ground truth Segmented	C1 %			C2 %			C3 %			M_I	$\mathcal{H}$
	C1	C2	C3	C1	C2	C3	C1	C2	C3		
Ground truth	100.0	0.0	0.0	0.0	100.0	0.0	0.0	0.0	100.0	100	100
K-means	74.5	1.2	23.9	0.5	88.9	10.6	8.0	23.7	68.3	68.8	60.3
FCM	74.8	1.2	23.5	0.5	89.2	10.3	8.1	24.3	67.6	68.9	60.08
SCFCM	75.7	2.4	21.9	0.6	92.4	7.0	8.4	35.0	56.6	70.06	63.75
GeodesicSCFCM	74.3	1.8	24.0	0.4	82.0	17.6	10.0	11.7	78.3	78.63	68.05
MVLS-random	77.9	1.1	20.9	0.8	89.3	9.8	10.8	22.0	67.2	73.6	66.35
MVLS-SCFCM	77.8	0.8	21.4	0.6	86.6	12.9	9.0	17.5	73.5	75.9	69.45

more sensitive to initialization (should start either completely inside or outside of the regions of interest). They suffer from contour leaking on weak edges (i.e., some nuclei edges) and early stopping on background edges (i.e., cytoplasm texture). Due to these problems, some significant nuclei regions are missed in Fig. 3d. The vector multiphase level sets (MVLS) are used with two different initialization concepts: uniformly distributed circles and SCFCM segmented regions. As expected, MVLS combined with SCFCM (Fig. 3f) produces better results specially in the lumen regions.

For quantitative region segmentation analysis (Table 1), we used confusion matrix based on the overlap between regions in segmented image and a reference image, and our more detailed evaluation measures *matching index M_I* and $\mathcal{H}$ introduced in [32] that takes into account localization, spatial coherence in terms of position, shape, size, etc., and over- and under-segmentation. While Geodesic-SCFCM results in higher values of M_I, when the over-under segmentation penalty is introduced with $\mathcal{H}$ criterion, MVLS-SCFCM outperforms all the tested methods.

5.2 Nuclei Detection Performance

Our proposed segmentation algorithm, spatial constraint for fuzzy C-means followed by multiphase vector-based level sets (SCFCM + MVLS), is further evaluated using nuclei point set matching. This module compares nucleus center detection results to the ground truth (GT) annotated by the expert. Detected centers are matched to ground truth centers using our automatic correspondence analysis algorithm. This point correspondence algorithm supports not only one-to-one matches but also many-to-one, one-to-many, one-to-none, or none-to-one matches that result from fragmentation, merge (under-segmentation of nuclei clusters), false detection, and missed center, respectively.

Figure 4 shows sample images of the four tissue types and their final SCFCM initiated multiphase vector-based level set segmentation. Merged nuclei regions are further processed to locate individual nuclei centers using the iterative voting (Sect. 4). Table 2 shows nuclei detection performance. Over-segmentation

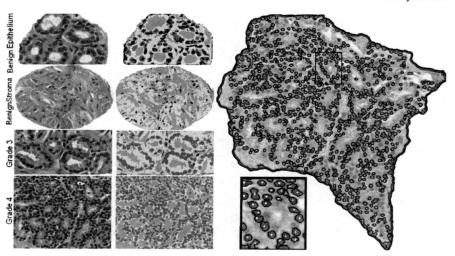

Fig. 4 *Left*: multiphase vector-based level sets (MVLS) segmentation of four tissue types into three categories: nuclei (dark-gray/*red and black*), pseudo-lumen (mid-gray/*green*), epithelial cytoplasm (light-gray/*yellow*). *Right*: region segmentation and nuclei detection result for a sample Grade 4 image, detected nuclei centers (white) and nuclei boundaries obtained using marker controlled watershed segmentation (black). Nuclei center recall: 81%, precision: 96%

Table 2 Statistics of comparison between the automatic nuclei center detection (DT) and the ground truth (GT)

Category	#GT	#DT	#Match (1-to-1)	#Under-segmentation	#Over-segmentation	#False negatives	#False positives	Recall	Precision
Benign epithelium	281	240	194 (69%)	66 (23%)	18 (6%)	3 (1%)	1 (0%)	78%	92%
Benign stroma	286	357	243 (85%)	27 (9%)	13 (5%)	3 (1%)	14 (5%)	90%	72%
Grade 3	553	630	451 (82%)	76 (14%)	15 (3%)	11 (2%)	9 (2%)	87%	77%
Grade 4	1,425	1,282	1,136 (80%)	228 (16%)	55 (4%)	6 (0%)	8 (1%)	85%	95%

(fragmentation) refers to the case where multiple detected centers match to a single ground truth center, under-segmentation (merge) refers to the case where a single detected center corresponds to multiple ground truth centers. False positives correspond to detected center which do not exist in the ground truth. False negative correspond to missed ground truth centers. Table 2 shows an average nuclei recall rate of 85% and an average precision rate of 84% across the four tissue types shown in Fig. 4. It should be noted that the proposed technique works extremely well (in terms of both recall and precision) even for Gleason Grade 4 images (row 4 in Fig. 4) with the highest number and density of cell nuclei. There is also some degree of inconsistency in the quality of the ground truth across experts in identifying indistinct nuclei; thus some of the false positives detected by the algorithm may indeed be correct. Overall, the false negative rate and the false positive rate are less than 5.5% excluding the benign stroma case.

6 Conclusion

In this chapter, we have described a robust fully automatic segmentation system for histopathology imaging derived from H&E-stained cancer tissue biopsies to be used as a first step to cancer grade classification. The system consists of three main modules. Spatial constraint fuzzy C-means developed previously by our group provides an accurate unsupervised initialization. An active contour algorithm that combines multispectral edge and region information through a vector multiphase level set framework and Beltrami color metric tensors refines the region segmentation. The process results in accurate identification of nuclei, lumen, and epithelial cytoplasm regions. Nuclei regions are further processed with an improved iterative kernel filtering approach to detect individual nuclei centers and to decompose densely clustered nuclei structures. Future extensions that we are exploring include incorporating a learning process to accommodate tissue variability as well as using human annotation to improve the overall segmentation accuracy.

References

1. Stotzka, R., Manner, R., Bartels, P., Thompson, D.: A hybrid neural and statistical classifier system for histopathologic grading of prostatic lesions. Analytical and Quantitative Cytology and Histology **17**(3) (1995) 204–218
2. Wetzel, A.: Computational aspects of pathology image classification and retrieval. The Journal of Supercomputing **11**(3) (1997) 279–293
3. Doyle, S., Hwang, M., Shah, K., Madabhushi, A., Feldman, M., Tomaszeweski, J.: Automated grading of prostate cancer using architectural and textural image features. In: IEEE International Symposium Biomedical Imaging: From Nano to Macro ISBI 2007. (April 2007) 1284–1287
4. Wittke, C., Mayer, J., Schweiggert, F.: On the classification of prostate carcinoma with methods from spatial statistics. IEEE Transactions on Information Technology in Biomedicine **11**(4) (2007) 406–414
5. Tabesh, A., Teverovskiy, M., Pang, H., Kumar, V., Verbel, D., Kotsianti, A., Saidi, O.: Multifeature prostate cancer diagnosis and gleason grading of histological images. IEEE Transactions on Medical Imaging **26**(10) (2007) 1366–1378
6. Yang, L., Tuzel, O., Chen, W., Meer, P., Salaru, G., Goodell, L., Foran, D.: PathMiner: A Web-Based tool for Computer-Assisted diagnostics in pathology. IEEE Transactions on Information Technology in Biomedicine **13**(3) (2009) 291–299
7. Huang, P., Lee, C.: Automatic classification for pathological prostate images based on fractal analysis. IEEE Transactions on Medical Imaging **28**(7) (2009) 1037–1050
8. Tosun, A.B., Kandemir, M., Sokmensuer, C., Gunduz-Demir, C.: Object-oriented texture analysis for the unsupervised segmentation of biopsy images for cancer detection. Pattern Recognition **42**(6) (2009) 1104–1112
9. Sertel, O., Kong, J., Catalyurek, U., Lozanski, G., Saltz, J., Gurcan, M.: Histopathological image analysis using Model-Based intermediate representations and color texture: Follicular lymphoma grading. Journal of Signal Processing Systems **55**(1) (April 2009) 169–183
10. Hafiane, A., Bunyak, F., Palaniappan, K.: Fuzzy clustering and active contours for histopathology image segmentation and nuclei detection. Lecture Notes in Computer Science (ACIVS) **5259** (2008) 903–914
11. Hafiane, A., Bunyak, F., Palaniappan, K.: Evaluation of level set-based histology image segmentation using geometric region criteria. In: IEEE International Symposium on Biomedical Imaging: From Nano to Macro, Boston, MA (Aug. 2009) 1–4

12. Hafiane, A., Bunyak, F., Palaniappan, K.: Clustering initiated multiphase active contours and robust separation of nuclei groups for tissue segmentation. In: IEEE International Conference on Pattern Recognition, Tampa, FL (Dec. 2008) 1–4
13. Hafiane, A., Zavidovique, B., Chaudhuri, S.: A modified fuzzy FCM with Peano scans to image segmentation. In: IEEE International Conference on Image Processing, Genova, Italy (Sept. 2005) 840–843
14. Kass, M., Witkin, A., Terzopoulous, D.: Snakes: Active contour models. International Journal of Computer Vision **1** (1988) 321–331
15. Malladi, R., Sethian, J.A., Vemuri, B.: Shape modelling with front propagation:A level set approach. IEEE Transactions on Pattern Analysis and Machine Intelligence **17**(2) (Feb. 1995) 158–174
16. Caselles, V., Kimmel, R., Sapiro, G.: Geodesic active contours. International Journal of Computer Vision **22**(1) (1997) 61–79
17. Sethian, J.A.: Level Set Methods and Fast Marching Methods: Evolving Interfaces in Computational Geometry, Fluid Mechanics, Computer Vision, and Materials Science. Cambridge University Press, Cambridge, UK (1999) ISBN 0-521-645573-3
18. Chan, T., Vese, L.: Active contours without edges. IEEE Transactions on Image Processing **10**(2) (Feb. 2001) 266–277
19. Vese, L., Chan, T.: A multiphase level set framework for image segmentation using the Mumford and Shah model. International Journal of Computer Vision **50**(3) (2002) 271–293
20. Chan, T., Sandberg, B., Vese, L.: Active contours without edges for vector-valued images. Journal of Visual Communication and Image Representation **11** (2000) 130–141
21. Yan, P., Zhou, X., Shah, M., Wong, S.: Automatic segmentation of high-throughput rnai fluorescent cellular images. IEEE Transactions on Information Technology in Biomedicine **12**(1) (January 2008) 109–117
22. Goldenberg, R., Kimmel, R., Rivlin, E., Rudzsky, M.: Fast geodesic active contours. IEEE Transactions on Image Processing **10**(10) (Oct 2001) 1467–1475
23. Malpica, N., de Solórzano, C., Vaquero, J., Santos, A., Vallcorba, I., Garcia-Sagredo, J., del Pozo, F.: Applying watershed algorithms to the segmentation of clustered nuclei. Cytometry **28**(4) (Aug. 1997) 289–297
24. Ersoy, I., Palaniappan, K.: Multi-feature contour evolution for automatic live cell segmentation in time lapse imagery. In: IEEE Engineering in Medicine and Biology Society (Aug. 2008) 371–374
25. Li, G., Liu, T., Nie, J., Guo, L., Malicki, J., Mara, A., Holley, S., Xia, W., Wong, S.: Detection of blob objects in microscopic zebrafish images based on gradient vector diffusion. Cytometry Part A **71**(10) (Oct 2007) 835–845
26. Yang, F., Jiang, T.: Cell image segmentation with kernel-based dynamic clustering and an ellipsoidal cell shape model. Journal of Biomedical Informatics **34**(2) (2001) 67–73
27. Yang, Q., Parvin, B.: Harmonic cut and regularized centroid transform for localization of subcellular structures. IEEE Transactions on Bio-Medical Engineering **50** (2003) 469–475
28. Byun, J., Verardo, M.R., Sumengen, B., Lewis, G.P., Manjunath, B.S., Fisher, S.K.: Automated tool for the detection of cell nuclei in digital microscopic images: Application to retinal images. Molecular Vision **12** (2006) 949–960
29. Schmitt, O., Hasse, M.: Morphological multiscale decomposition of connected regions with emphasis on cell clusters. Computer Vision and Image Understanding **113**(2) (2008) 188–201
30. Parvin, B., Yang, Q., Han, J., Chang, H., Rydberg, B., Barcellos-Hoff, M.H.: Iterative voting for inference of structural saliency and characterization of subcellular events. IEEE Transactions on Image Processing **16**(3) (March 2007) 615–623
31. Schmitt, O., Hasse, M.: Radial symmetries based decomposition of cell clusters in binary and gray level images. Pattern Recognition **41** (2008) 1905–1923
32. Hafiane, A., Chabrier, S., Rosenberger, C., Laurent, H.: A new supervised evaluation criterion for region based segmentation methods. Lecture Notes in Computer Science (ACIVS) **4678** (2007) 439–448

Chapter 42
A Dynamically Masked Gaussian Can Efficiently Approximate a Distance Calculation for Image Segmentation

Shareef M. Dabdoub, Sheryl S. Justice, and William C. Ray

1 Introduction

The human system of vision allows for perception and comprehension of the surrounding environment. Similarly, we use computational methods to reproduce and improve upon biological vision, allowing for automation of tasks normally requiring humans and the performance of otherwise impossible tasks. This is the domain of computer vision and image analysis. However, affording computers the ability to "perceive" is not an easy task. Most methods of image acquisition produce a two-dimensional projection of our three-dimensional world, thus losing a great deal of information [3]. In addition, the digital capture of images necessarily converts a continuous signal to a discrete one, resulting in further information loss. Where possible, both causes of loss require additional algorithmic processing to overcome.

1.1 Motivation

Our research includes the study and visualization of uropathogenic *Escherichia coli* (UPEC). These bacteria are the major causative agent of acute and recurring urinary tract infections (UTIs), mainly affecting women, and costing billions of dollars yearly in medical care and lost time. Unfortunately, UTIs resulting from such infections are difficult to treat because the bacteria possess highly evolved methods of evading both host defenses and exogenous treatments such as antibiotics. A major component of their evasion strategy is the formation of structured intracellular bacterial communities (IBCs) post-invasion, as seen in Fig. 1.

Very little is known about the structural architecture of the IBCs. In fact, it was only recently discovered that UPEC progresses through three developmental stages during intracellular growth [1]. These stages involve distinct changes in

---

S.M. Dabdoub (✉)
The Biophysics Program, The Ohio State University, Columbus, OH 43210, USA
e-mail: dabdoub.2@buckeyemail.osu.edu

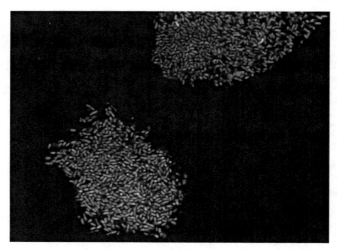

Fig. 1 Fluorescent microscope image of uropathogenic *Escherichia coli* (UPEC) colonization of a superficial bladder epithelial cell of a mouse. Images have 2D resolution of 0.102 μm per pixel and z-resolution of 0.6 μm per slice

colony structure, bacterial morphology, and individual growth rate. However, little is known quantitatively regarding these changes due to the fact that colonies can contain thousands of bacteria, making manual quantitation an impossibility. We therefore require an automated method of structural and compositional colony analysis. The laboratory strain we work with, UTI89, includes a plasmid with a GFP gene under control of a constitutive promoter. This results in significant amounts of GFP within the cytoplasm and, correspondingly, within each cell the emitted intensity decreases radially outward from the center. This, coupled with excellent image capture capabilities (Fig. 1), places us in the uniquely interesting position of having data relatively free from noise and error. In fact, this simplicity makes the data amenable to techniques that ironically remain largely unimplemented due to their very triviality, and the computational-side quest for "interesting" datasets and more refined analysis methods. Specifically, for the purpose of quantifying the different morphological types, it is not necessary to segment each pixel/voxel to a particular bacterium. Given the extreme regularity in the dimensions of these bacteria,[1] it is enough to identify what we have termed the "chewy center" (the region of greatest intensity) of each bacterium to assign a morphological class to a set of pixels.

Our first approach focused on identifying the innermost pixels of each bacterium by iteratively peeling layers of connected pixels starting from the outside toward the center. However, it soon became evident that this technique would be quite computationally expensive when applied to the large number of bacteria present in most IBCs. Therefore, we turned our attention to possible approximate alternatives.

---

[1] Each bacterium is approximately 3–70 μm × 1 μm × 1 μm (length × width × depth).

1.2 Shadowed Gaussian

An often indispensible preliminary step in computer vision algorithms is the application of a low pass filter to reduce detail and normalize pixel intensity variation. One of the most commonly used is the Gaussian blur, applied as a convolution between the image and a two-dimensional Gaussian function. Each pixel assumes the Gaussian-weighted average of its neighborhood, producing excellent smoothing of undesirable intensity variation. Additionally, when applied to objects with highly consistent dimensions, it acts as an approximate measure of distance in the same manner as the layer peeling: highlighting the inner sections of the bacteria. Of equal importance is that objects of interest maintain their original boundaries. Unfortunately, the radially symmetric nature of the Gaussian blur can have undesirable side effects. Namely, pixels internal to one bacterium may integrate intensity data from neighbors, resulting in false signal. Additionally, background pixels are altered by the intensity of the bacteria, causing a weakening of edges.

To address these problems, we have designed an algorithm called the Shadowed Gaussian that dynamically masks the convolution kernel to remove unwanted influences while maintaining the approximate distance measure of the original Gaussian. The essential idea is to identify regions of the kernel to exclude based on image intensity thresholds. As the Gaussian kernel passes over each image pixel, we identify surrounding image pixels matching the set of excluded intensities. All such pixels define barriers, and any kernel pixels that cannot "see" the central pixel due to blocked line of sight are excluded from the weighted average. In addition, we set our kernel size to match the average width of a single bacterium. This ensures that the centermost pixels of individual bacterium receive input from pixels spanning the entire bacterial width, while the masked Gaussian kernel ensures that the intensity contributed by a bacterium does not affect either the background intensity or any of its neighbors. In this manner, we provide maximum smoothing within the bacteria while maintaining the original perimeter, both of which optimize conditions for later image-based analysis techniques.

2 Methods

We define "background" pixels as a particular intensity value (or range) specified by the user. In normal convolution operations, the kernel is passed over the image, modifying the value of each pixel based on the contents of the kernel and the pixel neighborhood covered by it. By contrast, within each new kernel window the Shadowed Gaussian (Algorithm 1) determines the location of each pixel classified as background, and models it as an opaque object which blocks lines of sight from the central pixel to any pixels beyond it. In the image domain, these background pixels define edges (barriers), opposite to which are pixels belonging to background or other objects. Barrier pixels and the shadows cast behind them are removed from the kernel convolution operation.

Algorithm 1: Shadowed Gaussian
Input: image, kernel
Output: convolved image
foreach image pixel
foreach kernel pixel around image pixel
if pixel is background
determine shadow pixels and exclude all from kernel
else
convolve as normal

Algorithm 2: Determine pixel shadow
Input: kernel size, location of background pixel (BP) on kernel grid
Output: excluded pixels
Draw lines from center pixel to BP opposite facing corners;
Extend lines to kernel boundary;
Mark all pixels along lines;
Mark all pixels between previously marked pixels;

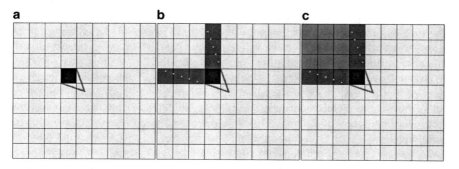

Fig. 2 The basic process of the shadow calculation procedure for one background pixel. (**a**) Rays of "light" are cast to the corners of the pixel. (**b**) Pixels along the *shadow lines* are marked. (**c**) All pixels between the *shadow lines* are filled in and marked. All pixels within the "shadow" of the *black* pixel are then removed from consideration during the kernel convolution process

The determination of the shadow cast by a single background pixel can be decomposed into four steps as set forth in Algorithm 2, and visualized in Fig. 2. This procedure is repeated for every background pixel within the kernel window until finally all shadows are calculated. All shadow pixels are marked for exclusion, and the result is a mask that has been overlaid on the Gaussian kernel. In this manner, the only pixels that contribute to the final value of the pixel under the kernel window are those within the object of interest.

The implementation of the masking algorithm can be greatly optimized by precalculating the shadow masks. The superset of pixels masked by any collection of barrier pixels is exactly the union of their individual contributions. This is amenable to precalculation of all single-pixel masks and run-time composition of full masked kernels by simple Boolean operations.

3 Discussion

Figure 1 represents the main motivation behind the development of this algorithm. We are concerned with quantitation and morphological classification of a sizeable number of highly regular, low noise objects. The bacteria produce a significant amount of fluorescent protein, but information is necessarily lost in digital capture, resulting in noise or false/attenuated signal. Thus preprocessing by some filter is highly desirable. When the Shadowed Gaussian is applied to meet this need, it performs the desired smoothing while confining the effect within the bounds of each bacterium, simultaneously identifying the "chewy center" necessary for morphological quantitation and classification as seen between Fig. 3a, c. Each of the bacteria have had their centermost regions smoothed and brightened, thus neatly approximating a distance function. In this way, the filtering process greatly eases later algorithmic image segmentation approaches.

In the next example, we have a selected region of a gray-scale image of a birch tree. The image in Fig. 3e is somewhat more complex in object shape, but it contains a definable background and provides a good example of the Shadowed Gaussian as applied to more common images. Here again, the dynamic masking of the Gaussian filter highlights in a distance-dependent manner while smoothing variations in intensity and preserving all the object boundaries of the original image.

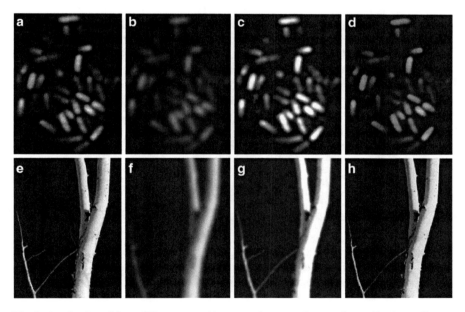

Fig. 3 Application of three different smoothing operations to an image of a small colony of bacteria and an image of a birch tree. (**a**) and (**e**), unprocessed images. (**b**) and (**f**), applied Gaussian blur with a 9×9 pixel kernel; (**c**) and (**g**) result of the Shadowed Gaussian applied with a 9×9 kernel and the background set as 10% (or less) of maximum fluorescence; (**d**) and (**h**), application of the bilateral filtering algorithm with $\sigma_s = 9.0$, $\sigma_r = 0.1$. A gray-scale image of a birch tree. A 9×9 kernel Gaussian blur applied to the birch tree image

3.1 Standard Gaussian

Previously, we mentioned the radially symmetric nature of the Gaussian blur. More precisely, each pixel is replaced with the sum of the weighted intensity of its neighbors according to the following equation [2]:

$$GB[I]_\mathbf{p} = \sum_{\mathbf{q} \in S} G_\sigma(||\mathbf{p} - \mathbf{q}||) I_\mathbf{q} \qquad (1)$$

where $\mathbf{p}$ is the current pixel, $\mathbf{q}$ is a pixel in the kernel neighborhood, I is the intensity of the pixel at $\mathbf{q}$, and G_σ is the Gaussian curve in 2D according to the equation:

$$G(x,y) = \frac{1}{2\pi\sigma^2} e^{-\frac{x^2+y^2}{2\sigma^2}} \qquad (2)$$

This establishes a radially weighted neighborhood with pixels closer in distance having more influence on the final intensity of the pixel in question. It is this distance-based radial weighting that makes the Gaussian blur so useful. However, the static nature of it causes undesirable results as is evident in Fig. 3b. Here the Gaussian blur has been applied to the familiar UTI89 colony. It does perform the smoothing function admirably, but causes a particularly undesirable result. The inclusion of pixels from the background and nearby bacteria in the final intensity value assignment leads to alteration of original object boundaries, resulting (in some cases) in the merging of objects. By contrast, the Shadowed Gaussian in Fig. 3c retains all the smoothing properties while rejecting the influence of pixels not belonging to individual bacteria. This results in much lower intensity variation within the objects while, at the same time, preventing weakening of object boundaries.

Similar problems can be seen when the Gaussian blur is applied to the birch tree image, as in Fig. 3f. The edges of the trunk are quite visibly feathered, and some smaller branches in the lower portion of the image are quite distorted or removed entirely from view. The Shadowed Gaussian applied to this image (Fig. 3g), however, preserves the original edges of the trunk as well as the more fine edges of the smaller branches.

3.2 Bilateral Filtering

The bilateral filter is an excellent overall method for image processing, but, as highlighted below, it is inappropriate for our image-processing needs. The algorithm attempts to preserve edges using the intuition that pixels should be treated as "close" if they are similar in the intensity domain as well as in the spatial domain. Tomasi and Manduchi [4] described the algorithm as a modification of the Gaussian blur (1) resulting in the following [2]:

$$BF[I]_\mathbf{p} = \frac{1}{W_\mathbf{p}} \sum_{\mathbf{q} \in S} G_{\sigma_s}(||\mathbf{p} - \mathbf{q}||) G_{\sigma_r}(I_\mathbf{p} - I_\mathbf{q}) I_\mathbf{q} \qquad (3)$$

where $W_\mathbf{p}$ is the normalization factor determined by:

$$W_\mathbf{p} = \sum_{\mathbf{q} \in S} G_{\sigma_s}(||\mathbf{p} - \mathbf{q}||) G_{\sigma_r}(I_\mathbf{p} - I_\mathbf{q}) \tag{4}$$

and G_{σ_r} is a Gaussian applied in the intensity domain. Returning to the microscope image of the UTI89 colony, we see that the bilateral filtering operation smooths the bacteria quite well while generally maintaining edge coherence. However, it produces a generalized diffusion of the white into the background, resulting in a halo effect around many of the bacteria. The Shadowed Gaussian (Fig. 3c) prevents the influence of non-background pixels outside of the object to which they belong. This explicitly prevents such a diffusion effect, resulting in a cleaner image for later processing.

Finally, we revisit the more common imagery represented by the birch tree in Fig. 3e. Again, the bilateral filtering (Fig. 3h) produces acceptable smoothing and edge cohesion, but within the trunk section only. The smaller branches in the lower part of the image are quite noticeably affected. Those that remain visible are quite reduced in size and, as a result, quite pixellated in some areas. The Shadowed Gaussian (Fig. 3g) improves upon this by preserving the original boundaries of the smaller branches.

4 Conclusion

Here we have presented a new algorithm that well approximates a distance function in images by smoothing and edge preservation, called the Shadowed Gaussian. This new method entirely replaces the normal Gaussian blur operation for situations in which blurring must be confined within certain areas of the image bounded by object edges. The Shadowed Gaussian also improves upon the excellent bilateral filtering algorithm when the important edges are relatively thin or weak.

The algorithm operates by dynamically masking the standard Gaussian kernel during the smoothing operation to entirely exclude pixels on the wrong side of an edge from influencing the pixels under examination. This has the effect of replacing the standard Gaussian weighting scheme by a simple 0/1 step function for edges and those pixels behind the edges. The original motivation for the Shadowed Gaussian was the domain of medical images, specifically that of epifluorescent microscopy of bacterial colonies. However, the algorithm works equally well to preserve edges in any gray-scale image for which a background is readily identifiable. In addition, we are currently investigating about extending the Shadowed Gaussian to the domain of color images by making use of luminosity as a nonbinary barrier with height approximately corresponding to brightness.

References

1. Justice, S.S., Hung, C., Theriot, J.A., Fletcher, D.A., Anderson, G.G., Footer, M.J., Hultgren, S.J.: Differentiation and developmental pathways of uropathogenic escherichia coli in urinary tract pathogenesis. Proceedings of the National Academy of Sciences of the USA **101**(5), 1333–1338 (2004)
2. Paris, S., Durand, F.: A fast approximation of the bilateral filter using a signal processing approach. International Journal of Computer Vision **81**(1), 24–52 (2009)
3. Sonka, M., Hlavac, V., Boyle, R.: Image processing, analysis, and machine vision second edition. International Thomson (1999)
4. Tomasi, C., Manduchi, R.: Bilateral filtering for gray and color images. IEEE International Conference on Computer Vision, p. 839 (1998)

Chapter 43
Automatic and Robust System for Correcting Microarray Images' Rotations and Isolating Spots

Anlei Wang, Naima Kaabouch, and Wen-Chen Hu

Abstract Microarray images contain a large volume of genetic data in the form of thousands of spots that need to be extracted and analyzed using digital image processing. Automatic extraction, gridding, is therefore necessary to save time, to remove user-dependent variations, and, hence, to obtain repeatable results. In this research paper, an algorithm that involves four steps is proposed to efficiently grid microarray images. A set of real and synthetic microarray images of different sizes and degrees of rotations is used to assess the proposed algorithm, and its efficiency is compared with the efficiencies of other methods from the literature.

1 Introduction

Microarray technique is a powerful new technology for large-scale gene sequence and gene expression analysis. It allows biologists to monitor thousands of genes in a single experiment simultaneously. This technique generates microarray images that contain a large volume of genetic data in the form of thousands of spots that need to be analyzed using digital image processing. The purpose of this processing phase is to extract each microscopic spot as well as to obtain its estimated background levels and intensity measures that are correlated with the gene expressions [1]. These estimates can be obtained mainly by performing three steps: (1) gridding and extraction of individual spots, (2) segmentation of spot images, and (3) quantification. Because of the large number of spots in each image to process as well as the noise and geometric rotations that affect microarray images, the processing is usually complex and time consuming. Other factors also increase the complexity of the analysis, including inaccuracies in spot printing, hybridization inconsistency, and environmental effects such as contamination [2]. Therefore, it is important that all steps are automated to save time, to remove user-dependent variations, and, hence, to obtain repeatable results.

N. Kaabouch (✉)
Electrical Engineering Department, University of North Dakota, ND 58202, USA
e-mail: naimakaabouch@mail.und.edu

Commercial systems – such as GenePix Pro, ScanAlyze, or GridOnArray – that perform some of these steps already exist. However, most of these systems use manual template matching [3] by adjusting spot size, spot spacing, and subarray location. Other software products incorporate automatic refinement searches for subarrays' locations given the size and spacing of spots. Although template matching is more robust to noise, it can fail to isolate spots if the image suffers from geometric rotations. Other techniques also have been proposed and used. These techniques are mostly based on the intensity projections of the image [4, 5, 8, 9] or combined with some morphological methods for grid segmentation [6]. However, the intensity projection-based approach is not only sensitive to noise but also sensitive to geometric rotations of the image, which also can make this approach fail to isolate spots.

In this paper, a technique that involves four major steps is proposed to efficiently grid microarray images. A set of real and synthetic microarray images of different sizes and degrees of rotations is used to assess the developed technique, and its efficiency is compared to the efficiencies of other methods from the literature.

2 Methodology

Microarray images are grouped into subarrays, with each subarray containing several hundred spots. An example is shown in Fig. 1, which contains 16 arrays and a large number of spots per subarray. One possible approach for extracting individual spots is, first, to divide the microarray image into subarrays and, then, grid each subarray to isolate each spot. However, in addition to the noise, most microarray

Fig. 1 Example of microarray image containing 16 subarrays

images suffer from geometric rotations and, hence, the gridding can fail if these rotations are not detected and corrected before spots' extractions.

In this research work, the main steps used to grid the microarray images are summarized as follows:

1. Correction of the global geometric rotation of the microarray image.
2. Division of the microarray image into subarrays.
3. Correction of each subarray geometric rotation.
4. Subarray gridding and spot extraction.

Each of these steps requires performing several substeps as follows.

2.1 Detection and Correction of the Global Geometric Rotation

To perform the correction of the global geometric deformation, the following steps are performed:

1. Enhance the microarray image using a power function given by

$$I_{out}(i,j) = cI_{in}(i,j)^{\gamma}, \tag{1}$$

where c and γ are constants, i and j represent the coordinates of the pixel located at the row i and column j in the image.
2. Convert the microarray image into a binary image.
3. Use a 3×3 median filter to remove the salt and pepper noise.
4. Find the centroids of the spots located in the first lines of the first row of subarrays in the microarray image.
5. Find the best straight line that contains these centroids using the least squares algorithm.
6. Compute the angle between the straight line and the edge of the microarray image.
7. Finally, use this angle to correct the original image, by rotating spots in the polar system.

2.2 Division of the Microarray Image into Subarray Images

To divide the microarray into subarrays, the following steps are performed:

1. Compute the profiles of the vertical and horizontal intensity projections of the microarray image. These profiles are given by:

$$P_h(i) = \frac{1}{M} \sum_{j=1}^{M} I(i,j) \tag{2}$$

$$P_v(j) = \frac{1}{N} \sum_{i=1}^{N} I(i,j) \tag{3}$$

Here, P_h and P_v represent the horizontal and vertical intensity projections of an $M \times N$ image.

2. Compute the first derivative of the vertical and horizontal profiles.
3. Derive the locations of the peaks in these profiles.
4. Compute the gap lengths between peaks in each profile.
5. Derive the maximum gap lengths in the horizontal and vertical profiles of step 2.
6. Use the maximum horizontal average gap lengths to perform the vertical gridding and the maximum vertical average gap lengths to perform the horizontal gridding of the original microarray image.

2.3 Detection and Correction of Each Subarray Geometric Rotation

Once the microarray image is divided into subarrays, each subarray's geometric rotation is corrected using the following steps:

1. Find the centroid of each spot located in the first row of the subarray.
2. Find the straight line that has the least squares for the centroids labeled in the last step.
3. Find the angle between the straight line and the edge of the subarray image.
4. Correct the rotation using the angle of the last step.

2.4 Subarray Gridding and Spot Extraction

The following steps illustrate the technique used to vertically and horizontally grid the subarray image to isolate each spot.

1. Compute the vertical and horizontal intensity projection profiles of the binarized subarray image.
2. Compute the first and second derivatives of the vertical profile and of the horizontal profile.
3. Use the following conditions to find the starting point of the coordinate i that needs to be used to perform the vertical gridding:
   - Point i must be located at the first quarter part of the horizontal profile
   - The first derivative at i is two times bigger than the first derivative is at $i-1$
   - The first derivative at i is two times smaller than the first derivative is at $i+1$
4. Use the following conditions to find the ending point of coordinate j that will be used to grid vertically:
   - j must be located at the last quarter part of the horizontal profile
   - The first derivative at j is two times smaller than the first derivative is at $j-1$
   - The first derivative at j is two times bigger than the first derivative is at $j+1$

5. Use the second derivative to differentiate between the beginning point and the ending point.
6. Use the starting and ending points to vertically divide the original subarray image.
7. Use the same steps (steps 3–6) to find the starting and ending points in the vertical profile to horizontally divide the original subarray image.

3 Results

To evaluate the robustness and performance of the proposed method, a set of real and synthetic microarray images of different sizes and qualities is used. An example of these images is illustrated in Fig. 1. This image, which represents a typical cDNA microarray image obtained from yeast, contains 16 subarrays. Each subarray holds 384 spots, with a total of 6,144 spots that have to be isolated and analyzed. As can be observed, this image suffers from noise and geometric rotations. In addition, the brightness of the spots is highly nonuniform. Most of the spots have intensity levels close to background levels, which make the manual gridding and spots extractions difficult and tedious.

Figure 2 shows the horizontal and vertical profiles of the intensity projections corresponding to the image shown in Fig. 1. The gaps between signals correspond to the gaps between subarrays. The detection of the middles of these gaps from the horizontal and vertical intensity projections allows the microarray image to be gridded into subarrays, as shown in Fig. 3.

Fig. 2 Profiles of the horizontal and vertical intensity projections of Fig. 1

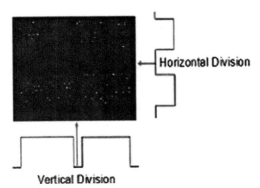

Fig. 3 Methodology for gridding an image consisting of four subarrays

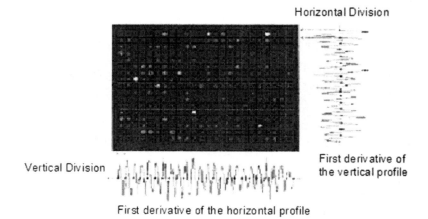

Fig. 4 Methodology for isolating spots in a subarray image

Fig. 5 Derivatives of the horizontal and vertical intensity projection profiles, respectively

To isolate each spot in each subarray image, the derivatives of the horizontal and vertical intensity projections are used to identify the width and height of each spot image. As shown in Figs. 4 and 5, the period of the horizontal signal gives the width of the spot image, while the period of the vertical signal gives the height of the spot image. Figure 6a, b gives the results of the gridded subarray #1 located on the top left of Fig. 1, using the intensity projection and the proposed technique, respectively. Comparing the two last rows in these two figures, one can see that the cells in the last row of Fig. 6a contain more than one spot per cell, while every cell in the last row of Fig. 6b contains only one spot per cell. The same results have also been observed in the other subarrays of this Fig. 1 image and in the subarrays of other microarray images that suffer from geometric rotations.

For every microarray image of the set, the efficiency of the developed system is achieved using the following criteria:

- The number of cells that contain more than one spot
- The number of cells that contain a partial spot or no spot
- The misalignment error that represents the total number of misaligned spots over the total number of pixels in all spots. This error is given by [7]:

$$E_{\text{Misalignment}} = \frac{\text{Number\_misaligned\_pixels}}{\text{Total\_number\_pixels}} \quad (4)$$

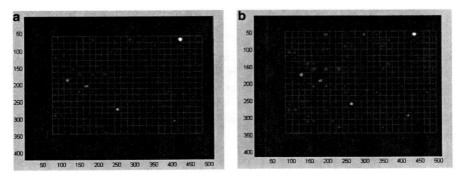

Fig. 6 Gridded subarray #1 of Fig. 1 using: (**a**) the intensity projection approach and (**b**) the proposed algorithm

Table 1 Results of Fig. 1 assessment

Criteria	Intensity projection			Proposed technique		
Number of cells that contain more or less than 1 spot	Original	Enhanced $\gamma > 1$	Enhanced $\gamma < 1$	Original	Enhanced $\gamma > 1$	Enhanced $\gamma < 1$
	194	96	72	0	48	0

We also investigated the impact of image enhancement on the efficiencies of the techniques, the proposed algorithm, the template matching approach, and the intensity projection approach.

Table 1 shows the results of the assessment corresponding to Fig. 1. As can be seen, the proposed algorithm isolated every spot in the original image, which resulted in no cell that contains less or more than one spot. However, the two other techniques, intensity projections and template matching, fail to isolate each spot. With the intensity projection technique, 194 cells contain more or less than one spot. The matching technique has the worst performance in this case, missing about 300 cells.

The impact of the enhancement depends on the quality of the image and the technique used for gridding. Table 1 also shows the enhancement's results corresponding to the image of Fig. 1. As can be seen, enhancing the image before using the intensity projection approach increases the efficiency of this technique. The number of cells drops from 194 cells for the original image to 96 if this image is enhanced using a power more than one ($\gamma > 1$), and 72 if this image is enhanced using a power less than 1 ($\gamma < 1$). On the one hand, the enhancement of the image before using the proposed technique decreases the efficiency of this technique if $\gamma > 1$, but does not affect the efficiency of the technique if $\gamma < 1$. On the other hand, the enhancement before using the matching technique does not affect the efficiency of this technique.

For the image of Fig. 1, the maximum misalignment error is about 6% when using the intensity projection, 9% when using the matching technique, and only 1.5% when using the proposed technique. In conclusion, it can be stated that the efficiency of the proposed technique is greater than the efficiencies of the other techniques in gridding and extracting spots.

4 Conclusion

In this paper, a simple and new approach to extract spots in microarray images is proposed. The robustness and efficiency of this technique is investigated and compared with other existing algorithms using a set of cDNA microarray images of different qualities and with different rotation angles. The results show that the proposed technique is the most accurate for gridding all types of microarray data images.

References

1. R. Nagarajan, "Intensity-based segmentation of microarrays images," IEEE Trans. Med. Imaging 22, 882–889, 2003.
2. V.M. Aris, M.J. Cody, J. Cheng, J.J. Dermody, P. Soteropoulos, M. Recce, and P.P. Tolias, "Noise filtering and nonparametric analysis of microarray data underscores discriminating markers of oral, prostate, lung, ovarian and breast cancer," BMC Bioinform., 5,185–193, 2004.
3. J. Buhler, T. Ideker, and D. Haynor, "Dapple: Improved Techniques for Find ing Spots on DNA Microarrays," UV CSE Tech. Rep. UWTR, 2000.
4. M. Steinfath, W. Wruck, H. Seidel, H. Lehrach, U. Radelof, and J. O'Brien, "Automated image analysis for array hybridization experiments," Bioinform., 17, 634–641, 2001.
5. A.N. Jain, T.A. Tokuyasu, A.M. Snijders, R. Segraves, D.G. Albertson, and D. Pinkel, "Fully automated quantification of microarray image data," Genome Res., 12, 2, 325–332, 2002.
6. R. Hirata, J. Barrera, and R. F. Hashimoto, "Segmentation of microarray images by mathematical morphology," Real-Time Imaging, 8, 6, 491–505, 2002.
7. Peter Bajcsy, "Gridline: Automatic Grid Alignment in DNA Microarray Scans," IEEE Transactions on Image Processing, 13, 1, 15–25, 2004.
8. Naima Kaabouch and Hamid Shahbazkia, "Automatic techniques for gridding cDNA microarray images," IEEE Electro/Information Technology Proceedings, Digital Object Identifier 10.1109/EIT.2008.4554300, pp. 218–222, 2008.
9. Anlei Wang, Naima Kaabouch, and Wen-Chen Hu, "A robust grid alignment system for microarray images," Applied Imagery Pattern Recognition 2009 Conference, IEEE Computer Society Technical Committee on Pattern Analysis and Machine Intelligence, 2009.

Chapter 44
Multimodality Medical Image Registration and Fusion Techniques Using Mutual Information and Genetic Algorithm-Based Approaches

Mahua Bhattacharya and Arpita Das

Abstract Medical image fusion has been used to derive the useful complimentary information from multimodal images. The prior step of fusion is registration or proper alignment of test images for accurate extraction of detail information. For this purpose, the images to be fused are geometrically aligned using mutual information (MI) as similarity measuring metric followed by genetic algorithm to maximize MI. The proposed fusion strategy incorporating multi-resolution approach extracts more fine details from the test images and improves the quality of composite fused image. The proposed fusion approach is independent of any manual marking or knowledge of fiducial points and starts the procedure automatically. The performance of proposed genetic-based fusion methodology is compared with fuzzy clustering algorithm-based fusion approach, and the experimental results show that genetic-based fusion technique improves the quality of the fused image significantly over the fuzzy approaches.

Keywords Registration · Fusion · Mutual information · Multi-resolution analysis · Genetic algorithm

1 Introduction

The process of combining complimentary information from two or more images of a particular organ into a single composite image is more informative and is more suitable for visual perception and computer processing. In medical imaging, different modalities of medical images carry different levels of complimentary information. In radiotherapy planning, the computed tomography (CT) data are very useful for imaging of bony structure, whereas magnetic resonance imaging provides the details

M. Bhattacharya (✉)
Indian Institute of Information Technology & Management, Morena Link Road,
Gwalior 474010, India
e-mail: mb@iiitm.ac.in; bmahua@hotmail.com

of soft tissue regions. PET images and SPECT imaging provide information of metabolic processes. In this chapter, we have attempted to describe a composite medical image registration [1–7] and fusion scheme [8–13] for detail visualization of the region of interest of body part for an improved clinical decision. We have already done experiment on registration process using MR (T_1 and T_2 weighted both) and CT imaging modalities of human brain for a patient having Alzheimer's diseases using shape theoretic approach. The control points on the concavities present in the contours are chosen to re-project ROI from the respective modalities in a reference frame [2]. Some image fusion methods have been introduced in the literatures [8–17]. The process of multimodal medical image registration as implemented in this work is the search strategy for maximizing the similarity metric using mutual information. Here the search strategy for maximizing the similarity metric is multi-resolution-based Genetic algorithm (GA) approach. Most of the optimization techniques are accurate when initial orientation is much close to the transformation that yields the best registration. To achieve the expected fusion accuracy, most of the state-of-the-art fusion methods have the typical requirement that a set of fiducial points must be manually identified and marked among the different modalities of the images to be fused. Clearly the marking of interested regions required typical anatomy-based gray-level information about the organs to be fused. The objective of this chapter is to introduce an *automatic* fusion system that can be used in clinical diagnosis and evaluated with accepted fusion accuracy. The overall image fusion scheme of the proposed algorithm involves the decomposition of registered images, computation of genetic algorithm-based selection technique, implementation of fusion rules and then reconstruction of the composite fused image. Section 1 of the chapter describes the introductory part, Sect. 2 explains the registration process of multimodal medical images, Sect. 3 relates the genetic algorithm-based selection techniques for fusion process implemented on registered images, Sect. 4 illustrates the experimental results, and Sect. 5 concludes the proposed scheme with a futuristic approach for further study.

2 Registration Process

Proposed methodology for registration of multimodal medical images implements correlation-based affine transformation of floating images as similarity metric based on mutual information and maximizes it to achieve appropriate registration of images. To validate the importance of the optimization method in registration problem, we have implemented the meta-heuristic search strategies–genetic algorithm in multi-resolution domain. In this chapter, we have proposed to use intensity-based mutual information (MI) or relative entropy to describe the dispersive behavior of the 2D histogram. If the images are geometrically aligned, MI of the corresponding pixel pairs is maximal. The main choices involved in image registration are interpolation technique, estimation of the probability distributions and the search strategy to optimize the similarity metric.

2.1 Affine Transformation

The transformation technique for registration of images can be categorized according to the degrees of freedom. Although elastic transformations are more realistic (as most body tissues are deformable to some degree), rigid body registration is preferred [18]. Rigid body registration for determination of global alignment is followed by local elastic registration. Other nonlinear registration methods align small blocks of the floating image to the reference image in a linear manner. The affine transformation preserves the parallelism of lines but not their lengths or their angles. It extends the degrees of freedom of the rigid transformation with a scaling factor and a shearing of images in each dimension. In the registration problem, the images to be registered are called *floating image* **F**, which is transformed into the *reference image* **R**. In this chapter, we have restricted the transformation **T** to *affine transform* only, which is much more practical in biomedical imaging. Let **T** denote the spatial transformation that maps features or coordinates from one image to another image. For 2D affine registration, the transformation matrix is:
$x' = a^*x + b^*y + c$ and $y' = d^*x + e^*y + f$

$$\begin{bmatrix} x' \\ y' \end{bmatrix} = T \begin{bmatrix} x \\ y \\ 1 \end{bmatrix} \quad T = \begin{bmatrix} a\,b\,c \\ d\,e\,f \end{bmatrix} \quad \text{where } T \text{ is a } 2 \times 3 \text{ matrix of coefficients.}$$

2.2 Computation of Mutual Information-Based Similarity Metric

Mutual information (MI) measures the statistical dependence between two variables or the amount of information that one variable contains about the other. Let two variables, A and B, with marginal probability distribution, $p_A(a)$ and $p_B(b)$ and joint probability distribution $p_{AB}(a,b)$. MI, $I(A,B)$, measures the degree of dependence of A and B by measuring the distance between the joint distribution $p_{AB}(a,b)$ and the distribution associated with the case of complete independence $p_A(a)\,p_B(b)$:

$$I(A,B) = \sum_{a,b} p_{AB}(a,b) \log \frac{p_{AB}(a,b)}{p_A(a)\,p_B(b)}$$

MI is related to entropy as $I(A,B) = H(A) + H(B) - H(A,B)$ with $H(A)$ and $H(B)$ being the entropy of A and B, respectively, $H(A,B)$ joint entropy. $H(A) = -\sum_a p_A(a) \log p_A(a)$

$$H(A,B) = -\sum_{a,b} p_{AB}(a,b) \log p_{AB}(a,b)$$

The feature space or joint histogram is a two-dimensional plot showing the combinations of gray values in each of the two images for all corresponding points.

2.3 Multi-Resolution Approach

A powerful but simple structure for representing the images at more than one resolution is the image pyramid. The base of the pyramid contains the highest resolution representation of the image; the apex contains a lowest resolution approximation. With moving up the pyramid, both size and resolution decrease. The base level J is size $2^J \times 2^J$ or $N \times N$, intermediate level j is size $2^j \times 2^j$, where $0 \leq j \leq J$. Fully populated pyramids are composed of $J+1$ resolution levels from $2^J \times 2^J$ to $2^0 \times 2^0$, but in practice most pyramids are truncated to $P+1$ levels, where $j = J-P, \ldots, J-2, J-1, J$ and $1 \leq P \leq J$, since a 1×1 pixel image is of little value. In the present work, we have applied Haar wavelet transform, and initial information of the affine transformation parameters are acquired from level $j-1$ approximate image.

2.4 Optimization Based on GA

To apply GA in our registration problem, we have encoded the six transformation parameters by the binary numbers to form the chromosomes in a particular generation and optimized them to achieve the best possible result. In the registration problem, let us consider the multimodality images to be registered are *Image A* and *Image B*. *Image A* is reference image and *Image B* is the affine transformed image which will be correctly registered with *Image A*. Now for 2D affine transformation, six parameters of *T* are required to transform an image.

3 Fusion Process

The overall image fusion scheme of the proposed algorithm involves the decomposition of input registered images, computation of genetic algorithm-based selection technique, implementation of fusion rules [8, 9, 19] and then reconstruction of the composite-fused image as shown in Fig. 1. The proposed methodology is developed with two registered input images as described in Sect. 2. First, the registered input images (A, B) are decomposed into high (D_A and D_B) and low (C_A and C_B) frequency sub-bands by HAAR wavelet transform. At a particular decomposition level, both sub-bands must carry out the information of that resolution. Genetic searching algorithm is applied for integrating maximum complementary information from the high frequency image sub-bands. The low frequency image sub-bands produced by DWT are averaged for accumulating the gross structure. Thus, the *selection rule* adapted for this algorithm can be described as:

$$C_F^j(u,v) = \text{mean}\{C_A^j(u,v), C_B^j(u,v)\}$$
$$D_F^j(u,v) = \text{max of }\{D_A^j(u,v), D_B^j(u,v)\}$$

The superscript j denotes the jth level of resolution.

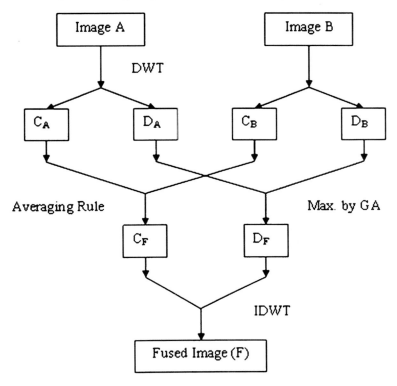

Fig. 1 Block diagram of image fusion scheme

Finally, by applying the inverse wavelet transform on the selected image sub-bands, the fused image (F) can be produced. It is noted that the proposed approach can be initiated without any fiducial points. It is an efficient, automatic and robust fusion technique using soft-computing approaches.

3.1 Averaging of LF Sub-band and Maximization of HF Sub-band Components

The fusion rule adapted in this chapter is based on *maximization* of high frequency components using genetic searching algorithms and averaging of low frequency sub-bands. The maximization and averaging process are independent of manual marking of fiducial points. More clearly, in the present approach, the fusion rules for complimentary feature selection are to average the approximate coefficients and maximize the details coefficients. GA is used for maximization of detail coefficients of the input images (A, B) at a particular resolution to achieve the improved fused images.

The approximate coefficients are selected by standard averaging method. Genetic searching algorithm is capable of preserving the important detail features into the fused image. GA is executed as an iterative process.

In the preset work, mutual information has been used for evaluating the quality of the fused image. It is also used for comparing the superiority of the proposed methodology over fuzzy-based fusion technique.

4 Experimental Results

In our study, first we execute the registration process as described in Sect. II between any two modalities CT, MR T1 and MR T2, and then the images are fused using the genetic searching technique, i.e. genetic-based maximization of HF sub-bands. The performance of each fusion methodology is measured with mutual information metric. To measure the superiority of GA-based fusion techniques, we have performed some comparative study of the fusion process considering fuzzy-based techniques. Tables 1–3 exhibit comparative study of genetic and fuzzy-based fusion method using MI as similarity measuring metric (Fig. 2). In Table 1, I1 & I2 are MR T1 and MR T2 images of the same patient and F1 & F2 are fusion algorithm using genetic and fuzzy fusion methodology respectively, Fig. 2.

In Table 2, I1 and I2 are MR T2 and CT images of the same patient and F1 and F2 are fusion algorithm using genetic and fuzzy fusion methodology, respectively (Fig. 3).

In Table 3, I1 and I2 are MR T1 and CT images of the same patient and F1 and F2 are fusion algorithm using genetic and fuzzy fusion methodology, respectively (Fig. 4).

Table 1 MR T1 vs. MR T2 fusion

MI	F1	F2
I1	1.8660	1.7177
I2	1.6590	1.5149
(I1 + I2)	3.525	3.2326

Table 2 MR T2 vs. CT fusion

MI	F1	F2
I1	1.8387	1.6509
I2	1.5618	1.3238
(I1 + I2)	3.4005	2.9747

Table 3 MR T1 vs. CT fusion

MI	F1	F2
I1	1.6772	1.5973
I2	1.5988	1.4824
(I1 + I2)	3.2760	3.0797

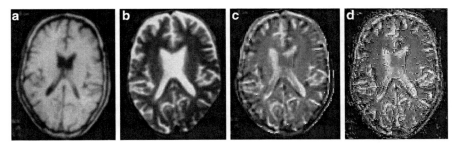

Fig. 2 (**a**) MR T1, (**b**) MR T2, (**c**) fused image using proposed genetic fusion algorithm (F1), (**d**) fused image using fuzzy clustering-based fusion technique (F2)

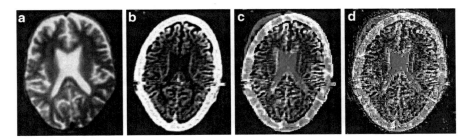

Fig. 3 (**a**) MR T2, (**b**) CT, (**c**) fused image using genetic-based proposed fusion algorithm (F1), (**d**) fused image using fuzzy clustering-based fusion technique (F2)

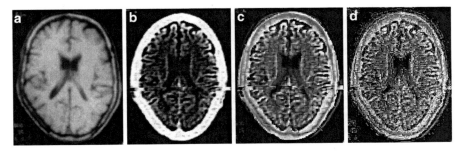

Fig. 4 (**a**) MR T2, (**b**) CT, (**c**) Fused image using proposed genetic fusion algorithm (F1), (**d**) fused image using fuzzy clustering-based fusion technique (F2)

5 Conclusion and Future Direction

This work describes combined computational approaches for medical image registration and fusion technique implemented on section of human brain using CT and MR T1 and MR T2 modalities. Our proposed technique shows an efficient and automated optimized process based on genetic algorithm to achieve a process for information combination for improved therapy planning by the physicians. In this work, we have presented an efficient mutual information (MI)-based image registration method combing GA search technique for non-rigid affine transformation. To overcome the influence of the existence of local maxima, GA mutation probability is adapted to 0.10. Presently all images to be fused are initially registered. The wavelet-based multi-resolution decomposition of the input images carries more complimentary information than the single resolution approach. Genetic-based automatic optimization technique maximizes the HF sub-bands according to the adapted fusion rule. The LF sub-bands produce the gross structure of the images. Thus, it is enough to average them to form the fused image. MI is again used as similarity measuring-metric that evaluates the dependency between fused and either of images A or B. Thus, it can be concluded that fused image must carry the information of both input images and more informative than either of single images. In this chapter, MI also evaluates the performance index (PI) for genetic and fuzzy-based fusion techniques and exhibits the better PI for GA-based fusion method. Most of the image fusion techniques developed so far are mostly application dependent; hence, a generalized fusion rule and algorithm is still needed in the field of medical image fusion.

Acknowledgements The authors gratefully acknowledge the help rendered by Prof. Rabindranath and other Scientists of National Brain Research Centre, Gurgao, India.

References

1. P. W. Josien, J. B. Pluim, A. Maintz and M. A. Viergever, "Mutual information based registration of medical images: a survey", *IEEE Trans. Med. Imaging*, vol. 22, pp. 986–1004, 2003
2. M. Bhattacharya and D. D. Majumder, "Registration of CT and MR images of Alzheimer's patient: a shape theoretic approach", *Pattern Recognit. Lett.*, vol. 21, no. 6–7, pp. 531–548, 2000
3. P. J. Slomka, J. Mandel, D. Downey and A. Fenster, "Evaluation of voxel-based registration of 3-D power Doppler ultrasound and 3-D magnetic resonance angiographic images of carotid arteries", *Ultrasound Med. Biol.*, vol. 27, no. 7, pp. 945–955, 2001
4. M. Bhattacharya and A. Das, "Affine registration by intensity and fuzzy gradient based correlation maximization", In: *Proceedings of IEEE 7th International Symposium on Bioinformatics & Bioengineering (IEEE BIBE-07)*, 14–17 October, Boston, MA, 2007
5. M. Bhattacharya and A. Das, "Registration of multimodality medical imaging of brain using particle swarm optimization", *LNCS* Springer-link, Intelligent Human Computer Interaction (IHCI), pp. 131–139, 20 to 23 January 2009
6. M. Jenkinson and S. Smith, "A global optimisation method for robust affine registration of brain images", *Med. Image Anal.*, vol. 5, no. 2, pp. 143–156, 2001
7. G. K. Matsopoulos, N. A. Mouravliansky, K. K. Delibasis and K. S. Nikita, "Automatic retinal image registration scheme using global optimization techniques", *IEEE Trans. Inform. Technol. Biomed.*, vol. 3, pp. 47–60, 1999
8. D. D. Majumder and M Bhattacharya, "Multimodal data fusion for Alzheimer's patients using dempster-shafer theory of evidence", 3rd Asian Fuzzy Systems Symposium 1998, Korea, *Proceedings* pp. 713–719, 1998
9. I. Bloch, "Information combination operators for data fusion: a comparative review with classification", *IEEE Trans. Syst. Man. Cybern. A Syst. Hum.*, vol. 26, no. 1, pp. 52–67, 1996

10. J. Zhang, X. Feng, B. Song, M. Li and Y. Lu, "Multi-focus image fusion using quality assessment of spatial domain and genetic algorithm", *Proceedings of IEEE Conference on Human System Interactions*, Krakow, Poland, pp. 71–75, May 25–27, 2008
11. Z. Wang, D. Ziou, C. Armenakis, D. Li and Q. Li, "A comparative analysis of image fusion methods", *IEEE Trans. Geosci. Remote Sens.*, vol. 43, no. 6, pp. 1391–1402, 2005
12. Y. Zheng, X. Hou, T. Bian and Z. Qin, "Effective image fusion rules of multi-scale image decomposition", *Proceedings of the 5th International Symposium on image and Signal Processing and Analysis*, 2007
13. D. Dutta Majumder and M. Bhattacharya, "Soft computing methods for uncertainty management in multimodal image registration and fusion problems", *Proceedings of the International Conference of Information Sciences JCIS -98*. North Carolina: Duke University, 1998
14. Q. Guihong, Z. Dali and Y. Pingfan, "Medical image fusion by wavelet transform modulus maxima", *Opt. Express*, vol. 9, pp. 184–190, 2001
15. J. A. Kennedy, O. Israel, A. Frenkel, R. Bar-Shalom and H. Azhari, "Improved image fusion in PET/CT using hybrid image reconstruction and super-resolution", *Int. J. Biomed. Imaging*, vol. 2007, p. 10, 2007
16. T. Z. Wei, W. J. Guo and H. S. Ji, "The wavelet transformation application for image fusion", In: *Wavelet Application VII*, H. H. Szu, ed., *Proceedings of SPIE 4056*, pp. 462–469, 2000
17. Y. Kirankumar and S. Shenbaga Devi, "Transform-based medical image fusion", *Int. J. Biomed. Eng. Technol.*, vol. 1, no. 1, pp. 101–110, 2007
18. R. Shekhar and V. Zagrodsky, "Mutual information-based rigid and nonrigid registration of ultrasound volumes", *IEEE Trans. Med. Imaging*, vol. 21, no. 1, pp. 9–22, 2002
19. G. Shafer, *A Mathematical Theory of Evidence*. Princeton: Princeton University Press, 1979

Chapter 45
Microcalcifications Detection Using Fisher's Linear Discriminant and Breast Density

G.A. Rodriguez, J.A. Gonzalez, L. Altamirano, J.S. Guichard, and R. Diaz

Abstract Breast cancer is one of the main causes of death in women. However, its early detection through microcalcifications identification is a powerful tool to save many lives. In this study, we present a supervised microcalcifications detection method based on Fisher's Linear Discriminant. Our method considers knowledge about breast density allowing it to identify microcalcifications even in difficult cases (when there is not high contrast between the microcalcification and the surrounding breast tissue). We evaluated our method with two mammograms databases for each of its phases: breast density classification, microcalcifications segmentation, and false-positive reduction, obtaining cumulative accuracy results around 90% for the microcalcifications detection task.

1 Introduction

Breast cancer is currently among the main causes of death by cancer. The American Cancer Society estimates that in 2009, 465,000 women around the world will die because of this type of cancer. Because every woman is under the risk of breast cancer, physicians recommend periodical clinical revisions (such as mammograms), with the idea of detecting any possible cancer symptom in an early phase, preventing it from getting to an advanced state in which the patient's life is in risk. Microcalcifications (small calcium deposits with 0.1–1.0 mm width) are the anomalies that tell us with more anticipation the beginning of a possible breast cancer, these can be detected in the mammogram and they look like small bright points of different shapes and sizes.

In the last two decades, several research groups have proposed different methods to detect microcalcifications making use of machine learning and computer vision

---

G.A. Rodriguez (✉)
National Institute for Astrophysics, Optics, and Electronics, Luis Enrique Erro No. 1,
Puebla 72840, Mexico
e-mail: g_rodriguez@inaoep.mx

techniques. If we focus on the segmentation step of the proposed methods, these can be classified into four types: regions based [1, 2], edges (borders) based [3, 4], clustering (thresholds) [5–8], and model based (supervised segmentation) [9, 10, 13]. The regions-based methods try to find a characteristic value of the pixels that belongs to a microcalcification, but due to the different tones between microcalcifications and their surrounding regions, finding this characteristic value is very difficult.

Another type of method to identify microcalcifications is based on borders detection. The main disadvantage of most of these methods is that they only detect those edges when there is a high contrast between the microcalcification and its surrounding region.

The most common segmentation technique used to detect microcalcifications is based on thresholds. Some of these methods assume that microcalcifications correspond to a high frequency in the image, so they first use a wavelets transformation to highlight the high frequencies of the image and then use a threshold segmentation method. One of the main disadvantages of these methods is that there is not always a high contrast between microcalcifications and their surrounding region as it happens in the case of breasts composed by a lot of glandular tissue (dense breasts). Then, the use of a single threshold for all types of breasts does not work properly, resulting in a bad detection of microcalcifications and the generation of a high number of false positives.

Finally, there are methods that use information about the objects that will be segmented. These methods are known as model based or supervised methods. Model-based methods are used when a geometric pattern can be found in the objects of interest, as it happens with microcalcifications images. It is important to note that model-based methods may have the problem that the identification results are highly dependent of the images used to learn the model. This problem usually happens when the descriptive characteristics are based on the images tones.

Analyzing the problems of the methods described before, we can notice that the difference in tone between the microcalcifications and their surrounding region is very important. For this reason, we propose to take into account the breast density, which causes differences in tone in the mammograms. This means that we need to identify the main tissue (fat or glandular) that composes the breast. With this information and using Fisher's Linear Discriminant (FLD) during the segmentation phase, our method is able to detect microcalcifications of any shape, size, and with different tonalities, even in breasts where there is not high contrast between the microcalcification and its surrounding breast tissue. After all possible microcalcifications have been detected and segmented, we obtain descriptive characteristics from them to train a classification algorithm to reduce the number of false positives generated in this step.

In Sect. 2 we describe how FLD works, in Sect. 3 we present the mammograms databases that we used to test our method, in Sect. 4 we show how the proposed method works, in Sect. 5 we present our results, and finally in Sect. 6, we show our conclusions.

2 Fisher's Linear Discriminant

The goal of any type of discriminant analysis is to find descriptive characteristics that allow the distinction of objects that belong to different classes. In the case of FLD, it is used to reduce the dimensionality of a space, keeping useful information to discriminate among classes. The main objective of FLD is to find a linear transformation U that maximizes the scatter rate between different classes projected into the new space and that minimizes the scatter rate within the projected classes.

In a more formal way, given a set of N examples, represented in an n-dimensional space, in which each of the examples belongs to one of k classes. The between-class scatter matrix S_b and the within-class scatter matrix S_w are defined as:

$$S_b = \sum_{i=1}^{k} N_i (\mu_i - \mu)(\mu_i - \mu)^T \tag{1}$$

$$S_w = \sum_{i=1}^{k} \sum_{j=1}^{N_i} (x_j - \mu_i)(x_j - \mu_i)^T \tag{2}$$

where μ is the general mean and μ_i is the mean of class i. Then, FLD chooses the linear transformation matrix U that maximizes the determinant of the S_b matrix of the projected examples over the determinant of the S_w matrix of the projected examples.

$$U_{\text{opt}} = \arg\max_{U} \frac{|U^T \cdot S_b \cdot U|}{|U^T \cdot S_w \cdot U|} \tag{3}$$

If S_w is a non-singular matrix, the optimum linear transformation U_{opt} corresponds to the m eigenvectors with higher eigenvalues of $\left(S_w^{-1}\right)(S_b)$. Once the optimum linear transformation is found, when we want to classify a new object, it is projected to the discriminant space and the Euclidean distance between it and each of the means of the different classes are calculated. The class assigned to the new object corresponds to the one closer to the object. In our case, there are only two classes: microcalcification and breast tissue. In Sect. 4 we describe how we use FLD to detect microcalcifications.

3 MIAS and ISSSTEP Databases

In this study, we used two mammograms databases, the MIAS and the ISSSTEP databases. The *Mammographic Image Analysis Society* (MIAS) is a public database that contains 322 images of medium lateral mammograms [11], and the size of each image is of 1024 × 1024 pixels and the images are stored in pgm format.

The second database that we used is known as the ISSSTEP database. This database was created with mammograms from the X-rays laboratory of the Social

Services and Security Institute of Employees to the Service of the State of Puebla. This database has 336 mammograms. The size of the images is of 1024 × 1024 pixels and the images are stored in the jpg format.

4 Proposed Method

The general scheme of the method proposed in this work (see Fig. 1) can be divided into three main phases. In the first phase, we recognize which is the predominant type of tissue found in the breast, either fat or glandular. From this classification, we select the discriminant space that better allows us to differentiate between microcalcifications and breast tissue images. In the second phase, the segmentation step, we use a fixed size window to analyze the breast looking for microcalcifications. The image inside the window is projected into the corresponding discriminant space, and there the image is analyzed to identify whether it corresponds to a microcalcification or to breast tissue. In case the image is identified as a microcalcification, this area of the mammogram is marked to indicate the existence of a possible microcalcification. This process is repeated until the window has analyzed the entire breast image.

In the final phase, we extract descriptive characteristics of the probable microcalcifications images, and we reduce the amount of false positives generated in the segmentation step using a classification algorithm. In the following subsections, we describe in a more detailed way each of the phases of the proposed method.

4.1 Breast Density Identification

In the first phase of this algorithm, we separate the breast from the background of the mammogram using fuzzy c-means. We only consider two groups, one of them contains all those pixels that belong to the breast and the other those pixels that are part of the background of the mammogram. We only keep the largest one that corresponds to the breast.

Considering that the main radiological characteristic that differentiates the fat from the glandular tissue is the regions tone, we calculate and analyze the breast

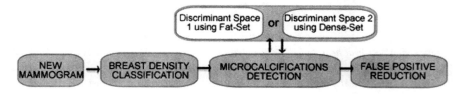

Fig. 1 Structure of proposed method

histogram; this histogram is divided into seven static intervals and for each of them we obtain its mean, frequency, and standard deviation. We selected these intervals in an experimental way, identifying that these intervals and attributes provided significant information for our task. In the final step of the breast density classification phase, we create a database using the characteristics extracted from each mammogram. This database is used in the training phase of the classifiers that allow us determining the breast category (fat or dense). In this work, we are interested in knowing the predominant breast tissue type; so we only take into account two classes (the cases classified as fat-glandular in the MIAS database are considered as dense).

4.2 Discriminant Spaces and Microcalcifications Detection Using FLD

The second phase of our method consists in the segmentation of microcalcifications using FLD. The goal of using FLD is to generate a discriminant space with a lower dimension in which we are able to distinguish between microcalcifications and breast tissue images. In the learning phase, we create two discriminant spaces, one for dense breasts and another for fat breasts. In order to generate these two spaces, we created two microcalcifications databases, one of them contains microcalcifications images from fat breasts and the other contains microcalcifications from dense breasts. Each database contains images of both classes (microcalcifications and breast tissue), and all of them are 12×12 pixels of size, stored in jpg format.

The process to generate a discriminant space with FLD (see code 1) starts with a set of M images, where m_1 of them belongs to the microcalcification class ($class_1$) and m_2 to the breast tissue class ($class_2$). After this, we calculate the general mean and the means of the two classes (lines 2–4). With these values, we calculate the between-class scatter matrix S_b and the within-class scatter matrix S_w (lines 9–10). We then find the linear transformation matrix U that is useful to project the data into a space in which we can easily discriminate between the two classes (see Sect. 2). If we directly apply FLD to our images, we fall into a problem; the matrix S_w is always singular. To solve this problem, the space is reduced using principal component analysis (PCA) before applying FLD [12] (lines 6–8). As the final step of the learning phase, all the images are represented as column vectors, and they are projected into the corresponding discriminant space (line 15).

Once we generated the two discriminant spaces, when we need to analyze a new mammogram looking for microcalcifications (see code 2), the mammogram is classified as a fat or dense (line 4). Then, in the segmentation phase, we use a 12×12 pixels window to analyze the breast area looking for microcalcifications. The region inside the window is projected into the corresponding discriminant space created with FLD (line 10).

Code 1. Microcalcifications Discriminant Space (image base)
1. Transform each image into a vector X_i
2. Mean_g = mean $(X_1, ..., X_m)$
3. Mean_c1 = mean $(X_i$ of class$_1)$
4. Mean_c2 = mean $(X_k$ of class$_2)$
5. matrix = $[X_1, .., X_m]$
6. M = matrix^T ∗ matrix
7. [eigevectors, eigenvalues] = eigen(M)
8. V = matrix∗ eigenvectors
9. S_B = calculateSb(Mean_g, Mean_c$_1$, Mean_c$_2$)
10. S_W = calculateSw(Mean_c$_1$, Mean_c$_2$)
11. Sb = $V^T * S_B * V$
12. Sw = $V^T * S_W * V$
13. [Eigenvec, Eigenval] = eigen(Sb, Sw)
14. U = V∗ Eigenvec
15. $(Y_i = U^T * X_i)$
16. Mean_class$_1$ = mean(images class$_1$)
17. Mean_class$_2$ = mean(images class$_2$)
18. Save U, Mean_class$_1$, Mean_class$_2$

Code 2. Detection of microcalcifications using FLD (Mammogram)
1. Mammo = Mammogram
2. breast = Fuzzy_cmeans (Mammo,2)
3. hist = histogram (breast)
4. density = density_breast (hist)
5. Load U, Mean_class$_1$, Mean_class$_2$ corresponding to the density
6. **While** (i belong to breast)
7. **While** (j belong to breast)
8. Image = Mammo $(i:1+11, j:j+11)$
9. VImage = vector (Image)
10. New_VImage = $U^T *$ VImage
11. Dist = Min_Euclid (New_VImage, Mean_class$_1$, Mean_class$_2$)
12. **if** (Dist == Mean_class$_1$)
13. Highlight(Mammo, $[i:i+11, j:j+11]$)
14. **end if**
15. j + + **end while**
16. i + + **end while**

After projecting the image, we use the Euclidean distance to identify whether the image is more similar to the microcalcification or to the breast tissue class (line 11). In case the region is classified as a microcalcification (line 12), it is highlighted in the mammogram and the window is moved one position. This process is repeated until the window has been moved to cover the whole breast image.

4.3 False-Positive Reduction

In the last phase of our method, we reduce the amount of false positives generated in the detection step. For this, we use descriptive characteristics (minimum and maximum gray levels, area, perimeter, convex area, orientation, minor axe length, major axe length, and solidity) of the probable microcalcifications. We trained different classifiers to distinguish between microcalcifications and breast tissue, thus reducing the number of false positives generated in the segmentation phase.

5 Experimental Results

We tested our method with the MIAS and the ISSSTEP databases described in Sect. 3. Each of the main phases was evaluated to measure its performance and then the global performance of the method was tested. We used as evaluation measures the proportion of true positives (TP) and the true negatives (TN), only for the detection phase we used the true-positives rate (TPR) and the number of false positive per image (FPI). For the evaluation of the breast density phase, we used a tenfold cross-validation technique; for this experiment, we used 295 mammograms (see Table 1). To evaluate the microcalcifications detection phase, we only used mammograms that contained microcalcifications (20 from the MIAS and 50 from the ISSSTEP). We tested our method with and without considering the breast density knowledge. We also compared our results with a wavelets and thresholds method [5] (see Table 2 and Fig. 2). Finally, the evaluation of the false-positive

Table 1 Results of breast density

Breast density identification phase				
	MIAS		ISSSTEP	
Classifiers	TP	TN	TP	TN
MLPerceptron	0.85	0.98	0.92	0.96
PART	0.81	0.94	0.92	0.97
IB1	0.85	0.96	0.9	0.97

Table 2 Results of microcalcifications detection

Microcalcifications detection phase								
	MIAS				ISSSTEP			
	Fat		Dense		Fat		Dense	
	TPR	FPI	TPR	FPI	TPR	FPI	TPR	FPI
FLD and breast density	91.8	1	90.9	4	94.3	6.1	92	7.4
FLD without breast density	88.9	18	70	21	65.1	2.3	83	5.3
Wavelets and thresholds [5]	81.2	16	74.1	20	85.7	19	76	20

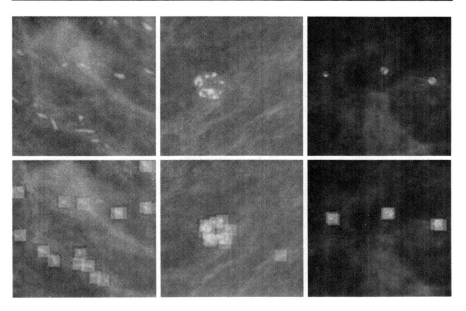

Fig. 2 Microcalcifications detected

Table 3 Results of false-positive reduction

False-positive reduction phase	MIAS				ISSSTEP			
	IB1		PART		IB1		PART	
	TP	TN	TP	TN	TP	TN	TP	TN
FLD with breast density	0.87	0.99	0.94	0.98	0.9	0.99	0.9	0.99
Wavelets and thresholds [5]	0.91	0.95	0.99	0.99	0.7	0.87	0.9	0.99

Table 4 Global results

	Global performance	
	TP	TN
MIAS	0.88	0.9
ISSSTEP	0.9	0.9

reduction phase was also performed with a tenfold cross validation technique, and we used two classifiers: IB1 and PART (see Table 3). The global performance of our method is shown in Table 4.

It is important to note that the TPR obtained with our method was higher than 90% for both databases, even with difficult breasts (dense breasts). Our method obtained a TPR 10% higher than when we do not take into account the breast density and higher than the method based in wavelets and threshold.

6 Conclusions and Future Work

In this work, we proposed a microcalcification detection method based on FLD that considers the breast density and allows reducing the amount of false positives generated during the segmentation phase. The proposed method obtained satisfactory results in the experiments, showing a performance higher than 90%. As future work, we propose to modify the density breast classification to identify which regions correspond to fat tissue and which ones to glandular tissue. Thus, we can use the discriminant spaces in an adaptively way instead of using one discriminant space for the entire breast.

References

1. Kim J and Park H (1997) Surrounding region dependence method for detection of clustered microcalcifications on mammograms. ICIP '97 Proceedings of the 1997 International Conference on Image Processing (ICIP '97), vol. 3, pp. 535–538
2. Morrow W, Paranjape R B, Rangayan R M and Desautels J E (1992) Region based contrast enhancement of mammograms. IEEE Transactions on Medical Imaging 11: 392–406
3. Fu J C, Lee S K, Wong S, Yeh J Y, Wang A H, and Wu H K (2005) Image segmentation feature selection and pattern classification for mammographic microcalcifications. Computer Medical Imaging and Graphics. doi:10.1016/j.compmedimag.2005.03.002

4. Lee Y and Tsai D (2004) Computerized classification of microcalcifications on mammograms using fuzzy logic and genetic algorithm. Proceedings of SPIE Medical Imaging, pp. 952–957
5. Flores B and Gonzalez J (2004) Data mining with decision trees and neuronal networks for calcifications detection in mammograms. Lecture Notes in Computer Science, pp. 232–241
6. Mousa R, Qutaishat M and Moussa A (2005) Breast cancer diagnosis system based on wavelet analysis and fuzzy neural. Expert Systems with Applications – Elsevier 28: 713–723
7. Wang T and Karayiannis N (1994) Detection of microcalcifications in digital mammograms using wavelets. IEEE Transactions on Medical Imaging 17: 498–509
8. Wu Y, Huang Q, Peng Y H and Situ W (2006) Detection of microcalcifications based on dual threshold. Lecture Notes in Computer Science 4046/2006: 347–354
9. Nishikawa R, El-Naqa I, Yang Y, Wernick M and Galatsanos N (2002) Support vector machine learning for detection of microcalcifications in mammograms. IEEE Transactions on Medical Imaging. doi:10.1109/ISBI.2002.1029228
10. Zhang W, Yoshida H, Nishikawa R and Doi K (1998) Optimally weighted wavelet transform based on supervised training for detection of microcalcifications in digital mammograms. Medical Physics 25: 949–956
11. Suckling J, Parker J, Dance D, Astley S, Hutt I, Boggis C, Ricketts I, Stamatakis E, Cerneaz N, Kok S, Taylor P, Betal D and Savage J (1994) The Mammographic Image Analysis Society digital mammogram database. Exerpta Medica: 375–378
12. Belhumeur P, Hespanha J, and Kriegman D (1997) Eigenfaces vs Fisherfaces: Recognition using class specific linear projection. IEEE Transactions on Pattern Analysis and Machine Intelligence. doi:10.1109/34.598228
13. Freixenet J, Oliver A, Martí R, Lladó X, Pont J, Pérez E, Denton E R E, and Zwiggelaar R (2008) Eigendetection of masses considering false positive reduction and breast density information. Medical Physics. doi:10.1118/1.289950

Chapter 46
Enhanced Optical Flow Field of Left Ventricular Motion Using Quasi-Gaussian DCT Filter

Slamet Riyadi, Mohd. Marzuki Mustafa, Aini Hussain, Oteh Maskon, and Ika Faizura Mohd. Nor

Abstract Left ventricular motion estimation is very important for diagnosing cardiac abnormality. One of the popular techniques, optical flow technique, promises useful results for motion quantification. However, optical flow technique often failed to provide smooth vector field due to the complexity of cardiac motion and the presence of speckle noise. This chapter proposed a new filtering technique, called quasi-Gaussian discrete cosine transform (QGDCT)-based filter, to enhance the optical flow field for myocardial motion estimation. Even though Gaussian filter and DCT concept have been implemented in other previous researches, this filter introduces a different approach of Gaussian filter model based on high frequency properties of cosine function. The QGDCT is a customized quasi discrete Gaussian filter in which its coefficients are derived from a selected two-dimensional DCT. This filter was implemented before and after the computation of optical flow to reduce the speckle noise and to improve the flow field smoothness, respectively. The algorithm was first validated on synthetic echocardiography image that simulates a contracting myocardium motion. Subsequently, this method was also implemented on clinical echocardiography images. To evaluate the performance of the technique, several quantitative measurements such as magnitude error, angular error, and standard error of measurement are computed and analyzed. The final motion estimation results were in good agreement with the physician manual interpretation.

1 Introduction

An early detection of coronary heart diseases can be performed by analyzing the motion of left ventricular. The abnormalities of its motion indicate the existence of ischemic segments of myocardium. The motion abnormalities may be weak-motion

M.M. Mustafa (✉)
Department of Electrical, Electronic and Systems Engineering, Faculty of Engineering and Built Environment, Universiti Kebangsaan Malaysia, Bangi, 43600 Selangor, Malaysia
e-mail: marzuki@eng.ukm.my

(hypokinetic), no-motion (akinetic), or not synchronic motion (dyskinetic) of myocardial segments. Currently, the myocardial motion is estimated manually based on the ultrasound cardiac video by doctor who has echocardiographic expertise.

The study of myocardial motion from 2D echocardiography has been an active research area in past decades. Various techniques have been proposed by researcher such as grids of intersection technique based on elastic nonrigid structure and Bayesian estimation framework [1–3]. Ledesma et al. developed a new spatiotemporal nonrigid registration to estimate the displacement fields from two-dimensional ultrasound sequence of the cardiac [4]. Another approach from Kazulinski (2001) detects the movement of myocardial using speckle tracking [5].

Other popular technique in terms of motion estimation, which is called optical flow technique, also has been implemented for myocardial. Loncaric and Macan introduced point-constrained optical flow to detect the left ventricular motion [6]. Tested on several synthetic images, the point constrained method reduced the optical flow error. However, the method failed to determine accurate flow fields of cardiac motion using real magnetic resonance images due to high variation in image brightness. Malpica et al. proposed a combined local and global regularization of the flow field for endocardial tracking [7]. The algorithm successfully tracked the cardiac wall on the myocardial contrast echocardiography sequences. Some wall determination error occurred when it was applied on the noisy images. In short, there are many opportunities to improve the performance of optical flow technique.

Generally, enhancement of left ventricular (LV) flow field can be performed by improving the quality of ultrasound images and the accuracy of motion estimation. In this chapter, we propose a new combination approach of Gaussian and discrete cosine transform (DCT) to reduce speckle noise and smoothen the computed optical flow field. On its own, these two techniques are very popular low pass filtering technique and video compression method, respectively. The proposed method, which we called quasi-Gaussian DCT (QGDCT) filter, is a quasi discrete Gaussian filter. Its coefficients are derived from a selected two-dimensional cosine basis function, and the Gaussian approach is used to suppress the speckle noise pattern while the selected DCT approach is used to preserve the image content.

2 Methodology

The methodology involves several tasks such as the collection of echocardiographic images, the reduction of speckle noise, the computation of optical flow field, and the improvement of flow field smoothness. In this work, the echocardiographic videos are obtained from patients with healthy and non-healthy cardiac at resting condition (baseline). The proposed filter is first applied to echocardiographic images to reduce their speckle noise. Then, the motion vector is computed by global regularization optical flow technique between two consecutive frames of echocardiographic images. To enhance the smoothness of motion vector, we implemented again the QGDCT on the computed flow field. Performance evaluation of the proposed technique has been performed by computing and analyzing several quantitative measurements such as magnitude error, angular error, and standard error of measurement. Before applying on clinical echocardiographic images, the algorithm was first validated on synthetic image that simulates a contracting myocardium motion.

3 Quasi-Gaussian DCT-Based Filter

The main drawback of Gaussian filtering is uniform blurring on the image region caused by the natural kernel of Gaussian. On the other hand, the DCT function was proven as a standard image coding technique which means that the DCT has a powerful function to represent an image. Our basic idea is to use the DCT function to create a new Gaussian kernel, called QGDCT-based filter.

Using 2D cosine basis function ($N = 8$) in Fig. 1, only 36 selected top-left functions are usually used for JPEG compression. These functions are not appropriate to be used to create the QGDCT kernel due to large computing time reason. According to our previous observation, the down-right function, indicated by red circle in Fig. 1, is more suitable since it represents the high frequency function.

In order to understand how to create the QGDCT kernel, it is easier by considering 1D basis function ($N = 8$) with $u = 7$ as shown in Fig. 1c. This function consists of positive and negative coefficients. A quasi-Gaussian model is created by removing the negative coefficients from the function as shown in Fig. 1d. It can be seen that the model is similar to the natural discrete Gaussian kernel without even term coefficients. The 2D QGDCT kernel is obtained by the similar steps applied on the down-right function of 2D DCT basis.

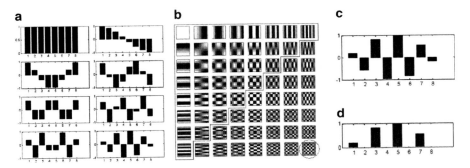

Fig. 1 Cosine basis function with $N = 8$ in (**a**) 1D, (**b**) 2D [8], (**c**) selected 1D DCT function, and (**d**) a quasi-Gaussian model

4 Global Regularization Optical Flow

Optical flow field is computed by assuming that the intensity of the object remains constant from the initial point of $I(x, y, t)$ toward the final position of $I(x + \delta x, y + \delta y, t + \delta t)$. Thus, the intensity constraint of a point is represented by:

$$I(x,y,t) = I(x + \delta x, y + \delta y, t + \delta t) \quad (1)$$

Within small values of δx, δy, and δt, the intensity constraint can be represented using the first order Taylor series expansion shown below,

$$I(x + \delta x, y + \delta y, t + \delta t) = I(x,y,t) + \frac{\partial I}{\partial x}\delta x + \frac{\partial I}{\partial y}\delta y + \frac{\partial I}{\partial t}\delta t \quad (2)$$

Simplifying (1) and (2), the basic constraint of the optical flow equation can be expressed as:

$$I_x v_x + I_y v_y + I_t = 0 \quad \text{or} \quad I_x u + I_y v + I_t = 0, \quad (3)$$

The intensity constancy assumption is very sensitive to brightness changes that contrarily often appear in natural cases. Therefore, it is important to introduce other assumptions to solve the estimation problem. Horn–Schunck [9] proposed global smoothness error (E_S) and data error (E_D) defined as

$$E_S = \iint_D (u_x^2 + u_y^2)(v_x^2 + v_y^2) dx dy$$
$$E_D = \iint_D (I_x u + I_y v + I_t)^2 dx dy \quad (4)$$

To determine the optical flow vector (u, v), they computed an iterative solution to minimize both errors by following the equation $E_D + \lambda\, E_S$, where λ represents the determined smoothing factor.

5 Results and Discussion

5.1 Speckle Noise Reduction

The first main task of the proposed filter is to enhance the quality of ultrasound images. To evaluate the filter capability, Riyadi et al. has computed some quantitative performances which are mean square error (MSE), peak signal to noise ratio (PSNR), speckle suppression index (SSI), and speckle image statistical analysis (SISA) [10]. According to the evaluation in [10], it can be concluded that the filtered image of QGDCT yields the best speckle suppression result when compared to the Mean, Median and Frost filters since the QGDCT has the lowest MSE, the highest PSNR and the lowest SSI values. In addition, the proposed QGDCT filter is also effective in preserving the image content since the SISA value is closed to 1.0.

5.2 Validation of Optical Flow Field on Synthetic Motion Images

To validate the enhancement of optical flow field, the proposed technique was implemented on synthetic sequences for which the actual motion field is known. A quarter part of parasternal short axis (PSAX) view of clinical echocardiographic image was used as ultrasound image model (Fig. 2). This model is then simulated by applying a radial displacement to down-right corner with increasing magnitude. This kind of motion reflects the myocardial contraction, whereas endocardial has larger displacement compared to epicardial. The performance of the proposed technique is assessed by computing mean absolute magnitude error and mean angular error. Table 1 indicated that the proposed filter performed the best result compared to other filters with 2.31% of mean relative magnitude error and 6.35° of mean angular error.

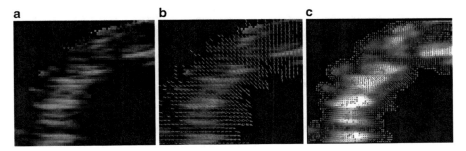

Fig. 2 (a) Original clinical ultrasound image; (b) synthetic motion of myocardial; and (c) computed optical flow using the proposed technique

Table 1 Mean relative magnitude error and mean angular error of computed optical flow using different filtering technique

Filter	Mean relative magnitude error (%)	Mean angular error (°)
Mean	38.2	9.14
Median	37.7	9.46
Frost	39.8	14.18
QGDCT	2.31	6.35

5.3 Application on Clinical Ultrasound Images

The proposed technique was implemented on clinical echocardiographic images. Figure 3a–d shows the myocardial movement sequences that indicate the diminishing motion of LV. These images were taken from a normal cardiac motion with PSAX view of the LV. The LV consists of a cavity in a centre surrounded by myocardial segments. The normal cardiac appearance is indicated by a uniform displacement of myocardial segments. They move in radial forward and backward directions with respect to the centre of cavity, which describe the profile of systole and diastole of a cardiac. In one period of time, the cavity size will shrink progressively during systole, and it will be followed by a brief instance of diastole that recovers the cavity [11].

Figure 4a shows the computed optical flow using the global regularization technique for myocardial motion of two consecutive images of Fig. 3a, b. The arrows represent the displacement of the pixels and their directions. The different arrows' colors represent the speed of displacements that change gradually from red to blue indicating a large to small displacement representation. The figure shows that the optical flow technique performed a right direction of LV motion. These arrows diminish as the move toward the centre of the cavity which is in agreement with the

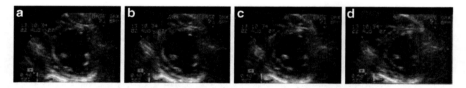

Fig. 3 (**a–d**) Clinical echocardiography images showing the myocardial movement sequences of the LV

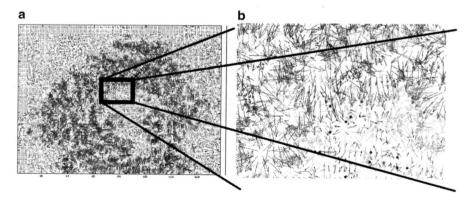

Fig. 4 (**a**) Computed optical flow of myocardial motion of two consecutive images of Fig. 3a, b; and (**b**) the zoom-view of error computed arrows

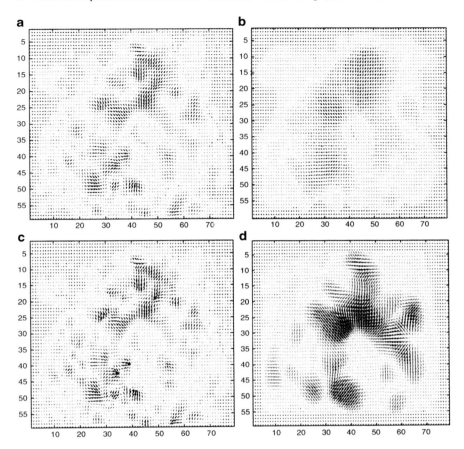

Fig. 5 The optical flow vector after using (**a**) mean, (**b**) median, (**c**) frost, and (**d**) the QGDCT filter

understanding of cardiac sequence image that during systole, the cavity size should diminish in a concentric manner. However, it can be seen that there are many error-computed arrows as shown in Fig. 4b.

To smoothen the arrows, the QGDCT filter is applied on the computed optical flow. A better result as shown in Fig. 5d is obtained. As a comparison, other result using mean, median, and Frost filtering are also presented in Fig. 5a–c, respectively. It can be seen that the QGDCT filtering has better smoothing property compared with the other filtering result.

The performance of the technique is also observed using the standard error of measurement (SEM):

$$\text{SEM} = \frac{\sum \sigma_\theta |(u,v)|}{\left(\sum |(u,v)|\right)^2} \tag{5}$$

Table 2 SEMs of enhanced flow field by different filters

Filter	SEM
Without filter	0.0880
Mean	0.0357
Median	0.0393
Frost	0.0380
QGDCT	0.0062

where σ_θ is the standard deviation of optical flow angle and $|(u,v)|$ is the magnitude of optical flow. The computed SEMs are tabulated in Table 2. According to the SEM, the QGDCT produces significant improvement compared with no filtering result and using other filters results. In this case, SEM represents the homogeneity of the arrow direction. It means that using the QGDCT, the computed arrows have more uniformity which complies with the true movement of myocardial segments.

The proposed technique for LV motion estimation has been implemented on more than 100 echocardiography sequences. The echo data consist of four standard views of echocardiography on baseline, low-dose, peak-dose, and recovery condition for healthy and non-healthy patients. According to our visual comparison on every sequence, the pattern of enhanced optical flow of myocardial motion agrees with the cardiologist manual interpretation.

6 Conclusion

In this work, new filter, called QGDCT, has been described. It is used to enhance the computed optical flow for myocardial motion estimation. The filter combines the concept of Gaussian filter and DCT technique. Selected function of 2D DCT is used to create a discrete quasi-Gaussian kernel. Our finding showed that the proposed filter can significantly improve the smoothness of optical flow field when compared to the initial result using global regularization technique and other filtering techniques. For future work, implementation on clinical echo images should be analyzed on each segment and quantified to obtain accurate result.

Acknowledgment The authors would like to acknowledge Universiti Kebangsaan Malaysia (Project code UKM-GUP-TKP-08-24-080) for the financial support awarded for this research.

References

1. Petland A. and Horowitz B. (1991) Recovery of non-rigid motion and structure. IEEE Transaction on Pattern Analysis and Machine Intelligent 13, pp. 730–742
2. Papademetris X., Sinusas A. J., Dione D. P., and Duncan J. S. (1999) 3D cardiac deformation from ultrasound images. Proceeding of MICCAI 99, pp. 420–429

3. Qazi M., Fung G., Krishnan S., Rosales R., Steck H., Rao R. B., Poldermans D., and Chandrasekaran D. (2007) Automated heart wall motion abnormality detection from ultrasound images using Bayesian networks. Proceedings of IJCAI 07, pp. 519–525
4. Ledesma-Carbayo M. J., Kybic J., Desco M., Santos A., Suhling M., Hunziker P., and Unser M. (2005) Spatio-temporal nonrigid registration for ultrasound cardiac motion estimation. IEEE Transcation on Medical Imaging 24(9), pp. 1113–1126
5. Kaluzyinski K., Chen X., Emelianov S., Skovoroda A., and O'Donnell M. (2001) Strain rate imaging using two-dimensional speckle tracking. IEEE Transactions on Ultrasonics, Ferroelectrics, and Frequency Control 40(2), pp. 1111–1123
6. Loncaric S. and Macan T. (2000) Point constrained optical flow for LV motion detection. Proceedings of SPIE 3978, p. 521
7. Malpica N., Garamedi J. F., Desco M., and Schiavi E. (2005) Endocardial Tracking in Contrast Echocardiography Using Optical Flow. Springer, Berlin, ISBMDA LNBI 3745, pp. 61–68
8. Richardson I. E. G. (2003) H.264 and MPEG-4 Video Compression: Video Coding for Next-generation Multimedia. Wiley, West Sussex
9. Horn B. K. P. and Schuck B. G. (1981) Determining optical flow. Artificial Intelligence 9, pp. 185–203
10. Riyadi S., Mustafa M. M., Hussain A., Maskon O., and Noh I. F. M. (2009) Quasi-Gaussian DCT Filter for Speckle Reduction of Ultrasound Images. Springer, Berlin, LNCS 5857. pp. 848–851
11. Anderson B. (2002) Echocardiography: The Normal Examination of Echocardiographic Measurements. Blackwell Publishing, Boston, Chapter 2

Chapter 47
An Efficient Algorithm for Denoising MR and CT Images Using Digital Curvelet Transform

S. Hyder Ali and R. Sukanesh

Abstract This chapter presents a curvelet-based approach for the denoising of magnetic resonance (MR) and computed tomography (CT) images. Curvelet transform is a new multiscale representation suited for objects which are smooth away from discontinuities across curves, which was developed by Candies and Donoho (Proceedings of Curves and Surfaces IV, France:105–121, 1999). We apply these digital transforms to the denoising of some standard MR and CT images embedded in white noise, random noise, and poisson noise. In the tests reported here, simple thresholding of the curvelet coefficients is very competitive with "state-of-the-art" techniques based on wavelet transform methods. Moreover, the curvelet reconstructions exhibit higher perceptual quality than wavelet-based reconstructions, offering visually sharper images and, in particular, higher quality recovery of edges and of faint linear and curvilinear features. Since medical images have several objects and curved shapes, it is expected that curvelet transform would be better in their denoising. The simulation results show the outperforms than wavelet transform in the denoising of both MR and CT images from both visual quality and the peak signal-to-noise (PSNR) ratio points of view.

Keywords Denoising · Discrete wavelet transform · Radon transform · Discrete curvelet transform · FFT · Thresholding

1 Introduction

The images usually bring different kinds of noises in the process of receiving, coding and transmission. The recent advent of digital imaging technologies such as CT and MRI has revolutionized modern medicine. To achieve the best possible diagnoses, it is important that medical images be sharp, clear and free of noise and artifacts.

S.H. Ali (✉)
Research Scholar, Anna University, Chennai, Tamil Nadu, India
e-mail: hyderfathima@yahoo.com

While the technologies for acquiring digital medical images continue to improve, resulting in images of higher and higher resolution and quality, noise remains an issue for many medical images. Removing noise in these digital images remains one of the major challenges in the study of medical imaging.

All noise elimination methods are divided into spatial and transformation domain. In spatial domain, the data operation is carried out on the original image, and the image grey values are processed to eradicate noise with methods like averaging and wiener filtering. In the transformation domain, the image is transformed, and the coefficients are processed to eliminate noise [1, 2]. If medical images are corrupted during capturing and transmission, it is impossible to rescue a human being from harmful effects. Recent wavelet thresholding-based denoising methods are capable of suppressing noise while maintaining the high frequency signal details. The wavelet thresholding scheme [3], which recognizes that by performing a wavelet transform of a noisy image, noise will be represented principally as small coefficients in the high frequencies. By setting these small coefficients to zero will eliminate much of the noise in the image.

In this chapter, we report initial efforts at image denoising based on curvelet transform – which have been proposed as alternatives to wavelet representation of image data [4]. The structural elements of curvelet transform include the parameters of dimension, location and orientation. Therefore, curvelet transform is superior to wavelet in the expression of image edge, such as geometry characteristic of curve and beeline. We exhibit higher peak signal-to-noise (PSNR) values on standard images such as CT and MRI, across a range of underlying noise levels. Curvelet denoising predicts that in recovering images which are smooth away from edges, and curvelets will obtain dramatically smaller asymptotic mean square error of reconstruction than wavelet methods.

2 Wavelet Denoising

The wavelet thresholding procedure removes noise by thresholding only the wavelet coefficients of the detail sub-bands, while keeping the low resolution coefficients unaltered. There are two thresholds frequently used, i.e. a hard threshold and a soft threshold [5], The hard-thresholding function keeps the input if it is larger than the threshold; otherwise, it is set to zero [1]. It is described as:

$$\eta_1(w) = wI(|w| > T) \qquad (1)$$

where w is a wavelet coefficient, T is the threshold and $I(x)$ is a function where the result is 1 when x is true and 0 vice versa.

The soft-thresholding function (also called the shrinkage function) takes the argument and shrinks it towards zero by the threshold. It is described as:

$$\eta_2(w) = [w - \text{sgn}(w)T]I(|w| > T) \qquad (2)$$

where sgn(w) is the sign of x. The soft-thresholding rule is chosen over hard-thresholding, for the soft-thresholding method yields more visually pleasant images over hard-thresholding. The equation shows the universal thresholding method in which N is the wavelet coefficients.

$$T = \sigma T = \sigma \sqrt{(2 \log N)} \quad (3)$$

3 Curvelet Transform

One type of signal representation is sparse, and it can represent a signal as superposition of a small numbers of components, the sparser the transformation, the more successful its use for signal denoising. Otherwise, the edges are prominent features in synthetic and real digital imagery. To address the problem of finding optimally sparse representations of objects with discontinuities along twice continuously differentiable (C^2) edge, a new tight frame of curvelets and fast digital curvelet transform have been introduced [6,7]. Recently, curvelet transform has been used on image denoising [8] and contrast enhancement [9]. Although the curvelet transform approaches can outperform other methods in image denoising, most shrinkage denoising approaches used are simply fixed thresholding. In image processing, edges are typically curved rather than straight, and ridgelets alone cannot yield efficient representations. However, at sufficiently fine scales, a curved edge is almost straight, and so to capture curved edges, one ought to be able to deploy ridgelets in a localized manner, at sufficiently fine scales. Curvelets are based on multiscale ridgelets combined with a spatial bandpass filtering operation to isolate different scales. Like ridgelets, curvelets occur at all scales, locations and orientations [10].

3.1 Curvelet Algorithm

The curvelet transform consists of an over complete representation of an image using a series of L2 energy measurements ranging across scale, orientation and position. Each curvelet consists of a tight frame constrained over a slice of the Fourier domain. In the spatial domain, the curvelet is a scaled and rotated Gabor signal along the width and is a scaled and rotated Gaussian signal along the length. One of the most important properties of the curvelet is length = width^2(length = 2^ − J, width = 2^ − (2*J), J = scale). This allows for the curvelet acting like a needle at fine scale representations. A brief overview of the mathematical framework from [11] is now presented to give the reader a formal representation of curvelets.

Each curvelet is defined by three parameters: scale J, orientation L and location K. Parabolic scaling matrix, rotation angle and orientation parameter are given by:

$$D_j = \begin{bmatrix} 2^{2J} & 0 \\ 0 & 2^J \end{bmatrix} \quad (4)$$

$$\Theta_J = 2\pi \times 2^{-J} \times L \tag{5}$$

$$K_g = (k1 \times \delta_1, k2 \times \delta_2) \tag{6}$$

where δ_1 and δ_2 are normalizing constants.

Curvelet basis function is given by:

$$\gamma(x_1, x_2) = \psi(x_1)\varphi(x_2) \tag{7}$$

where $\psi(x_1) = \text{Gabore}(x_1)$ and $\varphi(x_2) = \text{Gaussian}(x_2)$.

Finally, the curvelet parameterized by (J, K, L) can be defined as:

$$\Gamma_{(j,l,k)}(x_1, x_2) = 2^{3J/2} \times \Gamma(D_j \times R_{\theta J} \times (x_1, x_2) - K_g) \tag{8}$$

The algorithm of the curvelet transform of an image I can be summarized in the following steps [12]:

1. *Sub-band Decomposition*: The image f is decomposed into sub-bands using additive wavelet transform: $f \rightarrow (P_0, \Delta_1 f, \Delta_2 f, \Delta_3 f, \ldots)$
2. *Tiling*: Tiling is performed on the sub-bands $\Delta_1 f$, $\Delta_2 f$
3. *Ridgelet Analysis*: The Ridgelet transform is performed on each tile of the sub-bands $\Delta_1 f, \Delta_2 f$

The purpose of sub-band filtering is to decompose the image into additive components and each of which is a sub-band of that image. This step isolates the different frequency components of the image into different planes without down sampling as in the traditional wavelet transform. This sub-band filtering is performed by "*a trous*" algorithm.

Tiling is the process by which the image is divided into overlapping tiles. These tiles are small in dimension to transform curved lines into small straight lines in the sub-bands $\Delta_1 f$, $\Delta_2 f$. This improves the ability of the curvelet transform to handle curved edges.

3.2 Ridgelet Transform

A basic tool for calculating ridgelet coefficients is to view ridgelet analysis as a form of wavelet analysis in the Radon domain. We recall that the Radon transform of an object f is the collection of line integrals indexed by (θ, t) G $(0, 2\pi) \times R$ given by:

$$R_f(\theta, t) = \int f(x1, x2)\delta(x1\cos\theta + x2\sin\theta - t)dx1 dx2 \tag{9}$$

Hence, the ridgelet transform is precisely the application of a one-dimensional wavelet transform to the slices of the Radon transform where the angular variable θ is constant and t is varying. The ridgelet is similar to the 2D wavelet except that the point parameters (b_1, b_2) are replaced by the line parameters (b, θ).

4 Proposed Denoising Algorithm

The following is the various steps of our proposed algorithm:

1. Apply the a trous algorithm with J scales,
2. Set $B1 = Bmin$,
3. for $j = 1; ::::; J$ do,
   - partition the subband wj with a block size Bj and apply the digital ridgelet transform to each block,
   - if j modulo $2 = 1$ then $Bj + 1 = 2Bj$,
   - else $Bj + 1 = Bj$.
4. Perform Hard thresholding
5. Take Inverse Curvelet transform

5 Experimental Result

We took CT brain images embedded with Random and Gaussian noises and MR brain images embedded with Rician and Speckle noises with noise factor 30 for testing our proposed denoising algorithm in MATLAB. We applied curvelet method for denoising with curvelets at the finest scale with 4 decomposition level, complex block thresholding and cycle spinning to obtain better image denoising. We applied wavelet method for denoising with Debauchees wavelet at the finest scale with 4 decomposition level, and hard thresholding.

Table 1 shows the PSNR values of the denoised CT and MR images. As per the result, the differences between PSNR values of denoised images for curvelet and wavelet methods are positive values, and it is evident that curvelet-based denoising method is superior to the wavelet method in point of PSNR value.

Table 2 shows the mean and standard deviation values of denoised CT and MR images. As per the result, the SD values are lower for curvelet method than wavelet method, and it is evident that curvelet-based denoising method is superior to the wavelet method in point of SD value.

Table 1 PSNR values for denoised images

Image	Noise	CvT PSNR	WT PSNR	Difference (dB)
CT1	Random	31.78	28.25	3.53
	Gaussian	22.60	19.97	2.63
CT2	Random	30.00	25.05	4.95
	Gaussian	29.25	33.12	3.87
MR1	Speckle	35.35	25.10	10.25
	Rician	25.87	19.89	5.98
MR2	Speckle	31.64	24.96	6.68
	Rician	38.56	30.22	8.34

Table 2 Mean and standard deviation

Image	Noise	Mean		SD	
		CvT	WT	CvT	WT
CT1	Random	65.47	65.43	83.08	85.29
	Gaussian	72.36	72.37	75.38	78.54
CT2	Random	53.56	53.50	66.70	68.73
	Gaussian	60.73	60.61	61.50	65.38
MR1	Speckle	34.37	34.32	39.14	41.68
	Rician	41.60	41.63	34.41	41.31
MR2	Speckle	42.86	42.87	53.70	65.70
	Rician	51.10	51.47	57.43	61.32

Table 3 Entropy–curvelet transform

Image	Noise	Entropy		
		Original	Noised	Denoised
CT1	Random	3.97	18.00	4.79
	Gaussian	3.97	6.32	6.25
CT2	Random	3.17	18.00	5.78
	Gaussian	5.17	6.42	6.65
MR1	Speckle	4.85	18.00	5.31
	Rician	4.85	6.06	6.05
MR2	Speckle	4.16	16.00	4.84
	Rician	4.16	6.02	6.06

Table 4 Entropy–wavelet transform

Image	Noise	Entropy		
		Original	Noised	Denoised
CT1	Random	3.97	18.00	6.02
	Gaussian	3.97	6.32	7.06
CT2	Random	5.17	18.00	6.42
	Gaussian	5.17	6.42	7.05
MR1	Speckle	4.85	18.00	6.01
	Rician	4.85	6.06	6.64
MR2	Speckle	4.16	16.00	5.85
	Rician	4.16	6.02	6.79

Tables 3 and 4 show the entropy values for original, noised and denoised CT and MR images using curvelet and wavelet methods, respectively. As per the result, the entropy values are lower for curvelet method than wavelet method, and it is evident that curvelet-based denoising method is superior to the wavelet method in point of entropy value.

Figure 1a shows the MR brain image embedded with speckle noise with noise factor 30; Fig. 1b shows the wavelet denoised image with the PSNR value of

47 An Efficient Curvelet Denoising Algorithm for MR and CT Images

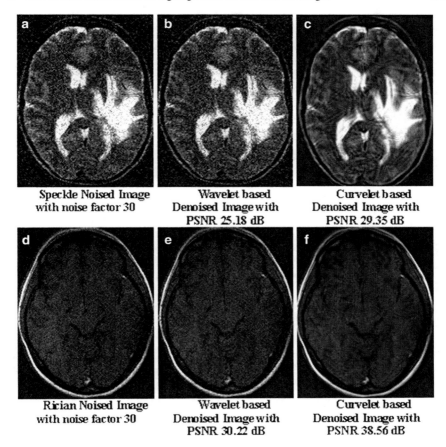

Fig. 1 Denoising of MR image – Speckle and Rician Noise

25.10 dB; Fig. 1c shows curvelet denoised image with improved PSNR value of 35.35 dB; Fig. 1d shows the MR brain image embedded with Rician noise with noise factor 30; Fig. 1e shows the wavelet denoised image with the PSNR value of 30.22 dB and Fig. 1f shows curvelet denoised image with improved PSNR value of 38.56 dB.

Figure 2a shows the CT brain image embedded with Random noise with noise factor 30; Fig. 2b shows the wavelet denoised image with the PSNR value of 28.25 dB; Fig. 2c shows curvelet denoised image with improved PSNR value of 31.78 dB; Fig. 2d shows the CT abdomen image embedded with Gaussian noise with noise factor 30; Fig. 2e shows the wavelet denoised image with the PSNR value of 29.25 dB and Fig. 2f shows curvelet denoised image with improved PSNR value of 33.12 dB.

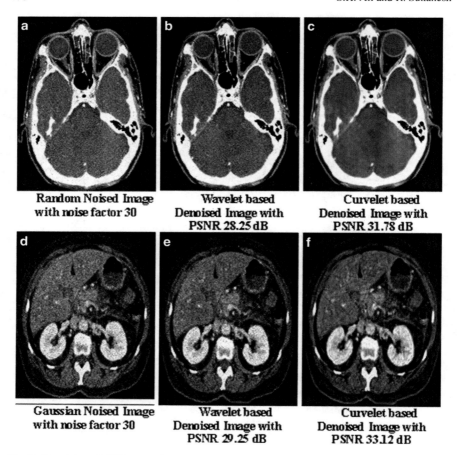

Fig. 2 Denoising of CT image – Random and Gaussian Noise

6 Conclusion

This chapter proposes the improvement algorithm in view of the wavelet transform denoising method's shortcoming. Take CT and MRI images as the examples. They are increased random noise separately with noise factor 30 and Gaussian noise with standard deviation 20. Then 2D wavelet transform and curvelet transform are applied to the noised images. After two-dimensional wavelet transform, edge and detailed information were mostly lost. After curvelet transform, detailed information is retained better than wavelet transform.

The curvelet method can achieve good denoising effect and has the widespread serviceability for CT and MRI images. Using curvelet transform denoising, random and Gaussian noises of CT and MRI images are small breakthrough to traditional method.

References

1. Donoho D.L., Johnstone I.M., 1994. Ideal spatial adaptation via wavelet shrinkage. Biometrika, 81, pp. 425–455
2. Quan P., Pan Z., Guanzhong D. et al., 1999. Two denoising methods by wavelet transform. IEEE Trans. Signal Process., 47(12), pp. 3401–3406
3. Zhing G., Xiaohai Y., 2004. Theory and application of MATLAB Wavelet analysis tools. National Defense Industry Publisher, Beijing, pp. 108–116
4. Candes E.J., Donoho D., 1999. Curvelet surprisingly effective nonadaptive Representation for object with edges. Proceedings of Curves and Surfaces IV. France, pp. 105–121
5. Donoho D.L., 1995. De-noising by soft-thresholding. IEEE Trans. Inform. Theory, 41(3), pp. 613–627
6. Candès E.J., Donoho D.L., 2004. New tight frames of curvelets and optimal representations of objects with C2 singularities. Commun. Pure Appl. Math., 57(2), pp. 219–266
7. Candès E.J., Demanet L., Donoho D.L., Ying L., 2006. Fast discrete curvelet transforms. SIAM Mult. Model. Simul., 5(3), pp. 861–899
8. Saevarsson B.B., Sveinsson J.R., Benediktsson J.A., 2004. Time invariant curvelet denoising. Proceedings of the 6th Nordic Signal Processing Symposium. Espoo, Finland, pp. 117–120
9. Starck J.L., Murtagh F., Candès E.J., Donoho D.L., 2003. Gray and color image contrast enhancement by the curvelet transform. IEEE Trans. Image Process., 12(6), pp. 706–716
10. Rakvongthai Y., 2008. Image denoising using uniform discrete curvelet transform and hidden markov tree model. University of Texas at Arlington, Tech. Rep., July 2008, EE5359. Online. Available: http://www-ee.uta.edu/dip
11. Starck J.L., 2001. Very high quality image restoration by combining wavelets and curvelets. Proc. SPIE, 4478, pp. 9–19
12. Wang Z., Bovik A., 2002. A universal image quality index, IEEE Signal Process. Lett., 9(3), pp. 81–84

(Mrs) R. Sukanesh is the Professor of Electronics and Communication Engineering in Thiagarajar College of Engineering, Madurai. She had received her bachelor's degree in ECE from Government College of Technology, Coimbatore, in 1982, PG Degree in Communication Systems from PSG College of Engineering and Technology in 1985, PhD in Biomedical Engineering from Madurai Kamaraj University in 1999 and PGDHE from IGNOU in 2001. Her area of research is Biomedical Engineering and Neural Networks. She has presented about 100 papers in International and National Conferences in home and abroad and published 50 research articles in International Journals of Engineering. She has to her credit a number of awards for her research papers from various academic and research organizations. To name a few, in the 12th International Conference on Biomedical Engineering held at Suntec City, Singapore, in 2005, she was lauded with the prestigious outstanding oral presentation award. She is also the recipient of The President of India's Prize and Jawaharlal Nehru Memorial Prize for her research contribution entitled "Genetic Algorithm Optimization of Fuzzy Outputs for Classification of Epilepsy Risk levels from EEG Signals", in the *21st Indian Engineering Congress* of the *Institution of Engineers*

[India] held at Guwhati in 2006. Madras Chapter of the Institution of Engineers awarded the woman Engineer award in the year 2008. She has been honoured with chairing the sessions of International conferences both in India and overseas. As a reviewer of technical papers for international conferences and journals, she has done a wonderful job in engineering research. She has also delivered more than 100 special and expert lectures and keynote and inaugural and valedictory addresses in various seminars, conferences and symposiums. She is a member of Board of Studies in Anna University and Mother Teresa Women's University, Kodaikanal, and a lead auditor for ISO 9001:2001 for educational institutions. She is a Fellow of Institution of Engineers and Life Member of Indian Society for Technical Education, Bio-Medical Engineering Society of India and Indian Association for Bio-Medical Scientists.

Prof. S. Hyder Ali received BE degree in ECE and ME in communication systems. He is now pursuing PhD in Biomedical Imaging at Anna University under the supervision of Dr (Mrs) R. Sukanesh. He is currently working as Associate Professor in School of Electrical Sciences, at VIT University, Vellore. His research interest includes Wavelet and time-frequency analysis, denoising, enhancement and segmentation of medical images. He is currently life member of three National Professional Societies. He has published several research papers in both national and international conferences.

Chapter 48
On the Use of Collinear and Triangle Equation for Automatic Segmentation and Boundary Detection of Cardiac Cavity Images

Riyanto Sigit, Mohd. Marzuki Mustafa, Aini Hussain, Oteh Maskon, and Ika Faizura Mohd. Nor

Abstract In this chapter, the computational biology of cardiac cavity images is proposed. The method uses collinear and triangle equation algorithms to detect and reconstruct the boundary of the cardiac cavity. The first step involves high boost filter to enhance the high frequency component without affecting the low frequency component. Second, the morphological and thresholding operators are applied to the image to eliminate noise and convert the image into a binary image. Next, the edge detection is performed using the negative Laplacian filter and followed by region filtering. Finally, the collinear and triangle equations are used to detect and reconstruct the more precise cavity boundary. Results obtained have proved that this technique is able to perform better segmentation and detection of the boundary of cardiac cavity from echocardiographic images.

1 Introduction

Cardiac cavity segmentation in two-dimensional short axis echocardiography images is very useful in helping doctors to diagnose patient cardiac condition. Various researches and methods have been conducted to detect cardiac cavity [1–9]. However, there is still room for innovation and development of methods and algorithms. To perform segmentation and detection of echocardiographic images, a number of researchers have used short axis images [1–6], while others [7,8] have used the long axis images. In [1,4,6], semi-automatic detection method was used, whereas in [3,7] a fully automated detection method was developed.

It is also interesting to note that in [5, 8], the active contour models or snakes algorithm has been used to perform segmentation of cardiac cavity. In [1], Klinger et al. applied mathematical morphology to carry out the segmentation of

---

R. Sigit (✉)
Department of Electrical, Electronic and Systems Engineering, Faculty of Engineering and Built Environment, Universiti Kebangsaan Malaysia, Bangi, Malaysia
e-mail: riyanto@eepis-its.edu

echocardiography images. Laine and Zong [2] presented border identification that depended on the shape modeling and border reconstruction from a set of images. Ohyama et al. [3] used ternary threshold method for detection of left ventricular endocardium. Maria et al. [1,4] applied semiautomatic detection of the left ventricular border. Chalana and David [5] presented multiple active contour model for cardiac boundary detection. Lacerda et al. [6] applied radial search for segmentation of cardiac cavity. Cheng et al. [7, 8] used watershed segmentation snakes for boundary detection and find the center point of the boundary.

In this study, we propose an alternative and computationally simpler method for segmenting and detecting the boundary of echocardiographic images automatically. The proposed method uses collinear and triangle equation. It involves applying a high boost filter along with the morphological and thresholding operators to eliminate noise and convert the image into a binary image. The use of a high boost filter is meant to enhance the high frequency component without affecting the low frequency component. This is followed by the implementation of a negative Laplacian filter meant for edge detection and consequently, the implementation of region filter. Finally, the collinear and triangle equations are used to detect and reconstruct the imprecise border.

2 The Proposed Technique

In this section, we describe the procedures involved to perform the automatic detection and reconstruction of the border based on estimation. The procedure which involves filter implementation, morphological, and thresholding operators is graphically described in a schematic diagram shown in Fig. 1.

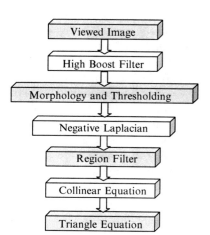

Fig. 1 Schematic diagram of the algorithm

2.1 Cardiac Image View

Typically, cardiac cavity images are presented using standard viewing method such as the short axis and long axis views. In this research, we process and analyze short axis images to test out our proposed segmentation and detection algorithm using the collinear and triangle equations. Figure 2 shows the short axis image of the left ventricular cardiac cavity from echocardiography that was used.

2.2 High Boost Filter

The first step to be implemented is the high boost filter algorithm [6], which uses the spatial mask as shown in Fig. 3. In this case, the high boost filter is used to enhance high frequency component while still keeping the low frequency components.

2.3 Morphology and Thresholding

After the high boost filter implementation, we subject the image to the morphological operation that basically deals with the opening and closing algorithm [1]. The main purpose of the opening and closing algorithm is to reduce speckle noise in the cardiac cavity image. The opening algorithm involves eroding image A by B and dilation by B. Mathematically this is achieved using the mathematical notation of the opening algorithm shown in (1).

$$A \circ B = (A \ominus B) \oplus B, \qquad (1)$$

where $\ominus$ and $\oplus$ denote erosion and dilation, respectively.

Fig. 2 The short axis image of the left ventricular of cardiac cavity from echocardiography

Fig. 3 Mask used for the high boost filter

−0.1111	−0.1111	−0.1111
−0.1111	9.89	−0.1111
−0.1111	−0.1111	−0.1111

Fig. 4 Kernel used for negative Laplacian

0	1	0
1	-4	1
0	1	0

The closing algorithm is implemented by dilating image A and eroding it by B. The mathematical notation of the closing algorithm is shown in (2).

$$A \bullet B = (A \oplus B) \ominus B. \tag{2}$$

2.4 Negative Laplacian Filter

Next, the negative Laplacian filter, which is a derivative filter, is realized to find areas of rapid change (edges) in the image. There are several ways to find an approximate discrete convolution kernel that approximates the effect of the Laplacian. A possible kernel is shown in Fig. 4 above.

2.5 Region Filter

Subsequently, we use region filter with the aim to eliminate the small contours. The region filter scans the contour and calculates the area of each contour. Regions with area that is smaller than the predetermined threshold are eliminated from the contour [6]. The threshold value was set to 25 pixels, and it was empirically determined.

2.6 Collinear Equation

The fifth step deals with the implementation of the collinear equation algorithm. The main purpose of such implementation is to optimize the number of contours by keeping and deleting some contours so that the resulted contour is closer to the actual boundary. This is achieved by finding the centroids of all existing contours using (3) as shown below:

$$\text{Centroid}(C) = \left(\frac{\sum_{k=1}^{n} Xk}{n}, \frac{\sum_{k=1}^{n}}{n} \right). \tag{3}$$

A collinear equation is then carried out from the center of the boundary to the centroids of each contour by finding the slope and intercept. The collinear equation used is as shown below in (4)–(6):

$$y = wx + b, \tag{4}$$

$$\text{Slope }(w) = \frac{n\Sigma xy - \Sigma x}{n\Sigma x^2 - (\Sigma)}, \tag{5}$$

$$\text{Intercept }(b) = \bar{y} - v. \tag{6}$$

Fig. 5 Triangle equation

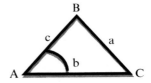

2.7 Triangle Equation

The final step is the application of the triangle equation. Figure 5 above shows a triangle, where A, B, and C are the corners, and a, b, and c are the distances between corners.

If a, b, and c are known, then the angle of the corner can be calculated using (7) and (8).

$$a^2 = b^2 + c^2 - 2bc(\cos A), \tag{7}$$

$$A = a\cos[(b^2 + c^2 - a^2)/2bc]. \tag{8}$$

The images obtained after the threshold operation of the cardiac cavity image result in images having closed and open borders or contours. Obviously, it is the closed contours image that we desire, and as such the problem of open contour images needs to be addressed. In this work, we use the small corner angle to reconstruct and closed the border or contour using the triangle equation.

To do so, we have used the OpenCV Library, which is a library for computer vision, and made modification to this library to get the boundary of cardiac activity contour and reconstruct it. A contour in this library that can be stored inside the memory storage is a sequence. A sequence in OpenCV is actually a linked list [7].

3 Experimental Results

The result of the high boost filter implementation is shown in Fig. 6a followed by the results after morphological operation with thresholding in Fig. 6b. In can be seen that high boost filter implementation yields an enhanced image of the cardiac cavity, which is further improved using morphological operations with thresholding. Result from the negative Laplacian filter implementation is shown in Fig. 6c, while result after region filtering is shown in Fig. 6d.

This method used collinear equation to keep and delete some contours; hence, the resulted contour is closer to the center boundary as shown in Fig. 7.

This method used triangle equation to reconstruct enclosed border. The images show that the method is finding the minimum small corner of boundary in the contour as in Fig. 8.

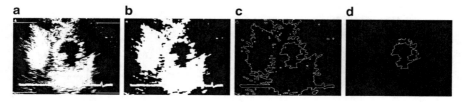

Fig. 6 Result from the high boost filter, morphological operation, Laplacian and region filter

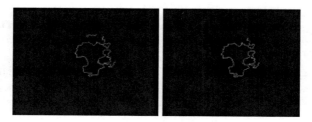

Fig. 7 The method allows one to keep and delete contour

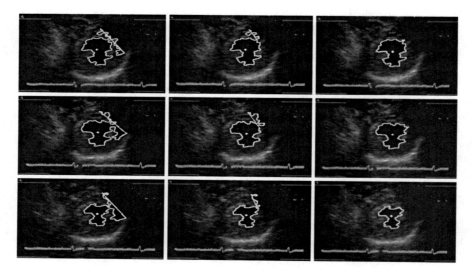

Fig. 8 The method is finding the minimum small corner of boundary

The testing images used in this research are extracted from a video recording that consists of nine repeated frames. The typical image size used is 320 pixels wide and 240 pixels high. It can be seen that the developed algorithm successfully detects and traces the boundary of the cardiac cavity in the video as it changed from large to small as depicted in Fig. 9.

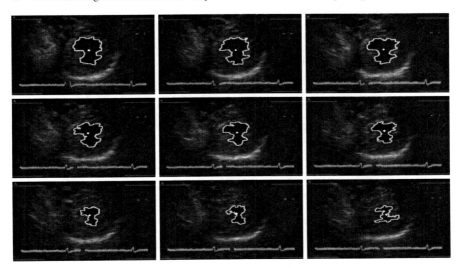

Fig. 9 Border detection of cardiac cavity image

4 Conclusion

It can be concluded that the proposed method using collinear and triangle equation can be used as an alternative solution to address problem in the area of computational biology and specifically in the segmentation and boundary detection problem in dealing with echocardiographic images.

Acknowledgment The authors would like to thank Universiti Kebangsaan Malaysia for the funding of this research through research grant contract number UKM-GUP-TKP-08-24-080.

References

1. J. W. Klinger, C. L. Vaughan, and T. D. Fraker, "Segmentation of echocardiographic images using mathematical morphology," *IEEE Trans. Biomed. Eng.*, 35, 1988, 925–934
2. A. Laine and X. Zong, "Border identification of echocardiograms via multiscale edge detection and shape modeling," *Proc. IEEE Int. Conf. Image Proc.*, 3, 1996, 287–290
3. W. Ohyama, T. Wakabayashi, F. Kimura, S. Tsuruoka, and K. Sekioka, "Automatic left ventricular endocardium detection in echocardiograms based on ternary thresholding method," in *15th International Conference on Pattern Recognition (ICPR'00)*, Barcelona, Spain, 2000, pp. 320–323
4. M. C. dos Reis, A. F. da Rocha, D. F. Vasconcelos, et al., "Semi-automatic detection of the left ventricular border," *30th Annual International IEEE EMBS Conference Vancouver*, British Columbia, Canada, August 20–24, 2008
5. V. Chalana, D. T. Linker, D. R. Haynor, and Y. Kim, "A multiple active contour model for cardiac boundary detection on echocardiography sequences," *IEEE Trans. Med. Imaging*, 15, 1996, 3

6. S. G. Lacerda, A. F. Da Rocha, D. F. Vasconcelos, et al., "Left ventricle segmentation in echocardiography using a radial search based image processing algorithm," *30th Annual International IEEE EMBS Conference Vancouver*, British Columbia, Canada, August 20–24, 2008
7. J. Cheng, S. W. Foo, and S. M. Krishnan, "Automatic detection of region of interest and center point of left ventricle using watershed segmentation," *IEEE Int. Symp. Circuits Syst.*, 1(2), 2005, 149–151
8. J. Cheng, S. W. Foo, and S. M. Krishnan, "Watershed-presegmented snake for boundary detection and tracking of left ventricle in echocardiographic images," *IEEE Trans. Inf. Technol. Biomed.*, 10(2), 2006, 414–416
9. M. Kass, A. Witkin, and D. Terzopoulos, "Snakes: Active contour models," Presented at the Int. Conf. Computer Vision, ICCV'87, London, UK, 1987

Chapter 49
The Electromagnetic-Trait Imaging Computation of Traveling Wave Method in Breast Tumor Microwave Sensor System

Zhi-fu Tao, Zhong-ling Han, and Meng Yao

Abstract Using the difference of dielectric constant between malignant tumor tissue and normal breast tissue, breast tumor microwave sensor system (BRATUMASS) determines the detected target of imaging electromagnetic trait by analyzing the properties of target tissue back wave obtained after near-field microwave radicalization (conelrad). The key of obtained target properties relationship and reconstructed detected space is to analyze the characteristics of the whole process from microwave transmission to back wave reception. Using traveling wave method, we derive spatial transmission properties and the relationship of the relation detected points distances, and valuate the properties of each unit by statistical valuation theory. This chapter gives the experimental data analysis results.

Keywords BRATUMASS · Near-field microwave · Traveling wave · Consistent estimate

1 Introduction

The dielectric constants of malignant tumor tissue ($\varepsilon = 50$) is larger than the dielectric constants of normal breast tissue ($\varepsilon = 10$). Hence, we can use this difference of electromagnetic condition to determine the location of the tumor in the breast area. We analyze the back wave, which is obtained by breast tumor microwave sensor system (BRATUMASS), to determine the characteristics of detected target. The characteristics of the whole near-field microwave conelrad transmission and back wave reception processing are the key points of obtained target properties relationship and reconstructed detected space. In this chapter, we deduce spatial transmission characteristics and the relationship of relation distances from detection points using traveling wave method from the ideal traveling wave transmission

M. Yao (✉)
East China Normal University, 3663 North Zhong-Shan Rd, Shanghai 200062, P.R. China
e-mail: myao@ee.ecnu.edu.cn

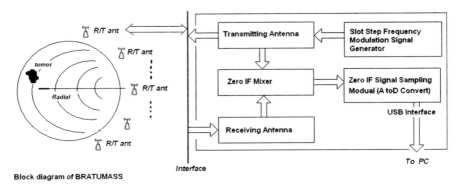

Block diagram of BRATUMASS

Fig. 1 Block diagram of BRATUMASS

Detected target center and the location of antennas. O is the location of transmission antenna, P is the location of receiving antenna, arrow for microwave direction

Fig. 2 Detected target and location of transceiver antenna

theory, and then evaluate each unit using statistical valuation theory. The block diagram of BRATUMASS is shown in Fig. 1 [1]. The detected target center and the location of antennas are shown in Fig. 2.

2 Description of Mathematical Model

2.1 The Relationship of Ideal Traveling Wave Transmission Characteristics in Homogeneous Medium Region

Considering the single detection point (as shown in Fig. 2), O is the location of transmission antenna, and sending signal is $f_1(t_1)$ at t_1 moment. P is the location of receiving antenna, and receiving signal is $f_2(t_2)$ at t_2 moment. Assume that O point is reference center, microwave transmits in homogeneous media region and ignores the transmission attenuation and measurement system factors. Then receiving signal at the P point obtained by traveling-wave method can be expressed as follows:
$f_2(t_2) = f_1(xp - xo, t_2) = f_1(xp, t_1 + \tau)$, where $t_2 - t_1 = \tau$, and τ is the transmission

delay. The source signal is $f_1(t_2) = f_1(x_0, t_1) = f_1(0, t_1)$. The analog signals before sampling of BRATUMASS can be expressed as:

$$llf(t) = [f_1(t_1) \times f_2(t_2)] \otimes h_L(t)$$
$$= [f_1(0, t_1) \times f_1(x_p, t_1 + \tau)] \otimes h_L(t) \quad (1)$$

where $h_L(t)$ is the impulse response of low-pass filter. Transmission antenna is the benchmark of uniform observation time, and then (1) becomes:

$$f(t) = [f_1(t) \times f_2(t + \tau)] \otimes h_L(t) = [f_1(0, t) \times f_1(x_p, t + \tau)] \otimes h_L(t) \quad (2)$$

Consider dual-domain Fourier transform [2] of traveling wave:

$$F(k, \omega) = \int_{-\infty}^{+\infty} \int_{-\infty}^{+\infty} f(x, t) e^{j(kx - \omega t)} dx dt \quad (3)$$

$$f(x, t) = \int_{-\infty}^{+\infty} \int_{-\infty}^{+\infty} F(k, \omega) e^{-j(kx - \omega t)} dk d\omega \quad \text{and} \quad f(0, t) = \int_{-\infty}^{+\infty} f(x, t) \delta(x) dx \quad (4)$$

From (4), we can get

$$f(0, t) = \int_{-\infty}^{+\infty} \int_{-\infty}^{+\infty} F(k, \omega) e^{-j(k \bullet 0 - \omega t)} dk d\omega = \int_{-\infty}^{+\infty} \left[\int_{-\infty}^{+\infty} F(k, \omega) e^{j \omega t} d\omega \right] dk \quad (5)$$

$$f(x_p, t) = \int_{-\infty}^{+\infty} \int_{-\infty}^{+\infty} F(k, \omega) e^{-j(k \bullet x_p - \omega t)} dk d\omega = \int_{-\infty}^{+\infty} \left(\int_{-\infty}^{+\infty} F(k, \omega) e^{j \omega t} d\omega \right) e^{-jk \bullet x_p} dk \quad (6)$$

Contrasting (5) and (6), the difference between source signal and receiving signal is just the phase change which is brought by the distance x_p in transmission direction. Considering Fourier transform of the BRATUMASS's mixer output $f(x)$ in (2):

$$F(\omega) = \int_{-\infty}^{+\infty} f(t) e^{-j\omega t} dt = \int_{-\infty}^{+\infty} f\left(\frac{x}{v}\right) e^{-j\frac{\omega x}{v}} d\left(\frac{x}{v}\right) = \int_{-\infty}^{+\infty} g(x) e^{-jkx} dx = G(k) \quad (7)$$

where $g(x) = f(x/v)/v$ is the microwave transmission characteristics along vector k direction, $k = \omega/v$, v is the microwave propagation velocity in medium region, and $G(k)$ is the Fourier transform of $g(x)$.

2.2 The Relationship of Transmission Characteristics in General Medium

Considering the impact of transmission medium and the measuring system [3], record signal is the cross-correlation function between received signal and sounding signal:

$$R_{12}(\tau) = f_1(t_1) \otimes f_2(-t_2) = \int_{-\infty}^{+\infty} b(t) R_{11}(\tau - t) dt \qquad (8)$$

where $R_{11}(t)$ is the autocorrelation function, $b(t)$ is the impulse characteristics of sounding target and measurement equipment's channel, and it can be expressed as the convolution of $h(t)$ and $w(t)$. For the convolution characteristics, (8) can be rewritten as:

$$R_{12}(\tau) = \int_{-\infty}^{+\infty} h(t) R_w(\tau - \tau) dt,$$

$$\text{where} \quad R_w(\tau) = \int_{-\infty}^{+\infty} w(t) R_{11}(\tau - t) dt \qquad (9)$$

Considering the characteristics of Fourier transform and convolution of (9), $R_{12}(\omega) = H(\omega) \bullet R_w(\omega)$, $R_{12}(\omega)$ is the Fourier transform of $R_{12}(\tau)$, $H(\omega)$ is the Fourier transform of $h(t)$, and $R_w(\omega)$ is the Fourier transform of $R_w(\tau)$. Then there are:

$$h(t) = \int_{-\infty}^{+\infty} H(\omega) e^{j\omega t} dt, \quad \text{and} \quad g(x) = \frac{h\left(\frac{x}{v}\right)}{v} = \int_{-\infty}^{+\infty} H(vk) e^{jkx} d\left(\frac{x}{v}\right). \qquad (10)$$

Medium characteristics distribution $g(x)$ can be obtained, in the direction of the transmission vector k.

2.3 Multi-Point Joint Estimation of Spatial Transmission Characteristic

Place the detection antenna according to the detection method of BRATUMASS [1]. Assume that there are N detection points and can be obtained N correlation functions, $g_1(x_1), g_2(x_2), \ldots, g_N(x_N)$, where $g_i(x_i)$ is the transmission characteristics function relative to i detection point, and x_i is the coordinates of wave vector k relative to i detection point. To make transmission characteristics function of detected region is $\mu(x, y)$, and any cell (x, y) in detected region has different detection distance relative to different detection point, $x_1, x_2, \ldots, x_N$. Measurement corresponding to each detection point can be seen as estimation of the cell spatial characteristics parameters under the detection condition:

$$\overline{\mu}(x, y) = \frac{1}{N} [g_1(x_1) + g_2(x_2) + g_3(x_3) + \ldots + g_N(x_N)] = \frac{1}{N} \sum_{i=1}^{N} g_i(x_i) \qquad (11)$$

Energy to reach the antenna is linear superposition, and then the signal from antenna is the power sum of all signals which have same delay. Hence, $g_i(x_i)$ is the sum of characteristics values which have the same transmission distance x_i relative to i detection point. However, for each detection point, any cell (x, y) in detected region has only one transmission distance relative to it.

Then satisfy the following relationship: [4]

$$A = \iint_\Omega \mu(x,y) dxdy = \int_{-\infty}^{+\infty} g_i(x_i) dx_i, i \in (1, 2, \ldots, N) \quad (12)$$

where Ω is the detected region. Consider equation:

$$g_i(x_i) = \iint_{(x,y) \in X(x_i)} \mu(x,y) dxdy \quad (13)$$

where $X(x_i)$ is point set, in which all points have the same transmission distance x_i relative to detection point i. Substitute (13) for (11):

$$\overline{\mu}(x,y) = \frac{1}{N} \sum_{i=1}^N g_i(x_i) = \frac{1}{N} \sum_{i=1}^N \iint_{(x,y) \in X(x_i)} \mu(x,y) dxdy \quad (14)$$

On the (14), $\mu(x, y)$ is double-counting N times. When $N \to \infty$, there is

$$\overline{\mu}(x,y) = \lim_{N \to \infty} \frac{N-1}{N} \mu(x,y) + \lim_{N \to \infty} \frac{1}{N} \iint_\Omega \mu(x,y) dxdy \quad (15)$$

That is $\overline{\mu}(x,y) \xrightarrow{N \to \infty} \mu(x,y)$, so the estimation is unbiased.

3 Simulation Experiment

3.1 Simulation Experiment 1

We use BRATUMASS to experiment according to the method in reference article [1]. There are 32 detection points uniform distribution in the boundary of circular region which radius is 80 mm. The detected region is filled with dielectric constant of 10 medium and embedded a ball with dielectric constant 50 and radius 7 mm, and the location of its center is 12 mm, −11.8 mm. It is shown in Fig. 3.

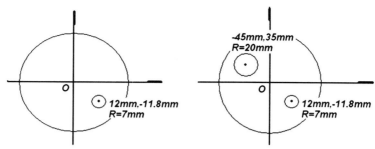

Fig. 3 Illustration of the detected region and targets location

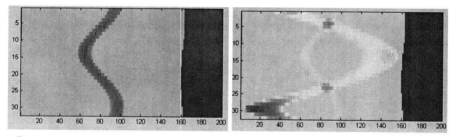

Fig. 4 Transmission characteristics data relative to single sampling point after mixer output

3.2 Simulation Experiment 2

There are 32 detection points uniform distribution in boundary of circular region which radius is 80 mm. The detected region is filled with dielectric constant of 10 medium and embedded two balls. One is the same as the ball in experiment 1, and the other one has dielectric constant 50, radius 20 mm, and the locations of its center −45 mm, 35 mm, also shown in Fig. 3 (right).

From the spatial characteristics distribution of above reconstruction region, this estimation method can be obtained transmission characteristics of detected region. Results of reconstruction were shown in Figs. 4 and 5.

Using traveling wave analysis to established transmission characteristics relationship based on sampling points is valid.

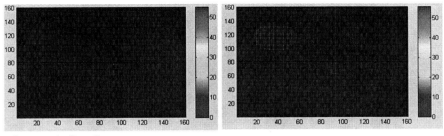

Reconstruction Target Space
Left: Target spatial characteristics distribution obtained by multi-points valuation in experiment one
Right: Target spatial characteristics distribution in experiment two

Fig. 5 Results of experiments

4 Conclusion

In this chapter, we give the relationship between spatial characteristics and transmission distance relative to single detection point of BRATUMASS and proposed a multi-point detection unbiased estimation method. Simulation results show that the method is effective.

Acknowledgment This work is supported by Shanghai Science and Technology Development Foundation under the project grant numbers 08JC1409200 and 03JC14026.

References

1. Meng Yao, et al. (2009) Application of Quantum Genetic Algorithm on Breast Tumor Detections with Microwave Imaging. GECCO 2009: 2685–2688. ISBN 978-1-60558-325-9
2. Ezzat G. Bakhoum and Cristian Toma (2009) Mathematical Transform of Traveling-Wave Equations and Phase Aspects of Quantum Interaction. August 23, 2009. http://www.hindawi.com/journals/mpe/aip.695208.pdf
3. Astanin, L.Y. and Kostylev, A.A. (2000) The Basis of Ultra-Wideband Radar Measurements. University of Defense Technology Press, Version 1, 126–128
4. Zhi-fu Tao, et al. (2009) Reconstructing Microwave Near-Field Image Based on the Discrepancy of Radial Distribution of Dielectric Constant. ICCSA 2009, 717–728

Chapter 50
Medical Image Processing Using Novel Wavelet Filters Based on Atomic Functions: Optimal Medical Image Compression

Cristina Juarez Landin, Magally Martinez Reyes, Anabelem Soberanes Martin, Rosa Maria Valdovinos Rosas, Jose Luis Sanchez Ramirez, Volodymyr Ponomaryov, and Maria Dolores Torres Soto

Abstract The analysis of different Wavelets including novel Wavelet families based on atomic functions are presented, especially for ultrasound (US) and mammography (MG) images compression. This way we are able to determine with what type of filters Wavelet works better in compression of such images. Key properties: Frequency response, approximation order, projection cosine, and Riesz bounds were determined and compared for the classic Wavelets W9/7 used in standard JPEG2000, *Daubechies8, Symlet8*, as well as for the complex Kravchenko–Rvachev Wavelets $\psi(t)$ based on the atomic functions $up(t)$, $fup_2(t)$, and $eup(t)$. The comparison results show significantly better performance of novel Wavelets that is justified by experiments and in study of key properties.

Keywords Wavelet transform · Compression · Atomic functions · Ultrasound images · Mammography images

1 Introduction

Different fields such as astronomy, medical imaging, and computer vision manage data of large volume. Hence, these data should be compressed to optimize the storage devices. There exist a lot of approaches in signal compression. Here, we present Wavelet-based techniques for compression procedures focusing on different threshold rules. The basic idea behind these techniques is to use Wavelets to transform data set into a different basis, where the non-important information can be eliminated. Also, we have tested as classical, as novel Wavelet algorithms based on atomic functions, which present excellent compression results. Below, decimated Wavelet transforms (WT) and the MAE fidelity criterion are used to evaluate the different

---

C. Juarez Landin (✉)
Autonomous University of Mexico State, Hermenegildo Galena No. 3, Col. Ma. Isabel, Valle de Chalco, Mexico State, Mexico
e-mail: cjlandin@yahoo.com.mx

compression methods for US and MG images. Investigating the key properties of the different Wavelets, we can justify the obtained experimental results in the compression of US and MG images.

1.1 Wavelet Transform and Filter Banks

The Discrete Wavelet Transform (DWT) is easy to realize using filter banks [1]. DWT can be implemented applying some equations, but it is usually made using filter bank techniques. The most popular scheme of the DWT for 2D signal applies only two filters for rows and columns, as in the symmetric filter bank.

In tests carried out previously, it was found that better results are obtained when compressing the ultrasound images with the Symlet Wavelet, and the mammography images with the Daubechies Wavelet [2]. Based on this fact, we realize an evaluation to compare the acting of three Wavelet families based on atomic functions with the Wavelets that presented better acting. We use the complex Kravchenko–Rvachev Wavelets $\psi(t)$ based on the atomic functions $up(t)$, $fup_2(t)$, and $eup(t)$ [3].

1.2 Compression by Wavelet Threshold

The three main steps of compression using the Wavelet coefficient and threshold technique are as follows:

(a) Calculate the Wavelet coefficient matrix applying WT to the original image
(b) Modify (threshold or shrink) the detail coefficients to obtain the reduced number of coefficients
(c) Encode the modified coefficients to obtain the compressed image

In the two-level sub-bands decomposition, the coefficients on the first level are grouped into the vertical details (LH_1), horizontal details (HL_1), diagonal details (HH_1), and approximations (LL_1) sub-bands. The approximations part is then similarly decomposed in second level sub-bands. The directions reflect the order, in which the high-pass (H) and low-pass (L) filters of the WT are applied along the two dimensions of the original image.

1.3 Compression by Wavelet Threshold

The thresholding functions [4] determine how the thresholds are applied to data. The most popular are four thresholds, and a single threshold ($\pm t$) is required for hard $\left[\delta_t^H(w)\right]$, soft $\left[\delta_t^S(w)\right]$, and garrote $\left[\delta_t^G(w)\right]$ functions, but for semisoft function $\left[\delta_{t_1,t_2}^{SS}(w)\right]$, there are require two thresholds ($\pm t_1$ and $\pm t_2$). Hard function does not modify the original data of the Wavelet coefficients; hence, for this reason we use only hard function in the experiments. The hard function is given below:

$$S = \begin{cases} x & si \quad |x| > t \\ 0 & si \quad |x| \le t \end{cases} \quad (1)$$

where x is the original signal, S is the thresholding signal, and t is threshold.

2 Wavelet Key Properties

The Wavelet decomposition algorithm uses two analysis filters: $\tilde{H}(z)$ (lowpass) and $\tilde{G}(z)$ (highpass). The reconstruction algorithm applies the complementary synthesis filters: H(z) (lowpass) and G(z) (highpass). These four filters constitute a perfect reconstruction filter bank. In the present case, the system is entirely specified by the lowpass filters H(z) and $\tilde{H}(z)$, which form a biorthogonal pair. The highpass operators are obtained by simple shift and modulation and shown below:

$$\tilde{G}(z) = zH(-z) \text{ and } G(z) = z^{-1}\tilde{H}(-z) \quad (2)$$

For all the Wavelet families used, the following properties were obtained to justify the results.

2.1 Frequency Response

This characteristic allows determining the behavior of the analysis and synthesis filters in a graphic way to appreciate the differences that there are among different Wavelet families used.

2.2 Approximation Order

This determines the number L of factors $(1+z^{-1})$ that divide the transfer function H(z). The approximation order plays a crucial role in Wavelet theory [5]. It implies that the scaling function $\varphi(x)$ reproduces all polynomials of degree lesser or equal to $n = L - 1$; in particular, it satisfies the partition of unity ($\Sigma_k \varphi(x-k) = 1$). They are also directly responsible for the vanishing moments of the Wavelet analysis: $\int x^n \tilde{\psi}(x)dx = 0$ for $n = 0, 1, 2, \ldots, L-1$. Finally, the order L also corresponds to the rate of decay of the projection error as a scale how it goes to zero [6].

The next point concerns the stability of the Wavelet representation and of its underlying multi-resolution bases. The crucial mathematical property is the translation of the scaling functions and Wavelets in Riesz bases [7]. Thus, one needs to characterize their Riesz bounds and other related quantities.

2.3 Riesz Bounds

The tightest upper and lower bounds, $B < 1$ and $A > 0$, of the autocorrelation filter of $\varphi(x)$ are the Riesz bounds of $\varphi(x)$ that are given by:

$$A, B = \inf_{c \in \ell^2}, \sup \frac{\left\| \sum_{k \in Z} c_k \varphi(x-k) \right\|_{L^2}}{\|c\|_{\ell^2}} \quad (3)$$

The existence of the Riesz bounds ensures that the underlying basis functions are in L^2, and that coefficients of transform are linearly independent in the ℓ^2 space. The Riesz basis property expresses equivalence between the L^2 norm of the expanded functions and the ℓ^2 norm of their coefficients in the Wavelet or scaling function basis. There is a perfect norm equivalence (Parseval's relation), if and only if $A = B = 1$, and in this case the basis is orthonormal.

2.4 Projection Cosine

The (generalized) projection angle θ between the synthesis and analysis subspaces V_a and $\tilde{V}_a$ is defined as [8]:

$$\cos \theta = \inf_{f \in \tilde{V}_a} \frac{\|P_a f\|_{L^2}}{\|f\|_{L^2}} = \frac{1}{\sup_{\omega \in [0, 2\pi]} \sqrt{a_\varphi(\omega) \cdot a_{\tilde{\varphi}}(\omega)}} \quad (4)$$

This fundamental quantity is scale independent, it allows comparing the performance of the biorthogonal projection $\tilde{P}_a$ with that of the optimal least squares solution P_a for a given approximation space V_a. Specifically, we have the following sharp error bound [9]:

$$\forall f \in L^2, \|f - P_a f\|_{L^2} \leq \|f - \tilde{P}_a f\|_{L^2} \leq \frac{1}{\cos \theta} \|f - P_a f\|_{L^2} \quad (5)$$

The projection angle θ between the synthesis and analysis subspaces should be 90° in orthogonal spaces. In other words, the biorthogonal projector $\tilde{P}_a$ will be essentially as good as the optimal one (orthogonal projector onto the same space) when the value $\cos \theta$ is close to 1.

2.5 Criteria of Fidelity

To evaluate in objective manner the fidelity of the compressed images, we apply the mean absolute error (MAE) and the compression rate (CR) criteria that characterize the difference between two images [10].

3 Simulation Results

We carried out numerous experiments to compare the performance of the compression algorithm. In these results, classical Wavelet filters are compared with three different families of Wavelet filters based en AFs. We use five decomposition levels in the compression procedure. Figures 1 and 3 present the obtained results in Ultrasound and Mammography respectively for MAE criterion. Figures 2 and 4 present the obtained results in Ultrasound and Mammography respectively for CR measure.

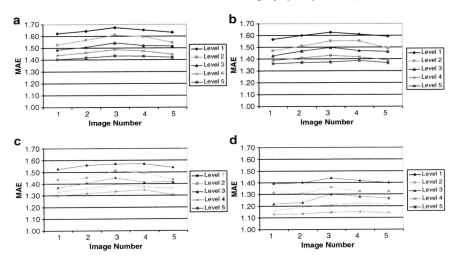

Fig. 1 MAE criterion for compressed ultrasound images with (**a**) Symlet filters, based on AF, (**b**) $up(t)$, (**c**) $fup_2(t)$, and (**d**) $eup(t)$ Wavelet filters

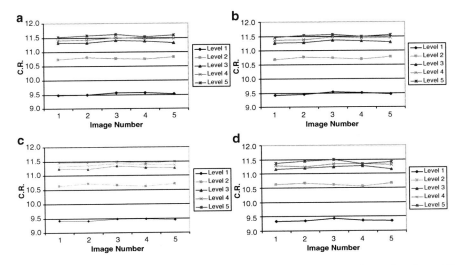

Fig. 2 CR criterion for compressed ultrasound images with (**a**) Symlet filters, based on AF, (**b**) $up(t)$, (**c**) $fup_2(t)$, and (**d**) $eup(t)$ Wavelet filters

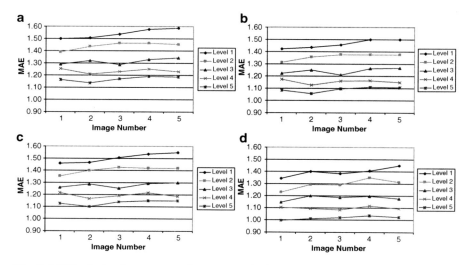

Fig. 3 MAE criterion for compressed mammography images with (**a**) Daubechies filters, based on AF, (**b**) $up(t)$, (**c**) $fup_2(t)$, and (**d**) $eup(t)$ Wavelet filters

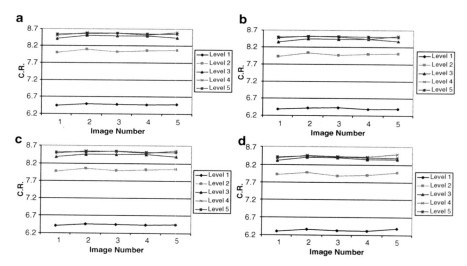

Fig. 4 CR criterion for compressed mammography images with (**a**) Daubechies filters, based on AF, (**b**) $up(t)$, (**c**) $fup_2(t)$, and (**d**) $eup(t)$ Wavelet filters

Finally, the Table 1 presents the key properties of the different Wavelets used in compression of the US and MG images.

It is known from statistical theory that the approximation property of estimation of random variable can be given in the form of relative error $\delta = 2(1 - r)$, where r is the correlation coefficient that is equal to projection cosine in this case. Hence, calculations of this error using Table 1 show that Wavelet based on $eup(x)$

Table 1 Summary of key properties for different Wavelet filters

Key properties for different Wavelet filters												
	Wavelet 9/7		Daubechies 8		Symlet 8		A.F.W. $up(t)$		A.F.W. $fup_2(t)$		A.F.W. $eup(t)$	
Type	Dec.	Rec.	Dec.	Rec.	Dec.	Rec.	Dec.	Rec.	Dec.	Rec.	Dec.	Rec.
Approximation order	4		4		4		4		4		4	
Projection cosine	0.98387		0.98879		0.98781		0.99176		0.99472		0.99769	
Riesz bounds	0.926	0.943	0.833	0.849	0.880	0.896	0.792	0.806	0.713	0.726	0.641	0.653
	1.065	1.084	1.267	1.290	1.273	1.295	1.514	1.542	1.802	1.834	2.145	2.183

can produce relative variance error of 0,00464 (6.8% in RMS value), but at the same time Wavelet Daubechies 8 gives value of 0,02242 (more than 15% in RMS value), and Wavelet 9/7 value of 0,03234 (more than 18% in RMS value). Hence, Wavelet based on $eup(x)$ function gives about three less a relative error in RMS values than Wavelet 9/7 used in JPEG2000 standard.

4 Conclusions

Numerous tests were carried out in comparing different Wavelets functions to choose the best one for compression of medical US and MG images. It is observed that for two modalities of images used in the tests, the Wavelets families based on atomic functions presented smaller levels of MAE, with relationship to their equivalent of the Daubechies and Symlet families. The compression level also decreased lightly but better image quality is conserved.

The Wavelet family based on the atomic function $eup(x)$ present better levels of MAE for two image modalities. It is observed in the carried out analysis that Daubechies and Symlet filters are most selective than the Wavelet 9/7 used in standard JPEG2000. The filters of the Wavelet based on the atomic function $eup(x)$ are presented the best frequency response. One can see that all the used filters have the same approximation order. This derives mainly in two things, the Wavelet filters have the same number of coefficients, and therefore, they imply the same computational complexity when being implemented in the compression algorithms, and the convergence of the error will be of the same order for all the filters.

The existence of the limits Riesz bounds demonstrates that the coefficients of the analysis and synthesis filters are lineally independent. The projection cosine measure shows that the families of Wavelets based on atomic functions are near to optimal ones; this implies that they are "better" orthogonal and "most independent." Also, the fact that they are lineally independent assures that errors are hardly presented to the decomposition/reconstruction procedure. Likewise, the different properties of the Wavelet filters found here demonstrates that the filters of the Wavelets families based on atomic functions can realize the approximation better in comparison with traditional Wavelets, potentially giving superior quality of compression, guaranteeing the same level of error.

Acknowledgments This work is supported in part by IPN, UAEM project 2703 and PROMEP.

References

1. Jähne, B., 2004. Practical Handbook on Image Processing for Scientific and Technical Applications. CRC Press, Boca Raton
2. Sanchez, J.L., Ponomaryov, V., 2007. Wavelet compression applying different threshold types for ultrasound and mammography images. GESTS Intern. Trans. Comput. Sci. Eng., vol. 39, no. 1, pp. 15–24
3. Gulyaev, Y., Kravchenko, V., Pustovoit, V., 2007. A new class of WA-systems of Kravchenko–Rvachev functions. Dokl. Math., vol. 75, no. 2, pp. 325–332
4. Jan, J., 2006. Medical Image Processing, Reconstruction and Restoration: Concepts and Methods. Taylor & Francis, CRC Press, Boca Raton
5. Vetterli, M., Kovacevic, J., 1994. Wavelets and Subband Coding. Prentice-Hall, Englewood Cliffs
6. Villemoes, L., 1994. Wavelet analysis of refinement equations. SIAM J. Math. Anal., vol. 25, no. 5, pp. 1433–1460
7. Meyer, Y., 1990. Ondelettes. Hermann, Paris
8. Unser, M., Aldroubi, A., 1994. A general sampling theory for nonideal acquisition devices. IEEE Trans. Signal Process., vol. 42, no. 11, pp. 2915–2925
9. Strang, G., Nguyen, T.Q., 1996. Wavelets and Filter Banks. Wellesley-Cambridge Press, Cambridge
10. Ponomaryov, V., Sanchez, J.L., Juarez, C., 2006. Evaluation and optimization of the standard JPEG2000 for compression of ultrasound images. Telecomm. Radio Eng., vol. 65, no. 11, pp. 1005–1017

Chapter 51
Cancellation of Artifacts in ECG Signals Using Block Adaptive Filtering Techniques

Mohammad Zia Ur Rahman, Rafi Ahamed Shaik,
and D.V. Rama Koti Reddy

Abstract In this chapter, various block-based adaptive filter structures are presented, which estimate the deterministic components of the electrocardiogram (ECG) signal and remove the noise. The familiar Block LMS algorithm (BLMS) and its fast implementation, Fast Block LMS (FBLMS) algorithm, is proposed for removing artifacts preserving the low frequency components and tiny features of the ECG. The proposed implementation is suitable for applications requiring large signal-to-noise ratios with fast convergence rate. Finally, we have applied these algorithms on real ECG signals obtained from the MIT-BIH database and compared its performance with the conventional LMS algorithm. The results show that the performance of the block-based algorithms is superior than the LMS algorithm.

1 Introduction

The electrocardiogram (ECG) is a graphical representation of heart's functionality and is an important tool used for diagnosis of cardiac abnormalities. The extraction of high-resolution ECG signals from recordings contaminated with background noise is an important issue to investigate [6–8]. Many approaches have been reported in the literature to address ECG enhancement using adaptive filters [1–4], which permit to detect time varying potentials and to track the dynamic variations of the signals. In [4], Thakor et al. proposed an LMS-based adaptive recurrent filter to acquire the impulse response of normal QRS complexes, and then applied it for arrhythmia detection in ambulatory ECG recordings. In these papers, the LMS algorithm operates on an instantaneous basis such that the weight vector is updated every new sample within the occurrence, based on an instantaneous gradient estimate.

There are certain clinical applications of ECG signal processing that require adaptive filters with large number of taps. In such applications, the conventional

---

M.Z.U. Rahman (✉)
Instrumentation Engineering, Andhra University, Visakhapatnam-530003, India
e-mail: mdzr55@gmail.com; mdzr_5@yahoo.com

LMS algorithm is computationally expensive to implement. The block processing of data samples can significantly reduce the computational complexity. By applying this strategy, a special implementation of the LMS algorithm is called the Block LMS (BLMS) algorithm. In a recent study, a steady-state convergence analysis for the LMS algorithm with deterministic reference inputs showed that the steady-state weight vector is biased, and thus the adaptive estimate does not approach the Wiener solution. To handle this drawback, another strategy was considered for estimating the coefficients of the linear expansion, i.e., the BLMS algorithm [9], in which the coefficient vector is updated only once every occurrence based on a block gradient estimation.

To the best of our knowledge, block-based adaptive filtering has not been considered previously within the context of filtering artifacts in ECG signals. In this chapter, we present BLMS and block normalized LMS (BNLMS) algorithms and their fast implementation to remove the artifacts from ECG. Such a realization is intrinsically less complex than its LMS-based counterpart. To study the performance of the proposed algorithm to effectively remove the noise from the ECG signal, we carried out simulations on MIT-BIH database for different artifacts. The simulation results show that the proposed algorithm performs better than the LMS counterpart to eliminate the noise from ECG.

2 Block-Based Adaptive Algorithms for Artifacts Removal from ECG Signal

The predominant artifacts present in the ECG includes: baseline wander (BW), power-line interference (PLI), muscle artifacts (MA), and motion artifacts (EM). These artifacts strongly affect the ST segment, degrade the signal quality, frequency resolution, produce large amplitude signals in ECG that can resemble PQRST waveforms, and mask tiny features that may be important for clinical monitoring and diagnosis. To allow doctors to view the best signal that can be obtained, we need to develop an adaptive filter to remove the noise to better obtain and interpret the ECG data.

Consider a length L LMS-based adaptive filter, depicted in Fig. 1, which takes an input sequence $x(n)$ and updates the weights as:

$$\mathbf{w}(n+1) = \mathbf{w}(n) + \mu \mathbf{x}(n) e(n) \tag{1}$$

where $\mathbf{w}(n) = [w_0(n)\, w_1(n) \cdots w_{L-1}(n)]^t$ is the tap weight vector at the nth index, $\mathbf{x}(n) = [x(n)\, x(n-1) \cdots x(n-L+1)]^t$, $e(n) = d(n) - \mathbf{w}^t(n)\mathbf{x}(n)$, with $d(n)$ being the so-called desired response available during initial training period and μ denoting so-called step-size parameter.

In order to remove the noise from the ECG signal, the ECG signal $s_1(n)$ with additive noise $p_1(n)$ is applied as the desired response $d(n)$ for the adaptive filter shown in Fig. 1. If the noise signal $p_2(n)$ possibly recorded from another generator

Fig. 1 Basic adaptive filter structure

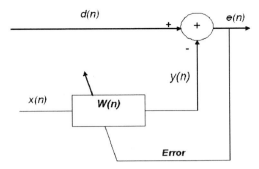

of noise that is correlated in some way with $p_1(n)$ is applied at the input of the filter, i.e., $x(n) = p_2(n)$, the filter error becomes $e(n) = [s_1(n) + p_1(n)] - y(n)$. The filter output $y(n)$ is given by:

$$y(n) = \mathbf{w}^t(n)\mathbf{x}(n) \tag{2}$$

Since the signal and noise are uncorrelated, the mean-squared error (MSE) is:

$$E[e^2(n)] = E\{[s_1(n) - y(n)]^2\} + E[p_1^2(n)] \tag{3}$$

2.1 The Block LMS Algorithm

In the proposed implementation, we considered a BLMS algorithm-based adaptive filter that takes an input sequence $x(n)$, which is partitioned into non-overlapping blocks of length P each by means of a serial-to-parallel converter [10]. With the j-th block, ($j \in Z$) consisting of $x(jP+r)$, $r \in Z_P = 0, 1, \ldots, P-1$, the filter coefficients are updated from block to block as:

$$\mathbf{w}(j+1) = \mathbf{w}(j) + \mu \Sigma_{r=0}^{P-1} \mathbf{x}(jP+r)e(jP+r) \tag{4}$$

where $\mathbf{w}(j) = [w_0(j)w_1(j)\ldots w_{L-1}(j)]^t$ is the tap weight vector corresponding to the j-th block, $\mathbf{x}(jP+r) = [x(jP-r)x(jP+r-1)\ldots x(jP+r-L+1)]^t$ and $e(jP+r)$ is the output error at $n = jP+r$, given by:

$$e(jP+r) = d(jP+r) - y(jP+r) \tag{5}$$

The sequence $d(jP+r)$ is the so-called desired response available during the initial training period and $y(jP+r)$ is the filter output at $n = jP+r$, given as:

$$y(jP+r) = \mathbf{w}^t(j)\mathbf{x}(jP+r) \tag{6}$$

The parameter μ, popularly called the step-size parameter, is to be chosen as $0 < \mu < \frac{2}{[P\,tr\mathbf{R}]}$ for convergence of the algorithm.

Table 1 A computational complexity comparison table ($M = L + P - 1, r = \log_2^M$)

Algorithm	MAC	ASC	Division	Shift
LMS	L	Nil	Nil	L
BLMS	$(L+1)P + L$	Nil	Nil	$(L+P)$
FBLMS	$10Mr + 10M$	$10Mr + 2M + P$	Nil	$35Mr + 18M + 2P$
BNLMS	$(L+1)P + 2L$	Nil	L	$(L+P)$
FBNLMS	$10Mr + 10M + L$	$10Mr + 2M + P$	L	$35Mr + 18M + 2P$

The convergence behavior of the BLMS algorithm is greatly improved if we replace the scalar step-size parameter μ by a diagonal matrix $\mu(n)$ in (4) whose diagonal elements are the set of normalized step-size parameters:

$$\mu(n) = \frac{\mu}{p + \mathbf{x}^t(n)\mathbf{x}(n)} \quad (7)$$

The resultant update equation corresponds to BNLMS algorithm.

2.2 Frequency Domain Implementation of Adaptive Filters

The fast block LMS (FBLMS) algorithm is nothing but a fast implementation of the BLMS algorithm in frequency domain. The element by element multiplication of the frequency domain samples of the input and filter coefficients is followed by an IDFT and a proper windowing of the result to obtain the output vector [5].

Finally, considering (4) for updating the tap-weight vector of the filter, it may correspondingly transform into the frequency domain as:

$$\mathbf{W}(j+1) = \mathbf{W}(j) + \mu FFT \begin{bmatrix} \mathbf{G}(j) \\ 0 \end{bmatrix} \quad (8)$$

Here $\mathbf{G}(j)$ is a matrix of first M elements of IFFT $[\mathbf{D}(j)\mathbf{U}(j)\mathbf{E}(j)]$, where $\mathbf{D}(j)$ is related to diagonal matrix of average signal power, $\mathbf{U}(j)$ is diagonal matrix obtained by Fourier transforming two successive blocks of input data, and $\mathbf{E}(j)$ is transform of error signal vector.

Similarly the frequency domain implementation of BNLMS is also considered and the resultant one is fast block normalized LMS (FBNLMS) algorithm. The computational complexity of these algorithms is summarized in Table 1.

3 Simulation Results

To show that block-based algorithms are really effective in clinical situations, we used the benchmark MIT-BIH arrhythmia database ECG recordings as the reference for our work, and real noise is obtained from MIT-BIH Normal Sinus Rhythm Database (NSTDB).

51 Cancellation of Artifacts in ECG Signals

For evaluating the performance of the proposed adaptive filter structures, we have measured the signal-to-noise (SNR) ratio's improvement and compared with LMS algorithm. For all the figures, *number of samples* are taken on x-axis and *amplitude* on y-axis, unless stated. For the evaluation of the performance of the algorithms, we have chosen first 4,000 samples of record number 100 from MIT-BIH database. Various adaptive filters structures are implemented based on BLMS, BNLMS, FBLMS, and FNBLMS algorithms.

3.1 Adaptive Baseline Wander Reduction

In this experiment, the original ECG signal (record no 100) is corrupted with real BW taken from the MIT-BIH NSTDB. The contaminated ECG signal is applied as primary input to the adaptive filter of Fig. 1.

Figure 2 gives the results of BW removal from the ECG record 100. From Fig. 2f, it can be observed that the BW has been effectively removed using FBNLMS

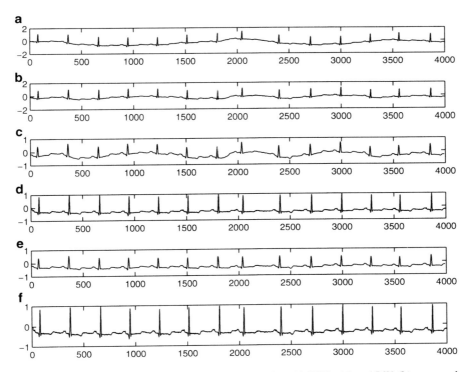

Fig. 2 Typical filtering results of baseline wander reduction. (**a**) ECG with real BW, (**b**) recovered signal using LMS algorithm, (**c**) recovered signal using BLMS algorithm, (**d**) recovered signal using FBLMS algorithm, (**e**) recovered signal using BNLMS algorithm, (**f**) recovered signal using FBNLMS algorithm

algorithm. From Fig. 2e, it is clear that the smoothing of the signal is good with BNLMS but because of the block-wise processing the signal attenuates more, as a result the SNR improvement is poor. For the BW removal, the SNR improvement is founded as 15.4984 and 16.8218 dB for FBLMS and FBNLMS, respectively, whereas the conventional LMS gets 2.1986 dB only.

3.2 Adaptive Power-Line Interference Canceler

The input to the filter is ECG signal corresponds to the data 100 corrupted with synthetic PLI with amplitude 1 mV and frequency 60 Hz, sampled at 200 Hz. The reference signal is synthesized PLI, and the output of the filter is recovered signal. These results are shown in Fig. 3.

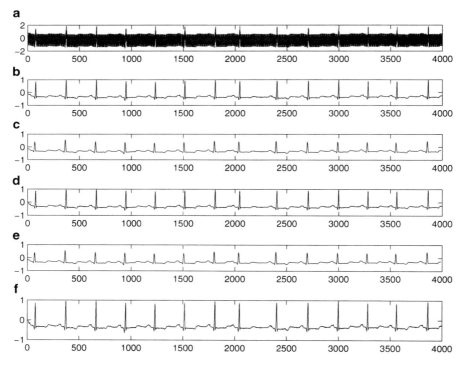

Fig. 3 Typical filtering results of PLI cancellation. (**a**) ECG with synthetic PLI, (**b**) recovered signal using LMS algorithm, (**c**) recovered signal using BLMS algorithm, (**d**) recovered signal using FBLMS algorithm, (**e**) recovered signal using BNLMS algorithm, (**f**) recovered signal using FBNLMS algorithm

The SNR improvement is found as 19.8458, 20.9911, and 18.5350 dB for FBNLMS, FBLMS, and LMS algorithms, respectively.

3.3 Adaptive Cancelation of Muscle Artifacts

To show the filtering performance in the presence of non-stationary noise, real MA is taken from the MIT-BIH Noise Stress Test Database. The MA originally had a sampling frequency of 360 Hz, and therefore they were anti-alias resampled to 128 Hz to match the sampling rate of the ECG. The original ECG signal with MA is given as input to the adaptive filter. MA is given as reference signal. These results are shown in Fig. 4.

The SNR improvement for FBNLMS is 16.8220 dB and that for FBLMS is 15.6777 dB, whereas the conventional LMS algorithm gets 3.5006 dB.

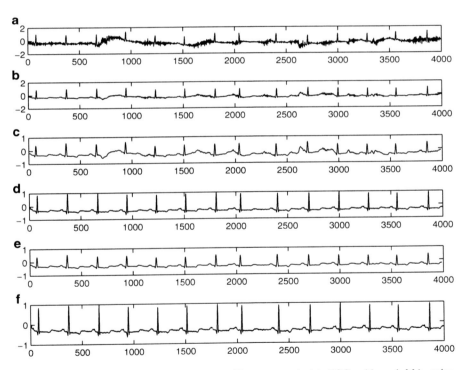

Fig. 4 Typical filtering results of muscle artifacts removal. (**a**) ECG with real MA noise, (**b**) recovered signal using LMS algorithm, (**c**) recovered signal using BLMS algorithm, (**d**) recovered signal using FBLMS algorithm, (**e**) recovered signal using BNLMS algorithm, (**f**) recovered signal using FBNLMS algorithm

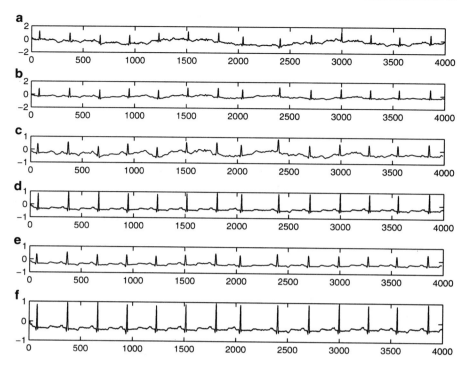

Fig. 5 Typical filtering results of motion artifacts removal. (**a**) ECG with real EM noise, (**b**) recovered signal using LMS algorithm, (**c**) recovered signal using BLMS algorithm, (**d**) recovered signal using FBLMS algorithm, (**e**) recovered signal using BNLMS algorithm, (**f**) recovered signal using FBNLMS algorithm

3.4 Adaptive Motion Artifacts Cancelation

To demonstrate this, we use MIT-BIH record number 100 ECG data with electrode EM added, where EM is taken from MIT-BIH NSTDB. The corrupted ECG signal with EM is given as input to the adaptive filter. The EM noise is given as reference signal. These results are shown in Fig. 5.

In this experiment, FBNLMS gets SNR improvement 15.2301 dB, FBLMS gets 13.9809 dB, and LMS algorithm gets 3.4948 dB.

4 Conclusion

In this chapter, the process of noise removal from ECG signal using block-based adaptive filters in time and frequency domain are presented. The various filter structures based on BLMS, FBLMS, BNLMS, and FBNLMS algorithms are implemented for noise cancellation. For this, the input and the desired response signals are properly chosen in such a way that the filter output is the best least squared estimate of the original ECG signal. The proposed treatment exploits the modifications in the weight update formula and thus pushes up the speed over the respective LMS-based realizations.

Our simulations, however, confirm that the SNR of the FLMS and FBNLMS based filters is better than that of LMS algorithm. The smoothing of the BNLMS is better than other algorithms, but signal gets attenuated. Also, the convergence rate of FLMS and FBNLMS algorithms is faster than LMS algorithm and computational complexity is less than their time domain implementation. Hence, these systems are well suited for wireless bio-telemetry applications, where high SNR and fast computations are needed.

References

1. B. Widrow, J. Glover, J. M. McCool, J. Kaunitz, C. S. Williams, R. H. Hearn, J. R. Zeidler, E. Dong, and R. Goodlin, "Adaptive noise cancelling: Principles and applications," *Proc. IEEE*, vol. 63, pp. 1692–1716, Dec. 1975.
2. O. Sayadi and M. B. Shamsollahi, "Model-based fiducial points extraction for baseline wander electrocardiograms," *IEEE Trans. Biomed. Eng.*, vol. 55, pp. 347–351, Jan. 2008.
3. Y. Der Lin and Y. Hen Hu, "Power-line interference detection and suppression in ECG signal processing," *IEEE Trans. Biomed. Eng.*, vol. 55, pp. 354–357, Jan. 2008.
4. N. V. Thakor and Y. -S. Zhu, "Applications of adaptive filtering to ECG analysis: Noise cancellation and arrhythmia detection," *IEEE Trans. Biomed. Eng.*, vol. 38, no. 8, pp. 785–794, 1991.
5. B. Farhang-Boroujeny, *Adaptive Filters-Theory and Applications*, Wiley, Chichester, UK, 1998.
6. P. E. McSharry, G. D. Clifford, L. Tarassenko, and L. A. Smith, "A dynamical model for generating synthetic electrocardiogram signals," *IEEE Trans. Biomed. Eng.*, vol. 50, no. 3, pp. 289–294, 2003.
7. L. Biel, O. Pettersson, L. Philipson, and P.Wide, "ECG Analysis: A new approach in human identification," *IEEE Trans. Instrum. Meas.*, vol. 50, pp. 808–812, Jun. 2001
8. A. K. Ziarani, A. Konrad, "A nonlinear adaptive method of elimination of power line interference in ECG signals", *IEEE Trans Biomed Eng*, vol. 49, no. 6, pp. 540–547, 2002.
9. S. Olmos , L. Sornmo and P. Laguna, "Block adaptive filter with deterministic reference inputs for event-related signals: BLMS and BRLS," *IEEE Trans. Signal Process.*, vol. 50, pp. 1102–1112, May. 2002.
10. B. Farhang-Boroujeny and K. S. Chan, "Analysis of the Frequency- Domain Block LMS Algorithm," *IEEE Trans. Signal Processing*, vol. 48, no. 8, pp. 2332–2342, Aug. 2000.

Chapter 52
Segmentation of Medical Image Sequence by Parallel Active Contour

Abdelkader Fekir and Nacéra Benamrane

Abstract This paper presents an original approach for detecting and tracking of objects in medical image sequence. We propose a multi-agent system (MAS) based on NetLogo platform for implementing parametric contour active model or snake. In NetLogo, mobile agents (turtles) move over a grid of stationary agents (patches). In our proposed MAS, each mobile agent represents a point of snake (snaxel) and it minimizes, in parallel with other turtles, the energy functional attached to its snaxel. Then, these turtles move over the image represented by a grid of patches. The set of these agents is supervised by Observer, the NetLogo global agent. In addition, Observer loads successively the frames of sequence and initializes the turtles in the first frame. The efficiency of our system is shown through some experimental results.

Keywords Image processing in medicine and biological sciences · Multi-agent system · Parallel active contour · Segmentation · Object tracking

1 Introduction

Image segmentation consists of extracting symbolic entities that are the regions and the contours. Several segmentation algorithms are proposed in the literature, based on discontinuity or similarity [1]. Since the publication by Kass et al. [2], the deformable models have become among the most used techniques to extract the discontinuity of boundaries from the interest structures [3, 4]. They were largely used in the medical image processing because of their flexibility, which allowed their adaptation to the variability of biological structures and the incorporation of the medical expertise into their conception [5–7]. Moreover, for several years, these

A. Fekir (✉)
Mathematics and Computer Science Department, Mascara University,
BP 763, Mamounia Route, 29000, Mascara, Algeria
e-mail: aekfekir@gmail.com; aekfekir@univ-mascara.dz

applications used a monolithic, sequential system to perform complex tasks. Some of these operations can be performed in parallel, or in a distributed fashion. Therefore, new approaches for tackling image segmentation from other angles are needed. One of these approaches, which is getting more and more popular, is represented by multi-agent system (MAS) [8–13].

In this paper, we propose MAS for segmentation of medical image sequence. Our MAS, based on NetLogo platform, evolves and moves all the points of contour active in parallel. Therefore, this article is organized as follows: Section 2 describes briefly about NetLogo, the multi-agent platform used in this work. In Sect. 3, we present active contour or the snake. Then, the proposed MAS approach of image sequence segmentation is detailed in Sect. 4. Some experimental results are presented in Sect. 5. Conclusion is given in the last section.

2 NetLogo Environment

NetLogo is a programmable modeling environment for simulating natural and social phenomena. It was authored by Uri Wilensky in 1999 [14] and has been in continuous development at the Center for Connected Learning and Computer-Based Modeling [15]. In NetLogo, there are four types of agents: turtles, patches, links, and observer. Turtles are reactive agents that move around the world. The world is two dimensional and is divided into a grid of patches. Each patch is a square piece of "ground" over which turtles can move. Links are agents that connect two turtles. The observer does not have a location. All four types of agents can run NetLogo commands. All three (observer, turtles, and patches) can also run "procedures." A procedure combines a series of NetLogo commands into a single new command that one defines, and more details are given in [14].

3 Active Contour

An active contour, as introduced by Kass et al. [2], is a curve described as an ordered collection of points, which evolves from its initial position to some boundary within the image (see Fig. 1). The contour active evolution is formulated as an energy

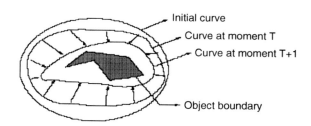

Fig. 1 Evolution of contour active

minimization; the snake energy is typically a linear combination of three terms: external energy, internal energy, and constraint energy:

1. External energy, which guides the snake toward the boundary of interest. It can be written as gradients of scalar potential functions in image I. This external energy is important in contour detection step.
2. Internal energy, which ensures that the segmented region has smooth boundaries. It defines the stiffness of the curve as well as the cohesion of the points. Then, it is intrinsic with the snake. Hence, this energy can be calculated using two forces: continuity force and curvature force.
3. Constraint energy, which provides a means for the user to interact with the snake. In this work, we have used two energies of context: balloon energy introduced by [16] and an improvement of directional energy proposed by [17].

4 The Proposed MAS Approach

In this proposed work, we introduce a cooperative approach for detecting and tracking of object in image sequence. The MAS proposed is used to implement the active contour.

4.1 Active Contour Used

The parametric model described in Sect. 2 is used in this approach. Our energy functional is given by the following formula:

$$E_{snake} = \Sigma_{i=1,N}(a\,E_{continuity}(Pi) + b\,E_{curvature}(Pi) + c\,E_{gradient}(Pi) + d\,E_{constraint}(Pi)) \quad (1)$$

$P_i = {}^t(x_i, y_i)\, i = 1..N$ are the points of snake and a, b, c, and d are coefficients attached to each energy.

In the first frame, we used the balloon energy as constraint energy. Then, we replaced this energy by directional energy in the other frames. For more detail about the utility of this change of the energy functional, see our paper [18].

4.2 Multi-Agent System

Our MAS is implemented using the NetLogo platform. We used three types of NetLogo agents: observer, turtles, and patches. The snake and its points (snaxels) are implemented using turtles; so, each turtle represents one point of snake. Similarly, each pixel of frame is represented by patch agent. These two types of agents (turtles and patches) are supervised by Observer The main stages of the proposed method are shown in Fig. 2.

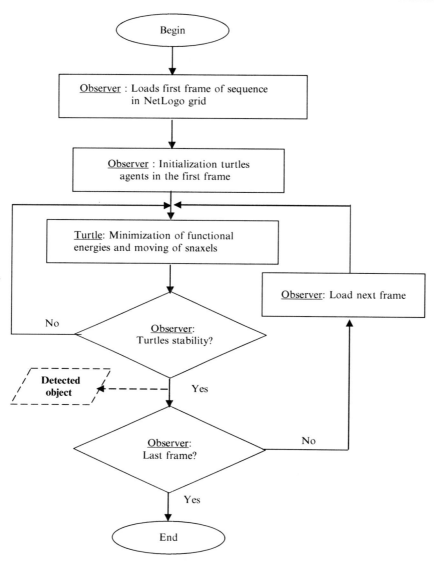

Fig. 2 Steps of proposed approach

4.2.1 Observer Roles

Observer performs several operations that are summarized as follows:

1. It loads the first frame of sequence in the NetLogo grid, and consequently it initializes patches by attaching the value of every image pixel to one patch. After object detection in the first frame, the Observer also loads other sequence frames in the order to track object.

2. Observer initializes the snake around the boundaries of object to be tracked. It creates turtles set by this initialization of snaxels.
3. Observer tests the turtle's stability as shown in Fig. 2. It is represented by percentage of fixed agents during a certain period.
4. It supervises the fusion of neighboring turtles if they are very close and it creates a new turtle between two neighboring turtles, which have high continuity energy.

4.2.2 Turtles' Roles

We have mentioned previously that these agents represent points of contour active. After its creation by Observer, each turtle has its own process to execute simultaneously with the other agents. This process is summarized in Algorithm 1.

Algorithm 1:

Repeat
- **For** all the neighbor pixels (patches) **do** calculate energies
- **For** all the neighbor pixels (patches) **do** Normalization
- Patch which had minimal energy functional represents the new position of turtle

Until stop Criterion

This algorithm is extracted from greedy algorithm [19]. The stop criterion in Algorithm 1 is the turtle's stability mentioned in Fig. 2. During this processing, each turtle needs several communications with other agents, which can be given as follows:

1. It contacts its two neighbor turtles to get their positions in the NetLogo grid. This information is necessary to calculate the energy and to evaluate the criterion of fusion with its neighbor agent. This criterion is evaluated by a comparison between the turtle continuity energy and the average distance between turtles.
2. The second type of turtle communication is established with Observer when the turtle wants to execute a fusion with its neighbor. This operation is achieved if Euclidian distance between neighbor turtles is minimal; two neighbor turtles are merged to obtain one turtle. Observer accepts or refuses this operation, and transmits its decision to communicant turtles.
3. The last type of communication consists to get from patch, the gray level of pixel, which is represented by this patch agent. Turtle agent use this value of gray level in evaluation of image energy.

5 Experimental Results and Discussions

In this section, we present some results in order to validate our proposed approach. We have used many sequences in experiments but we have given here only two types of image sequences: sequence of biological cell and echocardiographic sequence.

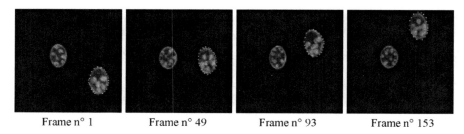

Frame n° 1　　Frame n° 49　　Frame n° 93　　Frame n° 153

Fig. 3 Detection and tracking object in biological images sequence

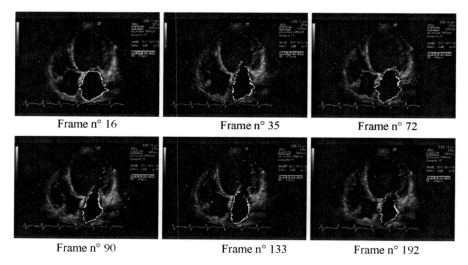

Frame n° 16　　Frame n° 35　　Frame n° 72

Frame n° 90　　Frame n° 133　　Frame n° 192

Fig. 4 Detection and tracking object in echocardiographic images sequence

In all the tests, we used a vicinity of 3 × 3. The criterion of stability (Fig. 2) is the stability of 90% of turtles (snaxels). Some results obtained are illustrated in Figs. 3 and 4. The turtles are represented by white pixels.

In the biological sequence, we have obtained good results. The object is detected in the first image and tracked in other frames of sequence. All turtles are stabilized in image sequences. This result is due to the simplicity of sequences and the absence of noise. In the second sequence also, we have obtained good results: The boundary object is tracked in all frames of sequence. We used 80 turtles to obtain continuous snake. But, we remark that there are some turtles, which did not have good positions. This is because of the following two reasons: First, this type of images is very noisy, and contains false areas (artifacts) around the boundaries of the object. Second, the percentage of stable turtles used in the test. If we increase this criterion, we will have a very important processing time.

6 Conclusion

In this paper, a novel MAS has been proposed, which is markedly different from previous cooperative approaches of image sequence segmentation. Parallel contour active is used to detect and track automatically an object in medical image sequence. This parallelism is implemented by our MAS based on NetLogo platform. The system has been applied to several image sequences. The results show good cooperation between agents in detection and tracking of object in sequence. After a temporal comparison between this cooperative approach and our pervious iterative approach [18], we propose an implementation of our MAS in C language in order to track the object in real time.

References

1. Coquerez JP, Philipp S. (1995) Analyse d'images: filtrage et segmentation. Masson, Paris.
2. Kass M, Witkin A, Terzopoulos D. (1988) Snakes: Active contour models. Int. J. Comput. Vis., 55:321–331.
3. McInerney T, Terzopoulos D. (1996) Deformable models in medical image analysis: A survey. Med. Image Anal., 1(2): 91–108.
4. He L, et al. (2008) A comparative study of deformable contour methods on medical image segmentation. Image Vis. Comput., 26:141–163.
5. Chang H, Valentino DJ. (2008) An electrostatic deformable model for medical image segmentation. Comput. Med. Imaging Graph., 32:22–35.
6. Xu J, Chutatape O, Chew P. (2007) Automated optic disk boundary detection by modified active contour model. IEEE Trans. Biomed. Eng., 54(3):473–482.
7. Fang W, Chan K, Fu S, Krishnan S. (2005) Incorporating temporal information for ventricular contour detection in echocardiographic image sequences. Eng. Med. Biol. 27th Ann. Conf., Shanghai, China, September 1–4, pp. 1099–1102.
8. Duchesnay E. (2001) Agents situés dans l'image et organisés en pyramide irrégulière: contribution à la segmentation par une approche d'agrégation coopérative et adaptative. Ph.D. thesis, Rennes-1 University, France.
9. Porquet C, Settache H, Ruan S, Revenu M. (2003) Une plate-forme multi-agent pour la segmentation d'images. Etude des stratégies de coopération contour-région ORASIS, 413–422.
10. Haroun R, Hamami L, Boumghar F. (2004) Segmentation d'images médicales par un système hybride flou – croissance de régions dans un système multi agents, JETIM, 21–30.
11. Bovenkamp EGP, Dijkstra J, Bosch JG, Reiber GHC. (2004) Multi-agent segmentation of IVUS images. Pattern Recognit. 37:647–663.
12. Benamrane N, Nassan S. (2007) Medical Image Segmentation by a Multi-Agent System Approach, MATES 2007, LNAI 4687, pp. 49–60, Springer-Verlag Berlin Heidelberg.
13. Chaib-Draa B, Jarras I, Moulin B. (2001) Systèmes multi-agents: principes généraux et applications. Hermès, Paris.
14. Wilensky U. (1999) NetLogo, Center for Connected Learning and Computer-Based Modeling, Northwestern University, Evanston, IL, http://ccl.northwestern.edu/netlogo/ June 2009
15. Tisue S, Wilensky U. (2004) "NetLogo: A Simple Environment for Modeling Complexity", Int. Conf. Complex Systems, Boston, MA.
16. Cohen L. (1991) On active contour models and balloons. Comput. Vis. Graph Image Process.: Image Underst., 53(2):211–218.

17. Lee B, Choi I, Jeon G. (2006) Motion-based moving object tracking using an active contour. IEEE Inter. Conf. Acoust. Speech Signal Process., pp. 649–651.
18. Fekir A, Benamrane N, Taleb-ahmed A. (2009) Segmentation d'une sequence d'images médicales par les contour actifs, ICSIP Guelma, Algeria.
19. Williams D, Sham M. (1992) A fast algorithm for active contour and curvature estimation. Comput. Vis. Graph. Image Process: Image Underst., 55(1):14–26.

Chapter 53
Computerized Decision Support System for Mass Identification in Breast Using Digital Mammogram: A Study on GA-Based Neuro-Fuzzy Approaches

Arpita Das and Mahua Bhattacharya

Abstract In the present work, authors have developed a treatment planning system implementing *genetic based neuro-fuzzy* approaches for accurate analysis of shape and margin of tumor masses appearing in breast using digital mammogram. It is obvious that a complicated structure invites the problem of over learning and misclassification. In proposed methodology, *genetic algorithm* (GA) has been used for searching of effective input feature vectors combined with *adaptive neuro-fuzzy model* for final classification of different boundaries of tumor masses. The study involves 200 digitized mammograms from MIAS and other databases and has shown 86% correct classification rate.

Keywords Breast cancer diagnosis system · Benignancy and malignancy · Fourier descriptors · Fuzzy c-means clustering · Genetic algorithm · Adaptive neuro-fuzzy

1 Introduction

Every year millions of women develop new cases of breast cancer around the world, and the detection procedures are based on clinical examination, breast imaging, and core biopsy. Mammographic images show signs of obstruction and many direct and indirect radiographic signs due to space occupying lesions in the tissue region of breast [1–3]. The conventional method is the localization of the lesion by needle [2] which is painful and costly. Many researchers have described potential approaches for breast cancer screening as reported in [4–20]. Benign tumors have clearly defined round or round to oval-shaped edges, whereas malignant masses spread quickly into surrounding breast tissues. They have irregular borders and multiple protrusions and made up of abnormally shaped cells [8, 21–23]. The present work

---

M. Bhattacharya (✉)
Indian Institute of Information Technology & Management, Morena Link Road, Gwalior 474003, India
e-mail: mb@iiitm.ac.in; bmahua@hotmail.com

is continuation of our earlier work based on *theory of shape* related to gradation of benignancy/malignancy of tumor in tissue region [8, 23, 24]. In the present chapter, authors have described a methodology for discrimination of tumor mass considering the type, size, and distribution into benign and malignant group. To describe the margin of masses very precisely, we introduce *Fourier descriptors* as shape-based features [25], and finally classifier has been designed using GA-based *neuro-fuzzy* techniques to discriminate the benignancy from malignancy growth of tumor lesion in breast identified by mammograms.

2 Proposed Method for Classification of Tumor Masses

Tumor masses have been extracted from surrounding normal breast tissues by fuzzy-based segmentation technique. Significant shape-based boundary features are selected. GA-based optimization technique has been implemented for reduction of feature space. Finally features are fed to an *adaptive neuro-fuzzy classifier* for a decision. The proposed method presented is shown in Fig. 1.

2.1 Segmentation of Tumor Mass

In the proposed technique, fuzzy c-means clustering algorithm used for intensity-based segmentation of masses. Total number of fuzzy cluster centers chosen is three as shown in Fig. 2. Cluster center *A* represents the healthy breast tissue. Second cluster *B* represents false presence of mass region and *C* represents actual mass region. The ultimate fuzzy partition membership functions have been shown in Fig. 2, which show that there is an overlapping between the membership functions *A*, *B*, and *C*.

In the present chapter, decision has been made on the basis, which is described as follows: if the possibility of a breast region regarding its belongingness to the

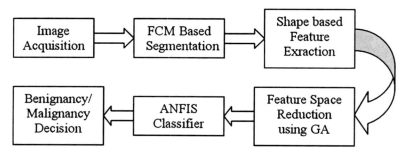

Fig. 1 Overview

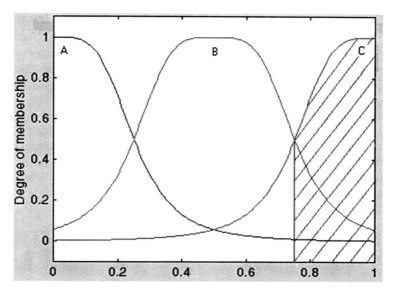

Fig. 2 Final fuzzy partition membership functions

calcification part is greater than 50%, i.e., the membership value of curve **C > 0.5**, decision may be taken that the particular region is belonging within the **calcified lesion**. According to the decision rule, the shaded region in Fig. 2 indicates the region of interest (ROI). Let $X = \{x_1, x_2, \ldots, x_n\}$ be a set of given data. A fuzzy *c-partition* of X is a family of fuzzy subsets of X, denoted by $P = \{A_1, A_2, \ldots, A_c\}$, which satisfies $\sum_{i=1}^{C} A_i(x_k) = 1$. The performance *index* of a *fuzzy partition P*, $J_m(P)$, is defined in terms of the cluster centers by the formula:

$$J_m(A, v_1, \ldots, v_c) = \sum_{k=1}^{n} \sum_{i=1}^{c} [A_i(x_k)]^m \|x_k - v_i\|^2$$

where $\|x_k - v_i\|^2$ represents the distance between x_k and v_i (v_i is the cluster centers). Clearly, the smaller the value of $J_m(P)$, the better the *fuzzy partition P*. Thus, the goal of fuzzy c-means clustering method is to find a fuzzy partition P that minimizes the *performance index* $J_m(P)$:

$$v_i = \frac{\sum_{k=1}^{N} [A_i(x_k)]^m x_k}{\sum_{k=1}^{n} [A_i(x_k)]^m} \quad \text{and} \quad A_i(x_k) = \frac{1}{\sum_{j=1}^{C} \left(\frac{d_{ik}}{d_{jk}}\right)^{2/(m-1)}}$$

2.2 Extraction of Boundary as Feature Using Fourier Descriptors

Presently, we introduce *Fourier descriptors* as the boundary features having the information about the shape and margin of the segmented masses. **Algorithm**: Let us consider a figure that describes K-points digital boundary in the x-y plane, staring at an arbitrary point (x_0, y_0) to the coordinate pairs (x_1, y_1), $(x_2, y_2), \ldots, (x_{k-1}, y_{k-1})$ along the boundary. These coordinates are represented by the form $x(k) = x_k$ and $y(k) = y_k$. Thus the boundary can be represented as, $s(k) = [x(k), y(k)] \ldots, k = 0, 1, 2, \ldots, K - 1$. Each co-ordinate pair can be treated as a complex number so that, $s(k) = x(k) + j*y(k) \ldots, k = 0, 1, 2, \ldots, K - 1$. The x-axis is treated as real axis and y-axis as the imaginary one. The Discrete Fourier Transform (DFT) of $s(k)$ is given below

$$a(u) = (1/K) \times \sum_{k=0}^{K-1} s(k) \times e^{-(j2\pi u(k)/K)} \ldots, \text{for } u = 0, 1, 2, \ldots, K-1.$$

The complex coefficient $a(u)$ is the *Fourier descriptor* of the edge points along the boundary. **Compactness** of a particular shape is another important feature used for boundary descriptor. It is defined as $(perimeter)^2/area$. *Compactness* is a dimensionless quantity and is minimal (nearly equal to 1) for round-shaped region. *Compactness* is insensitive to the orientation of the images. In the present work, computation of *compactness* has been considered as important shape feature.

2.3 Genetic Algorithm for Reduction of Feature Subspace

In proposed method, *genetic algorithm* (GA) [26–32] has been used to search two significant shape descriptors, which are able to represent the particular class of masses. Different image boundaries of tumor masses have been recognized on the basis of *Fourier descriptors* which play the role of payoff values (*objective function*) associated with individual strings. We encode the magnitude of *Fourier descriptors* as binary strings and optimize these strings to achieve best two features that are responsible for representing the class of the objects.

2.4 Classification of Significant Features

The proposed method uses *adaptive neuro-fuzzy network* [25, 33–36] for classification of features into benign and malignant groups Fig. 3. To adapt the network with ever-changing environments, hybrid-learning rule has been used.

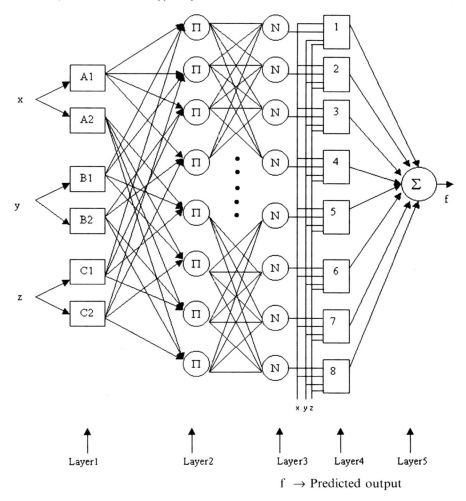

Fig. 3 Adaptive neuro-fuzzy model for final classification

Layer 1: Every node I in this layer is an adaptive node with a node function, $O_{1,i} = \mu_{Ai}(x)$, $O_{1,i} = \mu_{Bi}(y)$ for for $i = 1$, where x (or y) is the input to node i and A_i (or B_i) is a linguistic label (such as large or small) associated with this node. In other words, $O_{1,i}$ is the membership grade of fuzzy set A (A_1, A_2) or B (B_1, B_2). Here the membership function for A can be any appropriate parameterized membership function, such as generalized bell function:

$$\mu_A = \frac{1}{1 + \left|\frac{x-c_i}{a_i}\right|^{2b}}$$

where $\{a_i, b_i, c_i\}$ is the parameter set. As the values of these parameters change, the bell-shaped function varies accordingly. Parameters of this layer are referred to as *premise parameters*.

Layer 2: Every node in this layer is a fixed node labeled Π, whose output is the product of all the incoming signals:

$$O_{2,i} = w_i = \mu_{Ai}(x)\mu_{Bi}(y), \text{ for } i = 1, 2$$

In general, any T-norm operator that performs fuzzy AND can be used as the node function in this layer.

Layer 3: Every node in this layer is a fixed node labeled N. The i-th node calculates the ratio of the rule's firing strength to the sum of all rules' firing strengths:

$$O_{3,i} = \overline{w_i} = \frac{w_i}{w_1 - w_2}$$

For convenience, outputs of this layer are called normalized firing strengths.

Layer 4: Every node i in this layer is an adaptive node with a node function:

$$O_{4,i} = \overline{w_i} f_i = \overline{w_i}(p_i x + q_i y + r_i)$$

where w_i is a normalized firing strength from layer 3 and $\{p_i, q_i, r_i\}$ is the parameter set of this node. Parameters of this layer are referred to as *consequent parameters*.

Layer 5: The single node in this layer is fixed node labeled Σ, which computes the overall output as the summation of all incoming signals:

$$O_{5,i} = \Sigma \overline{w_i} f_i = \frac{\sum_i w_i f_i}{\sum_i w_i}$$

Limitation of *adaptive neuro-fuzzy* model is that the architecture is learned well only when the number of inputs is very small (3–4 only). In the present chapter, we set the significant input feature vector size to 3 only by GA, and there are two bell-shaped membership functions that are assigned for each input variable. Thus, number of fuzzy *if–then* rules for *adaptive neuro-fuzzy* learning are $2^3 = 8$ only.

2.5 Decision-Making Logic

A set of 200 different mammograms is used for a test of the proposed algorithms. In order to perform this comparison, it is necessary to define a Euclidean Distance function (μ_1) for the determination of roundness deviation, and the degree of malignancy is higher for higher value of μ_1. $\mu_1 = [(\mathbf{D_1} - \mathbf{O_1})^\mathbf{T}(\mathbf{D_1} - \mathbf{O_1})]^{0.5} = [(D_1 - O_1)^2]^{0.5}$ where, $\mathbf{D_1}$ = desired output value for benign microcalcification, $\mathbf{O_1}$ = output value of the test mammograms. The decision on the boundary of test masses is defined below: If $\mu_1 \leq 20$, the shape and margin of test masses are considered as: Almost Round or Round to Oval Shape & Smooth Boundary – Benign

Stage. If $20 \leq \mu_1 \leq 40$, the shape and margin of test masses are considered as: Lobulated Shape and Non-Circumscribed Boundary – tendency toward malignancy. If $\mu_1 \geq 40$, the shape and margin of the test masses are considered as: Irregular Shape and Ill-defined Boundary – possibly in malignant stage.

3 Experimental Results

We have applied the proposed algorithm to databases (MIAS and others) consisting of 200 images. Figure 4 shows some sample mammograms having tumor lesions, the enhanced ROI and the contour of the lesion. The classifier was first trained with obvious benign masses of *sixty dataset* as identified by the expert radiologists, and then other non-obvious cases have been tested and classified during the experiment. The training database consists of almost round or round to oval benign masses with circumscribed margins (examples of some of train data set are *data-1 and data-2*) as specified by the radiologists. On the basis of that *test image database* is evaluated. It has been noted that the desired value, assumed for *circumscribed benign masses* $(D_1) = 50$. Some examples of the non-obvious case studies are as follows:

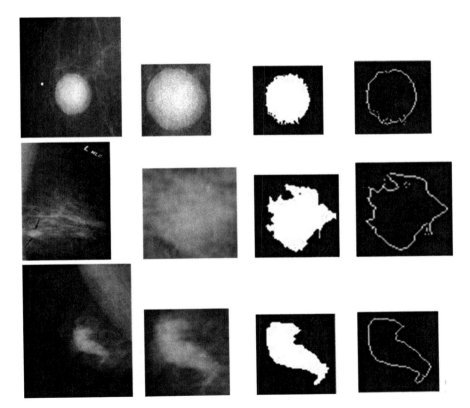

Fig. 4 Shows mammogram having lesion and the contour of lesion

3.1 Choice of Number of Generation in GA-Based Feature Subset Selection Model

In GA-based *feature subset selection* (FSS) model, the initial boundary feature space dimension (number of *Fourier descriptors*) is 90, and after four iterations (generations) the feature size converges into 6 significant points only. Among them only *first two maximum* value of the population is enough to represent the *shape and margin* of each mass. Third important feature to describe the shape of the mass is *measuring of compactness* of the mass. Thus, total number of input terminals for *adaptive neuro-fuzzy network* is kept into three (two for representing GA-based *boundary features*, one for representing *compactness*), and only with two membership functions on each would result in $2^3 = 8$ fuzzy *if–then rules* and offers best possible result also. The variation in average population and population size with different generation count has been graphically represented in Fig. 5a, b. Table 1 demonstrates the result for the detection of benignancy/malignancy of breast tumor lesion which are appearing in mammogram.

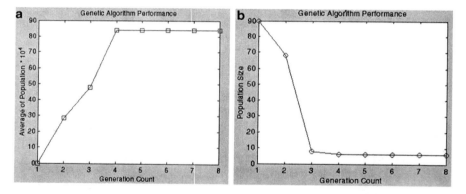

Fig. 5 (a) Variations in average population with different generation count. (b) Variations in population size with different generation count

Table 1 Results for benignancy/malignancy of few test database

Database	O_1	μ_1	Decision on shape	Final decision
Data 1	50.9258	0.9258	Round to oval	Benign stage
Data 2	50.1647	0.1647	Round to oval	Benign stage
Data 3	11.0055	38.9945	Non-circumscribed	Malignant stage
Data 4	17.4536	32.5464	Non-circumscribed	Malignant stage
Data 14	6.7601	43.2399	Irregular	Possibly malignant
Data 15	7.2857	42.7143	Irregular	Possibly malignant
Data 17	6.4994	43.5006	Irregular	Possibly malignant

4 Discussion

In proposed methodology for tumor classification using mammogram, authors have attempted to develop a technique based on GA-based *adaptive neuro-fuzzy* model by extracting the boundary of the lesion or ROI. This classification concerns the prediction regarding the prognosis of the disease toward either benignancy or malignancy using machine intelligence considering features such as shape and margin. *Adaptive neuro-fuzzy* is a robust soft-computing approach to take decision even in the presence of uncertainty.

Acknowledgment The authors would like to thank Dr S. K. Sharma of EKO X-ray and Imaging Institute, Kolkata, India, for his support.

References

1. M. McMaster, B.M. McCook, M.I. Shaff, A.K. Lamballe, A.C. Winfield, 'Dilated mammary veins as sign of superior vena cava obstruction', Special report: Breast Imaging Applied Radiology, November 1987
2. M.J. Homer, E.R. Pile-Spellman, "Needle localization of occult breast lesions with a curved end retractable wire", Applied Radiology 16, 547–548, 1986
3. S.A. Feig, R. McLelland, Breast carcinoma: current diagnosis and treatment. New York: Masson, 1983
4. S. Yu, L. Guan, "A CAD system for the automatic detection of clustered microcalcifications in digital mammogram films", IEEE Transactions on Medical Imaging, 19, 2, 115–126, 2000
5. M. Bhattacharya, A. Das, "Fuzzy logic based segmentation of microcalcification in breast using digital mammograms considering multiresolution", Proceedings of the International Machine Vision and Image Processing Conference (IMVIP-07), IEEE Computer Society, NUI Maynooth, Ireland, pp. 98–105, 2007
6. B. Verma, J. Zakos, "A computer-aided diagnosis system for digital mammograms based on fuzzy-neural and feature extraction techniques", IEEE Transactions on Information Technology in Biomedicine, 5, 1, 46–54, 2001
7. M. Bhattacharya, "A computer assisted diagnostic procedure for digital mammograms using adaptive neuro-fuzzy soft computing", IEEE Nuclear Science Symposium, Medical Imaging Conference, California, 29th Oct–4th Nov, 2006
8. D. Dutta Majumder, M. Bhattacharya, "Cybernetic approach to medical technology: application to cancer screening and other diagnostics", Millennium Volume of Kybernetes, International Journal of Systems & Cybernetics, 29, 7/8, 871–895, 2000
9. L.N. Mascio, J.M. Hermandez, C.M. Logan, "Automated analysis for microcalcifications in high resolution digital mammograms", Image processing, SPIE Proceedings, 1898, 472–479, 1993
10. D. Brzakovic, X.M. Luo, P. Brzakovic, "An approach to automated detection of tumors in mammograms", IEEE Transactions on Medical Imaging, 9, 3, 233–241, 1990
11. N.H. Eltonsy, G.D. Tourassi, A. Elmaghraby, "A concentric morphology model for detection of masses in mammography", IEEE Transactions on Medical Imaging, 26, 880–889, 2007
12. C.K. Abbey, R.J. Zemp, J. Liu, K.K. Lindfors, M.F. Insana, "Observer efficiency in discrimination tasks simulating malignant and benign breast lesions imaged with ultrasound", IEEE Transactions on Medical Imaging, 25, 2, 198–209, 2006
13. D. Gur, "Computer-aided detection performance in mammographic examinations of masses: assessment", Radiology, 223, 418–423, 2004

14. C. Varela, S. Timp, N. Karssemeijer, "Use of border information in the classification of mammographic masses," Physics in Medicine and Biology, 51, 2, 425–441, 2006
15. L.M. Bruce, R.R. Adhami, "Classifying mammographic mass shapes using the wavelet transform modulus-maxima method", IEEE Transactions on Medical Imaging, 18, 12, 1170–1177, 1999
16. M.A. Kupinski, M.L. Giger, "Automated seeded lesion segmentation on digital mammograms," IEEE Transactions on Medical Imaging, 17, 4, 510–517, 1998
17. N.R. Mudigonda, R.M. Rangayyan, J. Desautels, "Detection of breast masses in mammograms by density slicing and texture flow-field analysis," IEEE Transactions on Medical Imaging, 20, 12, 1215–1227, 2001
18. B. Sahiner, H.P. Chan, N. Petrick, M.A. Helvie, L.M. Hadjiiski, "Improvement of mammographic mass characterization using speculation measures and morphological features," Medical Physics, 28, 1455–1465, 2001
19. C. Klifa, J. Carballido-Gamio, L. Wilmes, A. Laprie, C. Lobo, E. DeMicco, M. Watkins, J. Shepherd, J. Gibbs, N. Hylton, "Quantification of breast tissue index from MR data using fuzzy clustering", Proceedings of the 26th Annual International Conference of the IEEE EMBS, San Francisco, September 1–5, 2004
20. L.M. Bruce, R.R. Adhami, "Classifying mammographic mass shapes using the wavelet transform modulus-maxima method", IEEE Transactions on Medical Imaging, 18, 12, 1170–1177, 1999
21. M. Set, J.T. Leith, "Dormancy, regression and recurrence: towards a unifying theory of tumor growth control", Journal of Theoretical Biology, 169, 327–328, 1994
22. N.P. Galatsanos, R.M. Nishikawa, I. El-Naqa, Y.Y.M.N. Wernick, "A support vector machine approach for detection of microcalcifications", IEEE Transactions on Medical Imaging, 21, 2, 1552–1563, 2002
23. M. Bhattacharya, D. Dutta Majumder "Knowledge based approach to medical image processing", in Pattern Directed Information Analysis (Algorithms, Architecture & Applications). New Age International Wiley, pp. 454–486, 2008, India
24. M. Bhattacharya, D. Dutta Majumder, "Breast cancer screening using mammographic image analysis", 16th International CODATA Conference, (8–12) November, Delhi, 1998
25. M. Bhattacharya, A. Das, "Object recognition using artificial neural network: case studies for noisy and noiseless images", Proceedings of the Irish Machine Vision and Image Processing Conference, pp. 52–59, 30th Aug–1st Sep., 2006, Dublin, Ireland
26. D.E. Goldberg, Genetic Algorithms in Search, Optimization and Machine Learning. Addison-Wesley, Reading, 1989
27. P. D'haeseleer, Context preserving crossover in genetic programming. In Proceedings of the 1994 IEEE World Congress on Computational Intelligence, Vol. 1, pp. 256–261, IEEE Press, Orlando, 1994
28. E. Burke, S. Gustafson, G. Kendall, "Diversity in genetic programming: an analysis of measures and correlation with fitness", IEEE Transactions on Evolutionary Computation, 8(1), 47–62, 2004
29. J. Kishore, L. Patnaik, V. Mani, V. Agrawal, "Genetic programming based pattern classification with feature space partitioning", Information Sciences, 131(1–4), 65–86, 2001
30. S. Luke, L. Spector, A revised comparison of crossover and mutation in genetic programming. In J. Koza, et al., editors, Proceedings of the Third Annual Genetic Programming Conference, pp. 208–213, San Francisco, CA. Morgan Kaufmann, 1998
31. J.D. Knowles, "Local-Search and Hybrid Evolutionary Algorithms for Pareto Optimization", PhD Thesis, University of Reading, Reading, UK, January 2002
32. H. Vafaie, K. De Jong, "Genetic algorithms as a tool for feature selection in machine learning", Proceeding of the 4th International Conference on Tools with Artificial Intelligence, Arlington, November, 1992
33. J.-S.R. Jang, "ANFIS: Adaptive-Network based Fuzzy Inference Systems", IEEE Transactions on Systems, Man and Cybernetics, 23, 3, 665–685, 1993

34. J.-S.R. Jang, C.T. Sun, E. Mizutani, Neuro-fuzzy and soft computing: a computational approach to learning and machine intelligent. Pearson Education, Upper Saddle River
35. C.-T. Lin, C.S.G. Lee, "Neural network based fuzzy logic control and decision system," IEEE Transactions on Computers, 40, 12, 1320–1336, 1991
36. L.-X. Wang, J.M. Mendel, "Back propagation fuzzy systems as nonlinear dynamic system identifiers," Proceedings of IEEE International Conference on Fuzzy Systems, San Diego, March 1992

Part VII
Computer-Based Medical Systems

Chapter 54
Optimization-Based Technique for Separation and Detection of Saccadic Movements and Eye-Blinking in Electrooculography Biosignals

Robert Krupiński and Przemysław Mazurek

Abstract Electrooculography (EOG) gives the possibility of eye tracking using biosignal measurements. Typical EOG signal consists of rapid value changes (saccades) separated by almost constant values. Additionally, the pulse shape from eyelid blinking is observed. The separation of them is possible using numerous methods, like median filtering. The proposed optimization method based on a model fitting using the variable number of parameters gives the possibility of features localization even for nearby saccades and blinking pulses.

1 Introduction

Oculography is the one of active research areas of biomeasurements due to advances in signal acquisition and processing driven by numerous applications [5]. The primary application of oculography is analysis of human vision system and is also applied to medical diagnosis. In the last decades, the new area of applications arose and the current research and applied systems are: the human–computer interfaces (e.g., virtual keyboard, the vehicle control, the wearable computers), the motion-capture systems (the direct animation of eyes of computer-generated character based on the real measurements of actor's eye-movement [3, 14]), the optimization of web services, the optimization of advertisement, the video compression driven by eye interest, and many others. A few methods of oculography were recently used [5] and the most important were video-oculography (VOG – based on the eye image analysis using a camera), infrared oculography (IROG – based on the corneal reflection of near infrared light source relative to the location of the pupil center, and electrooculography (EOG – based on the biopotentials measurements).

P. Mazurek (✉)
Department of Signal Processing and Multimedia Engineering, West Pomeranian University of Technology in Szczecin, 26-Kwietnia 10 St., 71-126 Szczecin, Poland
e-mail: przemyslaw.mazurek@zut.edu.pl

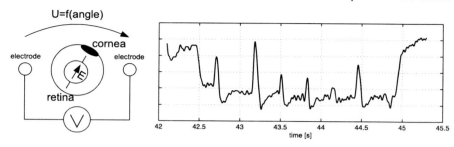

Fig. 1 EOG measurement model of the cornea–retina voltage (*left*); EOG and blinking signal waveforms (*right*)

The EOG method uses a simpler technique based on the measurements of human-generated biosignals with an amplitude modulated by eye movements. The eye voltage between a retina and a cornea has a quite large value about ± 1 mV [16]. The placement of electrodes around an eye (on the human skin) gives the possibility of eye's angular position estimation based on the voltage measurements (Fig. 1).

There are numerous variants of electrodes placements [4] and mostly three or seven electrode systems are typically used (another additional electrode for potential reference REF is also used, and so they are named also as four and eight electrode systems appropriately). The simplest placement (3/4 electrodes) assumes the symmetry of eye movements and eye-blinking [16]. Such configuration gives the possibility of two differential measurements (LEFT–UP and RIGHT–UP pairs) for the additional suppression of external disturbance sources (especially the 50/60 Hz power line interferences [15]) and also gives two almost orthogonal values corresponding to the orientation of eyes.

There are a few kinds of eye movements and the most significant is the saccadic movements [2, 5–7, 9]. The saccades are rapid changes of eye orientation and they are very fast; so the measurement system should process a signal with sufficient bandwidth. This signal is a step like; so a signal between saccades can be a constant value. The saccades are very important for the human vision analysis. The second signal is the smooth pursuit that exists during object movement tracking and is typically linear. The smooth pursuit occurs between the saccades; so the linear slopes may be observed instead of constant values. The next signal is a fixation related to the small correction of eye directions and are small eye movements. The last signal is the nystagmus – rapid movements of eyes. Additionally, the models of saccadic movements for typical person are discussed in [5].

2 EOG and Blinking Signal Separation Techniques

One of the most significant disturbances of the EOG signal is a blinking signal generated due to the eyelid movements. This signal has special interest in the EOG applications because it can be used in, e.g., the human–computer interfaces as an additional degree-of-freedom for interaction. In most applications, this signal should be separated from EOG.

The eye-blinking signal depends on the electrodes placements, and this signal is an additive pulse observed in the 3/4 electrode systems. The example of such signal is shown in Fig. 1.

The blinking signal depends on the speed of blinking, and the most typical blinking is a short pulse with comparable strength to the EOG values range. There are also the nonstandard blinking pulses, which are much more complicated to analysis and are not considered in this chapter. For example, the long-term blinking can be combined with the multiple saccades. Another example important for the 3/4 electrode system is related to the possibility of asymmetrical blinking signals. Humans can blink using a single eyelid and for such assumptions the 7/8 electrode systems should be used [17]. Such option extends the possibilities of human–computer interaction and is also important for the motion-capture systems.

The typical blinking signal and saccadic movements are time moment exclusive; so the blinking occurs between the saccades. Such assumption (true only for regular cases) is a basis for developing of separation algorithms.

The median filters are well-known filters for the removal of impulse noise. For a given input signal, the appropriate median filter removes blinking pulses, and so the clear EOG signal is obtained. The blinking signal is received using the subtraction operation of obtained signal from the median filter operation and the original signal. Two or more filters can be combined together for better results [10]. Numerous variants of separation techniques including median filtering (based on the single and multiple median filters with fixed and adaptive weights) are proposed [1, 8, 10, 12, 13]. One of the problems of median filters is the blinking signals located quickly after or before the saccade. Moreover, the blinking signal may occur during a saccade what is possible and is used intentionally by actors during motion-capture sessions. Such signals cannot be well separated by the median filters, which will be shown in comparison to the new proposed method based on a model and optimization.

3 Optimization Based on the EOG Signal Analysis Approach

The matching pursuit algorithm [11] uses the optimization technique for the set of prior defined bases, which is the motivation of our method based on the approximation of signal and the signal model. The parameters of model are subjects of optimization and they describe the EOG and blinking signal separately which gives the possibility of separate reconstruction of both signals.

The optimization efficiency (the time of calculations and obtained error) depends on the number of model's parameters; so the window-based approach is assumed. Only a small part of signal (the hundreds of samples – with up to a few saccades and blinkings) is processed at one time, and overall signal is processed in the separate steps according to the window position.

There are two important parts of algorithm: a model and an optimization technique.

3.1 EOG and Blinking Signal Model

The EOG signal is modeled by the sets of parameters, where t_i is the period time and s_i is the height of constant value between two saccades: $(t_i, s_i) \in S$.

The first aim of the algorithm is to estimate S for the assumed signal window. The second aim of the algorithm is to estimate the time position of blinking pulses b_j and the appropriate amplitude (the height of them) h_j, defined by the set: $(b_j, h_j) \in B$. The signal approximated reconstruction $\hat{Y}$ is given by the following formula:

$$\hat{Y} = \sum_i f_S(S,i) + \sum_j f_B(B,j) \qquad (1)$$

where f_S and f_B are the functions for the generation of interval between two saccades and for a blink pulse appropriately. The number of parameter pairs N_S, N_B is a variable, which is also a subject for the optimization process: $f: S \rightarrow N_S; f: B \rightarrow N_B$.

3.2 Optimization Method

There are numerous optimization techniques, and a nongradient oriented method should be used due to the variable number of blinking pulses B and the saccades S. The nongradient method based on the mutation of parameters is proposed for signal decomposition (separation).

A single iteration shown in Fig. 2 is divided into five optional steps, and optionality means that some steps cannot be processed during particular iteration. The selection of basic operation for every step is driven by uniform random generator. Such formulation allows the estimation of both signals and thereby separation, but results depend on the optimization algorithm, depending on a minima search method.

The assumed optimization technique permits using the knowledge-based mutation of parameters which is more convenient in comparison to the raw methods used in genetic programming. Moreover, in some cases a simple gradient search and a random value generation method are used locally for the convergence improvement. The optimization algorithm uses the minimization of mean squared error (MSE) criteria.

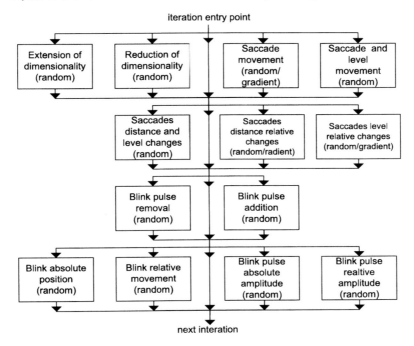

Fig. 2 Schematic of optimization algorithm

3.3 Dimensionality Changes and Balanced Movement of Saccade

The dimensionality changes are basic operations in first step and the changes of dimensionality of S. There are two kinds of changes of dimensionality – the extension and reduction.

The extension of dimensionality of S is possible by the random selection of constant value between two neighborhood saccades defined by (t_i, s_i) and splitting them into two random size lengths: $\{t_i, s_i\} \Leftarrow \{(t_i^*, s_i^*); (t_{i+1}^*, s_{i+1}^*)\}$, so the other values are not changed. A new height value is derived from the previous one: $s_i^* = s_{i+1}^* = s_i$. The reduction of dimensionality of S is possible by the random selection of two neighborhood constant values between three neighborhood saccades defined by: $\{(t_i, s_i); (t_{i+1}, s_{i+1})\}$. A substitute element is calculated using the following formula:

$$\{(t_i, s_i); (t_{i+1}, s_{i+1})\} \Leftarrow \left[t_i + t_{i+1}, \frac{(s_i + s_{i+1})}{2}\right] \quad (2)$$

so the new height of constant value is the mean of parent values.

The next operation is saccade movement – the modification of the time position of saccade movement responsible for a local optimization. For two neighborhood constant EOG values a saccade may exist. Using a random number generator or a gradient search method the length of both time periods is updated, so only the position of possible saccade is modified $t_i^* + t_{i+1}^* = t_i + t_{i+1}$. The next operation is related to the combined saccade and the level movement and is similar to the previous one, where, additionally, one of the randomly selected EOG levels are updated by a random value. Such operation improves the convergence because simple saccade movement modify less parameters.

3.4 Single Saccade Distance and Level Changes

The previously defined operation on the saccades influences only two neighborhood time intervals. The next operation influences the selected saccade, and it also influences the time position of following saccades. There are defined three operations: the first one is related to the random generation of completely new time interval and level. The next two operations are related to the time interval and level changes separately using the relative modifications. For the relative changes, the random generator or gradient approach for local minimization is used.

3.5 Dimensionality Changes of Blinking and Blinking Modification

The number of blinking pulses can be also modified and two operations are defined: the pulse addition and removal. As in the previous step, the no-change transition is possible. Using the random generator, the randomly selected pulse can be removed (the reduction of the dimensionality of blinking). Adding a new blinking pulse (the extension of the dimensionality of blinking) is also based on a random generator, but there is an additional constraint b_{min} related to the blink overlapping protection, which is used in the following formula:

$$\bigvee_j (b_j - b_{j+1} \geq b_{min}) \qquad (3)$$

A randomly selected blink pulse can be shifted to the new position using the absolute or relative movement with a previously defined position constraint. A blink amplitude can be also modified using the absolute or relative change. All operations in this step are driven by the random generators.

After all these steps, an error is calculated and the decision is made about the acceptance or rejection of a new proposed set of coefficients S and B.

3.6 Performance Comparison of Optimization and Median Filter Approaches

There are numerous methods proposed for the separation of EOG and blinking signal and in most cases the regular measurements are assumed, and most methodswork

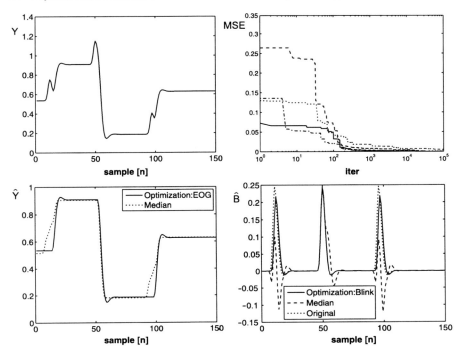

Fig. 3 Comparison of median and optimization methods for signal separation and reduction of error values during the optimization process (*right–up*)

quite well if the blinking pulses are between the saccades. For such assumption the median filters are sufficient, but if a blink pulse and a saccade are nearby, the performance of this method is reduced. In the tests, the synthetic generator of EOG and blinking signals for results comparison is used. In the test, the low-pass filtered signal and prior knowledge about this transmittance for signal fitting are assumed.

In Fig. 3, the convergence of optimization algorithm to the EOG and the blinking signals for a few drawn samples are shown.

For a median filter a different mask length can be used, and for the selection of optimal one the additional Monte Carlo test is used. The selected median filter with the length 29 is compared with the proposed optimization method, and for 150 samples the performance using the MSE criteria for the EOG and blinking separately is obtained. The MSE for the reconstructed EOG signal with the proposed optimization algorithm was ten times smaller than for the median method. Similarly, the MSE for the reconstructed blinking pulses with the proposed optimization algorithm was six times smaller than for the median method.

The example of signal separation into the EOG and blinking using the median and optimization method is shown in Fig. 3, and the optimization technique gives much better separation using a more sophisticated and time-consuming method.

4 Conclusions

The separation of blinking signal and the position of eye signal (the constant level values and saccades) are important for numerous applications. The proposed method based on a model fitting to two signals (the EOG and blinking) using the optimization algorithm gives the excellent results in comparison to the recent methods based on the median filters. Especially, the saccades and blinking can be separated even if they occur at almost the same time. For the known linear filter corresponding to the biological properties of eye and acquisition system, the proper signal can also be estimated.

Acknowledgements This work is supported by the UE EFRR ZPORR project Z/2.32/I/1.3.1/267/05 "Szczecin University of Technology – Research and Education Center of Modern Multimedia Technologies" (Poland).

References

1. Bankman, I.N., Thakor, N.V.: Noise Reduction in Biological Step Signals: Application to Saccadic EOG, Medical and Biological Engineering and Computing, 28(6):544–549, Nov(1990)
2. Becker, W.: Metrics. In: Wurtz, R.H. and Goldberg, M.E. (eds.) The Neurobiology of Saccadic Eye Movements, 13–67, Elsevier, Amsterdam (1989)
3. E.O.G Beowulf DVD 2'nd disc, Warner Brothers (2008)
4. Brown, M., Marmor, M., Vaegan, Zrenner, E., Brigell, M., Bach, M.: ISCEV Standard for Clinical Electro-oculography (EOG), Documenta Ophthalmologica, 113:205–212 (2006)
5. Duchowski, A.: Eye Tracking Methodology: Theory and Practice, Springer, Heidelberg (2007)
6. Fleming, B., Dobbs, D.: Animating Facial Features & Expressions, Charles River Media, Rockland (1999)
7. Gu, E., Lee, S.P., Badler, J.B., Badler, N.I.: Eye Movements, Saccades, and Multiparty Conversations. In: Deng Z., Neumann U. (eds.) Data-Driven 3D Facial Animation, 79–97, Springer, Heidelberg (2008)
8. Juhola M.: Median Filtering is Appropriate to Signals of Saccadic Eye Movements, Computers in Biology and Medicine, 21:43–49 (1991)
9. Krupiński, R., Mazurek, P.: Estimation of Eye Blinking Using Biopotentials Measurements for Computer Animation Applications. In: Bolc, L., Kulikowski J.L., Wociechowski, K. (eds.) ICCVG 2008. LNCS 5337, 302–310, Springer, Heidelberg (2009)
10. Krupiński, R., Mazurek, P.: Median Filters Optimization for Electrooculography and Blinking Signal Separation using Synthetic Model, In: 14–th IEEE/IFAC International Conference on Methods and Models in Automation and Robotics MMAR'2009, Miedzyzdroje, (2009)
11. Mallat, S.G., Zhang, Z.: Matching Pursuits with Time-Frequency Dictionaries, IEEE Transactions on Signal Processing, 3397–3415, Dec (1993)
12. Martinez, M., Soria, E., Magdalena, R., Serrano, A.J., Martin, J.D., Vila, J., Comparative Study of Several Fir Median Hybrid Filters for Blink Noise Removal in Electrooculograms, WSEAS Transactions on Signal Processing, 4:53–59, March (2008)
13. Niemenlehto, P.H.: Constant False Alarm Rate Detection of Saccadic Eye Movements in Electro-oculography, Computer Methods and Programs in Biomedicine, 96(2):158–171, November (2009) doi:10.1016/j.cmpb.2009.04.011
14. Sony Pictures Entertainment, Sony Corporation, Sagar, M., Remington, S.: System and Method for Tracking Facial Muscle and Eye Motion for Computer Graphics Animation, Patent US., International Publication Number WO/2006/039497 A2 (13.04.2006)

15. Prutchi, D., Norris, M.: Design and Development of Medical Electronic Instrumentation. Wiley, New Jersey (2005)
16. Schlgöl, A., Keinrath, C., Zimmermann, D., Scherer, R., Leeb, R., Pfurtscheller, G.: A Fully Automated Correction Method of EOG Artifacts in EEG Recordings, Clinical Neurophysiology 118:98–104 (2007) doi:10.1016/j.clinph.2006.09.003
17. Thakor, N.V.: Biopotentials and Electrophysiology Measurement. In: Webster, J.G. (ed.) The Measurement, Instrumentation, and Sensors Handbook, vol. 74, CRC Press, Boca Raton (1999)

Chapter 55
A Framework for Lipoprotein Ontology

Meifania Chen and Maja Hadzic

Abstract Clinical and epidemiological studies have established a significant correlation between abnormal plasma lipoprotein levels and cardiovascular disease, which remains the leading cause of mortality in the world today. In addition, lipoprotein dysregulation, known as dyslipidemia, is a central feature in disease states, such as diabetes and hypertension, which increases the risk of cardiovascular disease. While a corpus of literature exists on different areas of lipoprotein research, one of the major challenges that researchers face is the difficulties in accessing and integrating relevant information amidst massive quantities of heterogeneous data. Semantic web technologies, specifically ontologies, target these problems by providing an organizational framework of the concepts involved in a system of related instances to support systematic querying of information. In this paper, we identify issues within the lipoprotein research domain and present a preliminary framework for Lipoprotein Ontology, which consists of five specific areas of lipoprotein research: Classification, Metabolism, Pathophysiology, Etiology, and Treatment. By integrating specific aspects of lipoprotein research, Lipoprotein Ontology will provide the basis for the design of various applications to enable interoperability between research groups or software agents, as well as the development of tools for the diagnosis and treatment of dyslipidemia.

Keywords Lipoproteins · Lipoprotein ontology · Ontology · Classification · Metabolism · Pathophysiology · Etiology · Treatment

1 Introduction

Advances in lipoprotein research have led to an exponential growth in the generation of new experimental data. While large online databases exist as a universal pool of knowledge, researchers face challenges in navigating through overwhelming

M. Chen (✉)
Digital Ecosystems and Business Intelligence Institute, Curtin University of Technology, Enterprise Unit 4, De Laeter Way, Technology Park, Bentley, WA 6102, Australia
e-mail: m.chen@cbs.curtin.edu.au

amounts of information. The advent of semantic web technologies, specifically ontologies, targets these problems by enabling the assimilation of data and extracting relevant information into effective and efficient problem-solving tools. Ontologies are a medium of knowledge representation, which captures and conceptualizes a domain in terms of its associated concepts and instances. They provide a mechanism for sharing a common vocabulary in a domain to facilitate information exchange and are the basis for intelligent retrieval of information. Consequently, ontologies are becoming increasingly relevant in life sciences, as evident from the emergence of a number of biomedical ontologies [1, 2]. Some of these ontologies are the Gene Ontology, Protein Ontology, Lipid Ontology, among others [3–5]; however, an ontology specific to lipoproteins does not exist to date.

Clinical and epidemiological studies have identified lipoprotein dysregulation as a significant contributor to cardiovascular disease, the leading cause of death in the world today [6]. Dyslipidemia has also been found to be a central feature in disease states, such as diabetes and hypertension, which increases the risk of cardiovascular disease [7, 8]. Lipoproteins have been extensively researched since the first isolation of high-density lipoproteins (HDLs) in 1929 and low-density lipoproteins (LDLs) in 1950 [9]. In addition, the lipoprotein transport system has been well established since 1960s [9]. However, in spite of these advances, there is an increasing prevalence of dyslipidemia.

One of the major challenges that researchers face is the depth and breadth of disparate heterogeneous information in lipoprotein research. Correlation studies between various lipoproteins have indicated that the metabolism of plasma lipoproteins is complex and highly interrelated [10, 11]. This presents an opportunity for a successful implementation of an ontology-based model of lipoprotein pathways. In this paper, we identify issues within the lipoprotein research domain and present a preliminary framework for Lipoprotein Ontology. Lipoprotein Ontology provides a formal framework for lipoprotein concepts and relationships that can be used to annotate database entries and support the intelligent retrieval of information. Lipoprotein Ontology will provide the basis for the design of various computer-based applications to enable interoperability between research groups or software agents, as well as the development of tools for the diagnosis and treatment of dyslipidemia.

2 Ontologies in Biomedical Science

Biomedical research is increasingly becoming a data-driven endeavor, with large volumes of complex information derived from different sources, with different structures and different semantics. There have been efforts in the recent years in the organizing of biological concepts, such as controlled terminologies or ontologies [1, 2]. Terminologies promote a standardized way of naming these concepts. That is, preestablished hierarchies of terms are used to constrain selections made by users in annotating large document corpora. In contrast, ontologies provide an organizational framework of concepts involved in biological entities and processes in a system of hierarchical and associative relations that allows reasoning about biomedical knowledge. Other systems have also been developed to provide interoperability among different ontologies, such as the Unified Medical Language System [12], which provides a common frame of reference among the different research communities. The Open Biomedical Ontologies (OBO) Foundry hosts more than 70 open source ontologies associated with phenotypic and biomedical information [13].

The Gene Ontology (GO) project provides a set of dynamic, controlled vocabularies of gene products that can be applied in different databases to annotate major repositories for plant, animal, and microbial genomes [3]. GO is divided into three domains: cellular component, molecular function, and biological processes. The use of GO terms by several collaborating databases facilitates uniform queries among them.

Protein Ontology annotates terms and relationships within the protein domain and classifies that knowledge to allow reasoning [4]. Protein Ontology consists of seven generic concepts: Residues, Chains, Atoms, Family, AtomicBind, Bind, and SiteGroup. These generic concepts are placed into a class hierarchy of the Protein Ontology (ProteinComplex) to derive descriptions for any given protein, including: (1) protein sequence and structure information; (2) protein folding process; (3) cellular functions of proteins; (4) molecular bindings internal and external to proteins; and (5) external factors affecting final protein conformation [4].

Most recently, Lipid Ontology was developed to provide a structured framework for the effective derivation of lipid-related information [5]. Lipid Ontology mainly serves as a formal annotation for the classification and organization of information on lipids and to support navigation of text mining results from lipid literature. The ontology has been extended to describe the lipid nomenclature classification explicitly using description logics (OWL-DL) and to support reasoning and inference tasks. Currently, the Lipid Ontology has a total of 672 concepts and 75 properties. The ontology is the result of integrating schema components from existing biological database schemas, interviews with laboratory scientists, lipid and text mining experts.

To date, there does not exist an ontology for lipoproteins. Researchers have identified the role of certain lipoproteins in the Disease Database system [14]; however, this is limited to disease implications rather than a structured set of definitions and relations. The LOVD, a database of the LDL receptor (LDLR), contains 1,066 variations of the LDLR gene, which encodes the receptor for LDL cholesterol particles [15].

3 Issues Within the Lipoprotein Research Domain

Despite the abundance of research widely available on lipoproteins, a vast number of the world population suffers from dyslipidemia. Clearly, there exist a number of problems with the effectiveness of harnessing these data. Some of these are highlighted as follows:

Issues within the lipoprotein research domain

1. Lipoprotein dysregulation occurs as a consequence of alterations in the kinetics of lipoproteins. Correlation studies between various lipoproteins have indicated that the metabolism of lipoproteins is complex and highly interrelated [10, 11]. Currently, there are no tools which map the relationships between lipoprotein entities.
2. Treatment of dyslipidemia often involves the use of multiple lipid-regulating agents by targeting specific lipoproteins and utilizing the complementary mechanisms of action of different agents. However, drug interactions may increase the risk of adverse effects and/or morbidity in certain individuals [16]. Lipoprotein Ontology may potentially reduce this risk by mapping these interactions in a systematic way.

Data management problem

1. The amount of information. The size of the existing corpus of knowledge on lipoproteins is very large. The possibility of searching this information effectively is very low and in most cases some important information is neglected.
2. The dispersed nature of the information. Information regarding lipoproteins is dispersed over various resources and it is difficult to link this information, to share it and find specific information when needed.
3. The autonomous nature of the information sources. The broad spectrum of domain knowledge involved in studying lipoproteins leads to different representations, storage formats, access methods, and coverage of data. Information regarding lipoproteins does not interoperate, which makes collective analyses of data very difficult. In addition, many overlaps and redundancies are found in the data originating from different sources.

4 Lipoprotein Ontology Model

Lipoproteins are a water-soluble "lipid+protein" complex, which serves as a mode of transport for the uptake, storage, and metabolism of lipids. The basic lipoprotein structure comprises a hydrophobic core of triglycerides and cholesteryl esters, surrounded by a hydrophilic outer layer of phospholipids, cholesterol, and apolipoproteins (Fig. 1). Lipoproteins play a very crucial role in the regulation of biological and cellular functions in humans, and can be affected by a number of factors, including obesity, diet/nutrition, physical exercise, and other factors such as smoking and alcohol consumption.

Some preliminary work has been done on Lipoprotein Ontology, which was presented at the 22nd IEEE International Symposium on Computer-Based Medical Systems in August 2009 [17]. Lipoprotein Ontology covers the classification of lipoproteins, pathways of lipoprotein metabolism, pathophysiology of lipoproteins, causes of lipoprotein dysregulation as well as treatment of dyslipidemia.

55 A Framework for Lipoprotein Ontology

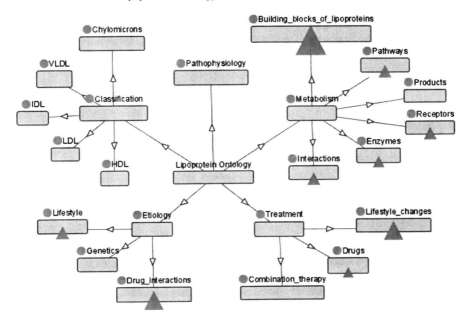

Fig. 1 Lipoprotein Ontology model consisting of five sub-ontologies and their subclasses

The lipoprotein transport system is critical for the supply, exchange, and clearance of essential lipids in the body. Various lipoproteins, apolipoproteins, enzymes, transporters, and receptors in this system constitute a delicate physiologic balance; disruption of one or more components of this system results in abnormal lipoprotein levels and increases the risk of cardiovascular disease. There have been considerable efforts by a multitude of different research groups working disparately to investigate different aspects of lipoprotein research. Thus, the need for a common information repository is warranted in order to fully appreciate the implications of lipoprotein dysregulation. By incorporating specific aspects of lipoprotein research into Lipoprotein Ontology, not only in terms of the classification of lipoproteins, but also understanding the metabolic pathways, pathophysiology, causes and treatment of abnormal lipoprotein levels, this impacts not only on identifying the risks, but also provides effective preventative measures. In this paper, we present a preliminary model for Lipoprotein Ontology using Protégé, which consists of five sub-ontologies:

1. Classification
2. Metabolism
3. Pathophysiology
4. Etiology
5. Treatment

An overview of Lipoprotein Ontology is shown in Fig. 1.

In our future work, Lipoprotein Ontology will be populated with concepts and instances from the literature. The database from which knowledge will be elicited is PubMed, which is a public online repository of peer-reviewed journal articles in the biomedical domain.

5 Significance of Lipoprotein Ontology

One of the most important benefits of ontologies is bridging the gap that exists between basic biological research and medical practice. Similarly, biological researchers also stand to benefit from being able to harness the clinical data and knowledge that are increasingly stored in computable forms. Lipoprotein ontology will have significant implications:

1. Annotation and integration of heterogeneous information sources into a single, organized body. Correlation studies between various lipoproteins have indicated that the metabolism of plasma lipoproteins is complex and highly interrelated. This presents an opportunity for a successful implementation of an ontology-based model of lipoprotein pathways.
2. The compilation of lipoprotein concepts under one framework will enable researchers to identify gaps in the literature, discover previously unknown links, and formulate new research questions.
3. Lipoprotein Ontology will form the basis for the design of various tools and applications to enable interoperability between research groups or software agents, thereby allowing for more efficient retrieval of information, saving time, and resources.
4. Diagnosis and development of treatments. By incorporating specific aspects of lipoprotein research into Lipoprotein Ontology, not only in terms of the classification of lipoproteins, but also understanding the metabolic pathways and pathophysiology, this impacts not only on identifying the risks, but also provides effective preventative measures.

6 Conclusion

The lipoprotein transport system is critical for the supply, exchange, and clearance of essential lipids in the body. Various lipoproteins, apolipoproteins, enzymes, transporters, and receptors in this system constitute a delicate physiologic balance; disruption of one or more components of this system results in abnormal lipoprotein levels and increases the risk of cardiovascular disease. Lipoprotein research is an extensive area with a multitude of different research groups working disparately to achieve different aims. Thus, the need for a common information repository is warranted in order to fully appreciate the implications of lipoprotein dysregulation. In this paper, we discuss issues that are present within the lipoprotein research domain and present a conceptual framework for Lipoprotein Ontology, which consists of classification, metabolism, pathophysiology, etiology, and treatment. Lipoprotein Ontology will be populated with concepts and instances derived from the literature. We believe that the design of Lipoprotein Ontology will form the basis for various computer-based applications that will enable interoperability between research groups or software agents, as well as the development of diagnostic tools and treatment of dyslipidemia. Our future work will focus on populating Lipoprotein Ontology with concepts and instances from the literature.

References

1. Y. Lussier, O. Bodenreider, "Clinical ontologies for discovery applications", in *Knowledge Discovery in the Life Sciences*, Springer Verlag, New York, 22, p. 450, 2007.
2. P. Lambrix, H. Tan, V. Jakoniene, L. Strömbäck, "Biological Ontologies", in Baker, Cheung (eds), in *Semantic Web: Revolutionizing Knowledge Discovery in the Life Sciences*, Springer, New York, pp. 85–99, 2007.
3. M. Ashburner, C.A. Ball, J.A. Blake, et al., "Gene ontology: tool for the unification of biology. The Gene Ontology Consortium", in *Nature Genetics*, vol. 25(1), pp. 25–29, 2000.
4. A.S. Sidhu, T.S. Dillon, E. Chang, "Integration of protein data sources through PO", in *Proceedings of the 17th International Conference on Database and Expert Systems Applications (DEXA 2006)*, Poland, pp. 519–527, 2006.
5. C.O. Baker, R. Kanagasabai, W.T. Ang, A. Veeramani, H.S. Low, M.R. Wenk, "Towards ontology-driven navigation of the lipid bibliosphere", in *BMC Bioinformatics*, vol. 9, pp. S5, 2008.
6. [AHA] American Heart Association. 2009. Heart disease and stroke statistics – 2009.
7. B. Isooma, P. Almgren, T. Tuomi, T.B. Forsén, K. Lahti, M. Nissén, et al, "Cardiovascular morbidity and mortality associated with the metabolic syndrome", in *Diabetes Care*, vol. 24, pp. 683–689, 2001.
8. S.M. Grundy, I.J. Benjamin, G.L. Burke, A. Chait, R.H. Eckel, B.V. Howard, et al, "Diabetes and cardiovascular disease: a statement for healthcare professionals from the American Heart Association", in *Circulation*, vol. 100, pp. 1134–1146, 1999.
9. R.E. Olson, "Discovery of the lipoproteins, their role in fat transport and their significance as risk factors", in *Journal of Nutrition*, vol. 128, pp. 439S–443S, 1998.
10. K.N. Frayn, *"Metabolic regulation: a human perspective"*, Blackwell Publishing, Oxford, 2003.
11. S. Eisenberg, "Lipoproteins and lipoprotein metabolism", in *Journal of Molecular Medicine*, vol. 61(3), pp. 119–132, 1982.
12. O. Bodenreider, "The Unified Medical Language System (UMLS): integrating biomedical terminology", in *Nucleic Acids Research*, vol. 32, pp. D267–D270, 2004.
13. B. Smith, M. Ashburner, C. Rosse, C. Bard, W. Bug, W. Ceusters, et al., "The OBO Foundry: coordinated evolution of ontologies to support biomedical data integration", in *Nature Biotechnology*, vol. 25, pp. 1251–1255, 2007.
14. Diseases Database. http://www.diseasesdatabase.com/content.asp. Updated 2009.
15. S.E. Leigh, A.H. Foster, R.A. Whittall, C.S. Hubbart, S.E. Humphreys, "Update and analysis of the University College London LDLR familial hypercholesterolemia database", in *Annals of Human Genetics*, vol. 72, pp. 485–495, 2008.
16. H.E. Bays, C.A. Dujovne, "Drug interactions of lipid-altering drugs", in *Drug Safety*, vol. 19, pp. 355–371, 1998.
17. M. Chen, M. Hadzic, "Lipoprotein ontology as a functional knowledge base", in *Proceedings of the 22nd IEEE International Symposium on Computer-Based Medical Systems*, New Mexico, USA, 2009.

Chapter 56
Verbal Decision Analysis Applied on the Optimization of Alzheimer's Disease Diagnosis: A Case Study Based on Neuroimaging

Isabelle Tamanini, Ana Karoline de Castro, Plácido Rogério Pinheiro, and Mirian Calíope Dantas Pinheiro

Abstract There is a great challenge in identifying the early diagnosis of the Alzheimer's disease, which has become the most frequent cause of dementia in the last few years, being responsible for 50% of the cases in western countries. The main focus of the work is the development of a multicriteria model for aiding in the decision making on the diagnosis of the Alzheimer's disease. It will be made by means of the Aranaú Tool, a decision support system mainly based on the ZAPROS method. The modeling and evaluation processes were conducted based on bibliographic sources, questionnaires, and on information given by a medical expert. The questionnaires analyzed were based mainly on patients' neuroimaging tests and were tried under various relevant aspects to the diagnosis of the disease.

1 Introduction

In the last few years, demographic studies have shown a progressive and significant increase in the elderly population of developed and developing countries [19]. Researchers agree that the advances in the medical area have significant importance on the increase of the life expectancy. Along with this fact, there is a major increase in the number of health problems among the elderly individuals, which, besides being of long duration, require skilled personnel, a multidisciplinary team, equipment, and additional high-cost tests.

Among the illness that occur especially in elderly people, we can say that the dementia is the one that deserves a major attention, since the chances of presenting the pathology increase exponentially as one gets older. Dementias are syndromes characterized by a decline in memory and other neuropsychological changes. It presents three main characteristics:

I. Tamanini (✉)
Graduate Program in Applied Computer Sciences, University of Fortaleza (UNIFOR),
Av. Washington Soares, 1321 – Bl J Sl 30, 60.811-905, Fortaleza, Brazil
e-mail: isabelle.tamanini@gmail.com

- Loss of memory, ranging from a simple oversight to a more severe case such as not remembering the own identity
- Behavior problems such as agitation, insomnia, tearfulness, inappropriate behavior and loss of normal social inhibition
- Loss of skills acquired throughout the life, such as getting dressed and cooking

The Alzheimer's disease is the most frequent cause of dementia and it is responsible (alone or in association with other diseases) for 50% of the cases in western countries [19]. According to [10], the disease was also recognized as the fifth leading cause of death in 2003 among those older than the age of 65. Besides, its incidence and prevalence double every 5 years, with estimated prevalence of 40% among people with more than 85 years of age [10]. Regardless of its high incidence, doctors fail to detect dementia in 21–72% of the cases [1]. Much of the disease characteristics and its evolution are still unclear, besides all the progress that has been made.

As it is known that the earliest the dementia is diagnosed, the greater are the chances of delaying its advance, and the greatest achievement of the researches nowadays is to find out a way to identify the disease on its earliest stage.

Even though the only way to have the definitive diagnosis of the disease is by the examination of the brain tissue (biopsy), a probable diagnosis can be made based on neurological and psychological tests, patient's clinical history, laboratory and neuroimaging tests, etc., being the latter the main focus of this chapter. Currently, a great variety of tests with this purpose are available, and one of the major challenges is to find out which test, or which characteristics of a test, would be more efficient to establish the diagnosis of dementia.

With the purpose of reducing the number of tests that an elderly person would have to do, the aim of this chapter is to determine which of the tests are relevant and would detect faster if the patient is developing the disease.

This way, the data given on the battery were analyzed and, considering the experience of the decision maker (the medical expert), a multicriteria model was structured based on the ZAPROS method to find out which questionnaires of a given set are more attractive to be applied to a patient to help on the diagnosis of the disease. The criteria were defined considering the characteristics of each questionnaire of the battery. The preferences were given through the analysis of the questionnaires results and the postmortem diagnosis of each patient, in a way to establish a relation between them. At the end, we had a ranking of the questionnaires, from the one that had the greatest importance on the diagnosis, to the one that had the least.

It was never thought of applying multicriteria methods in the health area, especially with the aim of assisting in the diagnosis of the Alzheimer's disease. The first model developed [2] was validated using small study cases presented on other papers [16]. Then, another model [3, 4] was validated with a Brazilian battery [20].

The model was extended later [7, 8] and was validated by the data provided on the battery of Consortium to Establish a Registry for Alzheimer's Disease (CERAD) [10]. This last model was based on three previous papers developed [5, 6, 18].

This work was based on two other works considering the Alzheimer's disease [23, 24], both applying multicriteria aiming the diagnosis of the disease.

2 The Alzheimer's Disease

First described in 1906, the Alzheimer's disease is characterized by the presence of senis plaques and tangled neurofibrillaries in the regions of the hippocampus and cerebral cortex, and the neurons appear atrophied in a large area of the brain (Fig. 1).

The Alzheimer's disease is a difficult diagnosed illness, since the initial symptoms are subtle and they progress slowly until they are clear and devastating. On the other hand, although the disease symptoms vary greatly, a pattern can be established.

There are still many aspects that remain unclear about the disease, although the greater advance that has been made in the last few years. Studies aiming at the identification of the fundamental points to the diagnosis and the posterior treatment of the pathology are the focus of many scientists nowadays.

There is a major importance in identifying the cases in which the risks of developing a dementia are higher, considering the few alternative therapies and the greater effectiveness of treatments when an early diagnosis is possible [12]. Besides, according to studies conducted by the Alzheimer's Association [13], the Alzheimer's disease treatments have significant resulting costs, and it is known that it is one of the costliest diseases, second only to cancer and cardiovascular diseases.

This way, the accuracy of the diagnosis is very important, and each day more treatments become available [9]. These efforts can help identifying conditions that are potentially reversible or treatable that have contributed to cognitive decline and

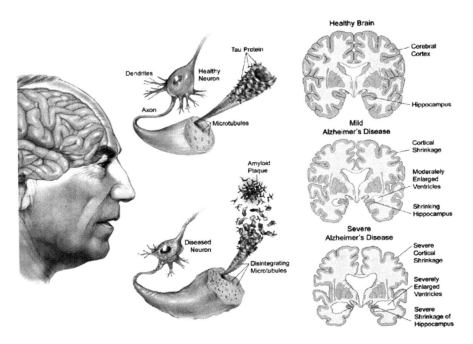

Fig. 1 The changes on a brain with Alzheimer's disease [11]

dementia. Based on the accuracy of the diagnosis, family members can plan the future needs of the patient and can also consider the implications of a particular diagnosis regarding their own future.

3 A New Approach Methodology

A methodology structured basically on the ZAPROS method [14] is proposed on [22]. It presents three main stages: *problem formulation, elicitation of preferences*, and *comparison of alternatives*, as proposed on the original version of the ZAPROS method. An overview of the approach is presented below.

3.1 Formal Statement of the Problem

The methodology follows the same problem formulation proposed in [14]:
Given:

1. $K = 1, 2, \ldots, N$, representing a set of N criteria
2. n_q represents the number of possible values on the scale of q-th criterion, $(q \in K)$; for the ill-structured problems, as in this case, usually $n_q \leq 4$
3. $X_q = x_{iq}$ represents a set of values to the q-th criterion, which is this criterion scale; $|X_q| = n_q (q \in K)$, where the values of the scale are ranked from best to worst, and this order does not depend on the values of other scales
4. $Y = X_1 * X_2 * \ldots * X_n$ represents a set of vectors y_i, in such a way that: $y_i = (y_{i1}, y_{i2}, \ldots, y_{iN})$, and $y_i \in Y$, $y_{iq} \in X_q$ and $P = |Y|$, where $|Y| = \prod_{i=1}^{i=N} n_i$
5. $A = \{a_i\} \in Y$, $i = 1, 2, \ldots, t$, where the set of t vectors represents the description of the real alternatives

Required: The multicriteria alternatives classification based on the decision maker's preferences.

3.2 Elicitation of Preferences

In this stage, the scale of preferences for quality variations (Joint Scale of Quality Variations – JSQV) is constructed. The elicitation of preferences follows the same proposed in [14]. However, instead of setting the decision maker's preferences based on the first reference situation and, then, establishing another scale of preferences using the second reference situation, we propose that the two substages be transformed in one. The questions made considering the first reference situation are the same as the ones made considering the second reference situation. Hence, both situations will be presented and must be considered in the answer to the question, not to cause dependence of criteria. The alteration reflects on an optimization of the process: instead of making $2n$ questions, only n will be made.

On the preferences elicitation for two criteria, the questions will be made dividing the QV into two items. For example, having the set of criteria $k = A, B, C$, where $n_q = 3$ and $X_q = q_1, q_2, q_3$, considering the pair of criteria A, B and the QV a_1 and b_1, the decision maker should analyze which imaginary alternative would be preferable: A_1, B_2, C_1 or A_2, B_1, C_1 (note that this answer must be the same to alternatives A_1, B_2, C_3 and A_2, B_1, C_3). If the decision maker answers that the first option is better, then b_1 is preferable to a_1, because it is preferable to have B_2 on the alternative instead of A_2.

3.3 Comparison of Alternatives

With the aim of reducing the number of incomparability cases, we apply the same structure proposed in [14], but modifying the comparison of pairs of the alternatives' substage according to the one proposed in [17].

Each alternative has a function of quality – V(y) [14], depending on the evaluations of the criteria that it represents. In [17], it is proposed that the vectors of ranks of the criteria values, which represent the function of quality, are rearranged in an ascending order. Then, the values will be compared to the corresponding position of another alternative's vector of values based on Pareto's dominance rule. Meanwhile, this procedure was modified for implementation because it was originally proposed to scales of preferences of criteria values, not for quality variation scales.

However, there are cases in which the incomparability of real alternatives will not allow the presentation of a complete result. These problems require further comparison. In such cases, we can evaluate all possible alternatives to the problem to rank the real alternatives indirectly. The possible alternatives should be rearranged in an ascending order according to their Formal Index of Quality (FIQ), and only the significant part will be selected for the comparison process (the set of alternatives presenting FIQ between the greatest and the smallest real alternatives' FIQ). After that, the ranks obtained will be passed on to the corresponding real alternatives.

3.4 Proposed Tool to the New Methodological Approach

In order to facilitate the decision process and perform it consistently, observing its complexity and with the aim of making it accessible, a tool was implemented in Java and it is presented by the following sequence of actions:

– *Criteria definition*: The definition of the criteria presented by the problem
– *Preferences elicitation*: Occurring in two stages: the elicitation of preferences for quality variation on the same criteria and the elicitation of preferences between pairs of criteria

- *Alternatives definition*: The alternatives can be defined only after the construction of the scale of preferences
- *Alternatives classification*: After the problem formulation, the user can verify the solution obtained to the problem. The result is presented to the decision maker so that it can be evaluated. The comparison based on all possible alternatives for the problem is possible, but it should be performed only when it is necessary for the problem resolution (for being an elevated cost solution).

4 CERAD: An Overview

The CERAD was funded in 1986 with the aim of establishing a standard process for evaluation of patients with possible Alzheimer's disease who enrolled in NIA-sponsored Alzheimer's Disease Centers or in other dementia research programs [15]. At that time, uniform guidelines were lacking as to diagnostic criteria, testing procedures, and staging of severity, despite the growing interest in clinical investigations of this illness. The following standardized instruments to assess the various manifestations of the disease were established: Clinical Neuropsychology, Neuropathology, Behavior Rating Scale for Dementia, Family History Interviews, and Assessment of Service Needs.

The battery of tests of the CERAD was used in this work. The neuropathological battery of CERAD, which is the center of interest of this work, is mainly structured on neuroimaging examinations of the patients, with the aim of verifying the existence of problems in the patients' brain, such as vascular lesions, hemorrhages, and microinfarcts, which could be relevant to the diagnosis of the disease.

5 The Multicriteria Model for the Diagnosis of the Alzheimer's Disease

In order to establish which of the tests are more likely to lead to a diagnosis faster, a multicriterion model was structured based on four tests of the CERAD's neuropathological battery: neuropathological diagnosis, cerebral vascular disease gross findings, microscopic vascular findings, and clinical history.

The criteria established were defined considering parameters of great importance to the diagnosis, based on the analysis of each questionnaire data by the decision maker and in her knowledge. The analysis of the questionnaires' data was carried out following some patterns, such that if a determined data are directly related to the diagnosis of the disease, and this fact can be found based on the number of its occurrences in the battery data, and this fact will be selected as a possible value

Table 1 Criteria for evaluation of the analyzed questionnaires of Consortium to Establish a Registry for Alzheimer's Disease (CERAD)

Criteria	Values of criteria
A – Verification of existence of other diseases	A1. There are questions about the severity of other brain problems
	A2. There are questions about the existence of brain problems
	A3. There are no questions about the existence of brain problems
B – Neuroimaging tests: MRI or CT	B1. One or more cerebral lacunes can be detected
	B2. Extensive periventricular white matter changes can be detected
	B3. Focal or generalized atrophy can be detected
C – Clinical studies	C1. Electroencephalogram
	C2. Analysis of cerebrospinal fluid
	C3. Electrocardiogram

of criteria. For being relevant to the diagnosis, one can notice that questionnaires that are able to detect this occurrence are more likely to give a diagnosis. Regarding the CERAD data, only the results of the tests of patients who had already died and on which the necropsy has been done were selected (122 cases), because it is known that necropsy is essential for validating the clinical diagnosis of dementing diseases.

This way, the criteria considered relevant to evaluate the CERAD's questionnaires aforementioned and the values identified for them are exposed in Table 1.

The preferences were elicited based on the analysis of the answers obtained on each questionnaire and the patients' final diagnosis with the medical expert. The facts that were more constant and extremely connected to the patient's final diagnosis were set as preferable over others. This way, it was possible to establish a order of the facts that had a direct relation with the patients' final diagnosis, and so, which of the questionnaires had the questions that were more likely to lead to it. The preferences scale obtained is given as follows: $b_1 \equiv c_1 \prec a_1 \prec c_2 \prec b_2 \prec b_3 \prec c_3 \prec a_2 \prec a_3$.

After the elicitation of preferences, the alternatives were defined. As the aim of this chapter is to give a preference order of application of the tests, the alternatives were formulated identifying which facts (the values of criteria) would be identified by each questionnaire, considering the tests it would require to be filled. Hence, we will have the questionnaires described as criterion values according to the facts verified by each one of them. For example: considering the CERAD's questionnaire *Cerebral Vascular Disease Gross Findings*, we can say that it has questions about the severity of the patients' brain problems, that a extensive periventricular white matter change can be detected by the tests required on answering it, and that, as clinical studies, a electroencephalogram would be analyzed to answer some of its questions. The model was applied to the Aranaú Tool [21], and the results presentation screen of the tool for this problem is presented in Fig. 2.

The order of preference to apply the tests shows that the questionnaire Cerebral Vascular Disease Gross Findings is more likely to detect a possible case of Alzheimer's disease than the others; hence, it should be applied first.

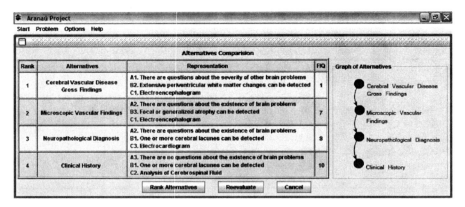

Fig. 2 Presentation of results

6 Conclusions

The main purpose of this work is to set up a order of questionnaires that were more likely to give the diagnosis. To do so, a multicriteria model was formulated based on the data available on CERAD. The criteria were formulated based on facts that have a direct relation to the diagnosis of the disease. These facts can be found on the data gathered on questionnaires, on which, most of the times, a neuroimaging examination of the patient was required. The preferences were based on information given by a medical expert that analyzed the data of each questionnaire, and, then, the alternatives were structured according to the facts that could be detected on answering each of these questionnaires.

The model was submitted to Aranaú tool [23], structured mainly on the ZAPROS method but with some modifications to improve the alternatives comparison process. The result obtained was the questionnaires order, from the most likely to lead to a diagnosis of disease, to the least one. This would enable a faster detection of patients that would develop the disease, increasing their chances of treatment.

As future works, we intend to improve the model proposed, involving more tests and questionnaires, to have a complete ranking. This would decrease the number of tests that patients have to go through and the number of questionnaires they would have to answer to get to the diagnosis. We also intend to verify which stage of the Alzheimer's disease would be detected based on the order established, trying to set up a order that would detect the disease in its earlier stages.

Acknowledgements The authors are thankful to the National Counsel of Technological and Scientific Development (CNPq) for all the support received and to the Consortium to Establish a Registry for Alzheimer's Disease (CERAD) for making available the data used in this case study.

References

1. Bassett, S.S.: Attention: Neuropsychological Predictor of Competency in Alzheimer's Disease. Journal of Geriatric Psychiatry and Neurology 12(4):200–205 (1999)
2. Castro, A.K.A. de, Pinheiro, P.R., Pinheiro, M.C.D.: Applying a Decision Making Model in the Early Diagnosis of Alzheimer's Disease. In: J. Yao, P. Lingras, W.-Z. Wu, M. Szczuka, N. Cercone and D. Slezak (Eds.): Rough Set and Knowledge Technology, 2nd International Conference (RSKT) Toronto, Canada, 2007 pp. 149–156. LNCS 4481 (2007)
3. Castro, A.K.A. de, Pinheiro, P.R., Pinheiro, M.C.D.: A Multicriteria Model Applied in The Early Diagnosis of Alzheimer's: A Bayesian Approach. VI International Conference on Operational Research for Development (ICORD) Fortaleza, Brazil, v.VI pp. 9–19 (2007)
4. Castro, A.K.A. de, Pinheiro, P.R., Pinheiro, M.C.D.: A Multicriteria Model Applied in the Diagnosis of Alzheimer's Disease. In: G. Wang, T. Li, J. W. Grzymsls-Busse, D. Miao, A. Skowron and Y. Y. Yao (Eds.): Rough Set and Knowledge Technology, 3rd International Conference (RSKT) Chengdu, China, 2008 pp. 612–619. LNCS 5009 (2008)
5. Castro, A.K.A. de, Pinheiro, P.R., Pinheiro, M.C.D.: A Hybrid Model for Aiding in Decision Making for the Neuropsychological Diagnosis of Alzheimer's Disease. In: C.-C. Chan, J. W. Grzymala-Busse and W. P. Ziarko (Eds.): 6th International Conference Rough Sets on and Current Trends in Computing (RSCTC) Akron, Ohio, 2008 pp. 495–504. LNCS 5306 (2008)
6. Castro, A.K.A.: Um Modelo Híbrido Aplicado ao Diagnóstico da Doença de Alzheimer. Master Thesis, Master Program in Applied Computer Sciences, University of Fortaleza (2009)
7. Castro, A.K.A. de, Pinheiro, P.R., Pinheiro, M.C.D.: An Approach for the Neuropsychological Diagnosis of Alzheimer's Disease. In: G. Wang, T. Li, J. W. Grzymsls-Busse, D. Miao, A. Skowron and Y. Y. Yao (Eds.): Rough Set and Knowledge Technology, 4th International Conference (RSKT) Gold Coast, Australia, 2009 pp. 216–223. LNCS 5589 (2009)
8. Castro, A.K.A. de, Pinheiro, P.R., Pinheiro, M.C.D.: Towards the Neuropsychological Diagnosis of Alzheimer's Disease: A Hybrid Model in Decision Making. In: M. D. Lytras, P. O. de Pablos, E. Damiani, D. E. Avison, A. Naeve and D. G. Horner (Eds.): Best Practices for the Knowledge Society. Knowledge, Learning, Development and Technology for All, Second World Summit on the Knowledge Society (WSKS) Chania, Crete. Communications in Computer and Information Science 49:(1):522–531. DOI: 10.1007/978-3-642-04757-2_57 (2009)
9. Daffner, K.R.: Current Approaches to the Clinical Diagnosis of Alzheimer's Disease. In: L. F. M. Scinto and K. R. Daffner (Eds.): Early Diagnosis of Alzheimer's Disease. Humana, New Jersey pp. 29–64 (2000)
10. Fillenbaum, G.G., van Belle, G., Morris, J.C., et al.: Consortium to Establish a Registry for Alzheimers Disease (CERAD): The First Twenty Years. Alzheimer's & Dementia 4(2):96–109 (2008)
11. How the Brain and Nerve Cells Change During Alzheimer's. Available via DIALOG. www.ahaf.org/assets/images/brain_and_nerve_cell_changes_border.jpg.Cited25Sep2009
12. Hughes, C.P., Berg, L.,Danziger, W.L., et al.: A New Clinical Scale for the Staging of Dementia. British Journal of Psychiatry 140(6):566–572 (1982)
13. Koppel, R.: Alzheimer's Disease: The Costs to U.S. Businesses in 2002. Alzheimers Association - Report (2002)
14. Larichev, O.: Ranking Multicriteria Alternatives: The Method ZAPROS III. European Journal of Operational Research 131(3):550–558 (2001)
15. Morris, J.C., Heyman, A., Mohs, R.C., et al.: The Consortium to Establish a Registry for Alzheimer's Disease (CERAD): Part 1. Clinical and Neuropsychological Assessment of Alzheimer's Disease. Neurology 39(9):1159–1165 (1989)
16. Morris, J.: The Clinical Dementia Rating (CDR): Current Version and Scoring Rules. Neurology 43(11):2412–2414 (1993)

17. Moshkovich, H., Mechitov, A., Olson, D.: Ordinal Judgments in Multiattribute Decision Analysis. European Journal of Operational Research 137(3):625–641 (2002)
18. Pinheiro, P.R., Castro, A.K.A. de, Pinheiro, M.C.D.: Multicriteria Model Applied in the Diagnosis of Alzheimer's Disease: A Bayesian Network. 11th IEEE International Conference on Computational Science and Engineering, São Paulo, Brazil, pp. 15–22 (2008)
19. Prince, M.J.: Predicting the Onset of Alzheimer's Disease Using Bayes' Theorem. American Journal of Epidemiology 143(3):301–308 (1996)
20. Silva, A.C.S: Métodos Quantitativos Aplicados a Políticas de Saúde Pública: Estudo de caso dos idosos. 202f Master Thesis, Instituto Tecnológico de Aeronáutica (ITA), São José dos Campos (2006)
21. Tamanini, I., Pinheiro, P.R.: Reducing Incomparability in Multicriteria Decision Analysis: An Extension of the ZAPROS Method. Pesquisa Operacional, v.3, pp. 1–16 (2010)
22. Tamanini, I., Pinheiro, P.R.: Challenging the Incomparability Problem: An Approach Methodology Based on ZAPROS. Modeling, Computation and Optimization in Information Systems and Management Sciences, Communications in Computer and Information Science. Springer, Heidelberg 14(1):344–353. DOI:10.1007/978-3-540-87477-5_37 (2008)
23. Tamanini, I., Castro, A.K.A., Pinheiro, P.R., Pinheiro, M.C.D.: Towards an Applied Multicriteria Model to the Diagnosis of Alzheimer's Disease: A Neuroimaging Study Case. 2009 IEEE International Conference on Intelligent Computing and Intelligent Systems (ICIS) v.3, pp. 652–656 (2009)
24. Tamanini, I., Castro, A.K.A., Pinheiro, P.R., Pinheiro, M.C.D.: Towards the Early Diagnosis of Alzheimer's Disease: A Multicriteria Model Structured on Neuroimaging. International Journal of Social and Humanistic Computing 1:203–217 (2009)

Chapter 57
Asynchronous Brain Machine Interface-Based Control of a Wheelchair

C.R. Hema, M.P. Paulraj, Sazali Yaacob, Abdul Hamid Adom, and R. Nagarajan

Abstract A brain machine interface (BMI) design for controlling the navigation of a power wheelchair is proposed. Real-time experiments with four able bodied subjects are carried out using the BMI-controlled wheelchair. The BMI is based on only two electrodes and operated by motor imagery of four states. A recurrent neural classifier is proposed for the classification of the four mental states. The real-time experiment results of four subjects are reported and problems emerging from asynchronous control are discussed.

1 Introduction

Brain machine interface (BMI) to control devices such as prosthetic arm, robots and wheelchair have been in the realm of neural rehabilitation research for the past decade or so [1–11]. BMI technologies can substantially improve the lives of people with devastating neurological disorders [11]. Initial demonstrations of the feasibility of such interfaces had relied on implanted electrodes on monkeys [3]. Although non-invasive electrodes suffer from measurement noise and spatial resolution, they are preferred for human electroencephalogram (EEG)-based studies [1]. BMIs are distinguished by the particular EEG feature they use to determine the user's intent. Two kinds of EEG-based BMI technologies (P300 Rebsamen et al. [7], and motor imagery [MI]) have been tested to control wheelchairs. Movement execution and its imagination induce event-related desynchronization and synchronization in the motor cortex, such MI can be an efficient strategy to operate a BMI [2]. BMI using MI is based on the recording and classification of circumscribed and transient EEG changes during different types of MI such as imagination of left hand, right hand and foot movement [2, 8, 10, 12, 13]. MI-based wheelchair control have been tested

C.R. Hema (✉)
School of Mechatronic Engineering, University Malaysia Perlis, 02600 Pauh, Perlis, Malaysia
e-mail: hema@unimap.edu.my

on simulated environments by two research groups [6, 10]. Our approach differs from [10] in three aspects, (1) we use only two electrodes and hand movement MI to achieve four mental states; (2) wheelchair automation is limited to obstacle avoidance only and (3) real-time wheelchair navigation is implemented in an indoor environment. The proposed BMI design employs mu, beta, lower gamma rhythms (8–40 Hz) and recurrent neural classifiers to spontaneously classify four MI tasks. The performance of the recurrent neural classifier is also compared with a conventional feed forward neural classifier.

2 Methods and Materials

2.1 Offline Synchronous Experiments

Synchronous experiments are conducted to collect data to train the neural classifiers. MI signals are recorded through a synchronous protocol for four motor tasks, relax forward, left and right and a stop control (NC); details of this paradigm is given in our earlier work in [14]. The experiment consisted of 10 runs with 40 trials each. Signals are recorded non-invasively for 3 s and amplified using an AD Instruments power lab EEG amplifier. Signals are recorded using two bipolar gold-plated cup electrodes from the C3 and C4 locations and an earth electrode at Fp1 as per the 10–20 International standards [15]; four volunteer healthy subjects (S1, S2, S3 and S4) participated in the experiments. All subjects had previous experience in BMI experiments. The signals are sampled at 200 Hz. The raw EEG signals are initially band-pass filtered between 8 and 40 Hz to obtain the mu, beta and lower gamma frequencies. Logarithmic values of six band power components (8–10, 11–12, 13–15, 16–18, 19–30 and 31–40 Hz) are extracted and used as the input features of the neural classifiers.

Recurrent neural networks (RNNs) have feedback connections, which add the ability to also learn the temporal characteristics of the data set [3, 16]. A three-layer RNN is modeled using six inputs, nine hidden and four output neurons. The choice of the hidden layer dimensionality was optimized to produce the best performing network with fewer elements [3]. To evaluate the performance of the proposed RNN, classification results are compared with a feed forward three layered network (FFNN) architecture. The FFNN is modeled using 6 input neurons, 25 hidden neurons, and 4 output neurons to represent the four mental states. Data samples are normalized using a binary normalization algorithm [17].

Almost 80% of the data samples are chosen experimentally to train the networks. The RNN and FFNN are tested with 20% of data samples. Random data samples are chosen for training and testing the networks. The networks are trained using a back propagation learning algorithm. Test results for task recognition are reported in Table 1 in terms of minimum, average and maximum accuracies (column 1) for all four subjects. The RNN model with lesser weight connections

57 Asynchronous Brain Machine Interface-Based Control of a Wheelchair

Table 1 Recognition performance of the neural classifier

Classification accuracy %	FFNN				RNN			
	S1	S2	S3	S4	S1	S2	S3	S4
Min	86.2	85.8	87.5	89.5	95	92.5	87.5	92.5
Ave	89.4	89.1	88.5	90	96.5	95.7	87.7	96.2
Max	90.8	89.9	90	97.5	97.5	97.5	92.5	97.5

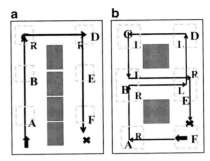

Fig. 1 Real-time experiment trajectory (**a**) protocol 1 and (**b**) protocol 2

(144) performs better task recognition in comparison with the FFNN (250). The RNN model with maximum performance is chosen for the asynchronous real-time experiments.

2.2 Real-Time Asynchronous Experiments

In the real-time experiments, subjects are seated on a BMI-wheelchair. The BMI is built on a Winner power wheelchair; the real-time control program runs on a HP Laptop and is designed to control the joystick of the wheelchair for the four states, stop, forward, left and right. Sensors are limited to proximity sensors mounted in front of the wheelchair for obstacle avoidance (shared control). Wheelchair is stopped when an obstacle is detected at a distance of 60 cm. The shared control algorithm is activated only when an appropriate signal from the subject is not received within 30 s after obstacle detection. The wheelchair speed is restricted to 0.5 m/s during a forward state and during a left or right state the wheelchair turns to 45° in the respective direction. Experiments are carried out in a 7 × 5 m indoor environment. The first protocol requires execution of three mental states (forward, right and stop) while the second protocol ensures that all the four states of the BMI are used during the wheelchair navigation as shown in Fig. 1a, b. S4 participated only in experiments for the first protocol. The task is to drive the BMI-wheelchair avoiding obstacles in an indoor environment using the two protocols. The subjects learned to drive and control the BMI wheelchair in 3 days, experimental sessions lasted for 2–4 h per day and each trial lasted for about 5–20 min, separated by 15-min breaks.

3 Results and Discussion

3.1 Protocol 1

In this protocol (Fig. 1a), subjects are asked to drive the wheelchair following a pre-specified path from the starting point through targets A–F and stopping at, a minimum of three targets chosen by the subject. Shared control is limited to stopping the wheelchair when obstacles are encountered, leaving the navigation control to the subjects. This protocol is designed to verify the performance of subjects to move from one point to another and to stop at a desired target. Experiments are conducted on day 1 and day 2, which were a week apart. Results are logged for time taken to reach each target and total distance traversed in each trial. Results of successful trials are shown in Table 2. Time in seconds to reach a target is indicated in columns 4–9 and hyphen indicates a missed target. Total time taken to complete the trial is shown in column 10. Performance (columns 11 and 12) is assessed based on targets reached and stops at targets; total distance covered in a trial is shown in column 13; the shaded cells indicate targets at which the subjects were able to generate stop signals.

On day 1, S1 was able to stop at two targets B and E, but missed target C (trial 3); on the same day in trial 5, S1 reached all six targets and stopped at targets A, C and D achieving 100% for target stops. S3 completed the protocol only on day 2 stopping at two targets A and B. S4 a 72-year-old subject who has experience in meditation, was able to stop at three targets (B, D and E) in trial 3 on day 1 and completed the protocol in the shortest time. The targets were missed when the subjects made a wrong turn before reaching it, this happened mostly at target F; except S3 all

Table 2 Time in seconds for protocol 1 for four subjects

| Subject | Day | Trial no. | Target locations | | | | | | Total A–F | Performance % | | Distance (m) |
			A	B	C	D	E	F		Target reached	Target stop	
S1	1	3	10	28	–	112	73	107	330	83.3	66.6	14.57
S1	1	5	10	32	78	205	39	46	410	100	100	14.57
S2	1	4	15	58	66	155	38	–	332	83.3	33.3	14.57
S2	1	6	12	85	46	285	34	54	516	100	66.6	14.57
S2	2	1	21	20	49	298	48	–	436	83.3	33.3	14.57
S2	2	2	7	80	37	145	71	65	405	100	33.3	14.57
S3	1	5	60	54	137	163	41	–	455	83.3	33.3	14.00
S3	1	6	43	71	121	102	–	57	394	83.3	33.3	14.57
S3	2	1	7	76	174	113	118	–	488	83.3	66.6	11.65
S3	2	3	27	106	198	32	175	50	588	100	66.6	14.57
S4	1	1	12	59	–	148	75	176	470	83.3	33.3	12.57
S4	1	3	9	48	29	52	154	40	332	100	100	14.57

subjects took more time to reach target D from C. Maximum stop time achieved at a target was 10.6 s at B by S4. S1 and S4 were able to achieve the goals on a single day with minimum target stop time of 5.6 s. S2 was able to reach all six targets on day 1 and day 2, but managed to stop only at two target locations.

3.2 Protocol 2

To further assess the performance of the BMI, three subjects (S1, S2 and S3) participated in experiments given in protocol 2 (Fig. 1b). Subjects drive the wheelchair following the pre-defined path avoiding two obstacles. This experiment requires execution of three left and three right turns in addition to forward signals to complete the protocol. Junctions B and E require significant effort from the subject to steer the wheelchair in the correct direction, as shared control is activated only when obstacles are encountered. Communication between the BMI and wheelchair are logged, the trajectory and shared control are manually logged. Samples of tra-

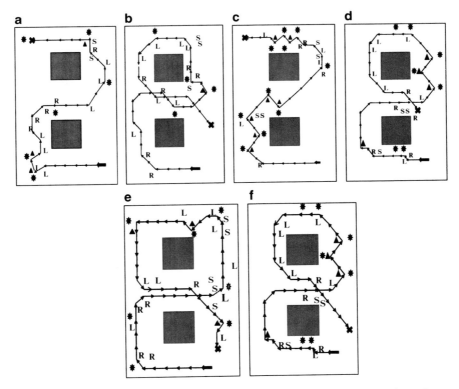

Fig. 2 (a) Protocol 2 – trajectory achieved by S2 (trial 5) on day 1, (b) Protocol 2 – trajectory achieved by S1 (trial 2) on day 1, (c) Protocol 2 – trajectory achieved by S3 (trial 3) on day 2, (d) Protocol 2 – trajectory achieved by S2 (trial 2) on day 2, (e) Protocol 2 – trajectory achieved by S1 (trial 4) on day 3, (f) Protocol 2 – trajectory achieved by S2 (trial 3) on day 3

jectories plotted from the logged results are shown in Fig. 2a–f, arrow indicates a forward signal, left, right and stop signals are represented by the L, R and S symbols, obstacle detection is shown using an asterisk (*) and the triangle (Δ), signifies the intervention of the control algorithm, which turns the wheelchair 90°, based on information from left and right proximity sensors. This intervention occurs only after a delay of 30 s and when the subject fails to generate the required left or right turn signal within the 30 s.

All subjects performed a minimum of four trials per day; Fig. 2a, b shows the best trajectory achieved by S2 and S1 on day 1. All subjects failed to complete the protocol on the first day. In most trials, S2 and S3 were unsuccessful in making a right turn at location B where shared control also failed. On the contrary, shared control significantly improved performance in the stretch between D and C. Navigation performance improved on second day of experiments (a week later). S3 was able to complete 50% of the protocol with assistance from the shared control as shown in Fig. 2c, d. S1 and S2 gained better control of the BMI on the next day (day 3) with minimal intervention from the shared control as depicted in Fig. 2e, f. S3 failed to complete the protocol even on day 3 because he could not make the required right or left turn at either B or E in all trials. When a subject executes more than two consecutive left or right turns, the wheelchair tends to rotates at a given location. Subjects learn to tackle such situations using the stop control. As is evident in Fig. 2c–f, all subjects are able to generate stop signals when corners or turns are required.

Significant improvement in the performance is observable for both protocols after sufficient training; however, subjects need to cope up with the need to generate stable EEG in the presence of obstacles. The proposed BMI wheelchair provides more flexibility to the user to control the direction and destination of the wheelchair compared to a P300 BMI wheelchair [7], which has fixed guided paths. P300 BMI have an advantage over MI BMI as they require less input from the users, while MI BMI users require more training in using the system, this can be a drawback for new users. However with sufficient training, MI-based BMI is more versatile and can adapt to new environments.

4 Conclusion

A BMI design using neural classifiers for a power wheelchair is presented; the studies described in this work suggest that MI-EEG acquired from two electrode locations C3 and C4 can be used to design a four-state BMI. The BMI system proposed is implemented in real-time using healthy subjects to control a wheelchair and allows individuals to navigate in an indoor environment using four mental states. In MI-based BMI, subjects require considerable training to switch between mental states efficiently; moreover, the users report fatigue when using the system for more than 30 min; the above drawbacks are to be further investigated, two possible approaches to minimize the first drawback is to improve the concentration skill of the users or increase the level of assistance through more automation to the wheelchair as suggested by Galan et al. [10] and Millan et al. [13], who have tested their approach on a simulated wheelchair. The BMI proposed in this study relies on MI from two electrodes only, which gives an advantage over the BMI systems using more electrodes [10]. The efficiency of the BMI could be further improved by using an online training algorithm for the neural classifier, which will be able to adapt to the mental state of the users. This is an area of our current research. While the current results are promising, experiments with patients (with motor neuron disorders) are essential to evaluate the feasibility of providing rehabilitation through a BMI wheelchair.

References

1. Wolpaw JR, Birbaumer N, McFarland DJ, Pfurtscheller G, and Vaughan TM, "Brain-computer interfaces for communication and control," *Clinical Neurophysiology*, vol. 113, pp. 767–791, 2002.
2. Pfurtscheller G and Neuper C, "Motor imagery and direct brain-computer communication," *Proceedings of the IEEE*, vol. 89, no. 7, pp. 1123–1134, 2001.
3. Sanchez JV, Erdogmus D, Nicolesis MAL, Wessberg J, and Principe JC, "Interpreting spatial and temporal neural activity through a recurrent neural network brain-machine interface," *IEEE Transactions on Neural Systems and Rehabilitation Engineering*, vol. 13, no. 2, pp. 213–219, 2005.
4. Borisoff JF, Mason SG, Bashashati A, and Birch GE, "Brain-computer interface design for asynchronous control applications: improvement to the LF-ASD asynchronous brain switch," *IEEE Transactions on Biomedical Engineering*, vol. 51, no. 6, pp. 985–992, 2004.
5. Pfurtscheller G, Muller-Putz GR, Pfurtscheller J, Gerner HJ, and Rupp R, "Thought control of functional electrical stimulation to restore hand grasp in a patient with Tetraplegia," *Neuroscience Letters*, vol. 351, no. 1, pp. 33–36, 2003.
6. Leeb R, Friedman D, Muller-Putz GR, Scherer R, Slater M, and Pfurtscheller G, "Self-paced (asynchronous) BCI control of a wheelchair in virtual environments: a case study with tetraplegic," *Computational Intelligence and Neuroscience*, no. 79642, 2007.
7. Rebsamen B, Burdet E, Guan C, Teo CL, Zeng Q, Ang M, and Laugier C, "Controlling a wheelchair using a BCI with low information transfer rate," *Proceeding of the IEEE 10th International Conference on Rehabilitation Robotics*, Netherlands, pp. 1003–1008, 2007.
8. Philips J, Millan JR, Vanacker G, Lew E, Galan F, Ferrez PW, van Brusel H, and Nutin M, "Adaptive shared control of a brain-actuated simulated wheelchair," *Proceedings of the IEEE 10th International Confeence on Rehabilitation Robotics*, Netherlands, pp. 408–414, 2007.
9. Bell CJ, Shenoy P, Chalodhorn R, and Rao RPN, "Control of a humanoid robot by a non-invasive brain-computer interface in humans," *Journal of Neural Engineering*, vol. 5, pp. 214–220, 2008.
10. Galan F, Nuttin M, Lew E, Ferrez PW, Vanacker G, Phillips J, and Millan JR, "A brain-actuated wheelchair: asynchronous and non-invasive brain-computer interfaces for continuous control of robots," *Clinical Neurophysiology*, vol. 119, pp. 2159–2169, 2008.
11. Daly J and Wolpaw J, "Brain–computer interfaces in neurological rehabilitation," *Lancet Neurology*, vol. 7, pp. 1032–1043, 2008.
12. Pfurtscheller G, Neuper C, Schlogl A, and Lugger K, "Separability of EEG signals recorded during right and left motor imagery using adaptive autoregressive parameters," *IEEE Transactions on Rehabilitation Engineering*, vol. 6, no. 3, pp. 316–325, 1998.
13. Millan JR, Renkens F, Mourino J, and Gerstner W, "Noninvasive brain-actuated control of a mobile robot by human EEG," *IEEE Transactions on Biomedical Engineering*, vol. 51, no. 6, pp. 1026–1033, 2004.

14. Hema CR, Yaacob S, Adom AH, Nagarajan R, and Paulraj MP, "Motor imagery signal classification for a four state brain machine interface", *International Journal of Biomedical Sciences*, vol. 3, no. 1, pp. 76–81, 2008.
15. Domenick DG, "International 10–20 Electrode Placement System for Sleep", 1998. http://members.aol.com/aduial/1020sys.html.
16. Engelbrecht AP, *Computational Intelligence an Introduction*, John Wiley and Sons Ltd. pp. 32–33, 2002.
17. Sivanandam SN and Paulraj M, *Introduction to Artificial Neural Networks* Vikas Publishing House, India. 2003, ch. 1, 5.

Chapter 58
Toward an Application to Psychological Disorders Diagnosis

Luciano Comin Nunes, Plácido Rogério Pinheiro,
Tarcísio Cavalcante Pequeno, and Mirian Calíope Dantas Pinheiro

Abstract Psychological disorders have kept away and incapacitated professionals in different sectors of activities. The most serious problems may be associated with various types of pathologies; however, it appears, more often, as psychotic disorders, mood disorders, anxiety disorders, antisocial personality, multiple personality and addiction, causing a micro level damage to the individual and his/her family and in a macro level to the production system and the country welfare. The lack of early diagnosis has provided reactive measures, and sometimes very late, when the professional is already showing psychological signs of incapacity to work. This study aims to help the early diagnosis of psychological disorders with a hybrid proposal of an expert system that is integrated to structured methodologies in decision support (Multi-Criteria Decision Analysis – MCDA) and knowledge structured representations into production rules and probabilities (Artificial Intelligence – AI).

Keywords Early diagnosis · Expert SINTA · Expert systems · Psychological disorders · Multicriteria

1 Introduction

Although psychological disorders have causes and specific symptoms, the establishment of diagnosis is not so easy to be determined by clinical and laboratory tests, but by analyzing the behavioral responses from a human being facing daily life events. Psychological disorders vary from person to person and also in the severity degree. In the same way as personality, consciousness, and intelligence, the term "abnormal behavior" is difficult to define because of subjectivity inherent to the subject. Major difficulties are encountered when trying to diagnose whether a

L.C. Nunes (✉)
Graduate Program in Applied Computer Sciences, University of Fortaleza,
Av. Washington Soares, 1321-Bl J, Sl 30, CEP: 60811-905, Fortaleza, Ceará, Brazil
e-mail: lcominn@uol.com.br; lcominn@bnb.gov.br

person is really suffering from a psychological disorder or not, since the symptoms will manifest small imperceptible signs to both the individual who is suffering the disorder and who lives together.

There are some aggravating factors, which include: first, the resistance offered by a person suffering from psychological disorder to accept the situation, and second, a healthy individual pretending to be suffering from the disorder to get some personal advantages [1].

Aiming to facilitate the understanding of the study given in this article, we present a hybrid model combining the methodology of multi-criteria decision support and expert system. This model will examine the main symptoms and causes of the following types of psychological disorders that have affected the economically active population in the world, covering various industries and professions: psychotic disorders, mood disorders, anxiety disorders, antisocial personality disorder, multiple-personality disorders, and addiction disorders. Afterward, the force relation between symptoms and causes will be established in order to create a hierarchy to set the rules that will be used as a base of knowledge to the expert system to help in the diagnosis in accordance with assigned value to variables representing the symptoms and causes. Important to note that hybrid models have been used to help in decision-making and the search for diagnoses in the health area and others areas, as referenced in previously published articles [2–7].

2 Psychological Disorders

2.1 General Information

For many years, mankind has thought "health" as the absence of hurt. Currently, the perception of "health" is associated with the absence of disease. The World Health Organization (WHO) defines *health* as a state of complete physical, mental and social well being, involving three areas that are interconnected with each other: physical health, psychological health, and social health.

The abnormal behavior consists in disordered emotion, disordered thinking or disordered action models identified by exceptions to a statistical standard distribution, by dysfunction or incapacity, by personal distress, and by the violation of social norms and cultural rules [8].

2.2 Psychological Disorders Discussed in This Article

The Mental Disorders Diagnostic and Statistical Manual, fourth edition (DSM-IV), published in 1994 by the American Psychological Association (APA) was decided to be used in this study, considering that it proportionates a better presentation structure

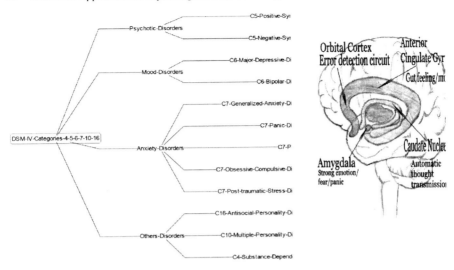

Fig. 1 Major categories of psychological disorders

and the best synthesis capacity without losing the quality and the details of the studied object. The map, described in Fig. 1, generated using the software HIVIEW, version 3.2.0.4, presents the most well-known psychological disorders. The illustration on the right side of Fig. 1 highlights some parts of the human brain where processed synapses are correlated with emotions and feelings evident in the behavior of a person.

2.3 Considerations About the Psychological Disorders Studied in This Article

One of the most drastic consequences for a person who suffers from psychological disorders is the image of incapable made by the society in which he or she lives. As much as the person strives to dispel the negative image formed, the damage is devastating and difficult to reverse, given the ease of lasting image. It is common to observe in a workplace how people react when a teammate suffers from a psychological disorder, even if he or she is being treated by specialists and taking medicines, most of the work team members observe this person with suspicion. A mutilation of a limb, or a severe infection controlled by medication, or even eradicated cancer at early stage provides chances for reinstatement of a person in the social life. However, the stigma of suffering from a severe psychological disorder usually invalidates the individual anywhere in the world, resembles the stigma of the leprosy in individuals, not long time ago. Currently, individuals suffering from leprosy are less stigmatized than individuals who suffer from psychological disorders.

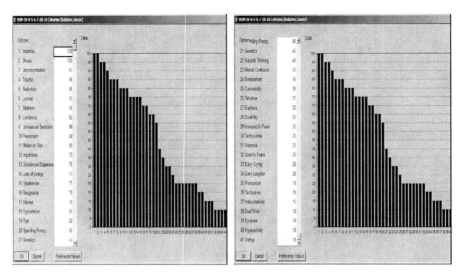

Fig. 2 Main events for the control of psycological disorders

One explanation for this discrimination is the misinformation from the great majority who make up the society, starting from the family of this individual. Based on the premise that it is possible to identify the causes and symptoms of psychological problems, we can try to diagnose these disorders as quickly as possible. For this, in this study were highlighted in Fig. 2 some of the main control events that are correlated with symptoms and causes that show some psychological disorders. For each event was assigned a confidence factor that indicates the weight on the outcome of the diagnosis. Note that these control events and confidence factors will be submitted to the methodology to support decision making with multiple criteria, described in item 3 below, and for the purpose of constructing an array of sense of values representing the differences in attractiveness between the events.

3 Model of Decision-Making Support Proposed

In order to establish a diagnosis in Psychology and Psychiatry, studies, experiments and careful observations of the symptoms that lead to the cause of evident disorder are necessary. A major difficulty in reaching a correct diagnosis is the complexity of highlighted factors, as well as the vast amount of information, that the expert must take into account including cultural, biological, psychosocial issues, information quality, and signs and symptoms common to many diseases. Because of this, the process becomes complicated and it is transformed into a difficult model. The methodology to support multicriteria decision-making has much to add to the Psychology of Health diagnosis processes.

3.1 Description of Multicriteria Methodology to Decision Making

There are three main stages in the decision support [9]: (a) Structuring the problem formulation and the goals identification. (b) Assessment of the alternatives and the criteria; (c) Recommendation: sensitivity and robustness analyses are made to check whether changes in the model parameters affect the final assessment outcome.

The Measuring Attractiveness by a Categorical-Based Evaluation Technique (MACBETH) method is an approach for multicriteria decision support. The research that initiated the method was performed by CA Bana e Costa and J.C. Vansnick in the beginning of the 1990s [10]. This methodology was in response to the question: *How to build a wide range of preferences from a range of options without forcing policy makers to produce their direct numerically preferences?* This approach allows assigning notes to each alternative through a paired comparison. Given the two alternatives, the decision which must be the most attractive in the case, has greater degree of confidence, and in which degree of attractiveness on a scale that has semantic correlation with ordinal scale. The program itself is the analysis of cardinal consistency (transitivity) and semantics (relations between the differences), suggesting in case of inconsistency how to resolve it. Linear programming is suggested a range of notes and intervals as they may change without making the problem inconsistent (PPL unfeasible). It also possibly adjusts the decision graphically and the value of the awarded marks, within the ranges allowed. According to the pioneers of this method, only after this adjustment, with the expert knowledge introduction, it gets characterized as the construction of a cardinal scale of values. The difference in attractiveness is very important in this methodology. For the MACBETH method is important the following reasoning: Given the impacts ij (a) and ij (b) of two potential actions a and b, according to a fundamental point of view FPVj, being judged a more attractive than b, the difference in attractiveness between a and b is judged as "null," "very weak," "weak," "moderate," "strong", "very strong," or "extreme." It introduced a scale formed by different semantic categories in attractiveness; the size is not necessarily equal to facilitate the interaction between the decision maker and analyst. The semantic categories, Ck, $k = 1 \ldots 6$, are represented as follows [11]:

- C1 very weak difference of attractiveness (or between zero and weak) → C1 = [s1, s2] and s1 = 0
- C2 weak difference of attractiveness → C2 =]s2, s3]
- C3 moderate difference of attractiveness (or between weak and strong) → C3 =]s3, s4]
- C4 strong difference of attractiveness → C4 =]s4, s5]
- C5 strong difference of attractiveness (or between strong and extreme) → C5 =]s5, s6]
- C6 extreme difference of attractiveness → C6 =]s6, +[

To facilitate the expression of absolute judgments of difference in attractiveness between the pairs of alternatives, it is useful to the construction of arrays of value judgments [12].

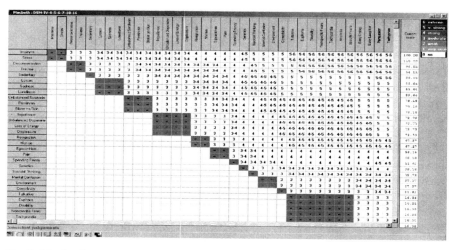

Fig. 3 Matrix of value-judgment and current scale

3.2 Results After Submitting the Events of Control

After submitting the events to the methodology to support decision making, implemented through software HIVIEW, which also runs MACBETH method, the control events cited in Fig. 2 were adjusted in order to compose the matrix of the main constant value in Fig. 3, which allows the visualization of the degree of attraction between the events and the Current Scale of confidence factors, and indicates that the results are consistent appearing "Consistent Judgments." The proposed model in this article suggests that the control events and confidence factors listed in Fig. 3 be exported to the Expert SINTA tool, to compose the knowledge base of an expert system to help the obsessive–compulsive disorder diagnosis.

4 Expert Systems

Expert systems are computer programs that offer solutions to certain problems in the same way that human experts offer solutions under the same conditions. The software Expert SINTA was created by a group of scholars at the Federal University of Ceará (UFC), and the State University of Ceará (UECE), called *Group SINTA* (Sistemas **INT**eligentes Aplicados or Applied Intelligent Systems) [13]. This is a computational tool that uses artificial intelligence (AI) techniques for automatic generation of expert systems. It uses a knowledge representation model based on production rules and probabilities, with the main objective of simplifying the construction work of expert systems through the use of a shared machine inference, the automatic construction of screens and menus, treatment of probabilistic rules production, and the use of sensitive explanations to the context of the knowledge base modeled. An expert system based on this type of model is very useful in classification problems. Some of the Expert SINTA main features are: use of backward chaining; use of confidence factors; tools for debugging; possibility to include on-line help for each knowledge base. As previously mentioned, the control events and confidence factors in psychological disorders are initially presented in Fig. 2, and subsequently analyzed and compared with each other, to calculate the difference in the degree of attractiveness for the construction and trial of the value matrix, generated by the MACBETH method, as shown in Fig. 3. After that, the information is compiled for, finally, feeding the Expert SINTA to the construction of expert system.

The process of building the expert system is composed of the following steps: (a) Registration of general information and goals of the knowledge base; (b) Registration for each event in Fig. 2 as a variable in the future specialist system; (c) Registration and definition to the objectives variables, which will point the diagnosis, with their level of confidence; (d) Construction of the user interfaces of the specialist system; (e) Construction of logical rules; (f) Definition to the best sequence for the logical rules; (g) Testing and execution of the specialist system; (h) Analysis of results, using log of execution provided by the expert system.

5 Conclusion and Future Works

Many intelligent techniques have been incorporated into the decision-making process. These techniques are based on technology of AI, such as expert systems, neural networks, intelligent agents, case-based reasoning, Fuzzy Logic and Genetic Algorithms. Despite these technological advances, much still needs to be done for the automation of decision-making processes, especially when it involves the multicriteria analysis. This article is part of a study intended to contribute to the development of an algorithm designed to automate the proposal for a connection between a methodology to support multicriteria decision-making and a program of construction of expert systems. Also provides a knowledge base expert system to aid in early diagnosis of psychological disorders. The information generated in the analysis of multicriteria methodology will be used in the algorithm in order to turn them into variables with the respective degrees of confidence, which will be processed by the inference engine of the expert system that will use this information to diagnose and incorporate them into its knowledge base or discard them. The test realized in this article attempts to show a hybrid model, using a connection manually. So it tries to present the feasibility of integration between the technologies mentioned: methodology to support multicriteria decision-making and expert system. With this proposal, we hope to help in the quality of automated diagnostics.

References

1. K. Huffman, M. Vernoy, J. Vernoy (2000) Psychology in Action, John Wiley & Sons, Inc., USA.
2. L.C. Nunes, P.R. Pinheiro, T.C. Pequeno (2009) An expert system applied to the diagnosis of psychological disorders. IEEE International Conference on Intelligent Computing and Intelligent Systems. IEEE PRESS, 363–367, China.
3. J.M.J. Aragonés, J.I.P. Sánchez, J.M. Doña, et al (2004) A neuro-fuzzy decision model for prognosis of breast cancer relapse. Current Topics in Artificial Intelligence, Springer, Berlin. 3040:638–645.
4. L. Qu, Y. Chen (2008) A hybrid MCDM method for route selection of multimodal transportation network. Advances in Neural Networks, Springer, Berlin. 5263:374–383.
5. L. Yu, S. Wang, K.K. Lai, L. Zhou (2008) An intelligent agent based multicriteria fuzzy group decision making model for credit risk analysis. Bio-Inspired Credit Risk Analysis, Springer, Berlin Heidelberg. Part IV:197–222.
6. A.K.A. de Castro, P.R. Pinheiro, M.C.D. Pinheiro (2009) An Approach for the Neuropsychological Diagnosis of Alzheimer's Disease: A Hybrid Model in Decision Making, Springer, Berlin. 5589:216–223.
7. A.K.A. de Castro, P.R. Pinheiro, M.C.D. Pinheiro (2008) A Hybrid Model for Aiding in Decision Making for the Neuropsychological Diagnosis of Alzheimer's Disease, Springer, Berlin. 5306:495–504.
8. R.O. Straub (2002) Helth Psychology. University of Michigan, Desrborn. Worth Publishers.
9. C.A. Bana e Costa, L. Ensslin, E.C. Correa, et al (1999) Decision support systems in action: integrates application in a multicriteria decision aid process. European Journal of Operational Research 133:315–335.
10. C.A. Bana e Costa, J.C. Vansnick (1999) "The MACBETH Approach: Basic Ideas, Software, and an Application", N. Meskens, M. Roubens (eds.), Advances in Decision Analysis, Book Series: Mathematical Modelling: Theory and Applications, Kluwer Academic Publishers, Dordrecht. 4:131–157.
11. C.A. Bana e Costa, J.C. Lourenço, M.P. Chagas, et al (2008) Development of reusable bid evaluation models for the Portuguese electric transmission company. Decision Analysis 5(1):22–42.
12. C.A. Bana e Costa, M.P. Chagas (2004) A career choice problem: An example of how to use MACBETH to build a quantitative value model based on qualitative value judgments. European Journal of Operational Research 153:323–331.
13. J.H.M. Nogueira, J.F.L. Alcântara, R.C. de Andrade, et al (1995) Grupo SINTA – Sistemas INTeligentes Aplicados. Federal University of Ceará (UFC).

Chapter 59
Enhancing Medical Research Efficiency by Using Concept Maps

Varadraj P. Gurupur, Amit S. Kamdi, Tolga Tuncer, Murat M. Tanik, and Murat N. Tanju

Abstract Even with today's advances in technology, the processes involved in medical research continue to be both time consuming and labor intensive. We have built an experimental integrated tool to convert the textual information available to the researchers into a concept map using the Web Ontology Language as an intermediate source of information. This tool is based on building semantic models using concept maps. The labor-intensive sequence of processes involved in medical research is suitably replaced by using this tool built by a suitable integration of concept maps and Web Ontology Language. We analyzed this tool by considering the example of linking vitamin D deficiency with prostate cancer. This tool is intended to provide a faster solution in building relations and concepts based on the existing facts.

1 Introduction

A typical researcher spends an inordinate amount of time in identifying, sorting, and analyzing already published research results [1]. We believe that this time can be cut down by several weeks with the use of appropriate tools and technologies [2]. The problem of inefficiency in medical research is because of the following reasons:

- There are inefficiencies when researchers use multiple databases with multiple interfaces, which may not all be known by a research assistant.
- Even when a research assistant is well versed in the search techniques, he or she may not be confident that all relevant research papers have been identified.
- When relevant research papers are identified, the next challenge is to efficiently extract information on materials, methods, and results illustrated in them.

The above-mentioned problems cause a significant time gap between the time results are obtained by one researcher and used by other researchers. There is also a

---

V.P. Gurupur (✉)
Department of Electrical and Computer Engineering, University of Alabama at Birmingham, Birmingham, AL 35294-1150, USA
e-mail: varad@uab.edu

significant time gap between the time most researchers form a consensus and the time such findings filter down to practicing physicians in the field. Even if a given research result is available, practicing physicians find it hard to keep up with such research because of time limitations [3, 4]. Assuming a physician jealously reserves a certain amount of his or her daily time to keep up with current research; he or she will face hundreds of articles on a given topic where he or she can only read a summary of only a handful of them. It takes many months, if not years, for research findings to filter down to practitioners. A case in point is the research on vitamin D [5], which has been going on for more than 30 years and has been gaining speed in the last few years. It is our opinion that many a research finding on vitamin D is simply out there waiting to be integrated.

1.1 Challenges in Finding the Right Research Material

It is usually observed that the medical researcher searches for an article on the Internet using an appropriate search engine and a keyword. The plethora of information available to researchers using this technique may not be a speedy way of integrating available information on a particular disease or disorder. What is needed is a search strategy based on the semantics of the information required. In other words, if a researcher searches for a particular disease or disorder, he or she must also be able to find the information related to it.

Normally, a researcher would spend a good deal of time reading various articles and construing necessary information from the article. The researcher will have to perform the following set of tasks to carry out the research activities:

- The researcher will have to identify a set of useful books and articles.
- The researcher will spend a good amount of time reading an article and construing necessary information from the article.
- The researcher will have to perform this task on many related and interesting articles to come up with a relation between the information presented in these articles [6].

The researcher will then build an idea or concept based on his or her understanding of the information construed by performing the above tasks. A given researcher or a practicing physician will also need to weigh the available evidence, which is discussed in the next section.

1.2 Sophisticated Tools Are Available for Medical Research

From Sect. 1.2, we can conclude that the present state of the medical research involves complicated tasks that could be very time consuming. In order to solve the foregoing problems, one has to address the underlying technological issues [7].

Multi-platforms with multi-interface requirements can be addressed with the concept of tools as services. Under such an approach, the researcher will use the tool that he or she always uses, such as an Excel spreadsheet. The software behind the tool automatically searches for the data requested by the researcher and populates the Excel spreadsheet with the required data. In this manner, a researcher does not waste time in learning unnecessary interfaces but continues to use the tool, which he or she has obtained familiarity.

The problem of determining whether all published information is identified can be alleviated by semantic search tools that perform "deep searches," which may not be accessible to commonly used tools such as Google. The problem of preparing comprehensive information, which can be used to bring research assistants or practicing physicians up to speed can be solved by concept maps discussed in the next section.

2 Using Concept Maps for Medical Research

A concept map is a graphical technique for organizing and representing knowledge, which is usually formed by networking-related concepts together. A research tool developed by the Institute for Human Machine Cognition [8] for this purpose is C-Map tools. "Concept maps are graphical tools for organizing and representing knowledge. They include concepts, usually enclosed in circles or boxes of some type, and relationships between concepts indicated by a connecting line linking two concepts." [9] A concept map allows an ordinary computer-literate individual to illustrate an idea, concept, or domain knowledge to be described in a graphical form [10, 11].

The interesting aspect of using a concept map tool is that it allows the domain expert to represent the knowledge in a graphical form, which can be easily processed by the human mind while its computability can be harnessed by easily converting the concept map into an XML-based format. The conceptual information thus obtained in the form of XML acquires the attribute of machine-actability and can be used to create semantic data in the form of Resource Description Framework (RDF) [12] and Web Ontology Language (OWL) [13].

3 Integrating Concept Maps and Ontology

The researchers can represent an idea or a concept in a graphical format, linking all the concepts together, resembling a growing neural network, effectively using the C-Map tools. The tool allows the conceptual information thus obtained to be automatically converted into OWL and RDF [14], as needed, in preparation for further processing. In fact, this robust domain-specific information stored in OWL can

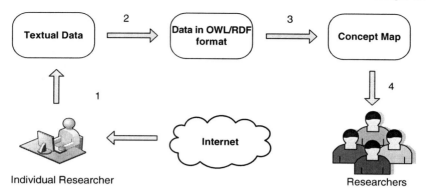

Fig. 1 The process of converting textual data into a concept map

be readily manipulated by the rule-based engine of the Jena Semantic Web Framework, as well as accessed by intelligent agents using Semantic Web technology for automated inference capability.

The greatest challenge for storing domain knowledge of a particular medical research area is to store this information in a human perceivable as well as machine-actable format. Concept maps can fulfill both of these requirements. While Concept maps can delineate a concept in the form of links and circles, it allows these concepts to be converted into XML-based ontology representation such as OWL and vice versa [15]. To achieve accumulation of domain knowledge in the form of concepts, the following steps will have to be performed:

1. Identify the electronic documents that pertain to the topic of medical research and its effects.
2. Convert the information contained in the document into an XML-based format.
3. Convert the information available in the XML-based format into a concept.

The concepts thus obtained can be further converted into OWL or other forms of Semantic Web-based format. This would allow an application program that can harness the versatility of Semantic Web-based information to extract the relevant information [2]. Overall, this process provides a path to conversion of natural language-based domain information into machine-actable information that can be readily used by an application program. The above-mentioned process helps us to achieve the following objectives:

1. It saves the amount of time required to decipher information available on a particular topic of medical research [16].
2. It automates the gathering of organized information and thereby provides a suitable path for linking its deficiency to diseases.

4 Relating Vitamin D Deficiency to Prostate Cancer

In order to provide a proof of concept, we will try to analyze the relationship between vitamin D deficiency and prostate cancer [17, 18]. Vitamin D is an important calcium homeostasis regulator and, moreover, modulates growth and differentiation. There seems to be a more than anticipated complex relationship between serum levels of vitamin D and the incidence of prostate cancer. Given the paucity and inconsistency of data relating vitamin D status to prostate cancer, as well as the complexity needed to achieve a meaningful conclusion for providing recommendations, building an unconventional method of research is inevitable. To achieve this, we propose the use of concept maps to delineate the complex research observations.

4.1 Using the Concept Map Alternative

A plausible solution to the above problem is to build a knowledgebase of vitamin D and its effect on prostate cancer. This can be achieved by using a combination of tools involving concept maps. By using this approach, the tasks mentioned in Sect. 1.1 could be obviated by building a set of concept maps from selected articles. This would also considerably reduce the amount of time and effort used to carry out research activities carried out by the individual researcher. The concepts that are built in the human mind can now be displayed in the form of a visual description available to the scientific community. To achieve this, we have performed the following set of tasks:

- Identify the electronic documents that pertain to vitamin D and its deficiency leading to prostate cancer.
- Convert the information contained in the document into an OWL document using a tool such as TopBraid Composer [19] developed by Franz Inc. [20].
- Build a concept map using the OWL document thus obtained using CMap Tools COE [14].

The concept map obtained by performing the above tasks can be easily used to depict the relationship between terminologies, concepts, and procedures related to vitamin D. Sometimes the research data can be populated on an Excel spreadsheet by the researcher, and it might be useful to convert this into a semantic model. The above mentioned steps can also be used to convert the data available on a spreadsheet. Overall, we are presenting an integrated tool to convert textual data into concept maps. While the CMap Tools software from IHMC has the ability to convert any document in OWL into a concept map, we have used TopBraid Composer to convert textual data into OWL. We would also like to mention that although this process provides a path to build concept maps from textual data, it is required to refactor the data in OWL to build a sensible concept map.

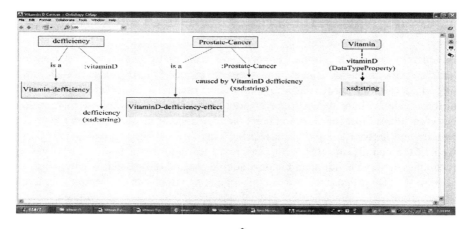

Fig. 2 Converting the information contained in an OWL document into a concept map

5 Summary and Conclusions

Today, the process of carrying out medical research involves time-consuming and labor-intensive tasks. There is a need to enhance the present scenario of medical research by using upcoming technologies such as concept maps and the Semantic Web. In this chapter, we have shown how a concept map can be used to depict a particular concept in medical research. We have also demonstrated how textual data can be converted into concept maps by integrating the required tools to achieve this task. In this way, we have provided a path for applying new visual tools to improve the area of medical research by reducing the amount of time and labor involved in carrying out complicated tasks.

This chapter illustrates a unique method for accelerating the process of medical research using a literature search on the subject of carcinogenesis prevention. There is a great need to improve the present day ad hoc ways of research using technologies such as concept maps and the Semantic Web. The proposed approach involves the usage of rapidly evolving cutting edge technologies such as CMap Tools COE that can revolutionize the area of medical research. We have provided a path for future generations of medical researchers to improve the techniques used to carry out research activities pertaining to a disease or disorder.

Acknowledgment This book chapter is a part of the Ph.D. dissertation of Varadraj Gurupur titled, "A Framework for Composite Service Development: Processs-as-a-Concept."

References

1. Calishain T (2007) Information Trapping: Real-Time Research on the Web. New Riders. Berkeley, California, USA.
2. Hu B (2008) Semantic web technologies can help save lives. Semantic Web Company Potal. http://www.semantic-web.at/1.36.resource.265.bo-hu-x22-semantic-web-technologies-can-help-save-lives-x22.htm. Accessed 11 August 2009.
3. Yamamoto K, Matsumoto S,et al (2008) A data capture system for outcomes studies that integrates with electronic health records: development and potential uses. Journal of Medical Systems. 32, 5:423–427.
4. Gurupur V, Tanju MN, et al (2009) Building Semantic Models Using Concept Maps for Medical Research. http://sdps.omnibooksonline.com/2009/. Accessed 05 January 2010.
5. Murray F (2008) Sunshine and vitamin D: a comprehensive guide to the benefits of the "sunshine vitamin". Basic Health Publications, Laguna Beach.
6. NIH. (2008) NIH roadmap for medical research: reengineering the clinical research enterprise. The NIH Common Fund. http://nihroadmap.nih.gov/clinicalresearchtheme/. Accessed 11 August 2009.
7. Hsieh SH, Hou IC, et al (2009) Design and implementation of a web-based mobile electronic medication administration record. Journal of Medical Systems. http://www.springerlink.com/content/e537234j8h7q0430/fulltext.pdf. Accessed 07 August 2009.
8. IHMC Official Website. http://www.ihmc.us/. Accessed 11 August 2009.
9. Novak JD, Cañas AJ (2006) The theory underlying concept maps and how to construct them. Technical report for Institute for Human and Machine Cognition. http://cmap.ihmc.us/Publications/ResearchPapers/TheoryCmaps/TheoryUnderlyingConceptMaps.htm. Accessed 21 August 2009.
10. Cañas AJ, Novak JD, et al (2004) Two layered approach to knowledge representation using conceptual maps and description logics. First Intl. Conference on Concept Mapping. http://cmc.ihmc.us/papers/cmc2004-205.pdf Accessed 21 August 2009.
11. Gurupur V (2010) A Framework for Composite Service Development: Process-as-a-Concept, Dissertation, Department of Electrical and Computer Engineering, UAB.
12. Allemang D, Hendler J (2008) Semantic web for the working ontologist: effective modeling in RDFS and OWL. Elsevier Inc, Burlington.
13. W3C (2004) OWL web ontology language overview. World Wide Web Consortium. http://www.w3.org/TR/owl-features/. Accessed 11 August 2009.

14. Padgett M (2008) CMAP tools for concept mapping and OWL authoring. Mike Padgett Website. http://www.mikepadgett.com/technology/technical/cmaptools-for-concept-mapping-and-owl-authoring. Accessed 11 August 2009.
15. Gurupur V (2008) Abstract software design framework: a semantic service composition approach, Dissertation Proposal, Department of Electrical and Computer Engineering, UAB.
16. Collett D (2009) Modeling survival data in medical research. CRC Press, Boca Raton.
17. Lim HW, Hönigsmann H, et al (2007) Photodermatology, Informa Healthcare, New York.
18. Zittermann A (2003) Vitamin D in preventive medicine: are we ignoring the evidence? British Journal of Nutrition. 89:552–572.
19. Franz Inc. (2009) TopBraid Composer. Franz Inc. Official Website. http://www.franz.com/agraph/tbc/. Accessed 11 August 2009.
20. Franz Inc. (2009) Franz Inc. Official Website. http://www.franz.com/. Accessed 11 August 2009.

Chapter 60
Analysis of Neural Sources of P300 Event-Related Potential in Normal and Schizophrenic Participants

Malihe Sabeti, Ehsan Moradi, and Serajeddin Katebi

Abstract The P300 event-related potential (ERP) is associated with attention and memory operations of the brain. P300 is changed in many cognitive disorders such as dementia, Alzheimer, schizophrenia, and major depression disorder. Therefore, investigation on basis of this component can help to improve our understanding of pathophysiology of such disorders and fundamentals of memory and attention mechanism. In this study, electroencephalography (EEG) signals of 20 schizophrenic patients and 20 age-matched normal subjects are analyzed. The oddball paradigm has been used to record the P300, where two stimuli including target and standard are presented with different probabilities in a random order. Data analysis is carried out using conventional averaging techniques as well as P300 source localization with low-resolution brain electromagnetic tomography (LORETA). The results show that the P300 components stem from a wide cerebral cortex network and defining a small definite cortical zone as its generator is impossible. In normal group, cingulate gyrus, one of the essential components of working memory circuit that was reported by Papez, is found to be the most activated area and it can be in line with the hypothesis that at least a part of the P300 is elicited by working-memory circuit. In schizophrenic group, frontal lobe is the most activated area that was responsible for P300 sources. Our results show that the cingulate gyrus is not activated in comparison with normal group, which is in line with previous results that dysfunction of the anterior cingulate cortex plays a prominent role in the schizophrenia disorder.

Keywords Experimental medicine and analysis tools · Biomedical engineering · Computer-based medical systems · P300 Event-related potential · LORETA

M. Sabeti (✉)
Department of Computer Science and Engineering, School of Engineering,
Shiraz University, Shiraz, Iran
e-mail: sabeti@shirazu.ac.ir

1 Introduction

One of the goals of cognitive neuroscience is to understand how the neural circuitry of the brain contributes to cognitive processes. It is too difficult to determine how specific populations of neurons operate on cognitive tasks. The best methods for measuring activity in distinct populations of neurons are invasive and usually cannot be applied to human subjects. Study of the many aspects of cognition is impossible in nonhuman subjects or the scientist cannot contribute the result of studies on animals directly to human being. Some modalities such as functional magnetic resonance imaging (fMRI) and positron emission tomography (PET) provide noninvasive measuring of changes in cerebral blood flow, which indirectly may be an indicator of neural activity, but changes in blood flow are so slow according to changes in brain neural activity, and they could not provide real-time measurement in most cognitive activities [1].

The P300 [2] is probably the most well-known component of the event-related potential (ERP) [1] that reflects attention/memory processes related to changes by new sensory inputs. Generally, ERP is a subgroup of electroencephalography (EEG) that directly measures the electrical charges in brain cortex in response to sensory, affective, or cognitive stimuli. ERP provides the requisite temporal resolution, it lacks the relatively high spatial resolution of PET and fMRI. However, ERP does provide some spatial information and many investigators are trying to use this spatial information to provide a measurement of the time course of neural activity in specific brain regions [1]. The first human studies on the normal origins of P300 focused on the hippocampal formation. Initial reports used depth electrodes that were implanted to help identify sources of epileptic foci in neurological patients. Their recordings suggested that at least some portion of the P300 is generated in the hippocampal areas of the medial temporal lobe [2, 3]. The fMRI achieving an acceptable compromise between spatial and temporal resolution has been used in several studies to investigate the neural basis of P300. Most of them have confirmed the involvement of the frontal, parietal, temporal, and cingulate areas to generate this ERP component [4, 5].

In recent years, different algorithms have been proposed for reconstructing the current source for a given scalp electrical distribution. Source localization based on scalp potentials requires a solution to an ill-posed inverse problem with many possible solutions. Selection of a particular solution often requires a priori knowledge acquired from the overall physiology of the brain and the status of the subject. A number of methods for localization of EEG sources have been investigated [6]. These methods can be divided into two main categories: equivalent current dipole model, in which the EEG signals are assumed to be generated by a relatively small number of focal sources, and the current distributed source model, in which all possible source locations are considered simultaneously.

Among the current distributed source methods, the low-resolution brain electromagnetic tomography (LORETA) [7] has been proved to have consistency with neuroimaging studies; therefore, LORETA might significantly improve knowledge on neural processes underlying the P300. Volpe et al. [8] studied the scalp topography and cortical sources of two subcomponents of P300 named P3a and P3b. Their findings on cortical generators are in line with the hypothesis that P3a reflects the automatic allocation of attention, while P3b is related to the effortful processing of task-relevant events. Jokisch et al. [9] used LORETA to analyze the neural processing of biological motion. They found that evidence for sources was located in right fusiform gyrus, right superior temporal gyrus, and areas associated with attentional aspects of visual processing. Saletu et al. [10] applied LORETA to identify brain regions involved in the processes of cognitive dysfunction in narcolepsy. They found that narcoleptic patients showed prolonged information processing, as indexed by N2 and P300 latencies and decreased energetic resources for cognitive processing.

In this study, an auditory oddball paradigm is used to record P300 ERP in 20 schizophrenic patients and 20 age-matched normal subjects. Whereas the precise neural origins of the P300 are as yet unknown, the main goal of this study is to determine the brain regions that are associated with P300 component. Another goal in this study is to determine how the schizophrenia affects the P300 neural sources. Therefore, LORETA is used to characterize the cortical distribution of P300 generators, and the results are evaluated statistically.

2 Data Acquisition

Schizophrenia is a mental disorder from which less than 1% of the whole world population suffers. According to the diagnostic criteria of the American Psychiatric Association (DSM-IV) [11], schizophrenic patients show some characteristic symptoms including delusions, hallucinations, or disorganized speech. About 20 patients with schizophrenia and 20 age-matched control subjects (all males) participated in this study. Age of control participants ranged from 18 to 55 years (33.4 ± 9.29 years; mean $\pm$ standard deviation (std)) and schizophrenic patients ranged from 20 to 53 years (33.3 ± 9.52 years; mean $\pm$ std). They were recruited from the Center for Clinical Research in Neuropsychiatry, Perth, Western Australia.

Each subject is participated in oddball paradigm, wherein two stimuli are presented with different probabilities in a random order. The participant is required to discriminate the infrequent target stimuli (high pitch beep) from the frequent standard stimuli (low pitch) by noting the occurrence of the target, by pressing a button. Electrophysiological data were recorded using a Neuroscan 24 Channel Synamps system, with a signal gain equal to 75 K ($150\times$ at the headbox). For EEG paradigms, 20 electrodes (Electrocap 10–20 standard system with reference to linked earlobes) were recorded plus left and right mastoids, VEOG and HEOG. According to the international 10–20 system, EEG data were continuously recorded from 20 electrodes (Fpz, Fz, Cz, Pz, C_3, T_3, C_4, T_4, Fp_1, Fp_2, F_3, F_4, F_7, F_8, P_3, P_4, T_5, T_6, O_1, O_2) with sampling frequency of 200 Hz.

3 P300 Event-Related Potential

The P300 is associated with attention and memory operation and probably originates from a distributed network of neurons. Although identifying a distinct explanation for this phenomenon has been so difficult, the P300 is produced in any tasks that requires stimulus discrimination – a basic psychological event that determines many aspects of cognition specially attention. The P300 component can be quantitatively characterized by amplitude and latency. Amplitude is defined as the difference between the mean pre-stimulus baseline voltage and the largest positive peak within a time window ranged in 250–500 ms. Latency is defined as the time that P300 is occurred after stimulus onset. P300 amplitude is thought to index brain activity that is required in the maintenance of working memory when the mental model of the stimulus context is updated. P300 latency is considered to be a measure of stimulus classification speed unrelated to response selection processes, and its timing is independent of behavioral reaction time. So P300 latency is an index of the processing time that occurs before response generation, and it provides a temporal measure of the neural activity underlying the processes of attention allocation and immediate memory [2].

4 P300 Source Localization

Low-resolution electromagnetic tomography algorithm [7] assumes current dipoles are distributed over a fixed lattice covering the brain tissue, the following linear relationship can be expressed:

$$\phi = A\mathbf{x} + \mathbf{n}, \qquad (1)$$

where $\phi = (\varphi_1, \varphi_2, \ldots, \varphi_M)^T$ is the vector of instantaneous EEG recordings, A is the lead field, $\mathbf{x} = (x_1, x_2, \ldots, x_{3 \times N})^T$ is the vector representing the strength of the dipole and n is the noise vector. M refers to the number of scalp recording electrodes and N refers to the number of current dipole sources within the solution space. Each column of A contains the potentials observed at the electrodes when the source vector has unit amplitude at one location and orientation and is zero at all others [6].

For 3D reconstruction methods, three dipole components should be placed in three directions of Cartesian coordinates to represent one dipole source at that location. Thus, the total number of dipole components is $3 \times N$. All possible sources in the LORETA method are in the 3D space of the whole brain and the orientation of the dipole at a specific location is decided by three orthogonal components. The computation of the lead field (A) was achieved by finite element method (FEM) or boundary element method (BEM) [12]. In such methods, the geometrical information about the brain layers and their conductivities has to be known.

The discrete inverse solution of LORETA is obtained by solving the problem:

$$\min ||B\mathbf{W}\mathbf{x}||^2 \text{ subject to } phi = A\mathbf{x}, \qquad (2)$$

where the definitions for ϕ, A, and $\mathbf{x}$ are the same as before. B is the discrete 3D Laplacian operator and imposes the smoothness constraint onto the ill-posed problem. $\mathbf{W}$ is a diagonal matrix, which consists of a column by column normalization of the lead field matrix A.

5 Experimental Results

Twenty schizophrenic patients and twenty age-matched normal subjects participated in this study. The neural generators of the P300 component is analyzed for each subject using the LORETA software [7], in the version providing current density values of 2,394 voxels in the cortical areas, modeled in the digitized Talairach atlas. Table 1 shows the neural generators of P300 component that is found for normal and schizophrenic groups, respectively. This method uses a Laplacian Weighted Minimum Norm algorithm without a priori assumption about a predefined number of activated brain regions, and thus it achieves a more open solution to the EEG inverse problem, which is closer to other brain imaging approaches. Figure 1 illustrates the LORETA localization result for a sample normal subject, and a sample schizophrenic patient.

Figure 1 shows that a specific point on the brain cortex cannot be defined, which can be responsible for generating the P300 sources in auditory oddball paradigm. Our results show that the LORETA localizes the neural activity from the cortical frontotemporoparietal network as the neural substrates of the scalp recorded P300. For statistical analysis, the frequencies of neural generators of P300 component for normal and schizophrenic groups according to cerebral cortex lobar division are shown in Table 2.

Using the Chi-square test, the topography of the P300 component of normal group was significantly different from that of the P300 component of schizophrenic group ($p = 0.006$). It is shown that in normal group, the most possible origin of P300 could be the limbic lobe (containing cingulate gyrus, parahippocampal gyrus, and uncus). The frequencies of general generators of P300 component for normal and schizophrenic groups according to right or left hemisphere are shown in Table 3.

Using the Chi-square test, there were no significant differences between each hemisphere for generating the P300 component of two mentioned groups ($p = 0.525$).

Table 1 The neural generators of P300 component for normal and schizophrenic participants

	Normal group	Schizophrenic group
Case 1	Posterior cingulate Limbic lobe	Superior temporal gyrus Temporal lobe
Case 2	Inferior frontal gyrus Frontal lobe	Superior frontal gyrus Frontal lobe
Case 3	Cingulate gyrus Limbic lobe	Medial frontal gyrus Frontal lobe
Case 4	Cingulate gyrus Limbic lobe	Paracentral lobule Parietal lobe
Case 5	Precuneus Parietal lobe	Precuneus Parietal lobe
Case 6	Anterior cingulate Limbic lobe	Precuneus Parietal lobe
Case 7	Precuneus Parietal lobe	Superior frontal gyrus Frontal lobe
Case 8	Anterior cingulate Limbic lobe	Orbital gyrus Frontal lobe
Case 9	Anterior cingulate Limbic lobe	Precuneus Parietal lobe
Case 10	Superior parietal lobule Parietal lobe	Superior temporal gyrus Temporal lobe
Case 11	Parahippocampal gyrus Limbic lobe	Inferior parietal lobule Parietal lobe
Case 12	Superior frontal gyrus Frontal Lobe	Middle frontal gyrus Frontal lobe
Case 13	Cingulate gyrus Limbic lobe	Superior frontal gyrus Frontal lobe
Case 14	Anterior cingulate Limbic lobe	Medial frontal gyrus Frontal lobe
Case 15	Medial frontal gyrus Frontal lobe	Precuneus Parietal lobe
Case 16	Uncus Limbic lobe	Orbital gyrus Frontal lobe
Case 17	Medial frontal gyrus Frontal Lobe	Middle temporal gyrus Temporal lob
Case 18	Cingulate gyrus Limbic lobe	Superior temporal gyrus Temporal lobe
Case 19	Superior frontal gyrus Frontal lobe	Superior frontal gyrus Frontal lobe
Case 20	Fusiform gyrus Temporal lobe	Postcentral gyrus Parietal lobe

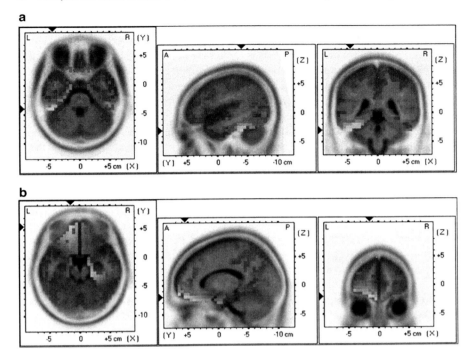

Fig. 1 The source localization result for (**a**) a sample normal subject, and (**b**) a sample schizophrenic patient

Table 2 The frequency of neural generators of P300 for different lobes

	Normal	Schizophrenic
Frontal lobe	5	8
Parietal lobe	3	7
Temporal lobe	1	4
Limbic lobe	11	1

Table 3 The frequency of neural generators of P300 for different hemisphere

	Normal	Schizophrenic
Left hemisphere	12	10
Right hemisphere	8	10

6 Discussion

In our study, a two-tone oddball paradigm is used to elicit the P300 component. Using LORETA algorithm, the neural sources of P300 component are compared for the two mentioned groups. Generally, our results show that finding a specific area of brain that could be responsible for generating P300 component is impossible. This result is substantially in line with previous studies, which have been carried out with different experimental techniques and paradigms that the P300 originates from a widespread neuronal network in brain and not from a specific region [13].

Table 1 shows that cingulate gyrus is the most common site that was responsible for P300 sources in normal group. As we know from previous studies, the P300 wave is generated by various circuits in brain that are responsible for working memory and attention. They suggest the neuroelectric events that underlie P300 generation stem from the interaction between frontal lobe and hippocampal/temporal–parietal function [2]. Our statistical analysis reveals the limbic lobe as the main source of P300, which is mainly in line with Desimone et al. [14] results. Previous fMRI studies, dealing with P300 components arising from an oddball auditory stimulation, reported that its generators were located in frontal, parietal, cingulate, and temporal areas [13]. Our results about the source activity underlying a target detection task are consistent with those reported in a recent studying by Mulert et al. [15], who also used a LORETA analysis. Cingulate gyrus is one of the essential components of working memory circuit that was reported by Papez and it has a widespread connection with most parts of cerebral cortex. Cingulate gyrus is found to be the most activated area and it can be in line with the hypothesis that at least a part of the P300 is elicited by working-memory circuit [16]. But we cannot show other parts of Papez circuit such as hippocampal formation to be a generator of P300, which was reported by McCarthy et al. [3].

As Table 1 shows, the frontal lobe is the most common site that was responsible for P300 sources in schizophrenic group. It confirms the previous study by Ska-Starzycka et al. [17] that cognitive impairment in schizophrenia is attributed to the functional hypofrontality. Our results show the cingulate gyrus is not activated in comparison with normal group. There is good evidence from neuroanatomic post-mortem and functional imaging studies that dysfunction of the anterior cingulate cortex plays a prominent role in the pathophysiology of schizophrenia [18].

We cannot define any significant dominancy between right or left hemispheres to generate the P300 components. The dominancy of hemispheres in attention or memory processes is so controversial. But many authors believed that we cannot absolutely mark one hemisphere as the main generator of P300. In conclusion, it seems that the P300 components stem from a wide cerebral cortex network and defining a small definite cortical zone as its generator is impossible. The LORETA source imaging method used in this study restricts the localization of electrical generators to cingulate gyrus (mainly) and frontoparietal areas.

Acknowledgments The authors of this paper present a special thanks to Dr. Gregory W. Price, who has given us his database including EEG of normal and schizophrenic participants.

References

1. Luck SJ (2005) An introduction to the event-related potential technique. References and further reading may be available for this article. To view references and further reading you must *purchase* this article. The MIT Press, USA.

2. Polich J (2007) Updating P300: an integrative theory of P3a and P3b. Clin Neurophysiol. 118: 2128–2148.
3. McCarthy G, Wood CC, Williamson PD et al (1989) Task-dependent field potentials in human hippocampal formation. J Neurosci. 9: 4235–4268.
4. Stevens AA, Skudlarski P, Gatenby JC et al (2000) Event-related fMRI of auditory and visual oddball tasks. Magn Reson Imaging 18: 495–502.
5. Clark VP, Fannon S, Lai S et al (2000) Responses to rare visual target and distractor stimuli using event-related fMRI. J Neurophysiol. 83: 3133–3139.
6. Sanei S, Chambers JA (2007) EEG signal processing. John Wiley & Sons, England.
7. Pascual-Marqui RD (2002) Standardized low-resolution brain electromagnetic tomography: technical details. Meth Find Exp Clin Pharmacol. 24(D): 5–12.
8. Volpe U, Mucci A, Bucci P et al (2007) The cortical generators of P3a and P3b: A LORETA study. Brain Res Bull. 73: 220–230.
9. Jokisch D, Daum I, Suchan B et al (2005) Structural encoding and recognition of biological motion: evidence from event-related potentials and source analysis. Behav Brain Res. 157: 195–204.
10. Saletu M, Anderer P, Saletu-Zyhlarz GM (2008) Event-related-potential low-resolution brain electromagnetic tomography (ERP-LORETA) suggests decreased energetic resources for cognitive processing in narcolepsy. Clin Neurophysiol. 119(8): 1782–1794.
11. Diagnostic and statistical manual of mental disorders, DSM-IV-TR (2000) Washington DC: American Psychiatric Association.
12. Hämäläinen MS, Sarvas J (1989) Realistic conductivity geometry model of the human head for interpretation of neuromagnetic data. IEEE Trans Biomed Eng. 36:165–171.
13. Brazdil M, Dobsik M, Mikl M et al (2005) Combined event-related fMRI and intracerebral ERP study of an auditory oddball task. Neuroimage. 26: 285–293.
14. Desimone R, Miller EK, Chelazzi L et al (1995) Multiple memory systems in the visual cortex. In: Gazzaniga MS (ed) Cogn Neurosci. 475–486.
15. Mulert C, Pogarell O, Juckel G et al (2004) The neural basis of the P300 potential. Eur Arch Psychiatry Clin Neurosci. 254: 190–8.
16. Squire LR, Kandel ER (1999) From short-term memory to long-term memory. In: Squire LR (ed) Memory From Mine to Molecules. New York: Scientific American Library 129–156.
17. Ska-Starzycka AB, Pascual-Marqui RD (2002) Dysfunctional prefrontal attention-related resources in schizophrenia localised by the low-resolution electromagnetic tomography (LORETA). International Congress Series (Recent advances in human brain mapping) 1232: 639–643.
18. Mulert C, Gallinat J, Pascual-Marqui R et al (2001) Reduced event-related current density in the anterior cingulate cortex in schizophrenia. NeuroImage. 13(4): 589–600.

Chapter 61
Design and Development of a Tele-Healthcare Information System Based on Web Services and HL7 Standards

Ean-Wen Huang, Rui-Suan Hung, Shwu-Fen Chiou, Fei-Ying Liu, and Der-Ming Liou

Abstract Information and communication technologies progress rapidly and many novel applications have been developed in many domains of human life. In recent years, the demand for healthcare services has been growing because of the increase in the elderly population. Consequently, a number of healthcare institutions have focused on creating technologies to reduce extraneous work and improve the quality of service. In this study, an information platform for tele- healthcare services was implemented. The architecture of the platform included a web-based application server and client system. The client system was able to retrieve the blood pressure and glucose levels of a patient stored in measurement instruments through Bluetooth wireless transmission. The web application server assisted the staffs and clients in analyzing the health conditions of patients. In addition, the server provided face-to-face communications and instructions through remote video devices. The platform deployed a service-oriented architecture, which consisted of HL7 standard messages and web service components. The platform could transfer health records into HL7 standard clinical document architecture for data exchange with other organizations. The prototyping system was pretested and evaluated in a homecare department of hospital and a community management center for chronic disease monitoring. Based on the results of this study, this system is expected to improve the quality of healthcare services.

1 Introduction

Population aging is a global trend. There were 600 million people aged 60 and over in 2000 and it is estimated that there will be 1.2 billion by 2025 and 2 billion by 2050 [1]. The need for tele-healthcare systems has increased in recent years. Many such systems have been developed by different vendors. If these vendors do not

---

E.-W. Huang (✉)
Department of Information Management, National Taipei University of Nursing and Health Sciences, 365, Ming-te Road, Peitou District, Taipei, Taiwan, R.O.C.
e-mail: huang@ntunhs.edu.tw

conform to standard methods of development, the data in the systems cannot be easily shared or exchanged with others. Health Level Seven (HL7) is an international standard that defines the format of health information exchange [2]. Tele-healthcare information systems may easily communicate with hospital information systems through HL7 standard. Following the development of extensible markup language (XML) and Simple Object Access Protocol (SOAP), web services have been created that can be executed by remote systems [3]. In this paper, we will share our experiences in the implementation of a tele-healthcare services system based on web services and HL7 standards.

2 Background and Literature Review

2.1 Aging Society

The United Nations has defined an "aging society" as a population in which the percentage of people over 65 years old is greater than 7%. When the percentage increases beyond 14%, the population is deemed an "aged society." The population of Taiwan is about 23.4 million. According to a government estimation report [4], the number of senior citizens over 65 years old accounted for 10.2% of the nation's total population in 2007. The report also predicted that the proportion would exceed 14% in 2017, meaning that the nation will become an aged society at that time. In 2008, each senior citizen is supported by 7.0 persons aged between 15 and 64. The number is forecasted to drop to 3.2 in 2026 and 1.4 by 2056. According to the statistical data from Bureau of National Health Insurance of Taiwan in 2006, the average spent by the elderly population for health insurance was about seven times that of other ages. The burden of supporting the healthcare expenses for the elderly population is significant.

2.2 Tele-Healthcare

In recent years, many countries have tried to apply information and communication technologies to the healthcare system. In the United States, Columbia University developed the Informatics for Diabetes Education and Telemedicine (IDEATel) project to bring healthcare into the homes of residents with diabetes. The system provided remote monitoring of physiologic measurement parameters, such as blood glucose and blood pressure. In addition, the project provided synchronous video conferencing and access to educational resources online [5].

Dorr et al. reviewed 109 articles of chronic care informatics systems. Most of the chronic diseases were diabetes, heart disease, and mental illness; 67% of the reviewed systems had a positive outcome [6]. Koch also researched the home tele-health of 578 publications between 1990 and 2003. Most of the publications dealt with the measurement of vital signs and audio/video consultations. Papers for improved information and communication technology tools and decision support for staff and patients are relatively spare [7]. Heberta et al. in Canada researched evidence supporting the effectiveness of home tele-healthcare for diabetes and proposed a decision framework to address complex issues of adopting home telehealth technology into practice [8]. In general, it has been concluded that home tele-health systems produce accurate and reliable data and capable of improving patient management [9].

3 Methods

The architecture of tele-healthcare services system in Fig. 1 was divided into client and remote server partitions. The client portion consists of a computer and vital sign measurement device that could be installed in a community service center or a patient's home. The remote server had six subsystems, including Case Information Management Subsystem (CIM), Vital Signs Capture and Transmission Subsystem (VCT), Personal Health Record Management Subsystem (PHM), Video Conference Subsystem (VCS), Care Staffs Information Subsystem (CSI), and Electronic Data Interchange Subsystem (EDI). Unified modeling language (UML) is a standard method to express the concept of the system components. We applied the architecture to describe the functions of each subsystem.

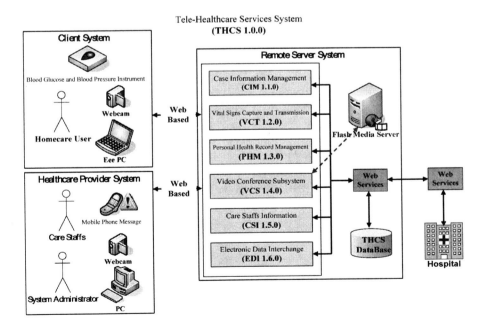

Fig. 1 The architecture of tele-healthcare services system

3.1 Case Information Management Subsystem

The subsystem had functions capable of adding, modifying, and querying the client's information. The subsystem contained the name, date of birth, and address of the individual.

It also provided an individual personal profile setting, which allowed the user to set the threshold values of systolic, diastolic, and blood glucose for the purposes of creating alerts. These functions allowed communication and collaboration between the medical staff and the users to achieve the goal of personal health management.

3.2 Vital Signs Capture and Transmission Subsystem

The function is to extract vital sign measurement data and to input them into the database. In this study, we used a two-in-one blood glucose blood pressure instrument manufactured by TaiDoc Corporation [10]. The client's computer extracted the measurement values through Bluetooth [11] or RS232, encoded the data to a HL7 message and finally called a web service to upload the message to the remote server. The web service parsed the HL7 message and saved the data into a database for later use. The smart web service also judged whether the measurement value was over the threshold. If so, a mobile phone message was sent to notify the medical staff or family immediately. The system also supported a history record query and trends through an Internet browser.

3.3 Personal Health Record Management Subsystem

The function of the PHM subsystem was to manipulate health records from the database. The system encoded the query condition to an HL7 message as a parameter and sent the message to the web services via the remote server. The web service then decoded the query message and converted into structure query language (SQL). After this, it extracted information from the database, encoded to a return message, and responded to the user. The system also provided the quality of life scale and Barthel index assessment for elderly people in long-term care. Through this system, patient's families, and caregivers were able to determine the condition of the patient anywhere and at anytime.

3.4 Video Conference Subsystem

In addition to communicating with family and medical staff by telephone, users may use the video conversation system to consult a medical expert about medical knowledge or access online health education systems. The conference was carried out with a dual-window screen on both the healthcare staff's and the patient's interfaces. The big window showed the image from the remote site while the small window showed the picture captured from a local camera. Before a conference, patient made a conference appointment by using the system to send a mobile phone message and an e-mail to his or her care staff.

3.5 Care Staffs Information Subsystem

In order to protect the privacy of the patients, only the assigned staffs were able to access the health records. Staffs can also search the health information on the Internet through this system.

3.6 Electronic Data Interchange Subsystem

It was possible to import data from another database and export data files in the HL7 V2.5 or CSV format. In order to create information that was readable by a human user, the system also generated HL7/CDA documents [12].

In this study, we used Microsoft Visual Web Developer to build the system. This software automatically generated codes supporting SOAP and XML of the web services and conformed to the standard. All of the transmission data were encoded to HL7 format as the web service parameter.

4 Results

Through the system analysis and design, a tele-healthcare information system was completed and included a web-based application server and client interface. The main functions are illustrated as follows:

4.1 Client Side System

The client interface system included instruments capable of measuring vital signs and a personal computer with a camera. Vital signs were transmitted to the personal computer through Bluetooth communication. The system was able to judge if the vital sign values were under a pre-specified normal range and presented the data on the screen and through speech by applying a text-to-speech (TTS) technology. For any abnormal measurement value, the system highlighted the value with red characters and sent a mobile phone message to the care provider. The user was then able to save the vital sign record to a local disk and upload the data to a remote server. Figure 2 shows the client portion of the system for a homecare user.

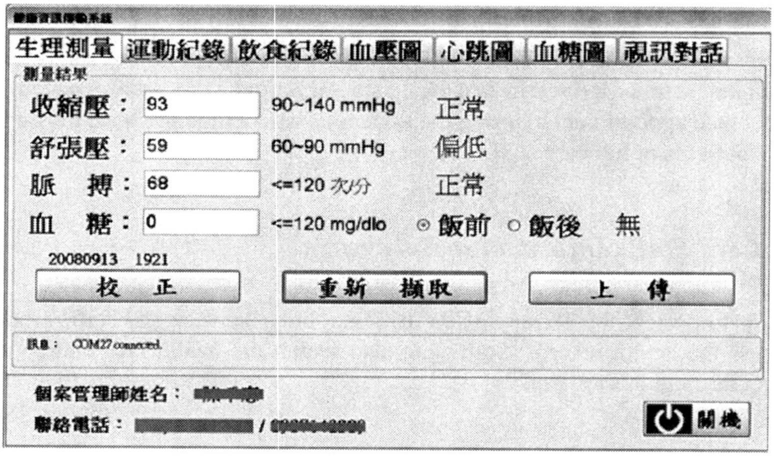

Fig. 2 The client side of the system for homecare user

For the homecare clients, a healthcare staff was able to communicate and partake in health education through video conferencing. Patients were able to make a conference appointment with the system. For this, the system notified the homecare nurse of the appointment through a mobile phone message and an email. The conference was carried out with dual-window screen on both the healthcare staff's and the patient's interfaces. The big window showed the image from the remote site while the small window showed the picture captured from local camera.

The system also provided the records of the Barthel index and quality of life scale as a means to evaluate the health condition of elderly people. The system supports the health records exchange function. The users were able to import and export data in HL7v2.x, CDA and CSV formats.

4.2 Web-Based Applications Server

There are three levels of system users including clients, caregivers, and system administrators. In order to protect the privacy of the patient, each user was provided with his or her own identification and password with corresponded to different access rights. The healthcare client could only read his or her own health records, while care givers were permitted to access the information of their specific clients. When a user logged in, the system automatically recognized the user identify and grouped him or her into a distinct interface with permissible functions.

The database stored blood pressure and glucose levels after each measurement. Caregivers were able to set the normal range of vital sign values for each client. The system notified the user if the measured value was out of the normal range. Users could search the stored records by establishing query conditions. The results of the query were then displayed on the screen. Figure 3 shows the systolic and diastolic blood pressures and pulses from a representative patient.

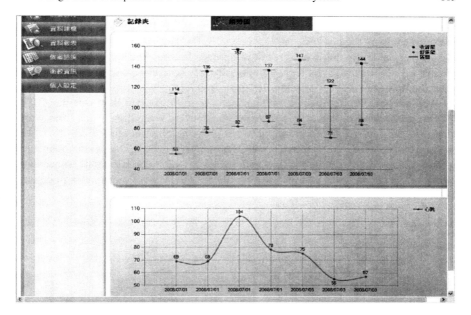

Fig. 3 The trend of systolic, diastolic blood pressures and pulses from a representative patient

5 Discussion and Conclusion

Most of the system users were elderly people who are not computer professionals. The user interface of the system should be designed to be exceptionally user friendly. All of the functions in the client system were condensed to four main categories included vital sign measurement, history records, user management, and system setting.

At the beginning, the system applied standard vital sign values as a threshold. The standards initially applied were that the systolic blood pressure was between 90 and 140 mmHg and the diastolic blood pressure was between 60 and 90 mmHg. When the measurement was less than or over than the standard value range, the system would alert and send a message to the caregiver. For some clients with hypertension, the blood pressure values were always higher than the standard, resulting in an alert each time. We have followed the caregiver's suggestion to modify system. The homecare nurses therefore set the threshold values for their clients, and each client may have different threshold values.

The system had the built-in function of sending a mobile phone message notification to the nurse for the purpose of monitoring homecare clients. The message was sent when the user uploaded a vital sign measurement that was over the threshold value. A nurse suggested that she needed to receive messages each time a user uploaded a measurement in order to monitor the changes for unstable patients. We have added a switch function to always, never, or conditionally send a message to accommodate the needs of the nurses and patients.

The system was designed to upload the measurement data to the server as soon as extracting it from the instruments. Occasionally, the data were lost because an unstable network at home that would result in uploads failure. The system was modified to save all of the data in the local computer before uploading and to delete the data after the upload.

We also believe that disease prevention is more important than treatment. The system may monitor the chronic diseases and prevent these conditions from worsening. In the future, we are planning to simplify the client interface portion of the system by replacing the computer with a mobile phone or PDA device connected to a server with 3G communication abilities. We hope the system will be accepted by more clients and will enhance the healthcare quality while promoting personal health.

Acknowledgments This work was supported in part by the National Science Council of Taiwan, R.O.C. under Grant NSC97-2221-E-227-001-MY3.

References

1. World Health Organization, 10 facts on ageing and the life course. http://www.who.int/features/factfiles/ageing/en/index.html.
2. HL7, Health Level 7 Version 2.5, Final Standard, An Application Protocol for Electronic Data Exchange in Healthcare Environment, 2003.
3. W3C, Web Services Glossary, http://www.w3.org/TR/ws-gloss/.
4. Council for Economic, Planning and Development. The estimates for future population in Taiwan area from 2008 to 2056. http://www.cepd.gov.tw/m1.aspx?sNo=0000455&key=&ex=%20&ic.
5. Starren J, Hripcsak G, Sengupta S, Abbruscato CR, Knudson PE, Weinstock RS, Shea S (2002) Columbia University's Informatics for Diabetes Education and Telemedicine (IDEATel) project: technical implementation. Journal of the American Medical Informatics Association 9 1:25–36.
6. David D, Bonner Laura M, Cohen Amy N, Shoai Rebecca S, Ruth P, Edmund C, et al (2007) Informatics Systems to Promote Improved Care for Chronic Illness: A Literature Review. Journal of the American Medical Informatics Association 14 2:156–163.
7. Koch S (2006) Home Telehealth-Current State and Future Trends. International Journal of Medical Informatics 75:565–576.
8. Marilynne A, Heberta, Korabek B, Scott RE (2006) Moving Research into Practice: A Decision Framework for Integrating Home Telehealth into Chronic Illness Care. International Journal of Medical Informatics 75:786–794.
9. Guy P, Mirou J, Claude S (2007) Systematic Review of Home Telemonitoring for Chronic Diseases: The Evidence Base. Journal of the American Medical Informatics Association 14 3:269–277.
10. TaiDoc Corporation, http://www.taidoc.com.tw/.
11. Wikipedia, Bluetooth http://en.wikipedia.org/wiki/Bluetooth.
12. Dolin RH, et al. HL7 Clinical Document Architecture. Release 2.0, HL7 V3 Ballot Document.

Chapter 62
Fuzzy Logic Based Expert System for the Treatment of Mobile Tooth

Vijay Kumar Mago, Anjali Mago, Poonam Sharma, and Jagmohan Mago

Abstract The aim of this research work is to design an expert system to assist dentist in treating the mobile tooth. There is lack of consistency among dentists in choosing the treatment plan. Moreover, there is no expert system currently available to verify and support such decision making in dentistry. A Fuzzy Logic based expert system has been designed to accept imprecise and vague values of dental sign-symptoms related to mobile tooth and the system suggests treatment plan(s). The comparison of predictions made by the system with those of the dentist is conducted. Chi-square Test of homogeneity is conducted and it is found that the system is capable of predicting accurate results. With this system, dentist feels more confident while planning the treatment of mobile tooth as he can verify his decision with the expert system. The authors also argue that Fuzzy Logic provides an appropriate mechanism to handle imprecise values of dental domain.

1 Introduction

This system is designed to help dentist decide treatment plan(s) for a mobile tooth. Tooth Mobility refers to the movability of a tooth resulting from loss of all or a portion of its attachment and supportive apparatus [3].

Medical decision making has always been a challenging task to perform since a physician has to rely on his intelligence, intuition, education, skills as well as experience in order to diagnose a problem. Decision making in dentistry is equally complicated and hence attract attention of researchers from mathematics, operation research, statistics and computer science. [1] describes various automated decision making systems in dentistry. The applicability of FL in medical domain was initially discussed in [14] and there after number of researchers have reported successful implementation of varied systems [2, 7–13].

V.K. Mago (✉)
Assistant Professor, DAV College, Jalandhar, India
e-mail: vijay.mago@gmail.com

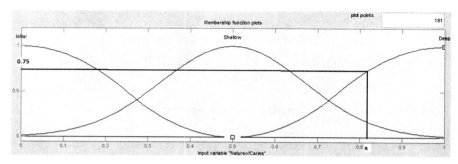

Fig. 1 Fuzzy set 'Deep' with smooth boundary

FL is a form of multi-valued logic derived from fuzzy set theory to deal with approximate reasoning [4]. A Fuzzy Expert System (FES) is a form of Artificial Intelligence that uses a collection of membership functions [5] and fuzzy rules [6]. The motivation behind FL is to imitate human reasoning capability that deals with *not well defined terms*. For instance, the statement: "A patient is having *deep* caries". The term *deep* is imprecise and FL deals with this imprecision using smooth boundary sets. This is shown in Fig. 1. On the horizontal axis, nature of caries is measured while on vertical axis, membership value of the input is quantified as membership value, denoted by a Greek symbol 'µ'. For instance, given the nature of caries as input 'x', its membership is denoted by $0.75 = \mu_{deep}(x)$.

1.1 Abstract View of the System

The sign symptoms are considered as input and corresponding treatment(s) as output. Dentist supplies the inputs through GUI. These are processed in FES. The resultant may be a single treatment or set of treatments. The FES has four main functional components: Fuzzifier, Inference Engine, Defuzzifier and Knowledge Base that is composed of if-then rules. This is shown in Fig. 2.

2 Materials and Methods

The treatment for tooth mobility depends upon several factors and includes a number of treatment processes. We have considered here four essential sign-symptoms (the inputs to the system) viz. Tooth Mobility (TM), Pain (P), Infection, and Occlusal Trauma (OT) and seven treatments (the output of the system) viz. Fixed Partial Denture (FPD), Splinting (Sp), Tooth Extraction (TE), Scaling (Sc), Occlusal Grinding (OG), Medication (Med) and Root Canal Treatment (RCT). They have their

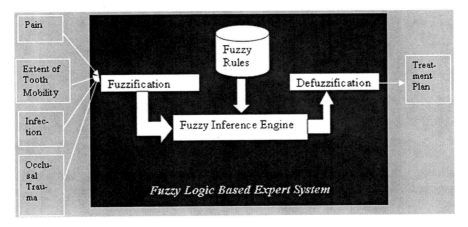

Fig. 2 The abstract view of the system

usual definitions. As discussed in Sect. 1.1, the whole system is divided into four functional components. The first step, fuzzification, involves designing membership functions for input and output variables. This is described below.

2.1 Fuzzification

With the help of an expert dentist, we determined the fuzzification of parameters. In a broader sense, fuzzification means designing the membership functions for variables. The membership function values for all the variables are mapped in the range of 0 to 1. We decided upon three linguistic variables: *Mild, Moderate, Severe* for mobility, three linguistic variables *Nopain, Onchewing and Moderate-to-Severe* for pain, three linguistic variables *NoInfection, GingivalInflammation and Periodontal_and_Periapical* for infection, two linguistic variables *Yes, No* for occlusal trauma as well as for each of the output parameters. The membership expressions defined for all the parameters are given below:

2.1.1 Membership Functions for Input Variable Mobility

For "mobility" value (say x), the fuzzy membership expression, using Z-shaped, Gaussian and S-shaped membership functions for linguistic variables *Mild, Moderate and Severe* respectively are:

$$\mu(\text{mild},[0.1,0.4]) = \begin{cases} 1, x \leq 0.1 \\ 1 - 2\left(\dfrac{x-0.1}{0.4-0.1}\right)^2, 0.1 \leq x \leq \dfrac{0.1+0.4}{2} \\ 2\left(0.4 - \dfrac{x}{0.3}\right)^2, \dfrac{0.1+0.4}{2} \leq x \leq 0.4 \\ 0, x \geq 0.4 \end{cases} \quad (1)$$

$$\mu(\text{moderate},[0.1699,0.5]) = e^{\frac{(x-c)^2}{2\sigma^2}} \quad (2)$$

where $c = 0.5$, the centre of the curve, $\sigma = 0.3$, the spread coefficient and e is the exponent.

$$\mu(\text{severe},[0.6,0.9]) = \begin{cases} 0, x \leq 0 \\ 2\left(\dfrac{x-0.6}{0.3}\right)^2, 0.6 \leq x \leq \dfrac{0.6+0.9}{2} \\ 1 - 2\left(\dfrac{x-0.9}{0.3}\right)^2, \dfrac{0.6+0.9}{2} \leq x \leq 0.9 \\ 1, x \geq 0.9 \end{cases} \quad (3)$$

2.1.2 Membership Functions for Input Variables Pain and Infection

For "pain" value, the fuzzy membership expressions for each linguistic variables *Nopain, Onchewing and Moderate-to-Severe* are similar to (1–3) respectively. For "infection" value (say x), the fuzzy membership expressions for each linguistic variable *NoInfection, GgingivalIinflammation and Periodontal_and_Periapical* are also analogous to (1–3) respectively.

2.1.3 Membership Functions for Input Variable Occlusal

For this variable value (say x), the fuzzy membership expressions are:

$$\mu(\text{no},0,0,0.4,0.6) = \begin{cases} 0, x \leq 0 \\ \dfrac{0.6-x}{0.2}, 0.4 \leq x \leq 0.6 \\ 0, x \geq 0.6 \end{cases} \quad (4)$$

$$\mu(\text{yes},0.4,0.6,1,1) = \begin{cases} 0, x \leq 0.4 \\ \dfrac{x-0.4}{0.2}, 0.4 \leq x \leq 0.6 \\ 0, x \geq 1 \end{cases} \quad (5)$$

2.1.4 Membership Functions for Output Variables

The output of the system can be one or set of treatments from: FPD, Sp, TE, Sc, OT, Med, RCT. Their membership functions are similar to (4–5).

2.2 Designing If-Then Rules

The decision procedure of the designed system is based upon the formation of fuzzy if-then rules. Out of four input parameters, Mobility, Pain, Infection have three linguistic variables each and Occlusal Trauma has two linguistic variables. Thus, a total of $54(=3^3*2^1)$ rule combinations are formed. During the evaluation of the rules, it was found that 12 rules generally do not exist due to certain medical restrictions. So, we excluded these rules from our knowledge base. Hence, we are left with a total of $54 - 12 = 42$ rules. These rules are defined in the rule editor of the Fuzzy Inference System supported by MATLAB [15].

2.3 Fuzzy Inference Mechanism

As an inference mechanism, we have used Mamdani algorithm [16] which maps the observation attributes of the given physical system into its controllable attributes, i.e. $U_1^*U_2^*\ldots^*U_r$ to $W(U_1, U_2, \ldots, U_r)$ are the universe of discourses for input variables and W is for the output variable. This fuzzy inference method is the most commonly used fuzzy methodology even though other methods like SAM, TSK [20] are also prominently used. A brief description of the working of this model is described below.

This model has been used as it is capable of handling fuzzy sets in the antecedent and the consequent part of the if-then rules. The basic structure of the rules is:

R_i: If x_1 is Ant_{i1} and x_2 is Ant_{i2} and .... and x_r is Ant_{ir}. Then y is Con_i where $i = 1, 2, \ldots, M$ and 'Ant' stands for the antecedent fuzzy set and 'Con' stands for the consequent fuzzy set.

$x_j (j = 1, 2, \ldots, r)$ are the input variables, y is the output variable. For simplicity sake, it is assumed that only one output is required. This is represented in Table 1.

Now assume an input: x_1 is Ant^*_1, x_2 is $Ant^*_2, \ldots, x_r$ is Ant^*_r where $Ant^*_1, Ant^*_2, \ldots, Ant^*_r$ are fuzzy subsets of $U_1, U_2, \ldots, U_r$, then the contribution of the $Rule_i$ is a fuzzy set whose membership function is computed by (6).

$$\mu_{Con^*i}(y) = (\alpha_{i1} \wedge \alpha_{i2} \wedge \ldots \wedge \alpha_{in}) \wedge \mu_{Con i}(y) \qquad (6)$$

where α_i is the matching degree of $Rule_i$ and α_{ij} is the matching degree between x_j and R_j's condition about x_j. This is computed by the (7).

$$\alpha_{ij} = \sup \left(\mu_{Ant^*j}(x_j) \wedge \mu_{Ant_{ij}}(x_j) \right) \forall x_j \qquad (7)$$

Table 1 Basic structure of if-then rules used in Mamdani model

Rule no.	If (antecedent)				Then (consequent)	Output of rule
	Variable x_1	Variable x_2	...	Variable x_r	Variable y	
R_1						
R_i	Fuzzy set Ant_{i1} (Output α_{i1}, Eqn7)	Fuzzy set Ant_{i2} (Output α_{i2}, Eqn7)	...	Fuzzy set Ant_{ir} (Output α_{ir}, Eqn7)	Fuzzy set Con_i	$\mu_{Con^*i}(y)$ (Eqn 6)
R_M						

Output of the Mamdani Model = Maximum of Last Column (8)

where '∧' denotes the min operator and 'sup' is supremum. The final output of the model is given by (8). It uses the clipping inference method [18].

$$\mu_{Con}(y) = \max\{\mu_{Con^*1}(y), \mu_{Con^*2}(y), \ldots, \mu_{Con^*M}(y)\} \quad (8)$$

2.4 Defuzzification

To convert fuzzy output of (8) to a crisp value, we apply centroid method of defuzzification [17]. Defuzzification is a process in which the fuzzy output is converted to a crisp output, i.e. non fuzzy output. There are seven distinct treatments. The defuzzification equation for one of the treatment, say RCT is given by (9).

$$RCT^* = \frac{\int \mu_{RCT}(x_i) \times x_i \, dx}{\int \mu_{RCT}(x_i) \, dx} \quad (9)$$

where RCT^* is the crisp output of the RCT fuzzy set, $\mu_{RCT}(x_i)$ is the membership value of x_i's in fuzzy set RCT (using (6)). x_i's are the values for four sign-symptoms {Tooth Mobility, Pain, Infection, Occlusal Trauma}.

3 Results and Discussion

The responses of the system and the dentist are collected. To verify the homogeneity of the two observations, Chi-square test of significance is applied. This is discussed below:

1. **Data:** See Table 2.
2. **Assumption:** It is assumed that we have a simple random sample from each one of the two populations of interest.
3. **Hypothesis:** H_0: The two populations are homogeneous with respect to treatments. H_A: The two populations are not homogeneous with respect to treatments. Let $\alpha = 0.05$.

Table 2 Aggregate of observations of Dentist's prediction and System's prediction

Observed data		Treatments							
		FPD	Sp	TE	Sc	OG	RCT	M	Total
Treatment suggested by	System	13	52	9	40	21	20	35	190
	Dentist	12	45	16	50	25	19	48	215
	Total	25	97	25	90	46	39	83	405

Table 3 Expected data of Dentist's prediction and System's prediction

Expected data		Treatments							
		FPD	Sp	TE	Sc	OG	RCT	M	Total
Treatment suggested by	System	11.72	45.50	11.72	42.22	21.58	18.29	38.93	190
	Dentist	13.27	51.49	13.27	47.77	24.41	20.70	44.06	215
	Total	25	97	25	90	46	39	83	405

4. **Test Statistics:** The test statistics is $\chi^2 = \Sigma \left[(O_i - E_i)^2 \big/ E_i\right]$. Data corresponding to expected values are shown in Table 3.
5. **Distribution of test statistics:** If H_0 is true, then data is homogeneous with $(7-1)^*(2-1) = 6$ degree of freedom.
6. **Decision Rule:** reject H_0 if the computed value of χ^2 is equal to or greater than 12.592 (Chi-square table value).
7. **Calculation of test statistics:** χ^2 value is 4.4998.
8. **Statistical decision:** Since 4.4998 is less than the critical value of 12.592, we are unable to reject the null hypothesis.
9. **Conclusion:** We conclude that the output produced by or predicted by the system is homogeneous to dentists' prediction.

An interesting observation: Dentists tend to provide excessive treatments mainly in the form of medication and extraction. This is not a new observation [19]. This problem can be checked with the system. We have noticed that a lot of research has been done in medical domain using the FL techniques or using combination of FL with other soft computing techniques. The immense potential of these techniques makes us believe that FL can prove to be very promising in dentistry too.

4 Conclusion

A FL based expert system is developed to assist dentist in clinical decision-making. It can be treated as an expert to the dentist during treatment of mobile tooth. Dentists often find it difficult to decide a treatment plan for the mobile tooth due to vagueness and imprecise values of the sign-symptoms. The availability of this system will enhance the confidence level of a dentist.

References

1. Khanna V, Karjodkar FR (2009) Decision Support Systems in Dental Decision Making: An Introduction. Journal of Evidence-Based Dental Practice 9(2):73–76
2. Mago VK, Prasad B, Bhatia A, Mago A (2008) A Decision Making System for the Treatment of Dental Caries. In: Bhanu Prasad (ed) Soft Computing Applications in Business. Studies in Fuzziness and Soft Computing, Vol 230. Springer, Germany
3. Thomas J, Zwemer (1993) Boucher's Clinical Dental Terminology: A Glossary of Accepted Terms in All Disciplines of Dentistry, 4th ed. Mosby, St. Louis, USA
4. Wikipedia.org (2009) http://en.wikipedia.org/wiki/Fuzzy_logic. Accessed October 2009
5. Ross TJ, Booker JM, Parkinson WJ (2002) Fuzzy Logic and Probability Applications: Bridging the Gap. Society for Industrial and Applied Mathematics, Philadelphia, and American Statistical Association, Alexandria, Virginia
6. Zadeh LA (1965) Fuzzy Sets. Information and Control 3:338–353
7. Fathi-Torbaghan M, Meyer D (1994) MEDUSA: A Fuzzy Expert System for Medical Diagnosis of Acute Abdominal Pain. Methods of Information in Medicine 33(5):522–529
8. Saritas I, Allahverdi N, Sert I U (2003) A Fuzzy Expert System Design for Diagnosis of Prostate Cancer. In: Rachev B, Smrikarov A (ed) Proceedings of the 4th International Conference Conference on Computer Systems and Technologies: E-Learning (Rousse, Bulgaria, June 19–20, 2003) ACM, New York
9. Allahverdi N, Torun S, Saritas I (2007) Design of a Fuzzy Expert System for Determination of Coronary Heart Disease Risk. In: Rachev B, Smrikarov A, Dimov D (ed) Proceedings of the 2007 International Conference on Computer Systems and Technologies (Bulgaria, June 14–15, 2007) ACM, New York
10. Watsuji T, Arita S, Shinohara S et al (1999) Medical application of fuzzy theory to the diagnostic system of tongue inspection in traditional Chinese medicine. In: IEEE International Fuzzy Systems Conference Proceedings (Seoul, South Korea, August 22–25, 1999) doi:10.1109/FUZZY.1999.793222
11. Kuo H-C, Chang H-K, Wang Y-Z (2004) Symbiotic Evolution-Based Design of Fuzzy-Neural Diagnostic System for Common Acute Abdominal Pain. Expert Systems with Applications 27(3):391–401
12. Wu M, Zhou C, Lin K (2007) An Intelligent TCM Diagnostic System Based on Intuitionistic Fuzzy Set. In: Proceedings of the Fourth international Conference on Fuzzy Systems and Knowledge Discovery. Doi: http://doi.ieeecomputersociety.org/10.1109/FSKD.2007.169
13. Schuh Ch, Hiesmayr M, Kaipel M et al (2004) Towards an intuitive expert system for weaning from artificial ventilation. In: Proceedings of IEEE Annual Meeting of the Fuzzy Information. doi: 10.1109/NAFIPS.2004.1337445
14. Zadeh LA (1988) Fuzzy Logic. Computer. doi:10.1109/2.53
15. Mathworks.com (2009) http://www.mathworks.com Accessed October 2009
16. Merer R, Nieuwland J, Zbinden, AM et al (1992) Fuzzy Logic Control of Blood Pressure During Anesthesia. IEEE Control Systems Magazine. 12(9):12–17
17. Bouchon-Meunier B (1995) Fuzzy Logic and Soft Computing. World Scientific Publishing Co, New Jersey
18. Dualibe C, Verleysen M, Jespers PGA (2003) Design of Analog Fuzzy Logic Controllers in Cmos Technologies. Kluwer Academic Publisher, The Netherlands
19. Brennian TA (1992) An Empirical Analysis of Accidents and Accident Law: The Case of Medical Malpractices Law. St. Louis University Law Journal 36:823–878
20. Yen J, Langari R (1999) Fuzzy Logic: Intelligence, Control, and Information. Prentice-Hall, New Jersey

Chapter 63
A Microcomputer FES System for Wrist Moving Control

Li Cao, Jin-Sheng Yang, Zhi-Long Geng, and Gang Cao

Abstract A portable close-loop control FES system, whose controller is a microcomputer, was proposed in this chapter. Considering the time-varying nonlinear of the muscle system, a self-adaptive PI control strategy was used. It included a neural network identifier (NNI) and a PI controller. NNI could get the variability of muscle working condition to identify muscle model online. Parameters of PI would be optimized by the results of NNI. Some tracking experiments had been done on able-bodied volunteers and the precision were under 4%.

Keywords Control · Wrist Moving · Neural Network · Functional Electrical Simulation (FES) · Muscles movement

1 Introduction

Functional electrical stimulation (FES) uses the low level current to stimulate patients' muscles and has been used to clinical treatments for many years [1]. Although the open-loop stimulations are still widely used in clinic, they are found having poor robustness properties and unsatisfactory for accurate movement control [2]. The main reason is that human muscle system is complications, nonlinear, and time varying [3, 4]. Recently, an important goal of FES research is to develop neuroprostheses that are useful for grasping, standing, and walking. Some of these functional activities require higher locomotion accuracy, resulting in potentially different ideal modeling strategies [5].

One of the recent attempts on FES were to develop clinical stimulators for a broad range of FES applications. Simcox et al. [6] developed an eight-channel portable transcutaneous stimulator for paraplegic muscle training. A separate PDA

L. Cao (✉)
College of Civil Aviation, Nanjing University of Aeronautics & Astronautics,
Nanjing 210016, China
e-mail: caoli@nuaa.edu.cn

provided the control strategy. Control and feedback signals could communicate with the stimulator. As we knew, these PDA- or PC-stimulator structure devices were really inconvenient for either doctors or patients, especially in maneuverability. Hart et al. [7] used a stimulator of novel design to explore the value of including doublet pulses in stimulation trains used for the correction of drop foot. The stimulator adopted PIC16F84 as the microcontroller. The stimulator took the several factors, such as memory space, portability, program complexity and reliability, into account.

Considered neuromuscular-skeletal systems existing the muscle fatigue, inherent time variance, time delay, and strong nonlinearities [2], nonlinear adaptive control strategies were the most suitable candidates to tackle the FES control problem [8]. Ferrarin et al. [9] presented a mixed strategy based on a closed-loop PI controller and an open-loop nonlinear inverse model of the plant. Tong et al. [10] used a three-layer structure ANN to clone the paraplegic expertise to walk with hand-switches. Kostov et al. [11] used Adaptive Logic Networks for real-time control of walking subjects with incomplete spinal cord injury that limited their ability to walk. The learning of rules was further improved by adding the history of the sensory data as suggested by Lan et al. [12]. Sepulveda et al. [13] developed a version of ANN by joint angles, and vertical ground reaction forces as input, and electrical activity of leg muscles as output. Another popular control is model reference adaptive control [14, 15]. It has a good performance of stability and robustness, but nonlinear recruitment characteristic is a serious problem [14].

The aim of this chapter is to develop a portable FES system, which can be used to train the paralyzed upper limbs easily. The main purpose is to control the arm movements under the given patterns. In order to make our FES system more robust and fitting to different situations, a potent adaptive control method has to be developed. It contains an adaptive PI (A-PI) controller and a neural network identifier (NNI). A-PI controller is used as an output-tuning unit, which parameters could optimize by the result of system output and NNI. NNI will be used as a system-apperceiving unit, which could feel the varieties of the control system. In addition, NNI is the online control model and is corrected by the system output.

2 Four-Channel Microcomputer Portable FES System

Figure 1 showed our four-channel portable FES system. It contained an adaptive controller, voltage converter, stimulation channel selector, MOSFET, HV power supply, reference stimulation parameter, and power supply unit. A MOSFET and a HV power supply unit were made up of a stimulation channel. The controller, we used here, was C8051F005. It is a 25MIPS with 32 kB of programmable flash memory, 2 kB of RAM, and two eight-channel on-board 12-bit ADC single chip processor. The power supply unit, which is a 9V battery, is used to supply the system power. Reference stimulation parameter was used to produce reference stimulation signal.

Stimulation channels produce the stimulation signal to drive the muscle contracting by surface electrodes. Once angle sensors detected the wrist movement, the

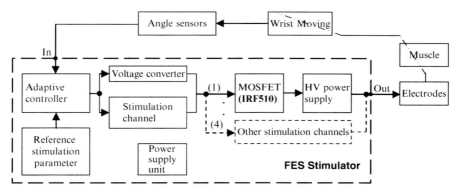

Fig. 1 Four-channel portable FES system

moving angle could be feeded back to the adaptive controller though it onboard ADC converters. The adaptive controller can modify the outputs according to reference parameters and feedback.

3 Adaptive PI Control by Neural Network Identifier

Figure 2 showed the control strategy we used here. The system is error feedback control and composed of two parts: PI controller and NNI. The input of the controller is desired joint angle (θ), variable (u) is the stimulation energy, and the system output (y) is joint angle of the stimulation wrist. $\hat{y}$ is the NNI output which is used to get the variable u. The PI controller parameters will be optimized online according to the error of $\hat{y}$ and θ to make the control strategy fit for the FES system-working environment.

Because the muscle system is a time-variance, time-delay, strong nonlinear system and every patient have different initialized state, NNI has two main functions here: to apperceive the characteristics of the different patients at the FES system initialized phase and to apperceive the changes of the muscle model during training online. To get higher convergence speed of NNI, a feedback three-layer NN structure with ten interim nodes was used by optimizing before. The error back-propagation algorithm, we used, was proposed in [16].

The error feedback control composed of two parts: PI controller and the training algorithm. The training algorithm could adjust the parameters of PI online by $\hat{y}$ and θ. It would be implemented as the parameter-adjusting algorithm in the model reference adaptive control. The inputs of the PI controller are feedback error $c_1(t)$ and total error $c_2(t)$. $c_1(t)$ is the system error at time t:

$$c_1(t) = \theta(t) - \hat{y}(t). \qquad (1)$$

And $c_2(t)$ is the total error:

$$c_2(t) = \Sigma c_1(t). \qquad (2)$$

Fig. 2 The control method of our portable system

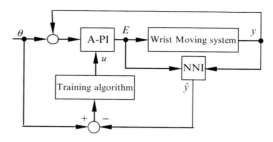

NNI identifies and feels the wrist musculoskeletal system continuously and guides the PI controller adjusting its parameters. The adaptive PI controller was organized as follows:

$$u(t) = k_1 c_1(t) - k_2 c_2(t), \qquad (3)$$

where k_1 and k_2 are control parameters. They can be corrected online according to NNI.

4 Wrist Tracking Experiment on Portable FES Device

In this experiment, our intention was to let the volunteer's right wrist tracking the desired trajectory produced by her left wrist moved at random (shown as Fig. 3). This will implement the paralyzed people self-training. The training time was 36 s to evaluate the performance of the control precision, robust, and response speed. The total error of NNI offline training was also limited within 0.15 during the initial phase and the iteration was within 150 times.

Figure 4 showed the training results by two situations: Adaptive-PI control (A-PI) and Fixed Parameter-PI (FP-PI) control. Figure 4a was the tracking result of FP-PI control. At the first 16 s, FP-PI control could track the desired trajectory well, but it could not get satisfactory tracking result during the left time. It showed that the tracking error would not be improved in control process although the control precision could be also accepted at the beginning if the control parameter being proper. Once the FES working, the muscle could feel it and the muscle conditions would change right away. Because P and I were fixed, the controller did catch up with this changing and made the tracking performance getting worse. Figure 4b showed its error (the total error of this time was 8.7%). The tracking error was increased gradually, and that showed FP-PI could adapt to the working condition of the muscle changed.

Figure 4c was the tracking result of A-PI method. To make these two experiments comparing more distinctly, we recorded the desired trajectory in FP-PI experiment and did A-PI control experiment in the next day at the same extern environment. We could see that A-PI could get better tracking result than FP-PI. For the first iteration, small output energy and large delay in response were observed. And as the iteration

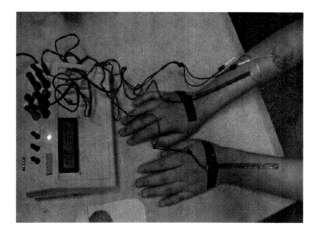

Fig. 3 FES experiment diagram

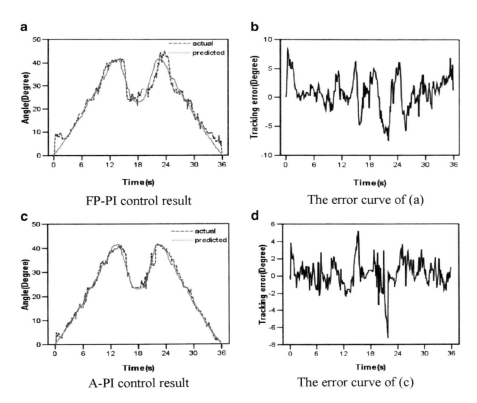

FP-PI control result   The error curve of (a)

A-PI control result   The error curve of (c)

Fig. 4 The performance of different PID controllers in our portable system

increases, the tracking error decreased gradually. It was evident that the tracking error became small when the NNI provides an initial model to PI controller. Furthermore, Fig. 4d showed that the tracking error decreased in whole control process. The results demonstrated that the control performance improved because our control strategy included an NNI. It could apperceive the real muscle system changes and revise the control parameters. The total control error of Fig. 4c was 3.6%.

5 Conclusion

A portable FES device had been developed in this chapter. Considering the system resource and the future demands of clinic training, a proper control strategy was proposed to realize the self-adaptive feedback control. Combining with the NNI and PI methods, the proposed control strategy has not only strong adaptive capability but also small computational cost. Some wrist tracking experiments on the ablebodied subject was done, and the results showed that the adaptive PI controller could perform the tests better than fixed PI controller. It has the good ability of decreasing average tracking error and response delay.

References

1. Rushton D.N. (2003) Functional electrical stimulation and rehabilitation—an hypothesis, Med. Eng. Phys., 25(1), 75–78
2. Jezernik S., Wassink R.G. and Keller T. (2004) Sliding mode closed-loop control of FES: controlling the shank movement, IEEE Trans. Biomed. Eng., 51(2), 263–272
3. Bobet J. and Stein R.B. (1998) A simple model of force generation by skeletal muscle during dynamic isometric contractions, IEEE Trans. Biomed. Eng., 45(8), 1010–1016
4. Frey Law L.A. and Shields R.K. (2006) Predicting human chronically paralyzed muscle force: a comparison of three mathematical models, J. Appl. Physiol., 100(3), 1027–1036
5. Rakos M., Freudenshu B., Girsch W., et al. (1999) Electromyogram-controlled functional electrical stimulation for treatment of the paralyzed upper extremity, Int. J. Artif. Organs, 23(5), 466–469
6. Simcox S., Davis G., Barriskill A., et al. (2004) A portable, 8-channel transcutaneous stimulator for paraplegic muscle training and mobility – A technical note. J. Rehabil. Res. Dev., 41(1), 41–52
7. Hart D.J., Taylor P.N., Chappell P.H., et al. (2006) A microcontroller system for investigating the catch effect: functional electrical stimulation of the common peroneal nerve, Med. Eng. Phys., 28(5), 438–448
8. Previdi F. and Carpanzano E. (2003) Design of a gain scheduling controller for knee-joint angle control by using functional electrical stimulation, IEEE Trans. Control Syst. Technol., 11(3), 310–324
9. Ferrarin M., Pavan E., Spadone R., et al. (2002) Standing-up exerciser based on functional electrical stimulation and body weight relief, Med. Biol. Eng. Comput., 40(3), 282–289
10. Tong K.Y. and Granat M.H. (1999) Gait control system for functional electrical stimulation using neural networks, Med. Biol. Eng. Comput., 37(1), 35–41
11. Kostov A., Andrews B.J., Popovic D.B., et al. (1995) Machine learning in control of functional electrical stimulation systems for locomotion, IEEE Trans. Biomed. Eng., 42(6), 541–551

12. Lan N., Feng H. and Crago P.E. (1994) Neural network generation of muscle stimulation patterns for control of arm movements, IEEE Trans. Rehabil. Eng., 2(4), 213–224
13. Sepulveda F., Granat M.H. and Cliquet A. Jr. (1998) Gait restoration in a spinal cord injured subject via neuromuscular electrical stimulation controlled by an artificial neural network, Int. J. Artif. Organs, 21(1), 49–62
14. Hatwell M.S., Oderkerk B.J., Sacher C.A., et al. (1991) The development of a model reference adaptive controller to control the knee joint of paraplegics, IEEE Trans. Automat. Contr., 36(6), 683–691
15. Bernotas L.A., Crago P.E. and Chizeck H.J. (1986) A discrete-time model of electrically stimulated muscle, IEEE Trans. Biomed. Eng., 33(9), 829–838
16. Rumelhart D.E., Hinton G.E. and Williams R.J. (1986) Learning representations by back-propagating errors, Nature, 323(9), 533–536

Chapter 64
Computer-Aided Decision System for the Clubfeet Deformities

Tien Tuan Dao, Frédéric Marin, Henri Bensahel, and Marie Christine Ho Ba Tho

Abstract A computer-aided decision system (CADS) based on ontology in pediatric orthopedics was developed to assess, without assumptions performed, the abnormalities of the musculoskeletal system of lower limbs. The CADS consists of four components. The first component is a diagnosis-based ontology, called Ontologie du Système Musculosquelettique des Membres Inférieurs (OSMMI). The second component is a database for collecting clinical observations, e.g., the birth classification of the clubfeet deformities. The third component uses statistical methods (principal component analysis and decision tree) for constructing an approach to evaluate new issues. The last component is an interactive module for managing the interaction between patients, experts, and the due CADS. Our system has been validated clinically with the real patient data obtained from the Infant Surgery Service in the hospital of Robert Debré in Paris. Our CADS is a good solution to compare the studies of the clubfeet deformities before and after the treatment using a universally scoring system. The assessment, conservative treatment, and monitoring were set up. Our system was developed to allow a better assessment for improving the knowledge and thus the evaluation and treatment of the musculoskeletal pathologies, e.g., the clubfeet deformities.

Keywords Computer-based medical systems · Biological data mining and knowledge discovery · Bio-ontologies + semantics · Medical informatics

1 Introduction

Clubfoot is one of the most common congenital orthopedic abnormalities for children [27]. The congenital deformities can be appeared with unilateral or bilateral abnormalities. The main strategy of treatment is the conservative treatment with

T.T. Dao (✉)
UTC - CNRS UMR 6600 Biomécanique et Bioingénierie, Compiègne, France
e-mail: tien-tuan.dao@utc.fr

different approaches such as the Ponseti technique [21], Ilizarov technique [15], or the French technique [3, 10, 25]. Each approach presents its robustness and also its disadvantages. Moreover, the comparison between different series is hardly realized due to the lack of common concepts. In order to allow the same language for pediatric orthopedists who are involved in clubfoot, an International ClubFoot Study Group (ICFSG) has been founded. A nomenclature and a rating system [4] were proposed for facilitating the comparison of the results of the various series of clubfoot. Furthermore, the better understanding of clubfoot deformities improves the assessment, the treatment, and the long-term follow-up of clubfoot. But the question is how to broadcast and share the knowledge skill in the medical and clinical research communities.

The computer-aided decision was used to model and understand the pathology. The expert systems, the first generation of the computer-aided decision system (CADS), have been studied since 1976 with Mycin as one of the older experts systems [24]. After that, many applications were developed in the medical diagnostic field [5, 12]. But these systems were not successful because they were black boxes. The diagnostic was done without the explanation about the pathological causes. Another ideal weakness of these experts systems is the interaction between patients (the users) and experts (surgeon, physiotherapist, etc.) is poor and not friendly.

The new generation of the expert system appeared with different terminologies such as computer-aided diagnosis system and CADS. These systems were developed to allow an interactive interface between the system and the user. In the literature, there are many studies concerning these approaches, but most of these systems were dedicated for one function in the diagnosis process, for example: mammography [6], imaging treatment [20], vocal cord diseases [30], otology [17]. Other limitations of these studies are that they did not take the aspect of post-diagnosis process as the treatment and follow-up into account. Nowadays, the web-based expert system was widely developed [19]. The software development technologies such as J2EE (Java 2 Platform, Enterprise Edition) and the expert system shell have many applications in real context [29]. These elements allowed the solutions to develop an online decision system managing the assessment, treatment, and follow-up of the pathologies.

The aim of this study was to develop a CADS based on the ontology in pediatric orthopedics to assess the abnormalities of the musculoskeletal system of the lower limbs. Our system was developed to allow a better assessment for improving the knowledge and thus the evaluation and treatment of the musculoskeletal pathologies, e.g., the clubfeet deformities.

2 Methods

The CADS allowed an interactive use of accumulated knowledge to reason, diagnose, and give appropriate decisions. Its operational scenario is illustrated in Fig. 1. Clinical experts will provide patient data input. The diagnosis algorithms are applied

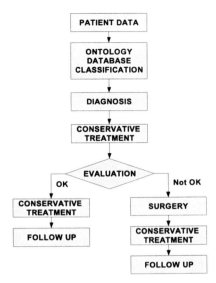

Fig. 1 Functional scenario of our CADS

to make a diagnosis decision. Then the conservative treatment is generated. After that, the evaluation is done to decide if the conservative treatment will be continued or the surgical decision is made. If the surgical decision is made, the conservative treatment will be done to maintain and improve the surgical quality. Finally, the follow-up is made to survey the patient until they cure their pathologies.

The CADS is resulting from a combination of different components (Fig. 2): a diagnostic-based ontology, named Ontologie du Systme Musculosquelettique des Membres Infrieurs (OSMMI); a database to collect the pathological data and a statistical classification. The interactive module is constituted of three parts consecutively: diagnosis, conservative treatment, and follow-up.

2.1 OSMMI Ontology

First, the conceptual structure of the OSMMI has to be established to have a general sight of ontology. Second, we start to create the OSMMI using the platform Protégé 2000.[1] The Ontology Web Language (OWL) was used as ontology format. This facilitates the mapping instance process into OSMMI ontology. Based on built ontology, the last component is the construction of the part of reasoning. The knowledge extraction process was developed by the following steps: (1) the enumeration of the important terms; (2) the definition of the classes and their hierarchy; (3) the

---

[1] www.protege.stanford.edu.

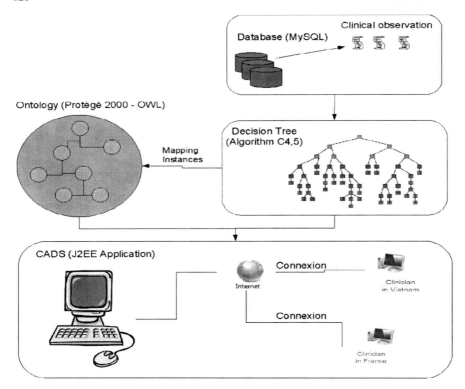

Fig. 2 Architecture of our CADS

definition of the properties of the classes – attributes; (4) the definition of the facets of the attributes: cardinality, types of value, domain, etc. and finally the creation of the instances.

2.2 Statistical Classification

First, the principal component analysis (PCA) was applied to study the inter-relationship and intra-relationship between all parameters of the clubfoot deformities. Second, a supervised clustering method was performed to classify three grades of the clubfoot deformities (moderate, severe, rigid). In this approach, each pathology corresponds a certain class known a priori and its symptoms, measurements, and parameters are considered as descriptive variables. Decision tree method was selected to generate the reasoning scheme of the clubfoot pathologies. The algorithm named C4.5 [2, 22] was used to generate the clubfoot decision tree. The C4.5 generates the rule set using C4.5 algorithm with strong generalization ability and strong comprehensibility [14, 18]. The rules generated are in conjunctive form such as if A and B then C where both A and B are the rule antecedents, C is the rule consequence. C4.5 uses a divide-and-conquer approach for growing decision tree. It is given below:

C4.5 Algorithm
Input: A learning set $S = \{s_1, \ldots, s_l\}$, a learning parameters set $\{x_i, i = 1, \ldots, n\}$, and a classifying set $C = \{w_1, \ldots, w_c\}$.
Output: A decision tree t.
Begin

(I) Check for all elements in the learning set S
(II) For each learning parameter x_i do:
   (II.1) Find the normalized information gain from splitting on x_i
   (II.2) Let x_{best} be the learning parameter with the highest normalized information gain
   (II.3) Create a decision node *node* that splits on x_{best} and recurse on the sublists obtained by splitting on x_{best} and add those nodes as children of *node*
End For;

End.

The information entropy and the information gain [22] were used to construct the decision tree above. Based on the rule base generated from the C4.5 algorithm, a universally online scoring algorithm of clubfeet deformities was developed as below:

Online Scoring Algorithm of Clubfeet Deformities
Input: The facts base of the clinical observations of the clubfoot deformities *FBC*. The rule-based knowledge base of the clubfoot deformities *RKBC*
Output: A grade of the clubfoot deformities G.
Begin

(0) Define the necessary function add for summarizing the total score
(1) Load the rule-based knowledge base *RKBC* into the inference system JESS (Java Expert System Shell)
(2) Enter the facts base *FBC* via the web-based interface
(3) Execute the RETE algorithm (Forgy 1982) to fire the corresponding rules and facts
(4) Search, store, and return the grade G of the clubfoot deformities
(5) STOP

End.

2.3 Experimental Data

A dataset of 1,000 (700 for the learning phase and 300 for the testing phase) clubfeet cases (105 moderate, 868 severe, 27 rigid) was used for constructing the pathological decision tree [3, 10, 25]. All informative parameters of clubfoot dataset are described in Table 1.

Table 1 Clubfeet deformities dataset: name, short description of all parameters

Name	Short description
Equinus deformity (eq)	Angle of the equinus in the sagittal plane
Equinus flexibility (eqf)	Level of the flexibility (stiffness, flexible, reducible) of the equinus angle
Varus deformity (vr)	Angle of the varus deviation in the frontal plane
Varus flexibility (vrf)	Level of the flexibility (stiffness, flexible, reducible) of the varus angle
Midfoot supination deformity (su)	Angle of the combination of the midtarsal foot supination and the medial rotation of the calcaneo-forefoot unit
Midfoot supination flexibility (suf)	Level of the flexibility (stiffness, flexible, reducible) of the midfoot supination angle
Calcaneo-midfoot unit deformity (de)	Angle of the derotation around the talus of the calcaneo-forefoot block
Calcaneo-midfoot unit flexibility (def)	Level of the flexibility (stiffness, flexible, reducible) of the calcaneo-midfoot unit angle
Cavus deformity (cv)	State of the cavus
Ankle dorsiflexors (adf)	State of muscles function in ankle dorsiflexors
Ankle plantar flexors (apf)	State of muscles function in ankle plantar flexors
Invertors (inv)	State of muscles function in foot invertors
Evertors (eve)	State of muscles function in foot evertors
Toe extensors (te)	State of muscles function in toe extensors
Toe flexors (tf)	State of muscles function in toe flexors

3 Computational Results

The architecture of our ontology includes 14 functional and anatomical structures of the musculoskeletal system of the lower limbs which are defined like classes of ontology: nervous system, ligament, muscle, tendon, cartilage, bone, limb, posture, support of load, diarthrosis joint, movement, articular contact, contact of environment and gait. The general architecture of the OSMMI ontology is illustrated in Fig. 3.

For the physiological and functional semantics of our ontology, what is considered as implicit on gait is defined by the following relations:

- Inform: the ligaments inform the nervous system if there are active signals
- Command: the nervous system command the muscles
- Attach: the muscles are attached to the bones through the tendons; the cartilages and the ligaments are attached to the bones
- Compose: the limb is composed by the different bones which correspond to a particular function in gait
- Act: the muscles, the support of load, the movement, and the ligaments act on the diarthrosis joint; the cartilage act the articular contact
- Influence: the contact of environment, the movement influence the gait
- Form: the limbs form the correspondent posture
- Support: the posture is supported by the support of load

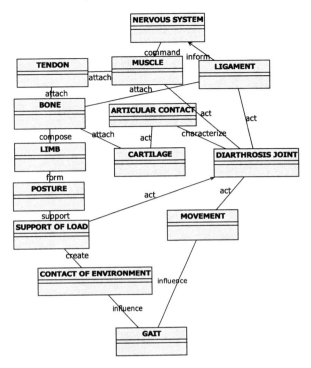

Fig. 3 General architecture of the OSMMI ontology

- Create: the support of load is created by the contact of environment
- Characterize: the articular contact characterize the diarthrosis joint

The PCA analysis was done and the distribution of all parameters of the clubfeet deformities is illustrated in Fig. 4. Based on the eigenvalues, all components showed the equal variance of the clubfeet parameters (7.6, 7.55, 7.34, 7.23, 7.03, 6.8, 6.73, 6.63, 6.58, 6.48, 6.31, 6.16, 6.02, 5.87, 5.69%, respectively).

The clubfoot decision tree of the clubfeet deformities was developed (Fig. 5). Each internal node tests a clubfoot parameter. Each branch corresponds to clubfoot parameter value. Each leaf node assigns a grade of the clubfeet deformities (moderate, severe, and rigid). The branch connecting the origin with a leaf node is a rule generated in conjunctive form. For example, the most right branch in the clubfoot decision tree demonstrates the following clauses (cavus = Yes and equinus deformality $\succ 0$ and varus flexibility $\succ 1$ and Calcaneo-MidFoot Unit deformity $\succ 0$ and Ankle Dorsiflexors = Non-Reactive and Evertors = Non-Reactive), then the grade of the clubfoot is rigid.

The rule set of the clubfoot deformities is generated from the pathological decision tree and based on the new clubfoot pretreatment evaluation form of the ICFSG. A total point of 20 is calculated [Grade I (moderate: 1–6), Grade II (severe: 7–14), Grade III (rigid: 15–20)]. An example of the equinus deformity is illustrated in

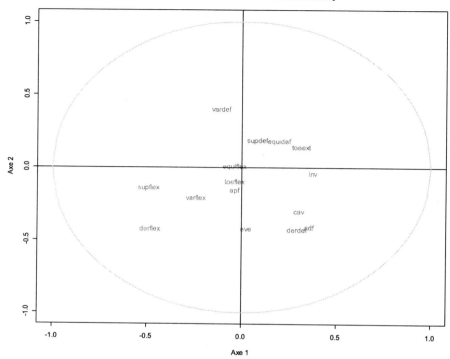

Fig. 4 2D plot parameter distribution using PCA

Table 2. In the rule-based knowledge of the clubfoot deformities, 50 rules are deduced and 24 facts are used for classifying the grade of the clubfoot deformities (moderate, severe, rigid).

The web-based interface of the scoring system is illustrated in Fig. 6.

4 Discussion

The originality of the present study is to propose a new modeling approach to integrate all components for comparing clubfeet deformities before and after treatment. A CADS was developed for better evaluation and then for treating clubfeet deformities. The individual development of each patient was taken into account to improve the treatment quality of his or her physical therapy or surgery.

The linear transformation of PCA showed that all parameters of the clubfeet deformities pathology are correlated and inter-dependant. This multidimensional statistical analysis permitted to preserve all parameters for developing clubfoot decision tree.

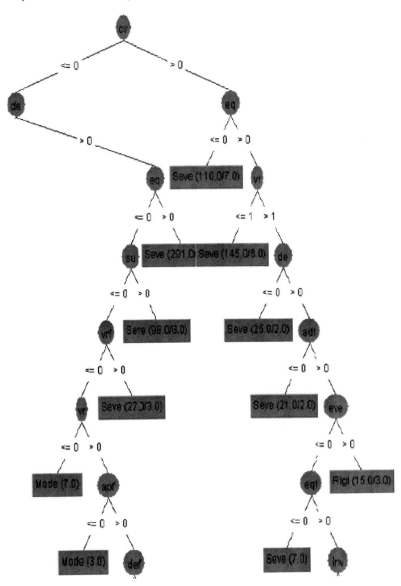

Fig. 5 Partial decision tree of the clubfoot deformities

The use of the diagnostic-based ontology for formalizing the rule-based knowledge is a new approach helping to improve the diagnosis decision. Therefore, our system was developed to broadcast and share the knowledge skill in the medical and clinical research communities. Approach of the connection of knowledge is necessary in all scientific fields, but diversity of representation of knowledge is an obstacle

Table 2 Rule-based knowledge of the clubfoot deformities: example of the equinus deformity

Function	Definition
R1	[defrule r1 (Deviation Equinus Sagittal-plane 90-45)] ⇒ [assert (Point-1 4)]
R2	[defrule r2 (Deviation Equinus Sagittal-plane 45-20)] ⇒ [assert (Point-1 3)]
R3	[defrule r3 (Deviation Equinus Sagittal-plane 20-0)] ⇒ [assert (Point-1 2)]
R4	[defrule r4 (Deviation Equinus Sagittal-plane 0-N20)] ⇒ [assert (Point-1 1)]
R5	[defrule r5 (Deviation Equinus Sagittal-plane N20)] ⇒ [assert (Point-1 0)]

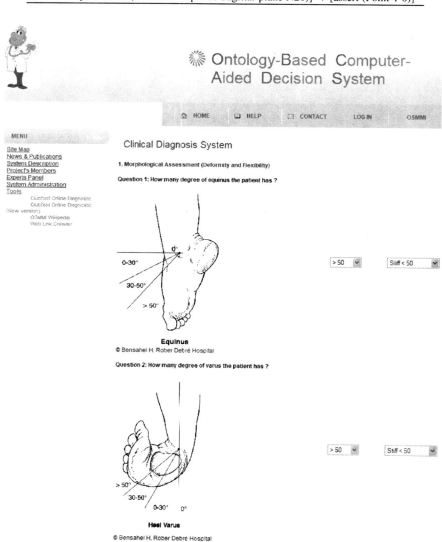

Fig. 6 Web-based interface of our CADS

Table 3 Rating score of real patients: clinician vs. our CADS

Subject	Clinician	Our CADS
Subject 1	13 (severe)	13 (severe)
Subject 2	17 (rigid)	17 (rigid)
Subject 3	20 (rigid)	20 (rigid)
Subject 4	16 (rigid)	16 (rigid)
Subject 5	13 (severe)	13 (severe)
Subject 6	13 (severe)	13 (severe)
Subject 7	8 (severe)	8 (severe)
Subject 8	8 (severe)	8 (severe)

to make shared knowledge possible. Ontology appeared as a good solution because it gives possibility to share the common comprehension of the structure of information between the researchers [26, 28]. Then it allows also the reuse of the knowledge in various systems. This approach is also applied to build the knowledge-based system using the accumulated knowledge to reason, diagnose, or give adequate decisions [7]. In the literature review, biomedical ontology is recently developed on breast cancer pathology [1], protein data models [23], and gene ontology [13]. But according to our knowledge ontology concerning biomechanics does not exist, we have initiated it recently for the musculoskeletal system [8].

In the literature, there are many approaches to construct the predictive medical model such as the artificial neural network [16], the support vector machine [14], and the decision tree. The comparison between these methods [11] was done in the previous study [9], and the decision tree method shows its robustness and its strong generalization ability. The same process can be applied for other pathologies such as Legg-Calvé-Perthes or cerebral palsy to construct their pathological decision tree. Furthermore, the CADS is an extensible tool, and other pathologies of the upper limbs can be easily added.

The conservative treatment and follow-up of the clubfeet deformities are set up via a user-friendly interface. The visualization of the conservative treatment and follow-up was well-defined throughout a step-by-step process. The technique and data used are guaranteed by the physiotherapists working over years at the hospital of Robert Debré in Paris, France [25]. Our system has been validated currently with the real patient data obtained from the Infant Surgery Service of this hospital. Some cases of the validation report are illustrated in Table 3. The feedback is very promising in terms of easy use and the accuracy of the clubfoot classification performed by our system as the classification performed by a clinician.

5 Conclusions and Perspectives

A CADS based on the ontology was developed. A universally scoring system and a remote treatment program were integrated. Different components and a multi-layers architecture were also addressed. The application of the clubfeet deformities was reported as a first case study. Rule-based knowledge of clubfoot deformities was deduced. The assessment, conservative treatment, and follow-up are set up. The remote access into our system is guaranteed through a dynamic web-based interface. Our system is developed to allow a better assessment for improving the knowledge and thus the evaluation and treatment of the pediatric orthopedics pathologies.

As perspective, this CADS could be extended to other children deformities such as Legg-Calvé-Perthes or cerebral palsy.

References

1. Abidi, S.R., Abidi, S.S., Hussain, S., Shepherd, M.: Ontology-based modeling of clinical practice guidelines: a clinical decision support system for breast cancer follow-up interventions at primary care settings. Studies in Health Technology and Informatics, **129**, 845–849 (2007).
2. Aitkenhead, M.J.: A co-evolving decision tree classification method. Expert Systems with Application, **34**, 18–25 (2008).
3. Bensahel, H., Dimeglio, A., Souchet, P.: Final evaluation of clubfoot. Journal of Pediatric Orthopaedics, Part B, **4**, 137–141 (1995).
4. Bensahel, H., Kuo, K., Duhaime, M., the International ClubFoot Study Group: Outcome evaluation of the treatment of clubfoot: the international language of clubfoot. Journal of Pediatric Orthopaedics, Part B, **12**, 269–271 (2003).
5. Botti, G., Fieschi, M., Fieschi, D., Joubert, M.: An expert system for computer-aided decision in diabetes therapeutic. Pathologie-Biologie, **2**, 101–106 (1985).
6. Christiane, M., Malich, A., Facius, M., Grebenstein, U., Sauner, D., Pfleiderer, S.O.R., Kaiser, W.A.: Are unnecessary follow-up procedures induced by computer-aided diagnosis (CAD) in mammography? Comparison of mammographic diagnosis with and without use of CAD. European Journal of Radiology, **51**, 66–72 (2004).
7. Daniel, L.R., Olivier, D., Yasser, B., David, G., Parvati, D., Mark, A.M.: Using ontologies linked with geometric models to reason about penetrating injuries. Artificial Intelligence in Medicine, **37**, 167–176 (2006).
8. Dao, T.T., Marin, F., Ho Ba Tho, M.C.: Ontology of the musculo-skeletal system of the lower limbs, 29th Annual International Conference of the IEEE EMBS, 386–389 (2007).
9. Dao, T.T., Marin, F., Ho Ba Tho, M.C.: Predictive mathematical models based on data mining methods of the pathologies of the lower limbs. 4th European Conference of the International Federation for Medical and Biological Engineering Antwerp, Belgium, November, 1803–1807 (2008).
10. Dimglio, A., Bensahel, H., Souchet, Ph., Mazeau, P.H., Bonnet, F.: Classification of clubfoot. Journal of Pediatric Orthopaedics, Part B, **4**, 129–136 (1995).
11. Dreiseitl, S., Ohno-Machado, L., Kittler, H., Vinterbo, S., Billhardt, H., Binder, M.: A comparison of machine learning methods for the diagnosis of pigmented skin lesions. Journal of Biomedical Informatics, **34**, 28–36 (2001).
12. Fieschi, M.: Sphinx : un systme expert d'aide la dcision en mdecine. Thèse Biologie Humaine, Fac. Mdecine Marseille (1983).
13. Hawkins, T., Chitale, M., Luban, S., Kihara, D.: PFP: Automated prediction of gene ontology functional annotations with confidence scores using protein sequence data. Proteins, **74**, 566–582 (2008).
14. He, J., Hu, H.J., Harrison, R., Tai, P.C., Pan, Y.: Transmembrane segments prediction and understanding using support vector machine and decision tree. Expert Systems with Applications, **30**, 64–72 (2006).
15. Jeffrey K.B., Raymond, S.: Correction of severe residual clubfoot deformity in adolescents with the Ilizarov technique. External Fixation Techniques for the Foot and Ankle, 571–582 (2004).

16. Kononenko I.: Machine learning for medical diagnosis: History, state of the art and perspective. Artificial Intelligence Medecine, **23**, 89–109 (2001).
17. Leigh, S.G., Robert, H.E., Marcus, D.A.: Clinical decision support systems and computer-aided diagnosis in otology. Otolaryngology Head and Neck Surgery, **136**, 21–26 (2007).
18. Lorena, A.C., Carvalho, A.C.P.L.F.: Protein cellular localization prediction with support vector machines and decision trees. Computer in Biology and Medecine, **37**, 115–125 (2007).
19. Moynihan, G.P., Fonseca, D.J.: PTDA2: A web-based expert system for power transmission design. Progress in Expert Systems Research, 1–24 (2007).
20. Pietka, E., Gertych, A., Witko, K.: Informatics infrastructure of CAD system. Computerized Medical Imaging and Graphics, **29**, 157–169 (2005).
21. Ponseti, I.V.: Congenital clubfoot. Fundamentals of treatment. Oxford: Oxford University Press (1996).
22. Quinlan, J.R.: C4.5 Programs for machine learning. San Francisco: Morgan-Kaufmann Publishers (1993).
23. Ratsch, E., Schultz, J., Saric, J., Lavin, P.C., Wittig, U., Reyle, U., Rojas, I.: Developing a protein-interactions ontology. Comparative and Functional Genomics, **4**, 85–89 (2003).
24. Shortliffe, E.H.: Computer based medical consultations: Mycin. New York: Elsevier (1976).
25. Souchet, P., Bensahel, H., Christine, T.N., Pennecot, G., Csukonyi, Z.: Functional treatment of clubfoot: a new series of 350 idiopathic clubfeet with long-term follow-up. Journal of Pediatric Orthopaedics, Part B, **13**, 189–196 (2004).
26. Stefan, S., Udo, H.: Towards the ontological foundations of symbolic biological theories. Artificial Intelligence in Medicine, **39**, 237–250 (2007).
27. Stephen, J.C., Birender, B., Cronan, C.K., Nigel, T.K.: Clubfoot. Current Orthopaedics, **22**, 139–149 (2008).
28. Thomas, B., Maureen, D.: Logical properties of foundational relations in bioontologies. Artificial Intelligence in Medicine, **39**, 197–216 (2007).
29. Tomic, B., Devedzic, V., Jovanovic, J., Andric, A.: Expert Systems Revisited: A practical approach. Progress in Expert Systems Research, 119–152 (2007).
30. Verikas, A., Gelzinis, A., Bacauskiene, M., Uloza, V.: Toward a computer-aided diagnosis system for vocal cord diseases. Artificial Intelligence in Medicine, **36**, 71–84 (2006).

Chapter 65
A Framework for Specifying Safe Behavior of the CIIP Medical System

Seyed Morteza Babamir

Abstract Adequate reliability of algorithms and computations of modern medical systems software is a matter of concern because the system software is in charge of satisfying *safety requirements* of the system environment, i.e., the patient. This chapter aims to present a framework for specifying the behavior of the Continuous Infusion Insulin Pump (CIIP) safety-critical medical system that satisfies diabetic's safety requirements.

1 Introduction

The diabetes medical system called the Continuous Infusion Insulin Pump (CIIP) system is a safety-critical one being responsible for the diabetic's safety. The system is used to take control of diabetics suffering from "type 1 diabetes," called youthful sugar. The disease emerges in all ages, even though children, the young people, and ages <30 are usually afflicted with the disease. The CIIP system intended to be continuously worn by the diabetic administers regular doses of insulin based on regular sampling of the wearer's blood sugar level. If the diabetes control system fails to satisfy requirement "normal blood sugar," the diabetic may be afflicted with cerebral, eye, heart, or kidney diseases. Requirement, "Diabetic's blood sugar should be normal" will be *violated* if the blood sugar is *low* or *high*. The violation of safety requirements may be verified by a monitor program in the CIIP system behavior; accordingly, we define *safety* as no violation. If we consider a violation as a bad event, safety will mean no bad event happens, as stated in [1]. Based on definition of safety, we define a *safety requirement* as the absence of the requirement violation. Therefore, for the CIIP system requirements, "Diabetic's blood sugar should not be *low*" and "Diabetic's blood sugar should not be *high*" are two safety requirements where *low* and *high* blood sugar levels are violations of the diabetic's requirement, "Diabetic's blood sugar should be normal."

S.M. Babamir (✉)
University of Kashan, Kashan, Iran
e-mail: babamir@kashanu.ac.ir

This chapter is meant to present a framework including three parts for specifying safe behavior of the CIIP system. The first part informally addresses *eliciting* patient's safety requirements stated in the physician natural language. The second part deals with *specifying* the system behavior through formal modeling interactions between the system and the diabetic. Finally, considering the first and the second part, the third part is intended for *specifying* the CIIP system behavior that should violate no diabetic's safety requirements.

2 Eliciting Informal Safety Requirements

The purpose of this section is the informal elicitation of the diabetic patient's safety requirements. According to the physician, blood sugar value varies from 1 to 20 where values: (1) below *safe-min* are *low*, (2) between *safe-min* and *safe-max* are *normal*, and (3) over *safe-max* are *high*. When the CIIP system is controlling the diabetic's blood sugar, the physician expects, (a) "Blood sugar never falls below *safe-min*," (b) "If blood sugar is above *safe-max*, the system have to deliver insulin," and (c) "The delivered insulin dose should be adequate, namely up to *max-single-dose* for one single dose and up to *max-daily-dose* for the daily dose." Considering the physician's expectations, safety requirements are as follows:

(SR_1) "Blood sugar never falls below *safe-min*," stated as (a)
(SR_2) "The diabetic should not overdosed on insulin," elicited from (b, c)
(SR_3) "The system should deliver no unnecessary insulin," elicited from (c)
(SR_4) "The system should deliver no short of insulin," elicited from (b, c)

Now, let us deal with the acuteness of the safety requirements violation; according to the physician, violation of SR_1 is *critical* because a sugar value below *safe-min* indicates a serious shortfall in the blood sugar and leads to the affliction that the system is not able to cope with. However, violation of other safety requirements is *unsafe* but remediable.

There are two requirements not being the main requirements posed by the physician but imposed by physical *constraints* of the system. The first constraint states that, on one hand, it is not feasible to measure blood sugar constantly and, on the other hand, the measurement should not be excessively delayed. Therefore, the physician suggests 10-min periods to measure the diabetic's blood sugar and expects that it should be satisfied by the system. The second constraint imposed by the system is that the reservoir system should consist of sufficient insulin at the time of dose delivery. Considering the physical constraints, three safety requirements are elicited:

(SR_5) "The system sensor should not be early in sampling"
(SR_6) "The system sensor should not be late in sampling"
(SR_7) "The system reservoir should not consist of insufficient insulin at the time of dose delivery"

Violation of SR_5 may lead to deliver dose early and this may lead to overdose; hence, the violation is critical. Violation of SR_6 and SR_7 may lead to deliver dose late, and this may lead to increase in the blood sugar level; hence, the violation is unsafe.

3 Modeling the CIIP System Behavior

The purpose of this section is to present the CIIP system behavior based on its *interaction* to the diabetic patient. For this purpose, an interaction model is represented where the CIIP system is segregated from the diabetic (Fig. 1). Then, the CIIP system is considered as a reactive system visiting the diabetic patient once every 10 min and reacts to him or her by a dose of insulin if necessary. Each system interaction shown in Fig. 1 includes a patient visit and a system reaction. *Attributes* Att_1 to Att_5 are considered to specify the safe system behavior. Attributes Att_2, Att_3, and Att_4 play the role of main attributes, and Att_1 and Att_5 play the role of auxiliary ones where Att_3 is an output attribute and the others are input ones. Att_2 indicating "blood parameters" is determined by the system sensor and Att_4 indicating "time passage" is determined by the system timer. Att_3 indicating "dose value" is determined by the system software and shows the system reaction.

Auxiliary *attributes* Att_1 and Att_5 indicating "the start time of sampling?" and "available insulin in reservoir?", respectively, are used to specify statuses of the system sensor and the system reservoir.

The system environment is the diabetic patient and the system components are input device (blood sensor), software, and output device (insulin delivery unit). The timer device sends an interrupt per 10 min to the input device to monitor the environment. The insulin delivery unit is an output device to control the diabetic and steered by the software component. Therefore, the system monitors and controls its environment by means of its components. The environment quantity to be monitored is a *sample* of the diabetic's blood. The system controls its environment through *insulin* whose dose value is determined by the software component and delivered by the output device. The sample quantity has two attributes: *start* of sampling (Att_1) and blood *sugar* (Att_2), and the insulin quantity has *dose* attribute (Att_4). While the

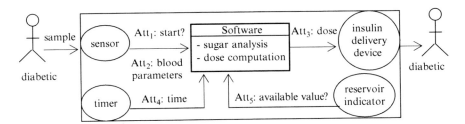

Fig. 1 The CIIP system interaction model

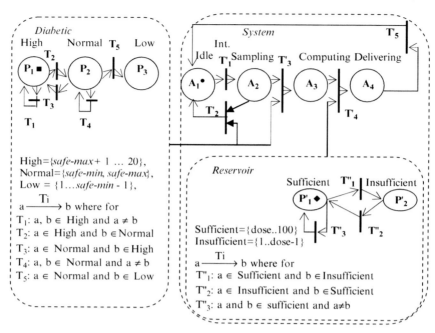

Fig. 2 The CIIP system behavioral model

system acquires Att_1 and Att_2 values from the patient, the patient acquires the Att_4 value from the system. The system mission (i.e., main task) is computation of a correct dose value timely.

Based on the interaction of the CIIP system with its environment depicted by Fig. 1, now we can deal with modeling system behavior using Petri nets [2]. Considering the system attributes and the interaction model (Fig. 1), the system behavior model is depicted in Fig. 2, consisting of the *system environment* behavior, i.e., diabetic patient behavior, the *system* behavior, and the *system reservoir* behavior. The first part constitutes the system input and output and the second and third parts constitute blood monitoring and controlling components of the system. As Fig. 2 shows, a Petri net consists of *places, tokens, arcs,* and *transitions* indicated by bubbles, bold marks, arrows, and bars, respectively.

Formally, a Petri net (PN) is defined by a 5-tuple (P, T, F, W, M0) [3], where $P = \{P_1, P_2, \ldots, P_n\}$ is a finite set of places; $T = \{T_1, T_2, \ldots, T_m\}$ is a finite set of transitions; $P \cup T \neq 0$; $F \subseteq (P \times T) \cup (T \times P)$ is a flow relation; $W : F \rightarrow N - \{0\}$ is a weight function, which associates a nonzero integer value to each flow relation; and $M_0 = \{m_0, m_1, \ldots, m_k\}$ is a initial number of tokens (called *initial marking*) in each place. The default value for a flow element is 1. Accordingly, we define the Petri net of the CIIP system (depicted by Fig. 2) by the 5-tuple (P, T, F, W, M_0) where:

$P = \{P_{(Diabetic)}, P_{(System)}, P_{(Reservoir)}\}$, where $P_{(Diabetic)} = \{P_1, P_2, P_3\}$,
$P_{(System)} = \{A_1, A_2, A_3, A_4\}$, and $P_{(Reservoir)} = \{P'_1, P'_2\}$,
$T = \{T_{(Diabetic)}, T_{(System)}, T_{(Reservoir)}\}$, where $T_{(Diabetic)} = \{T_1, T_2, T_3, T_4, T_5\}$,

$T_{(System)} = \{T'_1, T'_2, T'_3, T'_4, T'_5\}$, and $T_{(Reservoir)} = \{T''_1, T''_2, T''_3\}$, $W : F \to 1$, and
$M_0 = \{M_{0(Diabetic)}, M_{0(System)}, M_{0(Reservoir)}\}$, where $M_{0(Diabetic)} = \{1, 0, 0\}$, $M_{0(System)} = \{1, 0, 0, 0\}$, and $M_{0(Reservoir)} = \{1, 0\}$, i.e., $P_1 = 1$, $A_1 = 1$, and $P'_1 = 1$.

As Fig. 2 shows, to represent the diabetic and the system behaviors three concurrent Petri nets have been considered where places play the role of states. The states speaking of duration are considered: (1) as a *set of values*, used in the diabetic's behavior in Fig. 2 or (2) as an *action*, used in the system behavior. In case of the set of values, the range of the values has been stated in Fig. 2. A transition is called *enabled* if its input place has a mark (token); when an enabled transition fires, the mark of its input place is removed and its output place takes a mark.

Initially, the diabetic is in the "High" state, which may: (1) stay in the same state or (2) change to the "Normal" state. According to physician (Sect. 2), each diabetic's state is a representation of a range of values; therefore, a change in the diabetic's blood sugar value may lead to the change or no change of the diabetic's current state. Accordingly, each transition is considered as a *mapping* from its input arc denoting an input value (shown by "a") into its output arc denoting an output value (shown by "b"). The mappings associated with Petri nets transitions of the diabetic and the system reservoir have been expressed in Fig. 2. Transitions T_1 and T_4 in Fig. 2, for instance, indicate that although the sugar value increases or decreases, the value is still high/normal. However, transition T_2 in Fig. 2 indicates that a decrease in the blood sugar value leads to the change of the state. The "Low" state being a *critical state* cannot be managed by the system, and so it may not be changed to the "Normal" state by the system. This state will appear if the diabetic is overdosed or unnecessarily dosed up by the system.

The system is initially in the "Idle" state and will change to the "Sampling" state (i.e., T'_1 will fire) when the system timer generates an interrupt event (indicated by Int.). The system will return to the "Idle" state from the "Sampling" state (i.e., T'_2 will fire) when it identifies no increase in the diabetic's blood sugar and will change to the "Computing" state (i.e., T'_3 will fire) when it identifies an increase. The system will change from the "Computing" state to the "Delivering" state (i.e., T'_4 will fire) when the system reservoir has sufficient insulin. The system reservoir will stay in the "Sufficient" state while the computed dose is available to deliver.

4 Specifying the Safe Behavior of CIIP

The purpose of this section is to specify the CIIP behavior not risking the diabetic's safety requirements. As stated in Sects. 1 and 2, a safety requirement indicates that a bad event should not happen, i.e., a safety requirement should not be violated. Therefore, we should determine the CIIP behavior leading to the safety requirement's violation. To this end, we consider the system and the diabetic behaviors in Fig. 2 and specify those combinations of these behaviors leading to violation of the safety requirements informally stated in Sect. 2. In fact, these combinations show the hazardous behavior that its negation speaks of safety requirements.

Table 1 shows that the compound states risk the diabetic's safety requirements. P_1, P_2, and P_3 belong to $P_{(Diabetic)}$ (see Sect. 3), A_1, A_2, A_3, and A_4 belong to $P_{(System)}$, P'_1 and P'_2 belong to $P_{(Reservoir)}$, and X indicates a don't care state. We show each compound state in the $S_1 S_2 S_3$ form where S_1, S_2, and S_3 belong to $P_{(System)}$, $P_{(Diabetic)}$, and $P_{(Reservoir)}$, respectively. Initially, the system is in the "Idle" state. Having run out of the idle time, the system deals with sampling the diabetic's blood and identifies his or her blood sugar level, i.e., "High," "Normal," or "Low." In case of "High," indicated by "$A_2 P_1 X$," the system should make a transition to the "Computing" state indicated by "$A_3 P_1 X$." Also, in case of "Normal," the system should make a transition to the "Computing" state indicated by "$A_3 P_2 X$" if the blood sugar level has increased during this level; however, it should return to $A_1 XX$ if it has not.

There are three types of compound states, *critical, unsafe*, and *susceptible*. The hazard of running a critical state, such as "$A_2 P_3 X$" (showing the diabetic's "Low" state), and the hazard of running an unsafe state, such as "$A_3 P_1 P'_2$" (showing the shortage of insulin in the system reservoir) are self-evident; however, the hazard of running a susceptible state depends on the past diabetic's states. The susceptible state "$A_3 P_1 X$," for instance, may be subject to overdose computation. Table 1 describes the meaning of hazardous states, i.e., the state combinations that violate the safety requirements elicited in Sect. 2, and their acuteness. Each individual state

Table 1 The hazardous behaviors of the CIIP safety-critical medical system

#	Combination	Path sequence	System diabetic reservoir	Violated safety requirement	Description	Acuteness
1	$A_2 P_3 X$	$A_1 XX \rightarrow A2P3X$	√-√	SR_1	Fall of sugar level	Critical
2	$A_2 P_2 X$	$A_1 XX \rightarrow A_2 P_2 X$	√-√	SR_5/SR_6	Verifying sampling time	Susceptible
3	$A_2 P_1 X$	$A_1 XX \rightarrow A_2 P_1 X$	√-√	SR_5/SR_6	Verifying sampling time	Susceptible
4	$A_3 P_2 X$	$A_1 XX \rightarrow A_2 P_2 X$ $\rightarrow A_3 P_2 X$	√-√	$SR_2/SR_3/SR_4$	Verifying computed dose	Susceptible
5	$A_3 P_1 X$	$A_1 XX \rightarrow A_2 P_1 X$ $\rightarrow A_3 P_1 X$	√-√	SR_2/SR_4	Verifying computed dose	Susceptible
6	$A_3 P_1 P'_2$	$A_1 XX \rightarrow A_2 P_1 X$ $\rightarrow A_3 P_1 X$ $\rightarrow A_3 P_1 P'_2$	√-√-√	SR_7	Insufficient insulin in the reservoir	Unsafe
7	$A_3 P_2 P'_2$	$A_1 XX \rightarrow A_2 P_2 X$ $\rightarrow A_3 P_2 X$ $\rightarrow A_3 P_2 P'_2$	√-√-√	SR_7	Insufficient insulin in the reservoir	Unsafe

speaks of a set of values (for the diabetic and the system reservoir states) or an action (for the system states). Column "Matrices sequence" shows the path leading to a hazard situation.

5 Conclusions and Related Work

In constrast to some other methods, this chapter made some contributions. The *first* contribution is using the high-level specification of the safety requirement's violations expressed in the physician language and the low-level specification of the violations specified in the low-level language, i.e., the matrix sequences. On one hand, the high-level specifications are intelligible to the high-level user, i.e., the physician, but they are not susceptible for formal modeling because they are expressed in natural language. On the other hand, the low-level specifications are not intelligible to the high-level user but they are susceptible for formal modeling by the formal methods. Therefore, in the first step, this chapter dealt with bridging the gap between high-level specifications and low-level ones. In [4], Wang et al. have addressed modeling the insulin pump behavior where they have directly used the SOFL formal langue to specify the system behavior; hence, they have not applied the high-level specifications. In a similar work [5], Sommerville has directly used the Z formal language for specification of the insulin pump system and the safety requirements.

The *second* contribution is using a visual method by which two profits were made. The first is the more suitable understanding of the system behavior than the textual methods such as those applied in [4, 5], and the second is showing visual interactions of the medical system with its environment. The *third* contribution is that the proposed method has the capability to deal with verifying low-level specifications. In [6], Martinez-Sarriegui et al. have used visual collaboration diagram to specify and model the diabetes care, but this is not a formal method and they do not discuss the possibility of verification of their specifications. The Petri net visual method was used by the author of this chapter to specify another safety-critical system in [7].

5.1 Future Work

Having determined hazardous compound states (Table 1) leading to safety requirement's violation, one may compose a monitor program that should be executed in parallel with the CIIP system software to verify the system behavior. Figure 3 shows the combination of Fig. 1 with the monitor. One may compose the program monitor using the construction proposed by [8]. While the system is interacting with the diabetic, the monitor should get states of the diabetic, the system, and the system reservoir and deal with verifying them against the hazardous states. The monitor

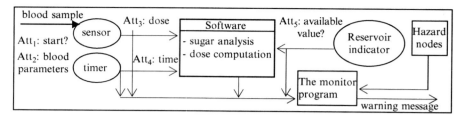

Fig. 3 The interaction of the CIIP system with the monitor program

program may be constructed: (1) as a distinct program running in parallel with the system software or (2) as an embedded code in proper places of the target program, i.e., the system software.

References

1. Pnueli A (1977) The temporal logic of programs. The 18th Annual of IEEE-CS Symposium on Foundations of Computer Science, 46–57
2. Jensen K and Kristensen L M (2009) Colored Petri-Nets, modeling and validation of concurrent systems. Springer, New York
3. Ghezzi C et al. (2003) Fundamentals of software engineering, 2nd edition. Prentice-Hall, Upper Saddle River
4. Wang J et al. (2007) Developing an insulin pump system using the SOFL method. The 14th Asia-Pacific Software Engineering Conference, 334–341
5. Sommerville I (2006) Software engineering, 8th edition. Addison Wesley, Reading
6. Martinez-Sarriegui I et al. (2009) Modeling and formal specification of a multi agent telemedicine system for diabetes care. The International Conference on Agents and Artificial Intelligence, ICAART2009, 507–512
7. Babamir S M and Babamir F S (2008). Behavioral specification of real-time requirements. The 15th Asia-Pacific Software Engineering Conference IEEE, 209–306. DOI: 10.1109/APSEC.2008.22
8. Delgado N et al. (2004) Taxonomy and catalog of runtime software-fault monitoring tools. IEEE Transactions on Software Engineering, 30 (22): 859–872

Part VIII
Software Packages and Other Computational Topics in Bioinformatics

Chapter 66
Lotka–Volterra System with Volterra Multiplier

Klaus Gürlebeck and Xinhua Ji

Abstract With the aid of Volterra multiplier, we study ecological equations for both tree system and cycle system. We obtain a set of sufficient conditions for the ultimate boundedness to nonautonomous n-dimensional Lotka–Volterra tree systems with continuous time delay. The criteria are applicable to cooperative model, competition model, and predator–prey model. As to cycle system, we consider a three-dimensional predator–prey Lotka–Volterra system. In order to get a condition under which the system is globally asymptotic stable, we obtain a Volterra multiplier, so that in a parameter region the system is with the Volterra multiplier it is globally stable. We have also proved that in regions in which the condition is not satisfied, the system is unstable or at least it is not globally stable. Therefore, we say that the three-dimensional cycle system is with global bifurcation.

Keywords Volterra Multiplier · Ecological Equations · Lotka–Volterra Tree Systems · 3D Predator–Prey Cycle System

1 Equations and Notions

The nonautonomous n-species Lotka–Volterra system is expressed as follows:

$$\dot{x}_i(t) = x_i(t)\left[b_i(t) - \sum_{j=1}^{n} a_{ij}(t)x_j(t)\right], \quad i \in \mathcal{N}, \tag{1}$$

where $\mathcal{N} = \{1,\ldots,n\}$, x_i represents the density of species i, b_i is the carrying capacity, and a_{ij} represents the effect of interspecific (if $i \neq j$) or intraspecific (if $i = j$) interaction. In vector form, system (1) is expressed as:

$$\dot{x} = X[b(t) - A_n(t)x], \tag{2}$$

X. Ji (✉)
Institute of Mathematics, Academy of Mathematics and Systems Science (AMSS), Chinese Academy of Sciences, Beijing 100190, China
e-mail: xhji@math.ac.cn

H.R. Arabnia and Q.-N. Tran (eds.), *Software Tools and Algorithms for Biological Systems*, Advances in Experimental Medicine and Biology 696,
DOI 10.1007/978-1-4419-7046-6_66, © Springer Science+Business Media, LLC 2011

where $x, b \in R^n$, $X = \text{diag}(x_1, \ldots, x_n)$, $A_n = (a_{ij})_{n \times n}$. The notations such as:

$$R^q_{+0} = \{u \in R^q : u_i \geq 0, \ i = 1, \ldots, q\}, \quad R^q_+ = \{u \in R^q : u_i > 0, \ i = 1, \ldots, q\},$$

are adopted, where q is any positive integer.

We consider arbitrary nonautonomous Lotka–Volterra tree systems with continuous time delay, where the growth rate of one species is not only decided by the density of all species at the time t but also influenced by its own density before the time t. The dynamical behavior of the total population of the species is described by the system of integro-differential equations:

$$\dot{x}_i(t) = x_i(t)\left\{b_i(t) - \sum_{j=1}^n a_{ij}(t)x_j(t) + \gamma_i \int_{-\infty}^t \alpha_i \exp[-\alpha_i(t-\tau)]x_i(\tau)\,d\tau\right\}, \quad (3)$$

where $\alpha_i > 0$, $\gamma_i \in R$, $i \in \mathcal{N}$. We will obtain sufficient conditions of ultimate boundedness to tree system (3). The criteria are applicable to cooperation model, competition model, and predator–prey model.

Definition 1.1. For fixed $i, j \in \mathcal{N}$ ($i \neq j$), species i and species j are said to be in interaction if $a_{ij}a_{ji} \neq 0$. With $a_{ii} > 0$, we define $a_{ij} < 0$ for cooperation model, $a_{ij} > 0$ for competition model, and $a_{ij}a_{ji} < 0$ for predator–prey model.

Definition 1.2. $A \in S_w$ means that there exists a positive diagonal matrix W such that $WA + A^T W$ is positive definite, and W is called a Volterra multiplier for A.

Definition 1.3. (e.g., [4]) The graph $G(A_n)$ consists of n vertices representing n species together with edges representing the interrelations between species. The graph $G(A_n)$ of system (1) is said to have a cycle if there are distinct indices $i_1, \ldots, i_m (m \geq 3)$ such that none of the elements $a_{i_1 i_2}, a_{i_2 i_3}, \ldots, a_{i_m i_1}$ vanish. A system is called a tree if its graph $G(A_n)$ is a connected graph without any cycle.

Definition 1.4. ([4]) The degree of i-th species in A_n is defined by the positive integer m_i that is the number of species connected with the i-th species in graph $G(A_n)$.

2 Main Results

We assume that the functions on the right-hand side of (1) are so smooth that for all initial conditions $(t_0, x_0) \in R \times R^n_+$, there exists a unique solution $x(t, t_0, x_0)(t \geq t_0)$ satisfying $x(t_0, t_0, x_0) = x_0$. We also assume the following hypotheses.

(H_1) $\sup_{t \in R} \| b(t) \| \leq \bar{b}$, $\bar{b} \in R_+$. (By $\| \cdot \|$, we mean the usual Euclidean norm.)
(H_2) The function matrix $A_n(t)$ has bounded entries $a_{ij}(t)$, $i, j \in \mathcal{N}$:

$$\hat{a}_{ij} = \inf_{t \in R} a_{ij}(t), \text{ for } i \neq j, \quad \hat{a}_{ii} = \inf_{t \in R} a_{ii}(t) > 0, \ i, j \in \mathcal{N}.$$

Definition 2.1. The solutions $x(t,t_0,x_0)$ of system (1) are said to be ultimately bounded with respect to the region R_+^n, if there exists a compact region $\Omega, \Omega \subset R_{+0}^n$, and a finite time $T = T(t_0,x_0)$ such that, for any $(t_0,x_0) \in R \times R_+^n$, we have $x(t,t_0,x_0) \in \Omega$ for all $t > T(t_0,x_0)$.

In order to obtain ultimate boundedness to system (3), we transform it into a system of ODEs using the usual linear-chain trick and introducing the supplementary nonnegative variables:

$$x_{n+i} = \int_{-\infty}^t \alpha_i \exp[-\alpha_i(t-\tau)] x_i(\tau) \, d\tau, \quad \alpha_i > 0, \quad i \in \mathcal{N}.$$

Substituting these new variables into (3), we obtain the following system:

$$\dot{x}_i = x_i \left[b_i(t) - \sum_{j=1}^n a_{ij}(t) x_j + \gamma_i x_{n+i} \right], \quad i \in \mathcal{N}, \tag{4}$$

$$\dot{x}_{n+i} = \alpha_i x_i - \alpha_i x_{n+i}, \quad i \in \mathcal{N},$$

together with the conditions

$$x_{n+i}(t)|_{t=-\infty} = 0, \quad \text{for} \quad i \in \mathcal{N}.$$

It is known (e.g., [8]) that systems (3) and (4) are equivalent in their qualitative dynamical behavior because of the exponential delay function in (3). Introducing vector $\tilde{x} = col(x_1,\ldots,x_{2n})$, we can write system (4) as:

$$\dot{\tilde{x}} = \begin{bmatrix} X & 0 \\ 0 & I_n \end{bmatrix} [\tilde{b}(t) - \tilde{A}(t)\tilde{x}], \tag{5}$$

where $\tilde{b} = col(b_1,\ldots,b_n,0,\ldots,0) \in R^{2n}$, I_n is the identity, and $\tilde{A}(t)$ is given by:

$$\tilde{A}(t) = [\tilde{a}_{ij}(t)]_{2n \times 2n} = \begin{bmatrix} A_n(t) & -\gamma \\ -\alpha & \alpha \end{bmatrix}, \tag{6}$$

where $\gamma = diag(\gamma_1,\ldots,\gamma_n)$ and $\alpha = diag(\alpha_1,\ldots,\alpha_n)$. We can easily conclude that the graph $G(\tilde{A})$ also has no cycle. Hence, (4) is a tree system. Let $2\mathcal{N} = \{1,\ldots,2n\}$,

$$\mathcal{I} = \{i \in 2\mathcal{N} : x_i = 0 \ \forall i \in \mathcal{I} \subseteq 2\mathcal{N}, \text{ and } x_j > 0 \ \forall j \in 2\mathcal{N} - \mathcal{I}\},$$

$$R_\mathcal{I}^{2n} = \{x \in R_{+0}^{2n} : x_j > 0 \ \forall j \in 2\mathcal{N} - \mathcal{I}\}.$$

Theorem 2.2. *Let (3) be tree system. Assume* $(H_1), (H_2)$ *and*

$$\hat{a}_{ii} > \tilde{m}_i \gamma_i, \quad \hat{a}_{ii}\hat{a}_{jj} > \tilde{m}_i \tilde{m}_j \hat{a}_{ij}\hat{a}_{ji}, \quad i \neq j; \ i,j \in \mathcal{N}, \tag{7}$$

where $\tilde{m}_i$ is the degree of i-th element in the graph of $G(\tilde{A}(t))$. Then the solutions of ODE system (4) are ultimately bounded with respect to R_I^{2n}.

Corollary 2.3. *Let (3) be tree system with $\gamma_i < 0$, $i \in \mathcal{N}$. Assume $(H_1), (H_2)$ and*

$$\hat{a}_{ii}\hat{a}_{jj} > m_i m_j \hat{a}_{ij}\hat{a}_{ji}, \quad i \neq j; \ i,j \in \mathcal{N}. \tag{8}$$

Then the solutions of ODE system (4) are ultimately bounded with respect to R_I^{2n}.

Stability in models with time delay has been studied by many authors. For predator–prey system, Corollary 2.3 is a nonautonomous correspondence to the result in [6]. To prove Theorem 2.2 a lemma is in order.

Lemma 2.4 ([4]). *For a tree system, if the inequalities*

$$\hat{a}_{ii}\hat{a}_{jj} > m_i m_j \hat{a}_{ij}\hat{a}_{ji}, \quad i \neq j \quad i,j \in \mathcal{N} \tag{9}$$

are fulfilled, then $\hat{A}_n \in S_w$ which is determined by $A_n(t)$ of system (1) and (H_2).

Applying a procedure in [4], we obtain a Volterra multiplier W for matrix $\hat{A}_n$ whose elements are given by induction from the formula

$$w_j = \frac{2\hat{a}_{ii}\hat{a}_{jj} - m_i m_j \hat{a}_{ij}\hat{a}_{ji}}{m_i m_j \hat{a}_{ji}^2} w_i$$

With $w_1 = 1$, obviously $W = \text{diag}(w_1, \ldots, w_n) \in R_+^n$ under (9). Thus, $\hat{A}_n \in S_w$.

Proof of Theorem 2.2: We can see that based on assumptions (H_1) and (H_2), we also have the following hypotheses.

$(\tilde{H}_1)$ $\sup_{t \in R} \| \tilde{b}(t) \| \leq \bar{\tilde{b}}$, $\bar{\tilde{b}} \in R_+$. (By $\|\cdot\|$, we mean the usual Euclidean norm.)
$(\tilde{H}_2)$ The function matrix $\tilde{A}(t)$ has bounded entries $\tilde{a}_{ij}(t)$, $i,j \in 2\mathcal{N}$.

$$\hat{\tilde{a}}_{ij} = \inf_{t \in R} \tilde{a}_{ij}(t), \text{ for } i \neq j, \quad \hat{\tilde{a}}_{ii} = \inf_{t \in R} \tilde{a}_{ii}(t) > 0, \ i,j \in 2\mathcal{N}.$$

Denote $\hat{\tilde{A}} = (\hat{\tilde{a}}_{ij}) \in R^{2n \times 2n}$, it is easy to see $\hat{\tilde{A}} \in S_{\widetilde{W}}$ with $\widetilde{W} = \text{diag}(\tilde{w}_1, \ldots, \tilde{w}_{2n}) \in R_+^{2n}$. Using a similar calculation to that in [3] which is about ultimate boundedness to nonautonomous tree systems without time delay, and taking $(\tilde{H}_1), (\tilde{H}_2)$ into account, we will be able to prove the theorem under conditions (7).

3 Special Systems and Example

Corollary 3.1. *Let (3) with $\gamma_i < 0$, $i \in \mathcal{N}$, be a chain system of competition (or cooperation) model. Assume $(H_1), (H_2)$ and*

$$\hat{a}_{11}\hat{a}_{22} > 2\hat{a}_{12}\hat{a}_{21},$$
$$\hat{a}_{ii}\hat{a}_{i+1,i+1} > 4\hat{a}_{i,i+1}\hat{a}_{i+1,i}, \quad 2 \leq i \leq n-2,$$
$$\hat{a}_{n-1,n-1}\hat{a}_{nn} > 2\hat{a}_{n-1,n}\hat{a}_{n,n-1}. \quad (10)$$

Then the solutions of ODE system (4) are ultimately bounded with respect to R_I^{2n}.

Corollary 3.2. *Let (3) with $\gamma_i < 0$, $i \in \mathcal{N}$, be a competition (or cooperation) system between one and multi-species. Assume $(H_1), (H_2)$ and*

$$\hat{a}_{11}\hat{a}_{jj} > (n-1)\hat{a}_{1j}\hat{a}_{j1}, \quad j \in \mathcal{N} - \{1\}. \quad (11)$$

Then the solutions of ODE system (4) are ultimately bounded with respect to R_I^{2n}.

Corollary 3.3. *Let (3) be a predator–prey tree system. Assume $\gamma_i < 0$, $i \in \mathcal{N}$, and assume $(H_1), (H_2)$ hold true. Then the solutions of the ODE system (4) are ultimately bounded with respect to R_I^{2n}.*

Example. In order to show the ease in using our criteria, we consider the system:

$$\frac{dx_1(t)}{dt} = x_1(t)\left\{2 + \sin \pi t - 2x_1(t) - x_2(t) - 3\int_{-\infty}^{t} 2\exp[-2(t-\tau)]x_1(\tau)\,d\tau\right\},$$
$$\frac{dx_2(t)}{dt} = x_2(t)\left\{4 + \cos \pi t - x_1(t) - 6x_2(t) - 8\int_{-\infty}^{t} 4\exp[-4(t-\tau)]x_2(\tau)\,d\tau\right\}.$$

It is easy to check that this system satisfies the conditions of Theorem 2.2. Thus, we conclude that the solutions of this system are ultimately bounded. If we consider the case without time delay, this system reduces into:

$$\frac{dx_1(t)}{dt} = x_1(t)\left[2 + \sin \pi t - 2x_1(t) - x_2(t)\right],$$
$$\frac{dx_2(t)}{dt} = x_2(t)\left[4 + \cos \pi t - x_1(t) - 6x_2(t)\right]. \quad (12)$$

which is given in [1], p. 176. It is easy to check that system (12) also satisfies the conditions of Theorem 2.2. Therefore, the solutions of system (12) are ultimately bounded. This result conforms to the solutions graphically illustrated in [1].

4 Three-Dimensional Predator–Prey Cycle System

We would also like to obtain Volterra multiplier for cycle system. Consider a three-dimensional Lotka–Volterra predator–prey cycle system:

$$\dot{x} = X(b - Ax), \qquad b \in R^3, \tag{13}$$

$$A = \begin{pmatrix} 1 & a & -a\varepsilon \\ -a\varepsilon & 1 & a \\ a & -a\varepsilon & 1 \end{pmatrix}, \quad a > 0, \quad \varepsilon > 0, \tag{14}$$

which has been investigated in [5]. We approach it in a different way and will get the same result.

Definition 4.1. Matrix $A = (a_{ij}) \in R^{n \times n}$ is said to be Liapunov stable if and only if every eigenvalue of A is with positive real part. Matrix $A = (a_{ij}) \in R^{n \times n}$ is said to be Volterra–Liapunov stable if $A \in S_w$.

As well known that a necessary condition for $A \in S_w$ is that A is Liapunov stable, and that (see [7]) for system (13) if $A \in S_w$, then it has a nonnegative and globally asymptotically stable equilibrium point for every carrying capacity $b \in R^3$.

Theorem 4.2. Matrix $A \in S_w$ if and only if its parameters a, ε satisfy condition

$$1 - \frac{2}{a} < \varepsilon < 1 + \frac{1}{a}, \quad a > 0, \quad \varepsilon > 0. \tag{15}$$

To obtain a Volterra multiplier for matrix A in (14), we need a lemma in [2].

Lemma 4.3 ([2]). Let $i, j \in \mathcal{N}$ be fixed and $i \neq j$. Let $a_{ii}, a_{jj} \in R_+$ and $a_{ij}, a_{ji} \in R - \{0\}$. If there is a positive real number μ satisfying the inequality:

$$a_{ii}a_{jj} > \mu a_{ij}a_{ji}, \quad i \neq j, \quad i, j \in \mathcal{N}, \tag{16}$$

then there exist positive constants $w_i, w_j \in R_+$ such that

$$w_i w_j a_{ii} a_{jj} > \frac{\mu}{4} \left(w_i a_{ij} + w_j a_{ji} \right)^2. \tag{17}$$

Applying above lemma we substitute $w = w_i/w_j$ into (17), which becomes

$$f(w) = w^2 \mu a_{ij}^2 - 2w(2a_{ii}a_{jj} - \mu a_{ij}a_{ji}) + \mu a_{ji}^2 < 0.$$

Two algebraical roots of $f(w) = 0$ are:

$$\xi_1^{ij}(\mu), \xi_2^{ij}(\mu) = \frac{2a_{ii}a_{jj} - \mu a_{ij}a_{ji} \mp 2\sqrt{a_{ii}a_{jj}(a_{ii}a_{jj} - \mu a_{ij}a_{ji})}}{\mu a_{ij}^2}$$

Under condition (16), we have:

$$f(w) < 0, \quad \text{for } \xi_1^{ij}(\mu) < w < \xi_2^{ij}(\mu),$$

where $0 < \xi_1^{ij}(\mu) < \xi_2^{ij}(\mu) < +\infty$. Applying same method as we did in preceding sections, we see that by choosing $\mu = 4$ [since $m_1 = m_2 = m_3 = 2$ for graph $G(A)$] we may get the positive definiteness for matrix $WA + A^T W$ provided the positive numbers w_1, w_2, w_3 satisfying

$$\frac{w_1}{w_2} \in \left(\xi_1^{12}, \xi_2^{12}\right), \quad \frac{w_1}{w_3} \in \left(\xi_1^{13}, \xi_2^{13}\right), \quad \frac{w_2}{w_3} \in \left(\xi_1^{23}, \xi_2^{23}\right) \quad (18)$$

simultaneously, where $\xi_1^{23} = \xi_1^{12}$, $\xi_2^{23} = \xi_2^{12}$ and

$$\xi_1^{12}, \xi_2^{12} = \frac{1 + 2ab \mp \sqrt{1 + 4ab}}{2a^2}; \quad \xi_1^{13}, \xi_2^{13} = \frac{1 + 2ab \mp \sqrt{1 + 4ab}}{2b^2} \quad \text{with } b = a\varepsilon.$$

Setting $w_3 = 1$ and taking note of (18), we have:

$$w_1 \in \left(\xi_1^{13}, \xi_2^{13}\right), \quad w_2 \in \left(\xi_1^{23}, \xi_2^{23}\right), \quad \frac{w_1}{w_2} \in \left(\frac{\xi_1^{13}}{\xi_2^{23}}, \frac{\xi_2^{13}}{\xi_1^{23}}\right). \quad (19)$$

Considering (18) and (19), we suppose:

$$\Xi = \left[\xi_1^{12}, \xi_2^{12}\right] \cap \left[\frac{\xi_1^{13}}{\xi_2^{23}}, \frac{\xi_2^{13}}{\xi_1^{23}}\right] \neq \emptyset \quad (20)$$

which implies there exists an element $w_1/w_2 \in \Xi$, or $\xi_1^{12} = \xi_2^{13}/\xi_1^{23} \in \Xi$. Hence, we have:

$$\frac{1 + 2ab - \sqrt{1 + 4ab}}{2a^2} = \left(\frac{1 + 2ab + \sqrt{1 + 4ab}}{2b^2}\right)\left(\frac{2a^2}{1 + 2ab - \sqrt{1 + 4ab}}\right). \quad (21)$$

Calculating (21), we reduce it into $b = a + 1$. Substituting $b = a + 1$ into:

$$\xi_1^{12} = \frac{1 + 2ab - \sqrt{1 + 4ab}}{2a^2} = \frac{w_1}{w_2} \in \Xi$$

we get $w_1/w_2 = 1$. Recalling $w_3 = 1$, we obtain $W = \text{diag}(1,1,1)$. We shall show that $W > 0$ is a Volterra multiplier for A under condition (15).

Proof of Theorem 4.2. In order to prove sufficiency, noting $\Delta_1 = 2 > 0$, we have only to find conditions under which:

$$\Delta_2 = \det\begin{pmatrix} 2 & a-b \\ a-b & 2 \end{pmatrix} > 0, \quad \Delta_3 = \det\begin{pmatrix} 2 & a-b & a-b \\ a-b & 2 & a-b \\ a-b & a-b & 2 \end{pmatrix} > 0.$$

By calculating, we get:

$$\{\Delta_2 > 0\} \cap \{\Delta_3 > 0\} = \left\{ (a,\varepsilon) \in R_+^2 \ : \ 1 - \frac{2}{a} < \varepsilon < 1 + \frac{1}{a} \right\}. \tag{22}$$

To show necessity, we obtain the eigenvalues of A which are $\lambda_1 = 1 + a - b$; $\lambda_2, \lambda_3 = 1 - (1/2)(a-b) \pm [(i\sqrt{3})/2](a+b)$. Thus, we have:

$$\{\lambda_1 > 0\} \cap \{\Re\lambda_2 > 0\} \cap \{\Re\lambda_3 > 0\} = \left\{ (a,\varepsilon) \in R_+^2 \ : \ 1 - \frac{2}{a} < \varepsilon < 1 + \frac{1}{a} \right\}. \tag{23}$$

The right-hand side of (22) and (23) is just the same as condition (15). □

Remark. This is a global bifurcation problem. The region given by (23) is a globally stable region which we denoted by Ω_s. Two curves $l_1, l_2 = \{\varepsilon = 1 \pm (1/a)\}$ divide R_+^2 into three regions $\Omega_s = \{1 - (2/a) < \varepsilon < 1 + (1/a)\}$, $\Omega_{u1} = \{\varepsilon > 1 + (1/a)\}$ and $\Omega_{u2} = \{\varepsilon < 1 - (2/a)\}$ for $(a,\varepsilon) \in R_+^2$. Inside Ω_s, the nonnegative equilibrium point of (13)–(14) is globally stable, and outside Ω_s, i.e., inside Ω_{u1} and Ω_{u2}, the system is unstable or at least it is not globally stable. Thus, we say that curves l_1, l_2 are two global bifurcation curves. And system (13)–(14) is with global bifurcation.

Acknowledgment This research was supported by the Deutsche Forschungsgemeinschaft.

References

1. K. Gopalsamy, Exchange of equilibria in two species Lotka–Volterra competition models, *Journal of Australian Mathematical Society Series B*, 24, 160–170, (1982)
2. Xinhua Ji, The existence of globally stable equilibria of n-dimensional Lotka–Volterra systems, *Applicable Analysis*, 62(1), 11–28, (1996)
3. Xue-Zhi Li, G. Gupur and Xin-Hua Ji, The criteria of ultimate boundedness for nonautonomous Lotka–Volterra tree systems, *Applied Mathematics Letters*, 14(4), 469–476, (2001)
4. Xue-Zhi Li, Chun-Lei Tang and Xin-Hua Ji, Global asymptotic stability for Lotka–Volterra tree systems, *Differential Equations and Dynamical Systems*, 8.2, 141–149, (2000)
5. Fen-guo Peng and Zhi-ming Zhou, Global stability for 3-dimensional prey–predator Volterra systems with loop, *Journal of Biomathematics*, 3(2), 159–170, (1988)

6. F. Solimano and E.Beretta, Existence of a globally asymptotical stable equilibrium in Volterra models with continuous time delay, *Journal of Mathematical Biology*, 18, 93–102, (1983)
7. Y. Takeuchi, *Global dynamical properties of Lotka–Volterra system*, World Scientific, Singapore (1996)
8. A. Wörz-Busekros, Global stability in ecological systems with continuous time delay, *SIAM Journal of Applied Mathematics*, 35, 123–134, (1978)

Chapter 67
A Biological Compression Model and Its Applications

Minh Duc Cao, Trevor I. Dix, and Lloyd Allison

Abstract A biological compression model, *expert model*, is presented which is superior to existing compression algorithms in both compression performance and speed. The model is able to compress whole eukaryotic genomes. Most importantly, the model provides a framework for knowledge discovery from biological data. It can be used for repeat element discovery, sequence alignment and phylogenetic analysis. We demonstrate that the model can handle statistically biased sequences and distantly related sequences where conventional knowledge discovery tools often fail.

1 Introduction

The genome stores all genetic information necessary for the development and functioning of all living organisms. It contains the instructions needed to construct other molecules such as RNA and proteins. The information is stored in a simple sequence over the alphabet of four nucleotides: adenine (A), cytosine (C), guanine (G) and thymine (T), and yet it accounts for the organism's inherited traits. This naturally leads to the very intriguing question: How much information is stored in a genome? It is also very interesting to be able to measure the information content of every symbol in the genomic sequence.

In information theory [17], the *information content* of an event is defined as the negative binary logarithm of the probability of the random variable representing the event. In other words, it measures how predictable the event is. The *mutual information* of two events measures the amount of information that can be obtained about one random variable by observing the other. This can be used to measure the relatedness of the two events. While the exact information content is not known, it can be approximated by compression which requires a compression model. The better

M.D. Cao (✉)
Clayton School of Information Technology, Monash University, Clayton, VIC 3800, Australia
e-mail: minhduc@monash.edu

the model compresses the data, the better it approximates the information content. However, compression of biological sequences is challenging [14]. Most general purpose text compression algorithms fail to compress DNA to below the naive 2 bits per symbol encoding. A number of special purpose compression algorithms for DNA have been developed recently. Most of these algorithms are slow and unable to compress long sequences. They, therefore, do not meet the demand of tools for analysing the large databases of sequences today.

This chapter presents the *expert model* [3], an algorithm for biological sequence compression. The expert model is shown to outperform existing biological compression algorithms in both compression performance and speed, and can compress whole genomes of eukaryotic species. Importantly, the model facilitates many knowledge discovery tasks: repeat element discovery, genome alignment and phylogeny inference.

2 The Expert Model Compression Algorithm

The expert model compresses a sequence by examining each symbol in turn. First, it estimates the probability distribution for the symbol. The actual symbol is then encoded with respect to the probability distribution. The encoding can be done efficiently by a coding scheme such as arithmetic coding [20] to produce code words that are arbitrarily close to the optimal theoretical length [17]: $L(x_i) = -\log_2 \Pr(x_i)$.

In order to form the probability distribution of a symbol, the algorithm maintains a set of *experts*, whose predictions of the symbol are blended to give a combined probability distribution. An expert is any model that can provide a probability distribution of the symbol at a position. An example is the *Markov expert* which uses a Markov model learnt from the statistics of the sequence up to the symbol. Since some biological sequences are highly repetitive, the expert model includes *repeat experts* each of which considers the current symbol to be copied from a symbol from a particular offset. Each repeat expert reviews its own performance over some recent history and accordingly builds a probability distribution for mutations. A hash table is used to suggest potential repeat experts. The hash table associates every position in the sequence with the hash key composed of k symbols preceding the position. It proposes the experts that suggest the current symbol is copied from the positions having the same hash key as the current position. The algorithm also has a parameter L, which specifies the maximum number of repeat experts to be employed at a time. The choice of the hash table size k and the expert limit L provides a tradeoff between compressibility and running time. Generally, a small k and a large L allow the model to search for repeats more exhaustively and thus give better compression at the cost of more time.

The reliability of each expert is continually evaluated. A reliable expert is given a high weight for combination, while an unreliable one has little influence on the final prediction or may be even ignored. An expert's weight is determined by the performance of the expert over a recent history. More specifically, the weight is proportional to the negative exponential of the length of the encoding message of symbols in the history window. A detailed description of the mathematics is presented in [3].

3 Compression Performance of the Expert Model

In [3] we show that the expert model outperforms existing compression algorithms on a standard dataset. The sequences in the standard dataset are relatively short, and thus do not demonstrate the intended full capability of the expert model. Here, we present the performance of the expert model on the human genome and the genomes of various species. The human genome was downloaded from Genbank Release 36. The expert model was configured with hash table size 11, and expert limits 200 and 10,000 in two runs, XM-F and XM-S, respectively. Table 1 shows the compression of the 24 human chromosomes by the expert model in the two

Table 1 Compression of the human genome

Chr	Len (Mb)	NML Rate	XM-F Rate	XM-F Time	XM-S Rate	XM-S Time
1	218.7	1.644	1.626	1h34	1.594	31h04
2	237.0	1.664	1.646	1h42	1.617	37h52
3	193.6	1.672	1.652	2h03	1.623	36h01
4	186.6	1.653	1.634	1h20	1.606	23h07
5	177.5	1.650	1.630	1h51	1.604	30h07
6	166.9	1.664	1.643	1h41	1.616	26h38
7	154.6	1.614	1.593	1h08	1.565	18h53
8	141.7	1.670	1.652	1h27	1.626	20h45
9	115.2	1.608	1.589	0h45	1.566	9h55
10	130.7	1.641	1.625	1h19	1.599	19h06
11	130.7	1.647	1.628	0h52	1.603	12h19
12	129.3	1.653	1.632	1h19	1.605	19h27
13	95.5	1.689	1.672	0h53	1.624	9h36
14	87.2	1.667	1.648	0h51	1.577	9h49
15	81.1	1.618	1.599	0h31	1.577	6h02
16	79.9	1.574	1.557	0h32	1.532	7h39
17	77.5	1.599	1.581	0h45	1.553	11h36
18	74.5	1.709	1.693	0h28	1.675	4h09
19	55.8	1.482	1.453	0h30	1.416	6h46
20	59.4	1.694	1.678	0h33	1.656	5h36
21	33.9	1.701	1.681	0h16	1.667	1h37
22	34.4	1.610	1.589	0h16	1.568	3h15
X	147.7	1.550	1.524	1h01	1.490	15h55
Y	22.8	1.149	1.121	0h07	1.113	0h26
Ave	2832	1.638	1.618	24h	1.591	368h

Table 2 Compression of various genomes

Species	Len (Mb)	XM Rate	XM Time
Thermoplasma volcanium	0.6	1.928	6s
Methanococcus jannaschii	1.7	1.813	11s
Haemophilus influenzae	1.8	1.878	11s
Bacillus licheniformis	4.2	1.914	37s
Escherichia coli	4.6	1.908	41s
Mycobacterium tuberculosis	4.4	1.834	58s
Mycobacterium avium	4.8	1.802	83s
Saccharomyces cerevisiae	12.2	1.818	3.5m
Dictyostelium discoideum	33.9	1.506	10.5m
Plasmodium falciparum	23.3	1.513	7.1m
Anopheles gambiae	225.0	1.748	114m
Drosophila melanogaster	120.3	1.833	20.1m
Caenorhabditis elegans	100.2	1.705	34m
Arabidopsis thaliana	119.0	1.659	43m
Vitis vinifera	290.2	1.404	103m
Oryza sativa	370.7	1.349	142.2m
Ciona intestinalis	141.2	1.397	55m
Tetraodon nigroviridis	302.3	1.775	108m
Gallus gallus	1042	1.757	726m

runs, and by another biological compression algorithm, the NML [9], which is the only algorithm found in the literature capable of compressing human chromosomes. The XM-F configuration took about 1 day on a 2.4 GHz processor to compress the 24 human chromosomes, while the NML was reported to take 3.5 h on a cluster of 12 workstations with 3.2 GHz processors [9]. Not only being faster, the XM-F also outperformed the NML by about 0.02 bps in every chromosome. The XM-S configuration performed even better. It made another improvement of 0.03 bps at the cost of more running time.

Table 2 shows the compression of a number of species' genomes. These species were selected from differing organisms, including bacteria, archea, single cell eukaryote, worm, plants and vertebrates. The lengths of these genomes are shown in the second column of the table. The expert model in this experiment was run with hash table size 11 and expert limit 200. The compression results in bits per symbol and the running times are presented in the third and the fourth columns, respectively.

4 Knowledge Discovery Using Expert Model

The preceding section has shown that DNA sequences are compressible, but only by a small ratio. For example, the best compression of the human genome from available tools is 1.591 bps, a saving of only 20% over the baseline *2 bits per symbol* encoding. Of the genomes selected, the genomes of *Vitis vinifera* (grape vine),

Oryza sativa (rice) and *Ciona intestinalis* (sea squirt) are the most compressible, and their saving rates are merely about 30%. The primary purpose of our research on biological compression, however, lies in the information theoretic approach for knowledge discovery. This section highlights some data mining tasks using the expert model.

4.1 Repeat Element Detection

Repeat DNA are abundant in eukaryotic genomes. As much as 55% of the human genome is made up of repeat elements [11]. Many human diseases are associated with repeat elements [2], and thus locating repeats in the genome is very important. However, it is very challenging. A long sequence over a small alphabet would contain many random matches while "genuine" repeat elements are approximate. For example, the ALU repeats in the human genome are about 13% different from each other. In other words, one would expect a mutation in every eight symbols in an ALU repeat element, whereas every 8-mer would be expected to occur randomly every $4^8 = 65536$ bases and thus would occur over 45,000 times in the human genome just by chance.

Unlike most other biological compression algorithms that rely on locating repeat DNA for compression, the expert model does not require the explicit identification of repeats from non-repeats. It instead blends the encoding of non-repeats and of repeats. A substantial repeat would result in a repeat expert that compresses significantly better than the Markov expert over some subsequence. Therefore, a subsequence is considered as a repeat if it is compressed by combining experts better than by the Markov expert alone. The compressibility of a symbol by a certain expert, or by the combination of experts, can be determined by examining the information content of the symbol [7]. By combining experts, the expert model can produce the sequence of information content from different sources of knowledge. The differences in information content with and without additional experts show their contribution.

Figure 1a shows the graph of information content along the HUMHBB sequence produced by the Markov expert. The expert compression performance is only slightly better than the baseline 2 bits per symbol. The information content of the sequence produced by combining the Markov expert with repeat experts is shown in Fig. 1b. The difference of the two information content sequences is shown in Fig. 1c. This amount of information is detected by repeat experts. Each spike of the plot corresponds to an area that is a repeat of an earlier subsequence. This is an example of finding repeats *ab initio* that is finding recurring patterns in a sequence.

The expert model can also be used to find repeat elements by comparing the sequence to a curated library of precomputed motifs. We obtained a collection of repeat elements in the human genome, the RepBase [8]. The HUMHBB sequence is compressed by combining the Markov expert with the *RepBase experts*, specialized repeat experts that base their prediction on the RepBase. Figure 1d shows the

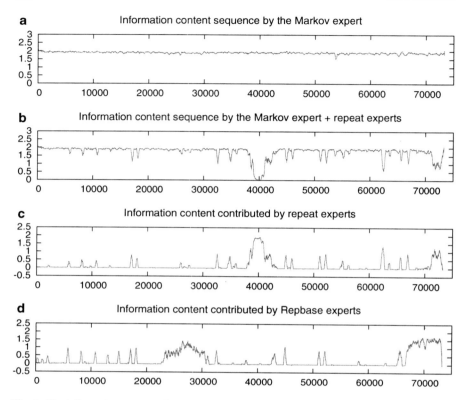

Fig. 1 The information content of sequence HUMHBB produced by various experts

information detected from the RepBase database. The spikes of the plot show the areas in HUMHBB that are similar to some elements in the RepBase. It is also noted that *ab initio* repeat finding locates a gene duplication around the position 40,000.

4.2 Genomic Local Alignment

A similar problem to repeat discovery is local alignment which attempts to identify regions of similarity between two or more biological sequences. Most existing alignment methods such as FASTA [15] and BLAST [1] are based on the dynamic programming paradigm. These methods generally search for short matches, called seeds, and extend seeds to include substitutions and gaps. These methods also depend on a scoring scheme and are insensitive when analysing distantly related sequences and sequences of biased composition.

A local alignment method called XMAligner [6] is developed based on the expert model. To align two sequences, the method compresses one sequence using

Table 3 Alignment performance of Promer, Nucmer and XMAligner

		Pf/Pk (%)	Pf/Pv (%)	Pf/Py (%)	Pf/Pg (%)
Nucmer	S_n	0.49	0.31	5.67	6.69
	S_p	51.20	38.26	91.98	90.49
Promer	S_n	43.15	39.76	48.98	46.23
	S_p	89.13	92.35	83.14	78.72
XM-Aligner	S_n	42.03	39.60	52.22	51.83
	S_p	90.81	91.65	89.45	86.50

the Markov expert learnt from this sequence, and repeat experts from the other sequence. A region is considered a potential homologue if there is a repeat expert whose prediction during its lifetime, if combined with the Markov expert's prediction, results in a lower conditional information content. The amount of reduction indicates the similarity of the two regions.

We applied the method to align the *P. falciparum* genome against the genomes of three other malaria parasites: *P. knowlesi*, *P. vivax* and *P. gallinaceum*. The length of each genome is about 23 megabases. We compared the performance of XMAligner with *Nucmer* and *Promer* from MUMmer package [10], which are the only available tools capable of aligning such long sequences. Each method is evaluated by comparing the alignment with the exon annotation of the *P. falciparum* genome. Table 3 shows the *sensitivity* (S_n) and the *specificity* (S_p) [6] of each method. Nucmer, which aligns at the nucleotide level, performs poorly on the dataset. Its sensitivity is less than 1% in aligning the genome of *P. falciparum* to the genomes of *P. knowlesi* and *P. vivax*. Promer aligns genomic sequences by translating potential coding regions to proteins and comparing at the protein level. Since these species are relatively distant and homologues in DNA show more diversity than at the protein level, Promer outperforms Nucmer as a result. XMAligner, despite aligning at the nucleotide level, is superior to both methods in both sensitivity and specificity.

4.3 Phylogenetic Sequence Analysis

The goal of molecular phylogenetics is to infer an evolutionary relationship among a set of species from some genetic sequence. Most phylogenetic methods can only be applied to short, homologous molecules such as genes or ribosomal RNA due to their intensive computational requirements. Phylogenetic analysis using different genes, however, may result in different trees due to the variation in evolutionary rates among genes [13]. Some genes may even have arisen from means other than inheritance from ancestors and thus are not available in all species involved.

It is suggested that phylogenetic analysis using whole genomes would overcome the inconsistency. However, there has been no available effective tool able to analyse large genomes. We propose using the *mutual information* of two genomes as an estimation of the genetic similarity between two species [4]. The expert model is

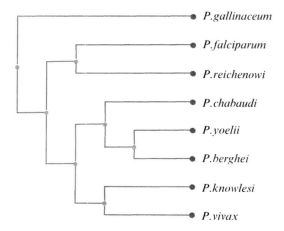

Fig. 2 The generated phylogenetic tree of the *Plasmodium* genus

used to estimate the information content of sequence X. The mutual information of X and Y can be estimated as the reduction in information content of X when Y is used as background knowledge to compress X. The more closely related the two species, the more genetic information they share and thus the more mutual information between their genomes. The phylogenetic tree of a number of species can be constructed from a matrix of pairwise distances between any two genomes by one of the distance techniques such as *neighbour joining* [16].

We applied the method to eight malaria parasites, namely *P. berghei, P. yoelii, P. chabaudi, P. falciparum, P. vivax, P. knowlesi, P. reichenowi* and *P. gallinaceum*. In order to adapt to the environment in the host and the mosquito vector blood, certain *Plasmodium* genes are under strong evolutionary pressure, which leads to variation in evolutionary rates among genes. Analyses using various loci show results that are inconsistent with each other [12]. Furthermore, the variation in composition distributions of these genomes is another challenge for statistical analysis. The AT content of the *P. falciparum* genome is as high as 80%, whereas the composition distribution of the *P. vivax* genome is more uniform even though both species parasitize on human. Conventional analysis tools are misled by such statistical bias [5]. Our method took just under 8 hours to process the 150 megabase dataset and generate a pairwise distance matrix of the sequences. The neighbour joining method was then applied to produce a tree which is shown in Fig. 2. The tree is consistent with most earlier work [12, 18].

5 Conclusion

We have presented the expert model a simple and effective algorithm for compression of biological sequences. The algorithm uses approximate repeats and statistical properties of biological sequences for compression. Our algorithm outperforms all published biological compressors to date while maintaining a practical running time. Furthermore, the algorithm is capable of compressing whole eukaryotic genomes. It has been used effectively to measure information content of the whole genomes of various species from different organism levels.

As a statistical compression method, the model is able to compute the information content of every symbol in a sequence which is useful in knowledge discovery [7, 19]. By examining the conditional information content and the mutual information content, patterns of interest can be found. The expert model is shown to facilitate discovering repeat elements in a sequence or finding instances of patterns from a given library. The model can also align genomic sequences with statistically biased compositions where traditional alignment tools are unable to perform.

The ability to compress long sequences provides an information theoretic approach to measure genetic distances between genomes for phylogenetic analysis. The approach can infer a reasonable phylogenetic tree for eight *Plasmodium* species, which has been known as a hard problem. The trees generated by our approach are largely consistent with the consensus trees from previous work.

References

1. S. F. Altschul, W. Gish, W. Miller, E. Myers, and D. Lipman. Basic local alignment search tool. *Journal of Molecular Biology*, 215:403–410, 1990
2. J. Buard and A. J. Jeffreys. Big, bad minisatellites. *Nature Genetics*, 15(4):327–328, 1997
3. M. D. Cao, T. I. Dix, L. Allison, and C. Mears. A simple statistical algorithm for biological sequence compression. *Proceedings of the 2007 Data Compression Conference*, 43–52, 2007
4. M. D. Cao, L. Allison, and T. I. Dix. A distance measure for genome phylogenetic analysis. *Lecture Notes in Computer Science*, 5866:71–80, 2009
5. M. D. Cao, T. I. Dix, and L. Allison. Computing substitution matrices for genomic comparative analysis. *Lecture Notes in Computer Science*, 5476:647–655, 2009
6. M. D. Cao, T. I. Dix, and L. Allison. A genome alignment algorithm based on compression. *BMC Bioinformatics*, 11:599, 2010
7. T. I. Dix, D. Powell, L. Allison, J. Bernal, S. Jaeger, and L. Stern. Comparative analysis of long DNA sequences by per element information content using different contexts. *BMC Bioinformatics*, 8(Suppl 2):S10, 2007
8. J. Jurka, V. V. Kapitonov, A. Pavlicek, P. Klonowski, O. Kohany, and J. Walichiewicz. Repbase update, a database of eukaryotic repetitive elements. *Cytogentic and Genome Research*, 110: 462–467, 2005
9. G. Korodi and I. Tabus. Normalized maximum likelihood model of order-1 for the compression of DNA sequences. *Proceedings of the 2007 Data Compression Conference*, 33–42, 2007
10. S. Kurtz, A. Phillippy, A. L. Delcher, M. Smoot, M. Shumway, C. Antonescu, and S. Salzberg. Versatile and open software for comparing large genomes. *Genome Biology*, 5(2), 2004
11. E. S. Lander, L. M. Linton, B. Birren, C. Nusbaum, M. C. Zody, J. Baldwin, and K. Devon. Initial sequencing and analysis of the human genome. *Nature*, 409:860–921, 2001
12. M. C. Leclerc, J. P. Hugot, P. Durand, and F. Renaud. Evolutionary relationships between 15 plasmodium species from new and old world primates (including humans): An 18s rDNA cladistic analysis. *Parasitology*, 129(16):677–684, 2004
13. E. Lerat, V. Daubin, and N. A. Moran. From gene trees to organismal phylogeny in prokaryotes:the case of the gamma-proteobacteria. *PLoS Biology*, 1(1):e19, 2003
14. C. G. Nevill-Manning and I. H. Witten. Protein is incompressible. *Proceedings of the 2007 Data Compression Conference*, 257–266, 1999

15. W. R. Pearson and D. J. Lipman. Improved tools for biological sequence comparison. *Proceedings of the National Academy of Sciences*, 85(8):2444–2448, 1988
16. N. Saitou and M. Nei. The neighbor-joining method: A new method for reconstructing phylogenetic trees. *Molecular Biology and Evolution*, 4(4):406–425, 1987
17. C. E. Shannon. A mathematical theory of communication. *The Bell System Technical Journal*, 27:379–423, 1948
18. M. E. Siddall and J. R. Barta. Phylogeny of plasmodium species: Estimation and inference. *The Journal of Parasitology*, 78(3):567–568, 1992
19. L. Stern, L. Allison, R. L. Coppel, and T. I. Dix. Discovering patterns in plasmodium falciparum genomic DNA. *Molecular and Biochemical Parasitology*, 118:175–186, 2001
20. I. H. Witten, R. M. Neal, and J. G. Cleary. Arithmetic coding for data compression. *Communications of the ACM*, 30(6):520–540, 1987

Chapter 68
Open Source Clinical Portals: A Model for Healthcare Information Systems to Support Care Processes and Feed Clinical Research

An Italian Case of Design, Development, Reuse, and Exploitation

Paolo Locatelli, Emanuele Baj, Nicola Restifo, Gianni Origgi, and Silvia Bragagia

Abstract Open source is a still unexploited chance for healthcare organizations and technology providers to answer to a growing demand for innovation and to join economical benefits with a new way of managing hospital information systems. This chapter will present the case of the web enterprise clinical portal developed in Italy by Niguarda Hospital in Milan with the support of Fondazione Politecnico di Milano, to enable a paperless environment for clinical and administrative activities in the ward. This represents also one rare case of open source technology and reuse in the healthcare sector, as the system's porting is now taking place at Besta Neurological Institute in Milan. This institute is customizing the portal to feed researchers with structured clinical data collected in its portal's patient records, so that they can be analyzed, e.g., through business intelligence tools. Both organizational and clinical advantages are investigated, from process monitoring, to semantic data structuring, to recognition of common patterns in care processes.

Keywords Healthcare management systems · Clinical portal · Reuse · Process monitoring · Semantic and syntactic interoperability · Clinical business intelligence

1 Introduction

The role of Information and Communication Technology (ICT) in the Italian healthcare sector has improved very much during the last years, with a growing trend among CIOs to consider ICT as a strategic lever to improve efficiency and effectiveness of care [1], despite healthcare ICT is often present usually in the diagnostic or administration area, in the form of isolated applications [2]. An important

---

N. Restifo (✉)
Fondazione Politecnico di Milano, via Durando 38/a, 20158 Milan, Italy
e-mail: Nicola.Restifo@fondazione.polimi.it; restifo@fondazionepolitecnico.it

contribution has been given by National and Regional rules, which are encouraging hospitals to invest in the digitalization of patients' electronic records. For example, the region of Lombardia recently established guidelines to foster the implementation of electronic enterprise-wide health records (EHR). These are intended to support daily activities of physicians and nurses, which are sources of a great amount of data, whose value in terms of clinical and scientific research could be enhanced if handled and stored in a structured way, according to common vocabularies to identify data inside the clinical process, also trying to reach alignment to international data classifications. Notwithstanding, the few cases in which clinical experimentations have been supported by ICT within a coherent approach [3] are being welcomed by researchers with a high satisfaction level. These showed that investments in the scientific area could bring huge benefits to research, with synergy on investments on ordinary clinical tools, since such functionalities share the same data.

2 Open Source in the Healthcare Sector: A Chance for Growth? The Clinical Portal at Niguarda Ca'Granda Hospital in Milan

Open source is a development method for software that harnesses the power of distributed peer review and transparency of the development process. The promises are better quality, higher reliability, more flexibility, easier application interoperability, lower costs, spur of innovation, and an end to predatory vendor lock-in. But even if supported by national rules (e.g., in Italy), solutions for the healthcare sector are not so widespread, and are mostly systems for radiology management, for organization of booking services, or basic EHRs [4]. First, the healthcare sector has special needs: most of the systems are "mission critical" and thus it is essential to ensure quality and reliability of the software, its compliance to regional standards, and continuity of service. Moreover, there is a need for healthcare organizations to support specific processes with vertical systems (use of speciality solutions).

2.1 A Real Case of Open Source: Niguarda's Clinical Portal

The A.O. Ospedale Niguarda "Ca'Granda" is the leading public hospital in Milan since 1939 and a national centre of excellence, hosting 26 centres of high specialization and is the only regional point of care qualified to perform any kind of tissue and organ transplant. It employs 740 doctors and 1,540 nurses, performing 54,000 admissions and over three million first aid treatments per year. Since year 2000, Niguarda has been revising the whole HIS and rethinking the traditional concept of patient record, conceiving it as a core instrument for supporting clinical and management processes. According to this point of view, Niguarda has developed a web enterprise Intranet portal (the clinical portal), which takes its place "above" the other components of the HIS. In fact, it is designed as an unique access point to vertical subsystems, networking them and thus linking clinical processes, allowing standardizing communication and process interaction between different hospital departments, also introducing automated workflows [5, 6]. The portal provides all patients' clinical data, including those relating to previous care events. Thus, each department can access or contribute to the creation of an EHR, in compliance with requirements of security and traceability according to JCI standards [7].

2.1.1 The Features of the Clinical Portal

Niguarda's portal allows clinical and administrative management both of inpatients and outpatients. For inpatients, the portal allows each ward, e.g., to manage directly its logistic, the pharmacy's inventory, and so on. As far as patient's management is concerned, the portal offers several features dealing with administrative data (e.g., DRG list and discharge summary for reimbursement by the NHS) and clinical activities. During hospitalization, the portal also allows physicians to request examinations or consult the repository of patients' medical reports, both generated by internal function of the portal (resignation letter, echocardiography, chest and digestive endoscopy, stress tests, etc.) or issued by other vertical subsystems (RIS, LIS, pathological anatomy). Recently, the portal has been redesigned by introducing a new functional module, the Clinical Dossier, conceived with a dual purpose: first, to experience a service of EHR that could enable the creation of completely paperless environment, and second, to allow the EHR's management also in mobility within the units through TabletPCs connected to the wireless network. The complete version of the Dossier implements almost all sheets of the paper record, including in nursing needs planning, body functions, and vital signs monitoring. Where a high need of mobility was needed, the Clinical Dossier has been implemented so as to provide a more intuitive interface and free-hand writing recognition. Digital signature of documents generated by the portal and system access through Radio Frequency Identification (RFId) badges are being implemented.

For outpatients, the clinical portal allows the computerized workflow management of the appointments, guiding staff through job planning, appointments' management, nurses' reception, examination, and reports issuing. Reception can be carried out by either a desktop PC or Personal Digital Assistant (PDA). Physicians have the possibility to access everyday's working list, updated in real time from the booking centre as patients arrive. Moreover, physicians can access the electronic prescription book and view previous reports from the hospital's central repository. At the end of each visit, digitally signed reports are sent in structured format to a common repository and can be consulted online by patients through the hospital's service of telemedicine or shared within the Regional SISS network, thus feeding an EHR at regional level in Lombardia.

2.1.2 The Architecture of the Clinical Portal

Niguarda's clinical portal is a pure web application, developed by the ICT unit, based on open source technologies and Oracle database, integrated with plugins acting as interfaces with external subsystems and rich clients. Since 2000, the portal migrated from a two-level (based on JSP) to a three-level architecture deployed as JBoss Web Application and characterized by a logical division of the three layers:

1. Presentation Layer: JSF (Java Server Faces), AJAX, and others
2. Business Logic: Java objects (POJO – Plain Old Java Object) for Business delegate and Model layer
3. Data: Data Access Object (DAO) interface, implemented by Hibernate persistence layer, supported by Oracle DBMS for data management

At the highest level, the user is provided with an advanced and capable interface, based on HTML, JSF, and AJAX technology. Some modules of the portal even belong to the RIA class of web applications: the Clinical Dossier itself is such. At this point, users, interacting with the presentation layer, handle a series of objects, managed by the front controller, which is also responsible for displaying proper feedback to users' actions. The controller then passes objects' instances to the layer of business delegate responsible for the business logic. After this, objects are passed to the DAO layer, which is responsible for data storage, abstracting the interaction with the database, and allowing to define standard methods. These can be implemented using a Persistence Manager like Hibernate [8]. Finally, data management is done by Oracle DBMS. The advantage of this type of architecture is that it can be easily integrated with external components, both at business level, using the Health Level Seven (HL7) messaging standard for the communication, and at data level, extending the methods implemented within the Persistence Manager. Today the portal manages via HL7 also integration procedures such as ADT flows, patients list synchronization, acquisition within the Clinical Dossier of structured data, and files from medical reports – e.g., LIS, order management, and so on.

The final step in the evolution path of the portal will be the transition to a fully compliant J2EE architecture [9]. Niguarda's target is an enterprise web architecture, with business logic implemented by Enterprise Java Bean (see Fig. 1). This kind of architecture allows software components to be implemented inside an application server (JBoss) that, as such, contains a series of integrated services including a logger, a persistence manager, and a web server. But above all J2EE also has important advantages in terms of performance and scalability of the developed system because it enables clustering. Moreover, an application server adherent to the J2EE patterns allows the use of connection pooling, load balancing, and caching services.

The adoption of such an architectural model leads to a deep change also in the organization of the IT Unit at Niguarda, supported by managerial and technology competences of Fondazione Politecnico di Milano, a research institute connected to the Politecnico di Milano Technical University. People working on the portal were

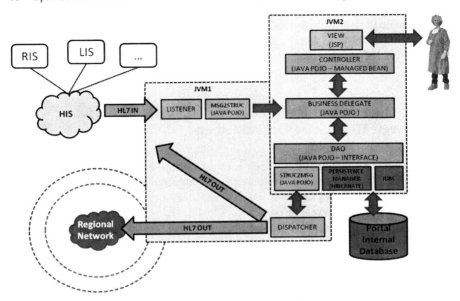

Fig. 1 The target J2EE architecture being implemented by Niguarda's portal

organized in a separate unit, divided into different areas dealing each one with a specific architectural layer. Moreover, new patterns were implemented for software development: Unified Process has become a standard, but some teams also experimented AGILE methodologies. Moreover, an internal roadmap has been developed to guide new releases of the system, and each new package gets tagged and its code commented before being allowed for release.

2.1.3 Benefits of Open Source as Experienced at Niguarda

The choice to use open source technologies instead of proprietary technologies for the realization of individual functional blocks allowed to obtain real benefits. First, the adherence to standard open languages, also for application servers and communication protocols, allows flexible integration with external components. This advantage is even stronger in the healthcare sector, where many vertical (proprietary) subsystems of the HIS must be integrated. In general, compliance to standards is an important aspect also in assuring application maintainability. The open source software used in the project are designed and maintained by Internet virtual communities (often with the support of organizations like the Apache Software Foundation) where continuous testing allows the hospital to access technologies that are evolving faster than proprietary, and that often prove being more stabile, efficient, and flexible. The use of open source languages also provides more opportunities for system customization and system enrichment, thanks to the possibility to edit the code of each program in the whole architecture. Finally, the fact that the architecture is made up of individual blocks, rather than a monolithic proprietary solution, allows interchangeability of components, lowering license costs.

3 Open Source and Reuse: Porting of a Clinical Portal Between Healthcare Organizations

Results gained at Niguarda have been recognized as a best practice in the region: other hospitals have started to evaluate the portal. Among these, the "Fondazione IRCCS Istituto Neurologico Carlo Besta" in Milan, an international centre of excellence for care and research on neurological diseases, has chosen the portal to implement its own clinical information system [10]. The whole project of reuse is based on a porting operation, which is "the process of adapting software so that it can be created for a computing environment that is different from the one for which it was originally designed" [11, 12]. Reusing an application means not "recycling": it involves design activities aiming at making the solution available to be customized in different settings, like Istituto Besta. Business analysts and developers from Niguarda and Fondazione Politecnico joined again to reach this goal. The process that droves the choice by Istituto Besta of porting Niguarda's application has followed several steps (see Fig. 2). A detailed analysis over required functionalities was paramount: Istituto Besta is a more specialized clinical centre, and so the portal needed some extensions.

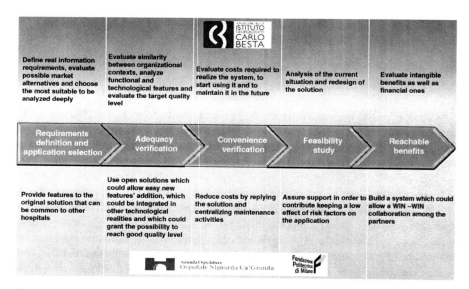

Fig. 2 A model to evaluate reuse opportunity [10]

Further analysis was done to evaluate some quality indicators: compliance to standards, maintainability, reliability, usability, support's quality, portability. Compliance to standard and technical consistency are strategic topics, since no communicating systems could affect the decision of installing them, and much more expensive actions should be taken to substitute already working systems. Maintainability is an aspect of real interest for both developers and users, since applications built with standard technologies are expected to require standard actions to be maintained; from Niguarda's point of view, this aspect is reinforced by the fact that maintenance made for one portal will help preventing bugs also on the other.

Istituto Besta went digital after the first roll-out of the outpatients module in October 2008. Benefits gained are essentially linked to lower costs of the application than those of market packages, which can be translated in satisfaction expressed by clinicians and future savings on development costs. On the other side, Niguarda can face lower costs using open source and then, by selling the application already developed, it could cover the costs sustained to build a solution, and its extensions working not only for Istituto Besta but also for itself (products developed on open platform may turn more convenient than solutions based on proprietary components).

Niguarda monitors Istituto Besta activities to gather feedbacks from users of a different environment and to understand which criticalities could be faced also in the original environment. Bugs are checked by remote-control from Niguarda, and each new release updated both portals. Current challenges regard the organization of the IT structure that will support further evolutions: while system management and user support are organized in each hospital, evolution should follow a coherent strategy. The analysis of application management processes in the open source world can suggest a solution to this problem. Moreover, spreading the acknowledgement of the solution among the Region is a unique opportunity to position itself as a reference product, to offer wider features to customers, and to gain the attention of public Institutions, which could turn to chances to get involved in new funded research projects.

4 "Medical Tutorial": The Clinical Portal gets Customized to Feed Scientific Research and Improve Process Governance

A further step in the cooperation between Niguarda, Istituto Besta, and Fondazione Politecnico is that of evolving the portal storing relevant entries as structured data related to patients and processes, to be used for assessments on the quality of the clinical processes within the organization, for analyses of system's performance and its use by healthcare operators, for creating clinical databases about neurological diseases to be used for clinical research. This means leading clinical activities and research pipelines to convergence. This is supported by the Italian Ministry of

Health within the project "Medical Tutorial: Web Integrated Information System for the Management of Clinical and Research Activities in the Field of Neuroscience."

4.1 Data Structuring Within Outpatients and Inpatients Management

The first implementation of Niguarda's clinical portal inside the Istituto Besta was the module for outpatients' management in October 2008. Since then, more than 1.300 reports per month have been issued digitally, representing more than 50% of the visits. At the end of the appointment, physicians are given the possibility to express a strictly reserved diagnosis for the patient (with signalling of certainty or suspicion), choosing among a set of codified pathologies besides the standard ICD 9 – CM coding, which was considered too administration-oriented. The definition of a semantic vocabulary involved an "expert" multidisciplinary team and resulted in codifying specific pathologies for neuro-oncology, neuropathy, epilepsy, and neuromuscular disorders. The definition of a vocabulary was made by identifying each pathology inside a tree-like model, where a unique identifier was assigned to each definition. As regards neuro-oncology documents, also cancer main and primitive site have been codified, in order to give the opportunity to express a complete diagnosis.

The Clinical Dossier is the side of medical tutorial designed to assist physicians and nurses during ward activities, gone live on June 2009. Inside the Clinical Dossier, many sections can be targeted as access points to insert structured data to feed research. For instance, ICD 9-CM is used to express pathological diseases inside familiar anamnesis, while inside pathological anamnesis, specific neurological classifications can provide consistent data. Another source of scientific data is the ADT system, where patient data are structured to enable the classification of diagnosis and procedures, now with extra information like the entry evidences and the admission cause. As regards clinical placement, physicians have to fill in fixed forms, developed according to their clinical daily needs, with information related to anamnesis, objective examination, and neurological exams. In this way, it will be possible to merge data related to entrance conditions, results of preliminary exams and clinical informations about the hospitalization, allowing to build back the complete history of the patient (family and chronic codified diseases, growing up history, altered exams, etc.). The system also allows to build discharge letters by importing some of the already entered data from daily clinical sheets.

As a pilot experimentation, the tool will enable the management of data related to the four codified pathologies, with particular attention to epilepsy. Data gathered for this disease will be structured in a standard document format, XML – CDA2 (Clinical Document Architecture release 2), to be sent to "Epinetwork," the regional pathology network for epilepsy.

4.2 Business Intelligence Capabilities Enabled by Medical Tutorial: Data Exploitation for Scientific Research and Process Monitoring

At present, research activities are performed through separate management of proprietary databases, fed apart from clinical activities with data related to various neurological pathologies. The main common obstacle is the interest of researchers in keeping strong control over data used to feed research, so that they look with some mistrust at the possibility given to everyone to participate to data gathering, especially for specific areas in which they researched for many years.

The development of the inpatients' section will be accompanied by that of a tool which will enable more complex analysis operating directly on the clinical database. Being a research foundation, Istituto Besta can find the key information on care processes by following clinical trials, where patients are monitored to follow the evolution of their diseases and their responses to therapies. Indeed, digitalization of trials could allow retrospective analysis and to identify regular paths in the care process of neurological diseases, possibly using business intelligence to detect whether the care process is carried on differently from standard paths [13].

Summing up, Fig. 3 shows how the scientific research process could change with Medical Tutorial, shifting from a traditional approach, based on the separation between the clinical and scientific sides, to a clinical-integrated approach with Medical Tutorial as unique internal source for clinical and scientific data. The traditional approach is based on manual feeding of scientific databases. Administrative staff is in charge of interpreting and extracting unstructured data from documents and systems. The clinical-integrated approach shows a borderline between internal and external knowledge. Medical Tutorial works as a unique source where all clinical and scientific data are stored, but enlarging the knowledge base integrating structured reports produced inside departmental systems and, above all, inside other healthcare organizations, connected in network of pathologies.

Business intelligence tools will be provided by Medical Tutorial itself, so that results could be directly accessible during daily activities and researchers. This will ensure more quality to data, which will provide more consistent information to improve patients' safety and more certainty to scientific research results. Accurate process monitoring through key performance indicators shared with managers is another feature enabled by workflow management capabilities and structured data on process activities [13] (e.g., the percentage of reports over the number of patients visited, the percentage of properly closed visits, the percentage of reports sent to the regional repository, the match between prescribed and delivered services).

5 Conclusions

Niguarda, Istituto Besta, and Fondazione Politecnico di Milano developed an agreement defining a model to manage organizational, technical, and functional issues related to the porting and the evolution management of Niguarda's portal using an

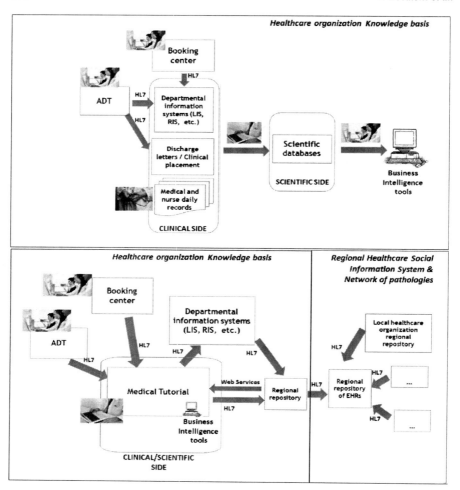

Fig. 3 Traditional and clinical-integrated approaches in clinical research [14]

open source framework. Niguarda's portal is conceived as a tool to manage clinical and administrative processes in the hospital, experienced by the user as a unique interface also to vertical subsystems of the HIS, also integrating a modular flexible EPR. The implementation at Istituto Besta shows how a health information system, developed to support clinical routine, could act as a collector of data to feed databases used for scientific, clinical, and governance purposes [15]. Data will be used for assessments on the quality of the clinical processes within the organization, for creating clinical databases about neurological diseases, and for feeding research business intelligence. This is enabled by a flexible architecture based on open source criteria, but above all by the definition of vocabularies to structure information and the use of interoperability standards (like HL7) for communication among applications and to the outside (e.g., pathology networks). Results show that this experience is a reference on how to pipeline clinical practice and research through a unique digital workspace.

References

1. Zakim D., Braun N., Fritz P., Alscher M.D., 2008. *Underutilization of information and knowledge in everyday medical practice: Evaluation of a computer-based solution.* BMC Medical Informatics and Decision Making 8: 50
2. Chauldry B., Wang J., Wu S. et al, 2006. *Systematic review: Impact of health information technology on quality, efficiency, and costs of medical care.* Annals of Internal Medicine 144 (10): 742–752
3. Tan J.K.H., 2001. *Health Management Information Systems: Methods and Practical Applications.* Sudbury: Jones & Bartlett Publishers
4. Apfelkraut.org, 2008. *Free Medical Software.* Last accessed October 18, 2008
5. Shepherd M., Zitner D., Watters C., 2000. *Medical portals: Web-based access to medical information.* Proceedings of the 33rd Hawaii International Conference on System Sciences
6. Baj E., Locatelli P., Gatti S., Restifo N., Origgi G., Bragagia S., 2009. *Open source: A lever for enhancing opportunities of healthcare information systems.* In: Proceedings of OSEHC 2009 – International Workshop on Open Source in European Health Care, T. Karopka & R. J. Cruz Correia, Eds., pp. 28–37, Porto: INSTICC Press
7. Joint Commission International on Healthcare Safety. *Guidelines and Standards.* Available on http://it.jointcommissioninternational.org/enit/Accreditation-Manuals/EBIAS400/1553/. For example ISBN: 978-1-59940-440-0
8. Sommerville I., 2007. *Software Engineering 8.* Upper Saddle River: Pearson Education
9. SUN, 2007. *Core J2EE Patterns.* Retrieved: http://java.sun.com/blueprints/corej2eepatterns/Patterns/. Last accessed October 17, 2008
10. Locatelli P., Baj E., Origgi G., Bragagia S., 2009. *Medical tutorial: Porting of a clinical portal between healthcare organizations.* In: HEALTHINF 2009 – International Conference on Health Informatics, L. Azevedo & A. R. London, Eds., pp. 375–380. Porto: INSTICC Press
11. Mooney J.D., 1997. *Bringing Portability to the Software Process.* West Virginia University, Dept. of Statistics and Computer Science
12. Garey A., 2007. *Software portability: Weighing options, making choices.* The CPA Journal 77 (11): 3
13. Protti D. J., 2005. *How business intelligence is making healthcare smarter.* NHS Connecting for Health World View Reports, http://www.connectingforhealth.nhs.uk
14. Baj E., Locatelli P., Restifo N., Origgi G., Bragagia S., 2009. *A clinical information system to feed scientific research and allow process monitoring.* Proceedings of the 2009 World Congress in Computer Science – WORLDCOMP, 14 July 2009, Las Vegas
15. Tang P.C., Fafchamps D., Shortliffe E.H., 1994. *Traditional Hospital Records as a Source of Clinical Data in the Outpatient Setting.* Knowledge Systems Laboratory

Chapter 69
Analysis and Clustering of MicroRNA Array: A New Efficient and Reliable Computational Method

Luca Sterpone, Federica Collino, Giovanni Camussi, and Claudio Loconsole

MicroRNAs (miRNAs) play determinant roles in gene expression and several cellular processes of mammalian. These processes include differentiation, development, apoptosis, and cancer pathomechanisms. In detail, miRNAs are known to regulate the expression of keys genes relevant to cancer and potentially to other diseases [1–3]. MiRNAs are short noncoding RNAs having a central function in gene expression, since miRNAs are involved in cellular differentiation processes [4,5], organism developments [6], and apoptosis. Furthermore, recent studies provide growing evidence of the contribution of miRNAs in cancer cellular mechanisms. Therefore, the analysis and the correct identification of miRNAs expression levels play a key role for better understanding and classifying cancer and other diseases. In parallel with the drastic growing of interest into the study of miRNAs, the support of systems and tools that allow an efficient and more robust analysis of miRNA's expression levels become increasing mandatory to have a better disease identification.

The miRNA's analysis flow consists of two phases: the PCR analysis and the clustering. Nowadays, the expression levels of miRNAs are evaluated by reverse transcription and quantitative real-time polymerases chain reaction (qRT-PCR). RT-PCR is a consolidate technique [7] which is performed on a biological target using plastic cards containing a series of miRNAs probes also called *plate*. Following the RT-PCR analysis, miRNAs expression levels are clustered to distinguish biologically meaningful information that can be used to classify and identify specific pathways for a given disease [8,9].

Modern probes, thanks to their high specificity and thanks to the possibility of monitoring different fluorochromes, allow to simultaneously observe more than one sequence [10]. Several methods have been proposed to analyze the fluorescence curve provided by RT-PCR analysis using array [11]. All methods rely on a mathematical model of the fluorescence curve signal assuming an ideal and optimal behavior of the RT-PCR amplification. The proposed computational method provides a relevant support in finding miRNAs role in specific biological

---

L. Sterpone (✉)
Dipartimento di Automatica e Informatica, Politecnico di Torino, Turin, Italy
e-mail: luca.sterpone@polito.it

pathways. We evaluated the effectiveness of the developed method on a set of miRNA experiments performed at the Molecular Biotechnology Center (MBC) of the University of Turin. The obtained results demonstrated the effectiveness and efficiency of the computational method. In particular, the proposed RT-PCR analysis method outperforms of nine times commercial software generally used for the RT-PCR data analysis.

1 Background

The polymerase chain reaction in real-time is a technique based on the simultaneous amplification and quantification of DNA strands. During the last decades, the PCR technique based on reverse transcription (RT-PCR) has become increasingly relevant as method for the analysis of gene expression levels, thanks to technology improvements resulting in drastic reduction of the execution time and in progressive usage of material less toxic and dangerous for the human operator that control the RT-PCR process. In order to evaluate the expression levels of miRNAs coming from various tissues, after their transformation in cDNA, thanks to reverse transcription, it is necessary to perform the RT-PCR process. The PCR is a biomolecular technique, which allows the exponential amplification of nucleic acid sequences *in vitro* [12]. The real-time PCR allows to measure in real time the concentration of a target sequence into a biological sample.

The cycle can be repeated up to 40 times, bringing to an ideal exponential amplification of the DNA target according to the formula: $N = 2^n$, where n is the cycle number, while N is the DNA target number of amplified copies, as illustrated in Fig. 1. However, in real condition not all the DNA target sequences are repli-

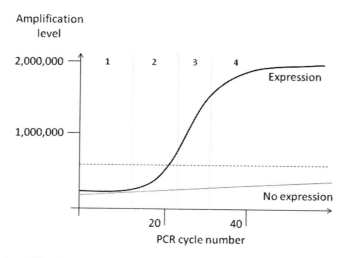

Fig. 1 Ideal amplification curve (PCR cycle with respect to fluorescence)

cated during a PCR cycle; therefore, to model more accurately the amplification, a coefficient comprised between 0 and 2 is used. The amplification formula is expressed by: $N = (1 + E)^n$, where E is the amplification efficiency and has a value comprises between 0 and 1.

The fluorescence curve emitted during an RT-PCR analysis has a sigmoid progress decomposable into four parts:

1. *Linear phase*: during this phase the fluorescence value is almost constant and it grows very slowly, since during the first PCR cycles the amplified product is not detected by the probes due to the minimal quantity.
2. *Exponential phase*: during this phase the fluorescence has a rapid growth, since there is an exponential growth of the target sequences.
3. *Second linear phase*: during this phase the fluorescence diminishes the rapid growth of the exponential phase and tends to have linear characteristics, since the target sequence growth is progressively reaching the quantity saturation.
4. *Plateau*: during this phase the fluorescence remains constant, since the target sequence has reached the saturation.

2 Previous Methods

Several methods have been proposed to analyze real-time PCR fluorescence curves and to support the clustering. Previous developed PCR analysis methods determine the threshold cycle number with derivative methods, searching the maximum second derivates without defining a specific baseline value. In general, nevertheless the capability to provide a CT value, one of the weak aspects of these methods is to not provide any indication of the reaction efficiency or the goodness of the PCR curve. In this section, we provide an overview of the previously developed methods underlining the basic concepts and the advantages presented by our proposed method with respect to these techniques.

A mathematical model for analyzing the relative quantification of real-time PCR has been proposed by Michael W. Pfaffl in [13]. The proposed mathematical method is based on a computation of the relative expression ratio based on the single real-time PCR and from the deviation of the computed threshold cycle with respect to a control sample.

The CP quantification happens always during the exponential phase. The reader can notice how the model proposed by Pffafl, the relative expression ratio of a target gene, is normalized with the expression of the endogenous reference target gene (possibly unregulated) to compensate the intra-PCR variation between the various experiments. In case the CP coefficient of the gene selected as reference is the same as the control gene and of the sample gene, the $\Delta CP = 0$. In this condition, (1) is simplified into (1):

$$\text{Ratio} = (E_{\text{target}})^{\Delta CP_{\text{target}}(\text{Control} - \text{Sample})} \qquad (1)$$

Equation (1) shows a mathematical model of the relative expression ratio in the real-time PCR under constant reference target gene. The CP values of the sample and control genes are equal and represent ideal condition (i.e., $\Delta CP_{ref} = 0$ and $E_{ref} = 1$).

The principal characteristic of the *MaxRatio* method is the identification of a consistent point in proximity of the exponential region of the real-time PCR fluorescence curve without requiring the human intervention [14].

The MaxRatio method analyzes the fluorescence curve in relation to the amplification and computes the ratio at each PCR cycle using (2) given below:

$$\text{Ratio}_n = \frac{\text{Signal}_n}{\text{Signal}_{n-1}} - 1 \qquad (2)$$

The method adopts the following principal characteristics:

1. Since the equation has a fluorescence value both at the numerator and at the denominator, the ratio is a dimensionless and self-normalizing expression. The self-normalizing property is useful to increase the insensitivity related to the signal variability between each miRNA of the same biological experiment.
2. The ratio has a value comprises between 0 and 1.
3. The ratio is based on the cycle-by-cycle ratio of the signal intensity; therefore, it differs from the derivative method such as the mathematical model presented in the previous section.

The computation of the ratio transforms the amplification sigmoid curve into an approximated sigmoid curve which has a well-defined peak, as illustrated in Fig. 2, where it is possible to observe the two most important values: the MaxRatio (MR) which is the maximum value of the ratio curve, and the Fractional Cycle Number (FCN) which is the cycle number in correspondence of the MR.

The MaxRatio method has been successfully evaluated on real-time PCR experiments; however, the method still fails in computing the CT value since it considers only the relative growth of the PCR fluorescence curve without considering possible false positive induced by systematic experimental errors.

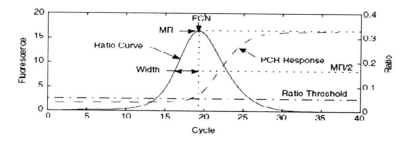

Fig. 2 PCR response and ratio curve for a positive reaction

3 The Proposed Method

The proposed method has the first goal of providing an automatic and robust analysis of data obtained from real-time PCR. This goal has been reached considering two fundamental needs. On one side, there is the need to uniform and quantify the evaluation of the fluorescence curve in order to obtain a certain confidential value with respect to the obtained curve; on the other side, the scientific community has the need to complement the analysis provided by the MaxRatio method that analyzes only the relative growth of the fluorescence curve without any consideration about the similarities between reactive real-time PCR analysis.

The proposed method achieves this goal since it is able to perform a robust analysis and able to avoid systematic errors due to the spectrographer or by the CCD camera that interferes with the correct reaction of the PCR amplification. The method introduces the concept of robustness inside of computational biology algorithm. The efficiency of the proposed method has been experimentally evaluated comparing with data obtained from instrumentation tools of the Applied Biosystem. It has been possible to observe how the proposed algorithm is able to identify false-positive analysis, such cases that are considered as normal amplification while they are not.

3.1 miRNA Expression Analyzer Tool

The miRNA expression analyzer consists of four modules for the analysis of the PCR fluorescence curves. The modules are as follows:

1. *Pre-elaboration*: it refines the data obtained from the real-time PCR monitoring system.
2. *Waveform evaluator*: it computes a waveform Index I_w.
3. *Relative analyzer*: it computes the maximum relative growth Index I_R.
4. *Evaluator*: it computes the final evaluation and whole quantification reporting a confidential Index I_C and the correspondent Threshold Cycle (CT).

The flow of the miRNA expression analyzer tool is depicted in Fig. 3.

The raw PCR fluorescence curves have an irregular distribution, as illustrated in an example in Fig. 4a, due to numerous oscillation of the measured signal among PCR cycles. Besides, the reader should note that each miRNA fluorescence spectrum consists of several curves. The pre-elaboration module normalizes the fluorescence values computing a balanced average on the whole values. The resulting waveform is illustrated in Fig. 4b.

After the pre-elaboration phase, the normalized fluorescence raw data are evaluated by the waveform evaluator and the relative analyzer.

The waveform evaluator assumes that a PCR fluorescence curve must contain the four characteristic phases (linear, exponential, second linear, and plateau). In case a

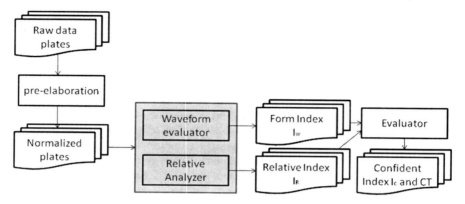

Fig. 3 The flow and the modules of the miRNA expression analyzer tool

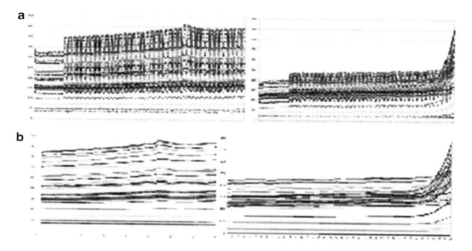

Fig. 4 Original PCR fluorescence raw data (**a**) Pre-elaborated PCR fluorescence raw data (**b**)

fluorescence curve did not have these characteristics, it biologically means that the extension phase of the considered miRNA is not expressed. The evaluation principle is based on the relative growth concept. In particular, the growth levels between PCR cycles are compared adopting the following rules:

- Given four consecutive points (a, b, c, and d) with coordinates expressed by cycle/level of fluorescence, as illustrated in Fig. 5. The following differences are executed: $\Delta_{ab} = b - a$, $\Delta_{bc} = c - b$, $\Delta_{cd} = d - c$.
- Considering the previous computed differences Δ_{ab}, Δ_{bc}, Δ_{cd}. The differences between two consecutive Δ are performed: $\Delta\Delta_{ac} = \Delta_{bc} - \Delta_{ab}$, $\Delta\Delta_{bd} = \Delta_{cd} - \Delta_{bc}$.

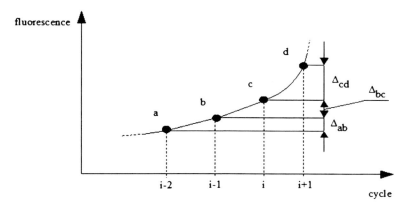

Fig. 5 Computation of the levels Δ_{ab}, Δ_{bc}, and Δ_{cd}

- The condition $\Delta\Delta_{bd} > \Delta\Delta_{ac}$ is verified to detect the beginning of the exponential growth.

The several characteristic phases are individuated with the similar condition: from linear to exponential $\Delta\Delta_{bd} > \Delta\Delta_{ac}$, from exponential to the second linear $\Delta\Delta_{bd} \leq \Delta\Delta_{ac}$, and finally from the second linear to the plateau $\Delta\Delta_{bd} < 0$. The waveform evaluator performs numerical computation on the normalized data containing discrete samples, and the following equation must be verified to identify a particular phase:

1. Linear phase: $y_l = kx$, $\frac{\partial^2 y_l}{\partial x^2} = 0$
2. Exponential phase: $y_e = he^{gx}$, $\frac{\partial^2 y_e}{\partial x^2} = g^2 he^{gx} > 0$
3. Second linear phase: $y_{l2} = wx$, $\frac{\partial^2 y_{l2}}{\partial x^2} = 0$
4. Plateau phase: $y_p = p$, $\frac{\partial^2 y_p}{\partial x^2} = 0$

Once a fluorescence curve is evaluated, the waveform evaluator indicates if the curve is correct or not, and in case it reports the individuated error. The waveform evaluator returns an index I_W that considers all the PCR fluorescence curves for a single miRNA.

On the other side, the *relative analyzer* evaluates the relative grown of the real-time PCR response. In particular, the algorithm computes the ratio curvature and it identifies the maximum and the correspondent cycle number. The relative analyzer returns for each of the real-time PCR fluorescence curve the correspondent index I_R. On the basis of the results of the waveform evaluator and of the relative analyzer, the evaluator will generate a final table containing the list of analyzed miRNA ordered in relation to the confidential indexes I_C computed as $I_W + I_R$.

4 Experimental Results

The proposed computational method has been evaluated on real-time PCR experimental analysis performed with a 7900HT Fast System manufactured by the Applied Biosystems. The PCR analyses have been performed on six different plates, each having 384 miRNAs.

The proposed computational method has been evaluated considering the results obtained by the miRNA expression analyzer algorithm. Please note that all the samples will be named in a unique alphabetical order since the biological data are currently protected by confidential agreement.

4.1 miRNA Expression Analyzer Results

The miRNA expression analyzer algorithm has been compared with the results obtained by the software algorithms provided by the Applied Biosystem tools.

The first analysis compared the threshold cycles computed by the miRNA expression analyzer and by the Applied Biosystem software. In particular, we calculated the average error while computing all the considered miRNAs. The results we obtained are reported in Tables 1 and 2. In Table 1 the average errors obtained considering all the miRNAs that are indicated as reactive by the 7900HT Fast Systems are reported, while in Table 2 the average errors without considering the miRNAs reported as incorrect by the proposed tool are reported.

As the reader can notice, the miRNA expression analyzer demonstrated a more efficient analysis, since the average error in computing the threshold cycle (CT) is nine times better than the commercial existing tool.

The second analysis of the miRNA expression analyzer algorithm has been focused on the waveform evaluation and in the computation of the consistency index I_C.

Table 1 Average error (in cycle) for all the miRNAs analyzed with the 7900HT Fast System

	Sample						
	A	B	C	D	E	F	Average sample
Average error	0.47	0.85	0.72	0.23	0.38	0.61	**0.54**

Table 2 Average error (in cycle) for the miRNAs analyzed with the 7900HT Fast System excluding the miRNAs reported as incorrect by the miRNA expression analyzer algorithm

	Sample						
	A	B	C	D	E	F	Average sample
Average error	0.02	0.01	0.11	0.00	0.10	0.13	**0.06**

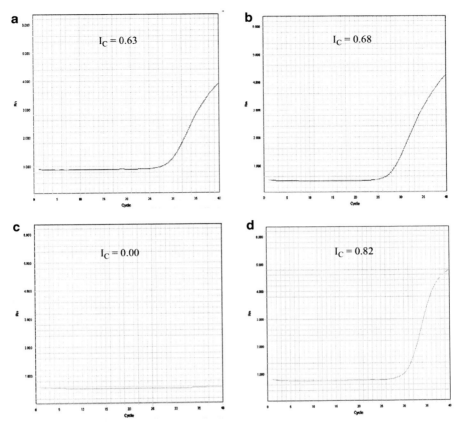

Fig. 6 Relevant fluorescence curves with the correspondent confidential index I_C

As illustrated in Fig. 6, it is possible to note the difference between the case 7.A and 7.B; they are reactive amplification but with a different growth. The case 7.C is a nonreactive amplification, while the case 7.D is the better amplification case, as it is possible to clearly distinguish all the four characteristic phases (linear, exponential, second linear, and plateau).

5 Conclusions

In this chapter, we present a novel method for the analysis and clustering of microRNA array. The main contribution of the proposed computational method is in providing a set of algorithms that can be used to improve the analysis of miRNA expression levels obtained by real-time PCR processes. The method allows to provide a global evaluation of the miRNAs expression profile and to improve the robustness and the reliability of the analysis. The method has been compared with existing software for the analysis of real-time PCR data. Experimental results demonstrated the efficiency and overall improvements on the miRNA expression analysis quality obtained by adopting the developed method.

References

1. He L., Hannon G., *MicroRNAs: smallRNAs with a big role in gene regulation*, Nat Rev Genet 2004, 5: 522, 531
2. Pasquinelli A.E., Hunter S., Bracht J., *MicroRNAs: a developing story*, Curr Opin Genet Dev 2005, 15: 200–205
3. Esquela-Kerscher A., Slack F., *Oncomirs-MicroRNAs with a role in cancer*, Nat Rev Cancer 2006, 6: 259–269
4. Chang S., Johnston R. Jr., Frokjaer-Jensen C., Lockery S., Hobert O., *MicroRNAs act sequentially and asymmetrically to control chemosensory laterality in the nematode*, Nature 2004, 430: 785–789
5. Lee R.C., Feinbaum R.L., Ambros V., *The C. elegans heterochronic gene lin-4 encodes small RNAs with antisense complementarity to lin-14*, Cell 1993, 75: 843–854
6. Reinhart B.J., Slack F.J., Basson M., Pasquinelli A.E., Bettinger J.C., Rougvle A.E., Horvitz H.R., Ruvkun G., *The 21-nucleotide let-7 RNA regulates developmental timing in Caenorhabditis elegans*, Nature 2000, 403: 901–906
7. Peltier H.J., Latham G.J., *Normalization of microRNA expression levels in quantitative RT-PCR assays: identification of suitable reference RNA targets in normal and cancerous human solid tissue*, RNA 2008, 14: 844–852
8. Deregibus M.C., Cantaluppi V., Calogero R., Lo Iacono M., Tetta C., Biancone L., Bruno S., Bussolati B., Camussi G., *Endothelial progenitor cell derived microvesicles activate an angiogenic program in endothelial cells by a horizontal transfer of mRNA*, Blood 2007, 110 (7): 2440–2448
9. Collino F., Revelli A., Massobrio M., Katsaros D., Schmitt-Ney M., Camussi G., Bussolati B., *Epithelial-mesenchymal transitino of ovarian tumor cells induce san angiogenic monocyte cell population*, Exp Cell Res. 15 October 2009, 315 (17): pp. 2982–2994, Epub 2009 June 16, PubMed PMID: 19538958
10. Whitcombe D., Theaker J., Guy S.P., Brown T., Little S., *Detection of PCR products using self-probing amplicons and fluorescence*, Nat Biotechnol 1999, 17 (8): 804–807
11. Sterpone L., *A novel dual-core architecture for the analysis of DNA microarray images*, IEEE Trans Instrum Meas 2009, 58 (8): 2653–2662
12. Bartlett J., Stirling D., *A short history of the polymerase chain reaction*, Springer Ser Meth Mol Biol 2003, 226: 3–6
13. Pfaffl M.W., *A new mathematical model for relative quantification in real-time RT-PCR*, Nucleic Acid Res 2001, 29 (9): 1–29
14. Shain E.B., Clemens J.M., *A new method for robust quantitative and qualitative analysis of the real-time PCR*, Nucleic Acid Res 2008, 36: 14

Chapter 70
Stochastic Simulations of Mixed-Lipid Compartments: From Self-Assembling Vesicles to Self-Producing Protocells

Kepa Ruiz-Mirazo, Gabriel Piedrafita, Fulvio Ciriaco, and Fabio Mavelli

Abstract The computational platform ENVIRONMENT, developed to simulate stochastically reaction systems in varying compartmentalized conditions [Mavelli and Ruiz-Mirazo: *Philos Trans R Soc Lond B Biol Sci* 362:1789–1802, 2007; *Physical Biology* 7(3): 036002, 2010], is here applied to study the dynamic properties and stability of model protocells that start producing their own lipid molecules (e.g., phospholipids), which get inserted in previously self-assembled vesicles, made of precursor amphiphiles (e.g., fatty acids). Attention is mainly focused on the changes that this may provoke in the permeability of the compartment, as well as in its eventual osmotic robustness.

Keywords Self-assembly · Vesicles · Protocells · Self-production · Lipid mixtures · Stochastic kinetics · Osmotic crisis

1 Introduction

Modeling the transition from prebiotic compartments (or "protocells") to biomembranes (i.e., membranes that can support – and be supported by – complex metabolic networks) can help us gain better understanding on the natural origins and dynamic behavior of the latter. Until recent years, most theories of the origin of life [1–3] considered compartmentation as a relatively late landmark, since it poses several problems. Apart from hindering free accessibility of substrates to the enclosed reaction domain, most phospholipids that constitute present day biomembranes are rather complex molecules, whose synthesis is enzymatically controlled [4] and, thus, highly improbable in prebiotic conditions.

K. Ruiz-Mirazo (✉)
Department of Logic and Philosophy of Science and Biophysics Unit (CSIC-UPV/EHU),
University of the Basque Country, Donostia-San Sebastian and Bilbao, Spain
e-mail: kepa.ruiz-mirazo@ehu.es

F. Mavelli (✉)
Chemistry Department, University of Bari, Bari, Italy
e-mail: mavelli@chimica.uniba.it

Nevertheless, alternative types of amphiphilic (lipid-like) compounds, such as simple isoprenoids [5] or fatty acids [6], have been proposed as a more tenable starting point and have actually been shown to self-assemble spontaneously into stable vesicles (closed bilayers). In particular, fatty acid vesicles are being extensively investigated (for a review, see [7]), and it is being found that their dynamic properties are quite different from those of standard liposomes (i.e., phospholipid vesicles). Among other things, it has been demonstrated that they can catalyze their own formation [8] and can be grown and reproduced in *in vitro* conditions [9-11], which is rather difficult to achieve with standard liposomes.

Another important feature of this type of vesicles is that they are much more permeable to different compounds, as compared to standard liposomes. This allows us to conceive the compartment of primitive protocells or vesicles as an easier barrier to cross, at least initially [12, 13]. Studies by Deamer and colleagues (see, e.g., [14]) had shown that the length of the hydrophobic chain of the lipid used had an important effect in the permeability of the bilayer (as expected, the longer the chain, the lower the permeability). But a series of experiments recently carried out in Szostak's lab [13] makes clear that there are other relevant features affecting permeability, as well, like the irregularities on the surface (caused by the different polar heads), or the fluidity and packing density of the hydrophobic tails.

All these *in vitro* experiments assume a scenario that contains some type of prebiotic lipid molecules, which spontaneously self-assemble into vesicles. Starting from that point, sooner or later, an important transition had to occur in which self-assembling membranes became also *self-produced*. In that transition, the naturally occurring lipid (e.g., a fatty acid) would be progressively substituted by an internally synthesized different one (e.g., a phospholipid), modifying the composition and properties of the membrane. In particular, its permeability to the different compounds coming in and out would certainly change, and this can have critical effects for the stability of the whole system.

The aim of this work is to develop a realistic model that addresses this problem, namely, the transformation of a self-assembling *vesicle* into a self-producing *protocell*, focusing on the consequences that the induced permeability changes may have in the viability of the latter. In order to do so, we make use of our platform ENVIRONMENT [15,16], which has been recently adapted to deal with membranes whose permeability is composition dependent, as explained below.

2 Methods: Mixed-Lipid Compartments in ENVIRONMENT

ENVIRONMENT is an object-oriented (C++) platform that has been developed to simulate stochastically (by means of a Monte Carlo algorithm: the Gillespie method [17, 18]) chemically reacting systems in globally nonhomogeneous conditions, in which diverse aqueous and lipidic domains are defined. Our general objective is to elaborate a computational tool that allows exploring the complex, self-organizing dynamics of systems where chemical reactions get coupled with compartment self-assembly and diffusion/transport processes. The general features of this platform have been already described elsewhere [15, 16, 19]; hence, here we will just briefly discuss its main novelty, i.e., the approach used to simulate solute permeability as a function of the membrane composition. In principle, the propensity density function related to solute exchange across a closed lipid bilayer, in our platform, is defined as follows:

$$p_X^{Tr} = D_X S_\mu \frac{\left|\left(C_X^{Env} - C_X^{Core}\right)\right|}{\lambda} \quad (1)$$

where D_X is the diffusion coefficient ($dm^2 s^{-1} mole^{-1}$), S_μ is the membrane surface (dm^2), λ is the membrane thickness (dm), and $\left(C_X^{Env} - C_X^{Core}\right)$ is the difference between the external and the internal concentration ($mole/dm^3$) of the solute X. $p_X^{Tr} dt$ gives the probability that one molecule of X crosses the membrane in the time interval $(t, t+dt)$ in the opposite direction of the concentration gradient. The deterministic flux of solute molecules, in turn, can be described by the differential equation:

$$\frac{1}{S_\mu} \frac{dn_X}{dt} = 10^{-1} P_X \left(C_X^{Env} - C_X^{Core}\right) \quad (2)$$

where n_X is the mole number of X in the vesicle aqueous core, while P_X is the solute macroscopic permeability usually expressed in cm/s. By comparing (1) and (2), it is easy to obtain the relationship:

$$D_X = 10^{-1} P_X \lambda N_A \quad (3)$$

which gives the molecular diffusion coefficient in terms of the macroscopic permeability, N_A being the Avogadro's number and 10^{-1} the conversion factor from cm to dm.

Now, (1) implies that a variation in the membrane composition can affect the propensity probability, since both the bilayer thickness λ and the solute diffusion coefficient D_X depend on the type of lipid or amphiphilic molecule involved. In order to account for this, ENVIRONMENT estimates λ by the ratio between the actual volume V_μ and surface S_μ of the membrane:

$$\lambda = \frac{V_\mu}{S_\mu} = \sum_j N_j v_j \bigg/ \frac{1}{2} \sum_j N_j a_j \quad (4)$$

where a_j and v_j are the surface head area and the molecular volume of the j-th species, respectively, and N_j is the respective number of lipid molecules present in the bilayer. Furthermore, for a binary mixture of two lipids, in this first approximation to the problem, we assume that the solute diffusion coefficient of the mixed membrane D_X^{Mix} will depend linearly on the membrane composition, according to the formula:

$$D_X^{Mix} = D_X^1 - \left(D_X^1 - D_X^2\right) \chi_2^S \quad (5)$$

where D_X^j ($j = 1, 2$) are the diffusion coefficients of solute X across the pure membrane of lipids 1 and 2, respectively, and $\chi_2^S = a_2 N_2 / (2 S_\mu)$ is the membrane surface fraction of the second amphiphile. We are aware that, in reality, things are much more complex (certainly nonlinear), but the lack of adequate experimental data on this question does not make easy a more accurate approximation, so far.

3 Protocell Scenario and Modeling Assumptions

In our model, even if the initial shape of vesicles/protocells will be taken as spherical (for the sake of standardizing initial conditions), it is not assumed that they must stay spherical all the time, or that they divide when they double their initial size (as it is so typically done). Instead, we consider that there is a relatively free relationship between volume and surface, within the following limits:

1. The actual surface of a protocell must be bigger than the theoretical spherical surface that corresponds to the actual volume at each iteration step. Otherwise the protocell will burst (in a simulated "osmotic crisis" – massive water inflow).
2. The actual surface of a protocell must be smaller than the theoretical surface that corresponds to two equal spheres of half the actual volume at each iteration step. Otherwise the protocell divides, giving two statistically equivalent ones.

So these are the conditions for system stability in the model, i.e., they establish the range of possible states in which our protocells will not break or divide (for more details, see [15, 16]). Defining the "reduced surface" of the system, Φ, as: $\Phi = S_\mu / \sqrt[3]{36\pi V_{core}^2}$ (i.e., the ratio between the actual surface S_μ and the surface of an ideal sphere of volume V_{core}), those conditions can be expressed as $1 \leq \Phi \leq \sqrt[3]{2}$. Besides, if one takes into account that the membrane is a relatively elastic structure, two additional parameters can be introduced in the following way:

$$1 - \varepsilon \leq \Phi \leq (1+\eta)\sqrt[3]{2} \qquad (6)$$

where ε and η are the burst and fission tolerance, respectively. Although these two parameters may change as functions of the membrane composition [19], in all simulation runs reported below they were fixed equal to 0.21 and 0.1, respectively. Thus, the actual stability range becomes $0.79 \leq \Phi \leq 1.386$.

Under these general conditions and modeling assumptions, in this work we explored a scenario where already self-assembled vesicles become self-producing protocells, thanks to an autocatalytic proto-metabolism (see Fig. 1). This internal metabolism cyclically transforms the component of the initial membrane: l (a single-tailed fatty acid precursor), into a more complex lipid molecule: L (a double-tailed phospholipid). The critical aggregation concentration (CAC) or relative solubility of this new lipid is much lower than that of the precursor, about two orders of magnitude. As a consequence of this, the freshly produced amphiphiles L are rapidly incorporated to the membrane, while l molecules are released from the bilayer to counterbalance the internal concentration depletion. As the amount of L increases, the membrane changes its composition and becomes less permeable to the different chemical species that can cross it (the nutrient molecules, X and Y, and, in particular,

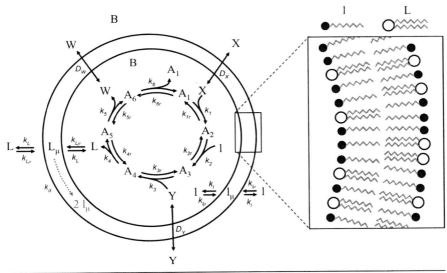

$k_i = 10$, $k_{ir} = 0.1$, for i={1-6}; $k_{l\mu} = k_{L\mu} = 7.6 \times 10^{19}$ s^{-1}M^{-1}dm^{-2}; $k_l = 4.56$ s^{-1}; $k_L = 7.6 \times 10^{-2}$ s^{-1}
$D_x = D_y = 4.62 \times 10^8$ dm^{2}s^{-1}mole^{-1}; $D_w^l = 1.93 \times 10^8$ dm^{2}s^{-1}mole^{-1}; $D_w^L = 5.40 \times 10^5$ dm^{2}s^{-1}mole^{-1}
$[X]_{env} = [Y]_{env} = 0.001$ M (constant); $[W]_{env} = 0$ (constant)
$[l]_{env} = [l]_{core} = 4$ mM; $[A_1]_{core} = 0.002$ M; $[B] = 0.2$ M (other initial concentrations set to zero)

Fig. 1 Reaction scheme through which a single chained lipid (l), with the addition of two externally fed molecules (X and Y) and the contribution of a series of metabolites (catalysts A_i; $I = 1-6$, at least one of whose initial concentration should be non-zero) is transformed into a more complex, double chained lipid (L), producing some waste (W) or leaving group, as well as one of the intermediary metabolites (to make the cycle properly autocatalytic [20]). A spontaneous decay process (L converting back into $2l$), assumed to occur within the membrane, was introduced to find conditions under which the final stationary state of the membrane is not pure. B is a non-reactive species (osmotic buffer), which is important to include for general stability reasons

the waste product W). The subsequent accumulation of a nonfunctional compound, like W, within the boundaries of the system could be a big threat to its stability, because of the natural tendency of water to enter the system (to compensate for the osmotic imbalance [15, 19]) and the eventual risk of bursting. In order to counterbalance the strength of the autocatalytic cycle in this sense, we introduced a decay process (back from L to $2l$) within the lipid bilayer (since in the aqueous core its effect would be minimal – the concentration of L is too low), searching for stationary states in which the membrane composition turns out to be mixed (see below).

4 Results

Indeed, we found that depending on the value of the decay constant (k_d) membranes with different final composition are obtained. As reported in Fig. 2, if k_d is above a relatively small value (e.g., $k_d = 1e - 3$, case A) the decay process has such a

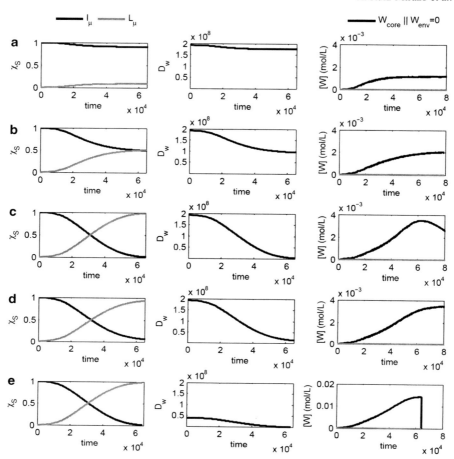

Fig. 2 Time course of membrane composition (*left*), waste diffusion coefficient (*centre*), and waste aqueous concentration in the vesicle core (*right*) depending on k_d values. $[W]_{env} = 0$ in the environment assuming a very rapid dilution of W. For more details, see the main text

strong effect that the conversion of the membrane (from *l* to *L*) is almost utterly suppressed. However, when k_d values are lower (e.g., $k_d = 1e-4$, as in B), a mixed membrane can surely be produced (1:1 ratio, in that case). If there is no decay at all (cases C and E), complete conversion occurs, as one would expect, since all *l* molecules are eventually used up by the proto-metabolic cycle and transformed into *L*. In those cases, if permeability values to W are low enough (as in E, where D_w^l is $3.85 \times 10^7 \, dm^2 s^{-1} \, mole^{-1}$), the system undergoes bursting by "osmotic crisis."

If the total amount of *l* molecules available is not limited (i.e., if the external environment is assumed to be a continuous source of them), the conversion will never be complete, but almost (see case D in Fig. 2 again). The advantage of this condition (for cases with no decay) is that it ensures the endless running of the proto-metabolic

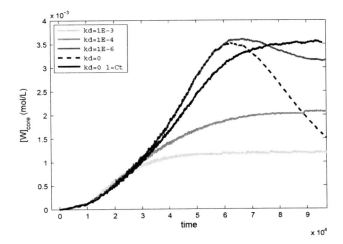

Fig. 3 Waste accumulation in the protocell core for different values of the decay constant k_d. The higher the values of k_d, the lower the W concentration levels, since permeability does not decrease so much. In addition, the proto-metabolic cycle deceleration effect should be taken into account, but this is more acute for smaller values of k_d (being due to scarcity of l)

cycle (which otherwise would cease operating, due to the lack of l). Nevertheless, the slowing down (or even temporary switching off) of the internal autocatalytic cycle can also have a positive effect as a way of regulating W production (and, therefore, as a way of avoiding an osmotic burst – see Fig. 3).

5 Discussion

The hypothesis that the first biosynthetically produced phospholipids were built from precursor molecules (e.g., fatty acids) available in prebiotic conditions may have, at least, two advantages: first, quite obviously, it does not require so many reaction steps (it would be like a tinkering strategy: the system makes use of what is already there); second, in line with the approach and results of this piece of work, the process could be facilitated by the presence of previously formed vesicles, which provide a hydrophobic compartment where the new, more complex lipids can be readily incorporated. Furthermore, if the first proto-metabolic network was autocatalytic, as we assumed here, the fact that the prebiotic lipid (l) is directly involved in the synthesis (of L) can help to regulate, as a limiting factor, an otherwise too rapid transition toward much more impermeable membranes, which would have to face strong osmotic problems. Thus, we shall continue exploring this type of scenario to gain better understanding of how membrane boundaries and reaction networks became so deeply complementary [21].

References

1. de Duve C (1991) *Blueprint for a cell: the nature and origin of life*. Neil Patterson Publishers, Burlington
2. Eigen M, Winkler-Oswatitsch R (1992) *Steps towards life: a perspective on evolution*. Oxford University Press, New York
3. Kauffman S (1993) *The origins of order: self-organization and selection in evolution*. Oxford University Press, Oxford
4. Peretó J et al (2004) Ancestral lipid biosynthesis and early membrane evolution. *Trends Biochem Sci* 29(9):469–477
5. Ourisson G, Nakatani Y (1994) The terpenoid theory of the origin of cellular life: the evolution of terpenoids to cholesterol. *Chem Biol* 1(1):11–23
6. Monnard PA, Deamer DW (2002) Membrane self-assembly processes: steps toward the first cellular life. *Anat Rec* 268:196–207
7. Morigaki K, Walde P (2007) Fatty acid vesicles. *Curr Opin Colloid Interface Sci* 12:75–80
8. Walde P, Wick R, Fresta M, Mangone M, Luisi PL (1994) Autopoietic self-reproduction of fatty acid vesicles. *J Am Chem Soc* 116:11649–11654
9. Berclaz N, Mueller M, Walde P, Luisi PL (2001) Growth and transformation of vesicles studied by ferritin labeling and cryotransmission electron microscopy. *J Phys Chem B* 105:1056–1064
10. Chen IA, Szostak JW (2004) A kinetic study of the growth of fatty acid vesicles. *Biophys J* 87:988–998
11. Hanczyc M, Fujikawa SM, Szostak JW (2003) Experimental models of primitive cellular compartments: encapsulation, growth, and division. *Science* 302:618–621
12. Deamer DW (2008) Origins of life: how leaky were primitive cells? *Nature* 454:37–38
13. Mansy S et al (2008) Template directed synthesis of a genetic polymer in a model protocell. *Nature* 454:122–126
14. Monnard PA, Deamer DW (2001) Nutrient uptake by protocells: a liposome model system. *Orig Life Evol Biosph* 31:147–155
15. Mavelli F, Ruiz-Mirazo K (2007) Stochastic simulations of minimal self-reproducing cellular systems. *Philos Trans R Soc Lond B Biol Sci* 362:1789–1802
16. Mavelli F, Ruiz-Mirazo K (2010) ENVIRONMENT: a computational platform to stochastically simulate reacting and self-reproducing lipid compartments. *Phys Biol* 7(3):036002. doi:10.1088/1478-3975/7/3/036002
17. Gillespie DT (1976) A general method for numerically simulating the stochastic time evolution of coupled chemical reactions. *J Comput Phys* 22:403–434
18. Gillespie DT (1977) Exact stochastic simulation of coupled chemical reactions. *J Phys Chem* 81:2340–2369
19. Ruiz-Mirazo K, Mavelli F (2008) On the way towards 'basic autonomous agents': stochastic simulations of minimal lipid-peptide cells. *Biosystems* 91:374–387
20. Ganti T (2002) On the early evolutionary origin of biological periodicity. *Cell Biol Int* 26:729–735
21. Varela FJ, Maturana H, Uribe R (1974) Autopoiesis: the organization of living systems, its characterization and a model. *Biosystems* 5:187–196

Chapter 71
A New Genetic Algorithm for Polygonal Approximation

Cecilia Di Ruberto and Andrea Morgera

Abstract In this chapter, the problem of approximating a closed digital curve with a simplified representation by a set of feature points containing almost complete information of the contour, i.e., dominant points, is addressed. We adopt an approach based on genetic algorithms (GAs) since they use parallel search and have good performance in solving optimization problems. The chromosome coincides with an approximating polygon and is represented by a binary string. Each bit, called gene, represents a curve point where dominant points have 1-value. The proposed algorithm enhances the selection and mutation phase avoiding the premature convergence issue. Our method is compared to other similar approaches and its efficiency is clearly demonstrated by experimental results giving a better approximation by lowering the error norm with respect to the original curves.

Keywords Digital curves · Dominant point · Genetic algorithm · Polygonal approximation · Shape representation

1 Introduction

Polygonal approximation consists in reducing the number of points of a digital closed curve. It leads to a higher speed during execution and a lower memory usage. Although the reduction of points could suggest a loss of information, the surviving ones contain the main features of the curve, less affected by noise and more suitable for classification purpose. Dominant points can suitably describe the curve for both visual perception and recognition [1, 2].

There are several approaches for dominant points detection: corner detectors [3, 4], sequential, iterative [5, 6], optimal [7–10], and meta-heuristic (tabu-search [9], ant colony [10, 11], and genetic algorithms [7, 12]).

A. Morgera (✉)
Department of Mathematics and Computer Science, University of Cagliari, Cagliari, Italy
e-mail: andrea.morgera@unica.it

H.R. Arabnia and Q.-N. Tran (eds.), *Software Tools and Algorithms for Biological Systems*, Advances in Experimental Medicine and Biology 696,
DOI 10.1007/978-1-4419-7046-6_71, © Springer Science+Business Media, LLC 2011

In this chapter we present a new method based on a genetic approach. The chromosome coincides with an approximating polygon and is represented by a binary string. Each bit, called gene, represents a curve point where dominant points have 1-value. The proposed algorithm enhances the selection and mutation phase avoiding the premature convergence issue. It is compared to other similar approaches and its efficiency is clearly demonstrated by experimental results giving a better approximation by lowering the error norm with respect to the original curves. The chapter is organized as follows. In the second section we describe the main aspects of GAs. In the third section we analyze the approximation polygonal problem and how it can be solved by a genetic approach. In the fourth section we describe our proposed method. Finally, we present some experimental results and comparisons.

2 Genetic Algorithms in Polygonal Approximation

In the last decade, GAs have found large space in the field of polygonal approximation. The starting point of GAs polygonal approximation is the modelling of the problem, which defines the structure of chromosomes to be used and the measure of comparison to use to quantify the goodness of the solutions. Generally, the selected chromosomes are binary, while the most used measure is the error of approximation. Given a closed curve consisting of n points and a value of v representing the number of points that the approximating polygon must have, each chromosome will be composed of n genes that can take the following values:

$$\text{gene} = \begin{cases} 0 \text{ the point is not part of the approximating polygon} \\ 1 \text{ the point is part of the approximating polygon} \end{cases}$$

From this we can see that the algebraic sum of the genes that make up a chromosome representing a permissible solution must be equal to v. Moreover, the points that make up the curve (and hence also those that make up the polygon) must be ordered (clockwise or counterclockwise) so that for every point you can determine its predecessor and its successor.

Consider a curve $C = \{p_1, p_2, \ldots, p_i, \ldots, p_j, \ldots, p_n\}$ consisting of n ordered points and identify with $\widehat{P_i P_j}$ and $\overline{p_i p_j}$, respectively, the arch and the chord of extremes p_i and p_j. The error obtained by approximating the arc $\widehat{P_x P_y}$ with the segment $\overline{p_x p_y}$ may be defined as:

$$L_{x,y} = \sum_{p_i \in \widehat{P_x P_y}} d^2(p_i, \overline{p_x p_y})$$

where $d(p_i, \overline{p_x p_y})$ represents the distance of point p_i from the chord $\overline{p_x p_y}$, which can be understood as the length of the perpendicular to the segment led from p_i to r, where r represents the line through the points p_x and p_y. Letting

$C' = \{v_1, v_2, \ldots, v_i, \ldots, v_j, \ldots, v_m\}$ with $m \leq n$, the list of vertices that make the polygonal approximation of C, the quantity

$$E_2(C, C') = L_{1,2} + L_{2,3} + \ldots + L_{m-1,m} + L_{m,1}$$

is defined as the sum of squared error (integral square error or *ISE*) and represents a quantification of the difference between the original curve and its approximation. In [12], a chromosome is used to represent a polygon and is represented by a binary string. A data reduction process is first applied to prevent unnecessary computation in the evolutionary process. The objective function is defined as the *ISE* between the given curve and the approximating polygon. Three genetic operators namely selection, crossover, and mutation have been constructed for this specific problem. The method proposed in [7] is different from the standard GAs in crossover and mutation operations to solve the specific application. It has good approximation results among several existing methods, but the computational load is relatively high because of the parallel search of GA.

3 The Proposed Method

In order to speed up the computational time, especially in the case of complex curves, we use the same technique described in [12]: in the preprocessing phase a matrix L, containing the values of quadratic errors of every possible segment, is created. In this way, we can compute quickly every single generation even if a time to build the matrix L is required. Taking into account that GAs tend to balance the exploration of the space research, it follows that a considerable part of the values contained in the array is never used during algorithm execution. Hence, we decided initially to create L empty and update it only during the steps demanding fitness control. Hence, when the algorithm needs to know the fitness value first check it: if the standard error of the required segment is present in the matrix we use it, otherwise the fitness computed value is stored in the matrix. By following this procedure, the first generations are computed slowly, while the speed is greatly increased as the matrix L is being filled.

The selection phase has been implemented, rather than with the roulette wheel technique, using the method of *simulated annealing* [13].

As for selection, even for the crossover was decided to use a completely different method than the one proposed by [12]. First of all we leave the elitism, because, after the introduction of Boltzmann selection, it was not considered influent in searching for solutions. We also eliminate the probability of crossover, preferring to process all the chromosomes with this operator; in fact we notice that the competition between parents and children is enough to keep an adequate number of chromosomes (in terms of balance) within the new population respect to the previous generation. Besides, we decide to increase the diversity of children by parents, through a decomposition of chromosomes characterized by a variable number of sub-strings.

To achieve this goal, we use a method inspired by the crossover adopted by [14]. Given two chromosomes X and Y involved in crossover process and represented as:

$$X = x_1 x_2 x_3 \ldots x_k$$

$$Y = y_1 y_2 y_3 \ldots y_k$$

starting from the first gene, we compute the value d so that:

$$\sum_{i=1}^{d} x_i = \sum_{i=1}^{d} y_i.$$

This allows to determine a sub-string that exchanged between the two genes permit to maintain two eligible solutions. After identifying the first group of genes, we repeat the procedure applying it from time to time to the portion of the chromosomes that have not been examined yet; the process ends when the last gene is reached. We decide to introduce an additional constraint: each sub-string should have a length not less than two. At this point, the method described in [14] uses an orthogonal array to determine the best combination of sub-strings suitable to generate the two best children. It implies that the second child differs from the first only for a sub-string: this could facilitate a low diversification of the population and lead to premature convergence. We therefore decide to build every time eight children randomly, where each i-th sub-string has a 50% chance of being taken from the first parent and 50% to be extracted from the second parent. This value is chosen because it was observed that using the L matrix, containing the errors of the segments, the construction of additional children does not involve significant worsening over time used; furthermore, we test that values greater than 8 do not produce significant enhancements. Thereafter eight children and two parents will compete: the two chromosomes with higher fitness are included in the next population.

During our experiments we have to deal with the premature convergence problem. We decide to observe in more detail the evolution of the population at different generations. This phase is characterized by analytical methods common to *numerical taxonomy*, a technique used in biology to classify organisms on the basis of similarity, without taking into account the evolutive relations. Premature convergence is due to the presence of chromosomes characterized by a value of fitness very high compared to other sectors of the population; this situation in turn leads to true direct cause of the problem, namely the reduction of diversity and consequently uniformity of the resulting chromosomes. So using the numerical taxonomy we compare the various chromosomes to identify the moment at which the features of some chromosomes begin to spread too intensely. The instrument used for the purposes of comparison is known as *index*, a statistical technique that allows to quantify the similarity between elements of a set. By considering the analysis of the population, to compare all possible pairs of chromosomes at the beginning of each generation the Jaccard index [15] is used. By experimental results, we also notice that if in the population had been placed two or more chromosomes identical with relatively high fitness values, convergence is reached very quickly, because in each generation the similar chromosomes increase exponentially. We decide then to modify identical chromosomes before applying the operators of selection, crossover, and mutation. We choose to apply a sort of deep and random mutation to the chromosomes in question, to increase the diversification of the population. For each set H_i of identical chromosomes, consisting of n_i elements, with $n_i > 1$, $n_i - 1$ are taken, and each is modified by a random shift of their genes. This modification brings a significant improvement in performance by reducing the variance between the values obtained in various tests on the same case and without an aggravation of the best results. Moreover, since the information offered by Jaccard index allow to identify not only the equal chromosomes but also those which are most similar to others, we decided to exploit them for further improvements. The similarity values are stored in a matrix where the rows and columns represent chromosomes. The row (or column) with the highest sum corresponds to the chromosome most similar to all others and therefore are more inclined to have children alike. By performing different actions in this direction, it is noted that the application of the shift to the first two random chromosomes, more similar to the others, results in additional improved performance. Since the application of the shift can be seen as a random mutation that can improve the spread of research within the space of solutions, we thought of making more profound operator intervention to the original mutation which, as written previously, tends to focus research toward specific pathways. Indeed, it is considered that the two approaches could offset each other and thus improve the behavior of the algorithm in such situations where we are close to a global optimum. Initially we increase the value of the probability of mutation, from 0.15 to 0.25; we also decide to use more operators into the final stages of processing. As mentioned above, the threshold t_e, representing the number of successive generations needed before the algorithm is stopped, is set, in the method of [12], to 10. Hence, we set $t_e = 18$ and apply the mutation, with a probability equal to 80%, to a portion of the population when the number of consecutive generations without improvement is greater than $t_e/2$. The number of chromosomes affected by this procedure must be equal to:

$$\frac{z}{t_e} - g_c - g_b + 2$$

where z is the population size, g_c is the current generation number, and g_b the number of generation in which the best chromosome has been discovered. As expected, the introduction of these changes has led in some cases to obtain better results than the previous ones, further reducing their variance. Initially, we decided to leave the preprocessing method proposed in [12] for the removal of collinear points. But if, on one hand, this produces a significant reduction in the workload of the algorithm, with benefits on execution speed, on the other hand, we observed that in the final results of more complex curves the positioning of some vertices was not entirely satisfactory. For example, there have been frequent cases where two (or more in the worst cases) vertices were close to each other, when it would be more appropriate to locate one of them in a different position. By eliminating the preprocessing phase the computational time is longer, but the results improved considerably. From the tests carried out, we infer that the removal of collinear points does not lead to significant losses of information only when we deal with simple digital curves; when we need to analyze more complex curves, it is preferable not to reduce points. For these reasons, we choose not to include this feature in the final version of the algorithm. We point out, however, that despite keeping the collinear point suppression step in preprocessing phase, global results obtained are better than those obtained by [12].

The proposed algorithm can be resumed as follows:

- **Input**: the original curve C and the number of segments m of the approximating polygon
- **Output**: List of vertices that approximate the given curve C

Steps:

1. Random generation of a population made by 60 chromosomes each representing possible polygonal approximation.
2. Localization of the best chromosome S_b; $g_b = 0$; $g_c = 1$; $t_e = 18$.
3. Generation computing. Random mutation through shift of any identical chromosome in the population and the three chromosome most similar to the others.
4. Boltzmann selection and mating pool construction.
5. Crossover : couple generation on chromosomes randomly choosen by the mating pool; for every couple random generation of eight children; competition between parents and children; insertion of the two best chromosomes into the new population.
6. New population chromosomes mutation with 25% of mutation probability for each gene.
7. New population best chromosome (S_{cb}) localization;
   if $F(S_{cb}) > F(S_c)$ **then** $S_b = S_{cb}$, $g_b = g_c$; **goto** 3
   else if $(g_c - g_b) = t_e$ **then goto** 7;
   else if $(g_c - g_b) > t_e/2$ **then** mutation of a new population chromosome part: we consider only those which have 80% probability of gene mutation; **goto** 3.
8. The algorithm returns S_b, representing the vertices list of the best C approximation found.

4 Experimental Results and Conclusions

We show now the results obtained by applying the algorithm to some images (Figs. 1–5). In the following tables, we indicate the parameter k, representing the number of segments of the polygonal approximation, the lowest and the average integral square error (Avg ISE), its variance (Var), and the average computational time expressed in seconds (Avg time). The average value is referred to 20 times execution tests. Other well-known polygonal approximation techniques are tested and compared to our algorithm.

From the results obtained, we can see that the algorithm implemented allows to obtain better results than other methods analyzed exploiting GAs. From Tables 6 and 7, we can also notice that for the examined cases the results appear to be equal or close to those reported in [16], using a method based on

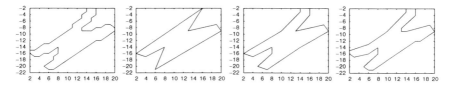

Fig. 1 Chromosome shape: Original (*60 points*), approximation by 8 segments, approximation by 15 segments, approximation by 22 segments

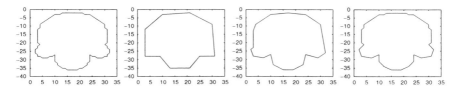

Fig. 2 Semicircle shape: Original (*102 points*), approximation by 10 segments, approximation by 18 segments, approximation by 30 segments

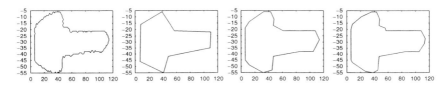

Fig. 3 Noise key shape: Original (*269 points*), approximation by 8 segments, approximation by 16 segments, approximation by 20 segments

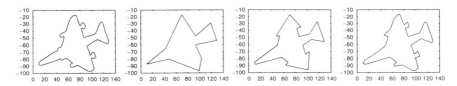

Fig. 4 F10 shape: Original (*409 points*), approximation by 10 segments, approximation by 20 segments, approximation by 35 segments

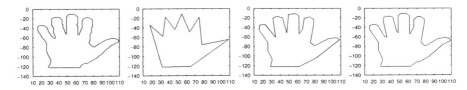

Fig. 5 Hand shape: Original (*528 points*), approximation by 11 segments, approximation by 28 segments, approximation by 42 segments

the elimination of dominant points which ensure to obtain the best result in 96% of cases. As for the computational time we can make some observations. It has been said that the preprocessing phase for the collinear points suppression was not included because, although it allows to reduce considerably the execution time, in most cases leads to greater errors on the polygonal approximations. However, we must remember that the results obtained, though lower quality, can be useful in many fields of application (e.g., in automated comparison of figures), which require not particularly close to the original curve figures. Hence, in these cases the preprocessing can be inserted to obtain a reduction of workload and hence the processing time. Another aspect that emerges from the Tables 1–5 is the processing time reduction as the number of segments k

Table 1 Results for Fig. 1 (60 points)

k	Best ISE	Avg ISE	Var	Avg time
8	13.43360	13.43360	0.00	1.91
9	12.07770	12.13250	0.01	1.86
10	8.06804	8.08098	0.00	1.98
12	5.81803	5.87695	0.03	1.90
14	4.16713	4.25055	0.02	2.01
15	3.79610	3.85910	0.01	1.99
17	3.12945	3.18453	0.00	1.94
18	2.82600	2.88708	0.00	2.11
20	2.33252	2.37767	0.00	2.08
22	1.93252	1.96881	0.00	2.16

Table 2 Results for Fig. 2 (102 points)

k	Best ISE	Avg ISE	Var	Avg time
10	38.9200	39.8390	16.89	2.82
12	26.0045	26.1159	0.10	2.69
14	17.3855	17.3855	0.00	2.85
15	14.3994	14.3994	0.00	2.79
17	12.2179	12.3572	0.04	2.97
18	11.1902	11.3016	0.01	2.85
19	10.0364	10.1940	0.04	2.86
20	9.0087	9.10742	0.03	3.07
22	7.0112	7.23467	0.02	3.22
25	4.6248	4.85342	0.06	3.18
27	3.7013	3.91912	0.04	3.08
30	2.6425	2.92136	0.02	3.04

Table 3 Results for Fig. 3 (269 points)

k	Best ISE	Avg ISE	Var	Avg time
8	550.1150	550.1150	0.00	16.83
10	237.1970	237.1970	0.00	14.09
13	107.7740	108.6160	2.40	11.71
15	83.3271	84.3572	1.15	11.57
16	73.3830	74.3279	0.45	11.20
18	57.7161	59.6446	1.64	11.00
20	51.2445	52.6971	3.67	10.42

Table 4 Results for Fig. 4 (409 points)

k	Best ISE	Avg ISE	Var	Avg time
10	2,603.6600	2,603.6600	0.00	37.42
20	628.25200	639.6410	482.33	22.73
30	143.19200	162.8280	703.49	19.07
35	94.9054	98.9640	14.09	18.58

Table 5 Results for Fig. 5 (528 points)

k	Best ISE	Avg ISE	Var	Avg time
11	3,177.9200	3,177.9200	0.00	57.91
19	402.6030	405.5330	171.72	37.18
28	148.6800	154.5570	71.83	29.45
35	88.5785	93.6038	17.38	27.87
42	51.3637	55.7249	12.74	26.63

Table 6 Comparative results between our method (OUR), Masood' one [16], and Ho-Chen's one [14]

	OUR		Masood	Ho-Chen	
k	Best ISE	Avg ISE	ISE	Best ISE	Avg ISE
Chromosome					
8	13.43360	13.43360	–	13.43	15.51
9	12.07770	12.13250	–	12.08	13.49
10	8.06804	8.08098	8.07	–	–
12	5.81803	5.81695	5.82	5.82	6.79
14	4.16713	4.25055	–	4.17	5.11
15	3.79610	3.85910	3.88	3.80	4.32
17	3.12945	3.18453	3.14	3.13	3.55
18	2.82600	2.88708	2.83	2.83	3.04
Semicircles					
10	38.9200	39.8390	–	38.92	44.07
12	26.0045	26.1159	–	26.00	29.53
14	17.3855	17.3855	–	17.39	20.14
15	14.3855	14.3994	14.40	–	–
17	12.2179	12.3572	–	12.22	14.58
18	11.1902	11.3016	–	11.34	12.86
19	10.0364	10.1940	10.04	10.04	11.52
22	7.0112	7.23467	7.01	7.19	8.52
27	3.7013	3.91912	3.70	3.74	5.03
30	2.6425	2.92136	2.64	2.84	3.57

increases. This phenomenon is caused by the operator of mutation. Please note in fact that if h_i is the gene to be changed and h_u e h_v are the previous and the next gene, respectively (both equal to 1), then the algorithm will process all the positions between (h_u, h_i) e (h_i, h_v) to find the best placement of h_i. As the number of polygons approximating segments increases, the space separating two consecutive genes equal to 1 within the chromosome reduces; therefore, the number of operations that

Table 7 Comparative results between our method (OUR), Huang and Sun'one [12], and Yin's one [7]

k	OUR Avg ISE	Avg time	Huang-Sun Avg ISE	Avg time	Yin Avg ISE	Avg time
Chromosome (Fig. 1)						
8	13.43360	1.91	13.9809	0.68	32.7134	8.61
12	5.81803	1.90	6.0960	0.64	20.7801	8.56
15	3.85910	1.99	4.3262	0.66	12.3058	8.63
18	2.88708	2.11	3.5422	0.68	7.71227	8.72
Semicircles (Fig. 2)						
10	39.8390	2.82	38.9200	1.18	93.4187	13.59
14	17.3855	2.85	21.8944	1.31	45.7951	13.36
19	10.1940	2.86	12.0241	1.35	27.9310	13.52
27	3.91912	3.08	6.5279	1.37	15.0895	13.65
Key (Fig. 3)						
8	550.1150	16.83	628.616	14.24	1,590.360	33.03
15	84.3572	11.57	125.144	14.62	387.210	33.11
18	59.6446	11.00	83.3598	14.71	300.167	33.21
20	52.6971	10.42	70.0745	14.73	246.844	33.19
F10 (Fig. 4)						
10	2,603.6600	37.42	2,746.4595	114.16	7,616.95	49.86
20	639.6410	22.73	841.5713	114.06	2,755.90	50.06
30	168.8280	19.07	322.9759	114.99	1,460.79	50.36
Hand (Fig. 5)						
11	3,177.9200	57.91	7,694.6000	25.51	14,532.80	64.65
28	154.5570	29.45	257.235	26.06	1,824.25	65.18
35	93.6038	27.87	169.99	26.38	1,163.73	65.29
42	55.7249	26.63	124.4370	26.28	907.716	65.65

the mutation operator must perform also decreases. In order to speed up the algorithm, we can consider to reduce the mutation probability; this is not suitable for two main reasons: this would lead to a worsening of final solutions and there would be a slowdown of convergence. This fact increases the number of generations to be produced and leads to the opposite effect to that intended, increasing the processing time required by the algorithm.

References

1. A. Carmona-Poyato, N.L. Fernández-García, R. Medina-Carnicer, F.J. Madrid-Cuevas. Dominant point detection: A new proposal. *Image and Vision Computing*, 23:1226–1236 (2005)
2. P. Cornic. Another look at dominant point detection of digital curves. *Pattern Recognition Letters*, 18:13–25 (1997)
3. C. Teh, R. Chin. On the detection of dominant points on digital curves. *IEEE Transactions on Pattern Analysis and Machine Intelligence*, 8:859–873 (1989)
4. W.Y. Wu. Dominant point detection using adaptive bending value. *Image and Vision Computing*, 21:517–525 (2003)

5. M. Marji, P. Siy. A new algorithm for dominant points detection and polygonization of digital curves. *Pattern Recognition*, 36:2239–2251 (2003)
6. B.K. Ray, K.S. Ray. Detection of significant points and polygonal approximation of digitized curves. *Pattern Recognition Letters*, 13:443–452 (1992)
7. P.-Y. Yin. A new method for polygonal approximation using genetic algorithms. *Pattern Recognition Letters*, 19:1017–1026 (1998)
8. P.Y. Yin. Genetic algorithms for polygonal approximation of digital curves. *International Journal of Pattern Recognition and Artificial Intelligence*, 13:1–22 (1999)
9. P.Y. Yin. A tabu search approach to the polygonal approximation of digital curves. *International Journal of Pattern Recognition and Artificial Intelligence*, 14:243–255 (2000)
10. P.-Y. Yin. Ant colony search algorithms for optimal polygonal approximation of plane curve. *Pattern Recognition*, 36:1783–1797 (2003)
11. C. Di Ruberto, A. Morgera. A new algorithm for polygonal approximation based on Ant Colony system. *ICIAP 2009 - 15th International Conference on Image Analysis and Processing, September 8-11, 2009, Vietri sul Mare - Salerno, Italy*, 633–641 (2009)
12. S.C. Huang, Y.-N. Sun. Polygonal approximation using genetic algorithms. *Pattern Recognition*, 32:1409–1420 (1999)
13. T. D. Mavridou, P. M. Pardalos. Simulated annealing and genetic algorithms for the facility layout problem: A survey. *Computational Optimization and Applications*, 7:111–126 (1997)
14. S.Y. Ho, Y.C. Chen. An effcient evolutionary algorithm for accurate polygonal approximation. *Pattern Recognition*, 34:2305–2317 (2001)
15. R. Real, J. M. Vargas. The probabilistic basis of Jaccard's index of similarity. *Systematic Biology*, 45:380–385 (1996)
16. A. Masood. Optimized polygonal approximation by dominant point deletion. *Pattern Recognition*, 41:227–239 (2008)

Chapter 72
Challenges When Using Real-World Bio-data to Calibrate Simulation Systems

Elaine M. Blount, Stacie I. Ringleb, and Andreas Tolk

Abstract Computer simulations allow us to gain insight into biological systems that would not be possible without destroying or changing the system in significant ways. To ensure that results are relevant, real-world bio-data should be used to calibrate simulations. Real-world data contain uncertainty due to the nature of how it is obtained. This chapter provides various sources on uncertainty and methods to cope with this challenge.

1 Introduction

Computer simulation has roots in mathematically based computation problems [1]. As modeling and simulation evolve, more difficult challenges are tackled with regard to gathering insight about how systems act and react to different situations and parameters [2]. New scenarios such as global warming or human behavior contain large numbers of variants with a great degree of uncertainty in their true value and interactions [3, 4]. It is due to these new problem sets that the representation, aggregation, propagation, and interpretation of results within a system containing uncertainty have become an important challenge to the modeling and simulation industry [5]. Being able to quantify the certainty of the input parameters and the certainty of M&S envisioned in this contribution is also known as an initial type of knowledge processing where computational activity builds the foundation for systemic activities and system-theory-based simulation [6]. Even more advanced applications are model-based and knowledge generation activities ultimately enabling knowledge processing.

Within disciplines such as computational engineering, computational biology, bioinformatics, computational systems biology, or computational drug discovery, use of computer simulation models has been established as a valid source for gaining

S.I. Ringleb (✉)
Old Dominion University, Norfolk, VA, USA
e-mail: sringleb@odu.edu

insight. To ensure validity and applicability, the system used must be valid for the context in which it was applied. *Validity* of a simulation system in this chapter is defined as reproducing the behavior of the simulated system within the boundaries of accuracy that can be tolerated for application or analysis of the simulation results. Calibration of the simulation system using data obtained from the real system is a challenge to researchers.

Input data may come from strong statistical information based on experimental data, sparse statistical information from a small data collection, intervals based on expert opinion, or expert qualitative descriptions [5,7,8]. Representing data can vary from a constant value, a randomly selected number from previously collected data, a randomly generated number from a statistical distribution, or a number and an uncertainty tolerance, depending upon the goals of the system and use of the particular parameter [5,7]. A constant value may not be the only possible value, necessitating use of several numbers from collected data or from a statistical distribution over many simulation runs. Using collected data limits the possible input values and may exhaust input in a short time period [7]. Using a statistical distribution requires fitting the data and can give numbers beyond expected boundaries within the simulation [7]. Often use of statistical distributions requires data be independent of each other, although this is not always the real life case [7].

Researchers are generally confronted with two categories of challenges when using real biological data to calibrate simulation systems. First, the biological system cannot be described regarding all of its parameters and dependencies. Even when a parameter is understood or defined as relevant, it may not be practical to include in the simulation due to computational expense. Challenges resulting from incomplete observability of the reference systems build the first category. Second, hidden assumptions and dependencies in the simulation need to be addressed. Simulation systems can be calibrated via predefined in input, but will also code parameters that are not accessible to the user. The transparency of the simulation may be limited due to reasons ranging from lack of documentation to intellectual property protection of the simulation developer. This chapter describes possible solution for the researchers for these two challenges.

When simulations are used for gathering insight, information tends to be gathered under conditions of high uncertainty with an incomplete model [4]. Within simulation validation, the modeler may analyze data and output from the simulation and compare it with the system modeled to ensure concurrence with respect to the answers and margins of error. Validation often includes determining uncertainty, sensitivity analysis, and model calibration. *Determining uncertainty* consists of identifying parameters and interactions within the system that are not known with certainty. *Sensitivity analysis* consists of calculating the degree of change to output with respect to changes in the input. Determining the optimum input parameter values that will give the system robustness is *model calibration*. With incomplete observability of the reference system these tasks can be difficult at best and uncertainty remains in the model. *Aleatory uncertainty* is due to random variations within the environment and has a stochastic nature [3–5,9]. Aleatory uncertainty is often represented as a probability distribution. *Epistemic uncertainty* is due to incomplete information about the model or system [4, 5, 9]. An increase in knowledge about the system reduces epistemic uncertainty. Epistemic uncertainty is particularly difficult to represent mathematically. To reduce uncertainty, it is important to gain insight into the particular model by gathering data and determining: how the parameters interact, the parameters of greatest influence, unexpected behaviors influenced by the parameters, and emergent behaviors.

2 Uncertainty from Incomplete Observability

Statistical techniques randomly predict potential values for use in a simulation. *Monte Carlo* simulation uses statistical distributions to generate input for repeated sampling in simulations [10]. Aleatory uncertainty is represented with first-order Monte Carlo techniques where random numbers are generated within a statistical distribution which is a best guess for the data. The input distributions for Monte Carlo can be *discrete* or *continuous* values [10]. If an input distribution cannot be identified, empirical data can be used for nonparametric bootstrapping. *Bootstrapping* consists of sampling data from real system and using it to achieve a viable simulation input distribution [10, 11]. Several variations of Monte Carlo techniques have evolved to model epistemic uncertainty. Second-order Monte Carlo techniques consist of two loops: an outer loop randomly generates the parameters that describe the mean and variance for the parametric distribution used to generate the numbers selected by the inner loop which generates the random input data for the simulation [12]. Other methods used to capture epistemic uncertainty include using randomly sampled sets of data parameters [13] or Markov chain Monte Carlo techniques [14] to estimate statistical parameters.

Monte Carlo approaches create input data based on specific statistical distributions, but do not propagate uncertainty through the model as does *Bayesian statistical methods*. Bayesian methods incorporate error bounds within the input and calculations [15, 16]. Six steps are given in [16] for validating Bayesian models: (1) specify inputs and parameters with uncertainties; (2) determine the evaluation criteria with respect to the domain of the input and the output, accounting for the context of how the model will be used, and adequate acquirement of field data and computer resources; (3) design validation experiments; (4) if needed, run computer model approximations for computationally expensive simulations; (5) analyze model output and compare with field data; (6) use output for current validation and for improving the model.

Bayesian networks use Bayesian statistics to reduce epistemic uncertainty in many areas of biological science [9]. A joint probability distribution is defined across important variables within the model where if A, B, and C are variables modeled as connected *nodes*, $p(a,b,c)$ is the probability($A=a$, $B=b$, $C=c$) via an iterative algorithm that matches data to an appropriate logical connection of variables. Serial, diverging, and converging connections can be modeled. A marginal likelihood solution is modeled as a probability distribution over the variables rather than as points. Bayesian analysis used in [17] infers phylogenetics for molecular sequences and traces virus dispersion to study the dynamics of epidemics. Bayesian statistics are used to model the discrete states of a finite number of genes as they change over time and space.

The *interval-based approach* [11] is a method of specifying input data when parameters of an input distribution are not known and propagating the uncertainty related to the interval through the simulation. Upper and lower bounds specify the distribution of the parameter: expo(2, 4) specifies exponential distribution with mean 3, where the width of the interval indicates the variability, not the actual variance. Interval arithmetic is used to compute the mean and variance based on the intervals. Three factors affect the level of confidence in the interval output: (1) interval width of the input parameters, (2) value of the input parameters with respect to the interval boundaries, and (3) number of experiment replications.

Fuzzy set theory, classification, and logic are alternatives to probability theory in modeling epistemic uncertainty. Objects are categorized into sets where fuzzy classifiers indicate the degree of membership [18]. A person may be short, medium, or tall, and the boundaries of the categories may overlap giving membership in more than one class. Degrees of membership are generally indicated in the interval [0, 1], and the values may be discrete or continuous. A person close to the midpoint of the category has a value close to 1; the outer edges are closer to 0. Fuzzy operators: union, intersection, inverse operators, and fuzzy inference structures operate on the fuzzy sets. Fuzzy classification was used for time of birth to assign unknown animal parents to several genetic groups using fuzzy membership in an animal model [27]. These results can be used to analyze genetic trend and mating selection of animals. A Fuzzy C-Means classification algorithm [19] classifies and compares simulated drug compounds into "C" categories based on their properties in [19] by minimizing a least squares function that estimates the distance to the center of clusters of data. Points were used to estimate values for parameters of the compounds used in a simulation of the hepatobiliary disposition of the drugs.

Ontological representations of uncertainty and probabilistic ontologies are described in more detail in [12]. The resulting frameworks have the potential to unify the aspects of the approaches summarized in this section.

3 Uncertainty Within Software

Data can be used as input for calculations, but some parameters are set within the software system. Many simulations use commercial software that does not give access to all parameters used by the simulation. This can make simulation validation, verification, and accreditation for a new scenario difficult. Parameters may be hidden in decision-making boundary conditions affecting logic rules. Experimental design techniques should expose the functionality of the system in unknown areas such as boundary conditions and important logic criteria that could affect the results of the project in addition to focusing on the simulation experiments. Input parameters can be varied *One Variable At a Time* (OVAT) to investigate the effect of factors. OVAT should be used on simulations with few parameters, or with regard to one particular parameter: it does not estimate interactions, is extremely timely, and can miss optimal settings of factors. *Sequential screening* is used to eliminate variables from future testing [20]. A tutorial on experimental design in [20] includes main effects, two way interactions, quadratic effects, and choosing values for input parameters. The full factorial approach varies each parameter by a set value within a group of experiments. The parameters are often varied in high/low settings: two values per parameter for three parameters would give $2^3 = 8$ runs. Sanchez provides information on many techniques derived from the full factorial method and a chart to help with experimental design selection.

Response surface methodology (RSM) [20–23] consists of determining the response of a system to changes in input parameters and is graphically displayed to ease interpretation. RSM incorporates experimental design techniques, but uses finer grids in areas of interest. Response surfaces can provide in sight when used in conjunction with many of the methods described in this chapter. Lehar [22] describes a response surface study of the interaction of biological systems and drug combinations to investigate the response of a particular biological system and provides a myriad of response visualizations.

Taguchi's robust parameter design categorizes input parameters into control and noise. *Control inputs* are changeable. *Noise inputs* are not, but happen as a matter of circumstance [23–25]. Several two step processes exist for determining control settings [2, 23]. The signal-to-noise ratio is minimized by adjusting control parameters that affect the output variation, then by adjusting parameters that affect output location, minimizing the least squares estimator of the difference between the output and real value. The effect of the control and noise factors on the location and dispersion can also be analyzed by looking at control-by-noise interaction plots and selecting control factor setting with flattest output. Ordering principles in parameter design help to ensure the most important data for making decisions is available [23]. Taguchi's methods contribute toward limiting model sensitivity and increasing parameter robustness. They have been used in food and industrial fermentations, molecular biology, wastewater treatment, and healthcare in addition to the manufacturing setting in which it was originally designed [25].

Dynamic data-driven applications offer new options by enabling new input while in execution [7, 26]. This is particularly helpful for expensive, long running simulations or situations where results are time critical. This feature would be useful in weather/climate prediction, geo-exploration, and bio-sensing and addresses data quality, uncertainty, time-dependent data, distributed systems, and streamed data requiring work in applications, mathematical algorithms, systems software, and measurements.

4 Schemas

All of the above have several criteria in common that they contribute toward successful analysis of uncertainty. The following steps can be used:

1. List all important input and system parameters. Identify importance of each variable with respect to influence on output, certainty of input, whether variable uncertainty is aleatory or epistemic in nature, and if the variable contributes toward the dispersion or location of the simulation results. Some data attributes may be unknown initially, but can be documented as they are illuminated.
2. Describe attributes of the model: expense of each run, predominant types of uncertainty, effects of a mistake due to incorrect results, tolerance needed in output, and type of model: dynamic system, agent based, gaming, etc.
3. Describe attributes important to the project that relate to the system. List obscurities or logical features important to the research with regard to software operation. Design and execute small sets of tests to demonstrate critical attributes to logic of simulation-mitigating research risks posed by the software.
4. Perform sensitivity analysis in areas of interest to enable results with minimal variance.
5. Design tests using experimental design techniques. Focus on time required for each run, data areas of interest, and plan methods for interpretation of results.
6. Run experiments, note problems or issues for evolution of your system and methodology. Document uncertain data used for future projects and note ideas for change. Analyze and interpret experiment results.

5 Summary

Simulation systems allow us to gain insight into biological systems that would not be possible without destroying or changing the biological system. The methods here contribute toward validity of simulation experiment results by taking into account uncertainty within the system. Methods such as Monte Carlo techniques, Bayesian statistics, data interval approaches, fuzzy logic, experimental design techniques, response surface analysis, and Taguchi's methods give researchers a variety of ways to represent aleatory and epistemic uncertainty. New methods are being investigated such as dynamic data-driven applications. Refinement and evolution continue in all areas related to dealing with uncertainty.

References

1. Hazelrigg, G. A. (1999) On the Role and Use of Mathematical Models in Engineering Design. Journal of Mechanical Design. 121:336–342
2. McAllister, M. L., Dockery, J., Ovchinnikov, S., Adlassnig, K. (1985) Tutorial on Fuzzy Logic in simulation. Proceedings of the 1985 Winter Simulation Conference. In: Gantz, D., Blais, G., Solomon, S. (eds). 40–44

3. Cipra, B. (2000) Revealing Uncertainties in Computer Models. Science. New Series. 287:960–961
4. Sokolowski, J. A., Banks, C. M. (2009) Principles of Modeling and Simulation: A Multidisciplinary Approach. Wiley, New York
5. Oberkampf, W. L., Helton, J.C., Joslyn, C. A., Wojtkiewica, S. F., Ferson, S. (2004) Challenge Problems: Uncertainty in System Response Given Uncertain Parameters, Reliability Engineering System Safety. 85:11–19
6. Oren, T (2009) Modeling and Simulation: A comprehensive and Integrative View. In: Ylmaz, L., Oren, T. (eds) Agent-directed Simulation and Systems Engineering, Wiley, Germany, pp 3–36
7. Kelton, W. D. (2007) Representing and Generating Uncertainty Effectively. Proceedings of 2007 Winter Simulation Conference. In: Henderson, S.G., Biller, B., Hsieh, M. H., Shortle, J., Tew, J. D., Barton, R. R. (eds). 38–42
8. Oberkampf, W. L., DeLand, S. M., Rutherford, B. M., Diegert, K. V., Alvin, K. F. (2000) Estimation of Total Uncertainty in Modeling and Simulation, Sandia Report, SAND 2000-0824
9. Needham, C. J., Bradford, J. R., Bulpitt, A. J., Westhead, D. R. (2007) A Primer on Learning in Bayesian networks for Computational Biology, pLoS Computational Biology. 3(8):1409–1416
10. Raychaudhuri, S. (2008) Introduction to Monte Carlo Simulation, Proceedings of the 2008 Winter Simulation conference, In: Mason, S. J, Hill, R. R., Monch, L, Rose, O., Jefferson, T, Fowler, J. W. (eds)
11. Bartarseh, O. G., Wang, Y. (2008) Reliable Simulation with Input Uncertainties Using an Interval-based Approach. Proceedings of the 2008 Winter Simulation Conference
12. Costa, P. C. G. (2005) Bayesian Semantics for the Semantic Web. Doctoral Dissertation. Department of Systems Engineering and Operations Research, George mason University, Fairfax
13. Kühn, C., Wierling, C., Kühn, A., Klipp, E., Panopoulou, G., Lehrach, H., Poustka, A. J. (2009) Monte Carlo Analysis of an ODE Model of the Sea Urchin Endomesoderm Network. BMC Systems Biology. 1–18
14. Wang, W., Sijn W., Symmans, W. F., Pusztai, L., Coombes, K. R. (2009) The Bimodiality Index: A Criterion for Discovering and Ranking Bimodel Signatures from cancer Gene Expression Profiling Data cancer Informatics 7:199–216
15. Bayarri, M. J., Berger, J. O., Paulo, R., Sacks, J., Cafeo, J. A., Cavendish, J., Lin, C. H., Tu, J. (2005) A Framework for Validation of Computer Models. National Institute of Statistical Sciences. Technical Report 162
16. Cavendish, J. C. (2003) A Framework for Validation of Computer Models. In: Ferguson, D. R., Peters, T. J. (eds). Mathematics for Industry–Challenges and Frontiers. A Process View: Practice and Theory. Cambridge University Press
17. Lemey, P., Rambaut, A. D., Alexei J., Suchard, M. A. (2009) Bayesian Phylogeograph Finds Its Roots. Computational Biology. 5:9:1–16
18. Merrick, J. R. W., Dinesh, V., Singh, A., van Dorp, J. R., Mazzuchi, T. A. (2003) Propagation of Uncertainty in a Simulation-based Maritime Risk Assessment Model Utilizing Bayesian Simulation Techniques. Proceedings of the 2003 Winter Conference
19. Sheikh-Bahaei, S., Hunt, C. A. (2006) Prediction of InVitro Hepatic Biliary Excreation using Stochastic Agent-based Modeling and Fuzzy Clustering, Proceedings of the 2006 Winter Simulation conference, IEEE
20. Sanchez, S. M. (2008) Better than a PetaFlop: The Power of Efficient Experimental Design, Proceedings of the 2008 Winter Simulation conference. 73–84
21. Kuehl, R. O. (2000) Design of Experiments: Statistical Principles of Research Design and Analysis, 2nd edn, Brooks/Cole, Pacific Grove
22. Lehar, J., Zimmermann, G. R., Krueger, A. S., Molnar, R. A., Ledell, J. T., Heilbut, A. M., Hort, L. F. III, Giusti, L. C., Nolan, G. P., Magid, O. A., Lee, M. S., Borisy, A. A., Stockwell, B. R., Keith, C. T. (2007) Chemical Combination Effects Predict Connectivity in Biological Systems. Molecular Systems Biology. 3:80:1–13
23. Wu, C. F. J., Hamada, M. (2002) Experiments – Planning, Analysis, and Parameter Design Optimization. Wiley, New York

24. Leon, R. V., Shoemaker, A. C., Kacker, R. N.(1987) Performance Measures Independent of Adjustment an Explanation and extension of Taguchi's Signal-to-Noise Ratios, Technometrics. 3:253–265
25. Rao, R. S., Kumar, C. G., Prakasham, R. S., Hobbs, P. J. (2008) The Taguchi Methodology as a Statistical Tool for Biotechnological Applications: A Critical Appraisal, Biotechnology Journal. 3:510–523
26. Darema, F. (2004) Dynamicd Data Driven Applications Systems; A New Paradigm for Application Simulation and Measurements. Computational Science. 4th International Conference. Krako, W. Poland, 3:662–669
27. Fikse, F. (2009) Fuzzy Classification of Phantom Parent Groups in Animal Model. Genetics Selection Evolution. 41:42

Chapter 73
Credibility of Digital Content in a Healthcare Collaborative Community

Wail M. Omar, Dinesh K. Saini, and Mustafa Hasan

Abstract With the increased number of new diseases that are appearing in the world, such as swine flu [influenza A(H1N1)], and the increased awareness of the importance of sharing medical ideas, information, experience, knowledge, and research results, there is an urgent demand for a collaboration framework. Such a framework depends on deploying, discovering, and using digital content. This inevitably leads to the generation of large amounts of digital content from different healthcare users, which requires massive resources to process, store, and retrieve them. Moreover, the digital content currently suffers from a lack of credibility, which is vital in healthcare applications. Thus, this chapter discusses briefly how grid computing can boost Web 2.0 communities. In addition, the chapter discusses a proposed scenario for offering a way of measuring the credibility of the published digital content.

1 Introduction

Collaborative frameworks leverage the sharing of ideas, information, experience, knowledge, resources, and feedback among people in a particular field of interest – as found for example in collaborative frameworks for physicians and healthcare advisors. The collaborative framework is useful in the development of the community[1] by sharing different types of digital content (text, audio, and video). The digital content is deployed through Web 2.0 interfaces (e.g., Wiki, blogs, RSS). A healthcare community is one of the digital communities that has started to grow rapidly with the development of Web 2.0 technologies. The healthcare community

---

[1] In this chapter, digital community, virtual community, and collaboration community have the same meaning. "Community" is used to refer to all of them.

W.M. Omar (✉)
Faculty of Computing and Information Technology, Sohar University,
P.O. Box 44, P.C. 311, Sohar, Sultanate of Oman
e-mail: w.omar@soharuni.edu.om

improves the awareness and responsibilities of patients toward their health: what type of food and exercise is useful for diabetes patients, what are the steps that need to be taken regarding swine flu (H1N1), and other examples. Moreover, the community facilitates the sharing of ideas, experience, and knowledge between physicians through exchanging medical experience, discussing challenging cases, sharing clinical insights, and obtaining results from the testing of new medicines. Healthcare management users also benefit from the community by getting the required information and experience from others in their management team.

As a result, a healthcare community assists in overcoming the shortage of specialists, the high patient load on hospitals, the cost of health services, and the difficulty in getting treatment in rural and remote locations. Many countries around the world, such as Australia, USA, and Canada, are suffering from these problems. The community provides attractive cost-effective solutions for the above-mentioned problems through offering advice to users, both physicians and patients. However, considering Web 2.0 as an advisory system is risky due to the quality of the published information. Therefore, the published contents should be benchmarked and tested according to trustees' content published by the relevant healthcare authority, such as a Ministry of Health or the World Health Organization (WHO).

However, this dream of having a healthcare collaborative framework, which deals with human lives, requires massive resources [1] and control. These resources include supercomputing, storage and backup systems, and communications, as well as health services such as remote medical monitoring systems, online medical testing labs, and video conferencing. The framework is deployed over heterogeneous systems, which require the adoption of an open standard format in exchanging medical information.

2 Web 2.0 and Healthcare Community

The use of Web 2.0 has increased significantly in recent years, with a focus on specific areas such as health, education, engineering, technology, and entertainment. Currently, the Web 2.0 framework is available in different formats, such as blogs, wikis, podcasts, tags, and social networking. These formats are merged with different enterprise applications to offer a base for user collaboration frameworks. Such applications are reviewed by Hogg et al. [2, 3], who conducted an in-depth investigation of 40 successful Web 2.0 applications. They summarized the range of characteristics to describe the phenomenon of Web 2.0 communities and to provide a systematic overview of current and emerging business models. Some of the best known Web 2.0 sites and communities are thehealthcareblog.com/, askdrwiki.com/, Patientslikeme.com, and cmepodcasting.com.

The reliability, quality of data, manageability, availability, and maintainability of the digital content still need to be improved and investigated further. Most people start to use the information available through different Web 2.0 platforms without awareness about the sources, quality, and validity of the information. For example, are we sure that X publishing correct information about H1N1? Is X really X or another person who is using X's name? Is the information that is published by X really up to date? There are many other similar questions that cannot be answered.

If we are discussing particular issues about a healthcare community and published information, such as information regarding swine flu, then the only trusted sources of medical information are those provided by healthcare authorities in different countries, official health organizations and research centers – e.g., the information provided by WHO regarding H1N1, the Canadian Ministry of Health (http://www.phac-aspc.gc.ca/alert-alerte/h1n1/index-eng.php), the Ministry of Health in Oman (http://www.moh.gov.om/a_h1n1/index.html), and other such bodies. Does this mean that we should only use information from an authorized source? If so, then we are not helping in developing the digital health community. The remainder of the chapter discusses how we can test the digital content.

Obviously, Web 2.0 community requires huge amount of resources. Omar et al. [1,4–7] proposed the use of Grid computing as a platform for offering a wide range of resources to Web 2.0 community.

Merging of Web 2.0 and grid computing will draw the future vision of how to support the next generation of the collaborative community. Grid computing is proposed as an attractive solution for offering different types of resources, and one that ensures the availability of resources for the operational framework.

3 Quality of Data

With the increasing involvement of users in providing digital resources to the digital community through Web 2.0, the lack of accuracy and quality of data are dramatically becoming a problem [8]. The importance of the accuracy of digital content that is published by users increases significantly for critical applications such as health, education, and geographic information systems [8]. Specifying the quality of data is essential for providing credibility for the published digital content. Credibility is defined as "the quality of being believed or trusted." Therefore, to reach to credibility, we need first to be sure about the quality of information.

A wide range of techniques exists to assess and improve the quality of data, such as record linkage, business rules, and similarity measures. Florian et al. [9] propose a system which determines trust relationships between community members automatically and objectively by mining communication data. The proposed model and system can improve the quality of the published digital content to a limit.

In order to discuss the quality of digital data, we need to have metrics that are used as a quality benchmark for the published digital contents. Olsina et al. [10] discuss how to model the different quality views by reusing and extending the ISO 9126–1 quality models for Web 2.0. Our proposed metrics for testing the quality of data will be the information that is published by a health authority, as described in the following section.

4 Scenario

The proposed scenario is projected to improve data quality, as shown in Fig. 1. The process is commenced by publishing information, experiences, data, etc., through Web 2.0 user interfaces (wikis, blogs, RSS, social networks, and others). The published information is stored as digital content in the system. The user requests, searches, or discusses subjects within the health community, e.g., swine flu. The system receives the request through the Web 2.0 user interface, discovers the required information, and then classifies it.

After categorizing the required digital contents, the system compares the requested materials from the users with the trust information that is available in the digital library. The sources of trust content in the digital library are provided by health authorities, official organizations, research centers, and in some cases the tested contents that have already passed the credibility process. Different mechanisms can be used to compare digital contents, starting with text comparison, mind map comparison, and ranking algorithms. The result of such a comparison will be given as a percentage that is attached to the tested digital content. The author of the digital content is considered as one of the parameters for the comparison process. For example, if the author is a physician, then a higher weight of trust would be given to that author than to a patient.

The following sections discuss the adopting of different classification algorithm for improving the classifying process of digital library.

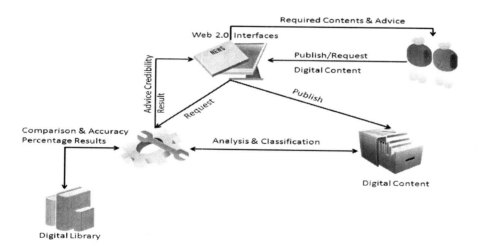

Fig. 1 Digital content credibility scenario

5 Classification Algorithms

In most health information systems, a multi-category classification is needed with better speed and accuracy [11]. Decision trees use concept space and category tests on word combinations in a document. The Repeating Pattern Algorithm is another method that can use repeating information in web pages to represent the semantic meaning. It can be used to improve the correct rate of classifying medical digital content. Gao et al. [12] analyzed and improved the traditional repeating patterns representation methods, and further proposed a new semantic representation of web information based on repeating patterns.

The semantic characters are divided into three levels: word, phrase, and sentence. The basic unit of our method is a word, which should be extracted from sentences. Words are kept in the correlative matrix and to find repeating patterns. Finally, the information on repeating patterns is used for categorizing the digital content.

In order to use the repeating pattern approach to represent semantic meaning of web information, the following definitions are used.

Word *Unit of Language denoted by w.*
Repeating pattern. *Strings a, b, c, and d are substrings of the string x.*
Repeating pattern Set. *Is a set of repeating patterns which is denoted by RPS $((name_1, weight_1), \ldots, (name_n, weight_n))$.*
γ-Approximating matching. *Suppose two strings*

$$x, y \in \sum, \text{ and } x = w_1 \ldots w_m, y = v_1$$

where x and y are Y-approximating matching, if and only if $|x| = |y|$, and $\sum_{i=1}^{|x|} |w_i - v_i| \leq \gamma$
When $w_i = v_i$, we say that $|w_i - v_i| = 0$; otherwise $|w_i - v_i| = 1$.

5.1 Algorithm for Repeating Pattern and Constructing Correlative Matrix and γ-Approximate Matching Algorithm

The algorithm for constructing the correlative matrix is shown in Figs. 2 and 3. After normalizing the repeating patterns, the method below is used to compute the weights of repeating patterns. From the analysis of semantics, all of the repeating patterns are categorized into two classes. From the analysis above, the weight computed equation of our approach is:

$$a_{ik} = \frac{\log(rf_{ik} + 1) \times \log\left(\frac{N}{cf_i}\right)}{\sqrt{\sum_{j=1}^{M} \log(rf_{ik} + 1) \times \log\left(\frac{N}{cf_i}\right)}}$$

Fig. 2 Pseudo-code for finding correlative matrix in repeating pattern

```
1.   exact-repetitions (t,matrix) {n=|t|}
2.   Begin
3.        for j←2 to n do
4.             If t[1] = t[j] then
5.             begin
6.                       matrix[1][j]←1
7.                       write 1j,1
8.             end
9.             else matrix [1][j]←0
10.       for i ←2 to n do
11.            for j←i+1 to n do
12.                 if t[i] then
13.                 begin
14.                      matrix [i][j] ← matrix [i-1][j-1]+1
15.                      put t[i-matrix[i][j]+1...i] into Repetition Set
16.                 end
17.                 else matrix [i][j]←0
18.  End
```

Fig. 3 Pseudo-code for γ-approximation matching

```
Y-approximate repetitions ,(t, Y, m) {n=|t|)
1.   begin
2.   a[0]=0
3.   for i←0 to n do
4.        for j←n-1 1 to 0 do
5.             a[j]←a[j-1]+|t[i]-t[j]|
6.             if i >m AND j>m then
7.             begin
8.                  a[j]←a[j]-|t[i-m]-t[j-m]|
9.             end
10.  end
```

In this equation, rf_{ik} is the repeating frequency of the i-th repeating pattern in digital content of class k, c_{fi} is the number of classes that the i-th repeating pattern appears in, N is the number of classes, M is the number of repeating patterns, and a_{ik} is the weight of the i-th repeating pattern in the digital contents of class k. The bigger the k is, the smaller is the reflection of the repeating pattern in the weight. Each class has its own repeating pattern set, denoted by $S[(sp_1, w_1), (sp_2, w_2), \ldots, (sp_n, w_n)]$; sp_i is the i-th repeating pattern, w_i is the weight of the i-th repeating pattern.

5.2 Digital Contents Categorization Algorithm

The γ-approximate matching algorithm is used for generating new repeating patterns, with the repeating pattern set denoted by T $[(sp_1, f_1), (sp_2, f_2), \ldots, (sp_n, f_n)]$. In order to categorize the testing set of digital content, the semantic distance of repeating patterns between known classes and testing set digital content should be calculated. The equation of semantic distance computing is:

$$\text{Distance}(C_i, D) = \sum_{i=1}^{n} w_i \times f_i$$

D is the testing set of digital content, C_i is the i-th class, w_i is the weight of repeating pattern sp_i which belongs to class C_i, and f_i is the frequency of repeating pattern sp_i in digital content D. From the equation above, the semantic distance between testing set with digital content D and all classes $(C_1, \ldots, C_N)$ can be calculated. Class i is chosen, corresponding to the maximum value of distance (C_i, D).

6 Conclusion

This chapter discusses the credibility of digital content that is provided by different types of users through the Web 2.0 framework. Such digital content is essential for the development of digital communities, such as in healthcare. The scenario that is presented in this chapter offers a way that can provide users with a method of improving their judgment on the published information, from the viewpoint of credibility. The work is still in early stage, and different algorithms of classification, decision making, and ontology could be introduced to augment the picture and improve the overall approach.

References

1. T. Solomonides, R. McClatchey, V. Breton, Y. Legré, S. Nørager. From Grid to Healthgrid: Prospects and Requirements in Healthgrid, 2005. 2005. Oxford, United Kingdom
2. C. Schoth, T. Janner. Web 2.0 and SOA: Coverging Concepts Enabling the Internet of Services. IEEE IT Professional, 2007. **9**(3): pp. 36–42
3. R. Hoegg, R. Martignoni, M. Meckel, K. Stanoevska-Slabeva. Overview of Business Models for Web 2.0 Communities. In Gemeinschaften in Neuen Medien. 2006
4. W. Omar, A. Taleb-Bindiab. Service Oriented Architecture for E-health Support Services Based on Grid Computing Over. In IEEE International Conference on Services Computing 2006. Chicago
5. W. Omar, A. Taleb-Bendiab, E-Health Support Services Based on Service Oriented Architecture. IEEE IT Professional, 2006. **8**(2): pp. 35–41
6. W. Omar, A. Abbas, A. Taleb-bindiab, SOAW2 for Web2 Framework. IEEE IT Professional, 2007. **9**(3): pp. 30–35
7. W. Omar, A. Abbas. Managing Web 2.0 Framework Based on Grid Computing Overlay. In 2nd IEEE International Conference on Digital Ecosystems and Technologies, 2008. DEST 2008. 2008. Phitsanulok, Thailand
8. M.F. Goodchild. Spatial Accuracy 2.0. In Eighth International Symposium on Spatial Accuracy Assessment in Natural Resources and Environmental Sciences. 2008. Liverpool
9. F. Skopik, H.-L. Truong, S. Dustdar. Trust and Reputation Mining in Professional Virtual Communities. In 8th International Workshop on Web Oriented Software Technology (IWWOST2009). 2009. San Sebastian: SpringerLink
10. L. Olsina, R. Sassano, L. Mich. Towards the Quality for Web 2.0 Applications. In 8th International Workshop on Web Oriented Software Technology (IWWOST2009). June, 2009. San Sebastian: SpringerLink

11. L. Liu, Z.Li, A. Wang, H. He. The Research of Decicsion Vector Machine in Web Information Classification. In 12th International Conference on Computer Supported Cooperative Work in Design (CSCWD). 2008. Xian: IEEE Xplore
12. K. Gao, B. Zhang, Y. Zhang, H. Wei, A. Ma. Study on Semantic Representation of Web Information Based on Repeating Patterns. In Fifth international Conference on Fuzzy Systems and Knowledge Discovery. 2008. Jinan, China, IEEE Computer Society, Washington, DC

Chapter 74
Using Standardized Numerical Scores for the Display and Interpretation of Biomedical Data

Robert A. Warner

Abstract The ability to review and analyze large amounts of data reliably and cost-effectively is important in both biomedical research and clinical care. We hypothesized that converting raw digital data to standard scores (Z scores) and gray-scale them based on their corresponding P values can be used to accomplish this. In Part 1 of the study, we recorded continuous digital electrocardiographic (ECG) and heart sound data from a subject undergoing acute anterior myocardial infarction (MI). We then computed Z scores of the digital data using the means and standard deviations of the data obtained during the pre-infarction period. In Part 2 of the study, we analyzed the digital ECG data from 576 subjects who had undergone coronary angiography and left ventriculography for the evaluation of possible coronary disease. We used the durations of Q waves in Lead aVF and of the initial R waves in Lead V2 as the ECG criteria of prior inferior and anterior MI, respectively. We calculated Z scores for these durations using the means and standard deviations of the subjects who had no angiographic evidence of coronary disease. Results show that in Part 1 of the study, the continuous recording of the gray-scale Z scores produced a highly intuitive display of the direction and statistical significance of simultaneous changes in five quantitative parameters known to be important for the detection and assessment of acute MI. In Part 2 of the study, the use of gray-scale Z scores revealed, in each of three subgroups, the distributions of subjects who met ECG criteria for inferior and anterior MI, respectively. Analyzing the Z scores to calculate diagnostic performances yielded results similar to those obtained using receiver-operating characteristic curves of the raw data. We conclude that the use of gray-scale Z scores is a highly efficient and statistically meaningful way to display diagnostic data produced by both continuous and individual recordings.

R.A. Warner (✉)
Laboratory for Logic and Experimental Philosophy, Simon Fraser University,
Burnaby, BC, Canada
e-mail: hillwarner@frontier.com

1 Introduction

The accurate and cost-effective analysis of the data used in biomedical research and clinical care is important. For example, it is often necessary to determine whether there is a statistically significant difference between the values of one or more clinical parameters for a subject and the values of the same parameters for a reference population. Similarly, it is also often important to detect significant temporal changes from the baseline values of relevant parameters in an individual, e.g., during clinical monitoring.

Since clinical evaluations and scientific studies often produce large volumes of data, the required analyses can be very labor intensive. Therefore, a method to permit the accurate and rapid review of these data would also be desirable.

In this study, we evaluated the use of a highly intuitive method for displaying and evaluating biomedical data accurately and efficiently. The method involves expressing the results of discriminative tests exhibited by a subject vs. the range of values of a baseline series or of a normal population using standard scores (Z scores), rather than the original numerical values.

2 Methods

2.1 Selection of Subjects

The present study has two parts:

1. We analyzed sequential diagnostic data that had been recorded from a 49-year-old woman in whom angioplasty of the left anterior descending coronary artery had been attempted. In the course of this procedure, there was inadvertent catheter-induced dissection of this vessel that resulted in arterial occlusion and subsequent nonfatal acute anterior myocardial infarction (MI). The presence of acute MI was documented by the combination of severe chest discomfort and marked elevation of the cardiac enzymes CK-MB and troponin.
2. We also analyzed electrocardiographic (ECG) findings obtained from 576 subjects (226 females), mean age 53, range 27–82 years, who were either angiographically normal (Group 1) or had either prior inferior (Group 2) or prior anterior (Group 3) MI proven by diagnostic coronary angiography and left ventriculography.

2.2 Diagnostic Data Obtained

In the subject who had undergone coronary angioplasty, simultaneous digital diagnostic data were recorded continuously both before and after the onset of the patient's chest discomfort. These data were obtained using an Audicor® (Inovise

Medical, Inc., Portland, OR, USA) device that records simultaneous digital 12-lead ECG and heart sound data from the surface of the body. This device has been extensively described elsewhere [1–4].

In the patients who had undergone diagnostic cardiac catheterization, we used digital 12-lead ECG data that had been obtained from each patient shortly before the catheterization, using standard ECG machines (Marquette Electronics, Milwaukee, WI, USA). To detect prior inferior and anterior MI, respectively, we measured the durations in milliseconds of the Q waves in standard ECG Lead aVF and of initial R waves in standard ECG Lead V2. Abnormally prolonged Q waves in Lead aVF are known to be associated with prior inferior wall MI [5, 6]. Abnormally abbreviated initial R waves in Lead V2 are known to be associated with prior anterior wall MI [7].

2.3 *Calculation, Significance, and Gray-Scale of Z Scores*

In all the subjects on whom we are reporting, we calculated standard (Z) scores of their diagnostic data. For each diagnostic parameter, the calculations use the individual value of interest and the mean and the standard deviation (SD) of the set of data being used for the comparison. For sequential data obtained from an individual subject, the means and SDs are calculated using the data recorded during the baseline period, i.e., before the suspected clinical event occurred. For diagnoses that use a single test applied to different subjects, the calculations use the data obtained from a group of patients who are known by independent criteria to be free of the abnormality of interest. The formula for calculating a Z score for a subject's value of a given parameter is:

$$Z \text{ Score} = (\text{Subject's Value} - \text{Mean Value})/SD$$

Thus, Z scores express in SDs the difference between a parameter's numerical value in an individual sample or patient vs. the mean value of the same parameter in the set of data that was used for the comparison. Whether the Z score has a positive or a negative value indicates whether the individual value is greater or less than the mean.

Z scores can be used to indicate the statistical significance of any difference between an individual value and the mean of the population to which the individual is being compared [8]. The greater the absolute value of a Z score, the lower is the P value associated with the comparison. To facilitate the interpretation of these comparisons, the Z scores can be gray-scale to indicate their corresponding P values. In Fig. 1, Z scores that do not differ significantly from the respective mean baseline values ($Z > -1.65$ and $Z < 1.65$; $P = NS$) are displayed as medium gray. Z scores that significantly exceed their respective baseline values ($Z \geq 1.65$; $P < 0.05$) are dark gray, and those that are significantly lower than their respective baseline values ($Z \leq -1.65$; $P < 0.05$) are light gray.

Gray-Scale Changes in Z Scores of ECG and Heart Sound Parameters
Acute Anterior Myocardial Infarction

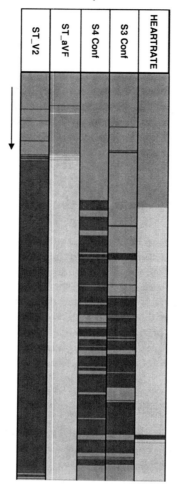

Conf = confidence (a measure of the strength of the recorded heart sounds), ECG = electrocardiogram, MI = myocardial infarction, S3 = third heart sound, S4 = fourth heart sound

Fig. 1 Gray-scale changes in Z scores of ECG and heart sound parameters. Acute anterior myocardial infarction

In the patients who underwent diagnostic cardiac catheterization, we used the distributions of Z scores in Group 1 to calculate specificities in Groups 2 and 3 and the distributions of Z scores in Groups 2 and 3 to calculate sensitivity for inferior and anterior MI, respectively.

To evaluate the accuracy of the diagnostic performances for prior inferior and anterior MI that we obtained using Z scores, we also determined these performances using standard receiver-operating characteristic (ROC) curves for Group 1 vs. Group 2 and for Group 1 vs. Group 3, respectively. These ROC curves were produced with commercially available statistical software (SPSS Version 14.0, SPSS, Inc., Chicago, IL, USA), used the raw measurement data from which the Z scores had been calculated and were used to determine the respective sensitivities of the Q and R wave durations at approximately 95% specificity. For all the statistical comparisons, we a priori chose an $\alpha < 0.05$ to indicate statistical significance.

3 Results

3.1 Sequential Diagnostic Data in the Same Subject

Figure 1 shows the continuous recording of ECG and sound data from the subject who sustained the acute anterior MI. The time course of the tracing proceeds from the top to the bottom of the figure as shown by the arrow's direction, and the arrow's point indicates the time at which the occlusion of the left anterior descending coronary artery occurred. The entire duration of the events portrayed in Fig. 1 is approximately 20 min. The figure shows that shortly after arterial occlusion, there were statistically significant changes in the following:

1. Increase (i.e., elevation) of the ST segments in ECG Lead V2 accompanied by a decrease (i.e., depression) of the ST segments in ECG Lead aVF. This pattern of ST segment changes typifies acute injury to the epicardial portion of the anterior left ventricular wall.
2. Increase in the acoustical evidence of an S4. This is characteristic of the abrupt reduction of left ventricular compliance that occurs with acute MI.
3. Subsequent increase in the acoustical evidence of an S3. This is characteristic of the onset of left ventricular systolic dysfunction that often accompanies MI.
4. Reduction in the heart rate, consistent with an increase in vagal tone that also commonly accompanies acute MI.

Thus, the changes in the values of the simultaneously displayed ECG and heart sound parameters are all directionally and temporally consistent with the known pathophysiology of acute anterior MI.

3.2 Diagnostic Data in Groups of Individual Subjects

There were 248 subjects in Group 1, 184 in Group 2, and 144 in Group 3. Determining the sensitivity and specificity for prior inferior MI of the Q wave durations in Lead aVF using Z scores vs. using ROC curves of the raw values yielded identical results: sensitivity = 72%, specificity = 95%. Similarly, determining the sensitivity and specificity for prior anterior MI of the initial R wave durations in Lead V2 using Z scores vs. using ROC curves of the raw values yielded identical results: sensitivity = 60%, specificity = 96%.

4 Discussion

Calculating Z scores using digital biomedical data can determine whether an individual value of a parameter differs significantly from the mean of a reference sample, whether that sample represents the baseline data in serial recordings or the values exhibited by groups of individual subjects. Depending on the purpose of a particular study, the means and standard deviations used to calculate the Z scores can be obtained from any subgroup of interest, e.g., selected on the basis of demographic features such as age, sex, and ethnicity. Thus, the comparisons can be as general or as specific as desired. The diagnostic findings for prior inferior MI show that the data on which Z scores are calculated need not have a Gaussian distribution (The distribution of Q wave durations in Lead aVF is skewed heavily to the right.). Also, if desired, raw data that have non-Gaussian distributions can be transformed to their squares or logarithms.

Figure 1 shows that an important advantage of Z scores is that they express the values of all parameters on the same scale, i.e., in units of the SD. This is useful in multiparameter monitoring in which different physiological parameters are usually expressed in different units that often have widely different scales and therefore cannot be meaningfully displayed on the same screen or printout. In Fig. 1, the directions and the time courses of the significant changes in each of the recorded parameters are all consistent with acute anterior MI, thereby increasing the likelihood that this diagnosis is correct. Thus, using Z scores to display changes in multiple parameters can increase both the ease and the accuracy of the interpretation of diagnostic data, especially if the diagnostic parameters are diagnostically orthogonal. Although Fig. 1 uses three shades of gray, other methods of visual coding can be used to demonstrate statistical significance at various levels of alpha, e.g., 0.05, vs. 0.01 vs. 0.001.

The second part of our study also shows that the proposed method can be used to analyze and display the distribution of the results of diagnostic tests that have been applied to large groups of subjects, e.g., in epidemiological investigations or drug studies.

As in any diagnostic endeavor, important caveats apply. Investigators should be mindful of Type I errors associated with multiple comparisons and remember that the assessment of all diagnostic information must include consideration of the prevalence of the abnormality of interest in the population being tested [9, 10].

The ubiquity of digital data in various biomedical fields makes the use of Z scores widely applicable. Of course, application of the proposed method of displaying and analyzing data need not be confined to biomedical disciplines, but can also be useful in other fields such as the physical and behavioral sciences, engineering, and business.

References

1. Zuber M, Kipfer P, Attenhofer Jost CH. Usefulness of acoustic cardiography to resolve ambiguous values of B-type natriuretic peptide levels in patients with suspected heart failure. *Am J Cardiol* 2007;100(5):866–869
2. Marcus GM, Gerber IL, McKeown BH, Vessey JC, Jordan MV, Huddleston M, McCulloch CE, Foster E, Chatterjee K, Michaels AD. Association between phonocardiographic third and fourth heart sounds and objective measures of left ventricular function. *JAMA* 2005;293(18):2238–2244
3. Gerber IL, McKeown BH, Marcus G, Vessy J, Jordan MV, Huddleston M, et al. The third and fourth heart sounds are highly specific markers for elevated left ventricular filling pressure and reduced ejection fraction. *J Card Fail* 2004;10(4):S36
4. Roos M, Toggweiler S, Zuber M, Jamshidi P, Erne P. Acoustic cardiographic parameters and their relationship to invasive hemodynamic measurements in patients with left ventricular systolic dysfunction. *Congest Heart Fail* 2006;12(4 Suppl. 1):19–24
5. Warner R, Hill NE, Sheehe PR, Mookherjee S, Fruehan CT, Smulyan H. Improved electrocardiographic criteria for the diagnosis of inferior myocardial infarction. *Circulation* 1982;66:422–428
6. Warner R, Hill N, Mookherjee S, Smulyan H. Electrocardiographic criteria for the diagnosis of combined inferior myocardial infarction and left anterior hemiblock. *Am J Cardiol* 1983;51:718–722
7. Warner R, Reger M, Hill N, Mookherjee S, Smulyan, H. Electrocardiographic criteria for the diagnosis of anterior myocardial infarction. Importance of the duration of precordial R waves. *Am J Cardiol* 1983;52:690–692
8. Warner RA, Olicker AL, Haisty WK, Hill NE, Selvester RH, Wagner GS. The importance of accounting for the variability of electrocardiographic data among diagnostically similar patients. *Am J Cardiol* 2000;86:1238–1240
9. Thatcher RW. Normative EEG databases and EEG biofeedback. *J Neurother* 1998;2(4):8–39
10. Berger JO, Strawderman WE. Choice of hierarchical priors: admissibility in estimation of normal means. *Ann Stat* 1996;24(3):931–951

Chapter 75
ImagCell: A Computer Tool for Cell Culture Image Processing Applications in Bioimpedance Measurements

Alberto Yúfera, Estefanía Gallego, and Javier Molina

Abstract This paper presents a computer tool for automatic analysis of cell culture images. The program allows the extraction of relevant information from biological images for pre- and postsystem analysis. In particular, this tool is being used for electrical characterization of electrode-solution-cell systems in which bioimpedance is the main parameter to be known. The correct modeling of this kind of systems enables both electronic system characterization for circuit design specifications and data decoding from measurements. The developed program allows cell culture image processing for geographic information extraction and generates cell count and equivalent circuit descriptions useful for system simulations.

1 Introduction

The impedance is a useful parameter for determining the properties of biological materials for several reasons: first, they are conductive [1], second, the impedance measurement represents a noninvasive technique, and third, it is a relatively cheap technique. Many biological parameters and processes can be sensed and monitored using its impedance as marker [2–5]. Impedance Spectroscopy (IS) of cell culture [6] and Electrical Impedance Tomography (EIT) in bodies [7] are examples of the impedance utility for measuring biological and medical processes and parameters. Classical real-time monitoring and imaging systems for biological samples are based on optical stimulation of samples, demanding bulky and expensive equipments. Embedded Complementary Metal-Oxide-Semiconductor (CMOS) sensors have been reported as an alternative for increasing the sensitivity to cell location and manipulation. The most popular are optical [8], capacitive [9], and impedance [10] based sensors. Despite the high number of papers with optical sensors over the

A. Yúfera (✉)
Electronic Technology Department, Computer Engineering School, Seville University,
Av. Reina Mercedes s/n, 41012, Spain
e-mail: yufera@us.es; yufera@imse-cnm.csic.es

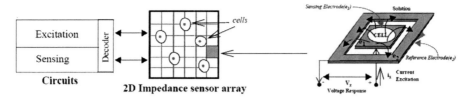

Fig. 1 Simplified system set-up: circuits and 2D electrode sensor array for bio-impedance measurement. Each sensor has e_1 and e_2 electrodes. Cell culture is done on electrode top

last years, they still need external lamps, optical fibers, etc., while capacitive and impedance-based detection do not rely on peripheral equipment.

This paper is related to a new method for impedance measurement with applications to cell culture systems. The system in Fig. 1 uses a two-dimensional electrode array as sensors [11, 12] together with CMOS circuits for impedance measurements [13]. Microelectronic circuits must be designed to work with constraints imposed by the electrode sensors. The whole system in Fig. 1 can be fully integrated in CMOS technologies [11]. When low concentration cell cultures are carried out on top of the electrode array, depending on the position of each cell, specific electrode-cell impedance will be measured, allowing cell detection. Electrical models reported for the electrode–cell interface description [12, 14] are the key for matching electrical simulations to real systems performance and hence decoding correctly the experimental results, usually known as a reconstruction problem. This kind of system can be used for real-time monitoring of cell cultures with the Electrical Cell Impedance Spectroscopy technique (ECIS) [6].

ImagCell is a computer tool that aids in cell culture image processing and reconstruction and helps in optimization of circuit design since it enables the emulation of biological loads. In the system shown in Fig. 1, the tasks of ImagCell are:

- To perform a pre-processing of a cell culture image to define the areas occupied by cells. Digital Image Processing (DIP) is focused on segmentation to discriminate the total area covered by cells.
- To incorporate the definition of the electrode area. This is important not only from the electrode-solution-cell system modeling and characterization point of view, but also because the electrode sensitivity of the impedance sensor will be dependent on its size and working frequency. The electrode-cell overlap area will be considered as the main parameter of the electrode-cell system.
- To deliver information to the electronic system design, including data files and an electrical description of the full system (electrode-solution-cells) necessary to reproduce confident electrical simulations. In our system, it measures the covered area, position, and cell number. ImagCell generates files in Analog Hardware Description Language (AHDL), required in functional and electrical simulations (SpectreHDL) for system design and validation.

The work is organized as follows: Sect. 2 describes the ImagCell interface. Program functions attached to the main menus are briefly detailed. Processing image algorithms are explained in Sect. 3. Electrode definition in the program is analyzed in Sect. 4. In Sect. 5, the measurements are scheduled and data can be extracted from images. Conclusions are highlighted in Sect. 6.

2 ImagCell Interface

The main functions developed by ImagCell are described from its interface. Input images are loaded and displayed in a historical register on the left panel: the *image panel* area. At the *center panel* area, the processed image is displayed. There are four main action modules described in the following.

Processing panel: ImagCell includes functions and algorithms for image processing. Their objective is to separate the background area from the cells by using segmentation, filtering, and morphological operations. This process can be done *automatically* or *manually* (by defining image processing functions and its parameters by the user). At the *electrode panel*, two actions are performed. First, a tool for scaling definition of the image size allows expressing a nondimensional image in micron units. Second, based on this scale, the electrode size is defined. The sensors are considered squared. ImagCell will show the resulting array of sensors on the main panel.

Measurement panel: The percent of cell coverage of each electrode can be obtained from the electrode-solution-cell overlapping area. This is the *fill factor* (ff) parameter. From the fill factor matrix, the parameterized electrical sensor (electrode) models are created using Analog Hardware Description Language (AHDL).

Advanced processing panel: For advanced users, this option allows to customize the segmentation process.

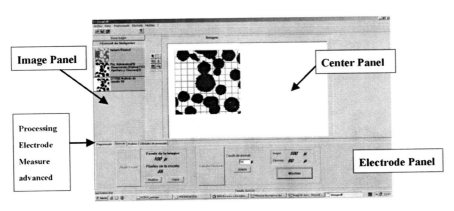

Fig. 2 ImagCell interface with: center panel, image panel, and the four functional panels: processing, electrode, measurements, and advanced processing. The electrode panel is being displayed at the figure

3 Image Processing

3.1 Processing Approach

The main objective of the image processing is to segment it, dividing into two areas: covered and noncovered (by cells). The proposed image processing is based on histogram information. This information is employed to define a threshold gray level. Figure 3 shows this process. When a threshold level is set at histogram, for example, the 160 gray levels, the image is easily binarized into two parts: with (black) and without (white) cells. However, not all images are directly binarized easily and some kind of preprocessing should be done before.

Two types of processing algorithms were considered: filter and morphological. They must enhance the original images, eliminating noise sources, detecting closed areas, smoothing images, etc., before binarization. Both are based on convolution functions between basic templates (kernel) and digital images.

The *filter algorithms* employed are median, mean, maximum, minimum, and Sobel, while the *morphological algorithms* included are dilation, erosion, opening, and closing. For both erosion and dilation, it is used as structural element, a start in which the pixel number at the main diagonal defines its size or length.

3.2 Image Catalog

The image histogram changes strongly from one image to another. The histograms were classified into five categories. For each one, the processing function parameters were customized to increase the quality of segmentation. The parameter values are fixed, but can be modified by the user in manual mode. Figure 4 shows the corresponding histograms.

- *Cat 1*: The histogram has an absolute maximum much bigger than the rest.
- *Cat 2*: The histogram has only a main slope. All pixels are near zero, and there are only few gray levels.
- *Cat 3*: The histogram only has a main slope. All pixels are near 255 levels, and, there are few gray levels.

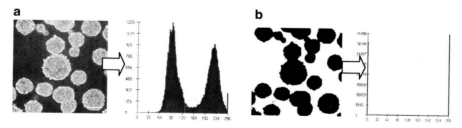

Fig. 3 (a) Original image and the corresponding histogram. (b) Image after binarization with the resulting histogram

- *Cat 4*: The image histogram has a nonuniform background.
- *Cat 5*: The histogram has two well separated peaks.

Based on this experience, the program has a proposal for image processing once it has been classified. If not possible, as in images cat 4, the manual mode is employed. Table 1 resumes the tasks required for each image category.

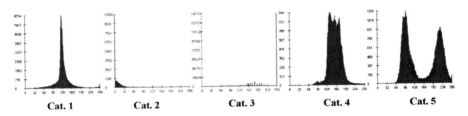

Fig. 4 Image histogram catalog

Table 1 Processing actions for each category of image histogram

Cat	Processing actions
1	1. Binarization plus Sobel. 2. B&W inversion for background. 3. Add images from points 1 and 2. 4. Closing (5). 5. Opening (3)
2	2. Binarization with threshold 0. 2. B&W inversion for background. 3. Closing (5). 4. Opening (3)
3	3. Binarization with threshold 254. 2. B&W inversion for background. 3. Closing (3). 4. Opening (6)
4	4. Histogram cannot be processed automatically
5	5. Automatic threshold 254. 2. B&W inversion for background. 3. Opening (3). 4. Closing (3)

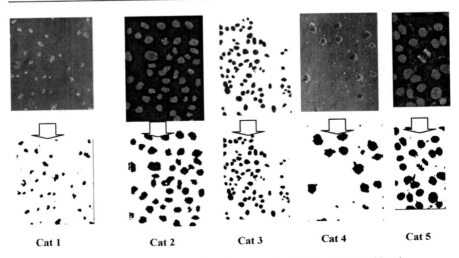

Fig. 5 Original and processed images for the five categories of histograms considered

4 The Sensor-Electrode

Two operations are developed from the electrode panel. First, a scale is defined, allowing the real dimensioning. The user can evaluate the involved dimensions and set the specific tool developed for this purpose. A 100 µm scale is set by default as input. Second, the sensor (squared) size is defined in the menu. Only squared shapes were considered for electrodes (Fig. 1), but this can be extended to any other shape. The main panel shows the final position for electrodes under the cell culture. In this paper, it has been considered that optimal electrodes must be sized similar to cell dimensions.

5 Measuring with ImagCell

Area Measurement can be done by detecting the cell-electrode area overlap. For each electrode, this process delivers the fill factor (ff) in the range [0, 1] representing the percent of electrode area covered by cells. These data are expressed in a matrix that can be displayed in the main panel. Also, information is stored in a data file compatible with other computer tools, such as MATLAB.

The cell number, *cell count*, at the image is approximated by defining the radius of a circular cell as a pattern. Other cell shapes can be easily considered.

One of the main motivations for ImagCell develop is to have a tool for rapid input processing of information from cell culture images. The circuit design in Fig. 1 is dependent on load to be sensed, in our case, the cells being necessary to adjust the circuit specifications to impedance values of electrode-solution-cell system and to select the optimum working frequency. ImagCell does a fast *generation of circuit model* from a cell culture image in AHDL [15] useful for SpectreHDL mixed-mode simulator. Figure 6 shows an example, where each electrode is described with its corresponding area and fill factor.

In the example, the input image is like a cat. 5 histogram. The automatic processing and performing opening and closing of length 3 were applied. The threshold found corresponds to level 157. The electrode/sensor size chosen is 50 µm^2. The center panel image displayed in Fig. 6 shows the sensor grid obtained. Each square represents an impedance sensor, as it is illustrated in Fig. 1. The measurement panel shows the fill factor matrix obtained, in which each number represents the cell-electrode overlapping area percent for a given electrode. Also, considering a radius of 35 µm for circular cells, 22 cells have been found in the image. Finally, Fig. 7 shows the file where the AHDL description of the electrode array is codified. Each line describes an electrode sensor in terms of its situation, area, and fill factor for electrical simulations. The area, in the last electrodes, considers the border effects derived from electrode size selection.

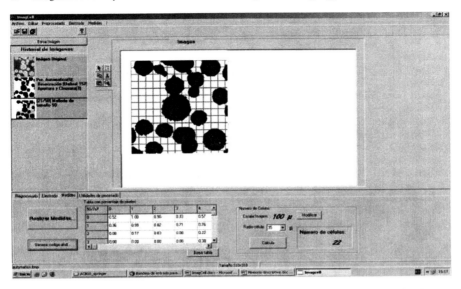

Fig. 6 Example of ImagCell interface: measuring area covered by cells and number of cells

```
module electrode_2D(e1,e2)
node [V,I] e1[15], e2[15]
{
    electrode_solution elec_1_1(e1[1],e2[1])(2500,0.52);
    electrode_solution elec_1_2(e1[1],e2[2])(2500,1.00);
    ......... .
    electrode_solution elec_14_13(e1[14],e2[13])(350,0.24):
    electrode_solution elec_14_14(e1[14],e2[14])(49,0.94); }
```

Fig. 7 SpectreHDL file for the electrode matrix: Area $(2,500\mu m^2)$ and fill factor (0.52 for sensor in position (0, 0))

6 Conclusions

This paper describes a tool for computer aid in cell culture image processing useful for bioimpedance measurement systems based on microelectrode sensors. The program performs image segmentation, focused on cell location, based on threshold algorithms. A wide number of cell culture images were analyzed and classified including them into the database. The bio-impedance sensor-electrode design was specified for optimum sizing on the basis of electrode-cells area overlap. Resulting data from image processing and electrode sizing allows cell count, electrode-cell area definition, and description of sensor-cell culture system in AHDL, useful for mixed-mode electrical simulations.

References

1. Ackmann, J., 1993. Complex Bioelectric Impedance Measurement System for the Frequency Range from 5Hz to 1MHz. *Annals of Biomedical Eng.*, 21, 135–146.
2. Beach, R. D., et al., 2005. Towards a Miniature *In Vivo* Telemetry Monitoring System Dynamically Configurable as a Potentiostat or Galvanostat for Two- and Three- Electrode Biosensors. *IEEE Trans. on Instrumentation and Measurement*, 54, 1, 61–72.
3. Yúfera, A., et al., 2005. A Tissue Impedance Measurement Chip for Myocardial Ischemia Detection. *IEEE Trans. on Circuits and Systems: Part I*, 52, 12, 2620–2628.
4. Radke, S. M., and Alocilja, E. C., 2004. Design and Fabrication of a Microimpedance Biosensor for Bacterial Detection. *IEEE Sensor Journal*, 4, 4, 434–440.
5. Borkholder, D., 1998. Cell-Based Biosensors Using Microelectrodes. *PhD Thesis*, Stanford University.
6. Giaever, I., et al., 1986. Use of Electric Fields to Monitor the Dynamical Aspect of Cell Behaviour in Tissue Culture. *IEEE Trans. on Biomedical Eng.*, BME-33, 2, 242–247.
7. Holder, D., 2005. Electrical Impedance Tomoghaphy: Methods, History and Applications", *Philadelphia: IOP*.
8. Manaresi, N., et al., 2003. A CMOS Chip for individual Cell Manipulation and Detection. *IEEE Journal of Solid Stated Circuits*, 38, 12, 2297–2305.
9. Romani, A., et al., 2004. Capacitive Sensor Array for Location of Bioparticles in CMOS Lab-on-a-Chip. *International Solid Stated Circuits Conference (ISSCC)*.
10. Medoro, G., et al., 2003. A Lab-on-a-Chip for Cell Detection and Manipulation. *IEEE Sensor Journal*, 3, 3, 317–325.
11. Hassibi, A., et al., 2006. A Programmable 0.18μm CMOS Electrochemical Sensor Microarray for Biomolecular Detection. *IEEE Sensor Journal*, 6, 6, 1380–1388.
12. Huang, X., et al., 2004. Simulation of Microelectrode Impedance Changes Due to Cell Growth. *IEEE Sensors Journal*, 4, 5, 576–583.
13. Yúfera, A., et al., 2008. A Method for Bioimpedance Measure With Four- and Two-Electrode Sensor Systems. *30th Annual Int. IEEE EMBS Conference*, 2318–2321.
14. Joye, N., et al., 2008. An Electrical Model of the Cell-Electrode Interface for High-density Microelectrode Arrays. *30th Annual Int. IEEE EMBS Conference*, 559–562.
15. Cadence Design Systems Inc (2010). *SpectreHDL Reference Manual*.

# Chapter 76
From Ontology Selection and Semantic Web to an Integrated Information System for Food-borne Diseases and Food Safety

Xianghe Yan, Yun Peng, Jianghong Meng, Juliana Ruzante, Pina M. Fratamico, Lihan Huang, Vijay Juneja, and David S. Needleman

Abstract Several factors have hindered effective use of information and resources related to food safety due to inconsistency among semantically heterogeneous data resources, lack of knowledge on profiling of food-borne pathogens, and knowledge gaps among research communities, government risk assessors/managers, and end-users of the information. This paper discusses technical aspects in the establishment of a comprehensive food safety information system consisting of the following steps: (a) computational collection and compiling publicly available information, including published pathogen genomic, proteomic, and metabolomic data; (b) development of ontology libraries on food-borne pathogens and design automatic algorithms with formal inference and fuzzy and probabilistic reasoning to address the consistency and accuracy of distributed information resources (e.g., PulseNet, FoodNet, OutbreakNet, PubMed, NCBI, EMBL, and other online genetic databases and information); (c) integration of collected pathogen profiling data, Foodrisk.org (http://www.foodrisk.org), PMP, Combase, and other relevant information into a user-friendly, searchable, "homogeneous" information system available to scientists in academia, the food industry, and government agencies; and (d) development of a computational model in semantic web for greater adaptability and robustness.

1 Introduction

Food-borne illness is an important public health concern in developing, as well as developed countries. Prevention of food-borne illness and outbreaks through effective interventions, availability of early warning systems, and reliable detection methods for food-borne pathogens is a critical issue worldwide. It is

X. Yan (✉)
U.S. Department of Agriculture, Agricultural Research Service,
Eastern Regional Research Center, Wyndmoor, PA 19038, USA
e-mail: xianghe.yan@ars.usda.gov

estimated that food-borne pathogens cause approximately 76 million cases of gastrointestinal illnesses, 325,000 hospitalizations, and 5,000 deaths in the United States annually [2, 5, 17].

Over the last three decades, remarkable advances in Information and Communication Technologies (ICTs), genomics, and other cutting-edge "omics" technologies have dramatically improved our ability to rapidly determine and interpret the mechanisms of survival and pathogenesis of human food-borne pathogens. Data collection, analysis, and the timely dissemination of these data are essential components for the planning, implementation, and evaluation of public health practices. There are numerous important mechanisms (Uniform Resource Locators [URLs]) for data sharing and accessing of food safety information, related to microbial and chemical contamination, pathogen characteristics and predictive microbiology, public health surveillance, risk assessment and risk analysis, inspection, management and regulation, recalls, violations, prevention and control, and others [25]. However, a centralized information system to handle the data flow from these information resources, to standardize the content of these resources and to integrate this information with data in public repositories is sorely lacking.

The purpose of this paper is to discuss how an integrated information system could be used to integrate data from heterogeneous resources to strengthen food-borne pathogen risk management, surveillance, and prevention systems and to lay the groundwork for a standard interoperable protocol that could serve as a nationwide food-borne pathogen-related warning system.

2 Challenges

There are many challenges associated with establishing a centralized Food Safety Information Reporting System (FSIRS), including data access issues, standards, and data format issues. One of the largest challenges when creating a FSIRS is accurate and reliable prediction of the combined effects of complex multifactorial factors on the growth and inactivation of food-borne pathogens. The development of predictive models involves conducting extensive scientific experiments to investigate the biological behaviors of microorganisms under a variety of conditions and fitting the experimental data into appropriate primary and secondary models. Data sharing is another large challenge facing the development of FSIRS. Data mining and large-scale statistical data analysis is a time-consuming process. The presentation of food-borne pathogen surveillance and prevention systems must be accurate and be ready to detect changes in complex heterogeneous data systems very quickly. This requires advanced algorithms, data structures, and dynamic communication tools (e.g., web) for detection and prediction of transmission patterns of food-borne pathogens. Advances in algorithms, data structures, and artificial intelligence allow for practical applications of data-driven outbreak detection methods, which can handle the complexity of the task at hand by learning from examples in historical data and from real or simulated recorded outbreaks. The Semantic Web first was considered by Tim Berners-Lee, inventor of the WWW, URIs, HTTP, and HTML in early 2001 "–semantics is considered to be the best framework to deal with the heterogeneity, massive scale, and dynamic nature of the resources on the Web" [22]. The semantic web services can search and evoke over thousands of Internet URLs quickly to embrace multiinputs and multioutputs for accurate and rapid food-borne pathogen emergency decision making. Several information technologies such as the extensible Markup Language (XML), the Web Ontology Language (OWL), and the Resource Description Framework (RDF) could be used to reduce the syntactic diversity and enable the system to manipulate and make inferences about the data, thereby defining meaningful relationships between the items. A deductive query language based on the semantic web ontology, OWL-QL, or other advanced technologies could facilitate user queries. The extended semantic markup language should be considered to represent both logical and probabilistic relations in web resources on a unified semantic basis. The developed reasoning methods are capable of resolving semantic differences between web resources, making probabilistic reasoning over the semantic web possible. Capability of resolving semantic differences is a significant advance in the area of distributed BN where existing methods all rely on exchanging beliefs via shared identical variables. This work also offers a solution to a persistent problem facing current semantic web ontologies, namely, their inability to represent and reason upon uncertain relations and inputs. In particular, treating probabilistically enhanced OWL ontologies as probabilistic resources, concept mapping between ontologies can be accomplished as evidential reasoning using the developed reasoning method. From both surveillance and prevention points of views, some critical challenges need to be addressed:

Heterogeneous Data Representation: There is a tremendous amount of online scientific literature, regulations, and pathogen profiling data. However, there is no standard language to represent semantics and heterogeneities of mined knowledge in the semantically heterogeneous scenario. Although manual translation is possible, it is time consuming and error prone if the size of knowledge is large.

The Correctness and Accuracy of Knowledge Prediction: The key problems associated with the correctness and accuracy of knowledge prediction is short of a unified data annotation ontology standard, corrected applied algorithms, and timely entry of public health data. Therefore, it is important to set up an ontology standard and develop a dynamic user interface by using semantic web technology.

Timing of Food Safety Emergency Responses: An important part of food-borne pathogen surveillance and prevention is the timing of food safety emergency responses. In some cases, determining the geographical scope of food safety emergency and/or investigating the course of the food safety "crisis" is difficult. Therefore, it is important to quickly compile multivariate data (ideally, in real time) through neural network analysis to maintain the data correlation ability of prediction hypotheses, regulatory threshold, and the individual cases. Advances in algorithms, data ontology analysis, and neural network allow for practical application of data-driven outbreak investigation methods.

3 Data Sources

Pathogen profiling of food-borne pathogens: The accurate classification of food-borne pathogens through integration of microbial genomics data with clinical and phenotypic observations is an important tool for the detection of patterns of food-borne pathogen transmission and the construction of epidemic trees.

PMP: The Pathogen Modeling Program (PMP), established and maintained at the USDA-ARS-Eastern Regional Research Center (ERRC), is a package of models that can be used to predict the growth, survival, and inactivation of food-borne bacteria, primarily pathogens, under various environmental conditions [16, 21, 26]. The PMP is periodically updated as new models and/or changes in the model-user interface become available.

Combase: This freely accessible Combined Database of Predictive Microbiology Information (ComBase) database was jointly developed by the USDA-ARS-ERRC, the Institute of Food Research, Norwich, U.K., Food Standards Agency, London, U.K., and the Food Safety Centre, Australia. ComBase [3, 4] contains data sets submitted by researchers and from publications that describe the rate of growth, survival, and inactivation of bacteria under diverse environments relevant to food processing operations.

PMIP portal: The Predictive Microbiology Information Portal (PMIP) contains three major sections [14], which provide access to predictive models for food-borne pathogens, relevant regulatory policies and guidelines, and microbial data related to pathogenic and spoilage microorganisms in a wide variety of food products.

Foodrisk.org: Foodrisk.org is an online resource managed by the Joint Institute for Food Safety and Applied Nutrition (JIFSAN), a joint Institute between the United States Food and Drug Administration (FDA) and the University of Maryland at College Park. Some of the content of Foodrisk.org is unique to the website and can only be found there.

Other online information resources: CDC's databases (PulseNet, FoodNet, and OutbreakNet) are already used to detect outbreaks early based on pathogen pulsed-field gel electrophoresis (PFGE) patterns [23, 24].

4 Roadmap to Integrated Information System of Food-borne Diseases and Food Safety

This FSIRS will be based on three different methodologies: Neural Network analysis, Bayesian Network Modeling, and Semantic Web (OWL/RDF) technology. The input to this system will include risk assessment information, document (rule)-based regulation policies, publications on experimental-based pathogen profiling, large-scale statistical data sets, and predictive modeling from real-time data

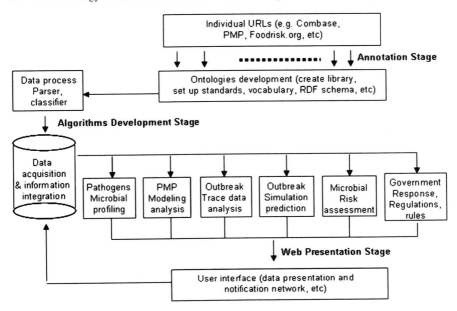

Fig. 1 Semantic ontology-driven information system architecture

and/or simulated outbreak analyses. The overall execution and management of this information system will be divided into three main substages as illustrated in Fig. 1.

Annotation Stage (Fig. 1): Ontology [27] development is the fundamental part of semantic web. Web search engines and the richest information resources such as Google using keywords, PubMed (indexed for MEDLINE) of NCBI, the European Molecular Biology Laboratory (EMBL), DNA Databank of Japan (DDBJ) through terminologies, and the most usable food safety knowledge via open URLs, such as Combase, foodrisk.org, PulseNet (http://www.cdc.gov/pulsenet/), Food Net (http://www.cdc.gov/FoodNet/), Outbreak Net (http://www.cdc.gov/OutbreakNet/), etc. to automatically retrieve text exemplars for each standardized ontology concept from the web could be used at this stage. Here, ontology is a formal, explicit description of concepts, their properties, and relationships among the terminology of food-borne pathogens. Note that the definition of terminology and standardized ontology will not be generated by Data Analysts alone, instead it will be a widely agreed-upon standard determined by experts in different fields. All these documents will then be classified into individual subcategories based on classifications related to food-borne pathogens and food safety.

Figure 2 describes the details of the food-borne pathogen ontology categories and subject descriptors. The flowchart of the categories and subject descriptors for food-borne pathogen ontology development in Fig. 2 not only includes the definition of classes, relationships, and attributes, but also defines a set of subcategories for further classification. To guarantee relevancy, the search will be guided by semantic information given in the ontology where the concept is defined. For example, instead

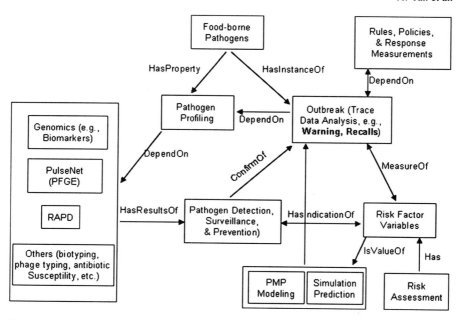

Fig. 2 The categories and subject descriptors for food-borne pathogen ontology development

of using "*Escherichia coli* O157:H7," the term for the concept, as search query, we can form the query by including all terms on the path from the root to this class in the ontology, e.g., "food-borne pathogen + bacteria + *Escherichia coli* O157:H7" for positive exemplars and "food-borne pathogen + bacteria − *Escherichia coli* O157:H7" for negative exemplars. Other ontological information can also be included, e.g., "nonpathogen," which is another super class of bacteria, "pink" which is the value for *Escherichia coli* O157:H7 colony color cultured on the agar medium called sorbitol MacConkey agar (SMAC) (a property), and "toxB," which is a property inherited from "*Escherichia coli* O157:H7."

Algorithms Development Stage (Fig. 1): Research in this direction could focus on methods to form search queries that best utilize available semantic information. Questions that will be addressed include, for example, what semantic information should be included; how to order the terms so that their semantic distances to the concept is reflected; and if and how to form alternative queries to represent disjunctive relations and synonyms [7–10, 12, 13, 15, 18, 20]. In addition, methods for processing returned pages to filter less relevant ones by, say, various data mining techniques such as clustering will be investigated. Research methods will be empirical. Computer experiments will be conducted to assess if and how much improvement can be gained with a particular technique. To formally categorize and resolve the semantic heterogeneity problems in food safety information resources, an ontology mining framework will first be to extend to discover semantics from both relational data and semistructured data and represent the mined knowledge with ontologies, which have been used for the formal specification of conceptualization in traditional knowledge engineering and the emerging Semantic Web. The heterogeneities between different data resources will be represented as formal ontological mappings. The mappings could be specified by humans manually or discovered by mapping tools automatically; however, the automatically discovered mappings will have some uncertainty. It should provide a formal representation of uncertain heterogeneities as fuzzy and probabilistic mappings by distinguishing subjective perceptions from objective measurements. From the distributed computing point of view, engineers may need to design knowledge translation algorithms for distributed data mining (DDM) systems in a client–server model. The clients will be data analysts from a domain (e.g., food processing industries or government agencies) and a DDM server will connect to local data resources. A software engineer will also ensure that the server will include a meta-data (e.g., ontologies and mappings) repository. The principle for this knowledge translation algorithm design is that the only thing transferred from local individual information resources (local miners) to the user (server) side is the translated knowledge through the Ontology & Mapping Repository. Engineers should avoid transferring data back and forth to achieve communication efficiency and good scalability. The data mining tasks could run on local resources to achieve locality of the computation. The XML namespace (XML + NS), XML schema, and RDF schema will be developed at this stage (Fig. 3).

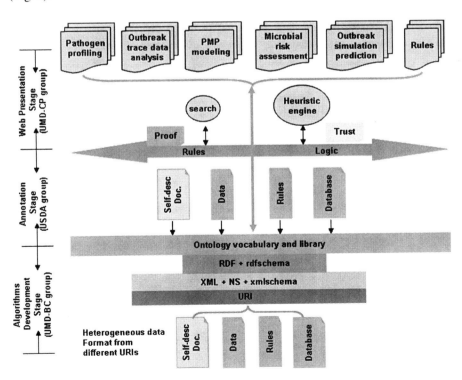

Fig. 3 Knowledge-based translation framework and the flowchart of research implementation plan (modified from Tim Berners Lee's vision of the Semantic)

Web Presentation Stage (Fig. 1): The semantic web (SW) extends the current web by specifying the semantics or meaning of information on the web using markup languages so that it can be understood and processed not only by humans but also by machines and programs [6]. Meanings of terms (or concepts) in web resources are defined unambiguously using ontologies stated at the "annotation stage," which are written using SW languages, to represent conceptualization of application domains. Existing SW languages are all logic based (e.g., RDF and OWL are based on description logic [DL], and ontologies [11] are based on first-order logic), and representation of meanings is expressed as logical sentences, and reasoning is done using logical inference. The SW languages are inadequate in specifying the semantics of probabilistic information. The final FSIRS web presentation can be classified into six different reporting categories, which will cover Pathogen Profiling, PMP Modeling Analysis, Microbial Risk Assessment, Outbreak Simulation Prediction, Rules, and Outbreak Trace Data Analysis as listed in Fig. 3. All the processed data in this system could be processed and presented for public reasoning with confidence, "trust" and "proof" with the help of dynamic advanced query algorithm "heuristic engine" (Fig. 3). As depicted in Figs. 1 and 3, this final FSIRS web presentation would consist of six modules:

Profiling Module: An expert, easily viewed annotated system, which makes formulation recommendations based on experimental data that include biomarker detection for identification; standardized pulsed-field gel electrophoresis (PFGE)-based molecular subtyping; and molecular profiling based on new technologies, such as microarray, biometrics, and next-generation sequencing technologies.

PMP Modeling Module: A back propagation neural network [1] which predicts dissolution rate of the recommended formulation using the mapping between formulation parameters and dissolution rates learned from samples of lab test data. To increase accuracy in this module, each model will consist of two sets of supporting prediction modeling: one predicted through Bayesian Network Modeling and the other obtained from actual laboratory experiments.

Outbreak Trace Data Analysis Module: A database-based tracking system for surveillance.

Outbreak Simulation Prediction Module: Based on simulation software to allow regulators and industry users to adjust variance factors when the predicted performance is not acceptable.

Risk Assessment Module: Food-borne pathogen risk assessment is the quantitative or qualitative value of risk related to public health and the estimated potential loss through prediction modeling.

Government Response, Regulations Module: A regulation-based expert system.

5 Concluding Remarks

This paper has discussed a solution to a persistent problem facing current semantic web ontologies, namely, their inability of representing and reasoning upon uncertain relations and inputs. In particular, treating probabilistically enhanced OWL ontologies as probabilistic resources, the concept mapping between ontologies can be accomplished as evidential reasoning using the developed reasoning methods. Once this system is created, it will cover all important aspects of food-borne pathogens and cost less [19] considering that resources are routinely underfunded, technically "isolated," or less comprehensive.

Robust, flexible, and extensible intelligent systems can be built in which both logical and probabilistic data of enormous quantity and variety in the web can be understood, utilized, and exchanged; knowledge can be continuously learned, integrated, and updated; and new and more complex problems can be solved. This means that food-borne pathogens are more likely to be detected and investigated comprehensively and sooner, whether they are in nonoutbreak or outbreak status. Many new applications, which are not possible to accomplish at present, may emerge in the near future. These include, for example, data mining from data sources with different semantics, and probabilistic semantic integration and interoperation of software systems.

This work is only a step toward bringing the ICTs into the food safety research area. Its success will likely inspire more research in this direction. Richer representation and information may be developed, and new methods (e.g., reasoning with distributions or other probabilistic relations embedded in food-borne pathogen surveillance data) may appear.

References

1. Adams BM, Saithanu K, and Hardin JM (2006) A neural network approach to control charts with applications to health surveillance. Invited talk at the 2006 Joint Statistical Meeting
2. Anonymous (1999). Health People 2000: Status Report - Food Safety Objectives. Food and Drug Administration, Food Safety and Inspection Service, and Centers for Disease Control and Prevention. September, 1999
3. Baranyi J and Tamplin M (2004) ComBase: A common database on microbial responses to food environments. J. Food Prot. 67:1834–1840
4. Baranyi J (2006) Using the ComBase database and associated software tools to predict microbial responses to food environments. Food Manufacturing Efficiency 1:9–13
5. Bean NH and Griffin PM (1990) Food borne disease outbreaks in the United States, 1973–1987: pathogens, vehicles, and trends. J. Food Prot. 53: 804–817
6. Berners-Lee T, Hendler J, and Lassila O (2001) The Semantic Web, Scientific American, May 2001
7. Ciccarese P, Wu E, Wong G, Ocana M, Konshita J, Ruttenberg A, and Clark T (2008) The SWAN biomedical discourse ontology. J. Biomedical Informatics 41:739–751
8. Cooper GF, Dash DH, Levander JD, Wong WK, Hogan WR, and Wagner MM (2006) Bayesian Methods for Diagnosing Outbreaks In: Wagner MM, Moore AW, and Aryel RM (ed) Handbook of Biosurveillance, Academic Press, Boston

9. Doan AH, Madhavan J, Domingos P, and Halevy H (2002) Learning to map between ontologies on the Semantic Web, In WWW2002, May 7–11, 2002, Honolulu, Hawaii, USA
10. Giugno R and Lukasiewicz T (2002) P-SHOQ(D): a probabilistic extension of SHOQ(D) for probabilistic ontologies in the Semantic Web, INFSYS Research Report 1843-02-06, Wien, Austria, April
11. Gruber TR (1993) A Translation Approach to Portable Ontology Specifications, Knowledge Acquisition 5: 199–220
12. Jiang X and Wallstrom GL (2006) A Bayesian network for outbreak detection and prediction, proceedings of the 21st national conference on Artificial intelligence, p. 1155–1160, July 16–20, 2006. Boston
13. Jiang X, Wallstrom GL, Cooper GF, and Wagner MM (2009) Bayesian prediction of an epidemic curve, J. Biomed. Inform. 42: 90–99
14. Juneja VK and Huang CA (2009) Predictive microbiology information portal (PMIP) with particular reference to the USDA-pathogen modeling program (PMP), Blackwell Publisher (Accepted, Book Chapter)
15. Lacher MS and Groh G (2001) Facilitating the exchange of explicit knowledge through ontology mappings. In the proceeding of 14th Int. Florida A.I. Research Society Conf., pp. 305–309. AAAI Press
16. McMeekin TA, Baranyi J, Bowman J, Dalgaard P, Kirk M, Ross T, Schmid S, and Zwietering MH (2006) Information systems in food safety management. Int. J. Food Microbiol. 112: 181–194
17. Mead PS, Slutsker L, Dietz V, McCaig LF, Bresee JS, Shapiro C, Griffin PM, Tauxe RV (1999) Food-related illness and death in the United States. Emerg. Infect. Dis. 5: 607–625
18. Mitra P, Noy NF, Jaiswal AR (2004) OMEN: A Probabilistic Ontology Mapping Tool. In Workshop on Meaning Coordination and Negotiation at the Third International Conference on the Semantic Web (ISWC-2004). Hisroshima, Japan
19. Peck M, Baranyi J, Belsten J (2003) Microbial database could cut costs. Food Manufacturer. June/2003
20. Prasad S, Peng Y, Finin T (2002) A tool for mapping between two ontologies using explicit information, In AAMAS'02 Workshop on Ontologies and Agent Systems. Italy, July 2002
21. Ross T, Baranyi J, McMeekin TA (1999) Predictive Microbiology and Food Safety. In: Robinson R, Batt C, Patel P (ed) Encyclopaedia of Food Microbiology, Academic Press, Boston
22. Sheth A (2004) From Semantic Search & Integration to Analytics, Dagstuhl Seminar on Semantic Interoperability and Integration, September 19–24, 2004. http://www.dagstuhl.de/04391/Materials/
23. Swaminathan B, Barrett TJ, Hunter SB, Tauxe RV (2001) PulseNet, the molecular subtyping network for foodborne bacterial disease surveillance, United States, Emerging Infectious Diseases 7:382–389
24. Swaminathan P, Gerner-Smidt LKNg, Lukinmaa S, Kam KM, Rolando S, Gutierrez EP, Binsztein N (2006) Building PulseNet International: an interconnected system of laboratory networks to facilitate timely public health recognition and response to foodborne disease outbreaks and emerging foodborne diseases, Foodborne Pathogens Diseases 3:36–50
25. Taylor MR, Batz M (2008) Harnessing knowledge to ensure food safety: opportunities to improve the nation's food safety information infrastructure. Gainesville, FL: Food Safety Research Consortium; (Report available on the FSRC website at http://www.thefsrc.org/FSII/)
26. Tamplin M, Baranyi J, Paoli G (2003) Software programs to increase the utility of predictive microbiology information. In: McKellar RC, Lu X (ed) Modelling Microbial responses in Foods, CRC, Boca Raton, Fla
27. Villaneuva-Rosales N, Dumontier M (2008) YOWL: An ontology-driven knowledge base for yeast biologists. J. Biomed. Inform. 41: 779–789

Chapter 77
Algebraic Analysis of Social Networks for Bio-surveillance: The Cases of SARS-Beijing-2003 and AH1N1 Influenza-México-2009

Doracelly Hincapié and Juan Ospina

Abstract Algebraic analysis of social networks exhibited by SARS-Beijing-2003 and AH1N1 flu-México-2009 was realized. The main tools were the Tutte polynomials and Maple package Graph-Theory. The topological structures like graphs and networks were represented by invariant polynomials. The evolution of a given social network was represented like an evolution of the algebraic complexity of the corresponding Tutte polynomial. The reduction of a given social network was described like an involution of the algebraic complexity of the associated Tutte polynomial. The outbreaks of SARS and AH1N1 Flu were considered like represented by a reduction of previously existing contact networks via the control measures executed by health authorities. From Tutte polynomials were derived numerical indicators about efficiency of control measures.

Keywords Algebraic epidemiology · Social network analysis · SARS · AH1N1 flu · Tutte polynomial · Graph theory · Computer algebra

1 Introduction

The numerical analysis of social networks has been applied in epidemiology, bio-surveillance, and bio-security [1, 2]. An alternative method of analysis was proposed in [3] using algebraic polynomials to represent the topological structures like graphs and networks observed in the epidemic patterns for SARS [4] in Taiwan [1, 3]. In this work, we show new examples in algebraic analysis of social networks applied to understand the epidemic patters exhibited by SARS in Beijing and AH1N1 flu in Mexico. The Tutte polynomials [5, 6] are the main tools for algebraic analysis of social networks, using the Maple package Graph-Theory [7]. In algebraic

J. Ospina (✉)
Logic and Computation Group, Physical Engineering Program,
School of Sciences and Humanities, EAFIT University, Medellín, Colombia
e-mail: jospina@eafit.edu.co

analysis, the evolution of a given social network is represented like an evolution for the algebraic complexity of the associated Tutte polynomial [3]. Reciprocally, the complexity reduction of a given social network is represented like an involution for the algebraic complexity of the associated Tutte polynomial [3]. This algebraic method is illustrated with simple examples in the next section. In the third section, the algebraic method is applied to more complex examples corresponding to the epidemic patterns exhibited by SARS-Beijing-2003 and AH1N1Flu-México-2009.

2 Algebraic Analysis of Social Networks: General Method

2.1 Algebraic Analysis for the Evolution of the Social Networks

A simple example of the evolution of a small social network is illustrated in Fig. 1 [8]. In Fig. 1a, a simple relation between two individuals is depicted and the corresponding graph is denoted G_1. In Fig. 1b, the graph G_1 is enlarged with a new member and a new graph is denoted G_2. Similarly in Fig. 1b the graph G_2 is modified and the resulting complete graph for three members is denoted G_3. In Fig. 1d, two new members are included in the Graph G_3 and then results the graph G_4. The evolution of the original graph G_1 continues through to Fig. 1e with the graph G_5; Fig. 1f with the graph G_6 and finally with the Fig. 1h and the graph G_7.

The evolution depicted in Fig. 1 can be analyzed using numerical indicators about the growing complexity of the network such as the mean degree and the number of spanning trees [1–3]. But it is interesting to note that the numerical analysis is not the only way to describe the evolution in the Fig. 1. The other way is the algebraic

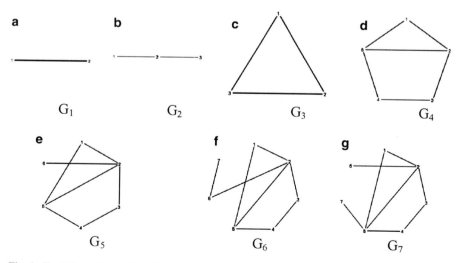

Fig. 1 Evolution of a small social network from the configuration G1 to G7 [8]

77 Algebraic Analysis of Social Networks for Bio-surveillance

method previously used in [3] which consists in to represent the original graphs (geometric or topological figures) by invariant polynomials (algebraic structures associated with topological structures). Specifically, we use the Tutte polynomial [3, 5, 6] which is a universal polynomial in computational graph theory and contains t topological information about the social networks. The corresponding Tutte polynomials for the graphs in Fig. 1 are respectively given by:

$$T(G_1, x, y) = x, \quad T(G_2, x, y) = x^2, \quad T(G_3, x, y) = y + x + x^2,$$
$$T(G_4, x, y) = y^2 + y + x + 2yx + 2x^2 + yx^2 + 2x^3 + x^4,$$
$$T(G_5, x, y) = x(y^2 + y + x + 2yx + 2x^2 + yx^2 + 2x^3 + x^4),$$
$$T(G_6, x, y) = x^2(y^2 + y + x + 2yx + 2x^2 + yx^2 + 2x^3 + x^4),$$
$$T(G_7, x, y) = x^2(y^2 + y + x + 2yx + 2x^2 + yx^2 + 2x^3 + x^4).$$

With these Tutte polynomials, the topological evolution of the social network in Fig. 1 is described in algebraic terms as a complexity of the Tutte polynomial for the network observed in the transition from the simple Tutte polynomials $T(G_1, x, y)$ and $T(G_2, x, y)$ to the more complex $T(G_6, x, y)$ and $T(G_7, x, y)$.

This algebraic analysis with Tutte polynomials reflects the evolution from the graph G_1 to the graph G_6, but the Tutte polynomials are not able to distinguish between the graphs G_6 and G_7 because the Tutte polynomials are the same for both graphs. In contrast when the algebraic analysis is executed using characteristic polynomials, we obtain the following results:

$$Ch(G_1, x) = x^2 - 1, \quad Ch(G_2, x) = x^3 - 2x, \quad Ch(G_3, x) = x^3 - 3x - 2,$$
$$Ch(G_4, x) = x^5 - 6x^3 - 2x^2 + 4x, \quad Ch(G_5, x) = x^6 - 7x^4 - 2x^3 + 7x^2 - 1,$$
$$Ch(G_6, x) = x^7 - 8x^5 - 2x^4 + 13x^3 + 2x^2 - 5x, \quad Ch(G_7, x) = x^7 - 8x^5 - 2x^4 + 11x^3 - 3x.$$

As we can see, the characteristic polynomials represent the evolution from G_1 to G_6 according to a process of enlarging the algebraic complexity of the polynomials. At the same time, the characteristic polynomial is able to discriminate between G_6 and G_7.

Another example of network evolution via the introduction of new members in a social network is illustrated in Fig. 2.

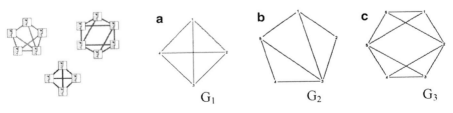

Fig. 2 Second example of network evolution from the configuration G1 to G3

The evolution in Fig. 2 contains three states represented by the graphs in the Fig. 2a–c. The corresponding Tutte polynomials are:

$$T(G_1, x, y) = y^3 + 3y^2 + 4yx + 2y + 2x + 3x^2 + x^3,$$
$$T(G_2, x, y) = y^3 + 2y^2 + 4yx + 2y^2x + 3yx^2 + y + x + 3x^2 + 3x^3 + x^4,$$
$$T(G_3, x, y) = 4x + 4y + 4y^3x + 5x^4 + 18yx^2 + 18y^2x + 11y^2 + 18yx + 11x^2 + 11y^3$$
$$+ 11x^3 + 5y^4 + y^5 + 2x^2y^2 + 4yx^3 + x^5.$$

We observe that the topological evolution in Fig. 2 is reflected in the algebraic evolution of the corresponding Tutte polynomials, namely the algebraic complexity of the Tutte polynomial $T(G_3, x, y)$ is evidently greater than the algebraic complexity of the Tutte polynomial $T(G_1, x, y)$.

From the Tutte polynomials, it is possible to derive numerical indicators about the topological complexity of the social networks. For example, the Tutte polynomial evaluated at $x = 1$ and $y = 1$ determines the number of spanning trees for the considered graph. In the case of the evolution in Fig. 2, we have the following numerical results: $T(G_1, 1, 1) = 16$, $T(G_2, 1, 1) = 21$, $T(G_3, 1, 1) = 128$. Then, we have the following ratios $T(G_2, 1, 1)/T(G_1, 1, 1) = 1.31$ and $T(G_3, 1, 1)/T(G_1, 1, 1) = 8$.

In words, these results indicate that the final graph G_3 has a relative complexity of eight respects to the initial graph G_1. In epidemiological terms, this means that the graph G_3 has eight times more probability to transmit infectious diseases than the graph G_1.

2.2 Algebraic Analysis of Complexity Reduction for Social Networks

The examples in Figs. 1 and 2 show cases when the network grows and its topological complexity is increased. Now, when the network reduces its topological complexity is also an interesting target for the algebraic analysis. For example, Fig. 3 shows a social network which is reduced via deletion of some links. The original network, denoted Phenol, is depicted in Fig. 3a; the reduced network, denoted DP, is depicted in Fig. 3b; and the network without links, denoted AP, is depicted in Fig. 3c.

For the graphs Phenol, DP, and AP in Fig. 3, the corresponding Tutte polynomials are given respectively by:

$$T(\text{Phenol}, x, y) = x(yx^6 + x^7 + x^8 + x^9 + x^{10} + x^{11}), \quad T(\text{DP}, x, y) = x^{12}, \quad T(\text{AP}, x, y) = 1.$$

With these polynomials, the transition from the graph Phenol to the graph AP is described in algebraic terms as a reduction of the algebraic complexity for the Tutte polynomials, given that the polynomial $T(\text{Phenol}, x, y)$ is more complex than the polynomial $T(\text{AP}, x, y) = 1$. From these polynomials, we obtain the following complexity ratios: $T(\text{DP}, 1, 1)/T(\text{Phenol}, 1, 1) = 1/6$, $T(\text{DP}, 1, 2)/T(\text{Phenol}, 1, 2) = 1/7$,

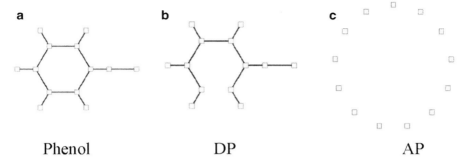

Fig. 3 First example in network reduction from the complex graph Phenol to the trivial graph AP

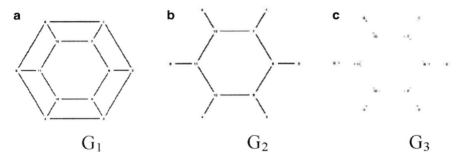

Fig. 4 Second example in network reduction from the complex graph G1 to the trivial graph G3

T(DP, 1, ∞)/T(Phenol, 1, 2) = 0. These ratios indicate that the topological complexity of the original graph Phenol is reduced to the seventh part when the original links are removed and the network is reduced to the ungraph AP. In epidemiological terms, the graph AP has a basic reproductive number equal to zero, and the graph DP has a basic reproductive number which is the seventh part of the basic reproductive number for the original network Phenol.

Another example with reduced networks is illustrated in Fig. 4. In this figure, three states are represented by the graphs in Fig. 4a–c. Figure 4a shows the original network denoted G_1; Fig. 4b shows the reduced graph denoted G_2 resulting when the more external edges in G_1 are deleted; Fig. 4c shows the final reduction denoted G_3 when all edges in G_2 are deleted.

The corresponding Tutte polynomials for the graphs in Fig. 4 are as follows:

$$\begin{aligned}T(G_1,x,y) =\ & 57y + 57x + 6yx^8 + 421x^5 + 36yx^7 + 39y^2x^5 + x^{11} + 176y^2 + 346yx \\ & + 227x^2 + 220y^3 + 122yx^6 + 696yx^2 + 417x^3 + 550xy^2 + 145y^4 \\ & + 748yx^3 + 487x^4 + 615x^2y^2 + 368xy^3 + 54y^5 + 7x^{10} + 168x^7 \\ & + 534yx^4 + 363x^3y^2 + 200x^2y^3 + 111xy^4 + 11y^6 + 28x^9 + y^7 \\ & + 292x^6 + 6y^2x^6 + 12y^5x + 18y^4x^2 + 48y^3x^3 + 141y^2x^4 + 288yx^5 \\ & + 6y^3x^4 + 78x^8 \end{aligned}$$

$T(G_2, x, y) = x^6(y + x + x^2 + x^3 + x^4 + x^5) : T(G_3, x, y) = 1.$

With these polynomials, the reduction from the graph G_1 to the graph G_3 is described in algebraic terms as a reduction of the algebraic complexity for the Tutte polynomials, given the polynomial $T(G_1, x, y)$ is more complex than the polynomial $T(G_3, x, y) = 1$.

3 Algebraic Analysis of Social Networks in SARS and AH1N1 Flu

The algebraic method previously introduced is applied here to SARS-Beijing-2003 [2] and AH1N1 flu-Mexico-2009. In the case of AH1N1 flu-Mexico-2009 is used a general model of social network was topologically reduced via the control measures prescribed by the public health authorities. In SARS-Beijing-2003, an observed and previously reported social network with geographical contacts is analyzed [1,2].

3.1 Algebraic Analysis of Social Networks in AH1N1 Flu-México-2009

A general model of social network which was exhibited during the recently occurred AH1N1 flu outbreaks in México is depicted in Fig. 5a [9]. Figure 5 shows a typical social network with a center composed of schools, hospitals, shopping centers, and work places, and a periphery composed of households with different numbers of members. This social network is able to transmit positive things like ideas and innovations, but the same social network is also able to transmit negative things like infectious diseases [10]. During the AH1N1 flu outbreak in Mexico, this general social network was topologically reduced to a simple cloud of points without any

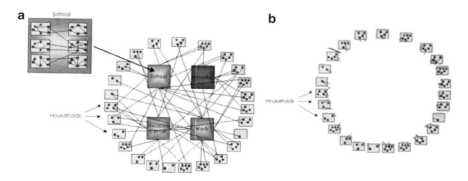

Fig. 5 General model of social network and its extreme reduction

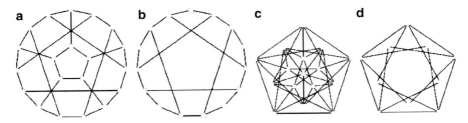

Fig. 6 Two simulations for the network reduction in Fig. 5

kind of links between the individuals such as that shown in Fig. 5b. This extreme reduction of social contacts was demanded by the public health authorities with the aim to stop the explosive propagation of flu.

Two simulations of the reduction depicted in Fig. 5 are shown in Fig. 6. The first simulation is given by Fig. 6a (original network represented by a flower snark graph denoted FS_5) and Fig. 6b (reduced network denoted RN). The second simulation is given by Fig. 6c (original network represented by a Clebsh graph denoted CG) and Fig. 6d (reduced network).

For the graphs in Fig. 6a, b, the corresponding Tutte polynomials are given respectively by:

$$\begin{aligned}
T(FS_5, x, y) = {} & 1145\,y^4x^6 + 4580\,yx^{11} + 3760\,y^3x^7 + 7878\,x + 7878\,y + 3220\,y^2x^9 \\
& + 114690\,x^3 + 191744\,y^2x^5 + 286\,x^{16} + 19\,y^{10} + 129158\,xy^2 \\
& + 10965\,y^5x^3 + 22700\,y^4x^4 + 47405\,y^3x^5 + 187515\,yx^2 + 97651\,y^2x^6 \\
& + 119792\,x^9 + 70904\,x^{10} + 273073\,x^6 + 293485\,x^4y^2 + 176070\,x^3y^3 \\
& + 90385\,x^2y^4 + 7767\,x^{13} + 26617\,y^2 + 3704\,y^7 + 215\,yx^{13} \\
& + 107135\,y^3x^4 + 198500\,x^2y^3 + 11938\,xy^6 + 27215\,y^5x^2 + 282743\,yx^6 \\
& + 261282\,x^5 + 336780\,x^3y^2 + 2715\,y^5x^4 + 36190\,y^4 + 950\,y^8 \\
& + 39815\,y^3 + 332265\,yx^3 + 39500\,y^2x^7 + 10624\,y^6 + 65\,y^3x^9 \\
& + 89925\,yx^8 + 86880\,xy^4 + 37697\,x^{11} + 2531\,xy^7 + 18052\,x^{12} \\
& + 80\,y^2x^{11} + 20\,y^5x^6 + 170\,y^9 + 173800\,yx^7 + 22832\,y^5 + 6285\,y^4x^5 \\
& + 38396\,xy^5 + 60\,y^7x^3 + 1240\,y^6x^3 + 15540\,y^3x^6 + 61874\,yx \\
& + 5335\,x^2y^6 + 125\,y^4x^7 + 268795\,x^2y^2 + 134515\,xy^3 + 200340\,x^4 \\
& + 12730\,y^2x^8 + 1150\,yx^{12} + 30\,y^8x^2 + 640\,y^3x^8 + 180402\,x^8 + 20\,y^9x \\
& + 55570\,y^4x^3 + 43135\,x^2 + 39505\,yx^9 + 14713\,yx^{10} + 379816\,yx^5 \\
& + 407935\,yx^4 + 610\,y^7x^2 + 140\,y^6x^4 + 369\,y^5x^5 + 239007\,x^7 \\
& + 330\,xy^8 + 615\,y^2x^{10} + y^{11} + 5\,y^4x^8 + 2977\,x^{14} + 66\,x^{17} \\
& + 1000\,x^{15} + 25\,yx^{14} + yx^{15} + 11\,x^{18} + 5\,y^2x^{12} + x^{19} + y^6x^5.
\end{aligned}$$

$$\begin{aligned}T(\text{RN}, x, y) =\ & 622x^7 + 27x + 27y + 320yx^6 + 195yx + 15y^2x^6 + 816x^5 + 402x^3 \\ & + 830yx^4 + 70y^3 + 170xy^3 + 475x^2y^2 + 290xy^2 + 220x^4y^2 \\ & + 65x^3y^3 + 20x^2y^4 + 782x^6 + 9y^5 + 422x^8 + 130yx^7 + y^6 + 152x^2 \\ & + 10y^3x^4 + 70y^2 + 35y^4 + 70y^2x^5 + 830yx^3 + 540yx^2 + 126x^{10} \\ & + 5yx^9 + 45xy^4 + 415x^3y^2 + 155x^2y^3 + 5xy^5 + 596yx^5 + 247x^9 \\ & + 35yx^8 + 672x^4 + 56x^{11} + x^{14} + 21x^{12} + 6x^{13}.\end{aligned}$$

From these polynomials, we obtain the following ratio between topological complexities: $T(\text{RN}, 1, 1)/T(\text{FS}_5, 1, 1) = 0.01632$; and this ratio indicates that the reduced network RN in the Fig. 6b has a topological complexity near to zero when RN is compared with the topological complexity of the original network FS_5 shown in Fig. 6a. In epidemiological terms, the reduction of the topological complexity when the contact network is reduced from FS_5 to RN is equivalent to the reduction of the infectiousness inside the reduced network RN with respect to the infectiousness inside the original network FS_5. In other words, the probability of an epidemic outbreak inside the reduced network RN is near to zero when such probability is compared with the probability of an epidemic outbreak inside the original network FS_5. In the second simulation of the reduction process in Fig. 5, the corresponding Tutte polynomials associated respectively with the graphs in the Fig. 6c, d are given by:

$$\begin{aligned}T(\text{CG},x,y) =\ & 1872172x + 34847530y^4 + 9112614y^2 + 21717820y^3 + 2678480y^{14} \\ & + 7870034x^2 + 43384468y^5 + 42011212y^7 + 35302105y^8 \\ & + 45431208y^6 + 105400y^{18} + 3958696x^7 + 15033470x^3 + 1872172y \\ & + 120y^{23} + y^{25} + 19759258y^{10} + 4907540y^{13} + 19283x^{11} \\ & + 27382885y^9 + 8367140y^{12} + 13306232y^{11} + 23866000y^8x^2 \\ & + 24896600y^6x^3 + 15110476yx + 140y^5x^8 + 93485328y^5x \\ & + 10555976yx^6 + 720y^2x^{10} + 93048150y^4x + 245520y^7x^5 \\ & + 886920y^8x^4 + 2446480y^{13}x + 275672y^5x^6 + 6205768y^2x^6 \\ & + 8920xy^{18} + 386480y^3x^7 + 367440y^{13}x^2 + 2514620y^7x^4 \\ & + 2691440y^9x^3 + 15520y^{13}x^3 + 5236520y^{12}x + 1680y^{19}x \\ & + 51332560yx^3 + 78503860y^2x^3 + 83435332y^5x^2 + 101400130y^4x^2 \\ & + 63725936y^7x + 45628390y^8x + 30079420y^9x + 60699616y^6x^2 \\ & + 78511920y^3x^3 + 60100y^{10}x^4 + 672y^{10}x^5 + 1360y^3x^9 \\ & + 13374520y^7x^3 + 14800y^5x^7 + 1082640y^{12}x^2 + 12724460y^5x^4 \\ & + 61328y^6x^6 + 43215300yx^4 + 40y^{13}x^4 + 10220y^{11}x^4 + 1152yx^{11} \\ & + 2783100y^{11}x^2 + 1033720y^{14}x + 12109800y^3x^5 + 6392880y^8x^3 \\ & + 15488yx^{10} + 259240y^9x^4 + 3860y^4x^8 + 54000y^8x^5 + 1451495x^8 \\ & + 2885x^{12} + 325x^{13} + 25x^{14} + x^{15} + 1355496y^{15} + 270930y^{17} \\ & + 632942y^{16} + 14236468x^5 + 36840y^{19} + 427155x^9 + 17576840x^4\end{aligned}$$

$+ 3044\,y^{21} + 680\,y^{22} + 8544936\,x^6 + 101355\,x^{10} + 11388\,y^{20} + 15\,y^{24}$
$+ 61562510\,y^4x^3 + 4 + 49009780\,y^2x^4 + 1305736\,y^2x^7 + 43880542\,y^2x$
$+ 946240\,y^4x^6 + 75108240\,y^3x + 79492384\,y^2x^2 + 2160\,y^{14}x^3$
$+ 41488560\,y^5x^3 + 39921392\,y^7x^2 + 10217568\,y^{11}x + 81408316\,y^6x$
$+ 103270060\,y^3x^2 + 3293168\,yx^7 + 305640\,y^{11}x^3 + 9200\,y^7x^6$
$+ 34000\,y^2x^8 + 2654560\,y^3x^6 + 18276422\,y^{10}x + 600\,y^{17}x^2$
$+ 78080\,y^{12}x^3 + 2443840\,y^5x^5 + 6070620\,y^6x^4 + 38438772\,yx^2$
$+ 12946480\,y^9x^2 + 983800\,y^{10}x^3 + 1080\,y^{12}x^4 + 37661120\,y^3x^4$
$+ 25320\,y^{15}x^2 + 16860\,y^2x^9 + 40\,y^{18}x^2 + 160\,y^{15}x^3 + 130088\,y^{16}x$
$+ 106320\,y^{14}x^2 + 5860600\,y^4x^5 + 6341100\,y^{10}x^2 + 37320\,y^{17}x$
$+ 859832\,y^6x^5 + 4680\,y^{16}x^2 + 1360\,y^6x^7 + 224\,y^{20}x + 130280\,yx^9$
$+ 90320\,y^4x^7 + 765300\,yx^8 + 390712\,y^{15}x + 187860\,y^2x^8 + 720\,y^8x^6$
$+ 20801316\,y^2x^5 + 25097376\,yx^5 + 23461820\,y^4x^4 + 40\,yx^{12}$
$+ 16\,y^{21}x + 8240\,y^9x^5.$

$T(RN, x, y) = 70y^2x^3 + 10y^4x^2 + 6x + 6y + 45y^3x + 96x^4 + 60y^2x + 95yx^4$
$+ 130yx^3 + 5yx^6 + 21\,x^7 + 6x^8 + 36yx^5 + 20y^2x^4 + 15y^3x^3 + 51x^6$
$+ 10y^4 + 15y^3 + 15y^2 + 40yx + 31\,x^2 + 4y^5 + y^6 + 100\,yx^2$
$+ 95\,y^2x^2 + 85x^5 + 71x^3 + 20y^4x + 5y^5x + 45y^3x^2 + x^9.$

From these polynomials we obtain the following ratio between topological complexities: $T(RN, 1, 1)/T(CG, 1, 1) = 0.5634 \times 10^{-6}$; and this ratio indicates that the reduced network RN in Fig. 6d has a topological complexity near to zero when RN is compared with topological complexity of the original network CG shown in Fig. 6c. In epidemiological terms, the reduction of the topological complexity when the contact network is reduced from CG to RN is equivalent to the reduction of the infectiousness inside the reduced network RN respect to the infectiousness inside the original network CG. In other words the probability of an epidemic outbreak inside the reduced network RN is near to zero when such probability is compared with the probability of an epidemic outbreak inside the original network CG. It is worthwhile to note that the ratio between RN and CG is minor than the ratio between RN and FS_5, and this indicates that the reduced network resulting from CG is more secure than the reduced network resulting from FS_5. The second simulation illustrates better the reduction process depicted in Fig. 5.

3.2 Algebraic Analysis of Social Networks for SARS-Beijing-2003

In [1] it was proposed that the inclusion of geographical contacts is very important to understand the epidemics and particularly to understand the SARS propagation. In [2] a weighted district network (WDN) was introduced with the aim to gain

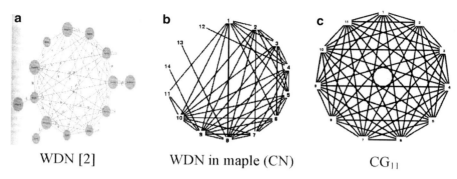

WDN [2]　　　　　WDN in maple (CN)　　　　　CG_{11}

Fig. 7 A weighted district network (WDN) relevant for Sars-Beijing-2003

insight about the SARS epidemic in Beijing during 2003. Figure 7a shows the WDN of [2] jointly with its maple implementation (Fig. 7b) and the complete graph with 11 vertices denoted CG_{11} (Fig. 7c).

In [2] numerical analysis of the WDN in the Fig. 7a, b was made and some information was obtained about the epidemic pattern of SARS. Here we make the algebraic analysis of the WDN using the Tutte polynomial. The topological complexity of the WDN is compared with the topological complexity of the complete graph CG_{11} shown in Fig. 7c. For the WDN denoted CN, the Tutte polynomial is very large but is a good algebraic representation of the WDN. To gain insight about the topological complexity of the WDN, we use the topological complexity of CG_{11} whose Tutte polynomial $T(CG_{11}, x, y)$ is easily computed but it is very large to be displayed here. In all case the relative complexity between WDN and CG_{11} is determined by $T(CG_{11}, 1, 1)/T(CN, 1, 1) = 12.789$.

4 Conclusions

Some examples of network evolution and network reduction were considered and analyzed using Tutte polynomials for graphs. The topological complexity of the social networks was mapped to the algebraic complexity of the corresponding Tutte polynomials. The cases of SARS-Beijin-2003 and AH1N1Flu-Mexico-2009 were considered, and simulations were applied to understand the effects of control measures imposed by the public health authorities. When the contact networks are reduced via the deletion of links between the individuals, the probability of transmission for infectious diseases is reduced appreciably with respect to the probability of transmission inside of very large and dense social networks. Using the Tutte polynomials, some factors for network reduction were derived and interpreted as efficiency indicators for the campaigns oriented to reduce the reproductive numbers for infectious diseases propagating in actual social networks. As a line for future research, it is proposed the application of ribbon graphs and Bollobás–Riordan polynomials in the algebraic analysis of social networks for epidemiology, bio-surveillance, and bio-security.

References

1. Yi-Da Chen, Chunju Tseng, Chwan-Chuen King, Tsung-Shu Joseph Wu, Hsinchum Chen. Incorporating Geographical Contacts into Social Network Analysis for Contact Tracing in Epidemiology: A Study on Taiwan SARS Data. D. Zeng et al. (Eds.): BioSurveillance 2007, *LNCS* 2007; 4506: 23–36
2. Xiaolong Zheng, Daniel Zeng *et al*. Network-Based Analysis of Beijing SARS Data. *LNCS* 2008; 5354: 64–73
3. Mario Vélez, Juan Ospina, Doracelly Hincapié. Tutte Polynomials and Topological Quantum Algorithms in Social Network Analysis for Epidemiology, Bio-surveillance and Bio-security. *LNCS* 2008; 5354: 74–84
4. Doracelly Hincapié, Juan Ospina.Bases para la modelación de Epidemias: el Caso del SARS en Canada. Revista de Salud Pública 2007; 9(1): 117–128. http://www.scielosp.org/scielo.php?pid=S0124006420070001000012&script=sci_arttext
5. Dorit Aharonov, Itai Arad, Elan Eban, Zeph Landau. Polynomial Quantum Algorithms for Additive approximations of the Potts model and other Points of the Tutte Plane. arXiv:quant-ph/070208. Presented at QIP 2007, Australia
6. F. Jaeger, D.L. Vertigan, and D.J.A. Welsh. On the Computational Complexity of the Jones and Tutte polynomials. *Mathematical Proceedings of the Cambridge Philosophical Society* 1990; 108: 35–53
7. GraphTheory Maple 12, www.maplesoft.com
8. Nicolas Rashevsky, Life, Information Theory and Topology. Bulletin of Mathematical Biophysics 1955; 17: 229–235
9. Lauren Ancel Meyers, Predicting the Path of Infectious Diseases [Internet]. www.utexas.edu/features/archive/2003/meyers.htm
10. Network Structure of Swine Flu Pandemic, TNT-the Network Thinker. [Internet]. www.thenetworkthinker.com/2009/04/network-str

Index

A

Active contour, 516–517
Adaptive logic networks, 616
Adaptive neuro-fuzzy model, 526–528
Affymetrix Analysis Suite v5.0 (Mas5), 148, 151–152
Affymetrix GeneChip® CustomExpress™, 4
Aging society, 600
Alzheimer disease, diagnosis, 556
   accuracy, 557–558
   brain changes, 557
   CERAD, 560–562
   diagnosis, 556
   incidence and prevalence, 556
   ZAPROS method
      comparison of alternatives, 559
      elicitation of preferences, 558–559
      Java, 559–560
      problem formulation, 558
ANN. *See* Artificial neural network
ARIMA (SARIMA), tuberculosis, 175
Artificial neural network (ANN)
   MHC class I binders/non-binders peptides (*see* MHC class I binders/non-binders peptides, ANN)
   tuberculosis, 175–176

B

Backpropagation (BP) and ELM training algorithm, 136–138
Batch cultivations. *See Streptomyces coelicolor* batch cultivations
Bayesian networks
   NeoMark project, 373
   PMP modeling module, semantic web, 748
   uncertainty, computer simulation, 711–712
Beltrami flow, real-time ET, 234–235, 237
Biclustering gene expression data
   coherent values, 123–124
   comparison of, 131–133
   datasets, 124
   K-means clustering algorithm, seed finding, 124–125
   residue score, 124
   seed-growing algorithms, 125
      ISIMSRDT, 129–131
      MSRDT, 126–128
      MSRT, 125–126
Binary decision diagrams (BDDs), 115
Bioimpedance measurement systems. *See* ImagCell
Bioinformatics
   algebraic analysis, social networks
      AH1N1 flu, 756–759
      complexity reduction, 754–756
      evolution, 752–753
      SARS flu, 759–760
      Tutte polynomials, 752–754
   computer simulations calibration
      challenges, 710
      uncertainty determination, 711–714
      validation, 710–711
   decision tree
      classification and regression tree, 193
      ID3, 192–193
   digital content credibility
      classification algorithm, 721–723
      Web 2.0, 718–720
   ensemble learning and decision tree
      bagging, 194
      boosting, 194
      cancer classification, 196–197
      genomics classification, 197–198
      stacked generalization, 194–195
   expert model
      algorithm, 658–659
      compression performance, 659–660
      knowledge discovery, 660–664

Food Safety Information Reporting System (FSIRS)
  architecture, 745
  challenges, 742–743
  data sources, food-borne pathogens, 744
  methodologies, 744
  semantic ontology-driven information system, 745–748
  Web Ontology Language (OWL), 743
gray-scale Z scores
  advantage, 730
  anterior myocardial infarction, 728
  calculation, 727
  digital diagnostic data, 726–727
  ECG and sound data changes, 729
  receiver-operating characteristic (ROC) curve, 729
  sensitivity and specificity, 729–730
  subjects selection, 726
ImagCell
  image catalog, 736–737
  image processing, 736
  interface, 735
  measurement, 738–739
  system set-up, 734
  tasks, 734
Lotka–Volterra system
  definitions, 648
  expression, 647–648
  hypotheses, 648
  lemma, 650
  predator-prey cycle system, 651–654
  theorem and corollary, 650
  ultimate boundedness, 649
miRNA expression analyzer, 683–687
mixed-lipid compartments, ENVIRONMENT
  membrane composition, vesicle core, 694
  methods, 690–692
  protocell scenario and assumptions, 692–693
  waste accumulation, protocell core, 695
  waste diffusion coefficient, vesicle core, 694
open source clinical portals, healthcare sector
  architecture, 670–671
  benefits, 671–672
  features, 669
  medical tutorial, 673–675
  quality indicators, 673
polygonal approximation
  genetic algorithms, 698–699
  procedure, 699–706
Biomedical systems
  adaptive systems, 378
  high-performance computing
    biological data mining, 380–381
    collaboration and multidisciplinary research, 382–383
    epigenetics, 382–383
    immune system models, 381
    limitation, 383
    model hybridisation, 381–382
    parallelism and large-scale computing complex systems, 379
    software and libraries, 379–380
  scale complexity, 378
  structural complexity, 378
Biovermiculations, patterned growth, 165–166
Block-based adaptive algorithms, ECG artifacts
  adaptive filter structure, 506–507
  artifacts, types, 506
  block LMS algorithm, 507–508
  electrocardiogram, 505–506
  fast block LMS (FBLMS) algorithm, 508
  noise removal, 506–507
  simulation results, 513
    adaptive baseline wander reduction, 509–510
    adaptive motion artifacts cancelation, 512
    adaptive power-line interference canceler, 510–511
    MIT-BIH arrhythmia database, 508, 509
    muscle artifacts, 511–512
Block LMS algorithm, 507–508
Bombyx mori, decision tree, 80
Boolean gene regulatory networks (BN), algebraic model checking
  Groebner bases for
    algorithm, 119–120
    polynomials, 118
  limitation of, 117, 121
  linear temporal logic specifications for, 116–117
  symbolic model checking, BDDs, 115
  singleton attractors in, 120–121
Boosting algorithm, 194
Bovine serum albumin (BSA), lysozyme binding, 276, 277
Brain machine interface (BMI), wheelchair navigation
  efficiency, 571

Index

MI-based BMI, 570–571
offline synchronous experiments
  feed forward three layered network, 566–567
  MI signals, 566
  recurrent neural networks, 566–567
  real-time asynchronous experiments, 567–570
Breast cancer, 53
  adaptive neuro-fuzzy model, 526–528
  benign tumors, 523–524
  decision-making logic, 528–529
  Fourier descriptors, boundary features, 526
  fuzzy-based segmentation
    calcification, breast, 524–525
    fuzzy cluster, 524
    fuzzy partition membership functions, 524, 525
    performance index, 525
  GA-based feature subset selection (FSS) model, 530
  genetic algorithm, Fourier descriptor reduction, 526
  mammograms, tumor lesions, 529
  microcalcification detection (*see also* Microcalcification detection, FLD)
    Fisher's Linear Discriminant (FLD), 453
    MIAS and ISSSTEP databases, 453–454
    segmentation steps, 452
Breast tumor microwave sensor system (BRATUMASS)
  detected target and transceiver antenna, 490
  simulation experiment
    detected region and targets location, 493–494
    transmission characteristics, 494–495
  traveling wave transmission characteristics
    general medium, 492
    homogeneous medium region, 490–491
    multi-point joint estimation, 492–493

C

Cancer classification studies, decision trees and ensemble tools, 196
Cardiac cavity segmentation
  boundary detection, 481–482, 487
  collinear equation, 484
    centroids, 484
    contour deletion, 485, 486
    high boost filter implementation, 485, 486
    Laplacian and region filter, 485, 486
  left ventricle, short axis image, 483
  morphological operation, 485, 486
  morphology and thresholding, 483–484
  negative Laplacian filter, 484
  region filter, 484
  triangle equation
    corner angle calculation, 485
    minimum small corner, 485, 486
    OpenCV Library, 485
Care staffs information (CSI) subsystem, 603
CART. *See* Classification and regression tree
Case information management subsystem, 602
Cells' tracking, image sequences, 255–256.
  See also High density cells' tracking, of image sequences
Cellular automaton (CA)
  and partial differential equations
    convergence maps and optimum convergence, 162–164
    Euler's methods, 160–161
    theoretical stability constraints, 161–162
  transition of, living cells, 159
Chemical-protein binding activity, graph patterns
  CCPC, 244
    classification accuracy of, 250, 251
    scores for, 247, 248
    training and testing phase, 248–249
  classification, 251
    accuracy, comparison of, 252
    approaches, 243–244
    common subgraph, 246
    contrast subgraph, 246
    COX2, classification accuracy of, 251
    dataset, 249
    experiment plan, 250
    isomorphism and subgraph isomorphism, 245, 246
    labeled graph, 245
    precision (recall) comparison, 252
    subgraph, 246
    support and growth rate, 247
    types of, 244–245
Chi-square test, 612–613
Chromosomal gene clusters, *Streptomyces coelicolor* batch cultivations
  expression profiles of, 9, 11, 12
  materials and methods, 5–6
  PhoP regulated genes, 12–13
Classification and regression tree (CART), 193
Clubfoot
  computer-aided decision system
    C4.5 algorithm, 626–627

dataset, 627–628
decision tree method, 626
online scoring algorithm of, 627
OSMMI ontology (*see* Computer-aided decision system (CADS))
partial decision tree, 629, 631
principal component analysis (PCA), 626, 629, 630
rule-based knowledge, 629–630, 632
conservative treatment, 623–624, 633
Clustering microarray data, normalization method
agglomerative hierarchical clustering, 147
background subtraction, 146
data processing methods
Mas5, 148
pre-normalized dataset, 147
RMA, 148
dendrogram, 149, 150
descriptive statistics, 148–149
pre-normalized dataset
descriptive statistics, 150–151
Mas5 scaling factors, 151–152
seeded data, subset of, 151, 152
Colon tumor, 52
Combined Database of Predictive Microbiology Information (ComBase), 744
Compactness, 526
Computer-aided biomarker discovery. *See* DiscoClini
Computer-aided decision system (CADS), 624
conservative treatment, 625
functional scenario, 624–625
OSMMI ontology
general architecture, 628, 629
knowledge extraction process, 625–626
Ontology Web Language (OWL), 625
statistical classification, 626–627
Web-based interface, 630, 632
Computer simulations calibration
challenges, 710
determining uncertainty
incomplete observability, 711–712
schemas, 714
software system, 712–713
validation, 710–711
Concept map
characteristics, 583
ontology, 583–584
vitamin D deficiency and prostate cancer, 585–586
XML-based format, 583

Consortium to Establish a Registry for Alzheimer's Disease (CERAD), 560–562
Continuous Infusion Insulin Pump (CIIP) safety-critical medical system
behavioral model, 640–641
hazardous behaviors, 642
high-level specification, 643
informal elicitation, 638
interaction model
main and auxiliary attributes, 639
patient visit and system reaction, 639
system environment, 639–640
monitor program, 643–644
physical constraints, 638
safety requirement's violations, 637
visual collaboration, 643
Contrast common pattern classifier (CCPC), 244
classification accuracy of, 250, 251
scores for, 247, 248
training and testing phase, 248–249
CT image denoising. *See* Curvelet-based approach, noise elimination
Curvelet-based approach, noise elimination
MR and CT brain images
entropy-curvelet transform, 476
entropy-wavelet transform, 476
mean and standard deviation, 476
PSNR values, 475
random and Gaussian noise, 478
Speckle and Rician Noise, 477
spatial domain, 472
transformation domain, 472
Cyanobacteria, patterned growth, 164–165

D

Data mining, 91–92
DcSymb language, 331
DcVisu language, 331
Decision support system, breast mass identification. *See* Breast cancer
Decision tree. *See also* Top Scoring Pair (TSP) decision tree
Bombyx mori, miRNA prediction, 80
classification and regression tree, 193
and ensemble classifiers, biological applications of
cancer classification, 196–197
genomics classification, 197–198
ID3
entropy of training set, 193
information gain (IG), 192

Rotation Forest, 213
Defuzzification, 612
Dementia, 555–556
Dendrogram, clustering microarray data, 149, 150
Description logics (DL) ontology, 359–360
Diabetic's safety requirements. *See* Continuous Infusion Insulin Pump (CIIP) safety-critical medical system
Digital content credibility classification algorithm
    correlative matrix and g-approximation matching, 721–722
    digital contents categorization algorithm, 722–723
    repeating pattern, definitions, 721
Web 2.0
    healthcare community, 718–719
    quality of data, 719
    scenario, 720
Digital mammogram, 529
DiscoClini
    dataflow steps, 329–330
    false discovery rate (FDR), 329
    goal, 329
    graphical languages, 331
    Hasse diagram, 330–331
    information visualization (IV), 329
    limitations, 333
    obesity research, 332–333
    visual data mining interface, 330
Disease vector population replacement
    insecticides, 335–336
    MEDEA
        limitations, 338
        parts, 337, 338
        Tribolium castaneum, 337–338
    simulation model, mosquito, 338–341
DNA chips, 27
DNA microarray classification, SLFN training algorithms
    BP and ELM training algorithm, 136–138
    cancer classification problems
        colon cancer data set, 141
        leukemia data set, 140
        performance comparison, 141, 142
        prostate cancer data set, 140–141
    RLS-ELM training algorithm, 138–139
    SVD-neural classifier, 139–140
Dyslipidemia, 548

E

Electromyographic (EMG) signals. *See* Surface electromyographic (SEMG) signals
Electronic data interchange (EDI) subsystem, 603
Electron tomography (ET), 233. *See also* Real-time ET, 3D noise reduction
Electrooculography (EOG)
    blinking signal
        3/4 electrode systems, 539
        human-computer interfaces, 538–539
        median filters, 539
    cornea-retina voltage, 538
    electrode placements, 538
    optimization technique
        algorithm, 541
        blinking, dimensionality changes and modification, 542
        blinking signal model, 540
        efficiency, 539–540
        vs. median filter approach, 542–543
        nongradient method, 540
        saccade, dimensionality changes and movement, 541–542
        single saccade distance and level changes, 542
    saccadic movements, 538
Ensemble learning, in bioinformatics
    bagging, 194
    boosting, 194
    and decision tree, biological applications of
        cancer classification, 196–197
        genomics classification, 197–198
    stacked generalization, 194–195
Euler's methods, cellular automaton
    Backward equation, 162
    Forward equation, 161
    transformation using, 160–161
Expert model, knowledge discovery
    genomic local alignment, 662–663
    phylogenetic sequence analysis, 663–664
    repeat element detection, 661–662
Expert system. *See* Fuzzy logic based expert system, tooth mobility
Extreme environments, patterned growth
    biovermiculations, 165–166
    cellular automaton (CA)
        dead center and live center histograms of, 167, 168
        and partial differential equations, 159–164
        transition of, living cells, 159

cyanobacteria, 164–165
dead center and live center histograms, 167–169
graphic of, 167, 168
Extreme learning machine (ELM) algorithm, 136–138

F

Fast block LMS (FBLMS) algorithm, 508
Feature selection (FS)
  dimensionality reduction for, 93
  filter and wrapper approach, 92
  heuristic function for, 93
  hybridized model for, 94
  interactive exploration approach
    algorithm, 96–98
    data set, plotting of, 96
    normalized data, plotting of, 97, 98
    variances, PCs, 97
  microarray gene expression data, 46
    BGS^3 best gene subset search strategy, 48
    entropy concepts, 47
    mutual information, 46–47
  principal component analysis
    definition, 95
    variable reduction procedure, 94
  relevancy and redundancy, 92
  rough set theory, 95
Feed forward three layered network (FFNN), 566–567
Fermentations, *Streptomyces coelicolor*, 6–7
Fick's law, 160
fills constructor, 359
Fisher's Linear Discriminant (FLD), 453. *See also* Microcalcification detection, FLD
FOLD-RATE method, PFD, 280–283
Food Safety Information Reporting System (FSIRS)
  architecture, 745
  challenges, 742–743
  data sources, food-borne pathogens, 744
  methodologies, 744
  semantic ontology-driven information system
    algorithms development stage, 746–747
    annotation stage, 745–746
    architecture, 745
    categories and subject descriptors, 746
    knowledge-based translation framework, 747
    web presentation stage, 748

Web Ontology Language (OWL), 743
Forecasting methods, 171–172. *See also* Tuberculosis (TB), forecasting methods
Four-channel portable FES system, 616–617
Fourier descriptors, boundary features, 526
Fraction of adsorbed viruses, 319–320
Functional data analysis, metabolomics
  Gaussian distribution, 311
  hydrazine toxicity, 314–315
  integrated NMR intensity, 313–314
  maximum likelihood (ML), 311
  mixed-effects model, 309–311
  observations features, 310
  standard t-test, 313
  temporal profiles comparison, 312
Functional electrical stimulation (FES) system
  adaptive logic networks, 616
  control strategy
    error feedback control, 617
    NNI, 617–618
    PI controller, 617
  four-channel portable FES, 616–617
  neuromuscular-skeletal systems, 616
  open-loop stimulations, 615
  PDA/PC-stimulator structure, 615–616
  wrist tracking experiment, 618–620
Fusion techniques
  composite-fused image, 444, 445
  genetic and fuzzy-based fusion method, 446–447
  HF sub-band components maximization, 445–446
  LF sub-band components, averaging process, 445–446
  selection rule, 444
Fuzzification
  linguistic variables, 609
  membership functions
    input variable mobility, 609–610
    input variable occlusal, 610
    input variables pain and infection, 610
    output variables, 611
Fuzzy clustering technique, SEMG signals, 203–206
Fuzzy c-means clustering, 523–524
Fuzzy logic based expert system, tooth mobility
  Chi-square test, 612–613
  components, 608, 609
  defuzzification, 612
  fuzzification
    linguistic variables, 609
    membership functions, 609–611

Index

if-then rules, 611
inference mechanism, 611–612
medical decision making, 607
signs and symptoms, 608–609
smooth boundary, 608

G
GeneChip® array, 13
Gene Ontology (GO) project, 549
Gene therapy
   human fetal mesenchymal stem cells (hfMSCs), 322–324
   intelligent agents, 318–319
   mathematical model
      fraction of adsorbed viruses, 319–320
      MOI, 321
      productively infected cells, 320
      transduction efficiency, 321
      viral replication index, 320
   pluripotent stem cells, 318
   regenerative process, 318
   transduction efficiency, hfMSCs
      experimental *vs.* simulation transduction, 322–324
      lentiviral vector, 323–324
      onco-retroviral vector, 322–323
      static transduction *vs.* centrifugational transduction, 322, 323
Genetic algorithm, 526
Genetic based neuro-fuzzy approaches. *See* Breast cancer
Gene tree, 288
Gene tree parsimony (GTP)
   Branch-and-Bound algorithm
      branching tree, 290–291
      components, 289
      consistent lower bound (CLB), 290
      criteria and minimum and maximum costs, 293
      duplication and loss costs, 292
      general principle, 291
      incremental forest, 291–292
      optimal cost and CPU time, 294
      properties, 292
      reconciliation costs, lowest common ancestor mapping, 289
Genomic and proteomic classification problems
   classification accuracy, 217–218
   DECORATE, 213
   ensemble building techniques, 212
   Kent ridge repository datasets, 217
   random forests, 212–213

ReliefF feature selection method, 214, 215, 218
Rotation Forest
   decision trees, 213
   elimination of, 214
   RRF, 214
   transformation of, 213–214
   single rotation forest decision tree, 214–216
   tenfold cross validation, 216
Genomics classification, decision trees and ensemble tools, 197–198
Genomics data
   correlation coefficient, 328
   DiscoClini
      dataflow steps, 329–330
      false discovery rate (FDR), 329
      goal, 329
      graphical languages, 331
      Hasse diagram, 330–331
      information visualization (IV), 329
      limitations, 333
      obesity research, 332–333
      visual data mining interface, 330
   linear correlation discovery, 328–329
   MFG data, 328
   ObeLinks, 328
Gini ratios. *See* Lorenz curves and Gini ratios
Goal-based agent, 319
Gouy–Chapman model, 274, 275, 277
Graphics processing units (GPUs)
   implementation, real-time ET, 237, 238, 241
Gray-scale Z scores
   advantage, 730
   anterior myocardial infarction, 728
   calculation, 727
   digital diagnostic data, 726–727
   ECG and sound data changes, 729
   receiver-operating characteristic (ROC) curve, 729
   sensitivity and specificity, 729–730
   subjects selection, 726
Groebner bases, algebraic model checking, 118–120
Gustafson–Kessel (GK) algorithm, 203, 204

H
Health Level Seven (HL7), 600
hfMSCs. *See* Human fetal mesenchymal stem cells
High density cells' tracking, image sequences
   cells' trajectories, three dimensions, 261

comparison of, 260
inactive and active cells, 258
overlap method, 258
segmentation, 256–257
size factor, 259
topological constraint
    cell image to graph, 260
    graph description of, 258
    similarity degree of, 259
    vertex matching, 259
High performance computing (HPC) implementations, real-time ET
    Beltrami flow, multithreaded implementation of, 237
    GPU, 237, 238
Hilbert spectrum (HS), 202–203, 205
Histone cluster, time-series microarray data, 61–63
Histopathology image analysis system
    nuclei center detection
        cluster decomposition, 418
        false negative and positive rate, 422
        individual nuclei centers, 418
        over and under segmentation, 421–422
        post-validation, 419
        pre-filtering, 418–419
        radial directions, 419
    spatial constraint for fuzzy C-means
        color image segmentation, 414
        edge-based geodesic active contours, 417–418
        geodesic active contours, 420–421
        Gleason grade 3 histopathology image, 420
        membership update function, 415
        objective function, 414–415
        quantitative region segmentation analysis, 421
        vector multiphase active contours, 416–417
Holt-Winter's method, tuberculosis, 174–175
Human fetal mesenchymal stem cells (hfMSCs), 322–324
Human papilloma virus (HPV) prevalence, health interventions
    age-gender discrete-time model, 185–187
    automatic sentiment classification, 184–185
    computational epidemiology, 183
    data, 183–184
    differential equation and Markov models, 182
    endemic prevalence, 186
    epidemic models, 184, 189
    public health, 181, 189
    temporal model, 188
    vaccine, 181–182

I

ICT enabled cancer reoccurrence prediction. *See* NeoMark project
ID3 algorithm, 192–193
If-then rules, 611
ImagCell
    image catalog, 736–737
    image processing, 736
    interface, 735
    measurement, 738–739
    system set-up, 734
    tasks, 734
Image compression. *See* Optimal medical image compression
Image denoising. *See also* Curvelet-based approach, noise elimination
    curvelet transform
        algorithm, 473–475
        ridgelet transform, 474
        structural elements, 472
    wavelet denoising
        hard-thresholding function, 472
        soft-thresholding function, 472–473
Image segmentation. *See* Optimal medical image compression
Incremental MSR difference threshold (ISIMSRDT)
    advantages of, 130–131
    algorithm, 129
    yeast and lymphoma datasets, 129–130
Informatics for Diabetes Education and Telemedicine (IDEATel) project, 600
International ClubFoot Study Group (ICFSG), 624
International Foldeomics Consortium, 280, 283
Intrinsic mode functions (IMF), 202–203

K

Kent Ridge Biomedical gene expression datasets, 30
K-means clustering algorithm, biclusters, 124–125
Knowledge discovery
    domain knowledge, 358
    expert model
        genomic local alignment, 662–663

Index

phylogenetic sequence analysis, 663–664
repeat element detection, 661–662
knowledge representation
  data mining method, 360
  description logics (DL) ontology, 359–360
  RAA method, 361–365
  SO-Pharm, 358
  transformations, 360
  levels, 358
  montelukast, 359
  phenotype features, 359
  problem, 358
Kolmogorov–Smirnov filter, miRNA recognition, 21–23
k-Top Scoring Pairs (k-TSP), 29, 32

L

Learning agent, 319
Left ventricular motion, QGDCT
  Gaussian filtering, 463
  global regularization optical flow, 464
  methodology, 462–463
  optical flow field validation, 465
  speckle noise reduction, 464–465
  ultrasound images
    computed optical flow, 466–467
    echocardiographic images, 466
    echo data, 468
    QGDCT filter, 467
    SEM, 467–468
Leukemia, 52–53
Leukocytes segmentation
  color features, 346–347
  evaluation, cells images, 350–351
  Markov random fields (MRF) model
    posterior probability, 350
    a priori probability, 349
    probability model, 349
  staining and cell population, real cells, 352
  texture analysis
    harmonic and generalized evanescent fields, 347–348
    Wold decomposition texture model, 347–348
Linear regression, tuberculosis, 173
Linear temporal logic (LTL) and model checking, 114–116
Lipoprotein ontology
  advantages, 548
  clinical and epidemiological studies, 548
  future aspects, 552

lipoproteins, 550
  research domain issues, 549–550, 552–553
  transport system, 551, 552
  significance, 552
  sub-ontologies and subclasses, 550, 551
Lorenz curves and Gini ratios
  bias
    first kind, order of classes, 84–85
    gene expression values, order of, 86
    second kind, 86
  experimentation, with reducing techniques
    Bayesian Net algorithm, 88
    consistency subset evaluation, 88–89
    correlation-based feature selection, 88
    libSVM algorithm, 89
    lung adenocarcinoma, 86, 87
    PCA, 87–88
  microarray data mining, 83
  SAM, 84
Lotka–Volterra system
  definitions, 648
  expression, 647–648
  hypotheses, 648
  lemma, 650
  predator-prey cycle system, 651–654
  theorem and corollary, 650
  ultimate boundedness, 649
Low-resolution brain electromagnetic tomography (LORETA), 590–591, 593
Lung cancer, 53
Lysozyme, binding to vesicles
  Hill plot of, 277
  myoglobin/bovine serum albumin-PC, 276, 277
  and myoglobin/PC/PG, 275, 276
  PC/PG, 274

M

Markov random fields (MRF) segmentation model
  posterior probability, 350
  a priori probability, 349
  probability model, 349
Maternal effect dominant embryonic arrest (MEDEA)
  limitations, 338
  parts, 337, 338
  simulation model, mosquito
    embryo survival and turnover, 339
    fitness effect, 341
    genetic mutation and reproduction, 338

percentage of insertion, 340
scenarios, 339
transposon control strategy, 340–341
Tribolium castaneum, 337–338
Mean squared residue threshold (MSRT), biclusters
advantages of, 126
yeast and lymphoma datasets, 125–126
MEDEA. *See* Maternal effect dominant embryonic arrest
Medical Functional Genomics (MFG), 327–328
Medical research
concept map
characteristics, 583
ontology, 583–584
vitamin D deficiency and prostate cancer, 585–586
XML-based format, 583
Excel spreadsheet, 583
inefficiency problem, 581
published information, 583
research material finding, challenges, 582
time gap, 582
Message-passing interface (MPI), 379–380
Metabolomics
short time series analysis
functional mixed-effects model, 309–311
Gaussian distribution, 311
hydrazine toxicity, 314–315
integrated NMR intensity, 313–314
maximum likelihood (ML), 311
observations features, 310
standard *t*-test, 313
temporal profiles comparison, 312
tasks and complications, 308
toxicology study, 308–309
Methionine genes, time-series microarray data, 62
MHC class I binders/non-binders peptides, ANN, 224
area under ROC curve (AROC), 227
epitope prediction, 224
evaluation parameters, 226–228
receiver operating characteristics (ROC), 226
SARS corona virus
alleles for, 227
data resources, 226
T-cell immune response, 224
vaccine designing, 224
variable learning rate, for training, 225–226
Microarray images, 435

estimated background levels, 433
global geometric deformation, 435
gridding, 435, 437
horizontal and vertical intensity projection profiles, 437
image enhancement, 439
intensity projections and template matching, 439
misalignment error, 438, 439
software products, 434
subarrays, 434
division, 435–436
geometric rotation, 436
gridding and spot extraction, 436–437
spot isolation, 438
Microarray techniques
biological evidence
breast cancer, 53
colon tumor, 52
leukemia, 52–53
lung cancer, 53
prostate cancer, 53
experimental methodology, gene expression
accuracy results, comparison of, 50
BGS^3, gene subsets selection, 49
visualization of, 51
feature selection (FS), 46
BGS^3 best gene subset search strategy, 48
entropy concepts, 47
mutual information, 46–47
mining and cell differentiation defects, in schizophrenia
differentiation process, 72
differentiation signature (DIF), 69–70
methods, 68–69
neural progenitor cells, 72
oligodendrocyte progenitor cells, 68, 71
percentages of common genes, 71
stemness/differentiation (S/D) set, 69
precision-reduced descriptions in
application of, 108–109
BIBE04, 104
cds28 dataset, expression behaviors in, 106
measured expression variation, 104–105
motivation, 102–103
rationale, 103
simplified data descriptions and subsequence-based clustering, 105–108
TESTGENES, 106, 107

Index 773

threshold clustering (TC) algorithm, 104
time-series, hierarchical signature clustering for
   clustering method, 58
   dendrogram, 60
   distance metric for, 64
   Euclidean distance, 58
   G. hirsutum, 63–64
   Hamming distance, 60
   k-means clustering, 60–61
   merging method, 60
   Saccharomyces cerevisiae, 61–63
   signature pattern, 59
Microcalcification detection, FLD
   breast density identification, 454–455
   discriminant spaces, 455–456
   false-positive reduction, 456, 458
   MIAS and ISSSTEP databases, 453–454
      breast density, 457
      false-positive reduction, 458
      microcalcifications, 457
MicroRNAs (miRNA)
   prediction, computational approach
      attribute measurement, 78–79
      attribute relevance analysis, 79
      classification of, 79
      decision tree for, *Bombyx mori*, 80
      filter-based approaches, 77
      gene finding, 76
      homology-based search, 77–78
      phases, 78
      secondary structure, 76
      target centered approaches, 77
      training dataset, 81
      Weka software, performance evaluation, 80
   recognition, *yasMiR* system
      using clustering, 22, 23
      $F1$ and $F2$ scores, 21
      features for, 18
      IE-NC and IE-M datasets, 20
      Kolmogorov–Smirnov filter, 21–23
      miPred datasets, prediction results for, 20–22
      miRBase 12.0, 19
      pivots, automatic selection of, 22–23
      Triplet-SVM and miPred, 18–19
      Triplet-SVM datasets, prediction accuracy for, 20
Minimum description length (MDL) approach. *See* Predictive minimum description length (PMDL)
Minimum Information About a Microarray Experiment (MIAME), 146
miPred datasets, miRNA recognition, 20–22
Mitoxantrone
   DNA binding mechanism, 386
   DNA tetramer sequences, 387
   molecular modeling methodology
      conformational changes, 392–394
      drug–DNA interactions, 390–392
      interaction energy, complexes, 388–389
      total energy, complexes, 389–390
   structure, 387
Model-based reflex agent, 319
Model checking, 113–114
MOI. *See* Multiplicity of infection
Molecular modeling, mitoxantrone
   conformational changes, 392–394
   drug–DNA interactions, 390–392
   interaction energy, complexes, 388–389
   total energy, complexes, 389–390
Molecular operating environment (MOE) software tool, 388
MR image denoising. *See* Curvelet-based approach, noise elimination
MSR difference threshold (MSRDT)
   advantages of, 128
   yeast and lymphoma datasets, 127–128
Multi-agent system (MAS), 515–516
   active contour, 517
   detection and tracking object
      biological images sequence, 519
      echocardiographic images sequence, 519
   NetLogo platform, 517
   observer roles, 518–519
   Turtles' roles, 519
Multimodality medical image registration
   affine transformation, 443
   computed tomography, 441–442
   GA, optimization, 444
   and image fusion scheme
      composite-fused image, 444, 445
      genetic and fuzzy-based fusion method, 446–447
      HF sub-band components maximization, 445–446
      LF sub-band components, averaging process, 445–446
      selection rule, 444
   multi-resolution approach, 444
   mutual information-based similarity metric, 443
   PET and SPECT imaging, 442
Multiplicity of infection (MOI), 321

Multi-resolution-based Genetic algorithm (GA) approach, 444
Myoglobin, lysozyme binding, 275–277

N
NeoMark project
  aim, 372
  data analysis
    disease evolution monitoring, 373–374
    early risk assessment, 373
  genomic data cleaning and filtering, 371
  image feature extraction, 370–371
  OSCC, 368
  qRT-PCR platform, 371–372
  system overview, 369
NetLogo, 516
Neural network identifier (NNI), 616–618
Noise cancellation. *See* Image denoising
Noise reduction. *See* Real-time ET, 3D noise reduction

O
ObeLinks, 328
Oculography, 537
Oligodendrocyte progenitor cells (OPCs), 68, 71
Ontology
  Gene Ontology (GO) project, 549
  lipid ontology, 549 (*see also* Lipoprotein ontology)
  Open Biomedical Ontologies (OBO), 549
  organizational framework, 548–549
  protein ontology, 549
  terminologies, 548–549
Open Biomedical Ontologies (OBO), 549
Open source clinical portals, healthcare sector
  architecture, 670–671
  benefits, 671–672
  features, 669
  medical tutorial, 673–675
  quality indicators, 673
Optimal medical image compression
  mammography images, 502
  ultrasound images
    CR criterion, 501
    MAE criterion, 501
  wavelet key properties
    approximation order, 499
    atomic function $eup(x)$, 503
    fidelity criteria, 500
    frequency response, 499
    projection Cosine, 500
    Riesz bounds, 500, 503
    simple shift and modulation, 499
    US and MG images, 502, 503
    wavelet threshold, 498–499
    wavelet transform and filter banks, 498
Oral squamous cell carcinoma (OSCC), 368

P
Pairwise statistical significance
  alignment score distribution, 298–299
  characteristics, 300
  function, 300
  P-value, 299–300
  vs. statistical significance, 299
Pathogen Modeling Program (PMP), 744
Peak signal-to-noise (PSNR) ratio, 475
Peritoneal antimicrobial pharmacokinetics model
  concentration profiles, 405
  dialysis schedule, 403
  dosing regimens, 403–404
  Monte Carlo implementation, 404
  schematic diagram, 403
  peritoneal cavity calculation
    computational framework, 409
    drain and refilling, 408–409
    dwell, 407–408
Personal health record management (PHM) subsystem, 602
Petri net (PN), 640–641
P300 event-related potential (ERP), schizophrenia
  amplitude, 592
  auditory oddball paradigm, 591
  cingulate gyrus, 596
  data acquisition, 591
  EEG sources, 590
  hippocampal formation, 590
  latency, 592
  LORETA, 590–591
  normal *vs.* schizophrenic participants
    Chi-square test, 593
    cingulate gyrus, 596
    frontal lobe, 596
    hemispheres, dominancy, 596
    neural generators, 593–595
    source localization, 595
  P300 source localization, 591–592
  scalp electrical distribution, 590
  spatial information, 590
Phosphatidylcholine (PC), lysozyme binding, 274–276

Index

Phosphatidylglycerol (PG), lysozyme binding, 274–276
Polygonal approximation
  genetic algorithms, 698–699
  procedure, 699–706
Predictive Microbiology Information Portal (PMIP), 744
Predictive minimum description length (PMDL)
  connectivity matrix, 39
  description length for, 39
  information theoretic models, 38
  mutual information (MI), 38, 39
  Saccharomyces cerevisiae data set, 40–42
    networks for, 41
  synthetic networks, simulation on, 39–40
    precision and recall values for, 40
  threshold sensitivity, 40
Principal component analysis (PCA)
  feature selection
    definition, 95
    variable reduction procedure, 94
  Lorenz curves and Gini ratios, 87–88
Prostate cancer, 53
Protein binding to vesicles, cooperativity
  cavitation energy, 273
  cluster analysis, 275
  lysozyme
    Hill plot of, 277
    myoglobin/bovine serum albumin-PC, 276, 277
    and myoglobin/PC/PG, 275, 276
    PC/PG, 274
  materials and methods, 272
  membrane-bound state, interfacial charge for, 274
  partition coefficient, 272–273
Protein Folding Database (PFD), 280–282
Protein folding kinetics modeling
  computational analysis and molecular modeling, 279
  correlation coefficients, 280, 282
  FOLD-RATE method, 280–283
  independent dataset in, 283
  mountain plot of, 282
  PFD, 280–282
Protein ontology, 549
Protein–Protein Interaction Prediction Engine (PIPE), 264, 269
Protein–protein interaction (PPI), string kernel (SK) approach
  amino acid sequencing, 264–266
  contingency table, 268
  experimental techniques, 267–268

method
  definition of, 266
  feature space, 266, 267
  mapping of, 267
  steps in, 265
  pairwise similarity (PPI-PS), 265
  performance of, weight decay factor, 269
  PIPE, 264, 269
  sequence similarity, pattern of, 264
  subsequence length values of, 269
Psychological disorders diagnosis
  abnormal behavior, 574
  aggravating factors, 574
  categories, 575
  control events, 576
  decision-making model, 576
    multicriteria methodology, 577
    value-judgment and current scale matrix, 578
  DSM-IV, 574–575
  expert systems, 578–579
  health, WHO, 574
  hybrid model, 574
  leprosy, 575
  signs and symptoms, 573–574

Q

Quasi-Gaussian discrete cosine transform (QGDCT)
  Gaussian filtering, 463
  left ventricular motion (*see* Left ventricular motion, QGDCT)

R

Random Forests, genomic and proteomic classification problems, 212–213
Real-time ET, 3D noise reduction
  geometric flow
    Beltrami flow, 234–235
    and feature preservation, 236
  HPC implementations, for modern platforms
    Beltrami flow, multithreaded implementation of, 237
    GPU, 237, 238
  processing time of
    GPU implementation, 241
    multithreaded implementation, 239, 240
  scalability, 239
  speedup factors of, 239, 240
Recurrent neural networks (RNNs), 566–567

Reduced minimal non-redundant rules
 (RMNR), 363
Regularized least-squares extreme learning
 machine (RLS-ELM) training
 algorithm, 138–139
Robust Multichip Average (RMA), 148
Role assertion analysis (RAA),
 pharmacogenomics
 assertion graphs exploration, 362
 binary table transformation, 362
 data mining, 362–363
 features, 361
 genotype–phenotype associations, 364–365
 implicit or hidden knowledge, 364
 RMNR interpretation, 363
 steps, 361–363
Role composition constructor, 360
Rotation Forest, genomic and proteomic
 classification problems
 decision trees, 213
 elimination of, 214
 RRF, 214
 transformation of, 213–214
Rotation of Random Forests (RRF), 214
Rough set theory, feature selection, 95
RT-PCR analysis
 amplification curve, parts, 680–681
 MaxRatio method, 681–682
 miRNA expression analyzer, 683–687

S

Saccharomyces cerevisiae
 data set, PMDL, 40–42
 time-series microarray data
 histone-related genes, 61–63
 methionine genes, 62
Sanger Protein Classification, 8
SARS coronavirus. *See* Severe acute
 respiratory syndrome (SARS)
 coronavirus, MHC class I
 binders/non-binders
Schizophrenia, cell differentiation defects
 differentiation process, 72
 differentiation signature (DIF), 69–70
 methods, 68–69
 neural progenitor cells, 72
 oligodendrocyte progenitor cells, 68, 71
 percentages of common genes, 71
 stemness/differentiation (S/D) set, 69
Seed-growing algorithms, biclusters, 125
 ISIMSRDT, 129–131
 MSRDT, 126–128
 MSRT, 125–126

Semantic ontology-driven food safety
 information system
 algorithms development stage, 746–747
 annotation stage, 745–746
 architecture, 745
 categories and subject descriptors, 746
 knowledge-based translation framework,
  747
 web presentation stage, 748
Sequence-comparison process
 database statistical significance
  BLAST programs, 303
  vs. pairwise statistical significance,
   300–301
 error per query *versus* coverage curves, 302
 orders, performance evaluation, 303
 pairwise alignment program
  alignment score distribution, 298–299
  characteristics, 300
  vs. database search programs, 300–301
  P-value, 299–300
  vs. statistical significance, 299
Severe acute respiratory syndrome (SARS)
 coronavirus, MHC class I
 binders/non-binders
 alleles for, 227
 data resources, 226
Simple reflex agent, 319
Single hidden layer feedforward neural
 networks (SLFNs). *See* DNA
 microarray classification, SLFN
 training algorithms
Singular value decomposition (SVD)-neural
 classifier, DNA microarray,
 139–140
Species tree, 288
Streptomyces coelicolor batch cultivations
 array design and samples
  materials and methods, 4
  validation of, 6
 chromosomal gene clusters, prediction of
  expression profiles of, 9, 11, 12
  materials and methods, 5–6
  PhoP regulated genes, 12–13
 data pre-processing, 5
 fermentations, 6–7
 MAYDAY, 13
 reproducibility
  fermenters, online and offline
   measurements for, 7
  materials and methods, 5
 variant genes
  categories, 9
  clustering of, 5

Index   777

QT clustering, 8, 10
Sanger Protein Classification, protein classes distribution, 8
String kernel (STRIKE) approach. *See* Protein–protein interaction (PPI), string kernel (SK) approach
Suggested Ontology for Pharmacogenomics (SO-Pharm), 358
Surface electromyographic (SEMG) signals database, 207
   fuzzy clustering technique, 203–204
   Hilbert–Huang analysis, 202–203
   recognition system
      feature extraction, 205
      fuzzy clustering classification, 205–206
      Hilbert spectrum generation, 205
      preprocessing, 204–205
      steps, 204
   supervised classification, 202
Synthetic networks simulation, PMDL, 39–40

T
Tele-healthcare services system
   aging society, 600
   chronic care informatics system, 600–601
   client interface system
      Barthel index and quality of life scale, 604
      homecare user, 603–604
      vital sign values, 603
   client portion, 601
   disease prevention, 606
   Health Level Seven (HL7), 600
   IDEATel project, 600
   mobile phone message notification, nurse, 605
   remote server, 601
      care staffs information subsystem, 603
      case information management subsystem, 602
      electronic data interchange subsystem, 603
      PHM subsystem, 602
      video conference subsystem, 602–603
      vital signs capture and transmission subsystem, 602
   unified modeling language, 601
   web-based applications server, 604–605
Tissue segmentation. *See* Histopathology image analysis system
Top Scoring Pair (TSP) decision tree, 29–30
   accuracy and model size, comparison with, 31
   algorithms, 28–29
   comparison of, and rule classifiers, 33
   ensembles of, 29
   gene labels and names, 32
   Kent Ridge Biomedical gene expression datasets, 30
   k-TSP methods, breast cancer dataset, 32
Transduction efficiency, 321
Triplet-SVM datasets, miRNA recognition, 20
TSP. *See* Top Scoring Pair (TSP) decision tree
Tuberculosis (TB), forecasting methods
   measure error, 177
   model development
      ARIMA (SARIMA), 175
      artificial neural network, 175–176
      decomposition method, 174
      Holt-Winter's method, 174–175
      linear regression, 173
      moving average, 173–174
   performance comparison of, 176, 177
   plot for, 176
   time series data for, 172
Turtles' roles, 519

U
Unified modeling language (UML), 601
Uropathogenic *Escherichia coli* (UPEC) visualization
   bilateral filtering, 430–431
   fluorescent microscope image, 425, 426
   innermost pixels, 426
   intracellular bacterial communities, 425–426
   shadowed Gaussian
      background pixel, 428
      barrier pixels, 427
      Gaussian blur, 427
      image intensity thresholds, 427
   smoothing operations, 429
   standard Gaussian, 430
   urinary tract infections, 425
Utility-based agent, 319

V
Video conference subsystem (VCS), 602–603
Viral replication index, 320
Vital signs capture and transmission subsystem (VCT), 602

W
Wavelet filters
   approximation order, 499

atomic function $eup(x)$, 503
fidelity criteria, 500
frequency response, 499
projection Cosine, 500
Riesz bounds, 500, 503
simple shift and modulation, 499
US and MG images, 502, 503
Web Ontology Language (OWL), 743
Wold decomposition texture model, 347–348
Wrist tracking experiment, 618–620

X
X3DNA software, 388

Y
Yeast and lymphoma datasets, biclusters
ISIMSRDT, 129–130
MSRDT, 127–128
MSRT, 125–126

Z
ZAPROS method
comparison of alternatives, 559
elicitation of preferences, 558–559
Java, 559–560
problem formulation, 558

CPSIA information can be obtained at www.ICGtesting.com
224139LV00003B/3/P